Plant Stress Tolerance

Plant Stress Tolerance: Molecular Mechanisms and Breeding Strategies, Volume Two explores methods of precise management of biotic stressors including pests and pathogens. This is based on advanced molecular technologies including mutagenesis, genetic engineering, genome-wide association study, marker-assisted selection, genomic selection, molecular marker-based platforms, functional genomics, multiple omics tools, high-throughput technologies, computational biology, epigenetic manipulation, and clustered regularly interspaced short palindromic repeats (CRISPR)-based genome editing.

This book proposes strategies involving immunity boosting through the release of genetic resources from naturally resistant plants, regulating defense systems by phytohormones, promoting disease tolerance by biostimulants, and nanotechnology such as nanocarriers for managing biotic stressors. These positive approaches help to advance and accelerate breeding programs for disease-tolerant crops against various pests and pathogens.

Plant Stress Tolerance: Molecular Mechanisms and Breeding Strategies, Volume Two is an ideal reference for the research fields of plant pathology, plant disease management, plant physiology, plant functional genomics, multiple omics, systems biology, and crop breeding. Regarding future research on disease-resilient crops, this book will inspire readers to face the challenge of global climate change and increases in population.

Plant Stress Tolerance

Molecular Mechanisms and Breeding Strategies

Volume Two

Edited by

Jen-Tsung Chen

CRC Press
Taylor & Francis Group
Boca Raton London New York

CRC Press is an imprint of the
Taylor & Francis Group, an **informa** business

Designed cover image: Shutterstock

First edition published 2025
by CRC Press
2385 NW Executive Center Drive, Suite 320, Boca Raton FL 33431

and by CRC Press
4 Park Square, Milton Park, Abingdon, Oxon, OX14 4RN

CRC Press is an imprint of Taylor & Francis Group, LLC

ISBN: 9781032895178 (hbk)
ISBN: 9781032895185 (pbk)
ISBN: 9781003543237 (ebk)

DOI: 10.1201/9781003543237

Typeset in Times
by codeMantra

Contents

Preface

In the modern world, biotic stress caused by pets and pathogens is still one of the main threats to global food and nutrition security. Volume Two of *Plant Stress Tolerance: Molecular Mechanisms and Breeding Strategies* aims to provide a series of ways for precisely managing these biotic stressors, highlighting the use of molecular and speed breeding strategies assisted by mutagenesis, genetic engineering, genomic selection, molecular marker-based platforms, functional genomics, multiple omics tools, high-throughput technologies, computational biology, and genome editing systems.

Phytopathogenic bacteria are responsible for mass damage to quality and yield losses of crops each year, and new strains emerge fast, consequently increasing the difficulty of managing. This book explores promising approaches to advancing and making efficient the development of bacterial-tolerant crops, covering the use of the genome-wide association study (GWAS), marker-assisted selection (MAS), genomic selection (GS), speed breeding (SB), and clustered regularly interspaced short palindromic repeats (CRISPR)-based genome editing. Additionally, strategies are proposed involving immunity boosting through the release of genetic resources from naturally resistant plants and nanotechnology such as targeted delivery of antimicrobial compounds via nanocarriers. The current achievements, limitations, and prospects were comprehensively discussed.

Genome editing technologies for developing viral disease-resistant crops are illustrated, focusing on the systems of transcription activator-like effectors (TALENs), zinc-finger nucleases (ZFNs), mega nucleases, and CRISPR/Cas9. Together with RNA interference and the application of viral vectors, these technologies are potent to improve the efficiency and precision of breeding toward the resilience of viral diseases in crops.

Among pests, insect herbivores can cause huge losses in global agricultural production. Under climate change, their polymorphism may increase and, therefore, bring new challenges for developing tolerant crops. The management and breeding strategies are proposed based on physiological and molecular approaches, exploring the roles of herbivory elicitors, octadecanoids, wound signals, wound-elicited responses, plant secondary metabolites, and so on, to improve insect resistance of crops.

To develop stress-resilient crops, scientists need to uncover one of the most important regulators for stress response, i.e., plant hormones. Therefore, this book refines the advances in crucial ones including ethylene, abscisic acid, salicylic acid, jasmonic acid, brassinosteroids, polyamines, and strigolactone, and illustrates their impacts on various stress conditions caused by biotic or abiotic factors. Among gene families of transcription factors, ATP-binding cassette transporters (ABC proteins) are recognized as crucial players involved in a range of cellular activities responsible for regulating plant stress responses. As components of plant immunity, ABC proteins participate in mechanisms including cuticle synthesis that protect plants against pathogens and pests. Also, some other members can mitigate environmental stresses by maintaining the cellular homeostasis of plants. They mediate the uptake, sequestration, and detoxification of various stressful factors such as ROS and heavy metals.

Plant cuticles play crucial roles in stress responses against drought, salinity, extreme temperatures, ultraviolet radiation, pollution, fungi, bacteria, viruses, and pests. Therefore, the modification of cuticle biosynthesis holds promise to advance the development of resilient crops. The impacts of stress conditions in the cellular processes of cutin matrix deposition and assembly, gene expression patterns of key players, and interactions between cuticle components, pests, and pathogens were comprehensively presented. The knowledge was proposed to advance breeding programs for organizing better quality and higher yield crops under either abiotic or biotic stress.

Biostimulants are beneficial microorganisms, metabolites, or substances, and their impacts on stress mitigation were explored in this book covering both biotic and abiotic stresses. Biostimulants possess multifunction and are promising in sustainable agriculture by enhancing quality and yield and reducing crop stress effects.

Approaches for managing nematodes in vegetables using beneficial microbes and controlling soil-borne pathogens by microbial consortia were presented, and these technologies are valuable for organizing precision and sustainable agriculture that supports the UN's SDGs.

An ideal reference for the research fields of plant pathology, plant disease management, plant stress physiology, plant functional genomics, and crop breeding, this book was collaboratively organized by diverse scientists and experts. The knowledge is expected to inspire novel ideas from the readers for designing their future investigations on disease-resilient crops to face the challenge of global climate change and an increase in human population. In the end, the editor would like to thank all contributors for their innovative chapters and the support from the publisher.

Editor

Dr. Jen-Tsung Chen is a professor of cell biology at the National University of Kaohsiung in Taiwan. He also teaches genomics, proteomics, plant physiology, and plant biotechnology. Dr. Chen's research interests include bioactive compounds, chromatography techniques, plant molecular biology, plant biotechnology, bioinformatics, and systems pharmacology. He became an active editor of academic books and journals in an effort to advance the exploration of multidisciplinary knowledge involving plant physiology, plant biotechnology, nanotechnology, ethnopharmacology, systems biology, and drug discovery. He serves as an editorial board member and a guest editor in several reputed journals. Dr. Chen published books in collaboration with international publishers on diverse topics such as drug discovery, herbal medicine, medicinal biotechnology, nanotechnology, bioengineering, plant functional genomics, plant speed breeding, CRISPR-based plant genome editing, and artificial intelligence. In 2023 and 2024, Dr. Chen was included in the World's Top 2% Scientists ranking by Stanford University.

Contributors

Licon Kumar Acharya
ICAR-National Research Centre for Integrated Pest Management
New Delhi, India

Shreni Agrawal
Department of Biotechnology, Parul Institute of Applied Science
Parul University
Vadodara, India

Abdul Razak Ahmed
Faculty of Agriculture, Plant Protection Department
Akdeniz University
Antalya, Turkey

Mohd Ali
Department of Zoology
Guru Nanak Dev University
Amritsar, India

Justice Frimpong Amankwah
China-ASEAN Joint Laboratory for Marine Aquaculture Technology Belt and Road Initiative (Shanghai)
Shanghai Ocean University
Pudong, China
and
Key Laboratory of Aquatic Genetic Resources Exploration and Utilization, Ministry of Education
Shanghai Ocean University
Pudong, China

Priscilla Amponsah
Key Laboratory of Integrated Pest Management on Crops in East China, Ministry of Agriculture
Nanjing Agricultural University
Nanjing, China

Prodipto Bishnu Angon
Institute of Agronomy
Hungarian University of Agriculture and Life Sciences
Gödöllő, Hungary

Augustine Antwi-Boasiako
National Center for Soybean Improvement, Key Laboratory of Biology and Genetics and Breeding for Soybean, State Key Laboratory of Crop Genetics and Germplasm Enhancement, Ministry of Agriculture
Nanjing Agricultural University
Nanjing, China
and
Council for Scientific and Industrial Research-Crops Research Institute (CSRI-CRI)
Fumesua, Kumasi, Ghana

Asad Azeem
Department of Plant Breeding and Genetics
University of Agriculture
Faisalabad, Pakistan

Anamika Barman
ICAR-Indian Agricultural Research Institute
Pusa, New Delhi, India

Zineb Belabess
Plant Protection Laboratory, Regional Center of Agricultural Research of Meknes
National Institute of Agricultural Research
Meknès, Morocco

Wissal Benaffari
Centre d'Agrobiotechnologie et Bioingénierie, Unité de Recherche labellisée CNRST (Centre AgroBiotech-URL-CNRST-05)
Cadi Ayyad University
Marrakesh, Morocco
and
Faculty of Science Semlalia, Laboratory of Agro-Food, Biotechnologies and Valorization of Plant Bioresources (AGROBIOVAL), Department of Biology
Cadi Ayyad University
Marrakesh, Morocco

Oumaima Benaissa
Phytopathology Unit, Laboratory of Functional Ecology and Environmental Engineering, Department of Plant Protection, Ecole Nationale d'Agriculture de Meknès
Sidi Mohamed Ben Abdellah University
Fez, Morocco

Khadija Benamar
Phytopathology Unit, Laboratory of Functional Ecology and Environmental Engineering, Department of Plant Protection, Ecole Nationale d'Agriculture de Meknès
Sidi Mohamed Ben Abdellah University
Fez, Morocco

Salma Benchlih
Phytopathology Unit, Department of Plant Protection
National School of Agriculture
Meknes, Morocco

Anurag Bera
Tea Research Association
Nagrakata, India

Indrani Bhattacharya
Department of Biotechnology, Parul Institute of Applied Science
Parul University
Vadodara, India

Ajanta Birah
ICAR-National Research Centre for Integrated Pest Management
New Delhi, India

Kaoutar Bourak
Plant Pathology Laboratory, Department of AgroBioSciences
Mohammed VI Polytechnic University
Hay Moulay Rachid Ben Guerir, Morocco

Hanane Boutaj
Centre d'Agrobiotechnologie et Bioingénierie, Unité de Recherche labellisée CNRST (Centre AgroBiotech-URL-CNRST-05)
Cadi Ayyad University
Marrakesh, Morocco
and
Faculty of Science Semlalia, Laboratory of Agro-Food, Biotechnologies and Valorization of Plant Bioresources (AGROBIOVAL), Department of Biology
Cadi Ayyad University
Marrakesh, Morocco
and
Faculty of Medicine and Pharmacy, Laboratory of Life and Health Sciences
Abdelmalek Essaâdi University
Tangier, Morocco

Harish Changotra
Department of Molecular Biology and Biochemistry
Guru Nanak Dev University
Amritsar, India

Abdelfattah Dababat
International Maize and Wheat Improvement Center (CIMMYT)
Ankara, Türkiye

Arpan Das
Department of Genetics and Plant Breeding
Bangladesh Agricultural University
Mymensingh, Bangladesh

Richa Das
Department of Biotechnology, Parul Institute of Applied Science
Parul University
Vadodara, India

Bishal Debnath
Punjab Agricultural University
Ludhiana, India

Juan Carlos Mateos del Amo
Consejo Superior de Investigaciones Científicas
Instituto de Hortofruticultura Subtropical y Mediterránea La Mayora
Dpt. Biología Molecular y Bioquímica
Universidad de Málaga-Consejo Superior de Investigaciones Científicas, and Universidad de Málaga
Universidad de Málaga
Málaga, Spain

Eva Domínguez
Consejo Superior de Investigaciones Científicas
Instituto de Hortofruticultura Subtropical y Mediterránea La Mayora
Universidad de Málaga-Consejo Superior de Investigaciones Científicas
Madrid, Spain

Suman Dutta
Department of Genetics
Ramakrishna Mission Vivekananda Educational and Research Institute
Howrah, India

Rafael Fernández-Muñoz
Consejo Superior de Investigaciones Científicas
Instituto de Hortofruticultura Subtropical y Mediterránea La Mayora
Universidad de Málaga-Consejo Superior de Investigaciones Científicas
Madrid, Spain

Fernando Gallardo
Dpt. Biología Molecular y Bioquímica
Universidad de Málaga
Málaga, Spain

Antonio Heredia
Dpt. Biología Molecular y Bioquímica
Universidad de Málaga
Málaga, Spain

Akbar Hossain
Division of Soil Science, Bangladesh Wheat, and Maize Research Institute
Dinajpur, Bangladesh

Chayma Ikan
Centre d'Agrobiotechnologie et Bioingénierie, Unité de Recherche labellisée CNRST (Centre AgroBiotech-URL-CNRST-05)
Cadi Ayyad University
Marrakesh, Morocco
and
Faculty of Science Semlalia, Laboratory of Agro-Food, Biotechnologies and Valorization of Plant Bioresources (AGROBIOVAL), Department of Biology
Cadi Ayyad University
Marrakesh, Morocco

Md. Inzamam-Ul-Haque
Faculty of Agriculture
Bangladesh Agricultural University
Mymensingh, Bangladesh

Md. Shafiul Islam
Department of Plant Pathology
China Agricultural University
Beijing, China

Lara Jadhav
Department of Microbiology, Parul Institute of Applied Science
Parul University
Vadodara, India

Yunfei Jiang
Faculty of Agriculture, Department of Plant, Food and Environmental Sciences
Dalhousie University
Nova Scotia, Canada

Faiz Ahmad Joyia
Centre of Agricultural Biochemistry and Biotechnology (CABB)
University of Agriculture
Faisalabad, Pakistan

Najwa Seddiqi Kallali
Phytopathology Unit, Department of Plant Protection and Environment
Ecole Nationale d'Agriculture de Meknès
Meknès, Morocco

Muhammad Sarwar Khan
Centre of Agricultural Biochemistry and Biotechnology (CABB)
University of Agriculture
Faisalabad, Pakistan

M. K. Khokhar
ICAR-National Research Centre for Integrated Pest Management
New Delhi, India

Sandeep Kour
Department of Zoology
Guru Nanak Dev University
Amritsar, India

Deepak Kumar
Department of Zoology
Guru Nanak Dev University
Amritsar, India

Pradeep Kumar
Department of Botany, MMV
Banaras Hindu University
Varanasi, India

Salah-Eddine Laasli
Phytopathology Unit, Department of Plant Protection and Environment
Ecole Nationale d'Agriculture de Meknès
Meknès, Morocco

Rachid Lahlali
Phytopathology Unit, Department of Plant Protection
Ecole Nationale d'Agriculture de Meknès (ENA)
Meknès, Morocco

B. S. Lakshmi
Department of Biotechnology
Anna University
Chennai, India

Amit Layek
Department of Agricultural Entomology
Bidhan Chandra Krishi Viswavidyalaya
Mohanpur, India

Ikram Leghrifi
Phytopathology Unit, Department of Plant Protection
National School of Agriculture
Meknes, Morocco
and
Bioactive Molecules, Sciences and Technologies Faculty, Laboratory of Microbial Biotechnology, Department of Biology, Sidi Mohamed Ben Abdellah University
Fez, Morocco

Dustin E. MacLean
Faculty of Agriculture, Department of Plant, Food and Environmental Sciences
Dalhousie University
Nova Scotia, Canada
and
Perennia Food and Agriculture Corporation
Nova Scotia, Canada

Durga Mahalanobish
Department of Agricultural Entomology
Bidhan Chandra Krishi Viswavidyalaya
Mohanpur, India

Debjyoti Majumder
Malda Krishi Vigyan Kendra
Uttar Banga Krishi Vigyan Kendra
Ratua, India

Swagata Mandal
Bidhan Charnda krishi Viswavidyalaya
Mohanpur, India

Suryakant Manik
ICAR-Indian Institute of Agricultural Biotechnology
Ranchi, India

Rajesh Kumari Manhas
Department of Microbiology
Guru Nanak Dev University
Amritsar, India

Abdelilah Meddich
Centre d'Agrobiotechnologie et Bioingénierie
Unité de Recherche labellisée CNRST (Centre AgroBiotech-URL-CNRST-05)
Cadi Ayyad University
Marrakesh, Morocco
and
Faculty of Science Semlalia, Laboratory of Agro-Food, Biotechnologies and Valorization of Plant Bioresources (AGROBIOVAL), Department of Biology
Cadi Ayyad University
Marrakesh, Morocco

Tatiana Minkina
Academy of Biology and Biotechnology
Southern Federal University
Rostov on Don, Russia

Sunil Kumar Mishra
Department of Pharmaceutical Engineering & Technology, Indian Institute of Technology
Banaras Hindu University
Varanasi, India

Fouad Mokrini
Biotechnology Unit, Regional Center of Agricultural Research
INRA
Rabat, Morocco

Shitosri Mondal
Department of Crop Botany
Faculty of Agriculture
Bangladesh Agricultural University
Mymensingh, Bangladesh

Monisha Mondol
Environmental Science Discipline, Life Science School
Khulna University
Khulna, Bangladesh

Sharah Jabeen Mou
Faculty of Agriculture
Bangladesh Agricultural University
Mymensingh, Bangladesh

Ghulam Mustafa
Centre of Agricultural Biochemistry and Biotechnology (CABB)
University of Agriculture
Faisalabad, Pakistan

Puja Ohri
Department of Zoology
Guru Nanak Dev University
Amritsar, India

Parkirti
Department of Zoology
Guru Nanak Dev University
Amritsar, India

Madhuri Pradhan
Genetic and Plant Breeding
CV Raman Global University
Bhubaneswar, India

Kaushik Pramanik
Department of Agricultural Entomology
Bidhan Chandra Krishi Viswavidyalaya
Mohanpur, India

Mohammed Radi
Phytopathology Unit, Department of Plant Protection
National School of Agriculture
Meknes, Morocco
and
Faculty of Sciences, Laboratory of Environment and Valorization of Microbial and Plant Resources
Moulay Ismail University
Meknes, Morocco

Vishnu D. Rajput
Academy of Biology and Biotechnology
Southern Federal University
Rostov on Don, Russia

Muhammad Khuram Razzaq
Faculty of Agriculture and Veterinary Sciences
Superior University
Lahore, Pakistan

Arpita Rani Roy
Faculty of Agriculture
Bangladesh Agricultural University
Mymensingh, Bangladesh

Debashis Roy
Dhaanya Ganga Krishi Vigyan Kendra
Ramakrishna Mission Vivekananda Educational and Research Institute
Sargachi, India

Tourya Sagouti
Phytopathology Unit, Department of Plant Protection
National School of Agriculture
Meknes, Morocco

Priyanka Saha
ICAR-Indian Agricultural Research Institute
Pusa, New Delhi, India

Gohar Sahakyan
Department of Biochemistry, Microbiology & Biotechnology, Research Institute of Biology
Yerevan State University
Yerevan, Armenia

Naira Sahakyan
Department of Biochemistry, Microbiology & Biotechnology, Research Institute of Biology
Yerevan State University
Yerevan, Armenia

Neha Sharma
ICAR-Indian Agricultural Research Institute
Pusa, New Delhi, India

Roohi Sharma
Department of Zoology
Guru Nanak Dev University
Amritsar, India

Amit Kumar Singh
Department of Pharmaceutical Engineering & Technology, Indian Institute of Technology
Banaras Hindu University
Varanasi, India

Fatima-Ezzahra Soussani
Centre d'Agrobiotechnologie et Bioingénierie, Unité de Recherche labellisée CNRST (Centre AgroBiotech-URL-CNRST-05)
Cadi Ayyad University
Marrakesh, Morocco
and
Faculty of Science Semlalia, Laboratory of Agro-Food, Biotechnologies and Valorization of Plant Bioresources (AGROBIOVAL), Department of Biology
Cadi Ayyad University
Marrakesh, Morocco
and
Faculty of Sciences, Laboratory of Plant, Animal, and Agro-Industry Productions
University Ibn Toufail
Kenitra, Morocco

Akivha Akter Sweety
Faculty of Agriculture
Bangabandhu Sheikh Mujibur Rahman Agricultural University
Gazipur, Bangladesh

Furqan Tayyab
Centre of Agricultural Biochemistry and Biotechnology (CABB)
University of Agriculture
Faisalabad, Pakistan

Kübra Budak Tek
Faculty of Agriculture, Plant Biotechnology Department
Akdeniz University
Akdeniz, Turkey

Mümin Ibrahim Tek
Faculty of Agriculture, Plant Protection Department
Akdeniz University
Akdeniz, Turkey

Kavindra Nath Tiwari
Department of Botany, MMV
Banaras Hindu University
Varanasi, India

Md. Shakil Uddin
Department of Soil, Water, and Ecosystem Sciences
University of Florida
Gainesville, FL, US

K. Varun
Department of Biotechnology
Anna University
Chennai, India

Vikram
Department of Zoology
Guru Nanak Dev University
Amritsar, India

Ian R. Willick
Kentville Research and Development Centre
Agriculture and Agri-Food Canada
Kentville, Canada

1 Plant Cuticle Synthesis and Tolerance to Stress

Juan Carlos Mateos del Amo, Rafael Fernández-Muñoz, Eva Domínguez, Antonio Heredia, and Fernando Gallardo

LIST OF ABBREVIATIONS

ABC	Adenosine triphosphate binding cassettes transporters
CER	ECERIFERUM gene family
CYP	Cytochrome P450 enzymes
DCF	Deficient in cutin ferulate gene product
DEWAX	Decrease wax biosynthesis transcription factor
ECR	Enoyl-CoA reductase
ER	Endoplasmic reticulum
FAE	Fatty acid elongase
GPAT	Glycerol-3-phosphate acyltransferases
HCD	3-Hydroxyacyl-CoA dehydratases
HD-ZIP IV	Homeodomain leucine zipper IV transcription factor
KCR	β-Ketoacyl-CoA reductase
KCS	β-Ketoacyl-CoA synthases
LACSs	Long-chain acyl-coenzyme A synthases
LTPs	Lipid transfer proteins
MAG	Monoacylglycerol
MYB	Myeloblastosis family of transcription factors
VLCFAs	Very-long-chain fatty acids
WDL1	Wilted Dwarf and Lethal 1 gene product
WIN/SHN	Wax inducer/shine family gene products

1.1 INTRODUCTION

The plant cuticle is a polymeric matrix that covers all aerial plant organs and is considered the primary barrier between the plant and the above-ground environment. It acts as an interphase not only preventing water loss but also protecting against potentially harmful UV radiation and pathogens, regulating gas exchange, and providing both mechanical support and controlling the expansion of organs (Domínguez et al., 2011). This complex matrix is deposited on the outer periclinal cell wall of the epidermal cells but also can be located on the anticlinal walls or boarding the stomata (Fernández et al, 2016).

The origin and evolution of the plant cuticle are linked to the colonization of the land by plants. It is widely thought that land plants appeared during the Cambrian Period (ca 470 Ma) when the Streptophyta algae evolved from the marine environment and started the conquest of the land. Fossil evidence of thin cuticle sheets appears in the geological record in the Late Ordovician (~450 Ma), which is considered one of the main features acquired to protect against desiccation. Early cuticles were thin and imperforated to facilitate gas diffusion in an elevated atmospheric CO_2 concentration and a more-refined feature for both gas and water regulations appeared with the evolution of stomata (~420 Ma; Willis and McElwain, 2014).

DOI: 10.1201/9781003543237-1

The cuticle functions described above rely on its hydrophobic nature. Although it was considered in the past as a free-polysaccharide structure and independent from the epidermal cells, today the plant cuticle is viewed as part of the outer cell wall of the epidermal cells, in other words, as a cutinized cell wall (Dominguez et al., 2011; Yeats and Rose, 2013).

With some variation between species, the plant cuticle corresponds to a cutin matrix with waxes, polysaccharides and other minor compounds. The cutin matrix is mainly formed by C_{16} and /or C_{18} hydroxy fatty acids (Philippe et al., 2020). Its composition varies among species as well as among organs and development stages within the same species (Fernández et al., 2016). In some species, an alternative polymer named cutan can be present as an alternative to cutin or in combination with it (Boom et al., 2005).

Waxes are present in the outermost cuticle surface as epicuticular waxes, but they also exist as intracuticular waxes within the cutin matrix. Waxes are composed mainly of alkanes, fatty alcohols, and fatty acids, and they can be found as morphological structures that differ among plant species shaping different wax crystal forms by self-assembly processes (Wu et al., 2018). Cyclic wax constituents (triterpenoids) are mostly found within the intracuticular layer (Buschhaus and Jetter, 2011). Minor compounds such as glycerol and phenolics may be found within the cuticle as well (Yeats and Rose, 2013). Phenolic compounds are secondary metabolites mainly produced by the shikimic acid route and include diverse defense-related compounds such as flavonoids and anthocyanins. Cinnamic acids and flavonoids found in plant cuticles (Cheynier et al., 2013) also contribute to their overall properties including mechanical support and UV protection (González et al., 2022).

The polysaccharide fraction in the plant cuticle is composed of cellulose, hemicellulose, and pectin (Heredia-Guerrero et al., 2016). The similar composition of the polysaccharide moiety of the cuticle and the primary cell walls and the physical interaction of these structures reflect that the cuticle is a modification of the outer epidermal wall (Domínguez et al., 2011). The polysaccharides present in the plant cuticle affect its elastic modulus, a key property in cuticle protective functions (López-Casado et al., 2007). Moreover, the importance of polysaccharides in water sorption and hydrogen bonding in the cuticle has been underlined (Tredenick and Farquhar, 2021).

1.2 CUTICLE SYNTHESIS AND BIOMECHANICS

Despite the origin of the plant cuticle is still viewed as a conundrum, the understanding of its biosynthesis has greatly risen during past decades thanks to major advances in studies conducted in model plants (Li and Chang, 2021). Phylogenetic analyses have suggested that cuticle biosynthesis machinery appeared in the last common ancestor of land plants (Kong et al., 2020) and evolved with gene family expansions also linked to selection pressures in the aero-terrestrial environment during the early stages of land plant evolution (Kong et al., 2020).

The plant cuticle components are synthesized in the epidermal cells and then secreted to the outermost surface (Yeats and Rose, 2013). The metabolic pathways of cuticle components have been revised in detail in several works (Yeats and Rose, 2013; Domínguez et al., 2015; Fich et al., 2016; Bhanot et al., 2021; Liu et al., 2022). Cuticular waxes derive from very-long-chain fatty acids (VLCFAs) with a chain length between C24 and C36 (Bhanot et al., 2021). Within the wax loads alkanes, aldehydes, ketones, esters, and primary and secondary alcohols are the foremost components, but other lipidic metabolites such as terpenoids can also be found (Jetter et al., 2006). During evolution, cuticular wax loads increased in seed plants compared to bryophytes (Kong et al., 2020). The wax assembling process starts with the de novo C_{16}–C_{18} fatty acid synthesis within the plastids (Lewandowska et al., 2020). Long-chain acyl-CoA synthetases (LACSs) associated with the endoplasmic reticulum (ER) activate C_{16}–C_{18} fatty acids with the addition of a CoA moiety. Within the ER, the fatty acid elongase (FAE) complex elongates activated C_{16}–C_{18} fatty acids to VLCFAs and converts them into primary alcohols and wax esters or alkanes, aldehydes, secondary alcohols, and ketones (Xue et al., 2017). Despite the intracellular transport of wax derivates from the ER to the plasma membrane is not fully understood, several hypotheses including transport associated with lipid transfer proteins

(LTPs) or vesicle-mediated transport have been proposed (Bernard and Joubès, 2013). Once the wax components have reached the plasma membrane, the transport to the cuticle matrix is postulated to be mediated by adenosine triphosphate binding cassettes (ABCs) transporters (Lewandowska et al., 2020) and glycosylphosphatidylinositol-anchored LTPs (Kim et al., 2012).

Cutin monomer biosynthesis occurs in plastids. Within the ER, the previously activated fatty acids undergo successive oxidations by enzymes from the cytochrome P450 family to form the cutin precursor monoacylglycerol (Fich et al., 2016). Some authors have highlighted the role of glycerol-3-phosphate acyltransferases in the formation of monoacylglycerol (Fich et al., 2016; Bhanot et al., 2021). Phenolic derivates, i.e., ferulic acid, can be esterified to monoacylglycerols due to the transferase activity of the enzyme encoded by the deficient in cutin ferulate (DCF) gene (Domínguez et al., 2015; Fich et al., 2016). These monomers are exported to the outer epidermal wall, where they are polymerized (Domínguez et al., 2015). Gas chromatography–mass spectrometry analyses have revealed that cuticular amounts of C_{18} and dicarboxylic acids are higher in seed plants than in mosses and ferns (Kong et al., 2020). It has been suggested that these differences could indicate an increased cutin matrix cross-linking in seed plants affecting some of their mechanical properties (Kong et al., 2020; López-Casado et al., 2007).

Most of the research on the genetic regulation of the cuticle synthesis process has been conducted in fleshy fruits (Hen-Avivi et al., 2014). The expression of cuticle biosynthesis core genes is regulated by a common set of transcription factors (Lee and Suh, 2015), including members of the APETALA2, homeodomain leucine zipper IV (HD-ZIP IV), and myeloblastosis (MYB) gene families, among others (Lee and Suh, 2015). A relation between epidermal cell differentiation and cuticle biosynthesis has been highlighted in some members of the MYB family (Lu et al., 2022). Moreover, coexpression analyses in biosynthetic, transporter, and regulatory genes conducted in model plants (i.e., *Arabidopsis thaliana*, *Solanum lycopersicum*, *Zea mays*, and *Oryza sativa*) indicated that cuticle synthesis is highly conserved among angiosperms (Kong et al., 2020).

An exocytosis mechanism was proposed as responsible for cuticle formation in the early stages of development (Heredia-Guerrero et al., 2008). The model is thought to be related to the procuticle onset and consists of the self-assembly of cutin monomers, resulting in electrodense particles named cutinsomes (Stępiński et al., 2020; Segado et al., 2020). It has been proposed that cutinsomes could be transported to the outer epidermal wall through the plasmalemma and the cell wall (Stępiński et al., 2020). Moreover, it has been highlighted that cuticle biosynthesis-related enzymes and cuticle monomers could be transported to the outermost surface attached to cutinsomes (Heredia-Guerrero et al., 2008; Stępiński et al., 2020).

The synthesis of the polysaccharides present in the cuticle is thought to occur in a similar way as described for those found in the primary cell wall. Cellulose, a major component of the cell wall, is synthesized in the plasma membrane and deposited directly on the outer surface (Lerouxel et al., 2006). In contrast, hemicelluloses and pectin are synthesized in the Golgi apparatus and then transported to the cell wall mediated by vesicles (Driouich et al., 2012). Most of the polysaccharide fraction belonging to the epidermal cell is not cutinized (Reynaud et al., 2021). Nevertheless, the potential role of a cutin:xyloglucan-transferase linking the polysaccharide fraction to the cutin matrix, contributing to the cuticle integrity and strengthening the epidermis, has been highlighted (Xin and Fry, 2021).

Finally, phenolic compounds present in plant cuticles are synthesized in the chloroplast by the shikimic acid pathway (Cheynier et al., 2013). They are usually accumulated in the vacuoles of epidermal, subepidermal, and guard cells, mostly as glycosylated forms (Li et al., 2021). To date, it is not clear how phenolics are incorporated into the cuticle. Some phenolics found in cuticles are linked by covalent bonds to the plant cell wall constituents, while others are linked to cutin oligomers or embedded between cuticular waxes as alkyl ester derivatives (Lattanzio et al., 2008).

Due to its composition, the plant cuticle conforms to a unique structure with a relevant resistance to deformation. When exposed to low stress, the cuticle behaves elastically, undergoing deformation to dissipate energy and, therefore, allowing the organs to adapt to normal plant growth. In contrast, cuticle stiffness and resistance to deformation reduce notably possible mechanical damage due to mechanical stress (Domínguez et al., 2011). A number of experiments conducted in tomato fruit (Petracek and Bukovac, 1995; López-Casado et al., 2007) and in other tissues of different species (Bargel et al., 2006) have underlined this viscoelastic and dual behavior of the cuticle. The incidence of agronomic issues such as fruit cracking and tolerance to biotic stress also relies on the mechanical properties of the plant cuticle (Petracek and Bukovac, 1995; Fernández-Muñoz et al., 2022).

Attending to cuticle composition, the cutin matrix is responsible for its extensibility properties, whereas its stiffness is provided mainly by polysaccharides (Domínguez et al., 2011). Both waxes and flavonoids act as matrix fillers, increasing the cuticle rigidity, hence reinforcing the role of polysaccharides (Domínguez et al., 2009; Khanal et al., 2013). Nevertheless, these rheological properties are also affected by environmental factors such as temperature and relative humidity (Khanal and Knoche, 2017).

1.3 CUTICLE AND ABIOTIC STRESS

Although plant cuticle and land plants evolved at 10–100 times the current levels of CO_2 concentration in the atmosphere, the current raising in atmospheric CO_2 levels due to anthropogenic activity may affect the synthesis of the cuticle. Gas emissions and their derived global warming effect produce a set of stressors for plant growth and productivity including increased temperature and UV radiation and reduction in water availability in some areas (Benevenuto et al., 2017; Ashraf et al., 2018; Raza et al., 2019). The plant cuticle, as an interphase between the plant and its environment, is considered a main factor in coping with plant stressors by limiting water loss when stomata are fully closed and providing to some extent isolation from changes in the external temperature (Schuster et al., 2016).

1.3.1 Tolerance to Drought Stress

Cuticle thickness or the total cuticular wax load does not correlate with reduced cuticular water permeability. Extraction of waxes with organic solvents in different species has shown that both intracuticular and epicuticular waxes contribute to water barrier properties of the cuticle, but the effect of the epicuticular wax layer depends on the whole composition of the cuticle in a species-specific manner (Jetter and Riederer, 2016). Similarly, the mechanical extraction of the epicuticular wax layer has highlighted that intracuticular waxes represent the primary transpiration barrier of the leaf (Zeisler-Diehl et al., 2018).

Changes in the expression of genes encoding enzymes of fatty acid synthesis may produce alterations in both cutin and wax accumulation (Yeats and Rose, 2013), Thus, mutants with impaired fatty acid biosynthesis have a less effective cuticle as evapotranspiration barrier (Weng et al., 2010; Rahman et al., 2021). Today, it is also known that changes in gene expression affecting cuticle synthesis are part of the plant response to different stress conditions (Liu et al., 2022). Water deficit increased by more than 60% the total cutin accumulation in *A. thaliana* leaves (Kosma et al., 2009), while other studies have described thicker wax layers in plants submitted to drought stress (Kosma et al., 2009; Xue et al., 2017). Modification in cutin and wax content has also been obtained by alteration of *LACS* gene expression. Overexpression of the *A. thaliana AtLACS2* homolog gene in soybeans resulted in higher levels of cutin but not in waxes (Ayaz et al., 2021). In contrast, increased accumulation of waxes in *Arabidopsis* overexpressing *LACS2* has been reported in response to submergence treatments (Xie et al., 2020). On the contrary, the *lacs2* single mutant and *lacs1 lacs2* double *Arabidopsis* mutant showed higher water loss rates (Weng et al., 2010). This evidence shows a clear role of LACS in wax content and derived cuticle properties. In addition, the spatial deposition of wax affects cuticle water permeability and water use efficiency as described by Park et al. (2010) in *wilted dwarf and lethal 1* (*WDL1*) rice mutants showing altered wax deposition on the leaf surface.

The accumulation of alkanes is enhanced in leaf cuticles after drought stress conditions (Chen et al., 2020). Expression studies of genes encoding relevant enzymes in alkane synthesis have received special attention, including β-ketoacyl-CoA synthases (KCSs), which catalyze the condensation of an acyl-CoA with malonyl-CoA, and enoyl-CoA reductase (ECR) that are involved in the elongation reactions of VLCFAs, and CER family genes, also described as *ECIFERUM* or *WAX2* genes encoding proteins with different activities (Chauhdhary and Panjabi, 2021). Thus, the overexpression of *KCS1*, *KCS2*, and *KCS6* in *Arachis hypogaea*, *Malus domestica*, and *Camellia sinensis*, respectively, resulted in enhanced both VLCFA and epicuticular wax accumulation (Lokesh et al., 2019; Guo et al., 2020; Liu et al., 2021), and the expression of *CERs* correlated positively with increased levels of alkanes in several species as *A. thaliana* (Chen et al., 2003; Bourdenx et al., 2011), cucumber (Wang et al., 2015), and tomato (Vogg et al., 2004; Karaba, 2007). Together, all these evidences outline the role of alkanes in the cuticle contribution to drought stress tolerance. Nevertheless, the integrity of the cutin matrix may have a pivotal role in cuticle relevance under water depletion conditions. Mutants with impaired expression in one or more genes related to cutin monomer synthesis showed substantial alteration in cuticle permeability (Yang et al., 2022a). A barley drought-hypersensitive mutant with a loss-of-function ABCG transporter presented a more brittle cuticle and the presence of lipid droplets in epidermal cells, attributed to untransported cutin monomers (Chen et al., 2011). *A. thaliana atabcg11* and *atabcg12* mutants accumulated cutin monomers due to their reduced transport to the outer epidermal cell wall, resulting in a fragile cuticle (Wu et al., 2022).

Our knowledge of gene networks governing cuticle biosynthesis is still fragmentary; nevertheless, there is increasing insight into the transcription factor families that will help to elucidate the regulation mechanisms of the synthetic core genes. *WAX INDUCER* (*WIN/SHN*) family genes including *SHN1*, *SHN2*, and *SHN3* have been proven to regulate not only the expression of wax biosynthesis genes in *Arabidopsis* (Aharoni et al., 2004), barley (Djemal and Khoudi, 2021), tomato (Al-Abdallat et al., 2014), wheat (Bi et al., 2018), and apple (Zhang et al., 2019b) but also with a clear effect plant tolerance to drought (Liu et al., 2022). Another widely known partner is the *MYB* family of transcription factors that are key regulators of cuticle biosynthesis, and their activity is linked to the modulation of cuticle properties and plant stress responses (Liu et al., 2022). Additionally, zinc-finger transcription factors, such as the *homeodomain leucine zipper IV* (*HD-ZIP IV*) family, upregulate wax biosynthesis genes, and wax transporters, thus increasing drought tolerance (Isaacson et al., 2009; Javelle et al., 2010; Wu et al., 2011; Wang et al., 2018b)

1.3.2 Tolerance to Salinity Stress

Similar to drought, high-salinity conditions inhibit plant growth and development through osmotic and ionic stress (Zhao et al., 2021). Impaired photosynthetic efficiency due to increased water depletion generates an overgrown amount of reactive oxygen species (ROS) damaging the plant (Miller et al., 2010). To avoid this situation, plants undergo different adaptations aiming to improve water use efficiency (Nabi et al., 2019). These adaptations not only include an effect on the growth of the root system and leaf area reduction or stomata closure but also increased cuticle thickness (Yavuz et al., 2023). As a layer regulating water loss, a hydrophobic cuticle helps plants to not desiccate under altered osmotic conditions.

Increased cuticle biosynthesis has been observed in model and crop plants in response to salinity stress (Liu et al., 2022). Flavonoids, which are present in plant cuticles as minor compounds, are well-known antioxidants (Yeats and Rose, 2013). Observations carried out in *Halophila johnsonii* leaves pointed out the potential role of flavonoids as ROS scavengers produced by high salinity (Gavin and Durako, 2012). However, additional research is needed on this topic, as its close relative *Halophila decipiens* does not present intracellular flavonoids in leaves, and a protective sunscreen role for cuticular flavonoids in these species is not clear (Gavin and Durako, 2011, 2012).

Arabidopsis plants subjected to sodium chloride (NaCl) treatment showed an increased amount of C_{16} and C_{18} fatty acids (Kosma et al., 2009). Some authors linked the synthesis of C_{18} species, mainly oleic (18:1), linoleic (18:2), and α-linolenic (18:3) acids, to counteract the effect of ROS production in lipid oxidation under salinity stress (He and Ding, 2020).

A number of genes encoding fatty acid synthesis enzymes and transcription factors are known to be involved in the general plant response to different abiotic stresses affecting also cutin and wax synthesis. Dehydration conditions are induced by drought and salinity through common plant response mechanisms. *Arabidopsis* plants overexpressing two *LACS* genes from apple, *MdLACS2* and *MdLACS4*, showed enhanced tolerance against drought and salinity stress conditions (Zhang et al., 2020a,b). Furthermore, some works have reported an increased expression in other *LACS* genes under salt stress (Zhang et al., 2018; Wang et al., 2023). Similarly, transgenic tobacco plants overexpressing SHN1-type transcription factor from *Hordeum vulgare* exhibited a higher tolerance against salt stress (Djemal and Khoudi, 2021). The overexpression of cotton cyclin-dependent kinases (CDK) in *Arabidopsis* also increased its tolerance to salinity (Magwanga et al., 2018). Promotor regions of *CDK* genes present dehydration-responsive element/C-repeat (DRE/CRT) elements, and it has been postulated that CDK genes could be a direct target of *HvSHN1* as it binds to DRE/CRT elements (Djemal and Khoudi 2021). Moreover, as described as regulators under drought stress, some *MYB* genes are involved in the response to salt stress. Thus, Zhang et al. (2020c) have highlighted that salt tolerance in *Arabidopsis* plants showed to be *AtMYB49*-expression dependent, a transcription factor that activates genes related to cutin and wax biosynthesis.

1.3.3 Tolerance to Extreme Temperature

At high temperatures, the plant cuticle can act as a heat sink since its heat capacity is higher than that of the air. This helps to mitigate the stresses generated by extreme temperatures on the aerial organs buffering internal organ temperature (Benítez et al., 2022). It has been observed that cuticular water permeability drastically increased at temperatures higher than 35°C in non-xerophytic plants (Schuster et al., 2016). These observations carried out in desert plants pointed out the biological role of the cuticle as part of the plant strategies to survive in such extreme habitats.

Water-spender *Citrullus colocynthis* and water-saver *Phoenix dactylifera* both dwell in the North African desert (Bueno et al., 2019). Water-saver plants close their stomata under stress conditions, while water-spenders increase their water uptake from the soil. Observations carried out by Lange (1959) and re-investigated by Bueno et al. (2019) demonstrated that water-saver species present increased leaf wax coverage than water-spender species. In both cases, the main compounds (around 90% of the total leaf wax load) were VLCFAs, while cyclic compounds comprised less than 5% (Bueno et al., 2019). Accordingly, these studies attribute the transpiration barrier properties of cuticles to VLCFAs, while triterpenoid accumulation has been reported to contribute negatively to this feature (Buschhaus and Jetter, 2012; Jetter and Riederer, 2016). However, contrary to this, Schuster et al. (2016) found that cyclic compounds were the main wax component in the desert plant *Rhayza stricta*, while the aliphatic fraction represented only a minor percentage of the total wax load. Similar observations by Busta et al. (2021) in the heat-tolerant cultivar *Sorghum bicolor* emphasized the high presence of triterpenoids in mature leaves. Altogether, these observations may suggest that mechanical strength provided by triterpenoids is required to maintain the efficiency of the water barrier in some species adapted to high temperatures (Schuster et al., 2016; Busta et al., 2021).

High temperature and also cold stress have been proven to enhance the expression of cuticle biosynthesis genes (Lee and Suh, 2015; Wang et al., 2018a; He et al., 2019). Several LACS genes were upregulated after treatment at 4°C in soybean (Wang et al., 2022). Single and double *LACS* knockdown mutants in *Arabidopsis* produced a lower wax load in the stem and an increment in cuticle permeability, water loss rate, and susceptibility to abiotic stresses (Jessen et al., 2015; Zhao et al., 2019). On the contrary, *A. thaliana* wax-overproducing mutant *dewax* showed improved tolerance to freezing temperatures compared to wild type (Rahman et al., 2021). The *cer3–6* mutant line produced fewer alkanes and showed reduced freezing tolerance, highlighting the key role of waxes in thermal stress tolerance (Rahman et al., 2021).

Studies on transcription factors related to cuticle synthesis and temperature stress are limited. Ectopic expression of barley *HvSHN1* in tobacco plants also enhanced heat tolerance, among other stresses (Djemal and Khoudi, 2021).

1.3.4 Tolerance to Ultraviolet Radiation

Ultraviolet (UV) radiation comprises UV-A (315–400 nm), UV-B (280–315 nm), and UV-C (280–200 nm), although only UV-A and UV-B spectrums reach the ground level (Robson et al., 2019). Increasing levels of UV-B arriving in the biosphere have been related to ozone layer depletion (Caldwell et al., 1998). Despite UV-B conforming to a small fraction of the total radiation reaching the Earth's surface, it can promote the formation of ROS, inducing stress in land plants (Robson et al., 2019). The first evidence of the cuticle's role against harmful UV radiation was highlighted by Steinmüller and Tevini (1985, 1986), who found increased alkane accumulation and total wax load in cucumber, bean, and barley under UV-B irradiation. Similar findings have been reported in sorghum (Grant et al., 1995), pea (González et al., 1996), sugar maple (Gordon et al., 1998), and coffee (Bernado et al.., 2021). Altogether, these results suggest that cuticle deposition or accumulation of certain VLCFAs has a role in decreasing UV-B penetration in aerial tissues (Robson et al., 2019).

Other molecules present in the plant cuticle, such as flavonoids and phenolics, have been identified as UV-screening compounds (Liu et al., 2022). Benítez et al. (2022) found a striking relation between cuticle thickness and these screening molecules alongside development stages in tomato fruit. The UV-A blockage capacity was positively correlated to cuticle thickness in immature fruits, while phenolic content remained low. However, the UV-A blockage increased during fruit ripening but not the cuticle thickness. This could be attributed to the accumulation of flavonoids in the tomato fruit cuticle (i.e., chalconaringenin) during ripening (España et al., 2014). Regarding UV-B radiation, phenolic compounds such as hydroxycinnamic acids present in plant cuticles have been shown to absorb it by a photostable and self-regenerating physicochemical mechanism as it was highlighted by González Moreno et al. (2022) in isolated cuticles from eight different plant species. Nevertheless, the genes underlying this stress resilience capacity have yet to be described.

1.3.5 Micropollutans and Cuticle Biosynthesis

Micropollutants are anthropogenic chemicals that occur in the environment at concentrations above natural background levels (Yang et al., 2022b). They include a wide variety of substances such as heavy metals, other inorganic chemicals, hormones, and pharmaceuticals that are released into the natural environment at low concentrations by natural processes and as byproducts of industrial activities (Abbasi et al., 2022). Despite their low emission levels, their low degradation rate produces accumulation in organs and tissues that can have grave consequences for plant, animal, and human life (Yang et al., 2022b).

Heavy metals penetrate the plant through roots and leaves (Shadid et al., 2017). While essential metals (Co, Cu, Fe, Mn, Mo, Ni, Se, and Zn) are needed for the proper functioning of enzymes, the presence of other metals such as Pb, Cd, and Hg are harmful to plants. The plant cuticle, as a physical barrier, regulates the intake of these external agents (Yeast and Rose, 2013). The adsorption of heavy metals on the leaf surface is determined by pollutants' atmospheric concentration and the features of the exposed leaf surface, i.e., cuticle composition. It has been observed that the load and morphology of epicuticular waxes are related to the retention levels of pollutants in leaves (Paull et al., 2019; Muhammad et al., 2020). Absorption of micropollutants to the inner tissues of leaves mainly occurs by passive diffusion through cuticle and stomata (Molina and Segura, 2021).

Plants able to survive in high polluted environments are suitable organisms to study tolerance to heavy metals. Observations realized on two species of *Avicennia* genus showed altered wax load, and the presence of heavy metals in the epicuticular wax layer of plant leaves exposed to a high level of heavy metals (Victório et al., 2020; Raju and Ramakrishna, 2021). Changes in cuticle composition in response to heavy metal presence have been reported in several works. Wax coverage of barley leaves increased by 50% in plants exposed to Cd (Hollenbach et al., 1997), and effects on suberin and wax biosynthesis have also been observed in ramie plants (*Boehmeria nivea*) under Cd treatment (Liu et al., 2015). In contrast, the application of fungicides in two coffee cultivars reduced the total wax load and altered the micromorphology of epicuticular waxes (Victório et al., 2020). Other authors have also reported reduced cuticle thickness after treatments with Cu (Marques et al., 2018). This effect could be because the excessive accumulation of heavy metals in the epicuticular waxes could impair their deposition and morphology (Victório et al., 2020).

Some authors highlighted similarities between salt and heavy metal stress tolerance, as both share common physiological mechanisms (Lucini and Bernardo, 2015; Victório et al., 2020). It has been observed that the overexpression of durum wheat *SHN1* transcription factor in tobacco plants conferred increased tolerance to Cd, Cu, and Zn (Djemal and Khoudi, 2022). Nonetheless, experiments conducted on cuticle-defective mutants exposed to heavy metal presence are lacking.

Li et al. (2020) found that cuticles of *Photinia serrulate* leaves showed a high retention capacity for Sr, but neither abaxial nor adaxial surface cuticles played a negative role in its uptake. The leaves of urban trees that presented a thinner wax layer accumulated a greater amount of particulate matter than those covered with a thicker layer (Corada et al., 2020). In addition, Zarinkamar et al. (2013) observed that Pb enhanced epicuticular wax deposition in *Hypericum perforatum* leaves. Altogether, the presence of micropollutants may lead to enhanced lipid synthesis pathways, increasing the epicuticular wax load. However, additional research is needed to elucidate how genes underlying the observed cuticle responses are regulated by micropollutants.

1.4 CUTICLE ROLE IN BIOTIC STRESS TOLERANCE

Plants cope with a wide range of pathogens, such as fungi, bacteria, viruses, and herbivore insects (Wang et al., 2020). These organisms may cause important losses to crop yield during plant growth and post-harvest (Singh et al., 2023). Global warming aggravates this problem, as warmer winters promote the activity of phytopathogens. Climate change may alter the distribution of pathogen vectors, mainly insects, leading them to colonize new territories and produce new environmental impacts on affected ecosystems.

The infection process starts in many pathosystems when the pathogen binds to the plant surface. Hence, epicuticular waxes could be considered the first point of interaction between the pathogen and the plant (Bhanot et al., 2021; Wang et al., 2020). Cuticle has been demonstrated to be crucial during pathogen attachment and recognition, not only due to its chemical composition but also for its role in the activation of defense responses (Bhanot et al., 2021; Lim et al., 2020). Pattern-triggered immunity occurs in the epidermal cell when surface receptors recognize in the apoplast certain molecules, named microbe-, pathogen-, or damage-associated molecular patterns (Ziv et al., 2018). Immunity is elicited when an effector protein secreted by the pathogen into the host cell is successfully recognized by the host, leading to a robust defense response. Recent studies have postulated crosstalk between pattern and effector-triggered immunity (Ngou et al., 2021; Nguyen et al., 2021; Yuan et al., 2021).

Pathogens usually penetrate the plant through natural openings such as stomata and hydathodes from leaves or fruit lenticels (Ziv et al., 2018). In addition, some organisms (i.e., fungi) use hydrolytic enzymes, mainly cutinases, to impair cuticle structure, hence contributing to hyphal penetration (Arya et al., 2021). The release of cutin monomers after cutinase function facilitates fungi infection, but it also may elicit plant defense responses (L'Haridon et al., 2011). Kolattukudy et al. (1995) already highlighted the role of cuticular lipids as messenger molecules during phytopathogen infection. Therefore, changes in cuticle thickness and permeability can modify to some extent the immune response against phytopathogens and plant susceptibility to infections (Bhanot et al., 2021).

Plants respond to challenges by bacterial, fungal, and viral pathogens in different ways depending on the biotrophic or necrotrophic lifestyle of the pathogens.

1.4.1 Fungi Infection

Fungi are among the main phytopathogens threatening crops and thereby compromising food security worldwide (Fones et al., 2020; Różewicz et al., 2021). Fungi have adopted different strategies to get through the cuticle, such as the excretion of cutinases, the formation of specialized infection organs (i.e., the appressorium), or in situ induction of spore germination (Arya et al., 2021). Pre-infection process always takes place at the cuticle level; hence, epicuticular wax composition is crucial to establish the plant–fungi interaction (Lewandowska et al., 2020; Arya et al., 2021).

There is actual evidence that fungi recognize epicuticular waxes to initiate the infection processes. Aldehydes, alkanes, and primary alcohols have been extensively described in the literature as signaling molecules in triggering the initiation of the infection (Hansjakob et al., 2011; Kong and Chang, 2018; Wang et al., 2020). Thus, spores of *Blumeria graminis* failed to germinate in plants where the cuticle layer was previously removed or presented impaired wax synthesis (Weidenbach et al., 2014; Kong et al., 2020; Arya et al., 2021). In line with these findings, overexpression of a decrease wax biosynthesis (DEWAX) transcription factor enhanced resistance to *Botryris cinerea* in *Arabidopsis* plants. Further analysis showed that DEWAX bound to promoter regions of different plant defense-related genes (Ju et al., 2017; Wang et al., 2020).

Conversely, other authors have highlighted a positive crosstalk between genes associated with cuticle synthesis and defense response. Overexpression of *MdMYB30*, which positively regulates *MdKCS1* expression, led to increased resistance to *Botryosphaeria dothidea* in apple (*Malus x domestica*) fruits (Zhang et al., 2019a). On the contrary, the knockdown of a cation co-transporter in *A. thaliana* produced the downregulation of cuticle biosynthesis genes and enhanced susceptibility to *Pseudomonas syringae* (Han et al., 2020).

As indicated above, cutinase activity has been reported to either stimulate fungi activity or facilitate plant defense mechanisms (Ziv et al., 2018). According to some authors, the accumulation of cutin monomers increases cutinase activity by facilitating the penetration of fungal hyphae into plant tissues (Woloshuk and Kolattukudy, 1986; Hansjakob et al., 2011; Ziv et al., 2018). Other authors have highlighted that a more permeable cuticle could enhance plant defense responses (L'Haridon et al., 2011; Aragón et al., 2017; Bhanot et al., 2021), Thus, numerous studies conducted on *A. thaliana* showed that cuticle-defective mutants with increased cuticle permeability presented enhanced resistance against *Botrytis cinerea*, but they were also susceptible to bacteria *Pseudomonas syringae* (Suh and Go, 2014; Survila et al., 2016; Ju et al., 2017; Ziv et al., 2018). Additional studies on cuticle role in fungal infection and a detailed list of genes involved in fungi–cuticle interaction have been described elsewhere (Ziv et al., 2018).

1.4.2 Bacteria Infection

As in the case of fungi, the plant cuticle plays a key role in regulating the interactions between plants and pathogenic and non-pathogenic bacteria (Schlechter et al., 2019). Recent research on bacteria–plant interactions at the cuticular level has focused on the intimate talk between plants and microorganisms at the leaf surface (Xia et al., 2019).

Cuticle permeation and wettability are key factors in ensuring epiphytes fitness; thus, bacteria synthesize biosurfactants to facilitate the permeation of compounds through the cuticle (Schlechter et al., 2019; Zeisler-Diehl et al., 2020). Along with the nutrient leaching from the apoplast, it has been theorized that epiphytic bacteria could degrade wax and/or cutin as energy sources, even though experimental evidence is lacking (Zeisler-Diehl et al., 2020). Surfactant synthesis may be necessary to ensure a suitable environment to degrade plant cuticle and to further increase its permeability (Pfeilmeier et al., 2016; Aragón et al., 2017). Most of the experimental evidence has been conducted on the interaction between *Pseudomonas syringae* and cuticle-defective plant mutants. Arabidopsis *lacs2* plants showed increased susceptibility to *P. syringae* pv *tomato* strain DC3000 (Pst DC3000) (Tang et al., 2007; Wang et al., 2020). MYB transcription factors family are thought to play a role in plant–bacteria interaction. Raffaele et al. (2008) observed that *MYB30* may control hypersensitive response via VLCFA biosynthesis. Ectopic expression of apple *MdMYB30* in *Arabidopsis* plants increased its resistance against the aforementioned Pst DC3000 strain. The same results were observed on *AtMYB96* overexpression (Seo et al., 2010; Zhang et al., 2019a).

Arabidopsis plants with an acyl-carrier protein mutation (*acp4*) failed to activate the systemic acquired resistance (SAR) in distal tissues. These plants presented a reduced $C_{16:0}$ pool, hence showing the importance of proper cuticle formation in SAR on pathogen recognition (Xia et al., 2019). Cuticle-impaired mutants with increased cuticle permeability accumulated more salicylic acid in epicuticular waxes rather than in apoplast, hence compromising SAR in distal tissues (Lim et al., 2020).

Favaro et al. (2020) inoculated a resistant and susceptible mandarin cultivar with *Xanthomonas* by spraying. Only relevant differences were found in the cuticle of the resistant cultivar after inoculation, which presented a higher wax load and a thicker cuticle during the early stages of leaf development. Moreover, the mechanical removal of epicuticular waxes makes the resistant cultivar susceptible to *Xanthomonas* infection. These data indicate that the early accumulation of waxes in leaves plays a pivotal role in the resistance to *Xanthomonas* (Favaro et al., 2020). In another study in tomato, Buxdorf et al. (2014) found that the overexpression of *SlSHN3*, a transcription factor related to wax biosynthesis, enhanced the resistance against *Xanthomonas campestris*. Despite these findings, the accumulation of epicuticular wax compounds can be also harnessed by bacterial and fungal pathogens, as it was highlighted by Bourdenx et al. (2011) in *Arabidopsis*.

1.4.3 Pests and Virus Infection

Insects are the most diverse animal taxon and have proceeded through a long coevolutionary process alongside plants (Whitney and Federle, 2013). Despite being necessary pollinators and seed dispersers for plants, they cause important yield losses to some of the most widely crops, such as wheat or soybean (Agrawal et al., 2011; Wang et al., 2020). Insect herbivory, from tissue chewers to phloem suckers, produces mechanical damage and also acts as a vector of viral diseases threatening crops worldwide (Rai et al., 2022).

Epicuticular waxes are the first components affecting plant–insect interaction. As the first layer upon plant–insect interaction, epicuticular waxes play a key role in regulating its ecology (Barbero, 2016). The slippery nature and morphology of the epicuticular wax crystals usually impair insect attachment and oviposition into the plant aerial organs (Salerno et al., 2020). Blenn et al. (2012) pointed out that oviposition by *Pieris brassicae* butterflies in *Arabidopsis* plants altered the epicuticular wax composition. The oxidation of wax compounds yields volatile molecules (i.e., nonanal) that could be related to defense purposes in *Zea mays* and poplar, according to Chen et al.'s (2023) findings. Some wax components could be exploited by insects to choose the most suitable place for their nutritional and reproductive purposes (Li and Ishikawa, 2006; Wang et al., 2020). These findings highlight the role of cuticular lipids as signaling chemicals in plant–insect interaction.

Cuticle-deficient mutants have shed light on the ecological interactions between insects and plants. A study using *Nicotiana glauca* mutants has shown how VLCFAs differently affect the growth of both chewing species and phloem feeders and the oviposition by *Bemisia tabaci*. In contrast, total wax load or alkane content did not affect chewer caterpillars, besides the fact that fatty alcohols ranging from C_{24} to C_{28} correlated negatively with insect growth (Negin et al., 2024). Similarly, aphid feeding was increased in mutants with impaired cuticle composition. While changes in wax composition did not affect insect behavior, epicuticular wax morphology was found to affect insect oviposition rate (Negin et al., 2024). Transcriptomic studies revealed the upregulation of cuticle biosynthesis genes in tea plants under tea green leafhopper attack (Zhao et al., 2021). Expression of a *CER1* homolog was markedly enhanced after insect attack, increasing C_{29} alkanes formation, while *Nicotiana glauca cer1* mutants with an extremely altered wax and cutin composition were more susceptible to aphid growth on plants (Negin et al., 2023, 2024).

Studies on viruses affecting cuticle features are limited. Most of the plant viruses are unable to overcome the cuticle so they rely on biological vectors for its transmission, including nematodes, fungi, bacteria, and insects (Xavier and Whitfield, 2023). Mechanical damage caused by the insect stylets constitutes the route of entry of viruses into the plant tissues. Maxwell et al. (2016) examined whether the viral transmission method could modify the epicuticular wax synthesis rate. While potato virus Y (PVY) and cucumber mosaic virus infection enhanced the transcript levels of *CER6* (a gene related to FA condensation) in *Nicotiana tabacum*, turnip vein clearing virus, which does not need a vector for its transmission, increased *CER9* expression, a negative regulator of wax synthesis, in *A. thaliana* (Maxwell et al., 2016). Virus-mediated alteration of the cuticle may modify its reflection index, acting as a visual cue to further attacks by other insects (Barbero et al., 2016; Maxwell et al., 2016).

1.5 CUTICLE IMPORTANCE IN BREEDING STRATEGIES FOR STRESS TOLERANCE

The growing atmospheric CO_2 levels aggravate the challenges plants must face, not only by increasing the stress conditions associated with a higher global temperature but also because of the rise in pathogen activity. The molecular mechanisms underlying the responses to both biotic and abiotic stress conditions have been a main topic in plant research (Pérez Méndez et al., 2022; Farooqi et al., 2022; Georgieva and Vassileva, 2023). The cuticle's role against stressor agents has been reviewed in the past years (Wang et al., 2020; Arya et al., 2021; Fernández-Muñoz et al., 2022; Liu et al., 2022).

Figure 1.1 depicts the main elements in cuticle formation with emphasis on the different stresses as described in the previous sections. As it also happens with other metabolic and signal transduction pathways, many enzymes and relevant transcription factors described here are also involved in the synthesis of other cell components and possibly related to other cell processes that are yet unknown. Nevertheless, due to its unique properties, cuticle modification may be considered for breeding strategy programs. In the past years, there have been a number of studies that provide insights into the identification of quantitative trait loci associated with cuticle deposition (Barraj et al., 2021; Yang et al., 2022a).

Within cuticle components, epicuticular waxes have received special attention regarding stress tolerance mechanisms. In a general sense, the accumulation of epicuticular waxes (i.e., alkanes) correlates positively with increased stress tolerance, which emphasizes that it may be a trait to be considered in breeding programs. Nevertheless, several fungi can explode the accumulation of alkanes to initiate pre-infection processes (Aragón et al., 2017). In turn, insects, the main vectors for plant–virus transmission, are affected by the slippery nature of epicuticular waxes and their morphology (Salerno et al., 2020). Further exploration of wax mixtures in different species may help in developing breeding programs to enhance crop tolerance against a wide range of stressors.

Cuticle integrity is important to avoid microcracking and the entry of pathogens into plant tissues along plant development but also in harvested fruit, which can be relevant to address the obtention of new cultivars producing high-quality fruits. During storage conditions, the cuticle can absorb water and reduce its mechanical resistance (Dominguez et al., 2011; Fernández-Muñoz et al., 2022). This property is mainly related to the cuticle polysaccharide fraction and may lead to fruit skin disorders such as fruit cracking, which can severely impair fruit shelf-life (Domínguez et al., 2011; Fernández-Muñoz et al., 2022). A study of the rheological properties of cuticles in tomato ripe fruits has been conducted (Benítez et al., 2022), but an investigation in other crops is needed to further assess cuticle mechanic protection role during storage.

The use of cuticle-deficient mutants has driven recent advances in studying the role of cuticles in stress tolerance. However, we are still far from fully understanding cutin matrix deposition and assembly processes, the gene expression patterns of transcription factors governing the biosynthesis of cuticles, or the complex crosstalk between cuticle components and organisms. Answering these questions may help researchers to develop robust breeding programs to improve plant stress tolerance and the gaining of knowledge to apply to other biotechnological approaches.

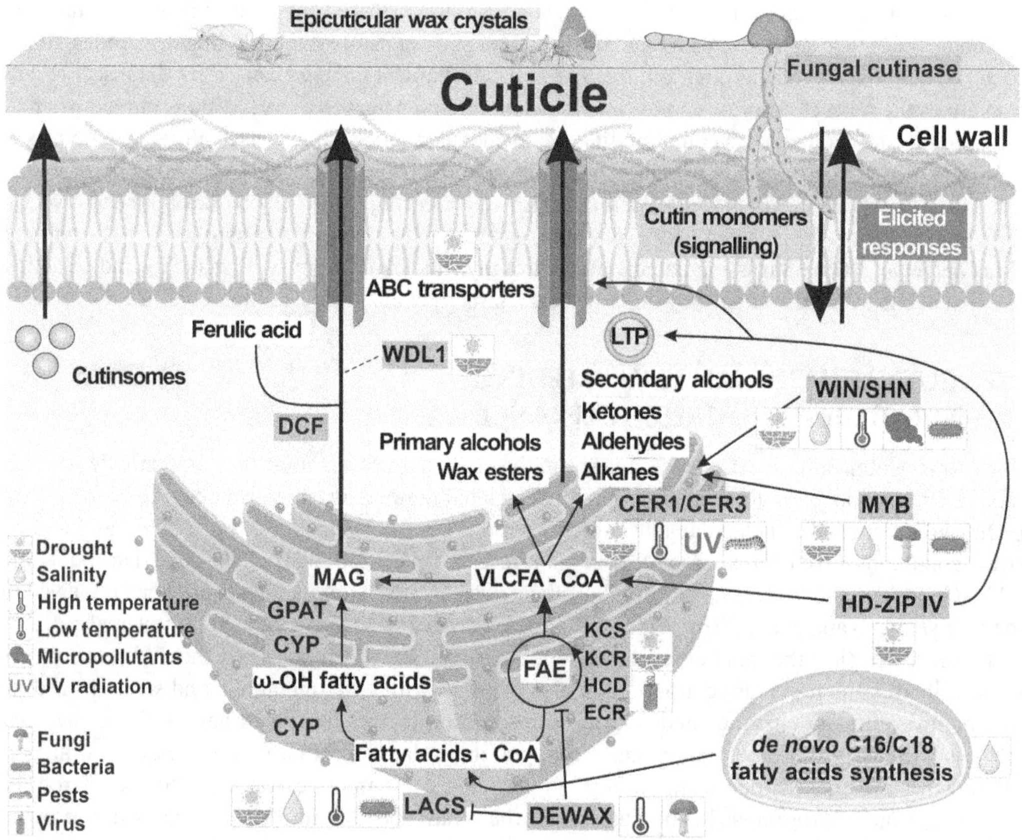

FIGURE 1.1 Overview of the effects of abiotic and abiotic stress on the biosynthesis and transport of cuticle components. See the text for a more detailed description of the effect of different stressors on gene expression and component content. ABC, adenosine triphosphate binding cassettes transporters; CER, ECERIFERUM gene family; CYP, cytochrome P450 enzymes; DCF, deficient in cutin ferulate gene product; DEWAX, decrease wax biosynthesis transcription factor; ECR, enoyl-coa reductase; FAE, fatty acid elongase; GPAT, glycerol-3-phosphate acyltransferases; HCD, 3-hydroxyacyl-CoA dehydratases; HD-ZIP IV, homeodomain leucine zipper IV transcription factor; KCR, β-ketoacyl-CoA reductase; KCS, β-ketoacyl-CoA synthases; LACS, long-chain acyl-coenzyme A synthases; LTPs, lipid transfer proteins; MAG, monoacylglycerol; MYB, myeloblastosis family of transcription factors; WDL1, wilted dwarf and lethal 1 gene product; WIN/SHN, wax inducer/shine family gene products. Figure created with BioRender (www.biorender.com).

ACKNOWLEDGMENTS

This work was supported by a grant from Agencia Estatal de Investigación, Ministerio de Ciencia e Innovation, Spain, co-financed by the European Regional Development Fund (project PID2021–126604OB).

REFERENCES

Abbasi, N. A., Shahid, S. U., Majid, M., & Tahir, A. (2022). Ecotoxicological risk assessment of environmental micropollutants. *Environmental Micropollutants, Advances in Pollution Research*, 331–337. https://doi.org/10.1016/B978-0-323-90555-8.00004-0.

Agrawal, A. A. (2011). Current trends in the evolutionary ecology of plant defence. *Functional Ecology*, *25*, 420–432. https://doi.org/10.1111/J.1365–2435.2010.01796.X.

Aharoni, A., Dixit, S., Jetter, R., Thoenes, E., van Arkel, G., & Pereira, A. (2004). The SHINE clade of AP2 domain transcription factors activates wax biosynthesis alters cuticle properties and confers drought tolerance when overexpressed in *Arabidopsis*. *The Plant Cell*, *16*, 2463–2480. https://doi.org/10.1105/TPC.104.022897.

Al-Abdallat, A. M., Al-Debei, H. S., Ayad, J. Y., & Hasan, S. (2014). Over-expression of SlSHN1 gene improves drought tolerance by increasing cuticular wax accumulation in tomato. *International Journal of Molecular Sciences*, *15*, 19499–19515. https://doi.org/10.3390/IJMS151119499.

Aragón, W., Reina-Pinto, J. J., & Serrano, M. (2017). The intimate talk between plants and microorganisms at the leaf surface. *Journal of Experimental Botany*, *68*, 5339–5350. https://doi.org/10.1093/jxb/erx327

Arya, G. C., Sarkar, S., Manasherova, E., Aharoni, A., & Cohen, H. (2021). The plant cuticle: An ancient guardian barrier set against long-standing rivals. *Frontiers in Plant Science*, *12*, 663165. https://doi.org/10.3389/FPLS.2021.663165.

Ashraf, M. A., Akbar, A., Askari, S. H., Iqbal, M., Rasheed, R., & Hussain, I. (2018). Recent advances in abiotic stress tolerance of plants through chemical priming: An overview. In A. Rakshit & H. Singh (Eds.), *Advances in seed priming* (pp. 51–79). Springer. https://doi.org/10.1007/978–981-13–0032–5_4.

Ayaz, A., Saqib, S., Huang, H., Zaman, W., Lü, S., & Zhao, H. (2021). Genome-wide comparative analysis of long-chain acyl-CoA synthetases (LACSs) gene family: A focus on identification evolution and expression profiling related to lipid synthesis. *Plant Physiology and Biochemistry*, *161*, 1–11. https://doi.org/10.1016/J.PLAPHY.2021.01.042

Barbero, F. (2016). Cuticular Lipids as a cross-talk among ants plants and butterflies. *International Journal of Molecular Sciences*, *17*, 1966. https://doi.org/10.3390/IJMS17121966.

Bargel, H., Koch, K., Cerman, Z., & Neinhuis, C. (2006). Structure-function relationships of the plant cuticle and cuticular waxes: A smart material? *Functional Plant*, *30*, 893–910. https://doi.org/10.1071/FP06139.

Barraj Barraj, R., Segado, P., Moreno-González, R., Heredia, A., Fernández-Muñoz, R., & Domínguez, E. (2021). Genome-wide QTL analysis of tomato fruit cuticle deposition and composition. *Horticulture Research*, *8*, 113. https://doi.org/10.1038/S41438-021-00548-5.

Benevenuto, R. F., Agapito-Tenfen, S. Z., Vilperte, V., Wikmark, O. G., van Rensburg, P. J., & Nodari, R. O. (2017). Molecular responses of genetically modified maize to abiotic stresses as determined through proteomic and metabolomic analyses. *PloS One*, *12*, e0173069. https://doi.org/10.1371/JOURNAL.PONE.0173069.

Benítez J. J., González Moreno, A., Guzmán-Puyol, S., Heredia-Guerrero, J. A., Heredia, A., & Domínguez E. (2022). The response of tomato fruit cuticle membranes against heat and light. *Frontiers in Plant Science*, *12*, 807723. https://doi.org/10.3389/FPLS.2021.807723.

Bernado, W. P., Rakocevic, M., Santos, A. R., Ruas, K. F., Baroni, D. F., Abraham, A. C., Pireda, S., Oliveira, D. S., da Cunha, M., Ramalho, J. C., Campostrini, E., & Rodrigues, W. P. (2021). Biomass and leaf acclimations to ultraviolet solar radiation in juvenile plants of *Coffea arabica* and *C. Canephora*. *Plants*, *10*, 640. https://doi.org/10.3390/PLANTS10040640.

Bernard, A., Joubès J. (2013). Arabidopsis cuticular waxes: Advances in synthesis export and regulation. *Progress in Lipid Research*, *52*, 110–129. https://doi.org/10.1016/J.PLIPRES.2012.10.002.

Bhanot, V., Fadanavis, S. V., & Panwar, J. (2021). Revisiting the architecture biosynthesis and functional aspects of the plant cuticle: There is more scope. *Environmental and Experimental Botany*, *183*, 104364. https://doi.org/10.1016/J.ENVEXPBOT.2020.104364.

Bi, H., Shi, J., Kovalchuk, N., Luang, S., Bazanova, N., Chirkova, L., Zhang, D., Shavrukov, Y., Stepanenko, A., Tricker, P., Langridge, P., Hrmova, M., Lopato, S., & Borisjuk, N. (2018). Overexpression of the TaSHN1 transcription factor in bread wheat leads to leaf surface modifications improved drought tolerance and no yield penalty under controlled growth conditions. *Plant Cell & Environment*, *41*, 2549–2566. https://doi.org/10.1111/PCE.13339.

Blenn, B., Bandoly, M., Küffner A., Otte, T., Geiselhardt, S., Fatouros, N. E., & Hilker, M. (2012). Insect egg deposition induces indirect defense and epicuticular wax changes in *Arabidopsis thaliana*. *Journal of Chemical Ecology*, *38*, 882–892. https://doi.org/10.1007/S10886-012-0132–8.

Boom, A., Sinninge Damsté, J. S., & de Leeuw, J. W. (2005). Cutan a common aliphatic biopolymer in cuticles of drought-adapted plants. *Organic Geochemistry*, *36*, 595–601. https://doi.org/10.1016/J.ORGGEOCHEM.2004.10.017.

Bourdenx, B., Bernard, A., Domergue, F., Pascal, S., Léger A., Roby, D., Pervent, M., Vile, D., Haslam, R. P., Napier, J. A., Lessire, R., & Joubès, J. (2011). Overexpression of *Arabidopsis* ECERIFERUM1 promotes wax very-long-chain alkane biosynthesis and influences plant response to biotic and abiotic stresses. *Plant Physiology*, *156*, 29–45. https://doi.org/10.1104/PP.111.172320.

Bueno, A., Alfarhan, A., Arand, K., Burghardt, M., Deininger, A. C., Hedrich, R., Leide, J., Seufert, P., Staiger, S., & Riederer, M. (2019). Effects of temperature on the cuticular transpiration barrier of two desert plants with water-spender and water-saver strategies. *Journal of Experimental Botany*, *70*, 1613–1625. https://doi.org/10.1093/jxb/erz018.

Buschhaus, C., & Jetter, R. (2011). Composition differences between epicuticular and intracuticular wax substructures: How do plants seal their epidermal surfaces? *Journal of Experimental Botany*, *62*, 841–853. https://doi.org/10.1093/JXB/ERQ366.

Buschhaus, C., & Jetter, R. (2012). Composition and physiological function of the wax layers coating *Arabidopsis* leaves: β-amyrin negatively affects the intracuticular water barrier. *Plant Physiology*, *160*, 1120–1129. https://doi.org/10.1104/PP.112.198473.

Busta, L., Schmitz, E., Kosma, D. K., Schnable, J. C., & Cahoon, E. B. (2021). A co-opted steroid synthesis gene maintained in sorghum but not maize is associated with a divergence in leaf wax chemistry. *Proceedings of the National Academy of Sciences of the United States of America*, *118*, e2022982118. https://doi.org/10.1073/PNAS.2022982118.

Buxdorf, K., Rubinsky, G., Barda, O., Burdman, S., Aharoni, A., & Levy, M. (2014). The transcription factor SlSHINE3 modulates defense responses in tomato plants. *Plant Molecular*, *84*, 37–47. https://doi.org/10.1007/S11103-013-0117-1.

Caldwell, M. M., Björn L. O., Bornman, J. F., Flint, S. D., Kulandaivelu, G., Teramura, A. H., & Tevini, M. (1998). Effects of increased solar ultraviolet radiation on terrestrial ecosystems. *Journal of Photochemistry and Photobiology B: Biology*, *46*, 40–52. https://doi.org/10.1016/S1011–1344(98)00184-5.

Chaudhary,, R. G., & Panjabi, P. (2021). Origin and diversification of *ECERIFERUM1 (CER1)* and *ECERIFERUM3 (CER3)* genes in land plants and phylogenetic evidence that the ancestral *CER1/3* gene resulted from the fusion of pre-existing domains. *Molecular Phylogenetics and Evolution*, *159*, 107101. https://doi.org/10.1016/j.ympev.2021.107101.

Chen, G., Komatsuda, T., Ma, J. F., Li, C., Yamaji, N., & Nevo, E. (2011). A functional cutin matrix is required for plant protection against water loss. *Plant Signaling & Behavior*, *6*, 1297–1299. https://doi.org/10.4161/PSB.6.9.17507.

Chen, J. Y., Kuruparan, A., Zamani-Babgohari, M., & Gonzales-Vigil, E. (2023). Dynamic changes to the plant cuticle include the production of volatile cuticular wax–derived compounds. *Proceedings of the National Academy of Sciences of the United States of America*, *120*, e2307012120. https://doi.org/10.1073/PNAS.2307012120.

Chen, M., Zhu, X., Zhang, Y., Du, Z., Chen, X., Kong, X., Sun, W., & Chen, C. (2020). Drought stress modify cuticle of tender tea leaf and mature leaf for transpiration barrier enhancement through common and distinct modes. *Scientific Reports*, *10*, 6696. https://doi.org/10.1038/s41598-020-63683-4.

Chen, X., Goodwin, S. M., Boroff, V. L., Liu, X., & Jenks, M. A. (2003). Cloning and characterization of the WAX2 gene of *Arabidopsis* involved in cuticle membrane and wax production. *The Plant*, *15*, 1170–1185. https://doi.org/10.1105/TPC.010926.

Cheynier, V., Comte, G., Davies, K. M., Lattanzio, V., & Martens, S. (2013). Plant phenolics: Recent advances on their biosynthesis genetics and ecophysiology. *Plant Physiology and Biochemistry*, *72*, 1–20. https://doi.org/10.1016/J.PLAPHY.2013.05.009.

Clark, J. W., Harris, B. J., Hetherington, A. J., Hurtado-Castano, N., Brench, R. A., Casson, S., Williams, T. A., Gray, J. E., & Hetherington, A. M. (2022). The origin and evolution of stomata. *Current Biology*, *32*, R539–R553. https://doi.org/10.1016/J.CUB.2022.04.040.

Corada, K., Woodward, H., Alaraj, H., Collins, C. M., & de Nazelle, A. (2021). A systematic review of the leaf traits considered to contribute to removal of airborne particulate matter pollution in urban areas. *Environmental Pollution*, *269*, 116104. https://doi.org/10.1016/J.ENVPOL.2020.116104.

Djemal, R., & Khoudi, H. (2021). The barley SHN1-type transcription factor HvSHN1 imparts heat drought and salt tolerances in transgenic tobacco. *Plant Physiology and Biochemistry*, *164*, 44–53. https://doi.org/10.1016/J.PLAPHY.2021.04.018.

Djemal, R., & Khoudi, H. (2022). The ethylene-responsive transcription factor of durum wheat TdSHN1 confers cadmium copper and zinc tolerance to yeast and transgenic tobacco plants. *Protoplasma*, *259*, 19–31. https://doi.org/10.1007/S00709-021-01635-Z.

Domínguez E., España L., López-Casado, G., Cuartero, J., & Heredia, A. (2009). Biomechanics of isolated tomato (*Solanum lycopersicum*) fruit cuticles during ripening: The role of flavonoids. *Functional Plant*, *36*, 613–620. https://doi.org/10.1071/FP09039.

Domínguez E., Heredia-Guerrero, J. A., & Heredia, A. (2011). The biophysical design of plant cuticles: An overview. *New Phytologist*, *189*, 938–949. https://doi.org/10.1111/J.1469–8137.2010.03553.X.

Domínguez E., Heredia-Guerrero, J. A., & Heredia, A. (2015). Plant cutin genesis: Unanswered questions. *Trends in Plant Science*, *20*, 551–558. https://doi.org/10.1016/J.TPLANTS.2015.05.009.

Driouich, A., Follet-Gueye, M. L., Bernard, S., Kousar, S., Chevalier, L., Vicré-Gibouin, M., & Lerouxel, O. (2012). Golgi-mediated synthesis and secretion of matrix polysaccharides of the primary cell wall of higher plants. *Frontiers in Plant Science*, *3*, 23540. https://doi.org/10.3389/FPLS.2012.00079.

España L., Heredia-Guerrero, J. A., Segado, P., Benítez J. J., Heredia, A., & Domínguez E. (2014). Biomechanical properties of the tomato (*Solanum lycopersicum*) fruit cuticle during development are modulated by changes in the relative amounts of its components. *New Phytologist*, *202*, 790–802. https://doi.org/10.1111/NPH.12727.

Farooqi, M. Q. U., Nawaz, G., Wani, S. H., Choudhary, J. R., Rana, M., Sah, R. P., Afzal, M., Zahra, Z., Ganie, S. A., Razzaq, A., Reyes, V. P., Mahmoud, E. A., Elansary, H. O., & El-Abedin, T. K. Z., & Siddique, K. H. M. (2022). Recent developments in multi-omics and breeding strategies for abiotic stress tolerance in maize (*Zea mays* L.). *Frontiers in Plant Science*, *13*, 965878. https://doi.org/10.3389/FPLS.2022.965878.

Favaro, M. A., Molina, M. C., Roeschlin, R. A., Gadea, J., Gariglio, N., & Marano, M. R. (2020). Different responses in mandarin cultivars uncover a role of cuticular waxes in the resistance to citrus canker. *Phytopathology*, *110*, 1791–1801. https://doi.org/10.1094/PHYTO-02–20–0053-R.

Fernández V., Guzmán-Delgado, P., Graça J., Santos, S., & Gil, L. (2016). Cuticle structure in relation to chemical composition: Re-assessing the prevailing model. *Frontiers in Plant Science*, *7*, 180208. https://doi.org/10.3389/FPLS.2016.00427.

Fernández-Muñoz R., Heredia, A., & Domínguez E. (2022). The role of cuticle in fruit shelf-life. *Current Opinion in Biotechnology*, *78*, 102802. https://doi.org/10.1016/J.COPBIO.2022.102802.

Fich, E. A., Segerson, N. A., & Rose, J. K. C. (2016). The plant polyester cutin: Biosynthesis structure and biological roles. *Annual Review of Plant Biology*, *67*, 207–233. https://doi.org/10.1146/Annurev-Arplant-043015–111929.

Fones, H. N., Bebber, D. P., Chaloner, T. M., Kay, W. T., Steinberg, G., & Gurr, S. J. (2020). Threats to global food security from emerging fungal and oomycete crop pathogens. *Nature Food*, *1*, 332–342. https://doi.org/10.1038/s43016-020-0075-0.

Gavin, N. M., & Durako, M. J. (2011). Localization and antioxidant capacity of flavonoids from intertidal and subtidal *Halophila johnsonii* and *Halophila decipiens*. *Aquatic Botany*, *95*, 242–247. https://doi.org/10.1016/J.AQUABOT.2011.07.004.

Gavin, N. M., & Durako, M. J. (2012). Localization and antioxidant capacity of flavonoids in *Halophila johnsonii* in response to experimental light and salinity variation. *Journal of Experimental Marine Biology and Ecology*, *416–417*, 32–40. https://doi.org/10.1016/J.JEMBE.2012.02.006.

Georgieva, M., & Vassileva, V. (2023). Stress management in plants: Examining provisional and unique dose-dependent responses. *International Journal of Molecular Sciences*, *24*, 5105. https://doi.org/10.3390/IJMS24065105.

González Moreno, A., de Cózar, A., Prieto, P., Domínguez, E., & Heredia, A. (2022). Radiationless mechanism of UV deactivation by cuticle phenolics in plants. *Nature Communications*, *13*, 1–11. https://doi.org/10.1038/s41467-022-29460-9.

Gonzalez, R., Paul, N. D., Percy, K., Ambrose, M., McLaughlin, C. K., Barnes, J. D., Areses, M., & Wellburn, A. R. (1996). Responses to ultraviolet-B radiation 280–315 nm of pea (*Pisum sativum*) lines differing in leaf surface wax. *Physiologia Plantarum*, *98*, 852–860. https://doi.org/10.1111/J.1399–3054.1996.TB06695.X.

Gordon, D. C., Percy, K. E., & Riding, R. T. (1998). Effect of enhanced UV-B radiation on adaxial leaf surface micromorphology and epicuticular wax biosynthesis of sugar maple. *Chemosphere*, *36*, 853–858. https://doi.org/10.1016/S0045–6535(97)10137-0.

Grant, R. H., Jenks, M. A., Rich, P. J., Peters, P. J., & Ashworth, E. N. (1995). Scattering of ultraviolet and photosynthetically active radiation by *Sorghum bicolor*: Influence of epicuticular wax. *Agricultural and Forest Meteorology*, *75*, 263–281. https://doi.org/10.1016/0168–1923(94)02210-B.

Guo, W., Wu, Q., Yang, L., Hu, W., Liu, D., & Liu, Y. (2020). Ectopic expression of cskcs6 from navel orange promotes the production of very-long-chain fatty acids (VLCFAs) and increases the abiotic stress tolerance of *Arabidopsis thaliana*. *Frontiers in Plant Science*, *11*, 564656. https://doi.org/10.3389/FPLS.2020.564656.

Han, B., Jiang, Y., Cui, G., Mi, J., Rob Roelfsema, M. G., Mouille, G., Sechet, J., Al-Babili, S., Aranda, M., & Hirt, H. (2020). Cation-chloride co-transporter 1 (CCC1) mediates plant resistance against *Pseudomonas syringae*. *Plant Physiology*, *182*, 1052–1065. https://doi.org/10.1104/PP.19.01279.

Hansjakob, A., Riederer, M., & Hildebrandt, U. (2011). Wax matters: Absence of very-long-chain aldehydes from the leaf cuticular wax of the glossy11 mutant of maize compromises the prepenetration processes of *Blumeria graminis*. *Plant Pathology*, *60*, 1151–1161. https://doi.org/10.1111/J.1365–3059.2011.02467.X.

He, J., Tang, S., Yang, D., Chen, Y., Ling, L., Zou, Y., Zhou, M., & Xu, X. (2019). Chemical and transcriptomic analysis of cuticle lipids under cold stress in *Thellungiella salsuginea*. *International Journal of Molecular Sciences*, *20*, 451. https://doi.org/10.3390/IJMS20184519.

He, M., & Ding, N. Z. (2020). Plant unsaturated fatty acids: Multiple roles in stress response. *Frontiers in Plant Science*, *11*, 562785. https://doi.org/10.3389/FPLS.2020.562785.

Hen-Avivi, S., Lashbrooke, J., Costa, F., & Aharoni, A. (2014). Scratching the surface: Genetic regulation of cuticle assembly in fleshy fruit. *Journal of Experimental Botany*, *65*, 4653–4664. https://doi.org/10.1093/JXB/ERU225.

Heredia-Guerrero, J. A., Benítez, J. J., & Heredia, A. (2008). Self-assembled polyhydroxy fatty acids vesicles: A mechanism for plant cutin synthesis. *BioEssays*, *30*, 273–277. https://doi.org/10.1002/BIES.20716.

Heredia-Guerrero, J. A., Benítez, J.J., Domínguez, E., Bayer, I. S., Cingolani, R., Athanassiou, A., & Heredia, A. (2016). Infrared spectroscopy as a tool to study plant cuticles. *Spectroscopy Europe*, *28*, 11–13.

Hollenbach, B., Schreiber, L., Hartung, W., & Dietz, K. J. (1997). Cadmium leads to stimulated expression of the lipid transfer protein genes in barley: Implications for the involvement of lipid transfer proteins in wax assembly. *Planta*, *203*, 9–19. doi:10.1007/s00050159.

Isaacson, T., Kosma, D. K., Matas, A. J., Buda,. G. J., He, Y., Yu, B., Pravitasari, A., Batteas, J. D., & Stark, R. E., Jenks, M. A., & Rose, J. K. C. (2009). Cutin deficiency in the tomato fruit cuticle consistently affects resistance to microbial infection and biomechanical properties but not transpirational water loss. *The Plant*, *60*, 363–377. https://doi.org/10.1111/J.1365-313X.2009.03969.X.

Javelle, M., Vernoud, V., Depège-Fargeix, N., Arnould, C., Oursel, D., Domergue, F., Sarda, X., & Rogowsky, P. M. (2010). Overexpression of the epidermis-specific homeodomain-leucine zipper IV transcription factor Outer Cell Layer1 in maize identifies target genes involved in lipid metabolism and cuticle biosynthesis. *Plant Physiology*, *154*, 273–286. https://doi.org/10.1104/PP.109.150540.

Jessen, D., Roth, C., Wiermer, M., & Fulda, M. (2015). Two activities of long-chain acyl-coenzyme a synthetase are involved in lipid trafficking between the endoplasmic reticulum and the plastid in *Arabidopsis*. *Plant Physiology*, *167*, 351–366. https://doi.org/10.1104/PP.114.250365.

Jetter, R., Kunst, L., & Samuels, A. L. (2006). Composition of plant cuticular waxes. In: M. Riederer & C. Müller (Eds.), *Biology of the plant cuticle* (pp. 145–181). Blackwell. https://doi.org/10.1002/9780470988718.ch4.

Jetter, R., & Riederer, M. (2016). Localization of the transpiration barrier in the epi- and intracuticular waxes of eight plant species: Water transport resistances are associated with fatty acyl rather than alicyclic components. *Plant Physiology*, *170*, 921–934. https://doi.org/10.1104/PP.15.01699.

Ju, S., Go, Y. S., Choi, H. J., Park, J. M., & Suh, M. C. (2017). DEWAX transcription factor is involved in resistance to *Botrytis cinerea* in *Arabidopsis thaliana* and *Camelina sativa*. *Frontiers in Plant Science*, *8*, 1210. https://doi.org/10.3389/FPLS.2017.01210.

Karaba, A. (2007). Improvement of Water Use Efficiency in Rice and Tomato using Arabidopsis Wax Biosynthetic Genes and Transcription Factors. PhD Thesis, Wageningen University.

Khanal, B. P., & Knoche, M. (2017). Mechanical properties of cuticles and their primary determinants. *Journal of Experimental Botany*, *68*, 5351–5367. https://doi.org/10.1093/JXB/ERX265.

Khanal, B. P., Grimm, E., Finger, S., Blume, A., & Knoche, M. (2013). Intracuticular wax fixes and restricts strain in leaf and fruit cuticles. *New Phytologist*, *200*, 134–143. https://doi.org/10.1111/NPH.12355.

Kim, H., Lee, S. B., Kim, H. J., Min, M. K., Hwang, I., & Suh, M. C. (2012). Characterization of glycosylphosphatidylinositol-anchored lipid transfer protein 2 (LTPG2) and overlapping function between LTPG/LTPG1 and LTPG2 in cuticular wax export or accumulation in *Arabidopsis thaliana*. *Plant & Cell Physiology*, *53*, 1391–1403. https://doi.org/10.1093/PCP/PCS083.

Kong, L., & Chang, C. (2018). Suppression of wheat TaCDK8/TaWIN1 interaction negatively affects germination of *Blumeria graminis* f.sp. *tritici* by interfering with very-long-chain aldehyde biosynthesis. *Plant Molecular*, *96*, 165–178. https://doi.org/10.1007/S11103-017-0687-4.

Kong, L., Zhi, P., Liu, J., Li, H., Zhang, X., Xu, J., Zhou, J., Wang, X., & Chang, C. (2020). Epigenetic activation of enoyl-coa reductase by an acetyltransferase complex triggers wheat wax biosynthesis. *Plant Physiology*, *183*, 1250–1267. https://doi.org/10.1104/PP.20.00603.

Kosma, D. K., Bourdenx, B., Bernard, A., Parsons, E. P., Lü S., Joubès J., & Jenks, M. A. (2009). The impact of water deficiency on leaf cuticle lipids of *Arabidopsis*. *Plant Physiology*, *151*, 1918–1929. https://doi.org/10.1104/PP.109.141911.

L'Haridon, F., Besson-Bard, A., Binda, M., Serrano, M., Abou-Mansour, E., Balet, F., Schoonbeek, H. J., Hess, S., Mir, R., Léon J., Lamotte, O., & Métraux, J. P. (2011). A permeable cuticle is associated with the release of reactive oxygen species and induction of innate immunity. *PLoS Pathogens*, *7*, 1002148. https://doi.org/10.1371/JOURNAL.PPAT.1002148.

Lattanzio, V., Kroon, P. A., Quideau, S., & Treutter, D. (2008). Plant phenolics: Secondary metabolites with diverse functions. In F. Daayf & V. Lattanzio (Eds.), *Recent advances in polyphenol research* (Vol. 1, pp. 1–35). https://doi.org/10.1002/9781444302400.CH1.

Lee, S. B., & Suh, M. C. (2015). Advances in the understanding of cuticular waxes in *Arabidopsis thaliana* and crop species. *Plant Cell*, *34*, 557–572. https://doi.org/10.1007/S00299-015-1772-2.

Lerouxel, O., Cavalier, D. M., Liepman, A. H., & Keegstra, K. (2006). Biosynthesis of plant cell wall polysaccharides - a complex process. *Current Opinion in Plant Biology*, *9*, 621–630. https://doi.org/10.1016/J.PBI.2006.09.009.

Lewandowska, M., Keyl, A., & Feussner, I. (2020). Wax biosynthesis in response to danger: Its regulation upon abiotic and biotic stress. *New Phytologist*, *227*, 698–713. https://doi.org/10.1111/NPH.16571.

Li, G., & Ishikawa, Y. (2006). Leaf epicuticular wax chemicals of the Japanese knotweed Fallopia japonica as oviposition stimulants for *Ostrinia latipennis*. *Journal of Chemical Ecology*, *32*, 595–604. https://doi.org/10.1007/S10886-005-9022–7.

Li, H., & Chang, C. (2021). Evolutionary insight of plant cuticle biosynthesis in bryophytes. *Plant Signaling & Behavior*, *16*, 1943921. https://doi.org/10.1080/15592324.2021.1943921.

Li, J., Halitschke, R., Li, D., Paetz, C., Su, H., Heiling, S., Xu, S., & Baldwin, I. T. (2021). Controlled hydroxylations of diterpenoids allow for plant chemical defense without autotoxicity. *Science*, *371*, 255–260. https://doi.org/10.1126/science.abe4713.

Li, Y., Luo, X., Bai, X., Lv, W., & Liao, Y. (2020). Adsorption of strontium onto adaxial and abaxial cuticle of *Photinia serrulata* leaf. *International Journal of Environmental Research and Public Health*, *17*, 1061. https://doi.org/10.3390/IJERPH17031061

Lim, G. H., Liu, H., Yu, K., Liu, R., Shine, M. B., Fernández J., Burch-Smith, T., Mobley, J. K., McLetchie, N., Kachroo, A., & Kachroo, P. (2020). The plant cuticle regulates apoplastic transport of salicylic acid during systemic acquired resistance. *Science Advances*, *6*, eaaz0478. https://doi.org/10.1126/SCIADV.AAZ0478.

Liu, L., Wang, X., & Chang, C. (2022). Toward a smart skin: Harnessing cuticle biosynthesis for crop adaptation to drought salinity temperature and ultraviolet stress. *Frontiers in Plant Science*, *13*, 961829. https://doi.org/10.3389/FPLS.2022.961829.

Liu, S., Tong, M., Zhao, L., Li, X., & Kunst, L. (2021). The ARRE Ring-type e3 ubiquitin ligase negatively regulates cuticular wax biosynthesis in *Arabidopsis thaliana* by controlling ECERIFERUM1 and ECERIFERUM3 protein levels. *Frontiers in Plant Science*, *12*, 752309. https://doi.org/10.3389/FPLS.2021.752309.

Liu, T., Zhu, S., Tang, Q., & Tang, S. (2015). Genome-wide transcriptomic profiling of ramie (*Boehmeria nivea* L. Gaud) in response to cadmium stress. *Gene*, *558*, 131–137. https://doi.org/10.1016/J.GENE.2014.12.057.

Lokesh, U., Venkatesh, B., Kiranmai, K., Nareshkumar, A., Amarnathareddy, V., Rao, G. L., Johnson, A. M. A., Pandurangaiah, M., & Sudhakar, C. (2019). Overexpression of ß-ketoacyl co-a synthase1 gene improves tolerance of drought susceptible groundnut (*Arachis hypogaea* L. cultivar K-6 by increased leaf epicuticular wax accumulation. *Frontiers in Plant Science*, *9*, 416357. https://doi.org/10.3389/FPLS.2018.01869.

López-Casado, G., Matas, A. J., Domínguez E., Cuartero, J., & Heredia, A. (2007). Biomechanics of isolated tomato (*Solanum lycopersicum* L.) fruit cuticles: The role of the cutin matrix and polysaccharides. *Journal of Experimental Botany*, *58*, 3875–3883. https://doi.org/10.1093/JXB/ERM233.

Lu, H. C., Lam S. H., Zhang, D., Hsiao, Y. Y., Li, B. J., Niu, S. C., Li, C. Y., Lan, S. Tsai, W. C., & Liu, Z. J. (2022). R2R3-MYB genes coordinate conicall cell development and cuticular was biosynthesis in Phalaenopsis Aphrodite. *Plant Physiology*, *188*, 318–331. https://doi.org/10.1093/plphys/kiab422

Lucini, L., & Bernardo, L. (2015). Comparison of proteome response to saline and zinc stress in lettuce. *Frontiers in Plant Science*, *6*, 240. https://doi.org/10.3389/FPLS.2015.00240.

Magwanga, R. O., Lu, P., Kirungu, J. N., Cai, X., Zhou, Z., Wang, X., Diouf, L., Xu, Y., Hou, Y., Hu, Y., Dong, Q., Wang, K., & Liu, F. (2018). Whole genome analysis of cyclin dependent kinase (CDK) gene family in cotton and functional evaluation of the role of CDKF4 gene in drought and salt stress tolerance in plants. *International Journal of Molecular Sciences*, *19*, 2625. https://doi.org/10.3390/IJMS19092625.

Marques, D. M., Veroneze Júnior V., da Silva AB,.Mantovani, J. R., Magalhaes, P. C., & Correa de Souza, T. (2018). Copper toxicity on photosynthetic responses and root morphology of *Hymenaea courbaril* L. (Caesalpinioideae). *Water Air Soil*, *229*, 138. https://doi.org/10.1007/s11270-018-3769-2.

Maxwell, D. J., Partridge, J. C., Roberts, N. W., Boonham, N., & Foster, G. D. (2016). The effects of plant virus infection on polarization reflection from leaves. *PLoS ONE*, *11*, e0152836. https://doi.org/10.1371/JOURNAL.PONE.0152836.

Miller, G., Suzuki, N., Ciftci-Yilmaz, S., & Mittler, R. (2010). Reactive oxygen species homeostasis and signalling during drought and salinity stresses. *Plant Cell & Environment*, *33*, 453–467. https://doi.org/10.1111/J.1365–3040.2009.02041.X.

Molina, L., & Segura, A. (2021). Biochemical and metabolic plant responses toward polycyclic aromatic hydrocarbons and heavy metals present in atmospheric pollution. *Plants*, *10*, 2305. https://doi.org/10.3390/PLANTS10112305.

Muhammad, S., Wuyts, K., & Samson, R. (2020). Immobilized atmospheric particulate matter on leaves of 96 urban plant species. *Environmental Science and Pollution Research*, *27*, 36920–36938. https://doi.org/10.1007/S11356-020-09246-6.

Nabi, R. B. S., Tayade, R., Hussain, A., Kulkarni, K. P., Imran, Q. M., Mun, B. G., & Yun, B. W. (2019). Nitric oxide regulates plant responses to drought salinity and heavy metal stress. *Environmental and Experimental Botany*, *161*, 120–133. https://doi.org/10.1016/J.ENVEXPBOT.2019.02.003.

Negin, B., Hen-Avivi, S., Almekias-Siegl, E., Shachar, L., Jander, G., & Aharoni, A. (2023). Tree tobacco (*Nicotiana glauca*) cuticular wax composition is essential for leaf retention during drought facilitating a speedy recovery following rewatering. *New Phytologist*, *237*, 1574–1589. https://doi.org/10.1111/NPH.18615.

Negin, B., Shachar, L., Meir, S., Ramirez, C. C., Rami Horowitz, A., Jander, G., & Aharoni, A. (2024). Fatty alcohols a minor component of the tree tobacco surface wax are associated with defence against caterpillar herbivory. *Plant Cell & Environment*, *47*, 664–681. https://doi.org/10.1111/PCE.14752.

Ngou, B. P. M., Ahn, H. K., Ding, P., & Jones, J. D. G. (2021). Mutual potentiation of plant immunity by cell-surface and intracellular receptors. *Nature*, *592*, 110–115. https://doi.org/10.1038/s41586-021-03315-7.

Nguyen, Q. M., Iswanto, A. B. B., Son, G. H., & Kim, S. H. (2021). Recent advances in effector-triggered immunity in plants: New pieces in the puzzle create a different paradigm. *International Journal of Molecular Sciences*, *22*, 4709. https://doi.org/10.3390/IJMS22094709.

Park, J. J., Jin, P., Yoon, J., Yang, J., Jeong, H. J., Ranathunge, K., Schreiber, L., Franke, R., Lee, I. J., & An, G. (2010). Mutation in wilted dwarf and lethal 1 (WDL1) causes abnormal cuticle formation and rapid water loss in rice. *Plant Molecular*, *74*, 91–103. https://doi.org/10.1007/S11103-010-9656-X.

Paull, N. J. Krix, D., Irga, P. J., & Torpy, F. R. (2020). Airborne particulate matter accumulation on common green wall plants. *International Journal of Phytoremediation*, *22*, 594–606. https://doi.org/10.1080/15226514.2019.1696744.

Pérez-Méndez N., Miguel-Rojas, C., Jimenez-Berni, J. A., Gomez-Candon, D., Pérez-De-luque A., Fereres, E., Catala-Forner, M., Villegas, D., & Sillero, J. C. (2021). Plant breeding and management strategies to minimize the impact of water scarcity and biotic stress in cereal crops under mediterranean conditions. *Agronomy*, *12*, 75. https://doi.org/10.3390/AGRONOMY12010075.

Petracek, P. D., & Bukovac, M. J. (1995). Rheological properties of enzymatically isolated tomato fruit cuticle. *Plant Physiology*, *109*, 675–679. https://doi.org/10.1104/PP.109.2.675.

Pfeilmeier, S., Caly, D. L., & Malone, J. G. (2016). Bacterial pathogenesis of plants: Future challenges from a microbial perspective: Challenges in bacterial molecular plant pathology. *Molecular Plant*, *17*, 1298–1313. https://doi.org/10.1111/MPP.12427.

Philippe, G., Geneix, N., Petit, J., Guillon, F., Sandt, C., Rothan, C., Lahaye, M., Marion, D., & Bakan, B. (2020). Assembly of tomato fruit cuticles: A cross-talk between the cutin polyester and cell wall polysaccharides. *New Phytologist*, *226*, 809–822. https://doi.org/10.1111/NPH.16402.

Raffaele, S., Vailleau, F., Léger A., Joubès J., Miersch, O., Huard, C., Blée E., Mongrand, S., Domergue, F., & Roby, D. (2008). A MYB transcription factor regulates very-long-chain fatty acid biosynthesis for activation of the hypersensitive cell death response in *Arabidopsis*. *The Plant*, *20*, 752–767. https://doi.org/10.1105/TPC.107.054858.

Rahman, T., Shao, M., Pahari, S., Venglat, P., Soolanayakanahally, R., Qiu, X., Rahman, A., & Tanino, K. (2021). Dissecting the roles of cuticular wax in plant resistance to shoot dehydration and low-temperature stress in *Arabidopsis*. *International Journal of Molecular Sciences*, *22*, 1554. https://doi.org/10.3390/IJMS22041554.

Rai, A., Sivalingam, P. N., & Senthil-Kumar, M. (2022). A spotlight on non-host resistance to plant viruses. *PeerJournal*, *10*, 12996. https://doi.org/10.7717/PEERJ.12996

Raju, K. A., & Ramakrishna, C. (2021). The effects of heavy metals on the anatomical structures of *Avicennia marina* (Forssk. *Vierh.*). *Revista Brasileira de Botanica*, *44*, 439–447. https://doi.org/10.1007/S40415-021-00698-9.

Raza, A., Razzaq, A., Mehmood, S. S., Zou, X., Zhang, X., Lv, Y., & Xu, J. (2019). Impact of climate change on crops adaptation and strategies to tackle its outcome: A review. *Plants*, *8*, 34. https://doi.org/10.3390/PLANTS8020034.

Reynaud, N., Petit, J., Bres, C., Lahaye, M., Rothan, C., Marion, D., & Bakan, B. (2021). The complex architecture of plant cuticles and its relation to multiple biological functions. *Frontiers in Plant Science*, *12*, 782773. https://doi.org/10.3389/FPLS.2021.782773.

Robson, T. M., Aphalo, P. J., Banaś A. K., Barnes, P. W., Brelsford, C. C., Jenkins, G. I., Kotilainen, T. K., Łabuz, J., Martínez-Abaigar, J., Morales, L. O., Neugart, S., Pieristè M., Rai, N., & Vandenbussche, F., Jansen, M. A. K. (2020). A perspective on ecologically relevant plant-UV research and its practical application. *Photochemical & Photobiological Sciences*, *18*, 970–988. https://doi.org/10.1039/C8PP00526E.

Różewicz, M., Wyzińska, M., & Grabiński J. (2021). The most important fungal diseases of cereals—problems and possible solutions. *Agronomy*, *11*, 714. https://doi.org/10.3390/AGRONOMY11040714.

Ryan, C. A., Lamb, C. J., Jagendorf, A. T., Kolattukudy, P. E., Kolaitukudy, P. E., Rogers, L. M., Li, D., Hwang, C. S., & Flaishman, M. A. (1995). Surface signaling in pathogenesis. *Proceedings of the National Academy of Sciences of the United States of America*, *92*, 4080–4087. https://doi.org/10.1073/PNAS.92.10.4080.

Salerno, G., Rebora, M., Piersanti, S., Gorb, E., & Gorb, S. (2020). Mechanical ecology of fruit-insect interaction in the adult mediterranean fruit fly *Ceratitis capitata* (Diptera: Tephritidae). *Zoology*, 139 125748. https://doi.org/10.1016/J.ZOOL.*2020*.125748.

Schlechter, R. O., Miebach, M., & Remus-Emsermann, M. N. P. (2019). Driving factors of epiphytic bacterial communities: A review. *Journal of Advanced Research*, *19*, 57–65. https://doi.org/10.1016/J.JARE.2019.03.003.

Schuster, A. C., Burghardt, M., Alfarhan, A., Bueno, A., Hedrich, R., Leide, J., Thomas, J., & Riederer, M. (2016). Effectiveness of cuticular transpiration barriers in a desert plant at controlling water loss at high temperatures. *AoB Plants*, *8*, plw027. https://doi.org/10.1093/AOBPLA/PLW027.

Segado, P., Heredia-Guerrero, J. A., Heredia, A., & Domínguez E. (2020). Cutinsomes and CUTIN SYNTHASE1 function sequentially in tomato fruit cutin deposition. *Plant Physiology*, *183*, 1622–1637. https://doi.org/10.1104/PP.20.00516.

Seo, P. J., Lee, S. B., Suh, M. C., Park, M. J., & Park, C. M. (2011). The MYB96 transcription factor regulates cuticular wax biosynthesis under drought conditions in *Arabidopsis*. *The Plant*, *23*, 1138–1152. https://doi.org/10.1105/TPC.111.083485.

Shahid, M., Dumat, C., Khalid, S., Schreck, E., Xiong, T., & Niazi, N. K. (2017). Foliar heavy metal uptake toxicity and detoxification in plants: A comparison of foliar and root metal uptake. *Journal of Hazardous Materials*, *325*, 36–58. https://doi.org/10.1016/J.JHAZMAT.2016.11.063.

Singh, B. K., Delgado-Baquerizo, M., Egidi, E., Guirado, E., Leach, J. E., Liu, H., & Trivedi, P. (2023). Climate change impacts on plant pathogens food security and paths forward. *Nature Reviews*, *21*, 640–656. https://doi.org/10.1038/s41579-023-00900-7.

Sivakumar, N., Sathishkumar, R., Selvakumar, G., Shyamkumar, R., & Arjunekumar, K. (2020). Phyllospheric microbiomes: Diversity ecological significance and biotechnological applications. *Plant Microbiomes for Sustainable Agriculture*, *25*, 113–172. https://doi.org/10.1007/978-3-030-38453-1_5.

Steinmüller D., & Tevini, M. (1985). Action of ultraviolet radiation (UV-B) upon cuticular waxes in some crop plants. *Planta*, *164*, 557–564. https://doi.org/10.1007/BF00395975.

Steinmüller D., & Tevini, M. (1986). UV-B-induced effects upon cuticular waxes of cucumber bean and barley leaves. In R. C. Worrest & M. M. Caldwell (Eds.), *Stratospheric ozone reduction, solar ultraviolet radiation and plant life*. NATO ASI Series (Vol. 8, pp. 261–269). Springer. https://doi.org/10.1007/978-3-642-70090-3_20.

Stępiński, D., Kwiatkowska, M., Wojtczak, A., Polit, J. T., Domínguez E., Heredia, A., & Popłońska K. (2020). The role of cutinsomes in plant cuticle formation. *Cells*, *9*, 1778. https://doi.org/10.3390/CELLS9081778.

Suh, M. C., & Go, Y. S. (2014). DEWAX-mediated transcriptional repression of cuticular wax biosynthesis in *Arabidopsis thaliana*. *Plant Signaling & Behavior*, *9*, 20463. https://doi.org/10.4161/PSB.29463.

Survila, M., Davidsson, P. R., Pennanen, V., Kariola, T., Broberg, M., Sipari, N., Heino, P., & Palva, E., T. (2016). Peroxidase-generated apoplastic ros impair cuticle integrity and contribute to DAMP-elicited defenses. *Frontiers in Plant Science*, *7*, 1945. https://doi.org/10.3389/FPLS.2016.01945.

Tang, D., Simonich, M. T., & Innes, R. W. (2007). Mutations in LACS2 a long-chain acyl-coenzyme A synthetase enhance susceptibility to avirulent *Pseudomonas syringae* but confer resistance to *Botrytis cinerea* in *Arabidopsis*. *Plant Physiology*, *144*, 1093–1103. https://doi.org/10.1104/PP.106.094318.

Tredenick, E. C., & Farquhar, G. D. (2021). Dynamics of moisture diffusion and adsorption in plant cuticles including the role of cellulose. *Nature Communications*, *12*, 5042. https://doi.org/10.1038/S41467-021-25225-Y.

Victório, C. P., dos Santos, M. S., de Mello, M. C., Bento, JPSP,, da Costa Souza, M., Simas, N. K., & do Carmo de Oliveira Arruda, R. (2020). The presence of heavy metals in *Avicennia schaueriana* Stapf & Leechman ex Moldenke leaf and epicuticular wax from different mangroves around Sepetiba Bay Rio de Janeiro Brazil. *Environmental Science and Pollution Research*, *27*, 23714–23729. https://doi.org/10.1007/S11356-020-08606-6.

Vogg, G., Fischer, S., Leide, J., Emmanuel, E., Jetter, R., Levy, A. A., & Riederer, M. (2004). Tomato fruit cuticular waxes and their effects on transpiration barrier properties: Functional characterization of a mutant deficient in a very-long-chain fatty acid β-ketoacyl-CoA synthase. *Journal of Experimental Botany*, *55*, 1401–1410. https://doi.org/10.1093/JXB/ERH149.

Wang, H., Ni, X., & Harris-Shultz, K. (2019). Molecular evolution of the plant ECERIFERUM1 and ECERIFERUM3 genes involved in aliphatic hydrocarbon production. *Computational Biology and Chemistry*, *80*, 1–9. https://doi.org/10.1016/J.COMPBIOLCHEM.2019.02.009.

Wang, J., Li, X., Zhao, X., Na, C., Liu, H., Miao, H., Zhou, J., Xiao, J., Zhao, X., & Han, Y. (2022). Genome-wide identification and characterization of the abiotic-stress-responsive lacs gene family in soybean *(Glycine max)*. *Agronomy*, *12*, 1496. https://doi.org/10.3390/agronomy12071496

Wang, W., Zhang, Y., Xu, C., Ren, J., Liu, X., Black, K., Gai, X., Wang, Q., & Ren, H. (2015). Cucumber ECERIFERUM1 (CsCER1) which influences the cuticle properties and drought tolerance of cucumber plays a key role in VLC alkanes biosynthesis. *Plant Molecular*, *87*, 219–233. https://doi.org/10.1007/S11103-014-0271-0.

Wang, X., C., Tian. X., Zhang, H., Li, H., Zhang, S., Li, H., & Zhu, J. (2023). Genome-wide analysis of the maize LACS gene family and functional characterization of the ZmLACS9 responses to heat stress. *Plant Stress*, *10*, 100271. https://doi.org/10.1016/J.STRESS.2023.100271.

Wang, X., Kong, L., Zhi, P., & Chang, C. (2020). Update on Cuticular Wax Biosynthesis and Its Roles in Plant Disease Resistance. *International Journal of Molecular Sciences*, *21*, 1–15. https://doi.org/10.3390/IJMS21155514.

Wang, Y., Sun, Y., You, Q., Luo, W., Wang, C., Zhao, S., Chai, G., Li, T., Shi, X., Li, C., Jetter, R., & Wang, Z. (2018). Three fatty acyl-coenzyme a reductases bdfar1 bdfar2 and bdfar3 are involved in cuticular wax primary alcohol biosynthesis in *Brachypodium distachyon*. *Plant and Cell Physiology*, *59*, 527–543. https://doi.org/10.1093/PCP/PCX211.

Wang, Z., Tian, X., Zhao, Q., Liu, Z., Li, X., Ren, Y., Tang, J., Fang, J., Xu, Q., & Bua, Q. (2018). The E3 ligase DROUGHT HYPERSENSITIVE negatively regulates cuticular wax biosynthesis by promoting the degradation of transcription factor roc4 in rice. *The Plant*, *30*, 228–244. https://doi.org/10.1105/TPC.17.00823.

Weidenbach, D., Jansen, M., Franke, R. B., Hensel, G., Weissgerber, W., Ulferts, S., Jansen, I., Schreiber, L., Korzun, V., Pontzen, R., Kumlehn, J., Pillen, K., & Schaffrath, U. (2014). Evolutionary conserved function of barley and *Arabidopsis* 3-ketoacyl-coa synthases in providing wax signals for germination of powdery mildew fungi. *Plant Physiology*, *166*, 1621–1633. https://doi.org/10.1104/PP.114.246348.

Weng, H., Molina, I., Shockey, J., & Browse, J. (2010). Organ fusion and defective cuticle function in a lacs1 lacs2 double mutant of *Arabidopsis*. *Planta*, *231*, 1089–1100. https://doi.org/10.1007/S00425-010-1110-4.

Whitney, H. M., & Federle, W. (2013). Biomechanics of plant–insect interactions. *Current Opinion in Plant Biology*, *16*, 105–111. https://doi.org/10.1016/J.PBI.2012.11.008.

Willis, K., & McElwain, J. (2013). *The evolution of plants*, 2nd ed. Oxford University Press.

Woloshuk, C. P., & Kolattukudy, P. E. (1986). Mechanism by which contact with plant cuticle triggers cutinase gene expression in the spores of *Fusarium solani* f. sp. pisi. *Proceedings of the National Academy of Sciences of the United States of America*, *83*, 1704–1708. https://doi.org/10.1073/PNAS.83.6.1704.

Wu, H., Liu, L., Chen, Y., Liu, T., Jiang, Q., Wei, Z., Li, C., & Wang, Z. (2022). Tomato SlCER1–1 catalyzes the synthesis of wax alkanes increasing drought tolerance and fruit storability. *Horticulture Research*, *9*, uhac004. https://doi.org/10.1093/HR/UHAC004.

Wu, R., Li, S., He, S., Waßmann F., Yu, C., Qin, G., Schreiber, L., Qu, L. J., & Gu, H. (2011). CFL1 a WW domain protein regulates cuticle development by modulating the function of hdg1 a class iv homeodomain transcription factor in rice and *Arabidopsis*. *The Plant*, *23*, 3392–3411. https://doi.org/10.1105/TPC.111.088625.

Wu, X., Yin, H., Shi, Z., Chen, Y., Qi, K., Qiao, X., Wang, G., Cao, P., & Zhang, S. (2018). Chemical composition and crystal morphology of epicuticular wax in mature fruits of 35 pear (*Pyrus spp*). cultivars. *Frontiers in Plant Science*, *9*, 679. https://doi.org/10.3389/FPLS.2018.00679.

Xavier, C. A. D., & Whitfield, A. E. (2023). Plant virology. *Current Biology*, *33*, R478–R484. https://doi.org/10.1016/J.CUB.2023.03.038.

Xia, Y., Gao, Q. M., Yu, K., Lapchyk, L., Navarre, D. R., Hildebrand, D., Kachroo, A., & Kachroo, P. (2009). An intact cuticle in distal tissues is essential for the induction of systemic acquired resistance in plants. *Cell Host & Microbe*, *5*, 151–165. https://doi.org/10.1016/J.CHOM.2009.01.001.

Xie, L. J., Tan, W. J., Yang, Y. C., Tan, Y. F., Zhou, Y., Zhou, D. M., Xiao, S., & Chen, Q. F. (2020). Long-chain acyl-coa synthetase LACS2 contributes to submergence tolerance by modulating cuticle permeability in *Arabidopsis*. *Plants*, *9*, 262. https://doi.org/10.3390/PLANTS9020262.

Xin, A., & Fry, S. C. (2021). Cutin:xyloglucan transacylase (CXT) activity covalently links cutin to a plant cell-wall polysaccharide. *Journal of Plant Physiology*, *262*, 153446. https://doi.org/10.1016/J.JPLPH.2021.153446.

Xue, D., Zhang, X., Lu, X., Chen, G., & Chen, Z. H. (2017). Molecular and evolutionary mechanisms of cuticular wax for plant drought tolerance. *Frontiers in Plant Science*, *8*, 621. https://doi.org/10.3389/FPLS.2017.00621.

Yang, H., Zou, Y., Li, X., Zhang, M., Zhu, Z., Xu, R., Xu, J., Deng, X., & Cheng, Y. (2022). QTL analysis reveals the effect of CER1-1 and CER1–3 to reduce fruit water loss by increasing cuticular wax alkanes in citrus fruit. *Postharvest Biology and Technology*, *185*, 111771. https://doi.org/10.1016/J.POSTHARVBIO.2021.111771.

Yang, Y., Zhang, X., Jiang, J., Han, J., Li, W., Li, X., Yee Leung, K., M, Snyder, S. A., & Alvarez, P. J. J. (2022). Which micropollutants in water environments deserve more attention globally? *Environmental Science and Technology*, *56*, 13–29. https://doi.org/10.1021/ACS.EST.1C04250.

Yavuz, D., & Rashid, B. A. R., & Seymen, M. (2023). The influence of NaCl salinity on evapotranspiration yield traits antioxidant status and mineral composition of lettuce grown under deficit irrigation. *Scientia Horticulturae*, *310*, 111776. https://doi.org/10.1016/J.SCIENTA.2022.111776.

Yeats, T. H., & Rose, J. K. C. (2013). The formation and function of plant cuticles. *Plant Physiology*, *163*, 5–20. https://doi.org/10.1104/PP.113.222737.

Yuan, M., Ngou, B. P. M, Ding, P., & Xin, X. F. (2021). PTI-ETI crosstalk: An integrative view of plant immunity. *Current Opinion in Plant Biology*, *62*, 102030. https://doi.org/10.1016/J.PBI.2021.102030.

Zarinkamar, F., Ghelich, S., & Soleimanpour, S. (2013). Toxic effects of Pb on anatomy and hypericin content in *Hypericum perforatum* L. *Bioremediation Journal*, *17*, 40–51. https://doi.org/10.1080/10889868.2012.751958.

Zeisler-Diehl, V. V., Barthlott, W. & Schreiber, L. (2020). Plant cuticular waxes: Composition function and interactions with microorganisms. In H. Wilkes (Ed.), *Hydrocarbons, oils and lipids, diversity, origin, chemistry and fate, handbook of hydrocarbon and lipid microbiology*. Springer. https://doi.org/10.1007/978-3-319-90569-3_7.

Zeisler-Diehl, V., Müller Y., & Schreiber, L. (2018). Epicuticular wax on leaf cuticles does not establish the transpiration barrier which is essentially formed by intracuticular wax. *Journal of Plant Physiology*, *227*, 66–74. https://doi.org/10.1016/J.JPLPH.2018.03.018.

Zhang, C. L., Hu, X., Zhang, Y. L., Liu, Y., Wang, G. L., You, C. X., Li, Y. Y., & Hao, Y. J. (2020). An apple long-chain acyl-CoA synthetase 2 gene enhances plant resistance to abiotic stress by regulating the accumulation of cuticular wax. *Tree Physiology*, *40*, 1450–1465. https://doi.org/10.1093/TREEPHYS/TPAA079.

Zhang, C. L., Mao, K., Zhou, L. J., Wang, G. L., Zhang, Y. L., Li, Y. Y., & Hao, Y. J. (2018). Genome-wide identification and characterization of apple long-chain acyl-coA synthetases and expression analysis under different stresses. *Plant Physiology and Biochemistry*, *132*, 320–332. https://doi.org/10.1016/J.PLAPHY.2018.09.004.

Zhang, C. L., Zhang, Y. L., Hu, X., Xiao, X., Wang, G. L., You, C. X., Li, Y. Y., & Hao, Y. J. (2020). An apple long-chain acyl-CoA synthetase MdLACS4 induces early flowering and enhances abiotic stress resistance in *Arabidopsis*. *Plant Science*, *297*, 110529. https://doi.org/10.1016/J.PLANTSCI.2020.110529.

Zhang, P., Wang, R., Yang, X., Ju, Q., Li, W., Shiyou, L., Tran, L. S. P., & Xu, J. (2020). The R2R3-MYB transcription factor AtMYB49 modulates salt tolerance in *Arabidopsis* by modulating the cuticle formation and antioxidant defence. *Plant Cell & Environment*, *43*, 1925–1943. https://doi.org/10.1111/PCE.13784.

Zhang, Y. L., Zhang, C. L., Wang, G. L., Wang, Y. X., Qi, C. H., Zhao, Q., You, C. X., Li, Y. Y., & Hao, Y. J. (2019). The R2R3 MYB transcription factor MdMYB30 modulates plant resistance against pathogens by regulating cuticular wax biosynthesis. *BMC Plant Biology*, *19*, 1–14. https://doi.org/10.1186/S12870-019-1918-4.

Zhang, Y. L., Zhang, C. L., Wang, G. L., Wang, Y. X., Qi, C. H., You, C. X., Li, Y. Y., & Hao, Y. J. (2019). Apple AP2/EREBP transcription factor MdSHINE2 confers drought resistance by regulating wax biosynthesis. *Planta*, *249*, 1627–1643. https://doi.org/10.1007/S00425-019-03115-4.

Zhao, H., Kosma, D. K., & Lü S. (2021). Functional role of long-chain acyl-coa synthetases in plant development and stress responses. *Frontiers in Plant Science*, *12*, 640996. https://doi.org/10.3389/FPLS.2021.640996.

Zhao, L., Haslam, T. M., Sonntag, A., Molina, I., & Kunst, L. (2019). Functional overlap of long-chain acyl-CoA synthetases in *Arabidopsis*. *Plant & Cell Physiology*, *60*, 1041–1054. https://doi.org/10.1093/PCP/PCZ019.

Zhao, X., Chen, S., Wang, S., Shan, W., Wang, X., Lin, Y., Su, F., Yang, Z., & Yu, X. (2020). Defensive responses of tea plants (*Camellia sinensis*) against tea green leafhopper attack: A multi-omics study. *Frontiers in Plant Science*, *10*, 1705. https://doi.org/10.3389/FPLS.2019.01705.

Ziv, C., Zhao, Z., Gao, Y. G., & Xia, Y. (2018). Multifunctional roles of plant cuticle during plant-pathogen interactions. *Frontiers in Plant Science*, *9*, 1088. https://doi.org/10.3389/FPLS.2018.01088.

2 Disease Resistance in Crops

Molecular Mechanisms and Breeding Strategies

Fatima-Ezzahra Soussani, Chayma Ikan, Hanane Boutaj, Wissal Benaffari, Rachid Lahlali, and Abdelilah Meddich

2.1 INTRODUCTION

The relationship between crops and their microbial enemies is a critical component in the quest for worldwide nutritional security (Fones et al. 2020). Plant diseases still present significant obstacles to agricultural output, necessitating a sophisticated comprehension of the molecular underlying resistance to infections in crops (Savary et al. 2019). Plant defense systems are a complex tapestry that includes molecular pathways of disease resistance (Deng et al. 2020). These systems' underlying theories, such as effector-triggered immunity (ETI), offer a theoretical foundation for understanding the plant's complex defense against pathogenic invasion (Han 2019; Remick et al. 2023). This chapter explores the molecular

vocabulary of genes responsible for resistance, with key players including nucleotide-linked repeats (NLRs), pattern-recognition receptors (PRRs), and antimicrobial peptides (El Kasmi 2021; Sirisha et al. 2024). We explore the intricate molecular ballet that coordinates the plant's defense against developing diseases, delving into the function of secondary metabolites and other defense responses. The emergence of new technologies has brought about a transformative period in the field of disease resistance research. Among these, CRISPR/Cas9 stands out as a powerful instrument that promises accuracy and versatility in modifying plant defense mechanisms (Kumar and Jain 2015; Sharma et al. 2023). Examining its potential to transform the genetic arms race between plants and diseases, this chapter evaluates the status of CRISPR/Cas9 applications in crop protection. Given our current position at the nexus of agricultural science and biotechnology, it is critical to explore the effective contribution of CRISPR/Cas9 to anticipate future disease-resistant crops (Zaidi et al. 2020; Shahriar et al. 2021). The deliberate planning of breeding programs is a critical component of crop disease defense (Nelson et al. 2018; Li et al. 2020; Hussain et al. 2023). However, several studies explored a variety of breeding strategies from traditional methods to the accuracy of marker-assisted selection (MAS) and the data-driven field of genomic selection. A detailed review of the advantages, as well as drawbacks of each approach, offers a road map for choosing the best approach for particular crop–pathogen interactions. Some studies have demonstrated effective breeding programs that produced disease-resistant crop types using actual case studies, demonstrating the palpable effects of strategic breeding on the agricultural landscape (Deng et al. 2020; Bettinelli et al. 2023; Misra et al. 2023). Theoretical knowledge and practical techniques must be combined as we set out on our scientific expedition into the domains of molecular mechanisms and breeding strategies that control crop disease resistance. This chapter seeks to advance the conversation on strengthening agricultural systems against the constant evolution of plant diseases by analyzing novel breeding tactics and deciphering their molecular complexities.

DOI: 10.1201/9781003543237-2

2.2 DISEASE RESISTANCE IN CROPS

2.2.1 The Importance of Disease Resistance in Crops for Global Food Security

Worldwide food security is a major topic as humanity's global population continues to increase. Producing disease-resistant crops is one of the most critical components in ensuring a consistent and ample supply of food (Lawrence et al. 2013). From fungal blights that decimate grains to viral mosaics that mottle vegetables, plant diseases pose a danger to agricultural productivity. According to Savary et al. (2012), these silent attackers steal between 20% and 40% of the world's crop production each year. The estimated yield for global disease damages in seven key crops from 2000 to 2023 varied from 2% to 5% for maize smut. White mold affects soybean, while bacterial blight and Fusarium wilt impact cotton. Additionally, the incidence of blast disease in rice and late blight in potatoes and tomatoes ranges from 10% to 25% (Table 2.1). Additionally, losses from fungal infections typically rise as crop production rises. This translates to millions of tons of lost food and causes malnutrition, hunger, and economic instability in areas that are already vulnerable. In a world striving to feed about 10 billion individuals by the decade 2050, this loss is simply unacceptable (Leung 2008). However, the development and cultivation of resistant crops to infectious diseases is a proactive and long-term strategy to lessen the effect of plant diseases on the world's food supplies (Stuthman et al. 2007). Disease-resistant cultivars are developed to possess genetic features that make them less susceptible to specific infections (Mores et al. 2021). These traits may include enhanced immunological responses, resistance to infection, and faster disease recovery. Disease-resistant crops contribute to higher and more dependable yields, particularly in resource-limited regions, by mitigating the impact of diseases on plant health (Savary et al. 2020). Farmers should anticipate higher yields and more abundant harvests when crops are less susceptible

TABLE 2.1
Expected Worldwide Yield Reductions Due to Diseases in Main Crops (2000–2023)

Crop	Disease Category	Average Annual Yield Loss (%)	References
Rice	Blast	10–30	Asibi et al. (2019)
	Sheath blight	17–33	Ahamad and Khan (2023)
	Bacterial leaf blight	15–30	Duy et al. (2013)
Wheat	Rusts	5–43	El-Orabey and Elkot (2020)
	Septoria leaf blotch	Up to 60	Takele et al. (2015)
	Fusarium head blight	Up to 35	Matny (2015)
Maize	Lethal necrosis disease	30–90	Mahuku et al. (2015)
	Rusts	20–50	Dey et al. (2002)
	Smut	22–46	Aydoğdu et al. (2015)
Potato	Late blight	10–25	Haverkort et al. (2016)
	Early blight	5–25	Leiminger and Hausladen (2012)
	Common scab	5–50	Shattock (2002)
Soybean	Rust	Up to 60	Bradley et al. (2021)
	Sudden death syndrome	5–15	Baetsen-Young et al. (2021)
	White mold	10–56	Mfuka et al. (2020)
	Fusarium wilt	2–15	Allen et al. (2015)
Cotton	Verticillium wilt	5–20	Bardak et al. (2021)
	Fusarium wilt	2–5	Hillocks and Kibani (2002)
	Bacterial blight	15–30	Blaise (2019)
Tomato	Late blight	10–25	Savary et al. (2012)
	Spotted wilt virus	12–26	Awondo et al. (2012)
	Bacterial wilt	10–20	Kurabachew et al. (2017)

to pathogens (Haq and Ijaz 2020). Additionally, disease-resistant crops generally exhibit improved quality traits, making them more desirable for both domestic consumption and international trade.

Cultivation of disease-resistant crops can reduce reliance on chemical pesticides, which are commonly employed to prevent crop diseases (Lamichhane et al. 2016). This shift toward more sustainable farming practices improves environmental health, reduces water contamination, and safeguards the health of farmworkers and surrounding ecosystems (Roberts et al. 2018). Additionally, it contributes to the development of agroecosystems that are more resilient and less prone to the negative consequences of chemical-intensive agriculture (Wang et al. 2021). On the opposing side, resistant crops to infectious diseases require less water and resources, making them essential tools for sustainable agriculture in a changing climate.

Disease-resistant crops provide farmers with a more secure and predictable income (Kahane et al. 2013). By reducing the risk of agricultural damages caused by diseases, farmers can better plan for the future, invest in their farms, and contribute to the overall economic stability of their communities (Li et al. 2020). This, in turn, helps to strengthen food supply chains and ensures a steady flow of produce to local and global markets. Adopting disease-resistant crops on a global scale has far-reaching implications for food security (Qaim 2020). As nations invest in research and development to create resistant varieties tailored to their specific agroclimatic conditions, a diverse array of crops can be safeguarded against prevalent diseases (Narayanasamy et al. 2011). This collective effort assists in building a more adaptable world food availability, making it better equipped for impending threats like climate change and new disease strains.

2.2.2 Triggers of Plant Disease Outbreaks

The disease triangle, proposed by Agrios (2005), serves as the foundation for understanding plant disease outbreaks. According to Madden et al. (2017), this conceptual model emphasizes the critical roles played by three essential components: an infectious pathogen, a vulnerable host, and a disease-promoting environment. Plant diseases can be better understood by looking at how these elements interact and how they start, spread, and worsen. The initiation and spread of diseases are substantially impacted by host density, which is influenced by monoculture techniques (Mundt 2014). To understand how agricultural practices like field aggregation, crop uniformity, and field size affect host vulnerability, researchers examine the genetic diversity of hosts and its consequences (Ekroth et al. 2019). The complicated link in genes and host density emphasizes how crucial it is to use robust and varied planting techniques to reduce the likelihood of disease outbreaks. Host density correlates positively with disease intensity, notably for foliar diseases caused by fungus (Rutten et al. 2021). The increased space between individual plants reduces the inoculum burden as the population density of the host plant decreases, which slows the transmission of pathogen spores. Furthermore, dense canopies produced by high plant densities raise internal humidity, which favors viruses that need moisture for infection (Sahni et al. 2021). Genetic diversity and host resistance mechanism adaptability are two important elements that affect pathogen virulence, a critical aspect in the formation of epidemics (Stukenbrock and McDonald 2008). Current research emphasizes how pathogens evolve in response to shifting environmental factors and host defense systems (Fisher et al. 2022). Another growing issue is the potential influence of global trade and the transfer of planting materials on the introduction of novel viruses to previously uninfected areas (Jones et al. 2021). In some cases, virulence genes in the virus match specific resistance genes in the host, leading to an 'arms race' that can accelerate the spread of epidemics (Flor 1955). Determining these relationships is essential for developing durable resistance strategies. Chakraborty and Newton 2011 reported that temperature, soil moisture, and climatic conditions are critical factors in disease development. Recent research has concentrated on the consequences of climate change for plant disease dynamics, emphasizing the necessity to adjust techniques for managing diseases under developing climatic conditions (Jiang et al. 2023). Predicting and controlling plant disease epidemics can be accomplished by examining how environmental factors foster the development of infectious agents.

Predicting the dynamics of disease requires an understanding of the intricate relationships that exist between virulent infections, susceptible hosts, and conducive environments (Engering et al. 2013). Investigating how changes to each element of the disease triangle affect the course of epidemics offers important information for creating successful management strategies. Although the disease triangle offers a fundamental framework, studies show complex interactions between these elements. Interactions between the host, environment, and pathogen further affect disease development. As per Kaweesi et al. (2014), some diseases, such as the cassava brown streak virus, flourish in particular conditions related to the host physiology and rainfall patterns. Beyond the individual elements of the disease triangle, Liliane and Charles (2020) emphasize the significance of global interconnection and shifting dynamics in determining the landscape of plant disease outbreaks. Plant material exchanges and increased international trade have facilitated the spread of pathogens into previously uninfected areas. Changes in farming methods, the emergence of novel diseases, and the aggressiveness of established pathogens can quickly tip the scales in favor of disease outbreaks in areas that were previously thought to be safe. Furthermore, Liliane and Charles underline the need for changing control strategies to keep pace with these dynamic developments. The rapid evolution of new virulent strains may overwhelm traditional techniques relying on static resistance genes. Therefore, to effectively handle the changing difficulties provided by plant diseases in a worldwide world, it will be imperative to embrace crop variety diversity, promote soil health and beneficial microbial communities, and create adaptable management strategies.

2.2.3 The Dynamics of Plant Disease in Natural Populations

Recent studies have investigated the occurrence and patterns of plant diseases in natural populations. Factors such as climate change, habitat fragmentation, and anthropogenic influences play pivotal roles in shaping disease prevalence (Jeger 2022). Research highlights the importance of monitoring disease occurrence as a key indicator of ecosystem health, emphasizing the need for comprehensive studies across diverse ecosystems to capture the complexity of disease dynamics (Meentemeyer et al. 2012). Recent research has focused on studying the changing patterns of interactions between hosts and pathogens in natural ecosystems. Wang et al. (2023) reveal intricate relationships between native plants and their pathogens, with natural selection contributing to shaping the process of tolerance to disease. The coevolutionary conflict between hosts and pathogens contributes to the diversity of resistance strategies observed in natural populations (Hall et al. 2011). Understanding these interactions provides insights into the resilience of natural ecosystems against disease threats. Environmental factors, including climate, soil conditions, and species composition, significantly influence disease interactions in isolated populations. Recent research emphasizes the impact of global warming in altering the distribution and severity of plant diseases, with shifts in temperature and precipitation patterns influencing pathogen life cycles (Pautasso et al. 2018). Exploring how these environmental drivers shape disease dynamics contributes to predicting and mitigating the impact of emerging plant diseases in natural ecosystems. Plant diseases can have cascading effects on biodiversity and ecosystem services. Recent studies highlight the connections between disease dynamics, species diversity, and ecosystem functions (Laine 2023). The loss of keystone plant species due to diseases can disrupt ecological interactions, affecting nutrient cycling, pollination, and overall ecosystem resilience (Gilbert and Parker 2016). Evaluating the broader ecological consequences of plant diseases in natural populations is essential for informed conservation strategies. In the face of emerging plant diseases, developing effective management strategies is crucial for preserving biodiversity. Recent research explores innovative approaches such as the employing of endophytes and microorganisms and ecological restoration to enhance the resilience of natural ecosystems to diseases (Mukherjee et al. 2021; Suman et al. 2022; Singh Rawat et al. 2023). Conservation efforts must consider the dynamic nature of plant diseases in natural populations to ensure the sustainable management of ecosystems in a changing environment.

2.3 MOLECULAR MECHANISM OF CROP DISEASE RESISTANCE

The plant defense system is based on its capacity to distinguish harmful substances, as well as to transduce signals while reacting offensively via processes involving multiple genomes. Pathogens aggressively attempt to dodge and disrupt response pathways, resulting in a decentralized, multicomponent immune system (Han et al. 2019). Recent advancements in molecular methods significantly improved the scientific comprehension of plant immunity, owing to the potential for agricultural applications (Martinelli et al 2015).

2.3.1 Overview of Fundamental Concepts of Plant Defense System

2.3.1.1 Plant Innate Immune System

Plant surface barriers such as cuticular lipids (Godwin et al. 2005), hard cell membranes, wax layers, secondary compounds (Cai et al. 2012), and antimicrobial peptides and enzymes (Underwood et al. 2012) prevent pathogen entry and afterward activation of stimulated defense mechanisms or disease symptom development. Plants have genetically processed innate immune systems that are responsive to molecular patterns associated with pathogen effectors and with microorganisms or pathogens (MAMP or PAMP) (Vidhyasekaran 2016). Pathogens that penetrate these protective layers are commonly encountered, and the defensive response is begun by two interdependent mechanisms: immunity triggered by effectors (ETI) and immunity induced by microbial-associated molecular patterns (MAMP/PAMP). Pathogens can produce various effectors that have been created as tools to aid in pathogen infection. However, some of them incorrectly activate plant immune receptors. The initial phase of inducible defenses, referred to as pattern-triggered immunity (PTI), is initiated by the extracellular identification of MAMPs or PAMPs, as well as DAMPs. PTI is critical for preventing unsuitable microorganisms from damaging the host's immune system and limiting disease by suitable pathogens in hosts that are susceptible (Hou et al 2019). In the initial defensive layer, plants use a vast number of PRRs such as receptor-like proteins (RLPs) and receptor-like kinases (RLKs) (Tang et al 2017). PTI presents the process whereby PRRs initiate an immune response. PRRs may also identify DAMPs, which are critical in innate immunity against non-adapted microbes or endogenous elicitors. Plant's DAMP signals can include oligogalacturonides and cellodextrins from plant cell walls, as well as cutin monomers (Haris et al. 2013). Additionally, after an assault, the activation of *Botrytis*-induced effector kinasel (BIK1), which is part of the PRRs complex, leads to a rise in cyclic nucleotide-gated channels (CNGC)-mediated cytosolic calcium (Ca^{2+}), which at first constitutes a crucial signal to immunity-triggered pathogen-associated molecular models (PAMPs), and the PRR complex, including botrytisl effector kinasel (BIK1), could have been triggered, resulting in a rise in CNGC-mediated cytosolic calcium (Ca^{2+}) (Leitao 2018). A mechanism of effector-triggered susceptibility is explained by the use of a series of pathogenicity effectors by pathogens to avoid PTI and develop a successful infection. In such circumstances, plants activate a second immune system, another mechanism termed ETI, which occurs mostly throughout the plant cells and provides an extensive and effective response. Accordingly, as a result, the stimulation of mitogen-activated protein kinase (MAPK) represents one of the processes of both the mechanisms of ETI and PTI that induce plant defense against infectious agents by altering the flow of ions throughout the barrier known as the plasma membrane; while increasing the levels of the apoplast ROS and cytosolic Ca^{2+}, they also trigger the accumulation of NO (nitric oxide) and the formation of phytohormones, which includes salicylic acid, ethylene, and jasmonic acid. Stomatal closure, transcriptional reprogramming, and callose deposition are among the other PTI and ETI processes. In response to (MAMP/DAMP) sensing, PRR stimulates intracellular signal transduction, which includes Ca^{2+} transmitting signals, CDPK, and numerous types of transcriptional modification precursors. Furthermore, eATP as a microbial apoplastic may stimulate the Ca^{2+} signaling. Pathogen colonization in plant tissues can alter crop transcriptional expression, potentially enhancing plant vitality (Figure 2.1).

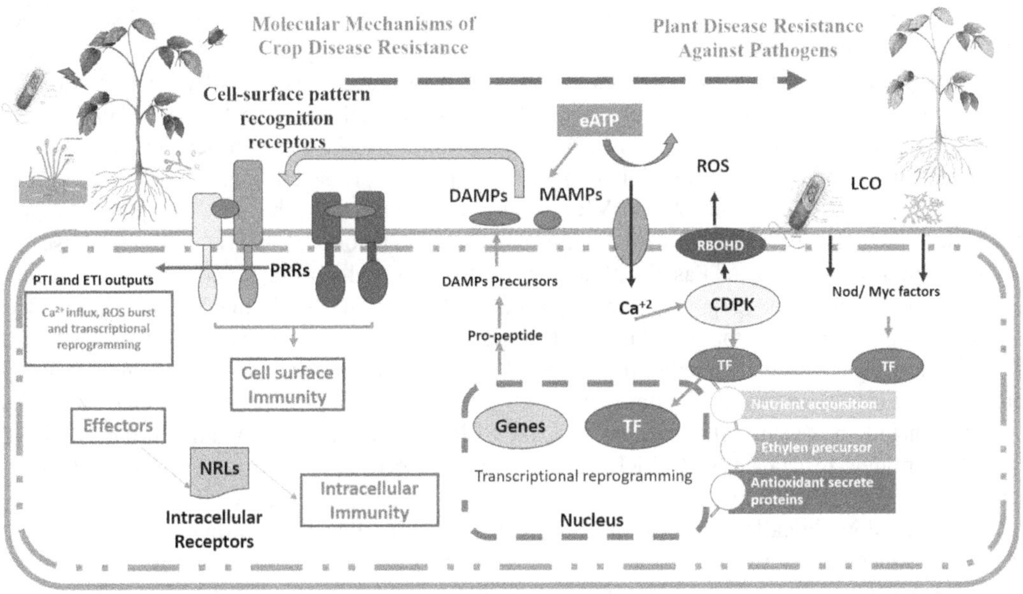

FIGURE 2.1 Understanding of the mechanism of plant immune recognition.

2.3.1.2 Cell Surface Immunity

RLKs/membrane-localized RLKs and RLPs represent both main immune system elements on plant cell surfaces that identify non-auto signatures as indicators of infection (Jones et al. 2006). RLPs/RLKs possess a variety of roles in crops, including managing plant immunity and disease resistance (Sanabria 2006). DAMPs (molecular patterns associated with damage) or pathogen/pest invasion patterns ligands called MAMPs are done by PRRs as cell surface immune receptors (Boller et al. 2009). Ligand detection often relies on co-receptors to convert non-self-perception throughout a response. Whereas receptors located on the cell surface represent the main actors in plant cell surface immunity, new research has revealed that membrane lipids play an increasing role in infection detection (Albert et al. 2020). Whatever their source, the invasion patterns identified by cell surface immune receptors are frequently progressively constrained ligands derived from pathogen or pest fitness factors. These critical components include cell wall constituents, bacterial flagellin subunits, and compounds released from proteins that are secreted into the host cytosol (Sharpee et al. 2016). At nanomolar concentrations, cell surface immune receptors detect these particular ligands, triggering signaling cascades such as MAPK activation, generation of ROS, cytosolic Ca^{2+} bursts, and alterations of numerous defense-related genome expressions (Macho et al. 2014). In general, cell surface immune responses are quite different from intracellular immunity; they are less volatile and don't lead to host cell death, thus helping to limit infection. Nevertheless, they represent a potent host defensive approach against infection, resulting in effective resistance (Boutrot et al. 2017). Pathogen/parasite effector proteins predominantly inhibit cell surface immune receptor signaling cascades.

2.3.1.3 Intracellular Immunity

Pathogens that may infect cells and cause disease must disrupt those mechanisms to establish infection. Further research into intracellular immunity will provide a fundamental comprehension of pathogenic processes and assist in the development of antimicrobial medicines that alert the immune system and target infectious agents. Plants employ NLRs to provide intracellular immunity. The other major kind of PRR is the NLR, which detects the activity of translocated effectors, triggering defensive processes to restrict infection transmission (Franceschetti et al. 2017).

In general, the effective immune response against pathogens is created by several distinct types of NRLs. The pattern-recognition hypothesis required the discovery of germline-encoded receptors (GRPs), which can recognize viruses as well as cell-invading substances from which they are naturally detected for immune activation without recourse to a selection mechanism. Plants rely on leucine-rich, NLR receptor proteins for intracellular immunity. The gene-for-gene paradigm states how plants must employ highly functional immune receptors expressed through R (resistance) genes to oppose and successfully control the AVR (avirulence) effectors of pathogens. This paradigm predates the molecular discovery of NLR receptors and effectors (Flor et al. 1971). NLRs are classed as AAA1 'signal-transducing ATPases with numerous domains', which share a core nucleotide-binding domain conserved in crops and fungal worlds (Takken et al. 2009) and include APAF1, the primary component of the mammalian apoptosome (Zou et al. 1999), as well as NLRP3 referred to NLR family pyrin domain-containing 3 and NLRC4 as NLR family CARD domain-containing protein 4, the metazoan immune system's best-characterized NLRs (Zhang et al. 2015). Plant NLRs typically include a C-terminal leucine-rich repeats (LRR) domain, a CC domain similar to RPW8 (powdery mildew resistance 8) (CC-RPW8), a variable N-terminal module with a TIR (toll/interleukin-1 receptor/resistance), coiled-coil (CC) domain (Dusxbury et al. 2016), and a central nucleotide-binding domain called NB-ARC (which is shared with APAF1), the programming language R DNA products, and CED4 (Qi et al. 2013). Furthermore, LRR domains might be found on the outermost layer of cells as well as in immune-mediated intracellular receptors, and they have been identified as ligand-binding motifs that regulate interactions between proteins throughout every domain of life. The LRR domain was previously linked to effector identification in some NLRs, but it may also be implicated in receptor auto-inhibition (Kanyuka et al. 2019). Van Ooijen et al. (2008) describe the NB-ARC domain as a molecular switch that communicates effector sensing via exchange of nucleotide (ADP for ATP). Immunity depends on RPW8-NLRs, CC-NLRs, and TIR-NLRs, which are differentiated by N-terminal domains (Maruta et al. 2022). In cultures, intermittent expression assays have revealed that N-terminal domains called integrated domains (IDs) can communicate directly with effectors and induce cell death in the absence of an effector (Chen et al. 2013). Interestingly, numerous NLR IDs have sequence/structure similarities to known virulence-associated host effector requirements, including transcription factors or proteins involved in cellular homeostasis (Cesari 2018).

2.3.2 Understanding of Fundamental Concepts of Plant Immunity: Molecular Mechanism

2.3.2.1 Pattern-Triggered Immunity Recognize PAMPs/DAMPs

Plant immunity requires the employing of a diverse spectrum of receptors and ligand types to provide complete protection against pathogen attack. PRRs on the cell surface identify PAMPs and DAMPs. PTI, or PRR-mediated immunity, is the ultimate final result. PTI is the initial component of the system's innate defenses, identifying PAMPs on the outermost layer of the cell's plasma membrane to avoid pathogen invasion. PTI employs PRRs on the surface of the membrane to recognize PAMPs generated by microorganisms. However, these infectious agents employ effector proteins to circumvent or damage the PTI system's immune response. Endogenous signals created during cell decay including PAMPs and DAMPs are detected by PRRs (Amarante-Mendes et al. 2018). DAMPs are important structures or components found in all kinds of infections (Patel et al. 2018). Oligogalacturonides (Galletti et al. 2008), Pep-13 (Brunner et al. 2002), ergosterol (Granado et al. 1995), bacterial flagellin, xylanase (Belien et al. 2006), lipopolysaccharides (LPS) (Dow et al. 2000), and cold-shock protein (Felix and Boller 2003) are examples. Plant species within a single family are the only ones that can recognize specific MAMPs and induce immune system responses. For example, factors of recognition have been observed solely within *Brassicaceae* crops (Kunze et al. 2004). Numerous MAMPs, such as chitin, LPS, and flagellin, elicit responses from the immune system in a variety of host species (Zipfel et al. 2006). Several plant diseases create lytic enzymes to break down plant tissue structural barriers. Consequently, peptides (Huffaker et al. 2006), cutin monomers, and cell wall fragments (Kauss et al. 1999) may act as endogenous elicitors known as DAMPs. Stimulation of the innate immunity system is affected by these DAMPs, which act as danger signals, as they often appear in the apoplast (Henry et al. 2012). Alkalinization of the propagation media sources is plant cell cultures' first line of defense against MAMPs. This event occurs across the plasma membrane, based on dramatic changes in the flux of K^+, Ca^{2+}, H^+, and Cl_2 ions. Elevated cytoplasmic Ca^{2+} levels are a critical stage in a reaction to PTI, and they are transmitted by Ca^{2+} buildup in chloroplast thylakoid lumens (Nomura et al. 2012). This is achieved by the Ca^{2+}-conducting channel via phosphorylation of PRR, a G-protein molecular switch, activation of a plasma membrane, and cyclic nucleotide channels (Chou et al. 2015). Calcium-binding proteins, such as calcineurin B-like and calcineurin-like proteins, calcium-dependent protein kinases, and calmodulin, identify and detect changes in [Ca^{2+}], and detection is enhanced in the presence of high Ca^{2+} in the cytoplasm. Ca^{2+} elevation in the cytosol is essential for other plant immune activities such as ROS regulation, closure of stomate, salicylic acid synthesis and the activation of defense-related genes. These proteins play a crucial role in signal transduction pathways, enabling plants to respond to various environmental stresses and pathogen attacks. (Ou et al. 2022). CDPKs, which serve as Ca^{2+}-sensing protein kinases, have recently been discovered as essential mediators of early immune signaling in PTI (Sardar et al. 2017).

2.3.2.2 Role of PRRs in Detection of PAMPs and DAMPs

Despite the discovery of PRRs, researchers have focused on studying the downstream responses elicited by protein-recognition receptors and the process of activating signaling components. The innate immune system contains PRRs, such as receptor kinases (RLKs) and RLPs, which are essential for recognizing pathogens and creating a protective immune response. PRRs may identify a diverse range of microbial components, including fungal polysaccharides (De Jonge et al. 2010), bacterial proteins (Sun et al. 2013), and viral nucleic acids (Niehl et al. 2016). Therefore, the knowledge of the process of PRRs is essential for managing pathogenic diseases, particularly against RNA viruses. Nucleic acids, such as DNA and RNA, have been identified as important PAMPs for diseases, particularly viruses. RNA is the main PAMP of RNA viruses, which comprise the majority of serious pathogenic viruses and create a major threat to plant health. PRRs include NOD_2, NLRP3, MDA5, RIG-I, TLR3, TLR7, TLR8, and several other minorities; they are categorized into numerous groups based on protein domain homology: RIG-I-like receptors (RLRs), C-type lectin receptors (CLRs), nucleotide-linked repeats (NLRs), toll-like receptors (TLRs), and AIM2-like receptors (ALRs) – the recently discovered cytosol DNA-sensing PRR and cyclic GMP–AMP synthase (cGAS). A broad spectrum of PAMPs from bacteria, viruses, fungi, and protozoa, including carbohydrates, nucleic acids, lipopolysaccharides, lipoproteins, and endogenous host DAMPs, which are linked to both homeostasis and autoimmune disorders, are sensed and detected by PRRs. If PRRs detect PAMPs or DAMPs, they activate intracellular cell signaling, resulting in transcriptional activation and the production of co-stimulatory molecules chemokines and cytokines (Chen et al. 2017). These receptors often contain LRRs that link to transmembrane domains for plasma membrane localization, extracellular ligands, and cytoplasmic kinase domains for signal transmission via phosphorylation (Zipfel et al. 2014). LRRs are very varied due to their capacity to bind to a diverse spectrum of elicitors. The main categories of PRRs are regulated by BAK1 and other SERKs referred to as somatic embryogenesis RLKs (Monaghan et al. 2012; Prince et al. 2014). Extensive signaling is not always begun; since when specific PRRs are engaged, they release kinase domains that penetrate the nucleus and stimulate transcriptomic remodeling (Park et al. 2014). PRRs detect a variety of molecules, including peptidoglycan (Park et al. 2012), fungal (chitin, xylanase), flagellin (Kaku et al. 2006), oomycete (β-glucan and elicitins) (Fliegmann et al. 2004; Du et al. 2015), viral (double-stranded RNA) (Niehl et al. 2016), and insect (aphid-derived elicitors) molecules. While much of this research has focused on particular molecular interactions in wheat, TaRLK-R1-3 has been related to resistance to rust (fungal organisms of the genus *Puccinia*) in *Arabidopsis* and PRRs such as TaLRK10 and TaRLP1.1 (Jiang et al. 2013). Unlike PRR, which identifies non-autopathogenic molecules during disease, additional receptors detect cellular components damaged by pathogenic enzymes, thereby initiating a more robust immune response to combat the infection.

2.3.2.3 Effector-Triggered Immunity

Intracellular receptors, such as NLRs, activate ETI in response to pathogen effectors, resulting in persistent resistance. ETI is widely utilized for crop disease management and resistance in agricultural production (Li et al. 2020). Individual cells in plant innate immunity are capable of sensing and detecting pathogen assaults (Muthamilarasan et al. 2013). Intracellular recognition mechanisms have developed to transform pathogen virulence factors (effectors) into immediate defense and identify their alterations at an early stage of the infection of pathogen's mechanisms (González et al. 2016). Successful resistance depends on diverse intracellular repeat receptors, NLRs, that recognize effector disruption throughout the cell. Activation of NLRs by effectors enhances transcriptional defense programs and interacts with a conserved basal resistance network (Jubic et al. 2019). ETI is an additional method of pathogen detection that is extremely durable in the face of pathogen disruption, mainly due to adaptive processes within the defense network. Triggered immunity involves the identification of virulence factors encoded by the pathogen, also known as effectors. Effectors are often detected indirectly through their pathogenic actions. ETI is a feasible approach to pathogen detection that is used in a variety of phyla, including plants; ETI signaling mechanisms are resilient to effector pathogen disruptions and are also more resistant to genetic alterations compared to PTI. However, the molecular mechanisms of ETI are more complex than simple PAMP receptor/ligand detection (Remick et al 2023).

2.3.2.4 Nucleotide-Linked/Leucine-Rich-Repeat Recognize Pathogen Effectors

Some pathogen effectors influence antimicrobial compound synthesis by interfering with protease secretion (Shen et al. 2007), vesicular trafficking via proteasome degradation (Nomura et al. 2006), and endocytosis. Intracellular leucine-rich repeat nucleotide-binding receptor proteins (NB-LRRs) stimulate NLR-mediated immunity. NLRs are categorized into three types: RPW8-type CC domain NLRs (RNLs); helical CC domain NLRs (CNLs); and toll/interleukin-1 receptor/resistance protein (TIR) NLRs (TNLs) (Jones et al. 2016). NLRs constitute one of the most rapidly developed protein families in cultures, as well as being sophisticated receptors that are able to detect a diverse array of alterations, including recognition of non-self and modified-self (Baggs et al. 2017). Compared to PRRs, NLRs are frequently found in the cytoplasm and contain LRRs at their C-terminus. Several plant-resistant diseases generate NLR receptors, which identify effector proteins released by infections, optionally in a direct manner or, indirectly, via host proteins targeted by the effector (Wu et al. 2017). When infectious effectors are recognized, their products change conformation, undergoing the transition from a compact ADP-bound state to an open ATP-bound state with exposed N-terminal domains, allowing downstream communication to begin (Cui et al. 2015). These are modular proteins classified into two types depending on their N-terminal domains: toll-interleukin 1 receptors and CC domains (Wu et al. 2017). The diversity of LRRs enables these receptors to recognize a variety of effector structures. Based on the conserved amino acids they contain, all motifs are named resistant nucleotide-binding site-A (RNBS-A), P-loop/Walker A, Kinase-2/Walker B, RNBS-D, RNBS-B, GLPL, RNBS-C, and MHDV, as exemplified by the rice CNL protein Pb1, which lacks a P-loop (Hayashi et al. 2010). It demonstrated that NLR receptors function in networks while revealing the distinction between sensor and helper NLRs, as well as NLRs essential for cell death (NRC). One of the most studied NLR–effector interactions is that of the NLRs RPM1 and RPS2, which sense the targeting of the negative resistance regulator RIN4 by the effectors AvrRpt2 and AvrRpm1 of *Pseudomonas syringae* (Marathe et al. 2003). Despite the remaining NLRs, RPS2 and RPM1 have locations in the cell membrane since their guardian, RIN4, is additionally located there (Axtell et al. 2003). After stimulation, RPM1 binds to the negative regulator AtTIP49a, which then binds to the promoter, triggering higher resistance responses (Holt et al. 2003). NLRs possess a variety of topologies, including integrated decoy regions that serve as effector constituents, including TF domains. The C-terminal WRKY TF domain of *Arabidopsis* resistance to *Ralstonia solanacearum* 1 (RRS1) binds directly to promoter sequences (Noutoshi et al. 2005). PopP2, a *Ralstonia solanacearum* effector, impacts this WRKY domain, thereby activating the NLR (Sarris et al. 2015). NLRs can additionally combine, as demonstrated by the CNL proteins RGA5 and RGA4 in rice, with RGA5 binding immediately to *M. oryzae*'s Avr-Pia and Avr1-Co39 effectors (Cesari et al. 2013). Further investigation is needed to understand how NLRs self-associate or disperse, especially in genes with financial significance, such as stem rust resistance genes in wheat (Casey et al. 2016). Similar to how PRRs rely on BAK1, NLRs convey signals via other proteins. CNLs and TNLs have been associated with non-race-specific disease resistance 1 (NDR1) and elevated disease vulnerability 1 (EDS1) proteins, correspondingly. NLRs have the ability to relocate particular areas of the cell, including endosomes and nuclei. When triggered by the *Phytophthora infestans* effector Avr3aKI, the potato CNL protein R3a migrates to endosomes and recruits additional effectors (Engelhardt et al. 2012). Barley CNL MLA proteins are concentrated in the nucleus, interfering with WRKY TFs and significantly reducing immunity (Shen et al. 2007).

2.3.2.5 Contributions of Secondary Metabolites to Disease Resistance

Plant metabolism and pathogenic infection are related (Fagard et al. 2014). Nutrition is a pathogen's major host need (Snoeijers et al. 2000), and plant metabolism regulates nutrition availability (Rojas et al.2014). The most significant feature of plant pathogen defense is energy availability, which is fulfilled by fundamental metabolic processes (Bolton et al. 2009). Secondary metabolites are widely recognized for their antibacterial properties, which present an effective contribution to plants' defense against a diversity of pathogenic interactions. Secondary compounds are usually thought to react with pathogenic organisms, and the involvement of secondary metabolites in infectious interactions between plants and pathogens has been the focus of extensive research on the immunity crop system response (Xu et al. 2022; Pusztahelyi et al. 2015; Maruta et al.2022; Bednarek et al. 2018). Several decades of scientific investigation have demonstrated that a varied range of secondary compounds contribute significantly to plant infection resistance. Their activities are significantly pooled in conserved frameworks, despite their significant structural variation, and many biosynthetic pathways are not conserved across the plant world (Humphry et al. 2010). The recognition of microorganisms by defense proteins, or MAMPs, contributes to generating and activating these secondary metabolites, as well as other conserved defensive systems for crops (Ahuja et al. 2012). Because of the wide spectrum of plant immunity, various classification criteria for secondary compounds have been created. A commonly employed categorization is based on how defense-related phytochemicals are generated and stored. Secondary plant metabolites are amino acids, chemical compounds, and nucleosides, resulting from primary metabolism by modifying specific enzymes and pathways; these metabolites include phenolic compounds, carotenoids, terpenes, and alkaloids (such as nicotine, morphine, caffeine, atropine, and quinine) (Katyal 2022; Pott et al. 2019; Jan et al. 2021). They assist plants in developing advanced responses to defend against pathogenic organisms that invade, increasing crop durability and creating resistance to disease metabolically (Abegaz and Kinfe 2019; Jan et al. 2021; Katyal 2022). Stimulation of different secondary compounds can be assisted by fundamental plant defense mechanisms, including the activity of defense proteins or MAMPs that are detected by PRRs (Ahuja et al. 2012; Ranf et al. 2018). Because of their diverse roles in plant immunity, they have been classified using a range of criteria, including the method of synthesis, defense-related phytochemicals, phytoalexin production, phytoanticipins, structures, modes of action, and common precursors (Piasecka et al. 2015). Phytoalexins are novel metabolites formed in reaction to infectious agents (Paxton et al. 1981), whereas phytoanticipins are metabolites created and generated in crop tissues for defense (Van Etten et al. 1994). Chemicals often disrupt cellular signaling pathways or target pathogenic components including enzyme inhibition, DNA alkylation, and the process of reproduction (Romirez-Gomez et al. 2019; Narayani and Srivastava 2017; Thakur et al. 2019). Recent research on medicinal plant secondary metabolites found to be effective against phytopathogens indicates that these metabolites could contribute to phytopathogenic fungi, including various *Fusarium* categories (Mafakheri et al. 2018). In this context, the most involved metabolite extracted from the aerial parts of plants, as established mainly in the essential oil, plays an important role against fungi, because the cell membrane consists primarily of ergosterol, the active fat-soluble chemical substances that may easily penetrate the fungal cell wall and function inside (Butassi et al. 2019). Infectious pathogen signals were discovered to promote *Arabidopsis thaliana* to create tryptophan-derived secondary metabolites (TDSMs). Immune-mediated activity via TDSMs is promoted by the resistance genes AIG2A and AIG2B. However, suppressor screening and heredity analysis have demonstrated that intentional coordination of TDSMs is essential for the regulation of the defense system in the absence of the AIG2B and AIG2A genomes (Wang et al. 2022).

2.4 MOLECULAR BREEDING STRATEGIES FOR IMPROVING DISEASE RESISTANCE OF CROPS

2.4.1 Different Types of Breeding Approaches

Plant breeding is divided into two strategies: conventional breeding, which involves the intentional crossing between closely associated individuals to produce novel species with desired characteristics, and molecular plant breeding, which refers to the process of developing or improving new varieties through molecular biology or engineering. Because it often takes several generations to identify and analyze the appropriate genotypes, traditional breeding may be restricted in its capacity to meet global food security (Tester and Langridge 2010). Molecular plant breeding progressed so fast that a diverse set of molecular markers remained established and employed for years. Molecular markers have strong heritability and may be tested by high-throughput techniques at a relatively early stage of growth, hence boosting selection gain. Currently, much work is being carried out using molecular markers, particularly where diagnostic markers are available (Li et al. 2020; Hasan et al. 2021; Xu et al. 2017; Yin and Qiu 2020).

2.4.1.1 Marker-Assisted Selection

The genome-wide investigation integrating primary genomes and QTLs underlying a variety of critical agricultural characteristics is assisting in the integration of biotechnology into conventional breeding practices. The MAS of characteristics throughout the selection procedure is a strategy created to circumvent the issues related to conventional plant breeding by moving the selection parameter from phenotypic to genetic factors, either directly or indirectly, using insights gathered from determining the geographic location of agronomically significant traits and associated molecular markers. MAS involves the employing of molecular markers to enhance phenotypic selections in the development of crops; thanks to the availability of a marker set it has become possible both for qualities determined by large genomes and for QTL. Nevertheless, scientific investigations using this approach have demonstrated that the effectiveness of MAS is determined by several variables, which include the level of interaction across the molecular marker and its target gene, the molecular basis that underlies the characteristic, the wide range of individuals that are eligible to be analyzed, and the historical context into which the gene of interest is to be transmitted. The effectiveness of a molecular marker is defined by its ability to exhibit polymorphisms in the nucleotide sequence, which allows various alleles of the molecular marker to be differentiated. MAS has been extensively utilized to create resistance to disease with desired agronomic characteristics. MAS for heritable characteristics is becoming increasingly important in breeding operations because it allows for quicker outcomes. Potential regular uses of MAS include traits associated with disease resistance and the quality of specific agricultural goods. To select varieties with long-term disease resistance, genetic markers closely connected to resistance genes are utilized for quick and effective indirect selection of target genes (Wang et al. 2005). MAS has been widely used to select rice varieties resistant to diseases of economic importance. The considerable economic losses caused by diseases have prompted scientists to apply molecular breeding methods and MAS to produce disease-resistant, high-yielding rice cultivars. According to Jeung et al. (2006), Xa4, Xa5, and Xa21 are three R genes that have been pyramided into tolerant cultivars in Korea, conferring resistance to dominant races. The MAS lines were then utilized to create mildew-resistant maize hybrids (Singh et al. 2009). Employing a molecular process, qualitative trait loci associated with sorghum late blight resistance have been discovered on chromosomes 6 and 3, which were utilized in the MAS approach to introgression into eight resistant maize varieties. QTLs that confer resistance to maize leaf and sheath anthracnose were also found. In rice, MAS has been utilized to effectively protect R genomes against blast and bacterial leaf blight. The R genes (xa5, xa13, and Xa21) for bacterial leaf blight were effectively produced in a resistant indica rice cultivar (PR106) using the MAS method, affording considerable resistance to the races of bacterial leaf blight found in India (Singh et al. 2001). Enhanced Pusa Basmati is the second successful product of India's MAS-based selection, which included the insertion of genes for bacterial leaf blight resistance (xa13 and Xa21). A sorghum blight-resistant maize line (CM139) and numerous QTL-NILs were developed and exploited to build disease-resistant maize cultivars (Prasanna et al. 2009). Hanson et al. (2016) succeeded in developing an array that is susceptible to five diseases: gray leaf spot *Fusarium* wilt race 2, late blight, bacterial wilt, tomato yellow leaf curl, and tobacco mosaic virus. Kumar et al. (2014) created cultivars that are susceptible by utilizing the tolerant sources EC-538408 (Solanum chilense) and EC-520061 (*Solanum peruvianum*) by marker-assisted genetic assembling of tomato leafroll virus resistance genes. El Mohtar et al. (2007) employed a PCR-based marker to identify tomato cultivars that have the I-2 genome, which provides susceptibility to *Fusarium oxysporum* f. sp. *lycopersici* Race 2. In contrast, Prasanna et al. (2015) engineered a tomato cultivar containing the Ty-2 and Ty-3 genomes, increasing resistance to the tomato leafroll infection.

2.4.1.2 Marker-Assisted Backcrossing

The primary objective of marker-assisted backcrossing (MABC) is to integrate a specific gene to add one or more qualities to an adapted variety that possesses not only a few undesirable characteristics, such as disease resistance, but many other desirable ones (Kwabena Osei et al. 2022; Lema 2018). MABC is an accurate and effective way to keep the key features of the target variety while introgressing a singular genetic locus governing a noteworthy trait (Kwabena Osei et al. 2022; Lema 2018). It includes using markers to select target loci, shortening the segment from the donor that encompasses the chosen locus of interest, and/or hastening the backcrossing process' recovery of the genetic code of the adapted or targeted variety (Lema 2018; Yang et al. 2019; Ramayya et al. 2021). Utilizing molecular markers in plants expedites the process of backcrossing to get the intended phenotype. MABC is considered an effective approach for providing integrated disease resistance to multiple species (Kim et al. 2021; Ramayya et al. 2021). Chukwu et al. (2020) and Sabar et al. (2019) have improved the resistance of Basmati rice to biotic stress by MABC, incorporating into the Basmati rice genetic background genes for defenses against both bacterial blight and blast, which may result in a loss of productivity of as much as 50% in enormously rice planting regions under suitable environmental circumstances. The most damaging fungal diseases affecting wheat worldwide are rusts, which are mainly controlled by improving resistance, as chemical control is costly and harmful to the environment. Resistance enhancement of a wheat variety susceptible to multiple rust has been achieved using MABC linked to resistance genes for this disease (Mallick et al. 2015; Singh et al. 2018). The use of molecular markers in the backcrossing approach has made it feasible to accurately transfer desired genes from unadapted germplasm to elite lines with minimum linkage traces (Kim et al. 2022; Ramayya et al. 2021).

2.4.1.3 Marker-Assisted Recurrent Selection

The objective of marker-assisted recurrent selection (MARS) is to choose the genotype with the highest performance within one or more related populations by identifying and selecting several genomic regions influencing the expression of complex traits through the utilization of molecular markers (Lema 2018; Singh et al. 2023). It enables genotypic selection and outcrossing between selected individuals during the same growing season for a selection cycle. The MARS method is proposed for traits like grain yield and biotic and abiotic resistance (Haque et al. 2021; Ramayya et al. 2021). As it can be used for many QTLs regulating the expression of a complex trait – a limitation of MAS and backcrossing – MARS can be one of the paramount methods in a molecular selection process (Assefa et al. 2019; Henkrar 2021). Consequently, MARS, which has the potential to integrate many favorable genes associated with the selection of a complex trait, may prove useful to enhance various quantitative traits, namely, disease resistance in multiple crops (Assefa et al. 2019; Singh et al. 2023). Several applications of MARS have proved effective in disease resistance, particularly in wheat, corn, and rice such as powdery mildew, leaf rust, stem rust, yellow rust, cereal cyst nematode, Hessian fly, wheat streak mosaic, barley yellow dwarf virus, and several cereal mosaic viruses (Miedaner and Korzun 2012; Reddy et al. 2011). Rahman et al. (2020) and Rahman (2018) mentioned that the average level of resistance to crown rot in populations improved after MARS was significantly higher than that of the previous wheat population. In *Coffea arabica*, MARS was assessed to find genes that protect from coffee berry diseases and coffee leaf rust. The method used was effective in selecting coffee hybrids that can serve as a resistance source in various crosses (Saavedra et al. 2023). Therefore, boosting MASR's importance need not apply to all crops equally; in fact, it might be a top focus for high-value crops, especially those with interesting economic and biological potential. However, the large-scale application of MARS can present certain challenges, namely, high costs and the unavailability of suitable markers (Assefa et al. 2019; Saavedra et al. 2023).

2.4.1.4 Genome-Wide Selection

Genome-wide selection (GWS) stands as a potent tool for identifying natural variations that underlie complex traits in crops, due to the quick advancement of sequencing technologies and computational techniques (Uffelmann et al. 2021). In general, GWS in agriculture takes advantage of a persistent resource: an assortment of diverse (and ideally homozygous) cultivars which may be genotyped just once and rephenotyped for many traits. From there, unique mapping populations for individual traits or QTLs in crops can be generated (Mills and Rahal 2019; Uffelmann et al. 2021). Genomic selection offers possible ways for forecasting the performance of new genotypes, considerably improving the process of selection (Nadeem et al. 2018). It is clear that the GWS research has enhanced our understanding of the molecular mechanisms and genetic foundation of prevalent diseases. Numerous crops, including rice, sorghum, and maize, have had successful genome-wide association studies conducted on them (Robertsen et al. 2019; Singh et al. 2019). The two primary models for GWS studies in crops are rice and maize, given the range of materials that have already been created and published (Robertsen et al. 2019; Wang et al. 2020). Combining association mapping with genome-wide prediction yielded good predictive accuracy when weighted genomic selection was used for *Fusarium* head blight resistance in triticale (Galiano-Carneiro et al. 2019). Likewise, GWS worked effectively as a technique to boost wheat (*Triticum aestivum* L.) resistance to *Fusarium* head blight (Verges et al. 2021). McElroy et al. (2018) suggest that the recourse to all markers in a global breeding framework is very interesting for enhancing resistance in the cocoa to broom disease and frost rot of pods induced by *Moniliophthora perniciosa* and *Moniliophthora roreri*, respectively. Similarly, O'Connell et al. (2022) confirmed the effective use of several markers for the prediction of red rot resistance at any stage of the sugarcane genomic breeding program. Thanks to advancements in genomic methods, it is now feasible to fully expose genetic diversity in attributes among crop species. The application of major resistance genes to improve resistance to biotic stress is therefore considered an effective and technically feasible approach to achieving optimal resistance and grain productivity (Robertsen et al. 2019; Uffelmann et al. 2021).

2.4.2 New Breeding Techniques for Pathogen Resistance: CRISPR/Cas9 as a Tool for Manipulating Plant Defense Systems

In recent years, new breeding technologies have become more interesting in plant research and consist of a broad spectrum of subjects such as plant tolerance against diseases (Nelson et al. 2018). According to Lusser and Davies (2013), alternatives to conventional plant breeding for agricultural improvement, incorporating resistance to infectious diseases, have been provided. These breeding technologies incorporate the most advanced and powerful molecular tools for precisely modifying one or more target genes. Mohanta et al. (2017) proved how CRISPR/Cas9 modification used in plant pathology to manage resistance to microorganisms and to target the viral genomes S has become a powerful approach to improve agricultural traits while revolutionizing genome editing technology. Despite their ploidy complexity, the genomes of representative plants *A. thaliana* and *Nicotiana benthamiana* have been edited; CRISPR/Cas9 has effectively modified the genetic code of almost all biological systems, including many species of plants, from monocots to dicots (Brooks et al. 2014; Fan et al. 2015). The CRISPR/Cas9 genome editing approach was first considered to be a microbial class II adaptive immune mechanism that was originally identified in *Streptococcus pyogenes* (Langner et al. 2018). It comprises two components: the Cas9 nuclease and (gRNA) a customized single guide RNA (Khatodia et al. 2016). The system consisting of CRISPR and Cas9 is an enormous throughput technique used in current genome research, including fresh research proving Cas9's ubiquitous usage in genetic administering drugs, and knockout mutations via non-homologous joining (NHEJ)-mediated double-strand breaks (DSBs) (Bortesi et al. 2016). A protospacer adjacent motif (PAM) sequence (5′-NGG-3′) is required for generating a DSB at the target site. The Cas9 system's capacity to generate DSBs in DNA has rendered it a suitable tool for modifying genes (Jinek et al. 2016). Because of its error-prone nature, NHEJ produces tiny segments (insertions/deletions) at a particular location. Numerous investigations on the mechanism underlying the activity of this powerful nuclease have shown the existence of CRISPR locations, a collection comprising repeating nucleotides found within the immunity systems of bacterial and archaeal organisms (Lee et al. 2019). These loci feature a unique sequence that comprises operons expressing Cas9, transcription machinery, and sequential repeats that were integrated within the genome of bacterial cells through the introduction of a viral infection or other foreign aggressor in response to an immune-triggered response (Zhang et al. 2014); these repeats are derived from different viral DNA genomes isolated by intervening sequences. And, in the laboratory, it requires no debonair machinery other than (i) an 18–22 bp extended RNA-guide sequencing that is identical to the intended target area and (ii) Cas9 nuclease to cause DSB by following the PAM sequence (Chen et al. 2019). A potentially less-than-faithful method of NHEJ or a more specific system of homology-directed repair corrects apparent CRISPR-induced DSBs (Liu 2019). NHEJ's error-prone nature allows it to cut a targeted sequence with insertions or deletions, resulting in the loss of function and crops that have been genetically modified. Following the creation and recognition of the PAM sequence, the gRNA has been cloned into the expression vector of selection, transformed, screened, and then validated (Lone et al. 2018; Pickar-Oliver et al. 2019). It has been established that Cas9 may be 'manipulated' by inserting any foreign nucleotide sequence into the genome of bacterial cells (Yin et al. 2017). To delete a gene, the CRISPR/Cas9 system is constructed and applied to explants using a biolistic strategy *Agrobacterium* or electroporation technique. Most studies support *Agrobacterium*-mediated processing due to its efficacy and safety (Wang et al. 2018). Regenerated seedlings originate through modified callus and are subsequently transferred to planting soil (Zhang et al. 2014). The predominantly used genome editing system is genotype II, and the more regularly used approach is CRISPR/Cas9 from *S. pyogenes*. The strategy of CRISPR/Cas9 gene-silencing technology has four major properties: cleavage of Cas9 3–4 nucleotides after the adjacent PAM (around 50 NGGs identify the PAM sequence) (Jinek et al. 2012), selection of a suitable binary vector for gRNA cloning, synthetic guide RNA (approximately 18–20 nucleotides) attaches to the target DNA, and transformation of the construction within explants using *Agrobacterium* or microprojectile gene delivery. The CRISPR/Cas9 defense approach is marketed as an outstanding instrument for plant technological advances, features a sequence-specific RNA-mediated adaptive immunity system discovered in prokaryotes, and is considered an excellent microbial defense tool that protects against viruses and bacteriophages (Hu et al. 2019). The network of CRISPR molecules is then transcribed to form a complete cRNA. This is known as lncRNA biosynthesis. The following cRNA recognizes and kills viral DNA sequences that include sequences complementary to the cRNA (Ahmad et al. 2020). The Cas9 nuclease, expressed by Cas genes around the CRISPR network, helps to degrade DNA (Liu et al. 2016).

2.5 CONCLUSIONS

Developing crops for disease resistance requires the introduction of new R genes and an understanding of disease resistance information against phytopathogens. In addition, intracellular and surface immune systems work together to provide strong resistance to infection. Plants recognize a variety of chemicals, both their own and others, as triggers for disease resistance. Plants have two layers of protection against these plant pathogens, presented as PTI and ETI. PRRs at the cell membrane identify chemicals from pathogens or external damage, as well as apoplastic pathogenic effectors. Nucleotide-linked leucine-rich repeat (NLR) proteins identify pathogenic effectors. Understanding the revolution in high-throughput molecular approaches and plant–molecular interactions has paved the way for advances in pathogen-resistant plant growth. Although modern plant breeding is based on ancient procedures, the implementation of new technology will undoubtedly increase efficiency and performance. To ensure resistant variants in the future, genotyping technologies, plant production strategies, breeding strategies, and field trial methodology must be implemented. We may anticipate the rise of a variety of ways that employ interdisciplinary thinking to optimize the benefits of introducing new technology. Among breeding strategies, CRISPR/Cas9 has transformed molecular biology and its applications in agriculture, signaling a watershed moment for the development of plants and productivity.

CDPK: Ca^{2+}-dependent protein kinase; DAMPs: damage-associated molecular patterns; LCO: lipo-chitooligosaccharide; MAMPs: microbial-associated molecular patterns; NRLs: intracellular nucleotide-binding/leucine-rich-repeat; PRR: pattern-recognition receptors; RbohD: respiratory-burst NADPH oxidase homolog D; ROS: reactive oxygen species.

REFERENCES

Abegaz, B. M., & Kinfe, H. H. (2019). Secondary metabolites, their structural diversity, bioactivity, and ecological functions: An overview. *Physical Sciences Reviews*, *4*(6), 20180100.

Adams, M. D., Kelley, J. M., Gocayne, J. D., Dubnick, M., Polymeropoulos, M. H., Xiao, H., Merril, C. R., Wu, A., Olde, B., & Palanques, R. M. (1991). Complementary DNA sequencing: Expressed sequence tags and human genome project. *Science*, *252*(5013), 1651–1656.

Agrios, G. N. (2005). *Plant pathology* (5th ed.). Academic Press.

Ahamad, F., & Khan, M. R. (2023). Incidence of sheath blight in irrigated rice and associated yield losses in Northern India. *Plant Disease*, *107*(10), 2907–2915.

Ahmad, S., Wei, X., Sheng, Z., Hu, P., & Tang, S. (2020). CRISPR/Cas9 for development of disease resistance in plants: Recent progress, limitations and future prospects. *Briefings in Functional Genomics*, *19*(1), 26–39.

Ahmad, S., Khan, H., Shahab, U., Rehman, S., Rafi, Z., Khan, M. Y., Ansari, A., Siddiqui, Z., Ashraf, J. M., Abdullah, S. M., Habib, S., Uddin, M. (2017). Protein oxidation: An overview of metabolism of sulphur containing amino acid, cysteine. *Frontiers in Bioscience-Scholar*, *9*(1), 71–87.

Ahuja, I., Kissen, R., & Bones, A. M. (2012). Phytoalexins in defense against pathogens. *Trends in Plant Science*, *17*(2), 73–90.

Albert, I., Hua, C., Nürnberger, T., Pruitt, R. N., & Zhang, L. (2020). Surface sensor systems in plant immunity. *Plant physiology*, *182*(4), 1582–1596.

Allen, T. W., Bradley, C. A., Sisson, A. J., Byamukama, E., Chilvers, M. I., Coker, C. M., Collins, A. A., Damicone, J. P., Dorrance, A. E., Dufault, N. S., Esker, P. D., Faske, T. R., Giesler, L. J., Grybauskas, A. P., Hershman, D. E., Hollier, C. A., Isakeit, T., Jardine, D. J., Kelly, H. M., Kemerait, R. C., Kleczewski, N. M., Koenning, S. R., Kurle, J. E., Malvick, D. K., Markell, S. G., Mehl, H. L., Mueller, D. S., Mueller, J. D., Mulrooney, R. P., Nelson, B. D., Newman, M. A., Osborne, L., Overstreet, C., Padgett, G. B., Phipps, P. M., Price, P. P., Sikora, E. J., Smith, D. L., Spurlock, T. N., Tande, C. A., Tenuta, A. U., Wise, K. A., & Wrather, J. A. (2017). Soybean yield loss estimates due to diseases in the United States and Ontario, Canada, from 2010 to 2014. *Plant Health Progress*, *18*(1), 19–27.

Amarante-Mendes, G. P., Adjemian, S., Branco, L. M., Zanetti, L. C., Weinlich, R., & Bortoluci, K. R. (2018). Pattern recognition receptors and the host cell death molecular machinery. *Frontiers in Immunology*, *9*, 2379.

Asai, T., Tena, G., Plotnikova, J., Willmann, M. R., Chiu, W. L., Gomez-Gomez, L., Boller, T., Ausubel, F. M., Sheen, J. (2002). MAP kinase signalling cascade in Arabidopsis innate immunity. *Nature*, *415*(6875), 977–983.

Asibi, A. E., Chai, Q., & Coulter, J. A. (2019). Rice blast: A disease with implications for global food security. *Agronomy*, *9*(8), 451.

Assefa, T., Assibi Mahama, A., Brown, A. V., Cannon, E. K. S., Rubyogo, J. C., Rao, I. M., Blair, M. W., & Cannon, S. B. (2019). A review of breeding objectives, genomic resources, and marker-assisted methods in common bean (*Phaseolus vulgaris* L.). *Molecular Breeding*, *39*. https://doi.org/10.1007/s11032-018-0920-0.

Athiyannan, N., Abrouk, M., Boshoff, W. H., Cauet, S., Rodde, N., Kudrna, D., Mohammed, N., Bettgenhaeuser, J., Botha, K. S., Derman, S. S., Wing, R. A., Prins, R., & Krattinger, S. G. (2022). Long-read genome sequencing of bread wheat facilitates disease resistance gene cloning. *Nature Genetics*, *54*(3), 227–231.

Awondo, S. N., Fonsah, E. G., Riley, D., & Abney, M. (2012). Effectiveness of tomato-spotted wilt virus management tactics. *Journal of Economic Entomology*, *105*(3), 943–948. https://dx. doi.org/10.1603/EC11272.

Axtell, M. J., & Staskawicz, B. J. (2003). Initiation of RPS2-specified disease resistance in Arabidopsis is coupled to the AvrRpt2-directed elimination of RIN4. *Cell*, *112*(3), 369–377.

Aydoğdu, M., Boyraz, N., & Yüksel, K. A. Y. A. (2015). Effect on yield losses on maize (*Zea mays* L.) caused by smut disease (*Ustilago maydis* (DC) Corda). *The Journal of Turkish Phytopathology*, *44(1–2–3)*, 23–30.

Baetsen-Young, A. M., Swinton, S. M., & Chilvers, M. I. (2021). Economic impact of fluopyram-amended seed treatments to reduce soybean yield loss associated with sudden death syndrome. *Plant Disease*, *105*(1), 78–86.

Baggs, E., Dagdas, G., & Krasileva, K. V. (2017). NLR diversity, helpers and integrated domains: Making sense of the NLR IDentity. *Current Opinion in Plant Biology*, *38*, 59–67.

Bardak, A., Çelik, S., Erdoğan, O., Ekinci, R., & Dumlupinar, Z. (2021). Association mapping of Verticillium wilt disease in a worldwide collection of cotton (*Gossypium hirsutum* L.). *Plants*, *10*(2), 306.

Bednarek, P., & Schulze-Lefert, P. (2018). Role of plant secondary metabolites at the host–pathogen interface. *Annual Plant Reviews*, *34*, 220–260.

Beliën, T., Van Campenhout, S., Robben, J., & Volckaert, G. (2006). Microbial endoxylanases: Effective weapons to breach the plant cell-wall barrier or, rather, triggers of plant defense systems?. *Molecular Plant-Microbe Interactions*, *19*(10), 1072–1081.

Bent, A. F., Kunkel, B. N., Dahlbeck, D., Brown, K. L., Schmidt, R., Giraudat, J., Leung, J., & Staskawicz, B. J. (1994). RPS2 of *Arabidopsis thaliana*: A leucine-rich repeat class of plant disease resistance genes. *Science*, *265*(5180), 1856–1860.

Bernoux, M., Zetzsche, H., & Stuttmann, J. (2022). Connecting the dots between cell surface-and intracellular-triggered immune pathways in plants. *Current Opinion in Plant Biology*, *69*, 102276.

Bettinelli, P., Nicolini, D., Giovannini, O., Stefanini, M., Hausmann, L., & Vezzulli, S. (2023). Breeding for black rot resistance in grapevine: Advanced approaches for germplasm screening. *Euphytica*, *219*(11), 113.

Blaise, D., & Kranthi, K. R. (2019). Cotton production in India. In K. Jabran & B. Singh Chauhan (Eds.), *Cotton production* (1st ed., pp. 193–215). Wiley. doi:10.1002/9781119385523.ch10.

Böhm, H., Albert, I., Fan, L., Reinhard, A., & Nürnberger, T. (2014). Immune receptor complexes at the plant cell surface. *Current Opinion in Plant Biology*, *20*, 47–54.

Boller, T., & Felix, G. (2009). A renaissance of elicitors: Perception of microbe-associated molecular patterns and danger signals by pattern recognition receptors. *Annual Review of Plant Biology*, *60*, 379–406.

Boller, T., & He, S. Y. (2009). Innate immunity in plants: An arms race between pattern recognition receptors in plants and effectors in microbial pathogens. *Science*, *324*(5928), 742–744.

Bolton, M. D. (2009). Primary metabolism and plant defense: Fuel for the fire. *Molecular Plant-Microbe Interactions*, *22*(5), 487–497.

Bonardi, V., Tang, S., Stallmann, A., Roberts, M., Cherkis, K., & Dangl, J. L. (2011). Expanded functions for a family of plant intracellular immune receptors beyond specific recognition of pathogen effectors. *Proceedings of the National Academy of Sciences*, *108*(39), 16463–16468.

Bortesi, L., Zhu, C., Zischewski, J., Perez, L., Bassié, L., Nadi, R., Forni, G., Lade, S. B., Soto, E., Jin, X., Medina, V., Villorbina, G., Muñoz, P., Farré, G., Fischer, R., Twyman, R. M., Capell, T., Christou, P., & Schillberg, S. (2016). Patterns of CRISPR/Cas9 activity in plants, animals and microbes. *Plant Biotechnology Journal*, *14*(12), 2203–2216.

Boutrot, F., & Zipfel, C. (2017). Function, discovery, and exploitation of plant pattern recognition receptors for broad-spectrum disease resistance. *Annual Review of Phytopathology*, *55*, 257–28.

Bradley, C. A., Allen, T. W., Sisson, A. J., Bergstrom, G. C., Bissonnette, K. M., Bond, J., Byamukama, E., Chilvers, M. I., Collins, A. A., Damicone, J. P., Dorrance, A. E., Dufault, N. S., Esker, P. D., Faske, T. R., Fiorellino, N. M., Giesler, L. J., Hartman, G. L., Hollier, C. A., Isakeit, T., Jackson-Ziems, T. A., Jardine, D. J., Kelly, H. M., Kemerait, R. C., Kleczewski, N. M., Koehler, A. M., Kratochvil, R. J., Kurle, J. E., Malvick, D. K., Markell, S. G., Mathew, F. M., Mehl, H. L., Mehl, K. M., Mueller, D. S., Mueller, J. D., Nelson, B. D., Overstreet, C., Padgett, G. B., Price, P. P., Sikora, E. J., Small, I., Smith, D. L., Spurlock, T. N., Tande, C. A., Telenko, D. E. P., Tenuta, A. U., Thiessen, L. D., Warner, F., Wiebold, W. J., & Wise, K. A. (2021). Soybean yield loss estimates due to diseases in the United States and Ontario, Canada, from 2015 to 2019. *Plant Health Progress*, *22*(4), 483–495.

Brooks, C., Nekrasov, V., Lippman, Z. B., & Van Eck, J. (2014). Efficient gene editing in tomato in the first generation using the clustered regularly interspaced short palindromic repeats/CRISPR-associated9 system. *Plant Physiology*, *166*(3), 1292–1297.

Butassi, E., Svetaz, L. A., Zhou, S., Wolfender, J. L., Cortes, J. C., Ribas, J. C., Díaz, C., Pérez-del Palacio, J., Vicente, F., & Zacchino, S. A. (2019). The antifungal activity and mechanisms of action of quantified extracts from berries, leaves and roots of *Phytolacca tetramera*. *Phytomedicine*, *60*, 152884.

Casey, L. W., Lavrencic, P., Bentham, A. R., Cesari, S., Ericsson, D. J., Croll, T., Turk, D., Anderson, P. A., Mark, A. E., Dodds, P. N., Mobli, M., Kobe, B., & Williams, S. J. (2016). The CC domain structure from the wheat stem rust resistance protein Sr33 challenges paradigms for dimerization in plant NLR proteins. *Proceedings of the National Academy of Sciences*, *113*(45), 12856–12861.

Cesari, S. (2018). Multiple strategies for pathogen perception by plant immune receptors. *New Phytologist*, *219*, 17–24.

Cesari, S., Thilliez, G., Ribot, C., Chalvon, V., Michel, C., Jauneau, A., Rivas, S., Alaux, L., Kanzaki, H., Okuyama, Y., Morel, J. B., Fournier, E., Tharreau, D., Terauchi, R., & Kroj, T. (2013). The rice resistance protein pair RGA4/RGA5 recognizes the *Magnaporthe oryzae* effectors AVR-Pia and AVR1-CO39 by direct binding. *The Plant Cell*, *25*(4), 1463–1481.

Chakraborty, S., & Newton, A. C. (2011). Climate change, plant diseases and food security: An overview. *Plant Pathology*, *60*(1), 2–14.

Chanclud, E., & Morel, J. B. (2016). Plant hormones: A fungal point of view. *Molecular Plant Pathology*, *17*(8), 1289–1297.

Chaparro-Garcia, A., Schwizer, S., Sklenar, J., Yoshida, K., Petre, B., Bos, J. I., Schornack, S., Jones, A. M. E., Bozkurt, T. O., & Kamoun, S. (2015). *Phytophthora infestans* RXLR-WY effector AVR3a associates with dynamin-related protein 2 required for endocytosis of the plant pattern recognition receptor FLS2. *PLoS One*, *10*(9), e0137071.

Chen, K., Wang, Y., Zhang, R., Zhang, H., & Gao, C. (2019). CRISPR/Cas genome editing and precision plant breeding in agriculture. *Annual review of Plant Biology*, *70*, 667–697.

Chen, N., Xia, P., Li, S., Zhang, T., Wang, T. T., & Zhu, J. (2017). RNA sensors of the innate immune system and their detection of pathogens. *IUBMB Life*, *69*(5), 297–304.

Chen, S., Liu, X., Zeng, L., Ouyang, D., Yang, J., & Zhu, X. (2011). Genetic analysis and molecular mapping of a novel recessive gene xa34 (t) for resistance against *Xanthomonas oryzae pv. oryzae*. *Theoretical and Applied Genetics*, *122*, 1331–1338.

Chen, S., Songkumarn, P., Venu, R. C., Gowda, M., Bellizzi, M., Hu, J., Liu, W., Ebbole, D., Meyers, B., Mitchell, T., & Wang, G. L. (2013). Identification and characterization of in planta–expressed secreted effector proteins from *Magnaporthe oryzae* that induce cell death in rice. *Molecular Plant-Microbe Interactions*, *26*(2), 191–202.

Chinchilla, D., Bauer, Z., Regenass, M., Boller, T., & Felix, G. (2006). The Arabidopsis receptor kinase FLS2 binds flg22 and determines the specificity of flagellin perception. *The Plant Cell*, *18*(2), 465–476.

Chinchilla, D., Zipfel, C., Robatzek, S., Kemmerling, B., Nürnberger, T., Jones, J. D., Felix, G., & Boller, T. (2007). A flagellin-induced complex of the receptor FLS2 and BAK1 initiates plant defence. *Nature*, *448*(7152), 497–500.

Chu, Z., Fu, B., Yang, H., Xu, C., Li, Z., Sanchez, A., Park, Y. J., Bennetzen, J. L., Zhang, Q., & Wang, S. (2006). Targeting xa13, a recessive gene for bacterial blight resistance in rice. *Theoretical and Applied Genetics*, *112*, 455–461.

Chukwu, S. C., Rafii, M. Y., Ramlee, S. I., Ismail, S. I., Oladosu, Y., Muhammad, I., Musa, I., Ahmed, M., Jatto, M. I., & Yusuf, B. R. (2020). Recovery of recurrent parent genome in a marker-assisted backcrossing against rice blast and blight infections using functional markers and SSRs. *Plants 9*, 1–15. https://doi.org/10.3390/plants9111411.

Cui, H., Tsuda, K., & Parker, J. E. (2015). Effector-triggered immunity: From pathogen perception to robust defense. *Annual Review of Plant Biology*, *66*, 487–511.

Day, B., Dahlbeck, D., & Staskawicz, B. J. (2006). NDR1 interaction with RIN4 mediates the differential activation of multiple disease resistance pathways in Arabidopsis. *The Plant Cell*, *18*(10), 2782–2791.

De Jonge, R., Peter van Esse, H., Kombrink, A., Shinya, T., Desaki, Y., Bours, R., Van der Krol, S., Shibuya, N., Joosten, M. H. A. J., & Thomma, B. P. (2010). Conserved fungal LysM effector Ecp6 prevents chitin-triggered immunity in plants. *Science*, *329*(5994), 953–955.

Deng, Y., Ning, Y., Yang, D. L., Zhai, K., Wang, G. L., & He, Z. (2020). Molecular basis of disease resistance and perspectives on breeding strategies for resistance improvement in crops. *Molecular Plant*, *13*(10), 1402–1419.

Deslandes, L., Olivier, J., Theulières, F., Hirsch, J., Feng, D. X., Bittner-Eddy, P., Beynon, J., & Marco, Y. (2002). Resistance to *Ralstonia solanacearum* in *Arabidopsis thaliana* is conferred by the recessive RRS1-R gene, a member of a novel family of resistance genes. *Proceedings of the National Academy of Sciences*, *99*(4), 2404–2409.

Dey, U., Harlapur, S. I., Dhutraj, D. N., Suryawanshi, A. P., Badgujar, S. L., Jagtap, G. P., & Kuldhar, D. P. (2012). Spatiotemporal yield loss assessment in corn due to common rust caused by Puccinia sorghi Schw. *African Journal of Agricultural Research*, *7*(37), 5265–5269.

Dixon, M. S., Jones, D. A., Keddie, J. S., Thomas, C. M., Harrison, K., & Jones, J. D. (1996). The tomato Cf-2 disease resistance locus comprises two functional genes encoding leucine-rich repeat proteins. *Cell*, *84*(3), 451–459.

Dodds, P. N., & Rathjen, J. P. (2010). Plant immunity: Towards an integrated view of plant– pathogen interactions. *Nature Reviews Genetics*, *11*(8), 539–548.

Dods, R. (2019). The Amino Acid Sequences of Proteins Determine Folding and Non-folding. *Concepts in Bioscience Engineering*, *2*, 37–83.

Dow, M., Newman, M. A., & Von Roepenack, E. (2000). The induction and modulation of plant defense responses by bacterial lipopolysaccharides. *Annual Review of Phytopathology*, *38*(1), 241–261.

Du, J., Verzaux, E., Chaparro-Garcia, A., Bijsterbosch, G., Keizer, L. C., Zhou, J. I., Liebrand, T. W. H., Xie, C., Govers, F., Robatzek, S., van der Vossen, E. A. G., Jacobsen, E., Visser, R. G. F., Kamoun, S., & Vleeshouwers, V. G. (2015). Elicitin recognition confers enhanced resistance to *Phytophthora infestans* in potato. *Nature Plants*, *1*(4), 1–5.

Durrant, W. E., & Dong, X. (2004). Systemic acquired resistance. *Annual Review of Phytopathology*, *42*, 185–209.

Duxbury, Z., Ma, Y., Furzer, O. J., Huh, S. U., Cevik, V., Jones, J. D. G., & Sarris, P. F. (2016). Pathogen perception by NLRs in plants and animals: Parallel worlds. *BioEssays*, *38*, 769–781

Duy, P. N., Lan, D. T., Pham Thu, H., Thi Thu, H. P., Nguyen Thanh, H., Phuong Pham, N., Auguy, F., Bui Thi Thu, H., Tran Bao Manh, T., Cunnac, S., & Pham, X. H. (2021). Improved bacterial leaf blight disease resistance in the major elite Vietnamese rice cultivar TBR225 via editing of the OsSWEET14 promoter. *PloS One*, *16*(9), e0255470.

Ekroth, A. K., Rafaluk-Mohr, C., & King, K. C. (2019). Host genetic diversity limits parasite success beyond agricultural systems: A meta-analysis. *Proceedings of the Royal Society B*, *286*(1911), 20191811.

El Kasmi, F. (2021). How activated NLRs induce anti-microbial defenses in plants. *Biochemical Society Transactions*, *49*(5), 2177–2188.

El Mohtar, C. A., Atamian, H. S., Dagher, R. B., Abou-Jawdah, Y., Salus, M. S., & Maxwell, D. P. (2007). Marker-assisted selection of tomato genotypes with the I-2 gene for resistance to Fusarium oxysporum f. sp. *lycopersici race 2. Plant Disease, 91*(6), 758–762.

El-Orabey, W. M., & Elkot, A. F. (2020). Prediction of leaf rust severity and yield loss in wheat based on environmental factors. *Journal of Plant Diseases and Protection, 127*, 507–519.

Engelhardt, S., Boevink, P. C., Armstrong, M. R., Ramos, M. B., Hein, I., & Birch, P. R. (2012). Relocalization of late blight resistance protein R3a to endosomal compartments is associated with effector recognition and required for the immune response. *The Plant Cell, 24*(12), 5142–5158.

Engering, A., Hogerwerf, L., & Slingenbergh, J. (2013). Pathogen–host–environment interplay and disease emergence. *Emerging Microbes & Infections, 2*(1), 1–7.

Fagard, M., Launay, A., Clément, G., Courtial, J., Dellagi, A., Farjad, M., Krapp, A., Soulié, M. C., & Masclaux-Daubresse, C. (2014). Nitrogen metabolism meets phytopathology. *Journal of Experimental Botany, 65*(19), 5643–5656.

Fan, D., Liu, T., Li, C., Jiao, B., Li, S., Hou, Y., & Luo, K. (2015). Efficient CRISPR/Cas9-mediated targeted mutagenesis in *Populus* in the first generation. *Scientific Reports, 5*(1), 12217.

Feys, B. J., Wiermer, M., Bhat, R. A., Moisan, L. J., Medina-Escobar, N., Neu, C., Cabral, A., & Parker, J. E. (2005). Arabidopsis senescence-associated gene101 stabilizes and signals within an enhanced disease susceptibility1 complex in plant innate immunity. *The Plant Cell, 17*(9), 2601–2613.

Fisher, M. C., Henk, D. A., Briggs, C. J., Brownstein, J. S., Madoff, L. C., McCraw, S. L., & Gurr, S. J. (2022). Emerging fungal threats to animal, plant and ecosystem health. *Nature, 484*(7393), 186–194.

Fliegmann, J., Mithöfer, A., Wanner, G., & Ebel, J. (2004). An ancient enzyme domain hidden in the putative β-glucan elicitor receptor of soybean may play an active part in the perception of pathogen-associated molecular patterns during broad host resistance. *Journal of Biological Chemistry, 279*(2), 1132–1140.

Flor, H. H. (1955). Host-parasite interaction in flax rust - its genetics and other implications. *Phytopathology, 45*, 680–685.

Flor, H. H. (1971). Current status of the gene-for-gene concept. *Annual Review of Phytopathology, 9*, 275–296

Fones, H. N., Bebber, D. P., Chaloner, T. M., Kay, W. T., Steinberg, G., & Gurr, S. J. (2020). Threats to global food security from emerging fungal and oomycete crop pathogens. *Nature Food, 1*(6), 332–342.

Franceschetti, M., Maqbool, A., Jiménez-Dalmaroni, M. J., Pennington, H. G., Kamoun, S., & Banfield, M. J. (2017). Effectors of filamentous plant pathogens: Commonalities amid diversity. *Microbiology and Molecular Biology Reviews, 81*(2), e00066–16.

Fu, D., Uauy, C., Distelfeld, A., Blechl, A., Epstein, L., Chen, X., Sela, H., Fahima, T., & Dubcovsky, J. (2009). A kinase-START gene confers temperature-dependent resistance to wheat stripe rust. *Science, 323*(5919), 1357–1360.

Galiano-Carneiro, A. L., Boeven, P. H. G., Maurer, H. P., Würschum, T., & Miedaner, T. (2019). Genome-wide association study for an efficient selection of Fusarium head blight resistance in winter triticale. *Euphytica, 215*. https://doi.org/10.1007/s10681-018-2327-8.

Galletti, R., De Lorenzo, G., & Ferrari, S. (2009). Host-derived signals activate plant innate immunity. *Plant Signaling & Behavior, 4*(1), 33–34.

Gao, Z., Johansen, E., Eyers, S., Thomas, C. L., Noel Ellis, T. H., & Maule, A. J. (2004). The potyvirus recessive resistance gene, sbm1, identifies a novel role for translation initiation factor eIF4E in cell-to-cell trafficking. *The Plant Journal, 40*(3), 376–385.

Garcia-Brugger, A., Lamotte, O., Vandelle, E., Bourque, S., Lecourieux, D., Poinssot, B., Wendehenne, D., & Pugin, A. (2006). Early signaling events induced by elicitors of plant defenses. *Molecular Plant-Microbe Interactions, 19*(7), 711–724.

Gilbert, G. S., & Parker, I. M. (2016). The evolutionary ecology of plant disease: A phylogenetic perspective. *Annual Review of Phytopathology, 54*, 549–578.

Göhre, V., Jones, A. M., Sklenář, J., Robatzek, S., & Weber, A. P. (2012). Molecular crosstalk between PAMP-triggered immunity and photosynthesis. *Molecular Plant-Microbe Interactions, 25*(8), 1083–1092.

González, C., Brito, N., & Sharon, A. (2016). Infection process and fungal virulence factors. In: Fillinger, S., Elad, Y. (Eds.), *Botrytis: the fungus, the pathogen and its management in agricultural systems Springer, Cham.* (pp. 229–246).

Grover, A., & Sharma, P. C. (2016). Development and use of molecular markers: Past and present. *Critical Reviews in Biotechnology, 36*(2), 290–302.

Gururani, M. A., Venkatesh, J., Upadhyaya, C. P., Nookaraju, A., Pandey, S. K., & Park, S. W. (2012). Plant disease resistance genes: Current status and future directions. *Physiological and Molecular Plant Pathology, 78*, 51–65.

Hall, A. R., Scanlan, P. D., Morgan, A. D., & Buckling, A. (2011). Host–parasite coevolutionary arms races give way to fluctuating selection. *Ecology Letters*, *14*(7), 635–642.

Han, G. Z. (2019). Origin and evolution of the plant immune system. *New Phytologist*, *222*(1), 70–83.

Hanson, P., Lu, S. F., Wang, J. F., Chen, W., Kenyon, L., Tan, C. W., Tee, K. L., Wang, Y. Y., Hsu, Y. C., Schafleitner, R., Ledesma, D., & Yang, R. Y. (2016). Conventional and molecular marker-assisted selection and pyramiding of genes for multiple disease resistance in tomato. *Scientia Horticulturae*, *201*, 346–354.

Haq, I. U., & Ijaz, S. (Eds.). (2020). *Plant disease management strategies for sustainable agriculture through traditional and modern approaches* (vol. 13). Springer.

Haque, M. A., Rafii, M. Y., Yusoff, M. M., Ali, N. S., Yusuff, O., Datta, D. R., Anisuzzaman, M., & Ikbal, M. F. (2021). Recent advances in rice varietal development for durable resistance to biotic and abiotic stresses through marker-assisted gene pyramiding. *Sustain*, *13*, 1–25. https://doi.org/10.3390/su131910806.

Harris, C. J., Slootweg, E. J., Goverse, A., & Baulcombe, D. C. (2013). Stepwise artificial evolution of a plant disease resistance gene. *Proceedings of the National Academy of Sciences*, *110*(52), 21189–21194.

Hasan, N., Choudhary, S., Naaz, N., Sharma, N., & Laskar, R. A. (2021). Recent advancements in molecular marker-assisted selection and applications in plant breeding programmes. *Journal of Genetic Engineering and Biotechnology*, *19*, 1–26. https://doi.org/10.1186/s43141-021-00231-1.

Haverkort, A. J., Boonekamp, P. M., Hutten, R., Jacobsen, E., Lotz, L. A. P., Kessel, G. J. T., Vossen, J. H., & Visser, R. G. F. (2016). Durable late blight resistance in potato through dynamic varieties obtained by cisgenesis: Scientific and societal advances in the DuRPh project. *Potato Research*, *59*, 35–66.

Hayashi, N., Inoue, H., Kato, T., Funao, T., Shirota, M., Shimizu, T., Kanamori, H., Yamane, H., Hayano-Saito, Y., Matsumoto, T., Yano, M., & Takatsuji, H. (2010). Durable panicle blast-resistance gene Pb1 encodes an atypical CC-NBS-LRR protein and was generated by acquiring a promoter through local genome duplication. *The Plant Journal*, *64*(3), 498–510.

Henkrar, F., & Udupa, S. (2021). Marker Assisted Selection in Plant Breeding : *Future Opportunities*. *1*, 237–247.

Hillocks, R. A., & Kibani, T. H. M. (2002). Factors affecting the distribution, incidence and spread of Fusarium wilt of cotton in Tanzania. *Experimental Agriculture*, *38*(1), 13–27.

Holt, B. F., Boyes, D. C., Ellerström, M., Siefers, N., Wiig, A., Kauffman, S., Grant, M. R., & Dangl, J. L. (2002). An evolutionarily conserved mediator of plant disease resistance gene function is required for normal Arabidopsis development. *Developmental Cell*, *2*(6), 807–817.

Hou, S. Liu, Z., Shen, H., & Wu, D. (2019). Damage-associated molecular pattern-triggered immunity in plants. *Frontiers in Plant Science*, *10*, 646.

Hu, N., Xian, Z., Li, N., Liu, Y., Huang, W., Yan, F., Su, D., Chen, J., & Li, Z. (2019). Rapid and user-friendly open-source CRISPR/Cas9 system for single-or multi-site editing of tomato genome. *Horticulture Research*, *6*, 7.

Huang, X. Q., Hsam, S. L. K., Zeller, F. J., Wenzel, G., & Mohler, V. (2000). Molecular mapping of the wheat powdery mildew resistance gene Pm24 and marker validation for molecular breeding. *Theoretical and Applied Genetics*, *101*, 407–414.

Huang, X., & Han, B. (2014). Natural variations and genome-wide association studies in crop plants. *Annual Review of Plant Biology*, *65*, 531–551. https://doi.org/10.1146/annurev-arplant-050213-035715

Huffaker, A., Pearce, G., & Ryan, C. A. (2006). An endogenous peptide signal in Arabidopsis activates components of the innate immune response. *Proceedings of the National Academy of Sciences*, *103*(26), 10098–10103.

Huma, H., & Majid Fazili, K. (2007). Plant protease inhibitors: A defense strategy in plants. *Biotechnology and Molecular Biology Reviews*, *2*(3), 68–85.

Humphry, M., Bednarek, P., Kemmerling, B., Koh, S., Stein, M., Göbel, U., Stüber, K., Piślewska-Bednarek, M., Loraine, A., Schulze-Lefert, P., Somerville, S., & Panstruga, R. (2010). A regulon conserved in monocot and dicot plants defines a functional module in antifungal plant immunity. *Proceedings of the National Academy of Sciences*, *107*(50), 21896–21901.

Hussain, M., Shakoor, N., Adeel, M., Ahmad, M. A., Zhou, H., Zhang, Z., Xu, M., Rui, Y., & White, J. C. (2023). Nano-enabled plant microbiome engineering for disease resistance. *Nano Today*, *48*, 101752.

Iyer-Pascuzzi, A. S., & McCouch, S. R. (2007). Recessive resistance genes and the Oryza sativa-Xanthomonas oryzae pv. *oryzae pathosystem*. *Molecular Plant-Microbe Interactions*, *20*(7), 731–739.

Jacob, F., Vernaldi, S., & Maekawa, T. (2013). Evolution and conservation of plant NLR functions. *Frontiers in Immunology*, *4*, 297.

Jacob, P., Kim, N. H., Wu, F., El-Kasmi, F., Chi, Y., Walton, W. G., Furzer, O. J., Lietzan, A. D., Sunil, S., & Dangl, J. L. (2021). Plant "helper" immune receptors are Ca2+-permeable nonselective cation channels. *Science, 373*(6553), 420–425.

Jan, R., Asif, S., Asaf, S., Du, X. X., Park, J. R., Nari, K., Bhatta, D., Lee, I. J., & Kim, K. M. (2023). Melatonin alleviates arsenic (As) toxicity in rice plants via modulating antioxidant defense system and secondary metabolites and reducing oxidative stress. *Environmental Pollution, 318*, 120868.

Jeger, M. J. (2022). The impact of climate change on disease in wild plant populations and communities. *Plant Pathology, 71*(1), 111–130.

Jeung, J. U., Kim, B. R., Cho, Y. C., Han, S. S., Moon, H. P., Lee, Y. T., & Jena, K. K. (2007). A novel gene, Pi40 (t), linked to the DNA markers derived from NBS-LRR motifs confers broad spectrum of blast resistance in rice. *Theoretical and Applied Genetics, 115*, 1163–1177.

Jiang, N., Yan, J., Liang, Y., Shi, Y., He, Z., Wu, Y., Zeng, Q., Liu, X., & Peng, J. (2020). Resistance genes and their interactions with bacterial blight/leaf streak pathogens (*Xanthomonas oryzae*) in rice (*Oryza sativa L.*): An updated review. *Rice, 13*(1), 1–12.

Jiang, Z., Ge, S., Xing, L., Han, D., Kang, Z., Zhang, G., Wang, X., Wang, X., Chen, P., & Cao, A. (2013). RLP1. 1, a novel wheat receptor-like protein gene, is involved in the defence response against *Puccinia striiformis f. sp. tritici. Journal of Experimental Botany, 64*(12), 3735–3746.

Jinek, M., Chylinski, K., Fonfara, I., Hauer, M., Doudna, J. A., & Charpentier, E. (2012). A programmable dual-RNA–guided DNA endonuclease in adaptive bacterial immunity. *Science, 337*(6096), 816–821.

Johal, G. S., & Briggs, S. P. (1992). Reductase activity encoded by the HM1 disease resistance gene in maize. *Science, 258*(5084), 985–987.

Jonah, P. M., Bello, L. L., Lucky, O., Midau, A., & Moruppa, S. M. (2011). The importance of molecular markers in plant breeding programmes. *Global Journal of Science Frontier Research, 11*(5), 5–12.

Jones, D. A., Thomas, C. M., Hammond-Kosack, K. E., Balint-Kurti, P. J., & Jones, J. D. (1994). Isolation of the tomato Cf-9 gene for resistance to *Cladosporium fulvum* by transposon tagging. *Science, 266*(5186), 789–793.

Jones, H., Whipps, J. M., & Gurr, S. J. (2001). The tomato powdery mildew fungus Oidium neolycopersici. *Molecular Plant Pathology, 2*(6), 303–309.

Jones, J. D. G., & Dangl, J. L. (2006). The plant immune system. *Nature, 444*, 323–329

Jones, J. D. G., Vance, R. E., & Dangl, J. L. (2021). Intracellular innate immune surveillance devices in plants and animals. *Science, 354*(6316), aaf6395.

Jones, J. D., Vance, R. E., & Dangl, J. L. (2016). Intracellular innate immune surveillance devices in plants and animals. *Science, 354*(6316), aaf6395.

Jubic, L. M., Saile, S., Furzer, O. J., El Kasmi, F., & Dangl, J. L. (2019). Help wanted: Helper NLRs and plant immune responses. *Current Opinion in Plant Biology, 50*, 82–94.

Kachroo, A., & Robin, G. P. (2013). Systemic signaling during plant defense. *Current Opinion in Plant Biology, 16*(4), 527–533.

Kahane, R., Hodgkin, T., Jaenicke, H., Hoogendoorn, C., Hermann, M., Keatinge, J. D. H., (Dyno), d'Arros Hughes, J., Padulosi, S., & Looney, N. (2013). Agrobiodiversity for food security, health and income. *Agronomy for Sustainable Development, 33*, 671–693.

Kanyuka, K., & Rudd, J. J. (2019). Cell surface immune receptors: The guardians of the plant's extracellular spaces. *Current Opinion in Plant Biology, 50*, 1–8.

Katyal, P. (2022). Plant secondary metabolites: Functions in plants and pharmacological importance. In: Sharma, A.K., Sharma, A. (eds) *Plant secondary metabolites: Physico-chemical properties and therapeutic applications* (pp. 437–457). Springer Nature, Singapore.

Kauss, H., Fauth, M., Merten, A., & Jeblick, W. (1999). Cucumber hypocotyls respond to cutin monomers via both an inducible and a constitutive H_2O_2-generating system. *Plant Physiology, 120*(4), 1175–1182.

Kawchuk, L. M., Hachey, J., Lynch, D. R., Kulcsar, F., Van Rooijen, G., Waterer, D. R., Robertson, A., Kokko, E., Byers, R., Howard, R. J., Fischer, R., & Prüfer, D. (2001). Tomato Ve disease resistance genes encode cell surface-like receptors. *Proceedings of the National Academy of Sciences, 98*(11), 6511–6515.

Kaweesi, T., Kawuki, R., Kyaligonza, V., Baguma, Y., Tusiime, G., & Ferguson, M. E. (2014). Field evaluation of selected cassava genotypes for cassava brown streak disease based on symptom expression and virus load. *Virology Journal, 11*(1), 1–15.

Khatodia, S., Bhatotia, K., Passricha, N., Khurana, S. M. P., & Tuteja, N. (2016). The CRISPR/Cas genome-editing tool: Application in improvement of crops. *Frontiers in Plant Science, 7*, 506.

Kim, M. S., Yang, J. Y., Yu, J. K., Lee, Y., Park, Y. J., Kang, K. K., & Cho, Y. G. (2021). Breeding of high cooking and eating quality in rice by marker-assisted backcrossing (Mabc) using kasp markers. *Plants, 10*, 1–18. https://doi.org/10.3390/plants10040804

Kim, M. S., Yu, J. K., Ko, S. R., Kim, K. J., Ji, H., Kang, K. K., & Cho, Y. G. (2022). Marker-assisted backcrossing (MABc) to improve eating quality with thin seed coat and aleurone layer of non-glutinous japonica variety in rice. *Genes (Basel)*, *13*, 1–16. https://doi.org/10.3390/genes13020210.

Knepper, C., Savory, E. A., & Day, B. (2011). The role of NDR1 in pathogen perception and plant defense signaling. *Plant Signaling & Behavior*, *6*(8), 1114–1116.

Konieczny, A., & Ausubel, F. M. (1993). A procedure for mapping Arabidopsis mutations using co-dominant ecotype-specific PCR-based markers. *The plant journal*, *4*(2), 403–410.

Kumar, A., Tiwari, K. L., Datta, D., & Singh, M. (2014). Marker assisted gene pyramiding for enhanced Tomato leaf curl virus disease resistance in tomato cultivars. *Biologia plantarum*, *58*, 792–797.

Kumar, S., Korra, T., Thakur, R., Arutselvan, R., Kashyap, A. S., Nehela, Y., Chaplygin, V., Minkina, T., & Keswani, C. (2023). Role of plant secondary metabolites in defence and transcriptional regulation in response to biotic stress. *Plant Stress*, *8*, 100154.

Kumar, V., & Jain, M. (2015). The CRISPR-Cas system for plant genome editing: Advances and opportunities. *Journal of Experimental Botany*, *66*, 47–57.

Kunze, G., Zipfel, C., Robatzek, S., Niehaus, K., Boller, T., & Felix, G. (2004). The N terminus of bacterial elongation factor Tu elicits innate immunity in Arabidopsis plants. *The Plant Cell*, *16*(12), 3496–3507.

Kurabachew, H., & Ayana, G. (2017). Bacterial Wilt caused by Ralstonia solanacearum in Ethiopia: Statusa and management approaches: A Review. *International journal of Phytopathology*, *5*(3), 107–119.

Kushalappa, A. C., Yogendra, K. N., & Karre, S. (2016). Plant innate immune response: Qualitative and quantitative resistance. *Critical Reviews in Plant Sciences*, *35*(1), 38–55.

Kwabena Osei, M., Danquah, A., Adu-Dapaah, H., Danquah, E., Blay, E., Massoudi, M., & Maxwell, D. (2022). Marker assisted backcrossing of alcobaca gene into two elite tomato breeding lines. *Frontiers in Horticulture*, *1*, 1–10. https://doi.org/10.3389/fhort.2022.1024042

Laine, A. L. (2023). Plant disease risk is modified by multiple global change drivers. *Current Biology*, *33*(11), R574–R583.

Lamichhane, J. R., Dachbrodt-Saaydeh, S., Kudsk, P., & Messéan, A. (2016). Toward a reduced reliance on conventional pesticides in European agriculture. *Plant Disease*, *100*(1), 10–24.

Lander, E. S. (1996). The new genomics: Global views of biology. *Science*, *274*(5287), 536–539.

Lawrence, G. J., Finnegan, E. J., Ayliffe, M. A., & Ellis, J. G. (1995). The L6 gene for flax rust resistance is related to the Arabidopsis bacterial resistance gene RPS2 and the tobacco viral resistance gene N. *The Plant Cell*, *7*(8), 1195–1206.

Lawrence, G., Lyons, K., & Wallington, T. (2013). Introduction: Food security, nutrition and sustainability in a globalized world. In: Lawrence, G., Lyons, K. and Wallington, T. (eds) *Food security, nutrition and sustainability* (pp. 1–23). Routledge.

Lee, K., Zhang, Y., Kleinstiver, B. P., Guo, J. A., Aryee, M. J., Miller, J., Malzahn, A., Zarecor, S., Lawrence-Dill, C. J., Joung, J. K., Qi, Y., & Wang, K. (2019). Activities and specificities of CRISPR/Cas9 and Cas12a nucleases for targeted mutagenesis in maize. *Plant Biotechnology Journal*, *17*(2), 362–372.

Leiminger, J. H., & Hausladen, H. (2012). Early blight control in potato using disease-orientated threshold values. *Plant Disease*, *96*(1), 124–130.

Leitao, J. (2018). *Encoding of calcium signals in innate immunity and development* [Dissertation, University of East Anglia].

Lema, M. (2018). Marker assisted selection in comparison to conventional plant breeding: Review article. *Agricultural Research & Technology Open Access Journal*, *14*, 1–10. https://doi.org/10.19080/artoaj.2018.14.555914

Leung, H. (2008). Stressed genomics: Bringing relief to rice fields. *Current Opinion in Plant Biology*, *11*(2), 201–208.

Li, L., Li, M., Yu, L., Zhou, Z., Liang, X., Liu, Z., Cai, G., Gao, L., Zhang, X., Wang, Y., Chen, S., & Zhou, J. M. (2014). The FLS2-associated kinase BIK1 directly phosphorylates the NADPH oxidase RbohD to control plant immunity. *Cell Host & Microbe*, *15*(3), 329–338.

Li, W., Deng, Y., Ning, Y., He, Z., & Wang, G. L. (2020). Exploiting Broad-spectrum disease resistance in crops: From molecular dissection to breeding. *Annual Review of Plant Biology*, *71*, 575–603. https://doi.org/10.1146/annurev-arplant-010720-022215

Liebrand, T. W., van den Berg, G. C., Zhang, Z., Smit, P., Cordewener, J. H., America, A. H., P., Sklenar, J., Jones, A. M. E., Tameling, W. I. L., Robatzek, S., Thomma, B. P. H. J., & Joosten, M. H. (2013). Receptor-like kinase SOBIR1/EVR interacts with receptor-like proteins in plant immunity against fungal infection. *Proceedings of the National Academy of Sciences*, *110*(24), 10010–10015.

Liliane, T. N., & Charles, M. S. (2020). Factors affecting yield of crops. *Agronomy-climate Change & Food Security*, *9*, https://dx.doi.org/10.5772/intechopen.90672

Liu, D., Hu, R., Palla, K. J., Tuskan, G. A., & Yang, X. (2016). Advances and perspectives on the use of CRISPR/Cas9 systems in plant genomics research. *Current Opinion in Plant Biology*, *30*, 70–77.

Liu, M., Rehman, S., Tang, X., Gu, K., Fan, Q., Chen, D., & Ma, W. (2019). Methodologies for improving HDR efficiency. *Frontiers in Genetics*, *9*, 691.

Lohithaswa, H. C., Jyothi, K., Sunil Kumar, K. R., Puttaramanaik, & Hittalmani, S. (2015). Identification and introgression of QTLs implicated in resistance to sorghum downy mildew (Peronosclerospora sorghi (Weston and Uppal) CG Shaw) in maize through marker-assisted selection. *Journal of Genetics*, *94*, 741–748.

Lone, B. A., Karna, S. K. L., Ahmad, F., Shahi, N., & Pokharel, Y. R. (2018). CRISPR/Cas9 system: A bacterial tailor for genomic engineering. *Genetics Research International*, (1), 3797214.

Loo, E. P. I., Tajima, Y., Yamada, K., Kido, S., Hirase, T., Ariga, H., Fujiwara, T., Tanaka, K., Taji, T., Somssich, I. E., Parker, J. E., & Saijo, Y. (2022). Recognition of microbe-and damage-associated molecular patterns by leucine-rich repeat pattern recognition receptor kinases confers salt tolerance in plants. *Molecular Plant-Microbe Interactions*, *35*(7), 554–566.

Lusser, M., & Davies, H. V. (2013). Comparative regulatory approaches for groups of new plant breeding techniques. *New Biotechnology*, *30*(5), 437–446.

Ma, X., Chen, X., Jin, Y., Ge, W., Wang, W., Kong, L., Ji, J., Guo, X., Huang, J., Feng, X.-H., Fu, J., & Zhu, S. (2018). Small molecules promote CRISPR-Cpf1-mediated genome editing in human pluripotent stem cells. *Nature Communications*, *9*(1), 1303.

Macho, A. P., & Zipfel, C. (2014). Plant PRRs and the activation of innate immune signaling. *Molecular Cell*, *54*, 263–272

Madden, L. V., Hughes, G., & Bosch, F. van den. (2017). *The study of plant disease epidemics*. APS Press. https://doi.org/10.1094/9780890545058.

Mafakheri, H., & Mirghazanfari, S. M. (2018). Antifungal activity of the essential oils of some medicinal plants against human and plant fungal pathogens. *Cellular and Molecular Biology*, *64*(15), 13–19.

Mahuku, G., Lockhart, B. E., Wanjala, B., Jones, M. W., Kimunye, J. N., Stewart, L. R., Cassone, B. J., Sevgan, S., Nyasani, J. O., Kusia, E., Kumar, P. L., Niblett, C. L., Kiggundu, A., Asea, G., Pappu, H. R., Wangai, A., Prasanna, B. M., & Redinbaugh, M. G. (2015). Maize lethal necrosis (MLN), an emerging threat to maize-based food security in sub-Saharan Africa. *Phytopathology*, *105*(7), 956–965.

Mallick, N., Vinod Sharma, J. B., Tomar, R. S., Sivasamy, M., & Prabhu, K. V. (2015). Marker-assisted backcross breeding to combine multiple rust resistance in wheat. *Plant Breed*, *134*, 172–177. https://doi.org/10.1111/pbr.12242.

Manning, V. A., Chu, A. L., Steeves, J. E., Wolpert, T. J., & Ciuffetti, L. M. (2009). A host-selective toxin of Pyrenophora tritici-repentis, Ptr ToxA, induces photosystem changes and reactive oxygen species accumulation in sensitive wheat. *Molecular Plant-Microbe Interactions*, *22*(6), 665–676.

Marathe, R., & Dinesh-Kumar, S. P. (2003). Plant defense: One post, multiple guards?!. *Molecular Cell*, *11*(2), 284–286.

Maruta, N., Burdett, H., Lim, B. Y., Hu, X., Desa, S., Manik, M. K., & Kobe, B. (2022). Structural basis of NLR activation and innate immune signalling in plants. *Immunogenetics*, *74*(1), 5–26.

Maruyama, T., Dougan, S. K., Truttmann, M. C., Bilate, A. M., Ingram, J. R., & Ploegh, H. L. (2016). Correction: Corrigendum: Increasing the efficiency of precise genome editing with CRISPR-Cas9 by inhibition of nonhomologous end joining. *Nature Biotechnology*, *34*(2), 210–210.

Matny, O. N. (2015). Fusarium head blight and crown rot on wheat & barley: Losses and health risks. *Advances in Plants & Agriculture Research*, *2*(1), 00039.

McElroy, M. S., Navarro, A. J. R., Mustiga, G., Stack, C., Gezan, S., Peña, G., Sarabia, W., Saquicela, D., Sotomayor, I., Douglas, G. M., Migicovsky, Z., Amores, F., Tarqui, O., Myles, S., & Motamayor, J. C. (2018). Prediction of cacao (Theobroma cacao) resistance to moniliophthora spp. diseases via genome-wide association analysis and genomic selection. *Frontiers in Plant Science*, *9*, 1–12. https://doi.org/10.3389/fpls.2018.00343

Meentemeyer, R. K., Haas, S. E., & Václavík, T. (2012). Landscape epidemiology of emerging infectious diseases in natural and human-altered ecosystems. *Annual Review of Phytopathology*, *50*, 379–402.

Meyers, B. C., Kozik, A., Griego, A., Kuang, H., & Michelmore, R. W. (2003). Genome-wide analysis of NBS-LRR–encoding genes in Arabidopsis. *The Plant Cell*, *15*(4), 809–834.

Mfuka, C., Byamukama, E., & Zhang, X. (2020). Spatiotemporal characteristics of white mold and impacts on yield in soybean fields in South Dakota. *Geo-Spatial Information Science*, *23*(2), 182–193.

Miedaner, T., & Korzun, V. (2012). Marker-assisted selection for disease resistance in wheat and barley breeding. *Phytopathology*, *102*, 560–566. https://doi.org/10.1094/PHYTO-05-11-0157

Mills, M. C., & Rahal, C. (2019). A scientometric review of genome-wide association studies. *Communications Biology*, *2*. https://doi.org/10.1038/s42003-018-0261-x

Mindrinos, M., Katagiri, F., Yu, G. L., & Ausubel, F. M. (1994). The A. thaliana disease resistance gene RPS2 encodes a protein containing a nucleotide-binding site and leucine-rich repeats. *Cell*, *78*(6), 1089–1099.

Mishina, T. E., & Zeier, J. (2007). Pathogen-associated molecular pattern recognition rather than development of tissue necrosis contributes to bacterial induction of systemic acquired resistance in Arabidopsis. *The Plant Journal*, *50*(3), 500–513.

Misra, V., Pandey, H., & Mall, A. K. (2023). Exploring CRISPR/Cas9 gene editing applications for enhancing disease resistance in sugar beet. *ACS Agricultural Science & Technology*. *4*(1), 4–21. https://doi.org/10.1021/acsagscitech.3c00193

Monaghan, J., & Zipfel, C. (2012). Plant pattern recognition receptor complexes at the plasma membrane. *Current Opinion in Plant Biology*, *15*(4), 349–357.

Mores, A., Borrelli, G. M., Laidò, G., Petruzzino, G., Pecchioni, N., Amoroso, L. G. M., Desiderio, F., Mazzucotelli, E., Mastrangelo, A. M., & Marone, D. (2021). Genomic Approaches to identify molecular bases of crop resistance to diseases and to develop future breeding strategies. *International Journal of Molecular Sciences*, *22*, 5423. https://doi.org/10.3390/ijms22115423.

Mukherjee, A., Bhowmick, S., Yadav, S., Rashid, M. M., Chouhan, G. K., Vaishya, J. K., & Verma, J. P. (2021). Re-vitalizing of endophytic microbes for soil health management and plant protection. *3 Biotech*, *11*, 1–17.

Mundt, C. C. (2014). Durable resistance: A key to sustainable management of pathogens and pests. *Infection, Genetics and Evolution*, *27*, 446–455.

Murugan, K., Babu, K., Sundaresan, R., Rajan, R., & Sashital, D. G. (2017). The revolution continues: Newly discovered systems expand the CRISPR-Cas toolkit. *Molecular Cell*, *68*(1), 15–25.

Muthamilarasan, M., & Prasad, M. (2013). Plant innate immunity: An updated insight into defense mechanism. *Journal of Biosciences*, *38*, 433–449.

Nadeem, M. A., Nawaz, M. A., Shahid, M. Q., Doğan, Y., Comertpay, G., Yıldız, M., Hatipoğlu, R., Ahmad, F., Alsaleh, A., Labhane, N., Özkan, H., Chung, G., & Baloch, F. S. (2018). DNA molecular markers in plant breeding: Current status and recent advancements in genomic selection and genome editing. *Biotechnology & Biotechnological Equipment*, *32*, 261–285. https://doi.org/10.1080/13102818.2017.1400401

Narayanasamy, P. (2011). *Crop diseases management: principles and practices*. New India Publishing Agency.

Narayani, M., & Srivastava, S. (2017). Elicitation: A stimulation of stress in in vitro plant cell/tissue cultures for enhancement of secondary metabolite production. *Phytochemistry Reviews*, *16*, 1227–1252.

Navarro, L., Zipfel, C., Rowland, O., Keller, I., Robatzek, S., Boller, T., & Jones, J. D. (2004). The transcriptional innate immune response to flg22. Interplay and overlap with Avr gene-dependent defense responses and bacterial pathogenesis. *Plant Physiology*, *135*(2), 1113–1128.

Nelson, R., Wiesner-Hanks, T., Wisser, R., & Balint-Kurti, P. (2018). Navigating complexity to breed disease-resistant crops. *Nature Reviews Genetics*, *19*(1), 21–33.

Ngou, B. P. M., Ahn, H. K., Ding, P., & Jones, J. D. (2021). Mutual potentiation of plant immunity by cell-surface and intracellular receptors. *Nature*, *592*(7852), 110–115.

Niehl, A., Wyrsch, I., Boller, T., & Heinlein, M. (2016). Double-stranded RNA s induce a pattern-triggered immune signaling pathway in plants. *New Phytologist*, *211*(3), 1008–1019.

Nomura, H., Komori, T., Uemura, S., Kanda, Y., Shimotani, K., Nakai, K., Furuichi, T., Takebayashi, K., Sugimoto, T., Sano, S., Suwastika, I. N., Fukusaki, E., Yoshioka, H., Nakahira, Y., & Shiina, T. (2012). Chloroplast-mediated activation of plant immune signalling in Arabidopsis. *Nature Communications*, *3*(1), 926.

Nomura, K., DebRoy, S., Lee, Y. H., Pumplin, N., Jones, J., & He, S. Y. (2006). A bacterial virulence protein suppresses host innate immunity to cause plant disease. *Science*, *313*(5784), 220–223.

Noutoshi, Y., Ito, T., Seki, M., Nakashita, H., Yoshida, S., Marco, Y., Shirasu, K., & Shinozaki, K. (2005). A single amino acid insertion in the WRKY domain of the Arabidopsis TIR–NBS–LRR–WRKY-type disease resistance protein SLH1 (sensitive to low humidity 1) causes activation of defense responses and hypersensitive cell death. *The Plant Journal*, *43*(6), 873–888.

O'Connell, A., Deo, J., Deomano, E., Wei, X., Jackson, P., Aitken, K. S., Manimekalai, R., Mohanraj, K., Hemaprabha, G., Ram, B., Viswanathan, R., & Lakshmanan, P. (2022). Combining genomic selection with genome-wide association analysis identified a large-effect QTL and improved selection for red rot resistance in sugarcane. *Frontiers in Plant Science*, *13*, 1–16. https://doi.org/10.3389/fpls.2022.1021182

Offor, B. C., Mhlongo, M. I., Dubery, I. A., & Piater, L. A. (2022). Plasma Membrane-Associated Proteins Identified in Arabidopsis Wild Type, lbr2-2 and bak1-4 Mutants Treated with LPSs from Pseudomonas syringae and *Xanthomonas campestris*. *Membranes*, *12*(6), 606.

Ou, X., Li, T., Zhao, Y., Chang, Y., Wu, L., Chen, G., Day, B., & Jiang, K. (2022). Calcium-dependent ABA signaling functions in stomatal immunity by regulating rapid SA responses in guard cells. *Journal of Plant Physiology*, *268*, 153585.

Paran, I., & Michelmore, R. W. (1993). Development of reliable PCR-based markers linked to downy mildew resistance genes in lettuce. *Theoretical and Applied Genetics*, *85*, 985–993.

Park, C. J., & Ronald, P. C. (2012). Cleavage and nuclear localization of the rice XA21 immune receptor. *Nature Communications*, *3*(1), 920.

Parker, J. E., Holub, E. B., Frost, L. N., Falk, A., Gunn, N. D., & Daniels, M. J. (1996). Characterization of eds1, a mutation in Arabidopsis suppressing resistance to Peronospora parasitica specified by several different RPP genes. *The Plant Cell*, *8*(11), 2033–2046.

Parveen, S., Shahzad, A., & Yadav, V. (2016). Molecular markers and their application in plant biotechnology. In: Shahzad, A., Sharma, S., Siddiqui, S. (Eds.), *Biotechnological strategies for the conservation of medicinal and ornamental climbers*. Springer, Cham. (pp. 389–413).

Patel, S. (2018). Danger-associated molecular patterns (DAMPs): The derivatives and triggers of inflammation. *Current Allergy and Asthma Reports*, *18*, 1–12.

Pautasso, M., Dehnen-Schmutz, K., & Holdenrieder, O. (2018). Plant health and global change—some implications for landscape management. *Biological Reviews*, *85*(4), 729–755

Peart, J. R., Mestre, P., Lu, R., Malcuit, I., & Baulcombe, D. C. (2005). NRG1, a CC-NB-LRR protein, together with N, a TIR-NB-LRR protein, mediates resistance against tobacco mosaic virus. *Current Biology*, *15*(10), 968–973.

Piasecka, A., Jedrzejczak-Rey, N., & Bednarek, P. (2015). Secondary metabolites in plant innate immunity: Conserved function of divergent chemicals. *New Phytologist*, *206*(3), 948–964.

Pickar-Oliver, A., & Gersbach, C. A. (2019). The next generation of CRISPR–Cas technologies and applications. *Nature reviews Molecular Cell Biology*, *20*(8), 490–507.

Poklar Ulrih, N. (2017). Analytical techniques for the study of polyphenol–protein interactions. *Critical Reviews in Food Science and Nutrition*, *57*(10), 2144–2161.

Pott, D. M., Osorio, S., & Vallarino, J. G. (2019). From central to specialized metabolism: An overview of some secondary compounds derived from the primary metabolism for their role in conferring nutritional and organoleptic characteristics to fruit. *Frontiers in Plant Science*, *10*, 835.

Prasanna, B. M., Pixley, K., Warburton, M. L., & Xie, C. X. (2010). Molecular marker-assisted breeding options for maize improvement in Asia. *Molecular Breeding*, *26*, 339–356.

Prasanna, H. C., Sinha, D. P., Rai, G. K., Krishna, R., Kashyap, S. P., Singh, N. K., Singh, M., & Malathi, V. G. (2015). Pyramiding T y-2 and T y-3 genes for resistance to monopartite and bipartite tomato leaf curl viruses of India. *Plant Pathology*, *64*(2), 256–264.

Prince, D. C., Drurey, C., Zipfel, C., & Hogenhout, S. A. (2014). The leucine-rich repeat receptor-like kinase brassinosteroid insensitive1-associated kinase1 and the cytochrome P450 phytoalexin deficient3 contribute to innate immunity to aphids in Arabidopsis. *Plant Physiology*, *164*(4), 2207–2219.

Prince, D. C., Drurey, C., Zipfel, C., & Hogenhout, S. A. (2014). The leucine-rich repeat receptor-like kinase brassinosteroid insensitive1-associated kinase1 and the cytochrome p450 phytoalexin deficient3 contribute to innate immunity to aphids in Arabidopsis. *Plant Physiology*, *164*(4), 2207–2219.

Pusztahelyi, T., Holb, I. J., & Pócsi, I. (2015). Secondary metabolites in fungus-plant interactions. *Frontiers in Plant Science*, *6*, 573.

Qaim, M. (2020). Role of new plant breeding technologies for food security and sustainable agricultural development. *Applied Economic Perspectives and Policy*, *42*(2), 129–150.

Qi, D., & Innes, R. W. (2013). Recent advances in plant NLR structure, function, localization, and signaling. *Frontiers in Immunology*, *4*, 348.

Qin, C., Yang, S., Chu, Y. H., Zhang, H., Pang, X. W., Chen, L., Tian, D. S., & Wang, W. (2022). Signaling pathways involved in ischemic stroke: Molecular mechanisms and therapeutic interventions. *Signal Transduction and Targeted Therapy*, *7*(1), 215.

Ragoussis, J. (2009). Genotyping technologies for genetic research. *Annual Review of Genomics and Human Genetics*, *10*, 117–133.

Rahman, M. M. (2018). Improving the crown rot resistance and tolerance of wheat using marker assisted recurrent selection (Doctoral dissertation).

Rahman, M., Davies, P., Bansal, U., Pasam, R., Hayden, M., & Trethowan, R. (2020). Marker-assisted recurrent selection improves the crown rot resistance of bread wheat. *Molecular Breeding*, *40*. https://doi.org/10.1007/s11032-020-1105-1.

Rajapakse, S., Belthoff, L. E., He, G., Estager, A. E., Scorza, R., Verde, I., Ballard, R. E., Baird, W. V., Callahan, A., Monet, R., & Abbott, A. G. (1995). Genetic linkage mapping in peach using morphological, RFLP and RAPD markers. *Theoretical and Applied Genetics*, *90*, 503–510.

Ramayya, P. J., Vinukonda, V. P., Singh, U. M., Alam, S., Venkateshwarlu, C., Vipparla, A. K., Dixit, S., Yadav, S., Abbai, R., Badri, J., Ram, T., Padmakumari, A. P., Singh, V. K., & Kumar, A. (2021). Marker-assisted forward and backcross breeding for improvement of elite Indian rice variety Naveen for multiple biotic and abiotic stress tolerance. *PLoS One 16*, 1–22. https://doi.org/10.1371/journal.pone.0256721.

Ranf, S. (2018). Pattern recognition receptors—Versatile genetic tools for engineering broad-spectrum disease resistance in crops. *Agronomy*, *8*(8), 134.

Reddy, V. D., Rao, P. N., Poduri, N., & Rao, K. V. (Eds.) (2011). Pests and pathogens: Management strategies. BS Publications.

Reina-Pinto, J. J., & Yephremov, A. (2009). Surface lipids and plant defenses. *Plant Physiology and Biochemistry*, *47*(6), 540–549

Remick, B. C., Gaidt, M. M., & Vance, R. E. (2023). Effector-triggered immunity. *Annual Review of Immunology*, *41*, 453–481.

Riesenberg, S., & Maricic, T. (2018). Targeting repair pathways with small molecules increases precise genome editing in pluripotent stem cells. *Nature Communications*, *9*(1), 2164.

Rietz, S., Stamm, A., Malonek, S., Wagner, S., Becker, D., Medina-Escobar, N., Vlot, A. C., Feys, B. J., Niefind, K., & Parker, J. E. (2011). Different roles of enhanced disease susceptibility1 (EDS1) bound to and dissociated from phytoalexin deficient4 (PAD4) in Arabidopsis immunity. *New Phytologist*, *191*(1), 107–119.

Roberts, D. P., & Mattoo, A. K. (2018). Sustainable agriculture—Enhancing environmental benefits, food nutritional quality and building crop resilience to abiotic and biotic stresses. *Agriculture*, *8*(1), 8. doi:10.3390/agriculture8010008.

Robertsen, C. D., Hjortshøj, R. L., & Janss, L. L. (2019). Genomic selection in cereal breeding. *Agronomy*, *9*, 1–16. https://doi.org/10.3390/agronomy9020095.

Rojas, C. M., Senthil-Kumar, M., Tzin, V., & Mysore, K. S. (2014). Regulation of primary plant metabolism during plant-pathogen interactions and its contribution to plant defense. *Frontiers in Plant Science*, *5*, 17.

Rowland, O., Ludwig, A. A., Merrick, C. J., Baillieul, F., Tracy, F. E., Durrant, W. E., Fritz-Laylin, L., Nekrasov, V., Sjölander, K., Yoshioka, H., & Jones, J. D. (2005). Functional analysis of Avr9/Cf-9 rapidly elicited genes identifies a protein kinase, ACIK1, that is essential for full Cf-9–dependent disease resistance in tomato. *The Plant Cell*, *17*(1), 295–310.

Ruffel, S., Dussault, M. H., Palloix, A., Moury, B., Bendahmane, A., Robaglia, C., & Caranta, C. (2002). A natural recessive resistance gene against potato virus Y in pepper corresponds to the eukaryotic initiation factor 4E (eIF4E). *The Plant Journal*, *32*(6), 1067–1075.

Rutten, G., Hönig, L., Schwaß, R., Braun, U., Saadani, M., Schuldt, A., Michalski, S. G., & Bruelheide, H. (2021). More diverse tree communities promote foliar fungal pathogen diversity, but decrease infestation rates per tree species, in a subtropical biodiversity experiment. *Journal of Ecology*, *109*(5), 2068–2080.

Saavedra, L. M., Caixeta, E. T., Barka, G. D., Borém, A., Zambolim, L., Nascimento, M., Cruz, C. D., de Oliveira, A. C. B., & Pereira, A. A. (2023). Marker-assisted recurrent selection for pyramiding leaf rust and coffee berry disease resistance alleles in Coffea arabica L. *Genes (Basel)*, 14. https://doi.org/10.3390/genes14010189.

Sabar, M., Akhter, M., Bibi, T., Riaz, A., Haider, Z., Khan, A. R., & Bibi, A. (2019). Basmati rice lines development carrying multiple bacterial blight resistance genes pyramided using the marker-assisted backcross breeding approach. *Molecular Breeding*, *39*. https://doi.org/10.1007/s11032-019-1047-7

Sahni, S., Prasad, B. D., Mishra, R. K., & Kumari, S. (2021). *Stem rot or white mold disease in grain legumes and its management. Diseases of pulse crops and their management.* Daya Publishing House® A Division of Astral International Pvt. Ltd

Salimath, S. S., Oliveira, A. C. D., Bennetzen, J. L., & Godwin, I. D. (1995). Assessment of genome origins and genetic diversity in the genus Eleusine with DNA markers. *Genome*, *38*(4), 757–763.

Sanabria, N., Goring, D., Nürnberger, T., & Dubery, I. (2008). Self/non self perception and recognition mechanisms in plants: A comparison of self-incompatibility and innate immunity. *New Phytologist*, *178*, 503–514

Sardar, A., Nandi, A. K., & Chattopadhyay, D. (2017). CBL-interacting protein kinase 6 negatively regulates immune response to *Pseudomonas syringae* in Arabidopsis. *Journal of Experimental Botany*, *68*(13), 3573–3584.

Sarris, P. F., Duxbury, Z., Huh, S. U., Ma, Y., Segonzac, C., Sklenar, J., Derbyshire, P., Cevik, V., Rallapalli, G., Saucet, S. B., Wirthmueller, L., Menke, F. L. H., Sohn, K. H., & Jones, J. D. (2015). A plant immune receptor detects pathogen effectors that target WRKY transcription factors. *Cell*, *161*(5), 1089–1100.

Savary, S., & Willocquet, L. (2020). Modeling the impact of crop diseases on global food security. *Annual Review of Phytopathology*, *58*, 313–341.

Savary, S., Ficke, A., Aubertot, J. N., & Hollier, C. (2012). Crop losses due to diseases and their implications for global food production losses and food security. *Food Security*, *4*(4), 519–537.

Savary, S., Willocquet, L., Pethybridge, S. J., Esker, P., McRoberts, N., & Nelson, A. (2019). The global burden of pathogens and pests on major food crops. *Nature Ecology & Evolution*, *3*(3), 430–439.

Seppänen, S. K., Syrjälä, L., Von Weissenberg, K., Teeri, T. H., Paajanen, L., & Pappinen, A. (2004). Antifungal activity of stilbenes in in vitro bioassays and in transgenic Populus expressing a gene encoding pinosylvin synthase. *Plant Cell Reports*, *22*, 584–593.

Shahriar, S. A., Islam, M. N., Chun, C. N. W., Rahim, M. A., Paul, N. C., Uddain, J., & Siddiquee, S. (2021). Control of plant viral diseases by CRISPR/Cas9: Resistance mechanisms, strategies and challenges in food crops. *Plants*, *10*(7), 1264.

Sharma, G., Ahalawat, S., Islam, A., Vignesh, S., Unnikrishnan, S., Kumar, P., Singh, A., Raghavan, M., & Singh, P. (2023). A comprehensive review of CRISPR/Cas9 based strategies in horticultural crop modification. *International Journal of Environment and Climate Change*, *13*(11), 310–323.

Sharpee, W. C. (2016). Identification and characterization of effectors from the filamentous fungal plant pathogen magnaporthe oryzae. Degree Doctor of Philosophy Discipline Plant Pathology Collections Dissertations.

Shattock, R. (2002). Compendium of Potato Diseases, Second Edition. W. R. Stevenson. *Plant Pathology*, *51*(4), 520–520. doi:10.1046/j.1365–3059.2002.06934.x

Shen, Q. H., Saijo, Y., Mauch, S., Biskup, C., Bieri, S., Keller, B., Seki, H., Ülker, B., Somssich, I. E., & Schulze-Lefert, P. (2007). Nuclear activity of MLA immune receptors links isolate-specific and basal disease-resistance responses. *Science*, *315*(5815), 1098–1103.

Shi, G., Zhang, Z., Friesen, T. L., Raats, D., Fahima, T., Brueggeman, R. S., Lu, S., Trick, H. N., Liu, Z., Chao, W., Frenkel, Z., Xu, S. S., Rasmussen, J. B., & Faris, J. D. (2016). The hijacking of a receptor kinase–driven pathway by a wheat fungal pathogen leads to disease. *Science Advances*, *2*(10), e1600822.

Singh Rawat, V., Kaur, J., Bhagwat, S., Arora Pandit, M., & Dogra Rawat, C. (2023). Deploying microbes as drivers and indicators in ecological restoration. *Restoration Ecology*, *31*(1), e13688.

Singh, A., & Shahi, J. P. (2012). Banded leaf and sheath blight: An emerging disease of maize (*Zea mays* L). *Maydica*, *57*(3), 215–219.

Singh, A., Jaiswal, J. P., & Badoni, S. (2018). Enhancing rust resistance in wheat through marker assisted backcross breeding. *Indian Journal of Genetics and Plant Breeding*, *78*, 19–25. https://doi.org/10.5958/0975-6906.2018.00003.2.

Singh, B., Mal, G., Gautam, S. K., & Mukesh, M. (2019). Whole-Genome selection in livestock. *Advances in Animal Biotechnology*, 349–364. https://doi.org/10.1007/978-3-030-21309-1_31.

Singh, G., Singh, N., Ellur, R. K. Balamurugan, A., Prakash, G., Rathour, R., Mondal, K. K., Bhowmick, P. K., Krishnan, S. G., Nagarajan, M., Seth, R., Vinod, K. K., Singh, V., Bollinedi, H., & Singh, A. K. (2023). Genetic enhancement for biotic stress resistance in basmati rice through marker-assisted backcross breeding. *International Journal of Molecular Sciences*, *24*, 1–18. https://doi.org/10.3390/ijms242216081.

Singh, S., Sidhu, J. S., Huang, N., Vikal, Y., Li, Z., Brar, D. S., Dhaliwal, H. S., & Khush, G. S. (2001). Pyramiding three bacterial blight resistance genes (xa5, xa13 and Xa21) using marker-assisted selection into indica rice cultivar PR106. *Theoretical and Applied Genetics*, *102*, 1011–1015.

Sirisha, V. L., Mitra, S., & Suprasanna, P. (2024). Plant immune system: Mechanisms and resilience. In: George Fink (ed) *Stress: immunology and inflammation* (pp. 9–21). Academic Press.

Snoeijers, S. S., Pérez-García, A., Joosten, M. H., & De Wit, P. J. (2000). The effect of nitrogen on disease development and gene expression in bacterial and fungal plant pathogens. *European Journal of Plant Pathology*, *106*, 493–506.

Sonah, H., Deshmukh, R. K., Sharma, A., Singh, V. P., Gupta, D. K., Gacche, R. N., Rana, J. C., Singh, N. K., & Sharma, T. R. (2011). Genome-wide distribution and organization of microsatellites in plants: An insight into marker development in Brachypodium. *Plos One*, *6*(6), e21298.

Song, W. Y., Pi, L. Y., Wang, G. L., Gardner, J., Holsten, T., & Ronald, P. C. (1997). Evolution of the rice Xa21 disease resistance gene family. *The Plant Cell*, *9*(8), 1279–1287.

Stukenbrock, E. H., & McDonald, B. A. (2008). The origins of plant pathogens in agro-ecosystems. *Annual Review of Phytopathology*, *46*, 75–100.

Stuthman, D. D., Leonard, K. J., & Miller-Garvin, J. (2007). Breeding crops for durable resistance to disease. *Advances in Agronomy*, *95*, 319–367.

Suman, A., Govindasamy, V., Ramakrishnan, B., Aswini, K., SaiPrasad, J., Sharma, P., Pathak, D., & Annapurna, K. (2022). Microbial community and function-based synthetic bioinoculants: A perspective for sustainable agriculture. *Frontiers in Microbiology*, *12*, 805498.

Sun, Y., Li, L., Macho, A. P., Han, Z., Hu, Z., Zipfel, C., Zhou, J. M., & Chai, J. (2013). Structural basis for flg22-induced activation of the Arabidopsis FLS2-BAK1 immune complex. *Science*, *342*(6158), 624–628.

Takele, A., Lencho, A., Getaneh, W., Hailu, E., & Kassa, B. (2015). Status of wheat septoria leaf blotch (septaria tritici roberge in desmaz) in south west and western shewa zones of oromiya regional state, Ethiopia. *Research in Plant Sciences*, *3*(3), 43–48.

Takken, F. L. W., & Tameling, W. I. L. (2009). To nibble at plant resistance proteins. *Science*, *324*, 744–746

Takken, F. L., & Goverse, A. (2012). How to build a pathogen detector: Structural basis of NB-LRR function. *Current Opinion in Plant Biology*, *15*(4), 375–384.

Tang, D., Guoxun, W., & Jian-Min, Z. (2017). Receptor kinases in plant-pathogen interactions: More than pattern recognition. *The Plant Cell*, *29*(4), 618–637.

Tester, M., & Langridge, P. (2010). Breeding technologies to increase crop production in a changing world. *Science*, *327*(5967), 818–822.

Thakur, M., Bhattacharya, S., Khosla, P. K., & Puri, S. (2019). Improving production of plant secondary metabolites through biotic and abiotic elicitation. *Journal of Applied Research on Medicinal and Aromatic Plants*, *12*, 1–12.

Thomas, C. M., Jones, D. A., Parniske, M., Harrison, K., Balint-Kurti, P. J., Hatzixanthis, K., & Jones, J. D. (1997). Characterization of the tomato Cf-4 gene for resistance to *Cladosporium fulvum* identifies sequences that determine recognitional specificity in Cf-4 and Cf-9. *The Plant Cell*, *9*(12), 2209–2224.

Thomas, C. M., Vos, P., Zabeau, M., Jones, D. A., Norcott, K. A., Chadwick, B. P., & Jones, J. D. (1995). Identification of amplified restriction fragment polymorphism (AFLP) markers tightly linked to the tomato Cf-9 gene for resistance to *Cladosporium fulvum*. *The Plant Journal*, *8*(5), 785–794.

Tian, W., Hou, C., Ren, Z., Wang, C., Zhao, F., Dahlbeck, D., Hu, S., Zhang, L., Niu, Q., Li, L., Staskawicz, B. J., & Luan, S. (2019). A calmodulin-gated calcium channel links pathogen patterns to plant immunity. *Nature*, *572*(7767), 131–135.

Torres, M. A., Dangl, J. L., & Jones, J. D. (2002). Arabidopsis gp91phox homologues AtrbohD and AtrbohF are required for accumulation of reactive oxygen intermediates in the plant defense response. *Proceedings of the National Academy of Sciences*, *99*(1), 517–522.

Tran, D. T., Chung, E. H., Habring-Müller, A., Demar, M., Schwab, R., Dangl, J. L., … & Chae, E. (2017). Activation of a plant NLR complex through heteromeric association with an autoimmune risk variant of another NLR. *Current Biology*, *27*(8), 1148–1160.

Tsuda, K., & Katagiri, F. (2010). Comparing signaling mechanisms engaged in pattern-triggered and effector-triggered immunity. *Current Opinion in Plant Biology*, *13*(4), 459–465.

Tsuda, K., Sato, M., Glazebrook, J., Cohen, J. D., & Katagiri, F. (2008). Interplay between MAMP-triggered and SA-mediated defense responses. *The Plant Journal*, *53*(5), 763–775.

Uffelmann, E., Huang, Q. Q., Munung, N. S., de Vries, J., Okada, Y., Martin, A. R., Martin, H. C., Lappalainen, T., & Posthuma, D. (2021). Genome-wide association studies. *Nature Reviews Methods Primers*. https://doi.org/10.1038/s43586-021-00056-9

Vallejos, C. E., Jones, V., Stall, R. E., Jones, J. B., Minsavage, G. V., Schultz, D. C., Rodrigues, R., Olsen, L. E., & Mazourek, M. (2010). Characterization of two recessive genes controlling resistance to all races of bacterial spot in peppers. *Theoretical and Applied Genetics*, *121*, 37–46.

van Esse, H. P., Reuber, T. L., & van der Does, D. (2020). Genetic modification to improve disease resistance in crops. *New Phytologist*, *225*, 70–86. https://doi.org/10.1111/nph.15967.

Van Ooijen, G., Mayr, G., Kasiem, M. M., Albrecht, M., Cornelissen, B. J., & Takken, F. L. (2008). Structure–function analysis of the NB-ARC domain of plant disease resistance proteins. *Journal of Experimental Botany*, *59*(6), 1383–1397.

Veloz-García, R., Marín-Martínez, R., Veloz-Rodríguez, R., Rodríguez-Guerra, R., Torres-Pacheco, I., González-Chavira, M. M., Anaya-López, J. L., Guevara-Olvera, L., Feregrino-Pérez, A. A., Loarca-Piña, G., & Guevara-González, R. G. (2010). Antimicrobial activities of cascalote (*Caesalpinia cacalaco*) phenolics-containing extract against fungus *Colletotrichum lindemuthianum*. *Industrial Crops and Products*, *31*(1), 134–138.

Verges, V. L., Brown-guedira, G. L.,& Van Sanford, D. A. (2021). Genome-wide association studies combined with genomic selection as a tool to increase *Fusarium* head blight resistance in wheat. *Crop Breeding, Genet Genomics*. https://doi.org/10.20900/cbgg20210007

Vidhyasekaran, P. (2016). *Switching on plant innate immunity signaling systems*. Springer.

Wang, B., Lin, Z., Li, X. Zhao, Y., Zhao, B., Wu, G., Ma, X., Wang, H., Xie, Y., Li, Q., Song, G., Kong, D., Zheng, Z., Wei, H., Shen, R., Wu, H., Chen, C., Meng, Z., Wang, T., Li, Y., Li, X., Chen, Y., Lai, J., Hufford, M. B., Ross-Ibarra, J., He, H., & Wang, H. (2020). Genome-wide selection and genetic improvement during modern maize breeding. *Nature Genetics*, *52*, 565–571. https://doi.org/10.1038/s41588-020-0616-3.

Wang, J. (1998). DNA biosensors based on peptide nucleic acid (PNA) recognition layers. A review. *Biosensors and Bioelectronics*, *13(7–8)*, 757–762.

Wang, X. Y., Gao, S., Chen, T., Wang, J., & Yu, F. H. (2023). Interactions between soil microbes and native species drive a diversity-invasibility relationship. *Biological Invasions*, *25*(5), 1461–1472.

Wang, X., Tu, M., Wang, D., Liu, J., Li, Y., Li, Z., Wang, Y., & Wang, X. (2018). CRISPR/Cas9-mediated efficient targeted mutagenesis in grape in the first generation. *Plant Biotechnology Journal*, *16*(4), 844–855.

Wang, Y. H., Liu, S. J., Ji, S. L., Zhang, W. W., Wang, C. M., Jiang, L., & Wan, J. M. (2005). Fine mapping and marker-assisted selection (MAS) of a low glutelin content gene in rice. *Cell Research*, *15*(8), 622–630.

Wang, Z., Yang, L., Jander, G., Bhawal, R., Zhang, S., Liu, Z., … & Hua, J. (2022). AIG2A and AIG2B limit the activation of salicylic acid-regulated defenses by tryptophan-derived secondary metabolism in Arabidopsis. *The Plant Cell*, *34*(11), 4641–4660.

Whitham, S., Dinesh-Kumar, S. P., Choi, D., Hehl, R., Corr, C., & Baker, B. (1994). The product of the tobacco mosaic virus resistance gene N: Similarity to toll and the interleukin-1 receptor. *Cell*, *78*(6), 1101–1115.

Wu, C. H., Abd-El-Haliem, A., Bozkurt, T. O., Belhaj, K., Terauchi, R., Vossen, J. H., & Kamoun, S. (2017). NLR network mediates immunity to diverse plant pathogens. *Proceedings of the National Academy of Sciences*, *114*(30), 8113–8118.

Xiao, S., Calis, O., Patrick, E., Zhang, G., Charoenwattana, P., Muskett, P., Parker, J. E., & Turner, J. G. (2005). The atypical resistance gene, RPW8, recruits components of basal defence for powdery mildew resistance in *Arabidopsis*. *The Plant Journal*, *42*(1), 95–110.

Xu, G., Moeder, W., Yoshioka, K., & Shan, L. (2022). A tale of many families: Calcium channels in plant immunity. *The Plant Cell*, *34*(5), 1551–1567.

Xu, Y., Li, P., Zou, C., Lu, Y., Xie, C., Zhang, X., Prasanna, B. M., & Olsen, M. S. (2017). Enhancing genetic gain in the era of molecular breeding. *Journal of Experimental Botany*, *68*, 2641–2666. https://doi.org/10.1093/jxb/erx135.

Yang, D., Tang, J., Yang, D., Chen, Y., Ali, J., & Mou, T. (2019). Improving rice blast resistance of Feng39S through molecular marker-assisted backcrossing. *Rice*, *12*. https://doi.org/10.1186/s12284-019-0329-3.

Yang, F., & Li, Y. (2017). The new generation tool for CRISPR genome editing: CRISPR/Cpf1. *Chinese Journal of Biotechnology*, *33*(3), 361–371.

Yang, H., Sun, Y., Wang, H., Zhao, T., Xu, X., Jiang, J., & Li, J. (2021). Genome-wide identification and functional analysis of the ERF2 gene family in response to disease resistance against *Stemphylium lycopersici* in tomato. *BMC Plant Biology*, *21*, 1–13.

Yang, K. Y., Liu, Y., & Zhang, S. (2001). Activation of a mitogen-activated protein kinase pathway is involved in disease resistance in tobacco. *Proceedings of the National Academy of Sciences*, *98*(2), 741–746.

Yin, K., & Qiu, J.-L. (2020). Genome editing for plant disease resistance. *Emerging Trends in Plant Pathology*, 577–590. https://doi.org/10.1007/978-981-15-6275-4_25.

Yin, X., Biswal, A. K., Dionora, J., Perdigon, K. M., Balahadia, C. P., Mazumdar, S., … & Bandyopadhyay, A. (2017). CRISPR-Cas9 and CRISPR-Cpf1 mediated targeting of a stomatal developmental gene EPFL9 in rice. *Plant Cell Reports*, *36*, 745–757.

Yun, B. W., Skelly, M. J., Yin, M., Yu, M., Mun, B. G., Lee, S. U., Hussain, A., Spoel, S. H., & Loake, G. J. (2016). Nitric oxide and S-nitrosoglutathione function additively during plant immunity. *New Phytologist*, *211*(2), 516–526.

Zaidi, S. S. E. A., Mahas, A., Vanderschuren, H., & Mahfouz, M. M. (2020). Engineering crops of the future: CRISPR approaches to develop climate-resilient and disease-resistant plants. *Genome Biology*, *21*(1), 1–19.

Zhang, F., Wen, Y., & Guo, X. (2014). CRISPR/Cas9 for genome editing: Progress, implications and challenges. *Human Molecular Genetics*, *23(R1)*, R40–R46.

Zhang, J., Li, W., Xiang, T., Liu, Z., Laluk, K., Ding, X., Zou, Y., Gao, M., Zhang, X., Chen, S., Mengiste, T., Zhang, Y., & Zhou, J. M. (2010). Receptor-like cytoplasmic kinases integrate signaling from multiple plant immune receptors and are targeted by a Pseudomonas syringae effector. *Cell Host & Microbe*, *7*(4), 290–301.

Zhang, L., Chen, S., Ruan, J., Wu, J., Tong, A. B., Yin, Q., Li, Y., David, L., Lu, A., Wang, W. L., Marks, C., Ouyang, Q., Zhang, X., Mao, Y., & Wu, H. (2015). Cryo-EM structure of the activated NAIP2-NLRC4 inflammasome reveals nucleated polymerization. *Science*, *350*, 404–409.

Zhang, S., & Klessig, D. F. (1998). Resistance gene N-mediated de novo synthesis and activation of a tobacco mitogen-activated protein kinase by tobacco mosaic virus infection. *Proceedings of the National Academy of Sciences*, *95*(13), 7433–7438.

Zhou, J. (2015). *Host target genes of the Xanthomonas oryzae pv. oryzae type III effectors for bacterial blight in rice*. [Doctoral dissertation, Iowa State University].

Zhou, N., Tootle, T. L., Tsui, F., Klessig, D. F., & Glazebrook, J. (1998). PAD4 functions upstream from salicylic acid to control defense responses in Arabidopsis. *The Plant Cell*, *10*(6), 1021–1030.

Zipfel, C. (2008). Pattern-recognition receptors in plant innate immunity. *Current Opinion in Immunology*, *20*(1), 10–16.

Zipfel, C. (2014). Plant pattern-recognition receptors. *Trends in Immunology*, *35*(7), 345–351.

Zou, H., Li, Y., Liu, X., & Wang, X. (1999). An APAF-1·cytochrome c multimeric complex is a functional apoptosome that activates procaspase-9. *Journal of Biological Chemistry*, *274*, 11549–1155.

3 Mutagenesis
A Promising Strategy for Elevating Plant Tolerance to Biotic Stress

Mohd Ali, Sandeep Kour, Deepak Kumar, Roohi Sharma, Parkirti Parkirti, Vikram, Rajesh Kumari Manhas, Harish Changotra, and Puja Ohri

3.1 INTRODUCTION

Plant life is continuously affected by a wide range of environmental conditions. These conditions are known to interfere with crop development, growth, reproduction, and overall productivity. Biotic and abiotic factors are the two main categories of environmental conditions that affect plant life. Abiotic factors include environmental cues such as heat, cold, drought, salt, heavy metals, ultraviolet (UV) radiation, and flood, while biotic factors include all pathogens and plant parasites such as fungi, bacteria, viruses, nematodes, and phytophagous insects and pests (Meena et al., 2017; Pramanik et al., 2023; Kumar et al., 2023; Ali and Ohri, 2023). Any kind of stress caused by biotic factors as a result of deterioration or damage is thought to pose a potential risk to food safety worldwide. Typically, the type of biotic stress that might affect the plants depends on the surrounding environmental factors (Sari et al., 2020; Gull and Kausar, 2019). To survive in such stressful environments, plants have intricate mechanisms in place that allow them to detect external cues and respond appropriately. To counteract the effects of these challenges, they set up a hierarchy of defense mechanisms. Additionally, plants possess mechanisms for damage repair in response to stress. These mechanisms often involve removing damaged tissue and recovering it (Chi et al., 2019). Plants naturally undergo genetic mutations to increase their resistance to pathogens, though this process takes many years to complete. Chromosome rearrangements or alterations in genes may result from mutations (Ramakrishnan et al., 2021). Thus, plant scientists developed the technique of using induced mutation to produce high-frequency genetic variety in plants at the start of the 20th century. Plant breeding techniques are widely used to induce mutations in plants through a variety of physical means (i.e., gamma rays, ionizing radiation, etc.), chemicals (sodium azide, ethyl methane sulfonate (EMS), diethyl sulfate, etc.), and genome editing methods (e.g., transcription activator-like effector nucleases (TALENs), zincfinger nucleases (ZFNs), and clustered regularly interspaced short palindromic repeats (CRISPR) (Sawarkar et al., 2023; Zhu et al., 2020). Desired genetic alterations, such as early maturation, disease resistance, and semidwarfism in plants, are brought about by induced mutation (Bhoi et al., 2022). Therefore, mutation breeding can be employed against genetically modified organisms and traditional breeding techniques and is a cost-effective and safe strategy.

3.2 MUTAGENESIS

Plant breeders have been using mutant breeding extensively in recent years. It is among the best methods for enhancing a variety of agronomic and commercially significant traits. It has been effectively used to raise crop plant productivity. Itis a widely held belief in mutation breeding that raising a mutagen dose would increase the frequency of mutations but at the expense of decreased survival and regeneration

DOI: 10.1201/9781003543237-3

rates (Khan et al., 2009). *Nicotiana tabacum* was the first commercially successful mutant cultivar, developed in 1934 (Oladosu et al., 2015). Later, a number of breeders were able to effectively create *Saccharum officinarum* mutants that are resistant to a variety of illnesses. Gamma radiation and chemical mutagen treatment resulted in the development of *S. officinarum* variation Co S 687, which is said to be resistant to red rot disease (Kaur et al., 2016). Crops given lesser doses of various radiation were found to be resistant to a range of illnesses. The rice blast disease, caused by *Pyricularia grisea* (the anamorph of *Maganaporthe grisea*), reduced the production of the crop globally by causing necrotic lesions on leaves, nodes, seedlings, and panicles. This lowered the yield of *Oryza sativa.* Using an induced mutation strategy, numerous attempts have been undertaken to create an *O. sativa* variety that is resistant to illness. A novel commercial variety with a unique agronomic trait, namely, resistance to blast (*Pyricularia oryzae*), was introduced as a result of a mutation that produced a new gene combination that had not previously existed in the germplasm pool (Miah et al., 2013).

3.3 MUTAGENS

Several mutagens can cause genetic diversity in plants. These consist of chemical, physical, and gene editing techniques (Figure 3.1).

3.3.1 Physical Mutagens

Physical mutagens in crops exhibit distinct, heritable traits resulting from spontaneous genetic changes (Suprasanna et al., 2015). These mutations can enhance or impair plant characteristics, influencing crop yield, resilience, and nutritional content. Scientists study and harness these mutants for agricultural advancements, aiming to create more robust and productive crop varieties (Bhat et al., 2021). Physical mutagens in crops include ionizing (X-rays, α-rays, β-rays, gamma rays, fast neutrons (FNs), and thermal neutrons) and non-ionizing radiations (ultraviolet light) (Gupta et al., 2019). Exposure to these mutagens induces genetic changes such as chromosome deletion, rearrangement, and loss of bases that ultimately change the structure and function of proteins, potentially leading to amended desirable traits in crops, and aiding in the development of new varieties with improved characteristics. The main ways that physical mutagens cause chromosomal abnormalities are by ionizing the target directly or indirectly through

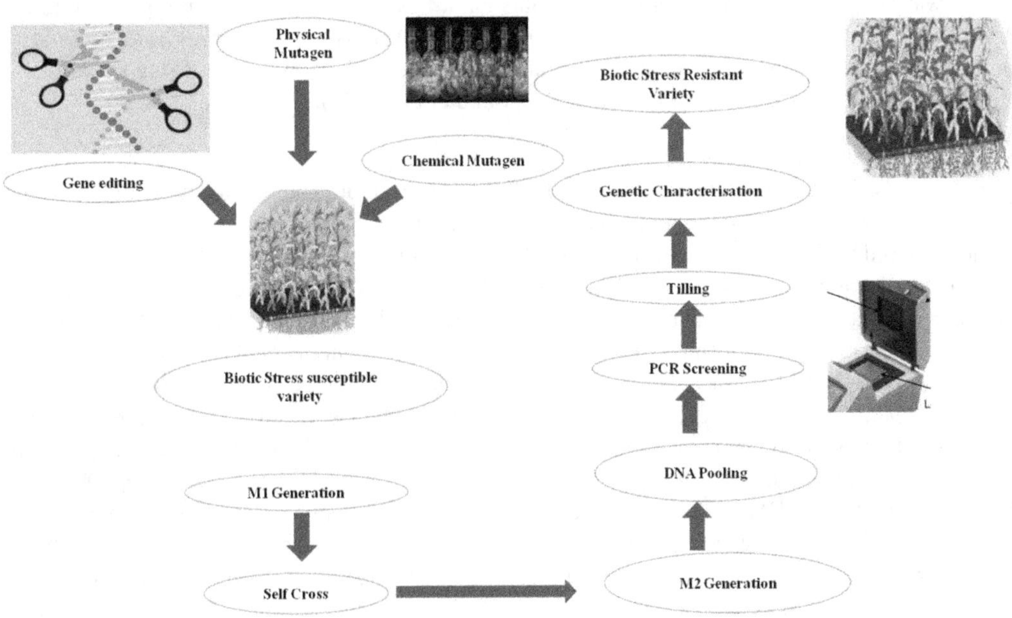

FIGURE 3.1 Diagrammatic illustration of mutagenesis techniques and identification methods for the development of plants resistant to biotic stress.

mutagenic free radicals that are produced when ambient facets are ionized (GMO EFSA panel, 2021). Over the years, about 3,200 mutant types of various agricultural plants have been produced for cultivation globally through the use of chemical and physical mutants (Mehta et al., 2018). By using ionizing radiation, plant breeders have created novel genetic combinations of plants with enhanced traits,for example, *Solanum tuberosum*'s production of tuber genotype PAU/RR-1501 γ-irradiation-derived mutants with desired processing features (Kaur et al., 2023). Furthermore, utilizing FN irradiation and gamma radiation, valuable mutants of plants have been created (Bhoi et al., 2022).

3.3.1.1 Gamma Rays

The most often utilized physical mutagen is gamma radiation. Muller employed it as a mutagenic agent for the first time in 1930 (Gulfishan et al., 2023). Using gamma radiation, a number of mutant *O. sativa*, *Cicer arietinum*, and *Solanum lycopersicum* cultivars have been created in India (Riviello-Flores et al., 2022). Physical mutagenesis, exemplified by gamma rays, offers superior penetration, yielding diverse plant variations. Its notable advantages over chemical treatments include reliable reproducibility and precision, attributed to the consistently high penetrating capacity of gamma rays. Irradiation techniques have been employed to induce diverse mutations, generating desired traits like semidwarfism, insect and disease resistance, and early maturity in plants (Mohamed et al., 2022). *S. officinarum* was successfully exposed to gamma radiation to increase its resistance to the sugarcane mosaic virus. Growing amounts of gamma radiation caused the germination rate to decrease, which was caused by a number of factors including decreased DNA synthesis, development inhibitor production, and obstruction of the auxin synthesis route (Kaur et al., 2016). Quite the reverse, 150 Gy of gamma radiation produced a mutant of *O. sativa* with a late maturity cycle, large grains, lodging resistance, and a modest resistance to blast disease (Schiocchet et al., 2014). After exposing the callus of *S. officinarum* to 0.05 Kr of gamma radiation, favorable effects were seen on a variety of significant characteristics, including yield, sugar content, plant height, number of green leaves per plant, and tiller count (Khan et al., 2009). Additionally, gamma radiation was utilized to create *C. arietinum* lines with high protein content and resistance to Ascochyta blight (produced by *Ascochyta rabii*) and mutants of *Nicotiana tabacum* that were resistant to blue mold disease (Kanouni et al., 2011). According to the study, gamma rays were the most potent mutagen. Gamma radiation, with its shorter wavelength and higher energy, penetrates plant tissues deeply, enhancing plant architecture through double-strand DNA breaks (Wang et al., 2022). This method serves as an alternative for breeding but lacks precision in genome mutation location. Moreover, identifying mutations is a time-consuming process (Olasupo et al., 2016; Sahu et al., 2020).

3.3.1.2 Fast Neutron

FN radiation is another alternate method of mutagenesis because of its ability to cause massive DNA sequence aberrations that result in malfunctions and knockouts of genes (Kumawat et al., 2019). DNA deletions caused by FN often vary in size from a few base pairs to megabases. FN technology enables the development of new crop varieties with improved traits such as higher yield, resistance to diseases, and enhanced nutritional content. This innovation enhances global food security by fostering the creation of resilient and high-performing crops. Precedence occurs in a variety of plants like *Pisum sativum* (Domoney et al., 2013), *Glycine soja* (Anderson et al., 2019), *Medicago truncatula* (Shen et al., 2023), and *Lotus japonicus* (Hoffman et al., 2007) for the use of FN mutagenesis in forward genetic screens. In *M. truncatula*, induced mutation has been observed by FN, which improved the genes for symbiosis, nodule formation, leaf patterns, and root architecture of the whole plant (Chen and Chen, 2018; Shen et al., 2023). Following EMS (Shamshad et al., 2023), gamma irradiation (Abdelnour-Esquivel et al., 2020), and FN (Malumpong et al., 2019) treatments, induced mutations relevant to biotic and abiotic stress in *O. sativa* have been observed. As a result of FN mutagenesis, structural alterations in the genomes of *O. sativa* (Malumpong et al., 2019) and *Glycine max* (Prenger et al., 2019) have previously been documented. FN technology has the benefit of being able to delete numerous genes at once, but it does not specify the target place within the genome (Kumawat et al., 2019).

3.3.1.3 Ultraviolet Rays

UV rays cannot deeply enter plant tissues because of their lower energy concentration, which makes them less invasive than gamma radiation. Because there is no insulating barrier, it is commonly used to treat calluses, cell suspension cultures, and protoplasts (Araujo et al., 2016; Blonstein and King, 2018). Exposing *Fragaria x ananassa* (Cirafine, Charlotte, and Candiss)to a UV-C radiation range of 0.85 and 1.70 kJ/m² (254 nm) effectively suppressed the production of gray molds by *Botrytis cinerea* by enhancing phenylalanine ammonia lyase (PAL) synthesis. Also, PAL activity led to the accumulation of phenylpropanoid compounds, promoting phytoalexin synthesis and plant defense (Forges et al., 2018; Zhang et al., 2021). However, excessive UV-C doses inhibited PAL synthesis and elevated the risk of black rot infection in some crops (Brown et al., 2001). Another study done on *Nicotiana tabacum* plants treated with ozone and ultraviolet light increased the production of pathogen-related (PR) proteins and salicylic acid (SA), which helped the plants become resistant to viral (such as tobacco mosaic virus (TMV)) infections (Kunz et al., 2008). After infection by a pathogen, a plant cannot instantly trigger every kind of defense mechanism. In addition to accumulating SA and PR-1 and aiding in the creation of the antibacterial chemical phytoalexins, radiation therapy can stimulate a variety of enzyme manufacturing pathways, increasing the plant's resistance to TMV. Similar results were obtained in the case of strawberry (Xu et al., 2019) and tomato (Escobar Bravo et al., 2019). UV rays are typically not favored by plant breeders as they cannot penetrate deep into plant tissues. This limitation hinders their effectiveness in inducing mutations for breeding purposes (Udage, 2021). Additionally, the time-consuming process of identifying mutant genes poses another drawback to this method, making it less practical for efficient and timely plant breeding efforts (Table 3.1).

3.3.2 Chemical Mutagens

Chemical mutagenesis is the most efficient and accessible strategy to create mutations with particular agronomic attributes in plants (Khan and Mayes, 2009). Chemical mutagens that cause mutation in plants include diethyl sulfate, 1-methyl-1-nitrosourea (MNU), and sodium azide (Suzuki et al., 2008). Many of these substances use reactive oxygen species to cause chromosomal damage in plants. Plant breeders utilize alkylating chemicals more than any other type of chemical mutagen. These chemicals cause mutations that result in substitutions in the bases of DNA, which alter the sequence of amino acids and further modify the function of proteins. According to the International Atomic Energy Association's database, alkylating chemical mutagenesis produces 80% of mutant plant species. On the contrary, 64% of mutants are produced by EMS, MNU, and 1-ethyl-1-nitrosourea (Oladosu et al., 2015).

3.3.2.1 Alkylating Agents

Alkylating chemicals react with purine and pyrimidine bases of DNA's phosphate. Another class of appropriate mutagens is base analogs. Since they are closely related to DNA bases, they can be incorporated into DNA in place of regular bases without impairing DNA replication. However, this may result in DNA pairing mistakes that cause mutations. One of the most powerful chemical mutation agents, EMS, has an alkylating characteristic that can result in low-frequency chromosomal abnormality and high-frequency gene mutation. Alkylating agents are a broad class of compounds that include diazo alkanes, alkyl phosphates, chloroethyl amines, alkyl sulfates, alkyl halides, sulfur mustards, ethylene imines, epoxides, alkyl methane sulfonates, alkyl nitrosoureas, alkyl nitrosoamines, alkyl nitrosoamides, alkyl sulfates, alkyl halides, alkyl phosphates, alkyl amines, and diazo alkanes. These are mutagens with alkylating properties that cause high-frequency gene mutation and low-frequency chromosomal aberrations (Oladosu et al., 2016).

3.3.2.2 Sodium Azide

These days, sodium azide plays a significant role in supporting crop plant agronomy. It is employed to improve crop plant resistance to damaging diseases, as well as to improve trait quality and yield. Numerous species are highly mutagenic to sodium azide. Sodium azide, on the contrary, has

TABLE 3.1
Disease-Resistant Mutant Plants Using Physical Mutagens

Stress Factor	Plant Species	Name of Disease	Mutagen Used	References
Rhizoctonia solani and Fusarium oxysporum	*Gossypium* spp.	Sheath blight and wilt diseases	Gamma irradiation	AbdulRahman et al. (2022)
Magnaporthe oryzae	*Trigonellafoenum-graecum*	Rice blast disease	Gamma irradiation	Gajbar et al. (2021)
Fusarium solani	*Helianthus annuus*	Fusarium rot disease	Gamma irradiation	Abu-Tahon et al. (2021)
Penicillium digitatum	*Citrus reticulate*	Green rots and molds	Gamma irradiation	Zhao et al. (2020)
Meloidogyne spp.	*Glycine max*	Root gall production and infestation by nematode	Gamma rays	Kankam et al. (2020)
Hypothenemus hampei Ferrari	*Coffea*spp.	Coffee berry borer disease	Gamma irradiation-induced DNA damage	Kiran et al. (2020)
Bactericera cockerelli via *Candidatus Liberibacter solanacearum*	*Solanum tuberosum* and *Solanum lycopersicum*	Zebra chip disease	Electron beam irradiation	Lei et al. (2020)
Mycosphaerella fragariae	*Fragaria x ananassa*	Leaf spot disease	Ultraviolet-C priming	Xu et al. (2019)
R. solani	*Solanum tuberosum*	Rhizoctonia stem canker and black scurf	Gamma rays	Kara and Arici (2019)
Frankliniella occidentalis	*Solanum lycopersicum*	Thrips infestation	Ultraviolet-enhanced salicylic acid	Escobar Bravo et al. (2019)
Magnaporthe oryzae triticum	*Triticum aestivum*	Blast disease	Gamma radiation	Harun-Or-Rashid et al. (2019)
Pseudomonas syringae DC3000	*Solanum lycopersicum*	Bacterial speck disease	Ultraviolet-enhanced jasmonicacid	Escobar Bravo et al. (2019)
Botrytis cinerea	*Fragaria x ananassa*	Graymolds	Uultraviolet-C radiation	Forges et al. (2018)
Penicillium expansum	*Pyrus pyrifolia* 'Niitaka'	Blue mold disease	Gamma irradiation	Jeong et al. (2017)
Sugarcane mosaic virus(SMV)	*Colletotrichum falcatum*	Red rot	Gamma rays	Kaur et al. (2016)
Puccinia melanocephala	*Saccharum officinarum*	Brown rust	gamma irradiation (^{60}Co)	Oloriz et al. (2011)
Fusarium graminearum	*Triticum aestivum*	Fusarium head blight	Fast neutron-induced chromosome deletion	Xiao et al. (2011)
Hyaloperonospora parasitica	*Arabidopsis thaliana*	Downy mildew	Ultraviolet-induced DNA damage	Kunz et al. (2008)
Sobemovirus	*Oryza sativa*	Rice yellow mottle virus	170, 201, and 240 Gray (Gy) gamma rays	Luzi-Kihupi et al. (2008)

minimal effects on a few other creatures. It is not, however, mutagenic to *Arabidopsis thaliana* or *Drosophila melanogaster*. The production of an organic azide metabolite controls sodium azide's mutagenic properties. In comparison to a plant that is not treated, the mitotic index decreases with increasing concentration and length of sodium azide treatment. The restraint sodium azide may have both mutagenic and genotoxic effects, based on its effect on the mitotic index. Similar results on *Hordeum vulgare* seeds with the addition of sodium azide have also been reported by other researchers (Table 3.1) (Ilbas et al., 2005). Furthermore, it has also been demonstrated to produce AT→GC base pair point mutations, transitions, and transversions (Bhoi et al., 2022).

3.4 GENE EDITING TECHNIQUES

A thorough grasp of the biological processes that must be included in the genome is necessary for the use of synthetic biology. Numerous techniques based on DNA, RNA, and proteins have been created to alter and insert the required agronomic features into crops. One of the most popular methods for gene targeting involves the random incorporation of genes into the existing genomes of target animals to produce a transgenic construct (Ahmar et al., 2021; Zhan et al., 2021). Therefore, transfer DNA insertion mutants were incorporated using transposons or retro-transposons, resulting in a random distribution of insertions, by plant biologists. There is a higher chance of developing mutant plants with incomplete functions, dominant-negative effects, or abnormal protein products when a random insertion occasionally fails to completely delete the open reading frame of a gene (Gebrie, 2023). These techniques cannot be used to successfully introduce single nucleotides into the genes or amino acids into proteins. To solve these issues, target-induced local genomic lesions and chemical mutagenesis techniques have been developed (Kumawat et al., 2019; Khalid et al., 2021). Even though crop improvement has been facilitated by induced mutagenesis using physical and chemical mutagens, these techniques are less preferred due to random mutation. Furthermore, it takes a lot of time and money (Chaudhary et al., 2019). Therefore, sequence-specific nucleases are becoming popular tools used in genome editing procedures in plant science. They have been applied to create high-yielding plants, make plants more resilient to environmental challenges, or increase the nutritional value of plants (Zafar et al., 2020; Li et al., 2022).

Homologous recombination, which was frequently utilized in bacteria, yeast, and mice for gene replacements or corrections, has been attempted numerous times in plants but has not yet been developed with a high degree of success (Mishra et al., 2020). The discovery of double-strand breaks by meganucleases, which have a recognition site of 18 bp and were originally discovered in yeast mitochondria, resulted in a paradigm change in the middle of the 1990s. Much more homologous recombination events could be seen with induced double-strand breaks caused by meganucleases, and double-strand breaks have also caused induced mutations due to improper repair processes (Puchta et al., 1993; Rinaldo and Ayliffe, 2015). Recognition sites are dispersed randomly throughout the genome and it is exceedingly time-consuming to redesign recognition motifs for certain target genes. ZFNs made it possible to target every gene for the first time (Khalil, 2020). The identification of TALEN proteins, which are transported to plant cells by bacterial pathogens of the genus *Xanthomonas* and activate plant genes to increase the pathogen's virulence, made it possible to target genome editing sites even more precisely (Liang et al., 2018; Gudeta, 2019). Another approach for gene editing in plants is oligonucleotide-directed mutagenesis (ODM) (Kamburova et al., 2017). Further, gene editing employing CRISPR/Cas9 and CRISPR-associated protein 9-based methods enables efficient and targeted alterations, creating a new opportunity to use this tool to enhance critical physiological and agronomic features (Dogar et al., 2022; Altaf et al., 2023).

The first sequence-specific nucleases to produce targeted double-strand breaks in eukaryotic genomes was a meganuclease. Meganucleases are naturally occurring proteins that are encoded by mobile genetic elements that are found in a wide variety of taxa, including bacteria, archaea, and eukaryotes (Pillay, 2020). *I-SecI*, the first meganuclease, was found in yeast (Faye et al., 1979). Because they can recognize DNA sequences between 14 and 40 bases long, meganucleases are the most specialized naturally occurring endonucleases. In comparison to type II restriction enzymes employed in recombinant DNA technologies, these enzymes have a broader recognition site (Hamdan et al., 2022). Meganucleases may effectively target and change any sequence of interest due to their lengthy recognition pattern and excellent specificity (Iqbal et al., 2020). For these causes, eukaryotic genomes have actively targeted double-strand breaks since the 1990s using meganucleases. In a seminal article from 1993, Puchta and colleagues showed that *I-SecI*-induced double-strand breaks in *Nicotiana tabacum* increased homologous recombination (Puchta et al., 1993). This signaled the start of precision genetic editing using sequence-specific nucleases in plants. Since then, numerous attempts to insert trait genes into plants have been made. Meganucleases have been effectively used for targeted gene editing in plants, but they have some drawbacks that prevent them from being widely used, such as limited enzyme catalytic activity, susceptibility to sequence degeneracy, and dearth of mature DNA-binding structure (Beyer and Iwaï, 2022; Bijlani et al., 2022).

ZFNs are collections of DNA recognition units built on the *Flavobacterium okeanokoites* I restriction enzyme's DNA-cleavage domain and zinc fingers. They can be used to modify the target genome and incorporate various genetic alterations, including insertions, deletions, inversions, translocations, and point mutations (Paschon et al., 2019; Saraswat et al., 2023). They have two domains: a nuclease domain on one side and a DNA-binding domain on the other. Nucleotide sequences ranging in length from 9 to 18 bases may be recognized by the DNA-binding domain's 3–6-zinc finger repeats (Yadav et al., 2023). The restriction enzyme *FokI*, required for DNA breakage, is found in the second domain (Patel et al., 2023). Three synthetic restriction enzymes, specifically ZFN-1, ZFN-2, and ZFN-3, are used in this procedure. ZFN-1: ZFN is now transported to the plant genome without requiring a repair template. When it reaches the plant genome, it causes double-stranded breaks in the host DNA, which causes non-homologous end joining of the DNA (Kumar et al., 2021; Lemarié and Marette, 2022). Non-homologous end joining either results in arbitrary mutations that are site-specific or a modest deletion or insertion of DNA. ZFN-2: In contrast to ZFN-I, the crop genome also receives a homology-directed repair coupled with a brief repair template. The template DNA and target DNA are homologous and connect to a particular sequence and cause a double-stranded rupture. An endogenous repair machinery that targets site-specific point mutations during homologous recombination kicks off the template's repair process.ZFN-3: The ZFN transcribing gene is known as ZFN3 once it has been inserted adjacent to the big repair template in the plant genome (Cai et al., 2009; Sprink et al., 2015). ZFN-mediated gene alteration has so far been documented in a number of different crops, including rice, wheat, maize, and soybean (Shukla et al., 2009; Jung et al., 2018; Ran et al., 2018; Bonawitz et al., 2019). According to Hilioti et al. (2016), the leafy-cotyledon1-like4 (L1L4) gene was effectively altered by the ZFN. The use of ZNFs has also been reported in *Nicotiana tabacum*, *Petunia hybrida*, *Arabidopsis thaliana*, *Ficus carica*, and *Malus domestica* to target endogenous and pre-reported reporter genes (Lu et al., 2016; Bhoi et al., 2022). However, their use as a crop editing tool is restricted due to the expensive and complex protein synthesis required for each targeted site, as well as the possibility of cytotoxicity effects caused by off-target breakage (Yadav et al., 2023) (Figure 3.2).

Similar to ZNFs, TALENs have become an alternate technique for genome editing (Salava et al., 2021). The DNA-cleavage motif of TALENs, like ZFNs, is the *FokI* domain, which cuts inside the 12–19-bp spacer sequence that divides each TALE binding site (Das andPaudel, 2021; González Castro et al., 2021). In contrast, the DNA-binding domain of TALENs is a tandem repeat of 33–35 amino acids, with significantly different amino acids at the 12th and 13th sites (Bhardwaj and Nain, 2021; Sakono et al., 2023). Repeat variable diresidues, which are specialized to recognize particular nucleotides, is the name given to these variable amino acids. The *Xanthomonas* bacteria's effectors, or tandem repeat proteins, are employed to recognize DNA (Shantharaj et al., 2023). Because of their straightforward interaction with DNA, TALENs, unlike ZFNs, are simple to target any DNA sequence. Off-target effects in TALENS could, however, result in double-strand breaks in other parts of the genome (Rhiel et al., 2023; Lei et al., 2023). Additionally, TALEN distribution into cells is difficult since they are far larger than ZFNs. However, TALENS is a useful method for producing plants with useful and commercially improved features (Becker and Boch, 2021). Successful applications of TALENS include *Brachypodium*, rice, tobacco, barley, and *Arabidopsis* (Cermaket al., 2011; Mahfouz et al., 2011; Christian et al., 2013; Shan et al., 2013; Wendt et al., 2013; Zhang et al., 2013). For instance, seven TALENS were used to target the five genes ADH1, MAPKKK1, DSK2B, TT4, and NATA2 in *Arabidopsis.* They discovered somatic mutations in the plants that were transgenic with an incidence of 2–15%, and they found that the mutations were passed on to the transgenic offspring with a frequency of 1.5–12% (Christian et al., 2013). In rice and barley, somatic mutagenesis was also reported. Rice was found to have a high mutation rate, which led to the creation of transgenic plants with plant pathogen resistance (Li et al., 2012; Wendt et al., 2013). Genomeediting techniques have been used to target the effector-binding elements found in the *SWEET* gene promoter

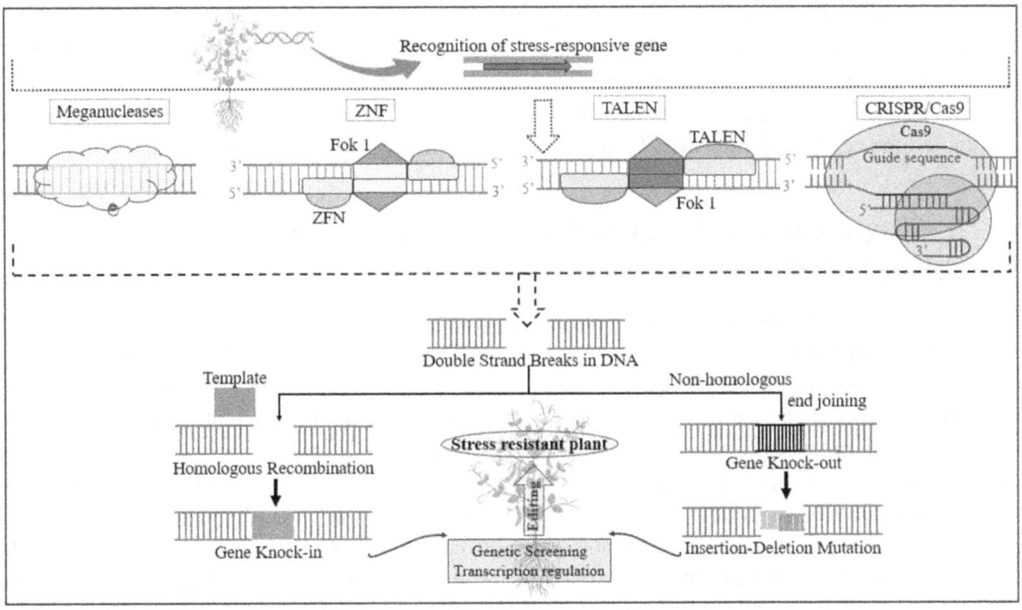

FIGURE 3.2 General illustration of sequence-specific nucleases-based gene editing in plants. Meganucleases, zinc finger nucleases (ZFNs), transcription activator-like effector nucleases (TALENs), and clustered regularly interspaced short palindromic repeats and CRISPR-associated protein 9 (CRISPR/Cas9) bring about double-strand breaks in DNA, which are further processed by homologous recombination and non-homologous end joining.

regions to make heritable changes to the genome that will prevent *S* gene activation, acting as a countermeasure to the *X. oryzae* infection method. Resistance to bacterial blight was demonstrated to be conferred by TALEN-induced mutations on overlapping effector-binding sites for PthXo3/AvrXa7 in the promoter of *OsSWEET14* (Li et al., 2012; Reschke et al., 2017). TALEN-mediated mutations in all three homologs of *MLO1* in a wheat line resulted in total defense against *Blumeria graminis*, the cause of powdery mildew in wheat (Wang et al., 2014). Recently, this line's phenotypic was thoroughly examined, and it was contrasted with TILLING mutants that offered some partial resistance to infection by pathogens. The effectiveness of powdery mildew resistance and vulnerability to the hemibiotrophic infectious agent *Magnaporthe oryzae*, which causes wheat blast disease, are correlated, according to the authors, while the results of infection with *Zymoseptoria tritici*, which causes Septoria leaf blotch, were unaffected. Additionally, it was discovered that altered plants do not exhibit the undesirable pleiotropic characteristics of early leaf senescence or callose accumulation that were seen in barley mlo mutants (Gruner et al., 2020; Matres et al., 2021).

ODM, which was originally successfully applied in mammalian systems, has since evolved into another cutting-edge method for plant gene editing (Sauer et al., 2016). To achieve site-directed editing of a gene or sequence of interest, ODM, a tool for targeted mutagenesis, uses a specific 20–100-base-long oligonucleotide, the sequence of which is identical to the target sequence in the genome but incorporates a single base pair modifications (with the goal mutation to be introduced in the genome) (Holme et al., 2019). These artificial oligonucleotides or repair templates bind to the targets and activate the cell's typical repair machinery, which detects the single inconsistency in the template and subsequently copies that mismatch or mutation into the target sequence by means of a repair technique (Mohanta et al., 2017; Sreedharan and Joseph, 2019). They are temporarily exposed to cells in plants using an array of specific means of delivery. As the plant cell breaks down the repair template oligonucleotide, the intended targeted single nucleotide or base editing that provides a unique function or characteristic is produced in the plant genome. Tissue culture techniques are used to reproduce cells with edited sequences, and traditional breeding is used to create novel genome-edited types with improved features (Abdurakhmonov, 2016; Wijerathna-Yapa et al., 2022).

3.5 CRISPR/CAS9 FOR GENOME EDITING IN MITIGATION OF BIOTIC STRESS

Biotic stressors, including phytopathogens such as fungi, bacteria, viruses, and herbivores, affect 11–30% of agricultural productivity every year by causing plant diseases and damaging plant commodities (van Esse et al., 2020). The challenge faced by plants in terms of biotic stress can be overcome by employing gene editing technologies. Advancements in gene editing techniques have led to the development of resistant crop varieties, thus, reducing losses in crop yields by enhancing plant defense against different pathogens (Wang et al., 2022). Among different gene editing techniques, the CRISPR/Cas system has gained much popularity by unraveling the utmost potential in developing resistance in plants against noxious disease-causing organisms. A one-stop solution to various diseases hampering crop productivity can be provided by creating a wide range of disease resistance toward different pathogens in staple crops (Xu et al., 2019). For instance, editing of the rice gene, *bsr-k1*, responsible for binding to and increasing the turnover of genes related to plant defense can provide a broad spectrum of disease resistance in rice (Zhou et al., 2018). Both bacterial leaf blight and leaf blight-resistant varieties of rice were created by 'switching off'*bsr-k1*defense-related genes (Zhou et al., 2018). Similarly, in tomatoes, by triggering gene editing at a single locus, wide-spectrum disease resistance was acquired (Thomazella et al., 2021). Resistance to *Xanthomonas*spp., *Pseudomonas syringae*, and *Phytophthora capsici* could be gained by mutating *SlDMR6-1* genes by CRISPER/Cas9-mediated gene editing that leads to enhanced levels of SA in plants, thereby reducing the abundance of pathogens and symptoms of diseases caused by them (Thomazella et al., 2021). Likewise, in barley, improved resistance to *Fusarium graminearum* and powdery mildew was developed by editing a defense-related gene, namely, *MORC1* with the help of CRISPR/Cas9 (Kumar et al., 2018).

Genome editing done with the aid of CRISPR/Cas9 has shown outstanding results in preventing diseases in rice plants. *SWEET* gene-encoded sucrose transporters are used by the majority of pathogens in several rice plants (Jiang et al., 2013). Bacterial leaf blight resistance was developed by targeting the promoter region of some *OsSWEET* genes by CRISPR/Cas9 (Xu et al., 2019; Olivia et al., 2019). The impact of leaf blight disease was decreased in rice, and its tolerance was enhanced by CRISPR/Cas9-mediated knocking out of the *OsERF922* gene, which was responsible for expressing ethylene response (Wang et al., 2016). Moreover, gene editing of *eIF4G*, an elongation factor in eukaryotes, using CRISPR/Cas9 led to the development of rice tungro virus-resistant rice plants (Macovei et al., 2018). Mildew-resistance locus (MLO) homologs were targeted in some plant species to boost immunity toward fungal infections. In wheat, resistance toward powdery mildew was enhanced by CRISPR/Cas9-facilitated simultaneous editing of MLO, TaMLO-A, TaMLO-B, and TaMLO-D homologs (Wang et al., 2014). Also, resistance toward the same disease in grapevine was enhanced by targeting the homologs of MLO, and a two-fold decrease in the sporulation of powdery mildew was observed in the gene-edited varieties of grapevine (Wan et al., 2020). CRISPR/Cas9 was also employed to target the *SlMlo1* gene to develop powdery mildew-resistant Tomelo,a transgene-free tomato (Nekrasov et al., 2017). Several studies have been conducted that demonstrate biotic stress tolerance in different plants by developing disease-resistant varieties through CRISPR/Cas9-mediated gene editing. Some of them are listed in Table 3.2.

TABLE 3.2
CRISPER/Cas9 Mediated Genome Editing for Biotic Stress Tolerance

Pathogen	Plant Species	Target Gene	Effect	References
Sphaerotheca fusca	Petuniacv. 'Mirage Rose'	*PhMLO1*	Decreased powdery mildew severity and symptoms in mutant plants	Xu et al. (2023)
Pseudomonas syringae pv. *tomato* (Pto) DC3000	Tomato	*SlHyPRP1*	Decreased bacterial growth in gene-edited plants	Tran et al. (2023)
Helicoverpa armigera and *Spodoptera litura*	Soybean	*GmUGT*	Increased resistance toward insect pests	Zhang et al. (2022)
S. litura	Soybean	*CALCIUM-DEPENDENT PROTEIN KINASE 38 (GmCDPK38)*	Regulated flowering time and enhanced insect resistance in soybean	Li et al. (2022)
Magnaporthe oryzae (*M. oryzae*) and *Xanthomonas oryzae* pv. *oryzae*	Rice	*Bsr-d1*, *Pi21*, and *ERF922*	Increased resistance to both rice blast and bacterial blast	Zhou et al. (2022)
Phytophthora infestans	Tomato	*miR482b* and *miR482c*	Reduced late blight disease symptoms, decreased reactive oxygen species levels, generated pathogen-resistant tomatoes	Hong et al. (2021)
Botrytis cinerea, *Rhizopus stolonifer*, and *Fusarium acuminatum*	Tomato	*PECTATE LYASE (SlPL)*	Reduced susceptibility and improved resistance to fungal diseases	Silva et al. (2021)
Erysiphe necator	Grapevine	*VvMOL3*	Boosted resistance to powdery mildew	Wan et al. (2020)
Oidium neolycopersici	Tomato	*POWDERY MILDEW RESISTANCE 4 (SlPMR4)*	Haustorial formation and hyphal growth decreased, reduced susceptibility to powdery mildew pathogen	Santillán Martínez et al. (2020)
Wheat dwarf virus	Barley	MP, CP, Rep/Rep, IR/virus genome	Increased disease resistance in plant	Kis et al. (2019)
P. syringae pv. *tomato* (Pto) DC3000	Tomato	*SlDMR6-1*	Developed disease-resistant plant varieties	Ortigosa et asl. (2019)
M. oryzae	Rice	ALB1, RSY1	Enhanced resistance toward rice blight	Foster et al. (2018)
Plant hopper and stem borer	Rice	*CYTOCHROME P450 71A1 (OsCYP71A1)*	Provided defense against insect pests	Lu et al. (2018)
Blumeria graminis f. sp. *tritici*	Wheat	*TaEDR1*	Developed disease resistance and eradicated mildew-induced cell death	Zhang et al. (2017)
Cucumber vein yellowing virus (CVYV), zucchini yellow mosaic virus (ZYMV), and papaya ring spot mosaic virus-W (PRSV-W)	Cucumber	*eIF4e*	Developed immunity and virus resistance in cucumber	Chandrasekaran et al. (2016)

3.6 CONCLUSION

The main objective of a plant breeding program is to increase the production of superior crops. Numerous environmental contaminants either directly affect the plants or increase their vulnerability to biological stressors. Therefore, it is imperative to produce crop plants that are resistant to stress to lessen the negative consequences of environmental hazards. Plant breeders frequently employ the highly effective induced mutation approach to increase the quality of their crops. Additionally, a novel framework for the application of mutation breeding in crop development has been offered by molecular methods. Therefore, induced mutation is thought to open up new possibilities for biological research and crop development. These are probably common strategies used in crops to improve production and quality characteristics. Mutation-induced varieties can withstand a range of biotic and abiotic stress situations and can assist in reducing the need for poisonous pesticides when dealing with harmful diseases. Therefore, it needs to be administered to a range of crops that are vulnerable to pests and insects. Although mutant plant species have contributed greatly, however, to achieving food security and fast economic growth, a great deal of work needs to be undertaken.

REFERENCES

Abdelnour-Esquivel, A., Perez, J., Rojas, M., Vargas, W., & Gatica-Arias, A. (2020). Use of gamma radiation to induce mutations in rice (Oryza sativa L.) and the selection of lines with tolerance to salinity and drought. *In Vitro Cellular and Developmental Biology-Plant*, *56*, 88–97.

Abdul Rahman, A. A., Ali, H. Z., Saaed, F. H., & Mohamed, M. K. (2022). Using of gamma irradiation to induce resistant in cotton plant against *Fusarium oxysporum* fir 16, Rhizoctonia solani c2 and Rhizoctonia solani z. *Biochemical and Cellular Archives*, *22*(1), 3013–3018.

Abdurakhmonov, I. Y. (2016). Genomics era for plants and crop species: Advances made and needed tasks ahead. In I. Y. Abdurakhmonov (Ed.), *Plant genomics* (pp. 3–28). In Tech.

Abu-Tahon, M. A., Isaac, G. S., & Mogazy, A. M. (2021). Protective role of fat hen (Chenopodium album L.) extract and gamma irradiation treatments against fusarium root rot disease in sunflower plants. *Plant Biology*, *23*(3), 497–507.

Ahmar, S., Mahmood, T., Fiaz, S., Mora-Poblete, F., Shafique, M. S., Chattha, M. S., & Jung, K. H. (2021). Advantage of nanotechnology-based genome editing system and its application in crop improvement. *Frontiers in Plant Science*, *12*, 663849.

Ali, M., & Ohri, P. (2023). Deciphering the synergistic effect of Jasmonic acid and Spermine in mitigating root-knot nematode stress in tomato plants through enhancing growth and activity of antioxidant enzymes. *South African Journal of Botany*, *161*, 21–35.

Altaf, M. A., Mandal, S., Behera, B., Mangal, V., Naz, S., Kumar, R., Ghorai, M., Singh, B., Dey, A., & Aftab, T. (2023). Salinity stress tolerance in solanaceous crops: Current understanding and its prospects in genome editing. *Journal of Plant Growth Regulation*, *42*(7), 4020–4036.

Anderson, E. J., Ali, M. L., Beavis, W. D., Chen, P., Clemente, T. E., Diers, B. W., & Tilmon, K. J. (2019). Soybean [Glycine max (L.) Merr.] breeding: History, improvement, production and future opportunities. *Advances in Plant Breeding Strategies: Legumes*, *7*, 431–516.

Araujo, S. D. S., Paparella, S., Dondi, D., Bentivoglio, A., Carbonera, D., & Balestrazzi, A. (2016). Physical methods for seed invigoration: Advantages and challenges in seed technology. *Frontiers in Plant Science*, *7*, 646.

Becker, S., & Boch, J. (2021). TALE and TALEN genome editing technologies. *Gene and Genome Editing*, *2*, 100007.

Beyer, H. M., & Iwaï, H. (2022). Structural basis for the propagation of homing endonuclease-associated inteins. *Frontiers in Molecular Biosciences*, *9*, 855511.

Bhardwaj, A., & Nain, V. (2021). TALENs: An indispensable tool in the era of CRISPR: A mini review. *Journal of Genetic Engineering and Biotechnology*, *19*(1), 1–10.

Bhat, M. A., Mir, R. A., Kumar, V., Shah, A. A., Zargar, S. M., Rahman, S., & Jan, A. T. (2021). Mechanistic insights of CRISPR/Cas-mediated genome editing towards enhancing abiotic stress tolerance in plants. *Physiologia Plantarum*, *172*(2), 1255–1268.

Bhoi, A., Yadu, B., Chandra, J., & Keshavkant, S. (2022). Mutagenesis: A coherent technique to develop biotic stress resistant plants. *Plant Stress*, *3*, 100053.

Bijlani, S., Pang, K. M., Sivanandam, V., Singh, A., & Chatterjee, S. (2022). The role of recombinant AAV in precise genome editing. *Frontiers in Genome Editing*, *3*, 799722.

Blonstein, A. D., & King, P. J. (2018). Plant mutant isolation via protoplasts. In R. Tage Eriksson (Eds.), *Plant protoplasts* (pp. 151–168). CRC Press.

Bonawitz, N. D., Ainley, W. M., Itaya, A., Chennareddy, S. R., Cicak, T., Effinger, K., Jiang, K., Mall, T. K., Marri, P. R., Samuel, J. P., & Pareddy, D. R. (2019). Zinc finger nuclease-mediated targeting of multiple transgenes to an endogenous soybean genomic locus via non-homologous end joining. *Plant Biotechnology Journal*, *17*(4), 750–761.

Brown, J. E., Lu, T. Y., Stevens, C., Khan, V. A., Lu, J. Y., Wilson, C. L., & Droby, S. (2001). The effect of low dose ultraviolet light-C seed treatment on induced resistance in cabbage to black rot (Xanthomonas campestris pv. campestris). *Crop Protection*, *20*(10), 873–883.

Cai, C. Q., Doyon, Y., Ainley, W. M., Miller, J. C., DeKelver, R. C., Moehle, E. A., Rock, J. M., Lee, Y. L., Garrison, R., Schulenberg, L., & Petolino, J. F. (2009). Targeted transgene integration in plant cells using designed zinc finger nucleases. *Plant molecular biology*, *69*, 699–709.

Cermak, T., Doyle, E. L., Christian, M., Wang, L., Zhang, Y., Schmidt, C., Baller, J. A., Somia, N. V., Bogdanove, A. J., & Voytas, D. F. (2011). Efficient design and assembly of custom TALEN and other TAL effector-based constructs for DNA targeting. *Nucleic Acids Research*, *39*(12), e82–e82.

Chandrasekaran, J., Brumin, M., Wolf, D., Leibman, D., Klap, C., Pearlsman, M., Sherman, A., Arazi, T. & Gal-On, A. (2016). Development of broad virus resistance in non-transgenic cucumber using CRISPR/Cas9 technology. *Molecular Plant Pathology*, *17*(7), 1140–1153.

Chaudhary, J., Deshmukh, R., & Sonah, H. (2019). Mutagenesis approaches and their role in crop improvement. *Plants*, *8*(11), 467.

Chen, Y., & Chen, R. (2018). Physical mutagenesis in Medicago truncatula using fast neutron bombardment (FNB) for symbiosis and developmental biology studies. In L. Cañas, & J. Beltran (Eds.), *Functional genomics in Medicago truncatula: Methods and protocols* (pp. 61–69).

Chi, Y. H., Koo, S. S., Oh, H. T., Lee, E. S., Park, J. H., Phan, K. A. T., Wi, S. D., Bae, S. B., Paeng, S. K., Chae, H. B & Lee, S. Y. (2019). The physiological functions of universal stress proteins and their molecular mechanism to protect plants from environmental stresses. *Frontiers in Plant Science*, *10*, 750.

Christian, M., Qi, Y., Zhang, Y., & Voytas, D. F. (2013). Targeted mutagenesis of *Arabidopsisthaliana* using engineered TAL effector nucleases. *G3: Genes, Genomes, Genetics*, *3*(10), 1697–1705.

Das, B. D., & Paudel, N. (2021). A review on reliability and validity of CRISPR/Cas9 technology for gene editing. *Jordan Journal of Biological Sciences*, *14*(3) ,503–511.

Dogar, A., Ali, M., Riaz, Z., Ali, Q., Ahmad, S., & Javed, M. (2022). Role of CRISPR to improve stress tolerance in plants. *Biological and Clinical Sciences Research Journal*, *2021*(1), e022–e022.

Domoney, C., Knox, M., Moreau, C., Ambrose, M., Palmer, S., Smith, P., Christodoulou, V., Isaac, P. G., Hegarty, M., Blackmore, T., & Ellis, N. (2013). Exploiting a fast neutron mutant genetic resource in Pisum sativum (pea) for functional genomics. *Functional Plant Biology*, *40*(12), 1261–1270.

EFSA Panel on Genetically Modified Organisms (GMO), Mullins, E., Bresson, J. L., Dalmay, T., Dewhurst, I. C., Epstein, M. M., Firbank, L. G., Guerche, P., Hejatko, J., Moreno, F. J., & Rostoks, N. (2021). In vivo and in vitro random mutagenesis techniques in plants. *EFSA Journal*, *19*(11), e06611.

Escobar Bravo, R., Chen, G., Grosser, K., Van Dam, N. M., Leiss, K. A., & Klinkhamer, P. G. (2019). Ultraviolet radiation enhances salicylic acid-mediated defense signaling and resistance to Pseudomonas syringae DC3000 in a jasmonic acid-deficient tomato mutant. *Plant Signaling and behavior*, *14*(4), e1581560.

Faye, G., Dennebouy, N., Kujawa, C., & Jacq, C. (1979). Inserted sequence in the mitochondrial 23S ribosomal RNA gene of the yeast *Saccharomyces cerevisiae*. *Molecular and General Genetics MGG*, *168*, 101–109.

Forges, M., Vàsquez, H., Charles, F., Sari, D. C., Urban, L., Lizzi, Y., Bardin, M., & Aarrouf, J. (2018). Impact of UV-C radiation on the sensitivity of three strawberry plant cultivars (Fragaria x ananassa) against *Botrytis cinerea*. *Scientia Horticulturae*, *240*, 603–613.

Foster, A. J., Martin-Urdiroz, M., Yan, X., Wright, H. S., Soanes, D. M., & Talbot, N. J. (2018). CRISPR-Cas9 ribonucleoprotein-mediated co-editing and counterselection in the rice blast fungus. *Scientific Reports*, *8*, 14355.

Gajbar, T. D., Kamble, M., Adhikari, S., Konappa, N., Satapute, P., & Jogaiah, S. (2021). Gamma-irradiated fenugreek extracts mediates resistance to rice blast disease through modulating histochemical and biochemical changes. *Analytical Biochemistry*, *618*, 114121.

Gebrie, A. (2023). Transposable elements as essential elements in the control of gene expression. *Mobile DNA*, *14*(1), 9.

González Castro, N., Bjelic, J., Malhotra, G., Huang, C., & Alsaffar, S. H. (2021). Comparison of the feasibility, efficiency and safety of genome editing technologies. *International Journal of Molecular Sciences*, *22*(19), 10355.

Gruner, K., Esser, T., Acevedo-Garcia, J., Freh, M., Habig, M., Strugala, R., Stukenbrock, E., Schaffrath, U., & Panstruga, R. (2020). Evidence for allele-specific levels of enhanced susceptibility of wheat mlo mutants to the hemibiotrophic fungal pathogen *Magnaportheoryzae*pv.*Triticum*. *Genes*, *11*(5), 517.

Gudeta, D. (2019). Genome editing: Tools and application in plants. *Journal of Microbiology and Biotechnology*, *4*(1), 000135.

Gulfishan, M., Bhat, T. A., Mir, R. A., Jahan, A., Shabir, F., & Khan, S. A. (2023). Plant mutagenesis: Terms and applications. In T. A. Bhat, & K. R. Hakeem (Eds.), *Biotechnologies and genetics in plant mutation breeding: Revolutionizing plant biology* (pp. 157–171). Apple Academic Press.

Gull, M., & Kausar, A. (2019). Screening the variability in salt tolerance of Sorghum bicolor L. by nutrients uptake and growth analysis of four genotypes. *Pharmacophore*, *10*(2), 43–50.

Gupta, S., Datta, A. K., Pramanik, A., Biswas, J., & Karmakar, R. (2019). X-ray and gamma irradiation induced chromosomal aberrations in plant species as the consequence of induced mutagenesis-an overview. *Plant Archives*, *19*(2), 09725210.

Hamdan, M. F., Karlson, C. K. S., Teoh, E. Y., Lau, S. E., & Tan, B. C. (2022). Genome editing for sustainable crop improvement and mitigation of biotic and abiotic stresses. *Plants*, *11*(19), 2625.

Harun-Or-Rashid, M., Meah, M. B., Uddin, M. I., Ahmed, S., & Kashem, M. A. (2019). Gamma radiated wheat for combating devastating blast disease (Magnaporthe oryzae Triticum) in Bangladesh. *Agricultural Science*, *1*(1), 1.

Hilioti, Z., Ganopoulos, I., Ajith, S., Bossis, I., & Tsaftaris, A. (2016). A novel arrangement of zinc finger nuclease system for in vivo targeted genome engineering: The tomato LEC1-LIKE4 gene case. *Plant Cell Reports*, *35*, 2241–2255.

Hoffmann, D., Jiang, Q., Men, A., Kinkema, M., & Gresshoff, P. M. (2007). Nodulation deficiency caused by fast neutron mutagenesis of the model legume Lotus japonicus. *Journal of Plant Physiology*, *164*(4), 460–469.

Holme, I. B., Gregersen, P. L., & Brinch-Pedersen, H. (2019). Induced genetic variation in crop plants by random or targeted mutagenesis: Convergence and differences. *Frontiers in Plant Science*, *10*, 1468.

Hong, Y., Meng, J., He, X., Zhang, Y., Liu, Y., Zhang, C., Qi, H., & Luan, Y. (2021). Editing miR482b and miR482c simultaneously by CRISPR/Cas9 enhanced tomato resistance to Phytophthora infestans. *Phytopathology®*, *111*(6), 1008–1016.

Iqbal, Z., Iqbal, M. S., Ahmad, A., Memon, A. G., & Ansari, M. I. (2020). New prospects on the horizon: Genome editing to engineer plants for desirable traits. *Current Plant Biology*, *24*, 100171.

Jeong, R. D., Jeong, M. A., & Park, M. R. (2017). Gamma irradiation-induced disease resistance of pear (Pyrus pyrifolia "Niitaka") against Penicillium expansum. *Journal of Phytopathology*, *165*(9), 626–633.

Jiang, W., Zhou, H., Bi, H., Fromm, M., Yang, B., & Weeks, D. P. (2013). Demonstration of CRISPR/Cas9/sgRNA-mediated targeted gene modification in Arabidopsis, tobacco, sorghum and rice. *Nucleic Acids Research*, *41*, e188.

Jung, Y. J., Nogoy, F. M., Lee, S. K., Cho, Y. G., & Kang, K. K. (2018). Application of ZFN for site directed mutagenesis of rice SSIVa gene. *Biotechnology and Bioprocess Engineering*, *23*, 108–115.

Kamburova, V. S., Nikitina, E. V., Shermatov, S. E., Buriev, Z. T., Kumpatla, S. P., Emani, C., & Abdurakhmonov, I. Y. (2017). Genome editing in plants: An overview of tools and applications. *International Journal of Agronomy*, *1*, 7315351.

Kankam, F., Addai, I. A., Larbi-Koranteng, S., & Asamane, T. A. (2020). Root-knot nematodes (meloidogyne spp.) infestation and reproduction on soybean (Glycine max L. Merrill) treated with varying doses of gamma rays. *Agricultural and Food Science Journal of Ghana*, *13*, 1197–1207.

Kanouni, H., Taleei, A., & Okhovat, M. (2011). Ascochyta blight (Ascochyta rabiei (Pass.) Lab.) of chickpea (Cicer arietinum L.): Breeding strategies for resistance. *International Journal of Plant Breeding and Genetics*, *5*(1), 1–22.

Kaur, M., Thind, K. S., Sanghera, G. S., Kumar, R., & Kashyap, L. (2016). Gamma rays induced variability for economic traits, quality and red rot resistance in sugarcane (Saccharum spp.). *International Journal of Science, Environment and Technology*, *5*(2), 355–65.

Kaur, R., Sharma, S. P., Kalia, A., Kaur, N., & Manchanda, P. (2023). In vitro induction and selection of mutants obtained through gamma irradiation with improved processing traits in potato (Solanum tuberosum L.). *International Journal of Radiation Biology*, *100*(1), 139–149.

Khalid, M. N., Amjad, I., Nyain, M. V., Saleem, M. S., Asif, M., Ammar, A., & Rasheed, Z. (2021). A review: Tilling technique strategy for cereal crop development. *International Journal of Applied Chemical and Biological Sciences*, *2*(5), 8–15.

Khalil, A. M. (2020). The genome editing revolution. *Journal of Genetic Engineering and Biotechnology*, *18*(1), 1–16.

Khan, A., & Mayes, S. (2009). Transition to accrual accounting. In C. Cottarelli (Eds.), *Technical Notes and Manuals* (pp. 1–24). Washington, DC 20431USA.

Khan, S., Al-Qurainy, F., & Anwar, F. (2009). Sodium azide: A chemical mutagen for enhancement of agronomic traits of crop plants. *Environment & We an International Journal of Science & Technology*, *4*, 1–21.

Kiran, R., Shenoy, K. B., & Venkatesha, M. G. (2020). Gamma radiation-induced DNA damage in adults of the coffee berry borer, Hypothenemus hampei Ferrari (Coleoptera: Curculionidae). *International Journal of Tropical Insect Science*, *40*, 773–779.

Kis, A., Hamar, E., Tholt, G., Ban, R., & Havelda, Z. (2019). Creating highly efficient resistance against wheat dwarf virus in barley by employing CRISPR/Cas9 system. *Plant Biotechnology Journal*, *17*, 1004–1006.

Kumar, N., Galli, M., Ordon, J., Stuttmann, J., Kogel, K.-H., Imani, J. (2018). Further analysis of barley MORC1 using a highly efficient RNA-guided Cas9 gene-editing system. *Plant Biotechnology Journal*, *16*, 1892–1903.

Kumar, N., Verma, S., Kumar, A., Dasila, H., Suyal, D. C., Kumari, G., Rawat, S., Jeena, N., Singh, M., & Kumar, A. (2023). Molecular insights into stress-responsive genes in the mitigation of environmental stresses. In D. Chandra & P. Bhatt (Eds.), *Unravelling plant-microbe synergy* (pp. 87–117). Academic Press.

Kumar, R., Nizampatnam, N. R., Alam, M., Thakur, T. K., & Kumar, A. (2021). Genome editing technologies for plant improvement: Advances, applications and challenges. In A. Kumar, R. Kumar, P. Shukla, & M. K. Pandey (Eds.), *Omics technologies for sustainable agriculture and global food security* (pp. 213–240). Springer.

Kumawat, S., Rana, N., Bansal, R., Vishwakarma, G., Mehetre, S. T., Das, B. K., Kumar, M., Yadav, S. K., Sonah, H., Sharma, T. R., & Deshmukh, R. (2019). Expanding avenue of fast neutron mediated mutagenesis for crop improvement. *Plants*, *8*(6), 164.

Kumawat, S., Rana, N., Bansal, R., Vishwakarma, G., Mehetre, S. T., Das, B. K., Kumar, M., Yadav, S. K., Sonah, H., Sharma, T. R., & Deshmukh, R. (2019). Expanding avenue of fast neutron mediated mutagenesis for crop improvement. *Plants*, *8*(6), 164.

Kunz, B. A., Dando, P. K., Grice, D. M., Mohr, P. G., Schenk, P. M., & Cahill, D. M. (2008). UV-induced DNA damage promotes resistance to the biotrophic pathogen Hyaloperonospora parasitica in Arabidopsis. *Plant Physiology*, *148*(2), 1021–1031.

Lei, J., Meng, J., Chen, I. W., Cheng, W., Beam, A. L., Islam, M. S., Bailey, W. D., Pillai, S., & Zhu-Salzman, K. (2020). Deleterious effects of electron beam irradiation on development and reproduction of tomato/potato psyllids, Bactericera cockerelli. *Insect Science*, *27*(6), 1311–1321.

Lei, Z., Meng, H., Zhuang, Y., Zhu, Q., & Yi, C. (2023). Chemical and biological approaches to interrogate off-target effects of genome editing tools. *ACS Chemical Biology*, *18*(2), 205–217.

Lemarié, S., & Marette, S. (2022). The socio-economic factors affecting the emergence and impacts of new genomic techniques in agriculture: A scoping review. *Trends in Food Science and Technology*, *129*, 38–48

Li, T., Liu, B., Spalding, M. H., Weeks, D. P., & Yang, B. (2012). High-efficiency TALEN-based gene editing produces disease-resistant rice. *Nature Biotechnology*, *30*(5), 390–392.

Li, X., Hu, D., Cai, L., Wang, H., Liu, X., Du, H., Yang, Z., Zhang, H., Hu, Z., Huang, F., & Wang, H. (2022). Calcium-Dependent Protein Kinase38 regulates flowering time and common cutworm resistance in soybean. *Plant Physiology*, *190*(1), 480–499.

Li, Y., Wu, X., Zhang, Y., & Zhang, Q. (2022). CRISPR/Cas genome editing improves abiotic and biotic stress tolerance of crops. *Frontiers in Genome Editing*, *4*, 987817.

Liang, Z., Chen, K., Yan, Y., Zhang, Y., & Gao, C. (2018). Genotyping genome-edited mutations in plants using CRISPR ribonucleoprotein complexes. *Plant Biotechnology Journal*, *16*(12), 2053–2062.

Lu, H. P., Luo, T., Fu, H. W., Wang, L., Tan, Y., Huang, J., Wang, Q., Ye, G. Y., Gatehouse, A. M., Lou, Y. G., & Shu, Q. (2018). Resistance of rice to insect pests mediated by suppression of serotonin biosynthesis. *Nature Plants*, *4*, 338–344.

Lu, H., Klocko, A. L., Dow, M., Ma, C., Amarasinghe, V., & Strauss, S. H. (2016). Low frequency of zinc-finger nuclease-induced mutagenesis in Populus. *Molecular Breeding*, *36*, 1–13.

Luzi-Kihupi, A., Shao-Mwalyego, F., Zakayo, J. A., & Mkuya, M. (2008). Mwangaza–a new early maturing, RYMV resistant rice mutant released in the United Republic of Tanzania. *Plant Mutation Reports*, *2*(1), 13–5.

Macovei, A., Sevilla, N. R., Cantos, C., Jonson, G. B., Slamet-Loedin, I., Cerm ˇ ák, T., Voytas, D. F., Choi, I. R., & Chadha-Mohanty, P. (2018). Novel alleles of rice eif4g generated by CRISPR/Cas9-targeted mutagenesis confer resistance to rice tungro spherical virus. *Plant Biotechnology Journal*, *16*, 1918–1927.

Mahfouz, M. M., Li, L., Shamimuzzaman, M., Wibowo, A., Fang, X., & Zhu, J. K. (2011). De novo-engineered transcription activator-like effector (TALE) hybrid nuclease with novel DNA binding specificity creates double-strand breaks. *Proceedings of the National Academy of Sciences*, *108*(6), 2623–2628.

Malumpong, C., Youngkom, P., Vanavichit, A., & Siang-Liw, M. (2019). Variation in tolerance to salinity stress at the reproductive stage in a large fast neutron mutant rice (Oryza sativa L.) population. *International Journal of Agricultural Technology*, *15*(3), 445–464.

Matres, J. M., Hilscher, J., Datta, A., Armario-Nájera, V., Baysal, C., He, W., Huang, X., Zhu, C., Valizadeh-Kamran, R., Trijatmiko, K. R., & Slamet-Loedin, I. H. (2021). Genome editing in cereal crops: An overview. *Transgenic Research*, *30*, 461–498.

Meena, K. K., Sorty, A. M., Bitla, U. M., Choudhary, K., Gupta, P., Pareek, A., Singh, D. P., Prabha, R., Sahu, P. K., Gupta, V. K., & Minhas, P. S. (2017). Abiotic stress responses and microbe-mediated mitigation in plants: The omics strategies. *Frontiers in Plant Science*, *8*, 172.

Mehta, A. K., & Basha, M. H. (2018). Application of mutation breeding in plants. In D. N. Bharadwaj (Eds.), *Advanced molecular plant breeding: Meeting the challenge of food security* (pp. 139–234). CRC Press.

Miah, G., Rafii, M. Y., Ismail, M. R., Puteh, A. B., Rahim, H. A., Asfaliza, R., & Latif, M. A. (2013). Blast resistance in rice: A review of conventional breeding to molecular approaches. *Molecular Biology Reports*, *40*, 2369–2388.

Mishra, N., Tiwari, P. N., Sahu, V. K., & Singh, Y. (2020). Genome editing in plants using CRISPR for crop improvement. *Recent Trends in Molecular Biology and Biotechnology*, 1, 153–173.

Mohamed, E., Adham, N. E. S., Mohd Esa, N. A. F., & Abd Aziz, M. H. (2022). Mutation in plant: Key of successful agriculture industry. In M. A. Rahim & M. Radzuan (Eds.), *Chemical process and sustainability in agricultural biotechnology* (pp. 99–109). Penerbit UTHM.

Mohanta, T. K., Bashir, T., Hashem, A., Abd_Allah, E. F., & Bae, H. (2017). Genome editing tools in plants. *Genes*, *8*(12), 399.

Nekrasov, V., Wang, C., Win, J., Lanz, C., Weigel, D., & Kamoun, S. (2017). Rapid generation of a transgene-free powdery mildew resistant tomato by genome deletion. *Science Report*, *7*, 482.

Oladosu, Y., Rafii, M. Y., Abdullah, N., Hussin, G., Ramli, A., Rahim, H. A., Miah, G., &Usman, M. (2016). Principle and application of plant mutagenesis in crop improvement: A review. *Biotechnology and Biotechnological Equipment*, *30*(1), 1–16.

Olasupo, F. O., Ilori, C. O., Forster, B. P., & Bado, S. (2016). Mutagenic effects of gamma radiation on eight accessions of Cowpea (Vigna unguiculata [L.] Walp.). *American Journal of Plant Sciences*, *7*(2), 339–351.

Oliva, R., Ji, C., Atienza-Grande, G., Huguet-Tapia, J. C., Perez-Quintero, A., Li, T., Eom, J.-S., Li, C., Nguyen, H., Liu, B. & Auguy, F. (2019). Broad-spectrum resistance to bacterial blight in rice using genome editing. *Nat. Biotechnol. 37*, 1344–1350.

Oloriz, M. I., Gil, V., Rojas, L., Veitía, N., Höfte, M., & Jiménez, E. (2012). Selection and characterisation of sugarcane mutants with improved resistance to brown rust obtained by induced mutation. *Crop and Pasture Science*, *62*(12), 1037–1044.

Ortigosa, A., Gimenez-Ibanez, S., Leonhardt, N., & Solano, R. Design of a bacterial speck resistant tomato by CRISPR/Cas9-mediated editing of SlJAZ2. *Plant Biotechnology Journal*, 2019, *17*, 665–673.

Paschon, D. E., Lussier, S., Wangzor, T., Xia, D. F., Li, P. W., Hinkley, S. J., Scarlott, N. A., Lam, S. C., Waite, A. J., Truong, L. N., & Rebar, E. J. (2019). Diversifying the structure of zinc finger nucleases for high-precision genome editing. *Nature Communications*, *10*(1), 1133.

Patel, A., Miles, A., Strackhouse, T., Cook, L., Leng, S., Patel, S., Klinger, K., Rudrabhatla, S., & Potlakayala, S. D. (2023). Methods of crop improvement and applications towards fortifying food security. *Frontiers in Genome Editing*, *5*, 1171969.

Pillay, M. (2020). Genome editing technologies for crop improvement. In S. Manjit Kang (Eds.), *Quantitative genetics, genomics and plant breeding* (pp. 33–43). CABI.

Pramanik, B., Sar, P., Bharti, R., Gupta, R., Purkayastha, S., Sinha, S., Chattaraj, S., & Mitra, D. (2023). Multifactorial role of nanoparticles in alleviating environmental stresses for sustainable crop production and protection. *Plant Physiology and Biochemistry*, *201*, 107831.

Prenger, E. M., Ostezan, A., Mian, M. R., Stupar, R. M., Glenn, T., & Li, Z. (2019). Identification and characterization of a fast-neutron-induced mutant with elevated seed protein content in soybean. *Theoretical and Applied Genetics*, *132*, 2965–2983.

Puchta, H., Dujon, B., & Hohn, B. (1993). Homologous recombination in plant cells is enhanced by in vivo induction of double strand breaks into DNA by a site-specific endonuclease. *Nucleic Acids Research*, *21*(22), 5034–5040.

Ramakrishnan, M., Satish, L., Kalendar, R., Narayanan, M., Kandasamy, S., Sharma, A., Emamverdian, A., Wei, Q., & Zhou, M. (2021). The dynamism of transposon methylation for plant development and stress adaptation. *International Journal of Molecular Sciences*, *22*(21), 11387.

Ran, Y., Patron, N., Kay, P., Wong, D., Buchanan, M., Cao, Y. Y., Sawbridge, T., Davies, J. P., Mason, J., Webb, S. R., & Hayden, M. J. (2018). Zinc finger nuclease-mediated precision genome editing of an endogenous gene in hexaploid bread wheat (*Triticum aestivum*) using a DNA repair template. *Plant Biotechnology Journal*, *16*(12), 2088–2101.

Reschke, M., Auguy, F., Doucoure, H., Szurek, B., Meynard, D., Portefaix, M., Cunnac, S. Ã., Guiderdoni, E., Boch, J., & Koebnik, R. (2017). Targeted promoter editing for rice resistance to *Xanthomonas oryza-e*pv.*oryzae* reveals differential activities for SWEET14âÄêinducing TAL effectors. *Plant Biotechnology Journal*, *15*(3), 306–317.

Rhiel, M., Geiger, K., Andrieux, G., Rositzka, J., Boerries, M., Cathomen, T., & Cornu, T. I. (2023). T-CAST: An optimized CAST-Seq pipeline for TALEN confirms superior safety and efficacy of obligate-heterodimeric scaffolds. *Frontiers in Genome Editing*, *5*, 1130736.

Rinaldo, A. R., & Ayliffe, M. (2015). Gene targeting and editing in crop plants: A new era of precision opportunities. *Molecular Breeding*, *35*, 1–15.

Riviello-Flores, M. D. L. L., Cadena-Iñiguez, J., Ruiz-Posadas, L. D. M., Arévalo-Galarza, M. D. L., Castillo-Juárez, I., Soto Hernández, M., & Castillo-Martínez, C. R. (2022). Use of gamma radiation for the genetic improvement of underutilized plant varieties. *Plants*, *11*(9), 1161.

Sahu, P. K., Sao, R., Mondal, S., Vishwakarma, G., Gupta, S. K., Kumar, V., Singh, S., Sharma, D., & Das, B. K. (2020). Next generation sequencing based forward genetic approaches for identification and mapping of causal mutations in crop plants: A comprehensive review. *Plants*, *9*(10), 1355.

Sakono, M., Oya, T., & Aoki, M. (2023). Development of a transcriptional activator-like effector protein to accurately discriminate single nucleotide difference. *ChemBioChem*, *24*(3), e202200486.

Salava, H., Thula, S., Mohan, V., Kumar, R., & Maghuly, F. (2021). Application of genome editing in tomato breeding: Mechanisms, advances and prospects. *International Journal of Molecular Sciences*, *22*(2), 682.

Santillán Martínez, M. I., Bracuto, V., Koseoglou, E., Appiano, M., Jacobsen, E., Visser, R. G., Wolters, A. M. A., & Bai, Y. (2020). CRISPR/Cas9-targeted mutagenesis of the tomato susceptibility gene PMR4 for resistance against powdery mildew. *BMC Plant Biology*, *20*(1), 1–13.

Saraswat, P., Chaturvedi, A., & Ranjan, R. (2023). Zinc finger nuclease (ZFNs) and transcription activator-like effector nucleases (TALENs) based genome editing in enhancement of anticancer activity of plants. In D. Pandita & A. Pandita (Eds.), *Plant-derived anticancer drugs in the OMICS era: Biosynthesis, functions and applications*. CRC Press.

Sari, F. O., & Nazli, M. (2020). Exploring the effects of "excessive tourism growth" on public health and ecosystem. *Journal of Hospitality and Tourism Insights*, *4*(1), 1–17.

Sauer, N. J., Mozoruk, J., Miller, R. B., Warburg, Z. J., Walker, K. A., Beetham, P. R., Schöpke, C. R., & Gocal, G. F. (2016). Oligonucleotide-directed mutagenesis for precision gene editing. *Plant Biotechnology Journal*, *14*(2), 496–502.

Sawarkar, H. A., & Mansoori, M. H. (2023). Methodology for physical and chemical mutagenic treatments. In T. A. Bhat & K. R. Hakeem (Eds.), *Biotechnologies and genetics in plant mutation breeding*: *Mutagenesis and crop improvement* (pp. 15–44). Apple Academic Press.

Schiocchet, M. A., Noldin, J. A., Raimondi, J. V., Tulmann Neto, A., Marschalek, R., Wickert, E.,…, & Andrade, A. D. (2014). SCS118 Marques-New rice cultivar obtained through induced mutation. *Crop Breeding and Applied Biotechnology*, *14*, 68–70.

Shah, P., & Swiatlo, E. (2008). A multifaceted role for polyamines in bacterial pathogens. *Molecular Microbiology*, *68*(1), 4–16.

Shamshad, A., Rashid, M., Jankuloski, L., Ashraf, K., Sultan, K., Alamri, S., Siddiqui, M. H., Munir, T., & Uz Zaman, Q. (2023). Effect of ethyl methanesulfonate mediated mutation for enhancing morpho-physio-biochemical and yield contributing traits of fragrant rice. *Peer Journal*, *11*, e15821.

Shan, Q., Wang, Y., Chen, K., Liang, Z., Li, J., Zhang, Y., Liu, J., Voytas, D. F., Zheng, X., & Gao, C. (2013). Rapid and efficient gene modification in rice and *Brachypodium* using TALENs. *Molecular Plant*, *6*(4), 1365–1368.

Shantharaj, D., Minsavage, G. V., Orbović, V., Moore, G. A., Holmes, D. R., Römer, P., Horvath, D. M., Lahaye, T., & Jones, J. B. (2023). A promoter trap in transgenic citrus mediates recognition of a broad spectrum of *Xanthomonas citri*pv.*citri* TALEs, including in planta-evolved derivatives. *Plant Biotechnology Journal.* https://doi.org/10.1111/pbi.14109.

Shen, Y., Ma, Y., Li, D., Kang, M., Pei, Y., Zhang, R., Tao, W., Huang, S., Song, W., Li, Y., & Chen, Y. (2023). Biological and genomic analysis of a symbiotic nitrogen fixation defective mutant in Medicago truncatula. *Frontiers in Plant Science, 14*, 1209664.

Shukla, V. K., Doyon, Y., Miller, J. C., DeKelver, R. C., Moehle, E. A., Worden, S. E., Mitchell, J. C., Arnold, N. L., Gopalan, S., Meng, X., & Urnov, F. D. (2009). Precise genome modification in the crop species *Zea mays* using zinc-finger nucleases. *Nature, 459*(7245), 437–441.

Silva, C. J., van den Abeele, C., Ortega-Salazar, I., Papin, V., Adaskaveg, J. A., Wang, D., Casteel, C. L., Seymour, G. B., & Blanco-Ulate, B. (2021). Host susceptibility factors render ripe tomato fruit vulnerable to fungal disease despite active immune responses. *Journal of Experimental Botany, 72*(7), 2696–2709.

Sprink, T., Metje, J., & Hartung, F. (2015). Plant genome editing by novel tools: TALEN and other sequence specific nucleases. *Current Opinion in Biotechnology, 32*, 47–53.

Sreedharan, M., & Joseph, R. M. (2019). Modern techniques in plant genetic engineering. In: Sreedharan, M., & Joseph, R. M. (eds) *Modern trends in agriculture* (pp. 276–290). Scire Science.

Suprasanna, P., Mirajkar, S. J., & Bhagwat, S. G. (2015). Induced mutations and crop improvement. In B. Bahadur, M. V. Rajam, L. Sahijram & K. V. Krishnamurthy (Eds.), *Plant Biology and Biotechnology: Plant Diversity, Organization, Function and Improvement* (Vol. 1, pp. 593–617).

Suzuki, T., Eiguchi, M., Kumamaru, T., Satoh, H., Matsusaka, H., Moriguchi, K., Nagato, Y., & Kurata, N. (2008). MNU-induced mutant pools and high performance TILLING enable finding of any gene mutation in rice. *Molecular Genetics and Genomics, 279*, 213–223.

Thomazella, D, Seong, K., Mackelprang, R., Dahlbeck, D., Geng, Y., Gill, U. S., Qi, T., Pham, J., Giuseppe, P., Lee, C. Y., Qi, T., Pham, J., Giuseppe, P., & Lee, C.Y. (2021). Loss of function of a dmr6 ortholog in tomato confers broad-spectrum disease resistance. *Proceedings of the National Academy of Sciences (PNAS), 118*, e2026152118.

Tran, M. T., Son, G. H., Song, Y. J., Nguyen, N. T., Park, S., Thach, T. V., Kim, J., Sung, Y.W., Das, S., Pramanik, D., & Kim, J. Y. (2023). CRISPR-Cas9-based precise engineering of SlHyPRP1 protein towards multi-stress tolerance in tomato. *Frontiers in Plant Science, 14*, 1186932.

Udage, A. C. (2021). Introduction to plant mutation breeding: Different approaches and mutagenic agents. *Journal of Agricultural Sciences (Sri Lanka), 16*(3), 466–483.

van Esse, H. P., Reuber, T. L., & van der Does, D. (2020). Genetic modification to improve disease resistance in crops. *New Phytologist, 225*, 70–86.

Wan, D. Y., Guo, Y., Cheng, Y., Hu, Y., Xiao, S., Wang, Y., & Wen, Y. Q. (2020). CRISPR/Cas9-mediated mutagenesis of VvMLO results in enhanced resistance to powdery mildew in grapevine (Vitis vinifera). *Horticulture Research, 7*, 116.

Wang, F., Wang, C., Liu, P., Lei, C., Hao, W., Gao, Y., Liu, Y. G., & Zhao, K. (2016). Enhanced rice blast resistance by CRISPR/Cas9-targeted mutagenesis of the ERF transcription factor gene OsERF922. *PLoS ONEE, 11*, e0154027.

Wang, J., Zhang, Y., Zhou, L., Yang, F., Li, J., Du, Y., Liu, R., Li, W., & Yu, L. (2022). Ionizing radiation: Effective physical agents for economic crop seed priming and the underlying physiological mechanisms. *International Journal of Molecular Sciences, 23*(23), 15212.

Wang, Y., Cheng, X., Shan, Q., Zhang, Y., Liu, J., Gao, C., & Qiu, J. L. (2014). Simultaneous editing of three homoeoalleles in hexaploid bread wheat confers heritable resistance to powdery mildew. *Nature Biotechnology, 32*(9), 947–951.

Wendt, T., Holm, P. B., Starker, C. G., Christian, M., Voytas, D. F., Brinch-Pedersen, H., & Holme, I. B. (2013). TAL effector nucleases induce mutations at a pre-selected location in the genome of primary barley transformants. *Plant Molecular Biology, 83*, 279–285.

Wijerathna-Yapa, A., Ramtekey, V., Ranawaka, B., & Basnet, B. R. (2022). Applications of in vitro tissue culture technologies in breeding and genetic improvement of wheat. *Plants, 11*(17), 2273.

Xiao, J., Jia, X., Wang, H., Zhao, R., Fang, Y., Gao, R., Wu, Z., Cao, A., Wang, J., Xue, Z., & Wang, X. (2011). A fast-neutron induced chromosome fragment deletion of 3BS in wheat landrace Wangshuibai increased its susceptibility to Fusarium head blight. *Chromosome Research, 19*, 225–234.

Xu, J., Naing, A. H., Kang, H., Lee, S. Y., Li, W., Chung, M. Y., & Kim, C. K. (2023). CRISPR/Cas9-mediated editing of PhMLO1 confers powdery mildew resistance in petunia. *Plant Biotechnology Reports, 17*(5), 767–775.

Xu, Y., Charles, M. T., Luo, Z., Mimee, B., Tong, Z., Véronneau, P. Y.,…, & Rolland, D. (2019). Ultraviolet-C priming of strawberry leaves against subsequent Mycosphaerella fragariae infection involves the action of reactive oxygen species, plant hormones and terpenes. *Plant, Cell and Environment*, *42*(3), 815–831.

Xu, Z., Xu, X., Gong, Q., Li, Z., Li, Y., Wang, S., Yang, Y., Ma, W., Liu, L., Zhu, B., Yang, Y., Ma, W., Liu, L., Zhu, B. & Zou, L., (2019). Engineering broad-spectrum bacterial blight resistance by simultaneously disrupting variable tale-binding elements of multiple susceptibility genes in rice. *Molecular Plant*, *12*, 1434–1446.

Yadav, R. K., Tripathi, M. K., Tiwari, S., Tripathi, N., Asati, R., Chauhan, S., Tiwari, P. N., & Payasi, D. K. (2023). Genome editing and improvement of abiotic stress tolerance in crop plants. *Life*, *13*(7), 1456.

Zafar, K., Sedeek, K. E., Rao, G. S., Khan, M. Z., Amin, I., Kamel, R., Mukhtar, Z., Zafar, M., Mansoor, S., & Mahfouz, M. M. (2020). Genome editing technologies for rice improvement: Progress, prospects and safety concerns. *Frontiers in Genome Editing*, *2*, 5.

Zhan, X., Lu, Y., Zhu, J. K., & Botella, J. R. (2021). Genome editing for plant research and crop improvement. *Journal of Integrative Plant Biology*, *63*(1), 3–33.

Zhang, W., Jiang, H., Cao, J., & Jiang, W. (2021). UV-C treatment controls brown rot in postharvest nectarine by regulating ROS metabolism and anthocyanin synthesis. *Postharvest Biology and Technology*, *180*, 111613.

Zhang, Y., Bai, Y., Wu, G., Zou, S., Chen, Y., Gao, C., & Tang, D. (2017). Simultaneous modification of three homoeologs of Ta EDR 1 by genome editing enhances powdery mildew resistance in wheat. *The Plant Journal*, *91*(4), 714–724.

Zhang, Y., Guo, W., Chen, L., Shen, X., Yang, H., Fang, Y., Ouyang, W., Mai, S., Chen, H., Chen, S., Ouyang, W., Mai, S., Chen, H., Chen, S. & Hao, Q. (2022). CRISPR/Cas9-mediated targeted mutagenesis of GmUGT enhanced soybean resistance against leaf-chewing insects through flavonoids biosynthesis. *Frontiers in Plant Science*, *13*, 802716.

Zhang, Y., Zhang, F., Li, X., Baller, J. A., Qi, Y., Starker, C. G., Bogdanove, A. J., & Voytas, D. F. (2013). Transcription activator-like effector nucleases enable efficient plant genome engineering. *Plant Physiology*, *161*(1), 20–27.

Zhao, B., Hu, S., Wang, D., Chen, H., & Huang, M. (2020). Inhibitory effect of gamma irradiation on *Penicillium digitatum* and its application in the preservation of Ponkan fruit. *Scientia Horticulturae*, *272*, 109598.

Zhou, X., Liao, H., Chern, M., Yin, J., Chen, Y., Wang, J., Zhu, X., Chen, Z., Yuan, C., Zhao, W., Zhu, X., Chen, Z., Yuan, C., Zhao, W. & Wang, J. (2018). Loss of function of a rice TPR-domain RNA-binding protein confers broad-spectrum disease resistance. *Proceedings of the National Academy of Sciences*, *115*, 3174–3179.

Zhou, Y., Xu, S., Jiang, N., Zhao, X., Bai, Z., Liu, J.,…, & Yang, Y. (2022). Engineering of rice varieties with enhanced resistances to both blast and bacterial blight diseases via CRISPR/Cas9. *Plant Biotechnology Journal*, *20*(5), 876–885.

Zhu, H., Li, C., & Gao, C. (2020). Applications of CRISPR–Cas in agriculture and plant biotechnology. *Nature Reviews Molecular Cell Biology*, *21*(11), 661–677.

4 Enhancing the Resilience of Apple to Fire Blight and Winter Injury in a Warming World

Dustin E. MacLean, Ian R. Willick, Rachid Lahlali, and Yunfei Jiang

4.1 INTRODUCTION

Apple (*Malus domestica* Borkh.) is one of the most widely planted fruit crops in temperate regions with an annual production in 2022 exceeding 95 million tons (FAO STAT 2024). In cool-climate regions, commercial orchards face constant challenges in producing apples with optimal color, storage, and flavor characteristics. To meet these needs, growers look for novel breeding material with horticultural improvements that include tree size, fruit quality, and improved resistance against economic threats such as disease and temperature extremes. For example, Malling 9 (M.9) is a widely used rootstock in North America that when paired with a scion will produce a tree that is half the size of a 'standard' tree with decent fruit quality but comparatively higher susceptibility to fire blight and freezing injury. This is particularly concerning as the causal agent of fire blight disease, *Erwinia amylovora* (Burill) (Winslow et al. 1920), is a major economic threat to apple production worldwide and in the United States is estimated to result in $100 million in crop loss each year (Norelli et al. 2003). Van Der Zwet et al. (2012) reported losses of approximately US $27.5 million for a single significant incident in 2007 in Switzerland in an assessment of economic losses during the past several decades.

Manifestation of fire blight occurs when apple trees are infected with more than 104 bacterial cells per flower (Philion 2014). Relative humidity (RH) plays a vital role in the size of the *E. amylovora* population and therefore the likelihood of infection. Epiphytic growth at the stigmatic level can only achieve a concentration of over 106 bacterial cells per crab apple blossom when RH is above 55%, whereas the floral cup (hypanthium) requires an RH of more than 80% (Pusey 2000). Therefore, rainy weather is essential for producing a moist environment where *E. amylovora* cells can multiply. However, dew that forms over the night can also increase humidity within the flower and even a slight wind can impact infection (Slack et al. 2022). In addition to the quantity of the *E. amylovora* cell population, the critical virulence factor type 3 secretion system (T3SS) must be activated for disease development. Elevated RH may enhance the expression of T3SS-related genes, allowing *E. amylovora* to infect a host plant by injecting proteins into it (Cui et al. 2021). It is also important to remember that pH is a factor in a successful infection. According to Shrestha et al. (2005), the ideal pH for *E. amylovora* growth is approximately 7.5, and this was demonstrated to have an impact on the organism's ability for growth and chemotaxis (Raymundo and Ries 1980). Furthermore, Pester et al. (2012) demonstrated that acidic environments reduce the virulence of *E. amylovora*. According to Pester et al. (2012), pH has an essential role in *E. amylovora* pathogenicity since it no longer induces the expression of T3SS genes in acidic environments (pH 4).

Ideal temperatures are also necessary for attaining the high cell density needed for floral infection (MAAARO 2011). According to Santander and Biosca (2017), the ideal temperature for the growth of *E. amylovora* is between 23°C and 30°C (73.4°F–86°F), with a threshold temperature of 18.3°C (64.9°F) (Dewdney et al. 2007). This is also the ideal temperature for the development of

DOI: 10.1201/9781003543237-4

fire blight. However, the pathogenicity of *E. amylovora* was demonstrated to persist between 14°C (57.2°F) and 4°C (39.2°F), indicating that the low bacterial cell growth rates slowed the infection (Santander and Biosca 2017). The bacteria can easily spread during this stage into the open pores of green tissues. This can make the blooms of tree fruits particularly vulnerable to infection. The availability of nutrients determines how temperature affects the physiology of *E. amylovora*. In nutrient-poor conditions, the same temperatures that are ideal for development and symptomatology in host tissues are not ideal for survival. In the same way, in nutrient-limiting conditions, suboptimal growth temperatures (such as 14°C (57.2°F)) are ideal for pathogen survival (Santander et al. 2014; Santander and Biosca 2017). The mild spring and summer temperatures encourage bacterial growth in the bark during canker creation and development, while healthy plant tissues are still available. This results in the onset and spread of fire blight symptoms. However, the most likely reason *E. amylovora* cells trapped in necrotic tissues fail to reproduce is that they are starved for nutrients (Sobiczewski et al. 2017). Under these conditions, the same temperatures which enable the bacteria to propagate to other structures, reproduce, and infect healthy plant tissues are detrimental to the long-term survival of *E. amylovora* cells that have been trapped in the dead and nutrient-depleted plant material.

Advances in flowering phenology due to unseasonably warm temperatures have created conditions where buds earlier in the spring are close to full bloom. Long-term observational records indicate that mid-bloom dates in Japan have advanced by 0.35 d/yr (Fujisawa and Kobayashi 2010) and in New York, United States by 0.20 d/yr (Wolfe et al. 2005). Similar projections over the twenty-first century for apple-growing regions in New York (Wolfe et al. 2018) and Austria (Unterberger et al. 2018) predict a 15- to –16-day advancement in bloom date. Observed advances in flowering phenology are partially due to a greater cumulative exposure to winter temperatures from 1.5°C to 13.0°C known as chill units that are necessary for buds to break (Shaltout and Unrath 1983).

In addition to fire blight, the early onset of flower development due to warmer-than-average spring temperatures has led to a greater susceptibility to injurious frosts. In the last 15 years, destructive spring frost events in 2007, 2012, and 2014 in the northeast United States have resulted in billions of dollars in crop losses (Wisniewski et al. 2018). Spring frosts generally result from radiation freezes that occur when nighttime temperatures drop below freezing, and humidity and wind speeds remain low. Lethal freezing injury can manifest at different temperature thresholds depending on the development stage of the bud. For example, in apples, fruit tree blossoms experience a lethal injury at the green tip around –15°C or in full bloom at around –2°C (Proebsting and Mills 1978). Susceptibility to early flower development will vary significantly as each apple cultivar has a different chilling requirement.

There is also growing uncertainty about future winter conditions. Scenarios developed by the Intergovernmental Panel on Climate Change (IPCC) predict that by 2100 there will be a doubling of atmospheric CO_2 and a mean increase in global temperatures of 6°C at more northern latitudes (IPCC 2014). Rising minimum temperatures have already been reported over the past 40 years (Vincent and Mekis 2006), and disproportionate increases in annual average minimum winter temperatures (IPCC 2014) are expected to further disrupt fruit tree production. What is worrisome is that the general warming observed thus far has not mitigated the propensity for lethal freezing events. For example, the Canadian province of Nova Scotia experienced unseasonably warm temperatures in December 2022 and January 2023, with the average temperature reaching 0.5°C. This left overwintering fruit buds more susceptible to an advective frost on February 3 that reached temperatures of –26°C and resulted in the province-wide loss of the peach (*Prunus persica* L. Batsch), cherry (*Prunus avium* L. and *Prunus cerasus* L.), and half of its wine grape (*Vitis vinifera* L.) crop.

To further complicate our understanding of how fruit trees survive over winter, growth under elevated CO_2 initiates freezing in plant tissues at warmer temperatures (Lutze et al. 1998) and delays the acquisition of increased capacity to tolerate or avoid freezing injury (Loveys et al. 2006). A survey of the known literature identified a correlation between the intrinsic ice nucleation temperature in leaves and atmospheric CO_2. Extrapolation of these curves suggests that doubling our atmospheric CO_2 from 350 to 700 ppm will raise the average intrinsic ice nucleation temperature by 1–2°C (Woldendorp et al. 2008). While this may not seem like a large temperature shift, a few degrees difference in ice nucleation temperature during a spring frost can be the difference between a healthy summer crop or the field-level loss of production.

More intense seasonal and diurnal temperatures, precipitation events, and air humidity will lead to earlier spring apple flower development and may coincide with greater activity and virulence of *E. amylovora*. We should expect a greater prevalence of fire blight and freezing injury if climate scenarios predicting changed frequencies and severity of extreme weather events proceed as predicted (IPCC 2014). The goal of this discussion is therefore two-pronged. First, we will provide context on the current understanding of the epidemiology of *E. amylovora*, horticultural practices that we can use to manage levels of inoculum within an orchard, known genetic resistance within the *Malus* genus, and new technologies that are being employed to enhance tree fruit performance against fire blight. We will then discuss how the tree maintains and loses cold hardiness, the growth-dormancy cycle, interactions between freeze–thaw and fire blight injury, and our current advancements in enhancing apple resistance to freezing injury. Finally, we will propose strategies to enhance the resilience of apple trees to these combined abiotic and biotic stresses.

4.2 DISCOVERY AND EPIDEMIOLOGY OF FIRE BLIGHT

Fire blight (Baker 1971) was first reported in the United States by Denning (1974) and has since been reported in over 40 countries (Peil et al. 2009). For more than 200 years, fire blight has been recognized as a damaging disease that affects apples and pears (Baker 1971). *E. amylovora* is a gram-negative enterobacterium that can infect woody tissues, fruits, flowers, vegetative shoots, and rootstock crowns (Winslow et al. 1920). The disease has multiple distinct phases, such as rootstock blight, shoot blight, and blossom blight (Figure 4.1). Epidemics of fire blight have proven difficult to contain or hinder due to the variety of host tissues that can be infected and the lack of treatment options available to control the disease. According to Van Der Zwet et al. (2012), the greatest risk of infection is during warm,

FIGURE 4.1 Rootstock blight (a), blossom blight (b), shoot blight (c) rootstock blight on roots (d), and fire blight canker (e) caused by fire blight (*Erwinia amylovora)* on apple. (Photos courtesy of Michelle Cortens, Tree Fruit Specialist with Perennia Food and Agriculture Corporation, Nova Scotia, Canada.)

humid weather, which is typically experienced in the spring and early summer and is favorable for the growth of bacteria. According to Khan et al. (2012), bacterial ooze serves as a reservoir from which bacteria can be dispersed to different areas of the tree or neighboring trees by insects, birds, wind, rain, and hail. As *E. amylovora* has a T3SS, it can secrete bacterial effectors into plant cells to further the infection process. Nonetheless, the host can identify effectors and initiate resistance responses.

During a wetting event, bacteria enter the plant through natural nectary apertures and selectively colonize stigmatic surfaces. Blossom blight is a total necrotic blight of the pedicels and flower cluster caused by an infection. After infecting blooms and forming cankers, bacteria are further transmitted to young, developing shoots by insects, wind-driven rain, and machinery. Through wounds from mechanical trauma, the bacteria infect the shoot tissue, resulting in the spread of a necrotic blight known as shoot blight. When bacteria move through the vascular system to the areas of the plant that are actively growing, a shoot blight infection may result in systemic infection of the entire tree and eventually result in the death of the tree (Peil et al. 2009). These systemic infections can lead to catastrophic losses during a season and have a significant impact on the orchard's life. Tree death frequently results from the infection of rootstocks through suckers or by other channels (Russo et al. 2008).

A comprehensive approach involving several strategies is necessary to effectively manage fire blight. These methods should aim to: (1) decrease the amount of inoculum available to initiate new infections; (2) build barriers to the pathogen's successful establishment on the host; and (3) lessen the susceptibility of the host to infection (Aldwinckle and Beer 1979; Van Der Zwet and Beer 1999). Most fire blight management techniques developed in the twentieth century concentrated on lowering the inoculum in an orchard and incorporating antimicrobial treatments to stop infection. See Van Der Zwet and Beer (1999) for examples and references therein. Although it has been acknowledged that improving host resistance is an essential component of managing fire blight (Norelli et al. 2003a), the adoption of this approach is hindered by the absence of commercially viable resistant cultivars and inadequate management techniques. It is now possible to alter this paradigm by taking advantage of recent advances in germplasm. First, size-controlling (typically dwarfing) rootstocks resistant to fire blight are being developed by apple rootstock breeding projects and are now being made available for commercial use (Norelli et al. 2003). Second, transgenic apple trees are currently undergoing field experiments after it was shown that genetic engineering of commercial apple cultivars could enhance fire blight resistance (Reddy et al. 2000). Third, it has been reported that chemical treatments enhancing host resistance are effective in managing fire blight (Yoder et al. 1999; Brisset et al. 2000; Maxson-Stein et al. 2002).

4.3 CURRENT FIRE BLIGHT MANAGEMENT

The use of spray treatments to prevent infection, particularly blossom infections, and phytosanitary efforts to lower the orchard's inoculum are the foundation of the current fire blight control strategy. Applying antibiotics like streptomycin, kasugamycin, or oxytetracycline is an effective strategy to manage fire blight. While most European nations prohibit or rigorously control the use of antibiotics, North America (Canada and the United States) simply restricts the use of antibiotics in organic production systems (McManus et al. 2002; McManus 2014). Early in the twentieth century, it was determined that eliminating persistent cankers from the orchard during winter pruning was an essential part of fire blight control strategies for reducing primary inoculum (Schroth et al. 1974). Moreover, cutting early season infections after bloom reduces the level of inoculum available for shoot infection, and applying copper sprays during the silver- or green-tip growth stage lowers the amount of primary inoculum in the orchard (Van Der Zwet and Beer 1999).

Another crucial aspect of managing fire blight is preventing the disease's blossom blight phase, and in the latter half of the twentieth century, considerable advancements were made in this regard. Studies have shown that the most important phase in the infection of blossoms is the formation of epiphytic populations of *E. amylovora* on the stigma (Miller and Schroth 1972; Thomson 1986; Wilson et al. 1989). To better control this stage of the disease by utilizing antibiotic sprays on time, several models based on phenological and meteorological data were created to forecast the occurrence of fire blight infection periods during bloom (Thomson 1986; Steiner 1989; Billing 1998; Smith 1998). MARYBLYT and Cougarblight are two of the most well-known fire blight prediction models that generate infection predictions based on weather and apple or pear phenology. However, their methods for calculating risk numbers are different. The DOS-based computer program MARYBLYT was created for orchards in West Virginia and Maryland, which predicts infection events and the emergence of symptoms and contains algorithms for the developmental stages of fire blight, such as canker, blossom, shoot, and trauma blight (Steiner and Lightner 1992). Smith created Cougarblight in Washington State after observing that earlier models did not fit well in the Pacific Northwest (Smith 1998). Weather and phenology are also used by Cougarblight but exclusively to forecast blossom infection by *E. amylovora*.

Even with these advancements, fire blight is still a challenging disease to manage. Over 90% control against streptomycin-sensitive strains of the disease can be achieved by strategically timing streptomycin application during bloom, yet failure to control blossom blight persists for several reasons. Extended durations of precipitation or the requirement to safeguard sizable areas frequently impede the prompt application of streptomycin, which necessitates application 24–48 hours prior to or 24 hours following an infection period for maximum effectiveness. Sometimes the use of erroneous meteorological data or the occurrence of exceptional variables beyond the model's standard prediction parameters cause models to fail to effectively anticipate infection periods. In most of western North America, streptomycin is less effective because of the development of streptomycin-resistant *E. amylovora* strains. In *E. amylovora*, there are two methods by which streptomycin resistance may develop. First, a single base pair mutation in the bacterial protein binding site for streptomycin can lead to the development of resistance. According to McManus et al. (2002) streptomycin binds to proteins and prevents them from being synthesized, which kills the bacteria. A single base pair mutation was discovered in the *rpsL* gene by Chiou and Jones in 1995. The mutation prevents streptomycin from binding to the protein and killing the bacteria. Developing resistance through the acquisition of the gene pair *StrA* and *StrB* is a second possibility. A plasmid encodes both of these genes (Huang and Burr 1999). Resistance can develop from these genes being transferred from one bacterium to another. Apple blossoms are home to a large number of non-pathogenic bacteria, some of which may carry the genes that confer resistance to streptomycin and can be transferred to *E. amylovora* at any time.

There have been no adequate streptomycin substitutes found, and oxytetracycline, while useful in regions where streptomycin-resistant bacteria are present, has proven to be only partially effective. Biological control agents have failed to control the pathogen in comparison to highly effective products such as streptomycin. Furthermore, after several years without significant fire blight outbreaks, producers tend to become slack in applying expensive control measures due to the unpredictable nature of fire blight infections and epidemics. There have been no management strategies available to manage shoot blight and rootstock blight during the summer months if growers are unable to manage blossom blight.

4.4 INCREASED SUSCEPTIBILITY OF APPLE TO FIRE BLIGHT

The adoption of high-density orchard systems in the twentieth century brought about significant changes to the apple industry, and the recent planting of sensitive cultivars and rootstocks has raised the risk of fire blight in apple orchards to previously unheard-of levels. Several horticultural benefits are associated with high-density planting systems: higher yields per hectare; better fruit quality due to increased light penetration into the tree canopy; increased tree precocity, which leads to quicker returns on investment; and more efficient use of labor and chemical inputs due to smaller trees. The use of dwarfing rootstocks is essential to a high-density apple orchard system. Malling (M.) 9 and M.26, the two most widely used dwarfing apple rootstocks, are extremely vulnerable to *E. amylovora*, and fire blight infection always results in tree death by girdling the rootstock.

In a four-year-old high-density planting, a 10% incidence of rootstock blight can result in substantial losses per hectare when replacement tree expenses, wasted investment in tree establishment and maintenance, and yield reductions over several years are taken into consideration (Momol et al. 1997). Currently, rootstocks cannot be adequately safeguarded from infection using chemical treatments or cultural methods. In addition, many commercially successful apple cultivars compared to most older cultivars are very susceptible to fire blight (see Bost and Windham 2002; Beckerman 2006 and references therein). Red Delicious, Liberty, Enterprise, and Freedom are generally considered to be most resistant to fire blight, whereas Golden Delicious, Empire, Honeycrisp, McIntosh, Mutsu, Spartan, Summerred, GoldRush, and Nova Easygro are moderately resistant. In contrast, the following cultivars are susceptible to fire blight: Braeburn, Fuji, Gala, Ginger Gold, Idared, Jonagold, Rome, Ambrosia, Cortland, Granny Smith, Gravenstein, Pink Lady, and Scotia. The severe fire blight susceptibility of modern apple rootstocks and scion cultivars, along with the higher planting costs of high-density plantings, has left many apple growers with crippling financial losses because of fire blight.

4.5 GROWTH RETARDANTS

Gibberellin production is inhibited by the plant growth regulator prohexadione-calcium (Apogee), which decreases longitudinal shoot growth (Rademacher 2000). Prohexadione-calcium regulation of vegetative growth on apples reduces the severity and amount of fire blight shoot infection (Yoder et al. 1999). Prohexadione-calcium promotes host resistance by decreasing plant vigor but fails to demonstrate antibacterial action against *E. amylovora*. Furthermore, prohexadione-calcium treatment of apples alters the pathways involved in the production of phenylpropanoid compounds, which may potentially improve resistance (Evans et al. 1999). Using plant growth retardants to alter apple trees' sensitivity to fire blight affords more flexibility than trying to alter susceptibility through standard horticultural techniques like adjusting fertilizer levels, pruning, and fruit thinning. Reducing shoot lengths should also reduce the need for annual pruning and improve light penetration into the center of apple trees.

When the petals on the king (first, central) flowers start to dehisce and turn senescent, prohexadione-calcium is applied near the end of the bloom. The duration of the inhibition of shoot growth is contingent on both the vigor of the tree and the rate at which prohexadione-calcium is applied (Jones 2001a). Within two weeks of application, a single, strategically placed treatment will significantly reduce shoot elongation. Retarding shoot growth in apple trees and preventing the secondary spread of fire blight are two other benefits of using trinexapac-ethyl (Moddus, Palisade), a chemical that is known to block gibberellin production and is currently used to protect cereal crops against lodging (Jones 2001b).

Since *E. amylovora* has become resistant to streptomycin, the most effective treatment in many apple-growing regions, there is an urgent need for new, effective materials to manage blossom and branch infections. The demand for efficient chemicals is greater than ever. However, new antibiotics have not been approved for the control of fire blight in recent years (McManus et al. 2002), and biological control agents are most effective when used in combination with antibiotics (Stockwell et al. 1996). Certain biological control agents (e.g., *Bacillus subtilis* QRD-137, *Pseudomonas fluorescens* A506, and *Pantoea agglomerans* C-91) and synthetic chemicals (e.g., oxolinic acid) have demonstrated potential in specific field trials.

4.6 SYSTEMIC ACQUIRED RESISTANCE INDUCERS FOR FIRE BLIGHT CONTROL

Through the systemic acquired resistance (SAR) pathway, acibenzolar-S-methyl (ASM, actigard, BION) triggers the natural defense mechanisms of the tree. Tobacco (Friedrich et al. 1996), wheat (Görlach et al. 1996), and Arabidopsis (Lawton et al. 1996), as well as apple (Brisset et al. 2000; Maxson-Stein et al. 2002), all exhibit increased expression of *PATHOGENECY RELATED*

PROTEIN (*PR*) genes when exposed to ASM. Herbaceous plant genes linked to SAR, including *PR-1* (antifungal properties), *PR-2* (β-1,3-glucanase), *PR-5* (thaumatin-like protein), and *PR-8* (class III chitinase), were also reported in apples, and they exhibited a notable uptick in expression three to five days following ASM treatment.

To give resistance enough time to develop, SAR-type compounds must be used prior to the onset of infection. According to Norelli et al. 2003, most trials begin ASM treatment at the pink stage of bud growth, which is roughly one week prior to the potential application of the first streptomycin treatments. Compared to pathogens of certain herbaceous plants, the effectiveness of the resistance conferred by ASM against *E. amylovora* does not seem to last as long on apples. When ASM sprays were applied weekly as opposed to biweekly, the best results were obtained in terms of controlling blossom blight and terminal infections (Maxson-Stein et al. 2002).

4.7 FIRE BLIGHT-RESISTANT ROOTSTOCKS FROM CONVENTIONAL PLANT BREEDING

The development of fire blight-resistant rootstocks has become a priority in rootstock breeding due to the increased severity of fire blight in apple crops on M.9 and M.26 rootstock. Resistant rootstocks prevent tree losses due to rootstock infection; however, they do not stop fire blight from infecting susceptible scion cultivars. Conventional breeding has produced fire-blight-resistant apple rootstocks far more rapidly than apple scion cultivars because fruit quality has no bearing on rootstock performance. Recently, rootstocks selected for resistance to fire blight were released by the breeding programs in Geneva, United States, and Vineland, Canada (Cummins and Aldwinckle 1994; Cline et al. 2001). New apple rootstocks with favorable pomological traits along with resistance to *E. amylovora* infection may offer effective control for the fire blight rootstock phase. The Geneva © apple rootstock (G.) breeding program's goal has been to create superior pomological rootstocks that are resistant to biotic and abiotic stressors, such as fire blight. Due to their excellent resistance to *E. amylovora*, *Malus* × *robusta* 'Robusta 5' and *M.* × *sublobata* P1286613 ('Novole') have been utilized extensively as parents in the Geneva rootstock breeding program (Gardner et al. 1980). Later research revealed that 'Robusta 5' and 'Novole' were differentially susceptible to specific strains of *E. amylovora*, yet, when rootstock selections were inoculated with a mixture of highly virulent strains, it was possible to select rootstocks with resistance to most *E. amylovora* strains in a non-differential manner (Norelli et al. 1986, 1987, 2003; Travis et al. 1998). Resistance selected against a single strain should not be as durable as resistance effective against strains with varied virulence patterns (or races). By 1994, the Geneva program produced and assessed approximately 350,000 seedlings from controlled crosses (Cummins and Aldwinckle 1994). The previous Horticultural Research Institute of Ontario at Vineland Station's rootstock breeding program aimed to create rootstocks that were simple to grow, hardy in cold climates, and impervious to common pests such as fire blight (Cline et al. 2001). Despite the termination of the breeding program, the material produced is still being evaluated and developed commercially. It has been observed that several of the Vineland (V.) rootstocks, including V.1, V.2, V.4, and V.7, are either tolerant or resistant to fire blight (Ferree et al. 1995; Cline et al. 2001).

The three main ways in which rootstocks can become infected are by internal bacterial dissemination from scion infections, infection of rootstock suckers (vegetative shoots growing out of the rootstock), and infection through cracks in the rootstock bark brought on by growth or other damage (Momol et al. 1997). Compared to M.9 and M.26 rootstock clones, trees on G.16, G.30, and Budagovsky (B.) 9 exhibited a high level of resistance to rootstock infection, resulting in no tree mortality. Rootstock infection only slightly affected G.11, resulting in 25% tree death. Along with B.9, Ferree and colleagues observed that, in Ohio orchard experiments, a higher percentage of trees on Poland (P.) 1, P.2, V.1, V.4, V.7, B.491, and B.469 rootstocks survived natural fire blight infections than on M.9 and M.26 (Ferree et al. 1995). See Table 4.1 for a list of fire-blight-resistant rootstocks.

TABLE 4.1
Select Apple Rootstock Vigor, Relative Resistance to Fire Blight, Phytophthora, Woolly Apple Aphid and Freezing Injury

		Biotic Resistance[c]			Freezing Injury (°C)[d]		
Rootstock[a]	Vigour[b]	Fire Blight	Phytophthora	Woolly Apple Aphid	Xylem Winter	Xylem Spring	Root LT_{50}
A.2	Semi-vigorous	S	S	S	ND	ND	−11
B.9	Dwarf	S to I	R	S	ND	ND	−10 to −14
G.11	Semi-dwarf	I	R	S to I	ND	ND	−10 to −14
G.16	Dwarf	R	R	S	ND	ND	ND
G.30	Semi-vigorous	R	R	S	ND	−32	−15
G.41	Dwarf	R	R	S	−33	−21	−10 to −15
G.65	Semi-dwarf	R	R	S	ND	ND	ND
G.202	Semi-vigorous	R	R	R	ND	ND	ND
G.935	Semi-dwarf	I to R	I to S	S	−40	−25	−16 to −19
M.7	Semi-dwarf	I	I	S	−35	−20	−7 to −8
M.9	Dwarf	S	R	S	<−40	−32	−8 to −10
M.26	Semi-dwarf	S	I	S	ND	ND	−10 to −16
M.M.106	Semi-vigorous	S	S	R	ND	ND	−11 to −16
V.5	Dwarf	I to R	ND	ND	<−40	−24	ND
V.6	Dwarf	I	ND	ND	−36	−26	ND
V.7	Dwarf	I to R	ND	ND	−38	−25	ND

[a] Abbreviations: Alnarp (A), Budagovsky (B), Geneva® (G), Malling (M), Malling-Merton (M.M), Vineland (V), or not tested (ND).

[b] The vigor of a rootstock can be classified as a dwarf (1.6–2.5 m), semi-dwarf (2.6–4.0 m), and semi-vigorous (4.1–6.0 m).

[c] Biotic resistance is ranked based on whether the rootstock is susceptible (S), intermediate resistance (I), or resistance (R) to fire blight (Cline et al. 2001; Barritt et al. 2002; Webster and Wertheim 2003); *Phytophthora* (Choi et al. 2021); or woolly apple aphid (Webster and Wertheim 2003; Choi et al. 2021).

[d] Freezing injury in spring- or winter-collected woody tissue is based on xylem browning (Moran et al. 2018). Values denoted as <−40°C indicate that no injury was observed at the minimum available freezing temperature – the lethal temperature at which half the roots exhibit injury during recovery (Embree 1988; Hampson et al. 1997; Privé and Embree 1997; Moran et al. 2011).

4.8 RESISTANCE FOUND IN WILD *MALUS* SPECIES

It is essential to select a genotype with a high degree of resistance when introducing wild, fire blight resistance into domesticated germplasm. Furthermore, to potentially pyramid resistant genes for superior resistance, it is critical to identify sources of resistance that operate through several methods. Mapping significant quantitative trait loci (QTLs) to determine if they map on different linkage groups (LGs) is one technique to accomplish this (Peil et al. 2014). Moreover, the inoculation of wild species with various strains of *E. amylovora* offers insights into several resistance mechanisms as fire blight resistance is strain-dependent (Norelli et al. 1984; Khan et al. 2012; Vogt et al. 2013; Emeriewen et al. 2015, 2017; Wöhner et al. 2018). The fewer pseudo-backcrosses are needed to produce a new commercial variety of apples; the closer a wild species' kinship with the domesticated apple is. There have been reports of distinct levels of fire blight resistance among *Malus sieversii* accessions (Harshman et al. 2017). *M. sieversii*, the major progenitor of *M. domestica* (Cornille et al. 2012; Duan et al. 2017), is projected to have less genetic drag than, for example, *M. fusca*, the Oregon crab apple, because of the tight genetic link with *M. domestica* and greater fruit sizes. Due to the lengthy generation periods of apples (Hanke et al. 2007) and the number of pseudo-backcrosses needed, it may require decades to introduce resistance genes from wild species.

Breeding can be accelerated and enhanced by the development of molecular markers linked to resistance loci (Hanke et al. 2007; Peil et al. 2007b). Additionally, 'rapid cycle breeding' and 'fast track breeding' are two methods used to accelerate the generation cycle. While 'fast track breeding' (Volz et al. 2009; Baumgartner et al. 2010, 2013) depends on supplying young apple seedlings with ideal growth conditions in growth chambers or glasshouses so that these seedlings can transition as early as possible from the juvenile to the adult phase, where flower and fruit formation is possible, 'rapid cycle breeding' is a transgenic approach (Kellerhals et al. 2017).

The high resistance level that can be inherited is the main benefit of employing wild species in fire blight resistance breeding. In an 'Idared'×*M.* × *robusta* 5 (Mr5) population on LG 3 (Peil et al. 2007a), the first fire blight resistance QTL identified in *Malus* wild species was found, and it accounted for almost 80% of the phenotypic variation. This QTL explained 67% and 83% of the phenotypic variation, respectively, and was later verified in a second phenotyping of this population and a different genetic background of a New Zealand M9×Mr5 population (Peil et al. 2007a). Both *M. domestica* cultivars and wild *Malus* accessions have been reported to be resistant to fire blight. Many resistance QTLs were found to be using genetic mapping techniques, along with a functionally proven resistance gene (*FB_MR5*) from the wild *Malus* accession *M.* × *robusta 5* (*Mr5*) (Fahrentrapp et al. 2013; Broggini et al. 2014) and candidate resistance genes found in *M. fusca* (Emeriewen et al. 2018) and 'Evereste' (Parravicini et al. 2011). Furthermore, few gene-for-gene relationships have been reported for the *Malus* – *E. amylovora* host–pathogen system thus far (Emeriewen et al. 2019). According to Durel et al. (2009), *M. floribunda* 821 and the ornamental apple variety 'Evereste' co-localize on LG12 with two distinct yet significant QTLs for resistance to fire blight. Although 'Evereste' includes Mf821 in its lineage, Durel et al. (2009) proposed that the resistance gene originated with the father, most likely from an unidentified wild species accession, rather than Mf821. Approximately 50% of the phenotypic variation was explained by the major QTL on 'Evereste' LG12, and approximately 6.5% by the minor QTL on LG15. A little over 40% of the phenotypic diversity was accounted for by the main QTL on LG12 of Mf821. While serine/threonine kinase and nucleotide-binding site-leucine-rich repeat (NBS-LRR) genes were found through detailed mapping of the fire blight locus on LG12 of 'Evereste' (Parravicini et al. 2011), functional evidence for the resistance gene responsible is still lacking. Emeriewen et al. (2017a) reported on a third fire blight resistance donor in the same region of LG12. The primary QTL identified in *M. xarnoldiana* accession MAL0004 accounted for approximately 50% of the phenotypic variation. Unlike *Mr5*, the QTL persisted following the mapping population's inoculation with the *E. amylovora* strain Ea3049, which completely breaks *Mr5*'s resistance (Emeriewen et al. 2017a). Similar to *M. xarnoldiana* accession MAL0004, a substantial QTL on LG10 was observed in the American wild species *M. fusca*, accession MAL0045, explaining around 66% of the phenotypic variance (Emeriewen et al. 2014). Ea3049 altered the QTL, but it did not break down (Emeriewen et al. 2015). A single potential resistance gene was found using a genomic walking technique (Emeriewen et al. 2018).

4.9 BREEDING FOR RESISTANCE USING APPLE CULTIVARS

Several cultivars' reduced susceptibility to fire blight has a recognized genetic basis, and molecular markers have the potential to be utilized in breeding programs. However, it is unclear how these cultivars' reduced sensitivity to fire blight or how the minor to moderate resistance is controlled genetically. Because of its elite origin and moderate and stable effects, the fire blight resistance locus found on LG7 of the *M. domestica* cultivar 'Fiesta', which was identified from a domesticated background has been deemed the most promising for fire blight resistance breeding. The moderately susceptible 'Cox's Orange Pippin', an older dessert apple, is the source of this locus, which was subsequently dubbed FB_F7 and accounts for 34–46% of the variation (Khan et al. 2006, 2007). Apple breeders can take advantage of the molecular markers that flank this QTL, even though the gene(s) within this locus have not yet been identified. Breeders are interested in *M. sieversii*, a progenitor of the modern apple, because of its large, tasty fruit and various horticultural and fruit quality characteristics similar to cultivated apples, in addition to employing sources of resistance or decreased susceptibility already present in *M. domestica.* In field-inoculated experiments, more than 200 accessions of *M. sieversii* have been characterized. Of these, 12 accessions were suggested for use as parents in breeding programs after being found to be resistant to fire blight (Harshman et al. 2017). Several of these accessions also showed a distinct resistance response, characterized by a low infection incidence, but a high severity once infection occurred.

Concurrently, 13 strain-specific and environment-dependent minor QTLs associated with fire blight resistance were found in a QTL study on a hybrid between *M. domestica* Royal Gala as the female parent and an elite accession of *M. sieversii* (PI 613981) as the male parent (Desnoues et al. 2018). These QTLs are distributed over eight apple LGs, the bulk of which are new. QTLs that were previously identified in the cultivars 'Florina' (Le Roux et al. 2010) and 'Co-op16×Co-op17' mapping population (Khan et al. 2013) were also co-localized with minor QTLs on LG10 and LG15. Depending on the pathogen strain, it has been proposed that different apple cultivars respond differently genetically to fire blight infection.

The 'Empire' and 'Gala' cultivars used distinct responses and cellular mechanisms for pathogen sensing and control, according to RNA sequencing and transcriptome profiling. Despite the early similarity in infection levels between the two cultivars, 'Empire' appears to suppress or slow down bacterial development more rapidly than 'Gala' after a few days. Nevertheless, following 24 hours post-inoculation, variations in the gene expression patterns between 'Empire' and 'Gala' were already detectable (Silva et al. 2019). Compared to major R-gene-based resistance, where high selection pressure could result in mutation in pathogen effectors to overcome resistance, minor and moderate quantitative resistance is typically broad spectrum and more durable.

4.10 NEW GENETIC TECHNOLOGIES FOR FIRE BLIGHT RESISTANCE

A considerable proportion of the annual apple production can be attributed to a small number of cultivars. Varieties including Gala, Red Delicious, Fuji, Golden Delicious, Honeycrisp, and others are among the thousands of apple varieties that have been established. The attractiveness, quality, flavor, and storability of these cultivars are highly valued by supermarkets and consumers alike, and growers also benefit from the ready market these cultivars provide because of consumer demand. Due to their heterozygosity, long generation time, and self-incompatibility, it is impossible to employ standard breeding methods to introduce disease-resistant genes in apples while maintaining the beneficial traits of a fruiting cultivar. Genetic engineering provides a very compelling alternative as it has the potential to provide more rapid results, resistance genes can be obtained from many sources, the expression of native apple genes can be modified, and the desirable qualities of the transformed cultivar or rootstock can be preserved. Given the popularity of these widely grown cultivars, one of the best ways to improve apple genetics is to use the new breeding technology [clustered regularly interspaced short palindromic repeats (CRISPR)/CRISPR-associated protein 9 (Cas9) or cisgenesis] to increase the cultivars' resistance to biotic and abiotic stress, fruit quality attributes, and storage disorders (Campa et al. 2019).

The cisgenic method is the first innovative breeding technology. While transgenic and cisgenic plants share the same DNA transfer techniques, cisgenic approaches use DNA fragments that are sourced from the same species or sexually compatible species (Schouten et al. 2006). As a result, cisgenic plants are identical to plants produced through traditional breeding. Potential risks and fitness consequences of cisgenic plants' toxicity and allergic potential are comparable to those of conventional breeding. The development of multi-disease-resistant cultivars while preserving all the elite traits of the cultivar would be accelerated using recombinant DNA technologies to transfer disease-resistance genes into resistant and susceptible apple cultivars (Schouten et al. 2006). Only Kost et al. (2015) have succeeded in introducing fire blight resistance into susceptible Gala to date. This was accomplished by inserting the resistance gene, *FB_MR5*, flanked by its native 5′-UTR of 1995 bp and third UTR of 1547 bp. These cisgenic Gala lines are currently undergoing testing in a field study in Switzerland (SECB 2016).

The early flowering approach is an additional tool. By employing transgenic early flowering plants in intermediate generations, Flachowsky et al. (2007) created a method to abbreviate the apple's generation cycle. Only transgene-free seedlings are chosen following the last cross. Based on the constitutive overexpression of *FRUITFULL*-homolog of the silver birch (*Betula pendula*) *MADS4* transcription factor in apple, this 'early flowering approach' was subsequently categorized as 'rapid cycle breeding' (Flachowsky et al. 2009).

There have also been uses for genome editing, which may trigger gene mutations at specific genomic regions. These methods create targeted double-strand DNA breaks in the cell genome by using sequence-specific nucleases, such as Cas9 nuclease. These breaks are then repaired by homologous recombination or non-homologous end joining. The CRISPR/Cas9 system is the preferred technique for genetic modification due to its ease of programming, capacity to target many genes, high efficiency, and abundance of vectors (Baltes et al. 2015; Osakabe et al. 2018). Through the editing of the *DIPM4*, the CRISPR/Cas9 system has been effectively used in Golden Delicious and Gala to decrease the susceptibility to fire blight (Pompili et al. 2020).

4.11 THE CHANGING FACE OF WINTER INJURY IN FRUIT TREES

Freezing injury is a recurring problem in northern cool climates and the main climatic limitation to many fruit crops (Palonen and Buszard 1997). Fruit trees can prepare for winter by reducing growth rates and setting terminal buds in the late summer that will serve as the basis for the crop in the subsequent year. Therefore, the capacity for winter survival of these fruit buds will influence the profitability and subsequent distribution of a cultivated variety. Exposure to low non-freezing temperatures enhances the capacity of a bud to tolerate or avoid freezing injury. The physiological manifestation of this phenomenon is sometimes referred to as plant cold hardiness. Cold acclimation is a common adaptation in plants that transiently enhances cold hardiness (Palonen et al. 2021). Trees can lose (de-acclimate) and re-gain (re-acclimate) their capacity to tolerate or avoid freezing injury when exposed to either positive or sub-zero temperatures for as many as two days (Gusta et al. 2009). However, if de-acclimated, tree buds swell and then lose their capacity to significantly re-acclimate and tolerate freezing injury.

4.11.1 Ice Nucleation, Supercooling, and Freezing

Part of the issue in developing strategies to mitigate freezing injury has been the difficulty in determining how the freezing process propagates throughout the plant. Here we have included a brief overview of how plants freeze. This is by no means an exhaustive review of the topic, and for additional perspectives we recommend the following literature: Asahina (1956); Weiser (1970); Sakai and Larcher (1987); Uemura et al. (2006); Gusta et al. (2009); Wisniewski et al. (2014a, 2020); Arora (2018); Takahashi et al. (2021).

Most plant tissue water can supercool below 0°C to avoid freezing. The degree to which this can occur is based on the principle that, while pure water will melt at 0°C, freezing will occur at a range of sub-zero temperatures. Cooling removes the sensible heat from and lowers the kinetic energy of water. At temperatures below 0°C, water is more likely to form stable hydrogen bonds that result in the formation of metastable clusters of regularly ordered molecules. Latent heat is released in the form of an exotherm upon the phase transition from liquid water to a stable ice lattice (Wisniewski et al. 2014a). While pure water can supercool to –38°C (Rasmussen and MacKenzie 1972), it will form stable ice nuclei at temperatures as warm as –3°C in the presence of heterogeneous nucleators such as plant-derived substances, microorganisms, and soil particulates (see reviews by Wisniewski et al. 2014a; Bredow and Walker 2017; Takahashi et al. 2021).

Plants that tolerate freezing have developed a variety of unique adaptations to ice nucleation and the formation of ice in plant tissues. There are some commonalities among the different variations of freezing tolerance. If the cooling rate is rapid, then the supercooled water inside cells is susceptible to intracellular ice nucleation or the propagation of ice from the extracellular space. In most instances, intracellular freezing under natural field conditions results in lethal injury (Gusta et al. 2009). Under slow cooling rates observed in nature (<2°C/h), a gradual reduction in temperature promotes the diffusion of intracellular water to the extracellular ice until a new equilibrium is reached between frozen and unfrozen phases of water in the plant. Some plants have adapted to sub-zero temperatures by developing tissues and organs (Ishikawa and Sakai 1985) that act as ice sinks in a similar fashion as the extracellular space. Regardless of whether the ice sink forms at a cellular, tissue, or whole organ level, the loss of intracellular (or whole tissue) water reduces the volume of cytoplasm, concentrates symplastic osmolytes, and depresses the freezing point of the remaining liquid water (Siminovitch and Scarth 1938). Subsequent ice recrystallization occurs through a thermodynamically favorable process where larger ice crystals will slowly grow at the expense of smaller crystals (Bredow and Walker 2017; Wisniewski et al. 2020). Injury arises when the growing ice surface interacts through adhesive or dehydrative pressures against the plant cell wall and the adjoining plasma membrane (see review by Takahashi et al. 2021). Cellular dehydration proportionally increases with declining temperature (Siminovitch and Scarth 1938). Membrane injury occurs when freezing-induced dehydration collapses the cell (cytorrhysis) or alters membrane structure. As cells re-hydrate, freeze-induced modifications to membrane structure or the accumulation of reactive oxygen species reduce membrane and, by extension, cell viability (Arora 2018).

Overwintering buds of certain species and xylem tissues can avoid freezing by maintaining water in a liquid state below the homogenous nucleation temperature of pure water. This phenomenon known as deep supercooling is a particularly remarkable adaptation that has fascinated physiologists and biophysicists for more than half a century (Quamme et al. 1972; George and Burke 1977; Hong and Sucoff 1982; Wisniewski et al. 2014a; Fujikawa et al. 2018; Neuner et al. 2019). Modifications that increase the thickness and rigidity of the cell wall are believed to play a role in promoting supercooling, which has recently been reviewed by Takahashi et al. (2021). The existence of anti-ice nucleators at low concentrations has been shown to lower the freezing point of water in a non-colligative manner. Plant-derived supercooling-promoting substances have been characterized as flavonol glycosides (Kasuga et al. 2008), hydrolyzable tannins (Wang et al. 2012), and terpenoids (Kawahara et al. 2000). Many of these secondary compounds were found in the xylem harvested from boreal trees and promoted deep supercooling to temperatures below –40°C (Fujikawa et al. 2009).

4.11.2 Manifestation of Freezing Injury in Fruit Trees

Freezing injury is a major constraint to the production of many temperate-zone fruit-bearing trees. While some woody perennials avoid the formation of ice within plant tissues by deep supercooling to temperatures below –60°C (Wisniewski et al. 2014b; Fujikawa et al. 2018; Takahashi et al. 2021), many cultivated fruit trees accrue lethal injury following prolonged exposure to temperatures ranging from –20°C to –40°C (Palonen and Buszard 1997). Understanding how plants are injured by freezing is further complicated by the fact that the lethal freezing temperature varies throughout the year and individual cells within the plant do not possess the same degree of hardiness against sub-zero temperatures. In apple, the bark and xylem of the trunk are more susceptible to freezing injury (Quamme and Hampson 2004) compared with first-year wood, and overwintering flower buds (Sakai and Larcher 1987). Interestingly, the trunks of certain apple cultivars are several degrees less freezing tolerant compared with young shoots and scaffold branches (Quamme and Hampson 2004) and can be more vulnerable to freeze–thaw injury (McArtney and Obermiller 2011). Organs of individual trees are further influenced by their immediate microclimate. Apple cv. Honeycrisp buds collected in midwinter can survive exposure to temperatures as low as –42°C, but the same buds collected from the same tree in the early spring cannot survive temperatures below –25°C (Willick pers. observations). After bud break in the spring, even the hardiest of fruit tree buds are prone to injury at temperatures just below freezing.

Roots from most species in the *Prunus* genus have a limited capacity to survive at sub-zero temperatures and experience injury following prolonged exposure to temperatures below –8°C (Embree 1988). The complete or partial loss of snow cover due to a midwinter thaw can remove an insulating layer and in regions that experience winter air temperatures below –20°C can cause freezing injury to the root systems of fruit trees and subsequent yield reductions in the spring (Wildung et al. 1973). Root injury manifests following spring bud break where new growth wilts and there is a loss of new growth. Overall injury can manifest as stunted tree growth to significant losses wherein replanting is the only viable option. Incorporating rootstocks with greater freezing tolerance can help prevent greater economic losses.

Winter sunscald or southwest injury is observed in young trees of certain species and does not manifest on mature wood of thick-barked trees (Figure 4.2). Fruit trees in the *Prunus* genus that develop thin, dark-colored bark as they mature remain susceptible and, in some instances, become more susceptible to winter sunscald as they age. Injury from sunscald can therefore be a persistent issue for fruit trees. Sunscald develops over winter on the sunny side of branches when incident radiation warms the tissues during a sunny day and is followed by rapid cooling when shaded by obstructions, such as other trees or cloud cover, or after sunset. Cross-sectional observations of sunscald-afflicted branches reveal lethal injury among the living phloem and cambium cells located near the outer bark (Sakai and Larcher 1987). While cooling rates rapid enough to cause lethal intracellular freezing in woody perennials are rare, rapid fluctuations due to changing light conditions are common in cool-climate orchards. During the summer, the resultant wound can scarify but repeated freeze–thaw injury over the winter or successive years will discolor the bark, and lead to cracking, sunken areas due to a lack of growth, and a loss of surface bark (Kesner and Hansen 1976). Continual localized injury weakens the wood and reduces the overall hardiness of the tree.

4.11.3 Post-Freezing Injury Caused by Fire Blight

Infection of freeze-injured tissues by opportunistic pathogens such as *E. amylovora* will further injure the tree. Cracks in the bark, sometimes referred to as radial shakes or frost cracks, provide entry points for spring fire blight infection (Figure 4.3). Much like winter sunscalds, frost cracks manifest on the south or southwest side of a tree that experiences large day and nighttime

FIGURE 4.2 Southwest injury on apple tree trunk (1) and a trunk cutting demonstrating the 'zone of delineation' between healthy and damaged tissue where southwest injury begins (2). (Photo courtesy of Michelle Cortens, Tree Fruit Specialist with Perennia Food and Agriculture Corporation, Nova Scotia, Canada.)

FIGURE 4.3 Frost crack visible on apple tree trunk. (Photo courtesy of Michelle Cortens, Tree Fruit Specialist with Perennia Food and Agriculture Corporation, Nova Scotia, Canada.)

discrepancies in minimum temperature. A drop in temperature after dusk can result in a differential temperature gradient between the outer and inner layers of the trunk. The accumulation of extracellular ice increases the volume of individual cells and places mechanical strain on the cell walls leading to collapse (Ishida 1963). Vertical cracks can form along weak points in the trunk when the cool outer layer contracts at a faster rate than the comparatively warm inner layer. Frost cracks will develop annually on susceptible trees unless management strategies are employed to minimize future occurrences.

In the spring, de-acclimated woody stems, blossoms, and new shoots can be injured by an early spring frost. Lethal intracellular injury leaves these new tissues flaccid and with a water-soaked appearance (Sakai and Larcher 1987). Over time, the injured tissue will wither and desiccate. We can distinguish between spring freezing injury and fire blight based on when the injury arises. Freezing injury will manifest in the week after a spring frost while injury from fire blight will progress over the days or weeks following an infection window (Van Der Zwet and Beer 1999). However, even non-lethal tissue injury caused by a spring frost can play a significant role in the infection process. This is sometimes referred to as trauma blight. Severe outbreaks of fire blight can follow late-season frosts, hailstorms, hurricanes, or following mechanical or pneumatic defoliation. Small wounds made by freezing or mechanical injury caused by hailstones or wind serve as ideal infection sites. Similarly, fall or pneumatic defoliation that exposes fruit to more light and cooler nighttime temperatures to promote fruit coloration and maturity, can inadvertently injure leaves and allow for bacterial entry (Willick pers. observations).

4.11.4 The Regulation of Cold Acclimation

The capacity to tolerate or avoid freezing injury is not a constant property and will vary throughout the year in response to environmental conditions. Cold acclimation is an adaptation wherein a plant can enhance survival at sub-zero temperatures through exposure to declining temperature, photoperiod, and shifts in light quality (Weiser 1970; Kidokoro et al. 2022). However, exposure to low temperatures appears to be the primary factor in achieving maximum levels of freezing survival (Gusta et al. 2009). Cold acclimation in woody perennials can be divided into three distinct stages (Weiser 1970). The first stage of acclimation is triggered by short days of cool nights between 10°C and 20°C. This corresponds with a nominal increase in the capacity to tolerate and avoid freezing injury and allows fruit trees to withstand temperatures ranging from –15°C to –20°C (Hong and Sucoff 1982). Secondary cold acclimation is induced following exposure to temperatures ranging from 10°C to 0°C which can allow fruit trees to withstand temperatures ranging from –15°C to –30°C. Subsequent exposure of fruit trees to non-lethal sub-zero temperatures promotes further desiccation, and relative accumulation of cryoprotectant sugars promotes survival at –40°C (Quamme 1976). It is important to note that the lethal freezing temperatures will vary within a tree. Typically, the base of the trunk contains the most tender tissues as cold acclimation proceeds at the tissue level from the periphery to the trunk and from the top to the bottom of the tree (Palonen and Buszard 1997).

The molecular basis for cold acclimation depends on a combination of biophysical and biochemical factors that are driven by the up- and downregulation of specific sets of genes. These factors include but are not limited to modifications to the cell wall (Takahashi et al. 2021), plasma membrane (Uemura et al. 2006; Arora 2018), the accumulation of osmoregulatory sugars and proteins with anti-ice inhibitory properties (Wisniewski et al. 2020). Experimental work on cold acclimation has identified several *COLD RESPONSIVE* (*COR*) proteins, such as dehydrins and transcription factors that control the expression of multiple *COR* genes (Van Buskirk and Thomashow 2006). Research in *Arabidopsis* identified cis-acting *DEHYDRATION-RESONSIVE ELEMENTS (DRE)* with an A/GCCGAC core sequence that regulates the expression of dehydration and *COR* (Yamaguchi-Shinozaki and Shinozaki 1994). The *DRE BINDING PROTEIN 1/ C-REPEAT BINDING FACTORS (DREB1/CBF)* transcription factors were subsequently characterized as members of the *APETALA2/ETHYLENE-RESPONSIVE FACTOR (AP2/ERF)* family and contribute to 12%–20% of the cold-induced transcriptional changes (Vogel et al. 2005; Mizoi et al. 2012). Members of this transcription factor family bind to a promoter motif referred to as the drought-responsive element. There are three characterized *DREB1/CBF* in *Arabidopsis*: *DREB1B/CBF1*, *DREB1C/CBF2*, and *DREB1A/CBF3*. Identified *DREB1/CBF* in herbaceous and woody plants contain common AP2 DNA-binding domain motifs (Mizoi et al. 2012). Analysis of the apple genome indicates there are five *DREB1/CBF* and one pseudo- *DREB1/CBF* in apple that are scattered throughout the genome (Wisniewski et al. 2014b). Whereas in the same study, a search of the peach genome identified six *DREB1/CBF* located in a single LG.

Numerous transcription factors, light- and photoperiod-related proteins, hormones, and kinases have been implicated in the regulation of *Arabidopsis CBF/DREB1s* (Kidokoro et al. 2022). *PHYTOCHROME-INTERACTING FACTORS 3 (PIF)*, *PIF4* and *PIF7* are transcription factors (Lee and Thomashow 2012; Jiang et al. 2017). However, the role of the *PIFs* may be species-specific as a tomato *PIF4* homolog (*SlPIF4*) enhanced *DREB1* expression in plants exposed to cold acclimating conditions (Wang et al. 2020). In cold acclimating *Arabidopsis*, *INDUCER OF CBF EXPRESSION 1 (ICE1)* is a MYC-like transcription factor implicated in the expression of *DREB1A/CBF3* using an *ice1-1* knockout line (Chinnusamy et al. 2003). Subsequent studies have illustrated that other *ICE* is involved in the expression of *DREB1/CBF*. For examples, see Wisniewski et al. (2018) and references therein. However, recent experiments with *ice1-1* identified that *DREB1A/CBF3* repression was due to DNA methylation-mediated gene silencing triggered by a T-DNA insert (Kidokoro et al. 2020; Kim et al. 2020). For this reason, it is problematic to include *ICE* in our understanding of how plants perceive cold until additional research elucidates its role in cold acclimation.

There are several other cold-inducible transcription factors (Doherty et al. 2009; Dong et al. 2011; Kidokoro et al. 2021). In response to rapid declines in temperature, *CALMODULIN-BINDING TRANSCRIPTION ACTIVATOR 3 (CAMATA)* and *CAMATA5* were reported to activate the expression of *DREB1B/CBF1* and *DREB1C/CBF2* (Doherty et al. 2009; Kidokoro et al. 2017). Whereas circadian oscillators *LATE ELONGATED HYPOCOTYL (LHY)* and *CIRCADIAN CLOCK-ASSOCIATED 1 (CCA1)* appear to function as positive regulators of *DREB1A/CBF3* and

DREB1C/CBF2 when *Arabidopsis* are exposed to slow or rapid cooling (Dong et al. 2011; Kidokoro et al. 2017). Subsequent work by Kidokoro and colleagues further identified in cold acclimated *Arabidopsis* that MYB proteins *REVEILLE4/LHY-CCA1-Like 1* (*RVE/LCL1*) and *RVE8/LCL5* serve as transcriptional activators of *DREB1* expression (Kidokoro et al. 2021) and *RVE/LCLs* are co-activated when plants are exposed to low temperatures by *NIGHT LIGHT INDUCIBLE AND CLOCK REGULATED PROTEIN 3* (*LNK3*) and *LNK4* (Kidokoro et al. 2023). The activity of *LNK3* and *LNK4* appears to be regulated by low-temperature-induced phosphorylation, whereas *LHY* and *CCA1* repress *DREB1* under control conditions and are degraded under low temperatures (Kidokoro et al. 2021). The presence of two sets of regulators suggests that plants have adapted different mechanisms to enhance their overall cold hardiness in response to rapid diurnal temperature fluctuations and slowly in response to shifts in seasonal temperature.

Since responses among similar proteins, metabolites and genes have been observed in woody perennials (Wisniewski et al. 2014b; Artlip et al. 2019), it is often reported in the literature that woody perennials *DREB1/CBF*s are regulated similarly as *DREB1/CBF*s observed in *Arabidopsis*. This however is not necessarily the case. The expression of *DREB1/CBF*s in poplar (*Populus* spp.) differed in annual versus perennial tissues (Benedict et al. 2006) and the timing of *DREB1/CBF* expression varies prior to and after the onset of dormancy (Welling and Palva 2008). Birch (*Betula pendula* Roth.) trees grown under long days induced the expression of species-specific *DREB1/CBF*s in a similar fashion as *Arabidopsis* after 15 minutes of exposure to low, non-freezing temperatures (Welling and Palva 2008). However, the same birch grown under short-day conditions experienced a delayed induction of *DREB1/CBF*s after 2 hours of exposure to low, non-freezing temperatures and maintained expression for up to 2 days (Welling and Palva 2008). Interestingly, cooling dormant birch trees to –10°C did not induce the expression of *DREB1/CBF* or *COR* which suggests that these genes do not play a direct role in the third stage of cold acclimation.

4.11.5 Cessation of Growth, Bud Set, and Dormancy in Fruit Trees

Fruit trees in preparation for winter undergo a series of phenological changes. These involve the cessation of growth, cold acclimation, bud set, and the induction, maintenance, and release of dormancy (Faust et al. 1997; Campoy et al. 2011). The scope of this section is limited to understanding how temperature and photoperiod, or the length of day and night in 24 hours, influence the molecular mechanisms underpinning bud set and dormancy as it relates to growth cessation and cold acclimation. Knowledge pertaining to hormonal regulation of bud dormancy is an incredibly complex topic that has been reviewed extensively (Tanino 2004; Cooke et al. 2012; Liu et al. 2021). For this reason, the role of hormones in the induction, maintenance, and release of dormancy will not be a major topic covered in this section.

Dormancy manifests as the absence of growth in the form of cell expansion or division in meristem tissues (Rohde and Bhalerao 2007). While not based on the cessation of growth in tree bud meristem cells, the classifications established by Lang et al. (1987) widely used in the horticultural literature are defined based on the physiological manifestation of dormancy. To remain consistent with the literature, we will phrase our discussions around these three classifications of dormancy. Lang and colleagues defined paradormancy as the cessation of growth controlled by physiological factors within the plant but external to the affected tissue structure (Lang et al. 1987). A classic example of paradormancy is observed in peach trees where inhibition of growth in lateral buds is controlled by adjacent leaves and can be removed through mechanical defoliation resulting in a forced blooming event by lateral buds (Lloyd and Couvillon 1974). Endodormancy is defined as growth cessation controlled by physiological factors internal to the affected tissue or organ (Lang et al. 1987). Following endodormancy, buds on fruit trees can enter an ecodormant state. Growth cessation during ecodormancy is controlled by environmental factors external to the plant such as sub-zero temperatures in late winter and spring that are unfavorable for growth (Lang et al. 1987). Bud break following endodormancy or ecodormancy requires exposure to a genetically predetermined amount of cool (chill units)

(Richardson et al. 1974) or warm temperatures (heat units) (Charrier et al. 2011) respectively. Tree buds in an endodormant state are therefore less prone to de-acclimation compared with ecodormant tree buds (Kalberer et al. 2007). Sensitivity to freezing in ecodormant buds will rise proportionally with de-acclimation and floral development (Vitasse et al. 2014).

The cessation of growth and induction of endodormancy in most cool-climate woody perennials is stimulated through exposure to a short photoperiod (Garner 1923; Wareing 1956; Weiser 1970; Cooke et al. 2012). Critical day length to induce growth cessation in natural populations of woody perennials will change along a north-south latitudinal cline (Smithberg and Weiser 1968; Junttila 1980). Wherein the cessation of apical growth in more southern populations requires a shorter photoperiod compared with more northern populations. Interestingly, exposure of northern populations of woody perennials to diurnally alternating temperatures can induce growth cessation even when plants are exposed to a 24-hour photoperiod (Heide 1974; Junttila 1980). Whereas bud set in most woody perennials can be modulated by temperature (Kalcsits et al. 2009; Tanino et al. 2010), with more northern populations requiring warmer temperatures to induce bud set compared with southern populations. Similar trends are observed regarding the induction of cold acclimation in woody perennials (Smithberg and Weiser 1968) and perennial grasses (Willick and Lowry 2022). Willick and Lowry (2022) postulated that an earlier response to declining temperatures may provide more northern populations with additional time to prepare for winter and provide an advantage when these plants are subjected to injurious fall frosts.

Many fruit trees and members of the Rosaceae family do not require a shift in photoperiod to induce bud set (Garner 1923). This photoperiod insensitivity with respect to bud set and growth cessation has been reported in apples, pears, mountain ash, and stone fruit (Nitsch 1957; Heide 2011). Thresholds for temperature-induced growth cessation and subsequent induction of endodormancy vary by species (Heide and Prestrud 2005; Heide 2008). Heide and Prestrud (2005) demonstrated in Alnarp (A.) 2, B.9, M.9, and Malling-Merton 106 (MM.106) apple and Brokmal pear rootstocks that temperatures below 12°C resulted in the cessation of growth and induction of endodormancy regardless of a long or short-day photoperiod. Whereas the cessation of growth in sour cherry (*P. cerasus* L. cv. Gisela 5 and Weirroot), sweet cherry (*P. avium* L. wild), and plum (*P. insitia* L. cv. St. Julien A and Weito) required exposure to 9°C to display a similar insensitivity to photoperiod (Heide 2008). Regardless, all fruit trees formed terminal winter buds after growth cessation and exposure to 8 weeks at 9°C (Heide and Prestrud 2005; Heide 2008). Continuous exposure to low temperatures between 6°C and 9°C for approximately 1000 hours was required for endodormancy release (Heide and Prestrud 2005; Heide 2008).

Many genes in fruit trees are believed to play a role in the establishment and release of dormancy. The genetic regulation of chilling requirement in Rosaceae has been associated with homologs of the *Arabidopsis* floral repressor *SHORT VEGETATIVE PHASE (SVP)*, such as the *DORMANCY ASSOCIATED MADs BOX (DAM)* in apple (Wu et al. 2017), pears (Saito et al. 2013), apricot (*Prunus armeniaca* L.) (Sasaki et al. 2011) and peach (Bielenberg et al. 2008). The observation of *DAMs* in the evergreen peach mutant (Bielenberg et al. 2008)) or the co-suppression of Royal Gala apple *DAMs* with RNA interference (Wu et al. 2021) led to the inability of the transgenic trees to cease growth and induce dormancy. This subsequently resulted in significantly higher *GIBBERELLIN 20 OXIDASE-2* (*MdGA20ox-2*) and *FLOWERING LOCUS T* (*MdFT2*) expression, and greater concentrations of gibberellin and jasmonate in the terminal buds of RNAi lines compared with the control plants. *EARLY BUD BREAK 1 (EBB1)* encoding a transcription factor of the *AP2/ERF* family was identified in a photoperiod-sensitive poplar population to be a positive regulator of bud break (Yordanov et al. 2014). Subsequent work by Azeez and colleagues identified that *EBB1* negatively regulates *SVP-LIKE* (*SVL*) expression whereas a temperature-responsive and epigenetically regulated *EBB3* serves as a positive regulator of bud break (Azeez et al. 2021).

Bioinformatic analysis of apple *CBF* peach, and pear *DAM* and poplar *EBB* genes identified CBF-binding motifs in promoters (Wisniewski et al. 2015). Analysis of Japanese apricot (*P. Mume*) identified an interaction between *PmCBF1* and *PmDAM1* (Zhao et al. 2018). Following the bud break of almond flower buds, a reduction in *DREB1/CBF* expression corresponded with higher *MAD1* and *MAD3* expression (Barros et al. 2012). Wisniewski et al. (2015) used a transformed Malling 26 (M.26) apple rootstock with the overexpression of peach *PpCBF1* to identify links between *DAM*, *EBB*, and *RGL* (*DELLA* genes). Through this work, Wisniewski et al. (2015) identified that some members of these gene families contained C-repeat regions in their promoter regions and surmised that they served as target sites for CBFs. The importance of the M.26 rootstock transformation studies to freezing injury is discussed further in the following section.

4.11.6 Rootstocks and Their Role in Reducing Freezing Injury

Development and uptake of hardy rootstocks can enhance the production of fruit trees in cool-climate orchards (Table 4.1). For example, in the Canadian province of Quebec, many growers use winter-hardy rootstocks such as Ottawa 3 (O.3.), which was developed and released in 1970 by Agriculture and Agri-food Canada (Embree 1988). Most commercial orchards in North America are planted on M.9, M.26, B.9 and are comparatively more freezing tolerant compared with other commercially available options (Embree 1988; Hampson et al. 1997). The Malling and Budagovsky rootstocks have other weaknesses such as disease susceptibility. In particular, M.26 and M.9 are noted for their susceptibility to fire blight. The Geneva© rootstocks were selected for disease resistance (Autio et al. 2005) and some have been noted for their freezing tolerance (Moran et al. 2011). A controlled rootstock freezing study identified that comparatively fire blight tolerant Geneva© 935 developed roots that survived to between –16°C and –19°C, which was on average 4°C lower than root survival temperatures observed for the M.26 (Moran et al. 2011). New rootstocks need to be tested for disease resistance and cold hardiness in trunk tissues.

A series of transgenic studies further elucidated the role of *DREB1/CBF* overexpression in enhancing the freezing tolerance of apple rootstocks (Wisniewski et al. 2011; Artlip et al. 2013, 2014, 2016, 2019). Wisniewski and colleagues examined the functional role of peach *PpCBF1* overexpression in the rootstock M.26 of apple (Wisniewski et al. 2011). When grown under short days (8 hours), the resultant transgenic apple lines with constitutive overexpression of *PpCBF1* exhibited earlier growth cessation and leaf senescence compared with the M.26 control apple trees (Wisniewski et al. 2011). An LT_{50} was calculated to assess the temperature at which half a population survived freezing injury. Transgenic apple line T166 leaves had a non-acclimated and cold acclimated LT_{50} of –9.5°C and 13.0°C which was 4.2°C and 5.7°C cooler than the non-transformed M.26 apple trees (Wisniewski et al. 2011). Results from a three-year field study confirmed that overexpression of *PpCBF1* resulted in reduced growth, early leaf senescence, delayed bud break, enhanced LT_{50}, and changes in leaf pigment accumulation compared with the non-transformed M.26 rootstock controls (Artlip et al. 2014). These two studies highlight the importance of *DREB1/CBF* expression in enhancing apple freezing tolerance.

Overexpression of *PpCBF1* in the rootstock was later assessed to see if it was transmitted to the scion (Artlip et al. 2016). Royal gala scions were grafted to the non-transformed M.26 or transgenic M.26 (T166) rootstocks and assessed for phenological development over three field seasons. Interestingly, *PpCBF1* was not detected in the phloem or cambium of 'Royal Gala' grafted to T166 rootstocks which indicates that there was no graft transmission of *PpCBF1* mRNA. Growth as measured with a stem caliper and overall height was reduced in the Royal Gala/T166 trees compared with the Royal Gala/M.26 (Artlip et al. 2016). Characterization of gene expression in bark tissues at different states of dormancy identified the most differentially expressed genes in April when dormancy release and bud break occurred in M.26 but not T166 (Artlip et al. 2019). Not surprisingly, genes associated with growth cessation such as the inactivation of auxin, gibberellic acid, and cytokinin were upregulated in T166 while genes responsible for cambial growth, and cell division were downregulated in T166 relative to M.26.

Taken together the work of Artlip and colleagues (Artlip et al. 2016, 2019) suggests that the overexpression of *PpCBF1* influences growth hormone homeostasis to delay growth in the spring and promote leaf senescence in the fall. Growth cessation in trees often corresponds with greater overall freezing tolerance (Tanino et al. 2010) and delayed bud break would promote a greater capacity to re-acclimate following a late winter thaw. Future work is needed to identify whether the constitutive expression of other *DREB1/CBF* and *CORs* would be advantageous to mitigating the negative effects of winter warming. Care must be taken to evaluate whether the benefits of improved freezing tolerance are outweighed by any reductions in crop load.

4.12 CONCLUSIONS

Management practices aimed at mitigating the effects of fire blight and freezing-related stress limitations to increased productivity have demonstrated that we can achieve progress at a regional level. This has allowed production to extend into climatic areas that may have been considered marginal by previous metrics. An example in North America is how the timing of spray treatments is managed to prevent blossom infections by *E. amylovora*. Restrictions on the availability of effective products to control fire blight infections and the number of times that a block can be sprayed in a season have created the need for an improved system to inform spray schedule decisions. The development of models such as MARYBLYT and Cougarblight that generate infection predictions based on weather and fruit tree phenology have helped orchard managers identify more optimal spraying windows to reduce the level of inoculum available for shoot and blossom infection. Growers that do not have resistance in their orchards apply streptomycin based on these fire blight forecasting systems to minimize the number of streptomycin applications required to inhibit further resistance development to streptomycin. It has also been recommended in many regions to apply copper-based treatments during dormancy, prune affected branches as soon as possible, and treat fire blight with oxytetracycline in orchards where streptomycin resistance exists. One effective way to prevent the bacteria from further colonizing the tree and spreading to nearby trees is to remove the infected branch 8–12 inches below the symptomatic tissue as soon as a tree exhibits any signs of fire blight. These infected branches should be removed from the orchard immediately and tools used for cutting these branches should be sterilized between uses.

When establishing a new orchard, problems can be avoided by carefully choosing cultivars and rootstocks (see Tables 4.1 and 4.2) that are resistant to fire blight. Orchard sites should contain soils that drain properly, have a pH between 5.5 and 6.5, and have sufficient organic matter content. Applying excessive nitrogen should be avoided. It is necessary to establish integrated pest control measures for sucking-piercing insects, such as plant bugs, aphids, and leafhoppers. It is also essential to implement an annual pruning program to remove as many fire blight cankers as possible. The spread of disease can be greatly reduced by using control products (such as streptomycin for blossom blight treatment in conjunction with a prediction model to time bloom applications, or copper or Bordeaux mixture before buds open). If blossom blight is an issue in mature-bearing trees, using Apogee (prohexadione-calcium) may help reduce the tree's susceptibility to the disease.

The genetic potential for improved resistance to fire blight, de-acclimation, dormancy induction, and general freezing injury is likely already present in cultivated varieties and wild accessions. In most breeding programs, tree fruits are often selected under conditions of optimal fertility, temperature, moisture, disease, and pest control. The introduction of the Geneva® and potentially the Vineland series rootstocks have provided additional opportunities for winter hardiness and fire blight management. To take advantage of the untapped potential, novel scion x rootstock combinations need to be screened concurrently under conditions more likely to be encountered within a specific region, rather than at one centralized location.

Tree fruit breeders tend to rely on yield, flavor profile, and storability, when selecting new scion material and understandably cannot always wait for basic research identifying companion traits related to how the tree tolerates or avoids stress. With the onset of changing climatic conditions and increased biotic and abiotic pressures, there is now a greater need for phenotyping fruit trees to obtain the needed information on adaptations to multi-stress environments. The incorporation of hyperspectral, LiDAR, and thermal imaging via satellite and unmanned vehicles may inform precision horticultural management strategies. However, care must be taken to ground these imaging-based non-destructive observations within a mechanistic understanding of why and how the plant is responding to stress.

The increasing precision and affordability of molecular techniques to modify scion and rootstock germplasm opens considerable possibilities for future research. This work will continue to make meaningful improvements to our mechanistic understanding of how fruit trees respond to stress.

REFERENCES

Aldwinckle, H. S., & Beer, S. V. (1979). Fire blight and its control. *Horticultural Reviews*, https://doi.org/10.1002/9781118060742.ch11.

Arora, R. (2018). Mechanism of freeze-thaw injury and recovery: A cool retrospective and warming up to new ideas. *Plant Science*, *270*, 301–313. https://doi.org/10.1016/j.plantsci.2018.03.002.

Artlip, T. S., Wisniewski, M. E., & Norelli, J. L. (2014). Field evaluation of apple overexpressing a peach CBF gene confirms its effect on cold hardiness, dormancy, and growth. *Environmental and Experimental Botany*, *106*, 79–86. https://doi.org/10.1016/j.envexpbot.2013.12.008.

Artlip, T. S., Wisniewski, M. E., Arora, R., & Norelli, J. L. (2016). An apple rootstock overexpressing a peach CBF gene alters growth and flowering in the scion but does not impact cold hardiness or dormancy. *Horticulture Research*, 3. https://doi.org/10.1038/hortres.2016.6.

Artlip, T. S., Wisniewski, M. E., Bassett, C. L., & Norelli, J. L. (2013). CBF gene expression in peach leaf and bark tissues is gated by a circadian clock. *Tree Physiology*, *33*, 866–877. https://doi.org/10.1093/treephys/tpt056.

Artlip, T., McDermaid, A., Ma, Q., & Wisniewski, M. (2019). Differential gene expression in non-transgenic and transgenic "M.26" apple overexpressing a peach CBF gene during the transition from eco-dormancy to bud break. *Horticulture Research*, *6*, 1–16. https://doi.org/10.1038/s41438-019-0168-9.

Asahina, E. (1956). The freezing process of plant cell. *Contributions from the Institute of Low Temperature Science, Series B*, 83–148. https://eprints.lib.hokudai.ac.jp/dspace/bitstream/2115/20219/1/10_p83-126.pdf

Autio, W., Robinson, T., Barritt, B., Cline, J., Crassweller, R., Embree, C., Ferree, D., Garcia, M., Greene, G., & E. H. (2005). Performance of 'Fuji' and 'Mcintosh' apple trees after 5 years as affected by several dwarf rootstocks in the 1999 NC-140 apple rootstock trial. *Journal of The American Pomological Society*, *59*, 202–214.

Azeez, A., Zhao, Y. C., Singh, R. K., Yordanov, Y. S., Dash, M., Miskolczi, P., Stojkovič, K., Strauss, S. H., Bhalerao, R. P., & Busov, V. B. (2021). Early bud-break 1 and early bud-break 3 control resumption of poplar growth after winter dormancy. *Nature Communication*, *12*, 1123.

Baker, K. F. (1971). Fire blight of pome fruits: The genesis of the concept that bacteria can be pathogenic to plants. *Hilgardia*, *40*, 603–633. https://doi.org/10.3733/hilg.v40n18p603.

Baltes, N. J., Hummel, A. W., Konecna, E., Cegan, R., Bruns, A. N., Bisaro, D. M., & Voytas, D. F. (2015). Conferring resistance to geminiviruses with the CRISPR–Cas prokaryotic immune system. *Nature Plants*, *1*, 1–4.

Barritt, B. H., Konishi, B. S., Dilley, M. A., & Pusey, L. (2002). Resistance of apple rootstocks to fire blight caused by *Erwinia amylovora*. In *I International Symposium on Rootstocks for* Deciduous *Fruit Tree Species* (Vol. 658, pp. 387–390). Zaragoza, Spain. https://www.actahort.org/books/658/

Barros, P. M., Gonçalves, N., Saibo, N. J. M., & Oliveira, M. M. (2012). Cold acclimation and floral development in almond bud break: Insights into the regulatory pathways. *Journal of Experimental Botany*, *63*, 4585–4596.

Baumgartner, I. O., Patocchi, A., Franck, L., Kellerhals, M., & Broggini, G. A. L. (2010). Fire blight resistance from'Evereste'and Malus sieversii used in breeding for new high quality apple cultivars: Strategies and results. In *XII International Workshop on Fire Blight* (Vol. 896, pp. 391–397). Warsaw (Poland). https://www.ishs.org/symposium/146

Baumgartner, I. O., Patocchi, A., Lussi, L., Kellerhals, M., & Peil, A. (2013). Accelerated introgression of fire blight resistance from *Malus × obusta* 5 and other wild germplasm into elite apples. In *XIII International Workshop on Fire Blight*, vol. 1056, pp. 281–287. Zürich (Switzerland). https://www.ishs.org/symposium/323

Beckerman, J. (2006). Disease susceptibility of common apple cultivars. Purdue Extension BP-132-W. https://www.extension.purdue.edu/extmedia/bp/bp-132-w.pdf

Benedict, C., Skinner, J. S., Meng, R., Chang, Y., Bhalerao, R., Huner, N. P. A., Finn, C. E., Chen, T. H. H., & Hurry, V. (2006). The CBF1-dependent low temperature signalling pathway, regulon and increase in freeze tolerance are conserved in *Populus* spp. *Plant, Cell & Environment*, *29*, 1259–1272. https://doi.org/10.1111/j.1365-3040.2006.01505.x.

Bielenberg, D. G., Wang, Y., Li, Z., Zhebentyayeva, T., Fan, S., Reighard, G. L., Scorza, R., & Abbott, A. G. (2008). Sequencing and annotation of the evergrowing locus in peach [*Prunus persica* (L.) Batsch] reveals a cluster of six MADS-box transcription factors as candidate genes for regulation of terminal bud formation. *Tree Genetics & Genomes*, *4*, 495–507.

Billing, E. (1998). Fire blight risk assessment: Billing's integrated system (BIS) and its evaluation. In *VIII International Workshop on Fire Blight* (Vol. 489, pp. 399–406). Kuşadası, Turkey.

Bost, S., & Windham, A. (2002). *SP277-R-Fire blight*. The University of Tennessee Agricultural Extension Service. https://doi.org/10.5962/bhl.title.41918.

Bredow, M., & Walker, V. K. (2017). Ice-binding proteins in plants. *Frontiers in Plant Science*, *8*, 1–15. https://doi.org/10.3389/fpls.2017.02153.

Brisset, M.-N., Cesbron, S., Thomson, S. V, & Paulin, J.-P. (2000). Acibenzolar-S-methyl induces the accumulation of defense-related enzymes in apple and protects from fire blight. *European Journal of Plant Pathology*, *106*, 529–536.

Broggini, G. A. L., Wöhner, T., Fahrentrapp, J., Kost, T. D., Flachowsky, H., Peil, A., Hanke, M., Richter, K., Patocchi, A., & Gessler, C. (2014). Engineering fire blight resistance into the apple cultivar 'Gala'using the FB _ MR 5 CC-NBS-LRR resistance gene of *Malus* × *robusta* 5. *Plant Biotechnology Journal*, *12*, 728–733.

Campa, M., Piazza, S., Righetti, L., Oh, C.-S., Conterno, L., Borejsza-Wysocka, E., Nagamangala, K. C., Beer, S. V., Aldwinckle, H. S., & Malnoy, M. (2019). HIPM is a susceptibility gene of Malus spp.: Reduced expression reduces susceptibility to *Erwinia amylovora*. *Molecular Plant-Microbe Interactions*, *32*, 167–175.

Campoy, J. A., Ruiz, D., & Egea, J. (2011). Dormancy in temperate fruit trees in a global warming context: A review. *Scientia Horticulturae (Amsterdam)*, *130*, 357–372.

Charrier, G., Bonhomme, M., Lacointe, A., & Améglio, T. (2011). Are budburst dates, dormancy and cold acclimation in walnut trees (*Juglans regia* L.) under mainly genotypic or environmental control? *International Journal of Biometeorology*, *55*, 763–774.

Chinnusamy, V., Ohta, M., Kanrar, S., Lee, B., Hong, X., Agarwal, M., & Zhu, J.-K. (2003). ICE*1*, a regulator of cold-induced transcriptome and freezing tolerance in *Arabidopsis*. *Genes Development*, *17*, 1043–1054.

Chiou, C. S., & Jones, A. L. (1995). Molecular analysis of high-level streptomycin resistance in *Erwinia amylovora*. Phytopathology, 85, 324–328.

Choi, B. H., Kim, C. S., Jeong, Y. J., Park, I. H., Han, S. G., & Yoon, T. M. (2021). Resistance evaluation of g, cg, or m series apple rootstocks to soil-borne diseases (Phytophthora root rot, white root rot, and southern blight) and woolly apple aphid. *Horticultural Science and Technology*, *39*, 167–174. https://doi.org/10.7235/HORT.20210015.

Cline, J. A., Hunter, D. M., Bonn, W. G., & Bijl, M. (2001). Resistance of the Vineland series of apple root-stocks to fire blight caused by *Erwinia amylovora*. *Journal of The American Pomological Society*, *55*, 218.

Cooke, J. E. K., Eriksson, M. E., & Junttila, O. (2012). The dynamic nature of bud dormancy in trees: Environmental control and molecular mechanisms. *Plant Cell Environent*, *35*, 1707–1728.

Cornille, A., Gladieux, P., Smulders, M. J .M., Roldán-Ruiz, I., Laurens, F., Le Cam, B., Nersesyan, A., Clavel, J., Olonova, M., & Feugey, L. (2012). New insight into the history of domesticated apple: Secondary contribution of the European wild apple to the genome of cultivated varieties. *PLoS Genetics*, *8*, e1002703.

Cui, Z., Huntley, R. B., Schultes, N. P., Kakar, K. U., Yang, C.-H., & Zeng, Q. (2021). Expression of the type III secretion system genes in epiphytic *Erwinia amylovora* cells on apple stigmas benefits endophytic infection at the hypanthium. *Molecular Plant-Microbe Interactions*, *34*, 1119–1127.

Cummins, J. N., & Aldwinckle, H. S. (1994). Challenges to Malling 9, M. 26 and M. 7-new rootstocks from Geneva. *Compact Fruit Tree*, *27*, 25–27.

Denning, W. (1974). On the decay of apple trees. *Transactions of the Society for the Promotion of Agriculture, Arts and Manufactures*, *2*, 219–222.

Desnoues, E., Norelli, J. L., Aldwinckle, H. S., Wisniewski, M. E., Evans, K. M., Malnoy, M., & Khan, A. (2018). Identification of novel strain-specific and environment-dependent minor QTLs linked to fire blight resistance in apples. *Plant Molecular Biology Reporter*, *36*, 247–256.

Dewdney, M. M., Biggs, A. R., & Turechek, W. W. (2007). A statistical comparison of the blossom blight forecasts of MARYBLYT and Cougarblight with receiver operating characteristic curve analysis. *Phytopathology*, *97*, 1164–1176.

Doherty, C. J., Van Buskirk, H. A., Myers, S. J., & Thomashow, M. F. (2009). Roles for Arabidopsis CAMTA transcription factors in cold-regulated gene expression and freezing tolerance. *Plant Cell*, *21*, 972–984.

Dong, M. A., Farré, E. M., & Thomashow, M. F. (2011). Circadian clock-associated 1 and late elongated hypocotyl regulate expression of the C-repeat binding factor (CBF) pathway in Arabidopsis. *Proceedings of the National Academy of Sciences*, *108*, 7241–7246.

Duan, N., Bai, Y., Sun, H., Wang, N., Ma, Y., Li, M., Wang, X., Jiao, C., Legall, N., Mao, L., Wan, S., Wang, K., He, T., Feng, S., Zhang, Z., Mao, Z., Shen, X., Chen, X., Jiang, Y., Wu, S., Yin, C., Ge, S., Yang, L., Jiang, S., Xu, H., Liu, J., Wang, D., Qu, C., Wang, Y., Zuo, W., Xiang, L., Liu, C., Zhang, D., Gao, Y., Xu, Y., Xu, K., Chao, T., Fazio, G., Shu, H., Zhong, G.-Y., Cheng, L., Fei, Z., & Chen, X. (2017). Genome re-sequencing reveals the history of apple and supports a two-stage model for fruit enlargement. *Nature Communications*, *8*, 249. https://doi.org/10.1038/s41467-017-00336-7.

Durel, C.-E., Denance, C., & Brisset, M.-N. (2009). Two distinct major QTL for resistance to fire blight co-localize on linkage group 12 in apple genotypes 'Evereste'and *Malus floribunda* clone 821. *Genome*, *52*, 139–147.

Embree, C. (1988). Apple rootstock cold hardiness evaluation. *Compact Fruit Tree*, *21*, 99–105.

Emeriewen, O. F., Peil, A., Richter, K., Zini, E., Hanke, M.-V., & Malnoy, M. (2017a). Fire blight resistance of Malus× arnoldiana is controlled by a quantitative trait locus located at the distal end of linkage group 12. *European Journal of Plant Pathology*, *148*, 1011–1018.

Emeriewen, O. F., Richter, K., Hanke, M.-V., Malnoy, M., & Peil, A. (2015). The fire blight resistance QTL of *Malus fusca* (Mfu10) is affected but not broken down by the highly virulent Canadian *Erwinia amylovora* strain E2002A. *European Journal of Plant Pathology*, *141*, 631–635.

Emeriewen, O. F., Richter, K., Hanke, M.-V., Malnoy, M., & Peil, A. (2017). Further insights into *Malus fusca* fire blight resistance. *Journal of Plant Pathology*, *99*, 45–49.

Emeriewen, O. F., Richter, K., Piazza, S., Micheletti, D., Broggini, G. A. L., Berner, T., Keilwagen, J., Hanke, M.-V., Malnoy, M., & Peil, A. (2018). Towards map-based cloning of FB_Mfu*10*, identification of a receptor-like kinase candidate gene underlying the Malus fusca fire blight resistance locus on linkage group 10. *Molecular Breeding*, *38*, 1–14.

Emeriewen, O. F., Wöhner, T., Flachowsky, H., & Peil, A. (2019). Malus hosts–*Erwinia amylovora* interactions: Strain pathogenicity and resistance mechanisms. *Frontiers in Plant Science*, *10*, 551.

Emeriewen, O., Richter, K., Kilian, A., Zini, E., Hanke, M.-V., Malnoy, M., & Peil, A. (2014). Identification of a major quantitative trait locus for resistance to fire blight in the wild apple species *Malus fusca*. *Molecular Breeding*, *34*, 407–419.

Evans, J. R., Evans, R. R., Regusci, C. L., & Rademacher, W. (1999). Mode of action, metabolism, and uptake of BAS 125W, prohexadione-calcium. *HortScience*, *34*, 1200–1201.

Fahrentrapp, J., Broggini, G. A. L., Kellerhals, M., Peil, A., Richter, K., Zini, E., & Gessler, C. (2013). A candidate gene for fire blight resistance in *Malus× robusta* 5 is coding for a CC–NBS–LRR. *Tree Genetics & Genomes*, *9*, 237–251.

FAO STAT. (2024). License: CC BY-NC-SA 3.0 IGO. https://www.fao.org/faostat/en/#data/QCL.

Faust, M., Erez, A., Rowland, L. J., Wang, S. Y., & Norman, H. A. (1997). Bud dormancy in perennial fruit trees: Physiological basis for dormancy induction, maintenance, and release. *HortScience*, *32*, 623–629. .

Ferree, D. C., Hirst, P. M., Schmid, J. C., & Dotson, P. E. (1995). Performance of three apple cultivars with 22 dwarfing rootstocks during 8 seasons in Ohio. *Fruit Varieties Journal*, *49*, 171–178.

Flachowsky, H., Hanke, M., Peil, A., Strauss, S. H., & Fladung, M. (2009). A review on transgenic approaches to accelerate breeding of woody plants. *Plant Breeding*, *128*, 217–226.

Flachowsky, H., Peil, A., Sopanen, T., Elo, A., & Hanke, V. (2007). Overexpression of BpMADS4 from silver birch (*Betula pendula* Roth.) induces early-flowering in apple (*Malus × domestica* Borkh.). *Plant Breeding*, *126*, 137–145.

Friedrich, L., Lawton, K., Ruess, W., Masner, P., Specker, N., Rella, M. G., Meier, B., Dincher, S., Staub, T., & Uknes, S. (1996). A benzothiadiazole derivative induces systemic acquired resistance in tobacco. *Plant Journal*, *10*, 61–70.

Fujikawa, S., Kasuga, J., Takata, N., & Arakawa, K. (2009). Factors related to change of deep supercooling capability in xylem parenchyma cells of trees. In L.V. Gusta, M. E. Wisniewski, & K. K. Tanino (Eds.), *Plant Cold Hardiness, from the Laboratory to the Field* (pp. 29–42). CABI.

Fujikawa, S., Kuwabara, C., Kasuga, J., & Arakawa, K. (2018). Supercooling-promoting (anti-ice nucleation) substances. In M. Iwaya-Inoue, M. Sakurai, & M. Uemura (Eds.), *Survival strategies in extreme cold and desiccation, adaptation mechanisms and their applications* (pp. 289–320)

Fujisawa, M., & Kobayashi, K. (2010). Apple (*Malus pumila* var. domestica) phenology is advancing due to rising air temperature in northern Japan. *Global Change Biology*, *16*, 2651–2660.

Gardner, R. G., Cummins, J. N., & Aldwinckle, H. S. (1980). Inheritance of fire blight resistance in malus in relation to rootstock breeding1. *Journal of the American Society for Horticultural Science, 105*, 912–916.

Garner, W. W. (1923). Further studies in photoperiodism, the response of the plant to relative length of day and night. *Journal of Agricultural Research, 23*, 871–920.

George, M. F., & Burke, M. J. (1977). Cold hardiness and deep supercooling in xylem of shagbark hickory. *Plant Physiology, 59*, 319–325. https://doi.org/10.1104/pp.59.2.319.

Görlach, J., Volrath, S., Knauf-Beiter, G., Hengy, G., Beckhove, U., Kogel, K. H., Oostendorp, M., Staub, T., Ward, E., Kessmann, H., & Ryals, J. (1996). Benzothiadiazole, a novel class of inducers of systemic acquired resistance, activates gene expression and disease resistance in wheat. *Plant Cell, 8*, 629–643. https://doi.org/10.1105/tpc.8.4.629.

Gusta, L. V, Wisniewski, M. E., & Trischuk, R. G. (2009). *Patterns of freezing in plants: The influence of species, environment and experiential procedures* (pp. 214–225). CABI. https://doi.org/10.1079/9781845935139.0214.

Hampson, C., Quamme, H., & R, B. (1997). Performance of dwarfing apple rootstocks in five trials in British Columbia, Canada. *Fruit Varieties Journal, 51*, 183–191.

Hanke, M.-V., Flachowsky, H., Peil, A., & Hättasch, C. (2007). No flower no fruit—genetic potentials to trigger flowering in fruit trees. *Genes Genomes Genomics, 1*, 1–20.

Harshman, J. M., Evans, K. M., Allen, H., Potts, R., Flamenco, J., Aldwinckle, H. S., Wisniewski, M. E., & Norelli, J. L. (2017). Fire blight resistance in wild accessions of *Malus sieversii*. *Plant Disease, 101*, 1738–1745.

Heide, O. M. (1974). Growth and dormancy in Norway spruce ecotypes (*Picea abies*) I. Interaction of photoperiod and temperature. *Physiologia Plantarum, 30*, 1–12.

Heide, O. M. (2008). Interaction of photoperiod and temperature in the control of growth and dormancy of Prunus species. *Scientia Horticulturae (Amsterdam), 115*, 309–314.

Heide, O. M. (2011). Temperature rather than photoperiod controls growth cessation and dormancy in Sorbus species. *Journal of Experimental Botany, 62*, 5397–5404.

Heide, O. M., & Prestrud, A. K. (2005). Low temperature, but not photoperiod, controls growth cessation and dormancy induction and release in apple and pear. *Tree Physiology, 25*, 109–114.

Hong, S.-G., & Sucoff, E. (1982). Rapid increase in deep supercooling of xylem parenchyma. *Plant Physiology, 69*, 697–700. https://doi.org/10.1104/pp.69.3.697.

Huang, T., & Burr, T. J. (1999). Characterization of plasmids that encode streptomycin-resistance in bacterial epiphytes of apple. *Journal of Applied Microbiology, 86*, 741–751.

IPCC 2014. Climate Change 2014: Synthesis Report. Contribution of Working Groups I, II and III to the Fifth Assessment Report of the Intergovernmental Panel on Climate Change [Core Writing Team, R. K. Pachauri & L. A. Meyer (Eds)]. Geneva, Switzerland.

Ishida, S. (1963). On the development of frost cracks in Todomatsu trunks, Abies sachalinensis, especially in relation to their wetwood. *Research Bulletin of the College Experimental Forests, College of Agriculture, Hokkaido University, 22*, 273–373.

Ishikawa, M., & Sakai, A. (1985). Extraorgan freezing in wintering flower buds of *Cornus officinalis* Sieb. et Zucc. *Plant Cell Environment, 8*, 333–338. https://doi.org/10.1111/j.1365-3040.1985.tb01407.x.

Jiang, B., Shi, Y., Zhang, X., Xin, X., Qi, L., Guo, H., Li, J., & Yang, S. (2017). PIF3 is a negative regulator of the CBF pathway and freezing tolerance in Arabidopsis. *Proceedings of the National Academy of Sciences, 114*, E6695–E6702.

Jones, A. L. (2001a). Experience with Apogee in 2000, Rate versus concentration. In *Proceedings of Michigan State Horticultural Society 130*, 50–51.

Jones, A. L. (2001b). Challenges of controlling fire blight. *Compact Fruit Tree*, , *34*(3), 86–91.

Junttila, O. (1980). Effect of photoperiod and temperature on apical growth cessation in two ecotypes of Salix and Betula. *Physiologia Plantarum, 48*, 347–352.

Kalberer, S. R., Arora, R., Leyva-Estrada, N., & Krebs, S. L. (2007). Cold hardiness of floral buds of deciduous azaleas: Dehardening, rehardening, and endodormancy in late winter. *Journal of the American Society for Horticultural Science, 132*, 73–79.

Kalcsits, L. A., Silim, S., & Tanino, K. (2009). Warm temperature accelerates short photoperiod-induced growth cessation and dormancy induction in hybrid poplar (*Populus*× spp.). *Trees, 23*, 971–979.

Kasuga, J., Hashidoko, Y., Nishioka, A., Yoshiba, M., Arakawa, K., & Fujikawa, S. (2008). Deep supercooling xylem parenchyma cells of katsura tree (*Cercidiphyllum japonicum*) contain flavonol glycosides exhibiting high anti-ice nucleation activity. *Plant, Cell Environment, 31*, 1335–1348. https://doi.org/10.1111/j.1365–3040.2008.01835.x.

Kawahara, H., Masuda, K., & Obata, H. (2000). Identification of a compound in *Chamaecyparis taiwanensis* inhibiting the ice-nucleating activity of Pseudomonas fluorescens KUIN-1. *Bioscience, Biotechnology, and Biochemistry*, *64*, 2651–2656.

Kellerhals, M., Schütz, S., & Patocchi, A. (2017). Breeding for host resistance to fire blight. *Journal of Plant Pathology*, 37–43.

Kesner, C., & Hansen, C. (1976). Prevention of winter sunscald injury in Michigan Orchards. *Journal of the American Society for Horticultural Science*, *101*, 546–550.

Khan, M. A., Duffy, B., Gessler, C., & Patocchi, A. (2006). QTL mapping of fire blight resistance in apple. *Molecular Breeding*, *17*, 299–306.

Khan, M. A., Durel, C.-E., Duffy, B., Drouet, D., Kellerhals, M., Gessler, C., & Patocchi, A. (2007). Development of molecular markers linked to the 'Fiesta'linkage group 7 major QTL for fire blight resistance and their application for marker-assisted selection. *Genome*, *50*, 568–577.

Khan, M. A., Zhao, Y., & Korban, S. S. (2012). Molecular mechanisms of pathogenesis and resistance to the bacterial pathogen *Erwinia amylovora*, causal agent of fire blight disease in rosaceae. *Plant Molecular Biology Reporter*, *30*, 247–260. https://doi.org/10.1007/s11105-011-0334-1.

Khan, M. A., Zhao, Y., & Korban, S. S. (2013). Identification of genetic loci associated with fire blight resistance in Malus through combined use of QTL and association mapping. *Physiologia Plantarum*, *148*, 344–353.

Kidokoro, S., Hayashi, K., Haraguchi, H., Ishikawa, T., Soma, F., Konoura, I., Toda, S., Mizoi, J., Suzuki, T., & Shinozaki, K. (2021). Posttranslational regulation of multiple clock-related transcription factors triggers cold-inducible gene expression in Arabidopsis. *Proceedings of the National Academy of Sciences*, *118*, e2021048118.

Kidokoro, S., Kim, J.-S., Ishikawa, T., Suzuki, T., Shinozaki, K., & Yamaguchi-Shinozaki, K. (2020). DREB1A/CBF3 is repressed by transgene-induced DNA methylation in the Arabidopsis ice1-1 mutant. *Plant Cell*, *32*, 1035–1048.

Kidokoro, S., Konoura, I., Soma, F., Suzuki, T., Miyakawa, T., Tanokura, M., Shinozaki, K., & Yamaguchi-Shinozaki, K. (2023). Clock-regulated coactivators selectively control gene expression in response to different temperature stress conditions in Arabidopsis. *Proceedings of the National Academy of Sciences*, *120*, e2216183120.

Kidokoro, S., Shinozaki, K., & Yamaguchi-Shinozaki, K. (2022). Transcriptional regulatory network of plant cold-stress responses. *Trends Plant Science*, *27*, 922–935. https://doi.org/10.1016/j.tplants.2022.01.008.

Kidokoro, S., Yoneda, K., Takasaki, H., Takahashi, F., Shinozaki, K., & Yamaguchi-Shinozaki, K. (2017). Different cold-signaling pathways function in the responses to rapid and gradual decreases in temperature. *Plant Cell*, *29*, 760–774. https://doi.org/10.1105/tpc.16.00669.

Kim, J.-S., Kidokoro, S., Shinozaki, K., & Yamaguchi-Shinozaki, K. (2020). DNA demethylase ROS1 prevents inheritable DREB1A/CBF3 repression by transgene-induced promoter methylation in the Arabidopsis ice1-1 mutant. *Plant Molecular Biology*, *104*, 575–582.

Kost, T. D., Gessler, C., Jänsch, M., Flachowsky, H., Patocchi, A., & Broggini, G. A. L. (2015). Development of the first cisgenic apple with increased resistance to fire blight. *PLoS One*, *10*, e0143980.

Lang, G. A., Early, J. D., Martin, G. C., & Darnell, R. L. (1987). Endo-, para-, and ecodormancy: Physiological terminology and classification for dormancy research. *HortScience*, *22*, 371–377.

Lawton, K. A., Friedrich, L., Hunt, M., Weymann, K., Delaney, T., Kessmann, H., Staub, T., & Ryals, J. (1996). Benzothiadiazole induces disease resistance in Arabidopsis by activation of the systemic acquired resistance signal transduction pathway. *Plant Journal*, *10*, 71–82.

Le Roux, P.-M. F., Khan, M. A., Broggini, G. A. L., Duffy, B., Gessler, C., & Patocchi, A. (2010). Mapping of quantitative trait loci for fire blight resistance in the apple cultivars "Florina" and "Nova Easygro". *Genome*, *53*, 710–722. https://doi.org/10.1139/g10–047.

Lee, C.-M., & Thomashow, M. F. (2012). Photoperiodic regulation of the C-repeat binding factor (CBF) cold acclimation pathway and freezing tolerance in *Arabidopsis thaliana*. *Proceedings of the National Academy of Sciences*,*109*, 15054–15059.

Liu, J., Artlip, T., Sherif, S., & Wisniewski, M. (2021). Genetics and genomics of cold hardiness and dormancy (pp. 247–270). https://doi.org/10.1007/978-3-030-74682-7_12.

Lloyd, D. A., & Couvillon, G. A. (1974). Effects of date of defoliation on flower and leaf bud development in the peach (*Prunus persica* (L.) Batschl. *Journal of the American Society for Horticultural Science*, *99*, 514–517.

Loveys, B. R., Egerton, J. J. G., & Ball, M. C. (2006). Higher daytime leaf temperatures contribute to lower freeze tolerance under elevated CO_2. *Plant, Cell Environment*, *29*, 1077–1086. https://doi.org/10.1111/j.1365-3040.2005.01482.x.

Lutze, J. L., Roden, J. S., Holly, C. J., Wolfe, J., Egerton, J. J. G., & Ball, M. C. (1998). Elevated atmospheric [CO_2] promotes frost damage in evergreen tree seedlings. *Plant, Cell Environment*, *21*, 631–635. https://doi.org/10.1046/j.1365-3040.1998.00296.x.

MAAARO. (2011). *Lutte intégrée contre les ennemis du pommier*. Extrait du Publication 310F.

Maxson-Stein, K., He, S.-Y., Hammerschmidt, R., & Jones, A. L. (2002). Effect of treating apple trees with acibenzolar-S-methyl on fire blight and expression of pathogenesis-related protein genes. *Plant Disease*, *86*, 785–790.

McArtney, S., & Obermiller, J. (2011). Effect of dwarfing rootstocks on low temperature tolerance of "golden delicious" apple trees during winter 2008–2009. *Journal of The American Pomological Society*, *65*, 178–184.

McManus, P. S. (2014). Does a drop in the bucket make a splash? Assessing the impact of antibiotic use on plants. *Current Opinion in Microbiology*, 19, 76–82.

McManus, P. S., Stockwell, V. O., Sundin, G. W., & Jones, A. L. (2002). Antibiotic use in plant agriculture. *Annual Review of Phytopathoogy,l 40*, 443–465.

Miller, T. D., & Schroth, M. N. (1972). Monitoring the epiphytic population of *Erwinia amylovora*. *Phytopathology*, *62*, 1175–1182.

Mizoi, J., Shinozaki, K., & Yamaguchi-Shinozaki, K. (2012). AP2/ERF family transcription factors in plant abiotic stress responses. *Biochimica et Biophysica Acta*, *1819*, 86–96. https://doi.org/10.1016/j.bbagrm.2011.08.004.

Momol, M. T., Norelli, J. L., Breth, D. I., & Aldwinckle, H. S. (1997). *Rootstock blight of apple caused by Erwinia Amylovora, case studies from New York state. (Abstr.)Phytopathology*, *87*, S67.

Moran, R. E., Peterson, B. J., Fazio, G., & Cline, J. (2018). Genotypic variation in apple rootstock low temperature tolerance during spring and fall. *Journal of the American Society for Horticultural Science*, *143*, 319–332.

Moran, R. E., Sun, Y., Geng, F., Zhang, D., & Fazio, G. (2011). Cold temperature tolerance of trunk and root tissues in one- or two-year-old apple rootstocks. *HortScience*, *46*, 1460–1464. https://doi.org/10.21273/hortsci.46.11.1460.

Neuner, G., Kreische, B., Kaplenig, D., Monitzer, K., & Miller, R. (2019). Deep supercooling enabled by surface impregnation with lipophilic substances explains the survival of overwintering buds at extreme freezing. *Plant Cell Environment*, *42*, (2065–2074. https://doi.org/10.1111/pce.13545.

Nitsch, J.-P. (1957). Photoperiodism in woody plants. *Proceedings, American Society for Horticultural Science*, *70*, 526–544.

Norelli, J. L., Aldwinckle, H. S., & Beer, S. V. (1984). Differential host x pathogen interactions among cultivars of apple and strains of *Erwinia amylovora*. *Phytopathology*, *74*, 136–139.

Norelli, J. L., Aldwinckle, H. S., & Beer, S. V. (1986). Differential susceptibility of *Malus* spp. cultivars Robusta 5, Novole, and Ottawa 523 to Erwinia amylovora. *Plant Disease*, *70*, 1017–1019.

Norelli, J. L., Aldwinckle, H. S., Beer, S. V, and Lamb, R. C. (1987). The effects of virulence of *Erwinia amylovora* on the evaluation of fire blight resistance in Malus. *Phytopathology*, *77*, 1551–1555.

Norelli, J. L., Holleran, H. T., Johnson, W. C., Robinson, T. L., & Aldwinckle, H. S. (2003). Resistance of Geneva and other apple rootstocks to Erwinia amylovora. *Plant Disease*, *87*, 26–32.

Osakabe, Y., Liang, Z., Ren, C., Nishitani, C., Osakabe, K., Wada, M., Komori, S., Malnoy, M., Velasco, R., & Poli, M. (2018). CRISPR–Cas9-mediated genome editing in apple and grapevine. *Nature Protocol*, *13*, 2844–2863.

Palonen, P., & Buszard, D. (1997). Current state of cold hardiness research on fruit crops. *Canadian Journal of Plant Science*, *77*, 399–420. https://doi.org/10.4141/P96-013.

Palonen, P., Lettojärvi, I., Luoranen, J., Ruhanen, H., Rantanen, M., Haikonen, T., & Finni, S. (2021). Deacclimation and reacclimation of apple (*Malus × domestica* Borkh.), blackcurrant (*Ribes nigrum* L.) and raspberry (*Rubus idaeus* L.) shoots and buds under controlled conditions. *Scientia Horticulturae (Amsterdam)*. *289*, 110430. https://doi.org/10.1016/j.scienta.2021.110430.

Parravicini, G., Gessler, C., Denance, C., Lasserre-Zuber, P., Vergne, E., BRISSET, M., Patocchi, A., DUREL, C., & Broggini, G. A. L. (2011). Identification of serine/threonine kinase and nucleotide-binding site–leucine-rich repeat (NBS-LRR) genes in the fire blight resistance quantitative trait locus of apple cultivar 'Evereste.' *Molecular Plant Pathology*, *12*, 493–505.

Peil, A., Bus, V. G. M., Geider, K., Richter, K., Flachowsky, H., & Hanke, M.-V. (2009). Improvement of fire blight resistance in apple and pear. *International Journal of Plant Breeding*, *3*, 1–27.

Peil, A., Garcia-Libreros, T., Richter, K., Trognitz, F. C., Trognitz, B., Hanke, M., & Flachowsky, H. (2007a). Strong evidence for a fire blight resistance gene of *Malus robusta* located on linkage group 3. *Plant Breeding*, *126*, 470–475.

Peil, A., Hanke, M., Flachowsky, H., Richter, K., Garcia, T., & Trognitz, B. (2007b). Developing molecular markers for marker assisted selection of fire blight resistant apple seedlings. *Acta Horticulture*, *763*, 117–121.

Peil, A., Wöhner, T., Hanke, M.-V., Flachowsky, H., Richter, K., Wensing, A., Emeriewen, O., Malnoy, M., LeRoux, P.-M., Patocchi, A., & Kilian, A. (2014). Comparative mapping of fire blight resistance in Malus. *Acta Horticulture*, 47–51. https://doi.org/10.17660/ActaHortic.2014.1056.4.

Pester, D., Milčevičová, R., Schaffer, J., Wilhelm, E., & Blümel, S. (2012). *Erwinia amylovora* expresses fast and simultaneously hrp/dsp virulence genes during flower infection on apple trees. *PLoS One*, *7*, e32583.

Philion, V. (2014). Le feu bactérien: Biologie. *Fiche* 104. Institut de recherche et de développement en agroenvironnement.

Pompili, V., Dalla Costa, L., Piazza, S., Pindo, M., & Malnoy, M. (2020). Reduced fire blight susceptibility in apple cultivars using a high-efficiency CRISPR/Cas9-FLP/FRT-based gene editing system. *Plant Biotechnology Journal*, *18*, 845–858.

Privé, J.-P., & Embree, C. (1997). Freezing media affect cooling rate and regrowth of KSC 28 apple rootstock. *Canadian Journal of Plant Science*, *77*, 469–471.

Proebsting, E. L., & Mills, H. H. (1978). Low temperature resistance of developing flower buds of six deciduous fruit Species1. *Journal of the American Society for Horticultural Science*, *103*, 192–198.

Pusey, P. L. (2000). The role of water in epiphytic colonization and infection of pomaceous flowers by *Palonen, P., Lettojärvi. Phytopathology*, *90*, 1352–1357.

Quamme, H. A. (1976). Relationship of the low temperature exotherm to apple and pear production in North America. *Canadian Journal of Plant Science*, *56*, 493–500. https://doi.org/10.4141/cjps76–081.

Quamme, H. A., & Hampson, C. R. (2004). Winter hardiness measurements on 15 new apple cultivars. *Journal of The American Pomological Society*, *58*, 98–107.

Quamme, H., Stushnoff, C., & Weiser, C. J. (1972). The relationship of exotherms to cold injury in apple stem tissues. *Journal of the American Society for Horticultural Science*, *97*, 608–613. https://doi.org/10.21273/jashs.97.5.608.

Rademacher, W. (2000). Growth retardants: Effects on gibberellin biosynthesis and other metabolic pathways. *Annuual Review of Plant Biology*, *51*, 501–531.

Rasmussen, D. H., & MacKenzie, A. P. (1972). Effect of solute on ice-solution interfacial free energy. In H. H. G. Jellinek (Ed.), *Calculation from measured homogeneous nucleation temperatures BT - water structure at the water-polymer interface* (pp. 126–145). Springer.

Raymundo, A. K., & Ries, S. M. (1980). Chemotaxis of Erwinia amylovora. *Phytopathology 70*, 1066–1069.

Reddy, M. V. B., Norelli, J. L., & Aldwinckle, H. S. (2000). Control of fire blight infection of apple blossoms, 1999. Fungic. *Nematic Tests*, *55*, 22.

Richardson, E. A., Seeley, S. D., & Walker, D. R. (1974). A model for estimating the completion of rest for 'redhaven'and 'elberta'peach trees1. *HortScience*, *9*, 331–332.

Rohde, A., & Bhalerao, R. P. (2007). Plant dormancy in the perennial context. *Trends Plant Science*, *12*, 217–223.

Russo, N. L., Robinson, T. L., Fazio, G., & Aldwinckle, H. S. (2008). Fire blight resistance of Budagovsky 9 apple rootstock. *Plant Disease*, *92*, 385–391.

Saito, T., Bai, S., Ito, A., Sakamoto, D., Saito, T., Ubi, B. E., Imai, T., & Moriguchi, T. (2013). Expression and genomic structure of the dormancy-associated MADS box genes MADS13 in Japanese pears (*Pyrus pyrifolia Nakai*) that differ in their chilling requirement for endodormancy release. *Tree Physiology*, *33*, 654–667.

Sakai, A., & Larcher, W. (1987). *Frost survival of plants - responses and adaptation to freezing stress*. Springer.

Santander, R. D., & Biosca, E. G. (2017). *Erwinia amylovora* psychrotrophic adaptations: Evidence of pathogenic potential and survival at temperate and low environmental temperatures. *Peer Journal*, *5*, e3931.

Santander, R. D., Oliver, J. D., & Biosca, E. G. (2014). Cellular, physiological, and molecular adaptive responses of Erwinia amylovora to starvation. *FEMS Microbiology and Ecology*, *88*, 258–271.

Sasaki, R., Yamane, H., Ooka, T., Jotatsu, H., Kitamura, Y., Akagi, T., & Tao, R. (2011). Functional and expressional analyses of PmDAM genes associated with endodormancy in Japanese apricot. *Plant Physiology*, *157*, 485–497.

Schouten, H. J., Krens, F. A., & Jacobsen, E. (2006). Cisgenic plants are similar to traditionally bred plants: International regulations for genetically modified organisms should be altered to exempt cisgenesis. *EMBO Reports*, *7*, 750–753.

Schroth, M. N., Thomson, S. V, Hildebrand, D. C., & Moller, W. J. (1974). Epidemiology and control of fire blight. *Annual Review of Phytopathology*, *12*, 389–412.

SECB. (2016). Statement on the experimental release of cisgenic apple trees with improved resistance towards fire blight. https://www.efbs.admin.ch/en/statements/field-trials/cisgenic-apple-application

Shaltout, A. D., & Unrath, C. R. (1983). Rest completion prediction model for 'Starkrimson Delicious' apples. *Journal of the American Society for Horticultural Science*, *108*, 957–961.

Shrestha, R., Lee, S.-H., Hur, J.-H., & Lim, C.-K. (2005). The effects of temperature, pH, and bactericides on the growth of *Erwinia pyrifoliae* and *Erwinia amylovora*. *Plant Pathology Journal*, *21*, 127–131. https://doi.org/10.5423/PPJ.2005.21.2.127.

Silva, K. J. P., Singh, J., Bednarek, R., Fei, Z., & Khan, A. (2019). Differential gene regulatory pathways and co-expression networks associated with fire blight infection in apple (*Malus × domestica*). *Horticultural Research*, *6*. https://doi.org/10.1038/s41438-019-0120-z.

Siminovitch, D., & Scarth, G. W. (1938). A study of the mechanism of frost injury to plants. *Canadian Journal of Research*, *16c*, 467–481. https://doi.org/10.1139/cjr38c-043.

Slack, S. M., Schachterle, J. K., Sweeney, E. M., Kharadi, R. R., Peng, J., Botti-Marino, M., Bardaji, L., Pochubay, E. A., & Sundin, G. W. (2022). In-orchard population dynamics of *Erwinia amylovora* on apple flower stigmas. *Phytopathology*, *112*, 1214–1225.

Smith, T. J. (1998). Report on the development and use of cougarblight 98c-a situation-specific fire blight risk assessment model for apple and pear. In *VIII International Workshop on Fire Blight* (Vol. 489, pp. 429–436). Kuşadası, Turkey.

Smithberg, M. H., & Weiser, C. J. (1968). Patterns of variation among climatic races of red-osier dogwood. *Ecology*, *49*, 495–505.

Sobiczewski, P., Iakimova, E. T., Mikiciński, A., Węgrzynowicz-Lesiak, E., & Dyki, B. (2017). Necrotrophic behaviour of *Erwinia amylovora* in apple and tobacco leaf tissue. *Plant Pathology*, *66*, 842–855.

Steiner, P. W. (1989). Predicting apple blossom infections by *Erwinia amylovora* using the MARYBLYT model. In *V International Workshop on Fire Blight* (Vol. 273, pp. 139–148). Zürich (Switzerland).

Steiner, P. W., & Lightner, G. W. (1992). *MARYBLYT, a predictive program for forecasting fire blight disease in apples and pears*. University of Maryland.

Stockwell, V. O., Johnson, K. B., & Loper, J. E. (1996). Compatibility of bacterial antagonists of *Erwinia amylovora* with antibiotics used to control fire blight. *Phytopathology*, *86*, 834–840.

Takahashi, D., Willick, I. R., Kasuga, J., & Livingston, D. P. (2021). Responses of the plant cell wall to sub-zero temperatures: A brief update. *Plant Cell Physiology*, *62*, 1858–1866. https://doi.org/10.1093/pcp/pcab103.

Tanino, K. (2004). Hormones and endodormancy induction in woody plants. *Journal of Crop Improvment*, *10*, 157–199. https://doi.org/10.1300/J411v10n01_08.

Tanino, K. K., Kalcsits, L., Silim, S., Kendall, E., & Gray, G. R. (2010). Temperature-driven plasticity in growth cessation and dormancy development in deciduous woody plants: A working hypothesis suggesting how molecular and cellular function is affected by temperature during dormancy induction. *Plant Molecular Biology*, *73*, 49–65. https://doi.org/10.1007/s11103-010-9610-y.

Thomson, S. V. (1986). The role of the stigma in fire blight infections. *Phytopathology*, *76*, 476–482.

Travis, J. W., Rytter, J. L., & Hickey, K. D. (1998). The susceptibility of apple rootstocks and cultivars to *Erwinia amylovora*. In *VIII International Workshop on Fire Blight* (Vol. 489, pp. 235–242. Kuşadası, Turkey.

Uemura, M., Tominaga, Y., Nakagawara, C., Shigematsu, S., Minami, A., & Kawamura, Y. (2006). Responses of the plasma membrane to low temperatures. *Physiologia Plantarum*, *126*, 81–89. https://doi.org/10.1111/j.1399–3054.2005.00594.x.

Unterberger, C., Brunner, L., Nabernegg, S., Steininger, K. W., Steiner, A. K., Stabentheiner, E., Monschein, S., & Truhetz, H. (2018). Spring frost risk for regional apple production under a warmer climate. *PLoS One*, *13*, e0200201.

Van Buskirk, H. A., & Thomashow, M. F. (2006). Arabidopsis transcription factors regulating cold acclimation. *Physiologia Plantarum*, *126*, 72–80. https://doi.org/10.1111/j.1399–3054.2006.00625.x.

Van Der Zwet, T., & Beer, S. (1999). Fire blight--its nature, prevention, and control: A practical guide to integrated disease management. U.S. Department of Agriculture, Agriculture Information Bulletin No. 631, p. 97. http://dx.doi.org/10.22004/ag.econ.309584

Van Der Zwet, T., Orolaza-Halbrendt, N., & Zeller, W. (2012). *Fire blight, history, biology, and management*. APS Press.

Vincent, L. A., & Mekis, É. (2006). Changes in daily and extreme temperature and precipitation indices for Canada over the twentieth century. *Atmospheric Ocean*, *44*, 177–193. https://doi.org/10.3137/ao.440205.

Vitasse, Y., Lenz, A., & Körner, C. (2014). The interaction between freezing tolerance and phenology in temperate deciduous trees. *Frontiers in Plant Science*, *5*, 107326.

Vogel, J. T., Zarka, D. G., Van Buskirk, H. A., Fowler, S. G., & Thomashow, M. F. (2005). Roles of the CBF2 and ZAT12 transcription factors in configuring the low temperature transcriptome of Arabidopsis. *Plant Journal*, *41*, 195–211. https://doi.org/10.1111/j.1365-313X.2004.02288.x.

Vogt, I., Wöhner, T., Richter, K., Flachowsky, H., Sundin, G. W., Wensing, A., Savory, E. A., Geider, K., Day, B., & Hanke, M. (2013). Gene-for-gene relationship in the host–pathogen system *Malus* × *robusta* 5–*E rwinia amylovora*. *New Phytologist*, *197*, 1262–1275.

Volz, R. K., Rikkerink, E., Austin, P., Lawrence, T., & Bus, V. G. M. (2009). "Fast-breeding" in apple: A strategy to accelerate introgression of new traits into elite germplasm. *Acta Horticulture*, *814*, 163–168.

Wang, D., Kasuga, J., Kuwabara, C., Endoh, K., Fukushi, Y., Fujikawa, S., & Arakawa, K. (2012). Presence of supercooling-facilitating (anti-ice nucleation) hydrolyzable tannins in deep supercooling xylem parenchyma cells in Cercidiphyllum japonicum. *Planta*, *235*, 747–759. https://doi.org/10.1007/s00425-011-1536-3.

Wang, F., Chen, X., Dong, S., Jiang, X., Wang, L., Yu, J., & Zhou, Y. (2020). Crosstalk of PIF4 and DELLA modulates CBF transcript and hormone homeostasis in cold response in tomato. *Plant Biotechnology Journal*, *18*, 1041–1055.

Wareing, P. F. (1956). Photoperiodism in woody plants. *Annual Review of Plant Physiology*, *7*, 191–214.

Webster, A. D., & Wertheim, S. (2003). Apple rootstocks. In D. C. Ferree & I. J. Warrington (Eds.), *Apples, botany, production, and uses* (pp. 91–124). CABI.

Weiser, C. J. (1970). Cold resistance and injury in woody plants. *Science*, *169*, 1269–1278. https://www.jstor.org/stable/1729366.

Welling, A., & Palva, E. T. (2008). Involvement of CBF transcription factors in winter hardiness in birch. *Plant Physiology*, *147*, 1199–1211. https://doi.org/10.1104/pp.108.117812.

Wildung, K., Weiser, I. C. J., & Pellett, H. M. (1973). Temperature and moisture effects on hardening of apple roots. *HortScience*, *8*, 53–55.

Willick, I. R., & Lowry, D. B. (2022). Cold acclimation threshold induction temperatures of switchgrass ecotypes grown under a long and short photoperiod. *Physiologia Plantarum*, *174*, e13812.

Wilson, M., Epton, H. A. S., & Sigee, D. C. (1989). *Erwinia amylovora* Infection of Hawthorn Blossom II: The Stigma. *Journal of Phytopathology*. 127, 15–28.

Winslow, C.-E., Broadhurst, J., Buchanan, R. E., Krumwiede Jr, C., Rogers, L. A., & Smith, G. H. (1920). The families and genera of the bacteria final report of the committee of the society of American bacteriologists on characterization and classification of bacterial types. *Journal of Bacteriology*, *5*, 191–229.

Wisniewski, M., Gusta, L., & Neuner, G. (2014a). Adaptive mechanisms of freeze avoidance in plants: A brief update. *Environmental and Experimental Botany*, *99*, 133–140. https://doi.org/10.1016/j.envexpbot.2013.11.011.

Wisniewski, M., Nassuth, A., & Arora, R. (2018). Cold hardiness in trees: A mini-review. *Frontiers in Plant Science*, *9*, 1394.

Wisniewski, M., Nassuth, A., Teulières, C., Marque, C., Rowland, J., Cao, P. B., & Brown, A. (2014b). Genomics of cold hardiness in woody plants. *Critical Reviews in Plant Sciences*, *33*, 92–124. https://doi.org/10.1080/07352689.2014.870408.

Wisniewski, M., Norelli, J., & Artlip, T. (2015). Overexpression of a peach CBF gene in apple: A model for understanding the integration of growth, dormancy, and cold hardiness in woody plants. *Frontiers in Plant Science*, *6*, 129442.

Wisniewski, M., Norelli, J., Bassett, C., Artlip, T., & Macarisin, D. (2011). Ectopic expression of a novel peach (*Prunus persica*) CBF transcription factor in apple (*Malus* × *domestica*) results in short-day induced dormancy and increased cold hardiness. *Planta*, *233*, 971–983. https://doi.org/10.1007/s00425-011-1358-3.

Wisniewski, M., Willick, I. R., Duman, J. G., Livingston, D., & Newton, S. S. (2020). Plant antifreeze proteins. In Ramløv, H., & Friis, D. S. (Eds.), *Antifreeze proteins*, environment, systematics and evolution (Vol. 1, pp. 189–226). Springer.

Wöhner, T. W., Richter, K., Sundin, G. W., Zhao, Y., Stockwell, V. O., Sellmann, J., Flachowsky, H., Hanke, M., & Peil, A. (2018). Inoculation of *Malus* genotypes with a set of *Erwinia amylovora* strains indicates a gene-for-gene relationship between the effector gene eop1 and both *Malus floribunda* 821 and *Malus* 'Evereste.' *Plant Pathology*, *67*, 938–947.

Woldendorp, G., Hill, M. J., Doran, R., & Ball, M. C. (2008). Frost in a future climate: Modelling interactive effects of warmer temperatures and rising atmospheric [CO_2] on the incidence and severity of frost damage in a temperate evergreen (*Eucalyptus pauciflora*). *Global Change Biology*, *14*, 294–308. https://doi.org/10.1111/j.1365-2486.2007.01499.x.

Wolfe, D. W., DeGaetano, A. T., Peck, G. M., Carey, M., Ziska, L. H., Lea-Cox, J., Kemanian, A. R., Hoffmann, M. P., & Hollinger, D. Y. (2018). Unique challenges and opportunities for northeastern US crop production in a changing climate. *Climate Change*, *146*, 231–245.

Wolfe, D. W., Schwartz, M. D., Lakso, A. N., Otsuki, Y., Pool, R. M., & Shaulis, N. J. (2005). Climate change and shifts in spring phenology of three horticultural woody perennials in northeastern USA. *International Journal of Biometeorology*, *49*, 303–309.

Wu, R., Cooney, J., Tomes, S., Rebstock, R., Karunairetnam, S., Allan, A. C., Macknight, R. C., & Varkonyi-Gasic, E. (2021). RNAi-mediated repression of dormancy-related genes results in evergrowing apple trees. *Tree Physiology*, *41*, 1510–1523.

Wu, R., Tomes, S., Karunairetnam, S., Tustin, S. D., Allan, A. C., Macknight, R. C., & Varkonyi-Gasic, E. (2017). SVP-like MADS box genes control dormancy and budbreak in apple. *Frontiers in Plant Science*, *8*, 256625.

Yamaguchi-Shinozaki, K., & Shinozaki, K. (1994). A novel cis-acting element in an arabidopsis gene is involved in responsiveness to drought, low-temperature, or high-salt stress. *Plant Cell*, *6*, 251–264. https://doi.org/10.2307/3869643.

Yoder, K. S., Miller, S. S., & Byers, R. E. (1999). Suppression of fire blight in apple shoots by prohexadione-calcium following experimental and natural inoculation. *HortScience*, *34*, 1202–1204.

Yordanov, Y. S., Ma, C., Strauss, S. H., & Busov, V. B. (2014). EARLY BUD-BREAK 1 (EBB1) is a regulator of release from seasonal dormancy in poplar trees. *Proceedings of the National Academy of Sciences*, *111*, 10001–10006.

Zhao, K., Zhou, Y., Li, Y., Zhuo, X., Ahmad, S., Han, Y., Yong, X., & Zhang, Q. (2018). Crosstalk of PmCBFs and PmDAMs based on the changes of phytohormones under seasonal cold stress in the stem of *Prunus mume*. *International Journal of Molecular Sciences*, 19, 15.

5 Recent Advancements in Genetic Engineering Techniques in Disease Resistance of Crops

An Outlook

Licon Kumar Acharya, Suryakant Manik, Madhuri Pradhan, Ajanta Birah, M. K. Khokhar, Debjyoti Majumder, and Akbar Hossain

5.1 INTRODUCTION

Plant diseases have posed a substantial threat to food security since the beginning of agriculture, despite our best attempts to combat them. Pathogen attacks have recently grown in severity, and they are expected to worsen as weather patterns and environmental factors become more variable as a result of global climate change. The situation worsens as the global food trade facilitates the transfer of established and emerging pathogens (Bebber et al., 2014). The average global loss due to diseases and pests is estimated to be 11–30% or higher. Furthermore, crop losses are higher in places that are already experiencing food scarcity (Savary et al., 2019). Taking into account the growing worldwide population and the resulting requirement for increased world food supply, combating plant diseases has become an ethical imperative that necessitates immediate and effective action.

Chemical pesticides are predominantly used to tackle plant disease. However, some pesticides are rapidly losing efficiency as a result of pathogen evolution, and their use faces more stringent restrictions to minimize undesirable consequences (Berger et al., 2017). The best defense against pathogen invasion is plant disease resistance, and traditional crop breeding techniques have been crucial in creating resistant lines to specific diseases. Yet, the most difficult task is to maintain the robust agronomic qualities of elite cultivars while concurrently selecting for genetic resistance against several diseases (Van Esse et al., 2020). Conversely, the introduction of a single resistance through classical breeding facilitates pathogen adaptation to it. The fact that new disease races occasionally appear adds to the complexity of the issue, which needs to be resolved to maintain the stability of the global food supply. When conventional breeding and other management strategies fall short of options, genetic engineering (GE) comes to the forefront, which refers to the alteration of the genetic composition of plants, thereby enhancing resistance. When compared to traditional breeding, GE produces crops with desired agronomic features faster by altering just the desired gene of interest without changing the crop's entire genome. Furthermore, it permits the exchange of genetic material between genera and species, expanding the area of resistance source. This technology is highly beneficial for novel gene introduction into crops grown through vegetative propagation unlike the conventional breeding approach (Dong and Ronald, 2019). GE leads to sustainability by reducing farmers' dependence on chemical pesticides. The deployment of Cry proteins and the production of transgenic crops (Bt cotton, Bt brinjal, etc.) to manage damaging insects sets a great example in this regard (Klumper and Qaim, 2014; Smyth et al., 2015). Plant GE can be broadly

DOI: 10.1201/9781003543237-5

categorized into two categories: traditional transgenic methods and genome editing technologies. In the conventional transgenic method, plants undergo plant transformation to develop varieties with nonshelf DNA by inserting genes encoding agronomic features at random sites in their genomes (Lorence and Verpoorte, 2004). However, a more recent phenomenon called genome editing makes it possible to alter endogenous plant DNA by deleting, inserting, and replacing different lengths of DNA at specific locations (Barrangou and Doudna, 2016). In this chapter, an inventory of GE-based solutions currently in use against major plant pathogens has been highlighted (Figure 5.1).

Figure 5.1: GE based solutions to develop disease resistance in plants

5.2 DEPLOYMENT OF R GENES

In the field, disease resistance provided by a single R gene typically fails to last. Successful pathogens frequently mutate the matching Avr gene to circumvent detection by host R genes. Stacking is the practice of introducing many R genes at the same time, to increase durability and widen the resistance spectrum (Fuchs, 2017; Mundt, 2018). Since the formation of a pathogen strain that can overcome resistance given by numerous R genes at once is a rare occurrence, resistance conferred by stacked R genes is projected to be long-lasting. Marker-assisted selection-based breeding can accomplish this; however, it takes a long time to combine numerous loci using this method. Molecular stacking is the process of assembling numerous R gene cassettes onto a single plasmid and introducing this R gene cluster en bloc into a single genetic locus by plant transformation, an alternative to gene stacking by means of marker-assisted selection (Que et al., 2010). It simplifies the selection process, as all the *R* genes introduced this way are inherited as a single genetic locus. In crop species grown by vegetative means, where breeding stacks are impractical, molecular stacking is a highly effective means of engineering resistance to broad-spectrum diseases. Using *Agrobacterium*-based transformation of a sensitive potato cultivar, the triple wide-range potato late blight R genes, Rpi-sto1, Rpi-vnt1.1,

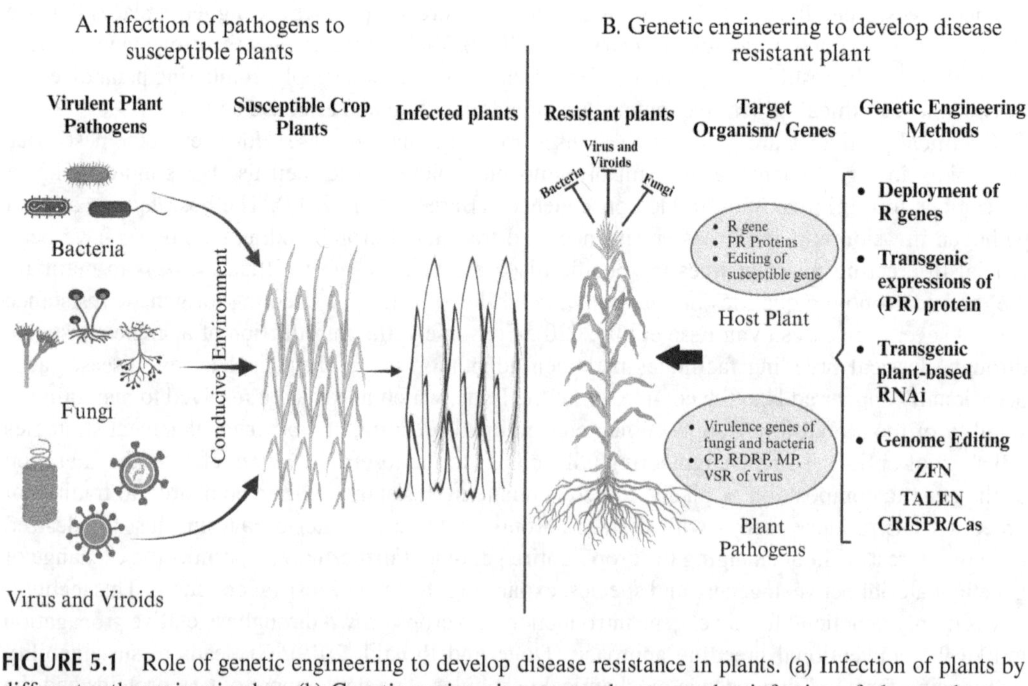

FIGURE 5.1 Role of genetic engineering to develop disease resistance in plants. (a) Infection of plants by different pathogenic microbes. (b) Genetic engineering approaches to combat infection of plant pathogens. CP, coat protein; MP, movement protein; PR, pathogenesis related; RdRP, RNA-dependent RNA polymerase; VSR, viral silencing suppressor.

and Rpi-blb3, were molecularly stacked. Under greenhouse circumstances, the resulting triple gene transformants showed widespread resistance equal to the total amount of the specific resistance imparted by the three different Rpi genes (Zhu et al., 2012). In related work, three distinct potato varieties were transformed using a single DNA fragment containing Rpi-vnt1.1 and Rpi-sto1. Remarkably, the introduction of the two R genes resulted in the R-gene-stacked lines, exhibiting a wide range of late blight resistance (Jo et al., 2014). Later, in the field, resistance to the late blight pathogen was established in both the triple- and double-gene-stacked potatoes by Haverkortet al. (2016). Ghislain et al. (2018) noticed that complete resistance was achieved in the field over multiple seasons by stacking triple resistance (*R*) genes from wild relatives [*RB*, *Rpi-blb2* from *Solanum bulbocastanum* and *Rpi-vnt1.1* from *S. venturii*] into African highland potato types.

5.3 TRANSGENIC EXPRESSIONS OF PATHOGENESIS-RELATED PROTEIN

Low-molecular-weight plant proteins known as pathogenesis-related (PR) proteins have a variety of structural characteristics. They are found in large quantities in plants, but when a pathogen attacks or when plants are under stress, their concentration increases (Van Loon et al., 2007). Currently, reports of 19 families of PR protein in various plants exist. Certain pathogenic fungi and oomycetes can exhibit direct antimicrobial activity through modifications to their structural integrity caused by PR proteins with chitinase and glucanase activity. Furthermore, the buildup of signaling phytohormones such as salicylic acid and the triggering of the hypersensitive response (HR) is strongly linked to the accumulation of PR proteins, which in turn triggers the development of SAR. Transgenic plants using PR genes are now being generated (Table 5.1) as a result of advancements in transformation techniques and the isolation of many PR genes. These plants have demonstrated efficacy in acquiring pathogen resistance in plants (Moosa et al., 2017; Nandi, 2016).

TABLE 5.1
An Illustration of a Transgenic Plant That Overexpresses PR Proteins to Combat Plant Diseases

PR Protein Family	Sources	Target Pathogen	Transgenic Plant	Reference
PR-2 (Glucanase)	Tobacco	*Cercospora arachidicola* *Aspergillus flavus*	Groundnut	Sundareshaet al. (2010)
	Rubber	*Phytophthora palmivora*	Rubber	Sunpao and Pornsuriya (2016)
PR-3 (chitinase)	Bitter gourd	*Magnaporthe grisea* *Rhizoctonia solani*	Rice	Li et al. (2009)
	Rice	*Verticillium dahliae* *Fusarium oxysporum*	Brinjal	Singhet al. (2015)
PR-4 (chitinase)	Rice	Magnaporthe grisea	Rice	Wang et al. (2011)
	Wild Chinese grape	*Uncinulanecator*	Common grapevine	Dai et al. (2016)
PR-5 (Thaumatin-like)	Wheat	*Puccinia triticina*	Wheat	Yanjun et al. (2017)
PR-12 (defensin)	Jatropha	*Rhizoctonia solani*	Tobacco	Wang et al. (2017)
PR-13 (thionin)	Barley	*Ceratocystis fimbriata*	Sweet potato	Muramoto et al. (2012)
PR-14 (lipid transfer protein)	Arabidopsis	*Fusarium graminearum*	Wheat	McLaughlin et al. (2020)
PR-15 (oxalate oxidase)	Wheat	*Sclerotinia sclerotiorum*	Soybean	Yang et al. (2019)

5.4 TRANSGENIC PLANT-BASED RNA INTERFERENCE (RNAi) STRATEGIES

Plants have a built-in defense system called RNAi against non-self nucleic acids. It starts with the sequence-specific identification of mRNAs and ends with the repression of their expression by translation inhibition or ribonuclease-directed destruction. The majority of viruses that cause disease in plants are single-stranded RNA viruses, and they use dsRNA intermediates to replicate their genomes in the cytoplasm of their hosts. In light of this, it is a useful tactic for managing a range of plant viruses by introducing an active RNAi mechanism into the plants. Since structured RNAs like hairpin RNAs and dsRNAs are potent inducers of RNA interference, most plants are now genetically modified (GM) to produce them. The adoption of GM 'Rainbow papaya,' which was created and deployed in Hawaii when no other means of controlling papaya ringspot virus (PRSV) were available, is among the most impressive success tales of RNA interference technology (Gonsalves, 2006). A novel tactic known as host-induced gene silencing (HIGS) is being investigated as a procedure to manage fungi, oomycetes, and nematodes in the wake of RNAi's success with plant viruses. The idea behind HIGS is to transgenically express short interfering RNAs (siRNAs) in the host plant, which the invader then ingests and silences the target pathogen gene. HIGS was initially seen in barley (*Hordeum vulgare*), where the effector gene Avra10 was silenced by the host due to HIGS, which decreased fungal development in the host plant (Nowara et al., 2010). Table 5.2 lists several instances of RNA interference techniques used to create transgenic plants resistant to fungi, viruses, and oomycetes. There are several negative effects of RNAi on the environment and plants. Reduced biodiversity could arise from RNA silencing's potential to alter the host genome and obstruct gene flow across related plants. Furthermore, some plant species have difficulty constructing RNAi. In summary, in situations when disease resistance is absent in wild populations, RNA interference (RNAi) ought to be employed as an additional control.

TABLE 5.2
RNAi Technology for Developing Disease-Resistant Transgenic Plant

Sl. No.	Pathogen	Mechanism	Transgenic Plant	References
1.	Potato virus Y (PVY)	Expressing the dsRNA obtained from the viral coat proteins' 3′-terminal end portion (CP)	Potato	Missiou et al. (2004)
2.	Beet necrotic yellow vein virus (BNYVV)	Targeting the viral coat protein	Tobacco	Andika et al. (2005)
3	Plum pox virus (PPV)	Targeting the viral coat protein	*Prunus domestica* and *benthamiana*	Hily et al. (2007)
4.	Cassava brown streak virus (CBSV) and the cassava brown streak Uganda virus (CBSUV)	Utilizing an RNAi construct made up of 397 nucleotides from the N-terminal and 491 nucleotides from the C-terminal of the virus's coat protein gene	Cassava	Patil et al. (2011)
5.	Tobacco streak virus (TSV)	Utilizing a coat protein gene that is 421 bp long and has sequences for both sense and antisense coat proteins	Tobacco and sunflower	Pradeep et al. (2012)
6.	Rice strip disease caused by the rice stripe virus (RSV)	RNAi construct composed of the CP gene and the disease-specific (SP) gene suppressed in two Japonica types susceptible to RSV	Rice	Zhou et al. (2012)
7.	Soybean mosaic virus (SMV)	Through the introduction of a Hc-Pro gene containing hairpin RNAi construct	Soybean	Kim et al. (2016)
8.	Tobacco streak virus	Utilizing TSV-coat protein-containing hairpin RNA	Peanut (*Arachis hypogaea* L.)	Senthilraja et al. (2018)

(*Continued*)

TABLE 5.2 (*Continued*)
RNAi Technology for Developing Disease-Resistant Transgenic Plant

Sl. No.	Pathogen	Mechanism	Transgenic Plant	References
9.	Rice tungro bacilliform virus (RTBV) and rice tungro spherical virus (RTSV)	By utilizing an RNAi construct with the highly conserved coat protein 3 (CP3) gene	*Oryza sativa*	Malathi et al. (2019)
10.	*Fusarium oxysporum* f. sp. *Cubense* (Foc)	By the expression of si-RNA mediated by intron hairpin RNA (ihp-RNA)	Banana	Ghag et al. (2014)
11.	*Verticillium dahliae*	Host-induced gene silencing (HIGS) of hydrophobin gene (*VdH1*)	Cotton	Zhang et al. (2016)
12.	*Bremia lactucae*	HIGS of cellulose synthase (*CES1*) genes	Lettuce	Govindarajulu et al. (2015)

5.5 GENOME EDITING TECHNOLOGY

The introduction of beneficial genes into crops using a trans-kingdom method has resulted in the evolution of GM crops, which have caused controversy. This is primarily because of fear-based agricultural policies that come from a lack of public knowledge and social and ethical issues. Because of this, researchers are now concentrating on creating genome editing instruments that can alter the genome without requiring the introduction of transgenes. It is feasible to accurately modify the genomic DNA of a cell or organism through genome editing. Sequence-specific nucleases are used as the core of genome editing technology to recognize specific DNA sequences and form double-strand breaks (DSBs) in the DNA at specific locations. DSBs are often repaired via two pathways: the homologous recombination process and the non-homologous end joining (NHEJ) pathway. Cells usually use the NHEJ pathway to fix DSBs. Nevertheless, NHEJ usually is inaccurate and often leads to mutations involving insertions or deletions. When a donor DNA template is available, DSBs are typically repaired through the homologous recombination pathway, leading to accurate base modifications or gene substitution. Currently, genome editing uses three main categories of sequence-specific nucleases. Among these are the CRISPR/Cas methods, transcription activator-like effector nucleases (TALENs), and zinc finger nucleases (ZFNs) (Yin and Qiu, 2019).

5.5.1 Zinc Finger Nuclease

Zinc finger DNA-binding domains and DNA-cleavage domains are combined to make ZFNs, synthetic restriction enzymes. Zinc finger domains can be designed to identify and attach to specific target DNA sequences. ZFNs may target specific regions inside complex genomes thanks to this customization (Ramirez et al., 2008). Two protein domains are present. The zinc finger is present in the DNA-binding domain, the first domain which consists of eukaryotic transcription factors. The nuclease domain, comprising the restriction enzyme *FokI*, is the second domain that catalyzes the breakage of DNA. Each of the three to six zinc finger repeats that make up a ZFN's DNA-binding domain may recognize anywhere from 9 to 18 base pairs. Even a pair of three-finger ZFNs, each of which recognizes a total of 18 base pairs, might theoretically target a single locus in a mammalian genome if these zinc finger domains show perfect specificity for their target site. ZFNs frequently use the cleavage domain of the type II restriction endonuclease *FokI*. To cleave DNA, the cleavage domain must dimerize, which requires a pair of ZFNs to target non-palindromic DNA locations. ZFNs typically join the cleavage domain to the terminus of each zinc finger domain. The two different ZFNs must attach to separate DNA strands with their C-termini spaced apart at a certain distance for the two cleavage domains to connect and cleave DNA (Carroll, D. 2011).

5.5.2 Transcription Activator-Like Effector Nucleases

TALENs is a genome editing technology that accurately modifies DNA sequences in live organisms. TALENs are artificial enzymes formed by combining a DNA-binding domain originating from transcription activator-like effectors (TALEs) with a DNA-cleaving domain, usually a nuclease domain. TALEs are naturally occurring proteins discovered in plant-infecting bacteria, particularly those from the genus *Xanthomonas*. These proteins are characterized by their capacity to bind to specific DNA sequences via a modular repeat structure. Each repeat in a TALE protein recognizes a single base pair in the DNA sequence, allowing for precise targeting. Once attached to their target location, TALENs' DNA-cleaving domain causes DSBs in the DNA (Christian et al., 2010). The DNA repair mechanism of a cell can then mend these breaks, allowing the appropriate genetic alterations to be introduced. TALEN technology has been utilized extensively in a variety of industries, including agriculture, biotechnology, and medicine, for gene editing, knockout, and regulation.

5.5.3 CRISPR/Cas System

Clustered regularly interspaced short palindromic repeats (CRISPR) is a revolutionary technology that allows scientists to accurately alter DNA sequences in live creatures. CRISPR systems evolved from bacterial and archaeal immune systems, serving as a defense mechanism against invading viruses and other foreign genetic material. The CRISPR system consists of two principal elements: a guide RNA (gRNA) and a CRISPR-associated (Cas) protein. The gRNA is designed to specifically match a target DNA sequence in the genome of the organism. The Cas protein, commonly Cas9, functions as a molecular scissor capable of cleaving DNA at the appropriate site. When the gRNA binds to the appropriate target sequence in the genome, it sends the Cas protein to that location. Afterward, the Cas protein slices the DNA, causing a DSB. The use of the cell's innate DNA repair systems enables the implementation of specific genetic alterations at the break position, such as insertions, deletions, or replacements of DNA sequences (Cong et al., 2013).

CRISPR technology has revolutionized genetics and molecular biology because of its simplicity, adaptability, and efficacy. It has enabled researchers to execute precise genome editing in a variety of taxa, including bacteria, plants, and animals, with remarkable ease and speed. CRISPR has numerous uses in fundamental research, biotechnology, agriculture, and medicine. However, its possible ethical and societal ramifications are hotly debated and scrutinized.

5.6 APPLICATION OF GENOME EDITING TOOLS IN THE DEVELOPMENT OF PLANT DISEASE RESISTANCE

Plant diseases caused by living organisms, including viruses, fungi, oomycetes, and bacteria, seriously jeopardize agricultural output and food security around the globe. Certain diseases cause significant losses in yield as well as the quality of field produce, fruits, and other plant parts that can be consumed. Traditionally, resistance to plant diseases has been established by introducing resistance genes (R genes) derived from wild relatives of cultivated crops. One typical resistance mechanism involves nucleotide-binding leucine-rich repeat (NB-LRR) proteins, which recognize particular pathogen avirulence (Avr) gene products. This detection causes a defensive response, which culminates in programmed cell death of the host cell, referred to as the hypersensitive reaction . Several traditional approaches have been established for producing disease-resistant plants. These include the pure line approach, which is designed for self-pollinating crops, the pedigree method, which is used to generate disease-resistant plants based on major genes, the backcross method, recurrent selection, and the development of interspecific hybrids. Moreover, disease-resistant plants have been effectively developed through mutation breeding. However, these old methods take a long period, and sometimes genes affecting plant development are linked to resistance alleles.

Targeting susceptibility (S) genes, which are essential for pathogen infection, is an additional strategy for achieving disease resistance. For many diseases to spread and proliferate, specific host genes are necessary. These host genes, also referred to as susceptibility genes or S genes, are necessary for pathogen identification, invasion, penetration, defense-evasion, and fulfilling the metabolic and structural requirements of the infection.

Mutations in S genes can provide wide-range, long-lasting recessive heritable resistance. Thus, S genes play a significant role in plant–pathogen interactions, regulating the host's susceptibility to infection. Three primary molecular mechanisms allow S genes to function: negative regulation of plant immune signals, persistent compatibility, and basic compatibility. Negative regulators of plant disease resistance, also known as host susceptibility (S) genes, are potential candidates for genome editing. Genome editing involves knowledge of the host genome sequences and molecular information about the target gene. Fortunately, components of plant innate immunity have been disclosed by extensive genetic and molecular research, and a rising number of plant species have been completely sequenced. This has provided a variety for dealing with a wide range of pest and disease preventive and control issues.

5.6.1 Genome Editing for Resistance against Phytopathogenic Bacteria

Bacterial infections provide significant challenges, particularly once outbreaks have commenced, due to the wide variety, rapid proliferation, and multimodality of bacterial pathogens. Nonetheless, advances in understanding the molecular underpinnings of host–bacterial interactions have resulted in the discovery of multiple host plant genes, including the susceptibility (S) genes connected to these complicated pathways.

Rice bacterial blight, caused by *Xanthomonas oryzae pv. oryzae* (*Xoo*), is a severe vascular bundle disease that kills rice plants. *Xoo* uses the type III secretion system to deliver TALE proteins to host cells. The TALE protein AvrXa7 from the Philippine strain PXO86 of *Xoo* interacts with the effector-binding element (EBE) in the promoter of *OsSWEET14*, while another TALE protein, PthXo2, targets the promoter of *OsSWEET13*, both boosting the virulence of particular *Xoo* strains. To counteract this, a technique involving two TALENs targeting EBEs in the *OsSWEET14* promoter was developed. This strategy attempted to impair *AvrXa7* binding while retaining *OsSWEET14*'s developmental function.

One more investigation used CRISPR/Cas9 to target the coding region and EBE of the CsLOB1 gene, an S gene implicated in the recognition of the effector PthA4 generated by the citrus canker-causing bacterium *Xanthomonas citri* sub sp. *citri*. Both studies found that *Xcc* resistance was conferred via CsLOB1 editing. Using the CRISPR/Cas9 system, null mutants were produced for the SlDMR6-1 gene, which is upregulated after *Pseudomonas syringae* infection in tomato plants. These mutants were resistant not only to *P. syringae* but also to the fungus *Phytophthora capsici* and several *Xanthomonas* species. Surprisingly, this resistance was achieved with no negative impact on tomato plant growth or development.

5.6.2 Genome Editing for Resistance against Phytopathogenic Fungi

Fungal infections pose a huge hazard to agriculture because they are the principal source of plant diseases that affect a wide range of crops around the world. Their capacity to adapt quickly, invade new hosts, overcome resistance conferred by R genes, and gain resistance to fungicides pose a significant challenge to disease control. To address this issue, genome editing has emerged as a promising strategy for modifying host susceptibility (S) genes.

Wheat Mlo genes have been altered using both TALEN and CRISPR technologies, and edited plants show resistance to the powdery mildew fungus *Blumeria graminis* f. sp. *tritici* (Bgt) when all six copies of TaMlo are mutated at the same time. This demonstrates the redundancy of TaMlo copies and the efficiency of genome editing in polyploid genomes. Similarly, targeted editing of

OsERF922 using CRISPR/Cas9 has resulted in the production of rice *oserf922* knockout mutants that demonstrate increased resistance to rice blast disease while leaving other significant agronomic features unchanged. In tomato, Mlo knockout mutants were created by targeting SlMlo1 with two sgRNAs at once. As expected, these tomato mutants are completely resistant to the tomato powdery mildew fungus, *Oidium neolycopersici*. This demonstrates the value of genome editing as a strategy for improving resistance to fungal diseases in a variety of crops.

5.6.3 Genome Editing for Resistance against Phytopathogenic Virus

Plant viral infections provide a substantial management problem due to viruses' fast development and interaction with insect vectors. Over the past 30 years, transgenic production of virus proteins or RNAs has been a popular method for increasing plant virus resistance. This approach, known as pathogen-derived resistance, has been widely used to attenuate the effects of viral infections on plants. Genome engineering methods like as ZFN and TALEN have seen little usage in the generation of virus-resistant plants.

Engineered tools, such as the non-nuclease artificial zinc finger protein (AZP), have been developed to target specific sections of viral genomes for crop enhancement. The AZP is designed to inhibit the replication origin of the beet severe curly top virus. Furthermore, ZFNs have been designed to target conserved areas within the Rep gene of tomato yellow leaf curl China virus (TYLCCNV) and tobacco curly shoot virus (TbCSV). These methods are critical in the production of transgenic crops that are more resistant to these viral diseases. TALENs have been used to create resistance in transgenic tobacco plants to TbCSV, TYLCCNV, and tomato leaf curl Yunnan virus (TLCYnV). This was accomplished by precisely targeting the Rep genes of these viruses, hence increasing the resistance capacity of transgenic tobacco plants. Currently, the growing of non-transgenic crops with these two methods is limited. However, genome editing tools such as the CRISPR/Cas system are increasingly being used to produce virus-resistant plants.

The majority of studies investigating CRISPR-mediated pathogen resistance in plants have primarily focused on conferring resistance against viruses. These studies have highlighted two primary strategies for engineering virus resistance:

1. Direct targeting of the virus genome
    1. Genome editing of plant DNA virus
    2. Genome editing of plant RNA virus
2. Targeting plant Susceptible (S) genes

5.6.3.1 Direct Targeting of the Virus Genome

5.6.3.1.1 Genome Editing of Plant DNA Virus

The earliest endeavors in engineered virus resistance employed the strategy of specifically attacking the viral genome in the plant cells. Introducing guide RNAs (gRNAs) aimed at the intergenic region (IR), coat protein (CP), or Rep genes in Cas9-expressing tobacco led to decreased accumulation of various significant DNA viruses, including tomato yellow leaf curl virus (TYLCV). Furthermore, these investigations unveiled the potential of a single gRNA targeting the conserved sequence (TAATATTAC) in the IR region to simultaneously aim for many Gemini viruses. A similar strategy was employed to target the endogenous banana streak virus (eBSV) within the genome of plantain banana (*Musa* spp.). Remarkably, 75% of the edited occurrences resulted in asymptomatic plants in comparison to their unaltered versions.

5.6.3.1.2 Genome Editing of Plant RNA Virus

Direct aiming of virus genomes through CRISPR offers a way to additionally build resistance against RNA viruses as well. This strategy involves targeting the viral RNA in place of DNA, utilizing Cas9 RNA-binding variants rather than DNA-binding Cas9. Zhang et al. (2018) developed *Nicotiana benthamiana* and *Arabidopsis* plants expressing FnCas9, targeting *cucumber mosaic virus* and *tobacco mosaic virus* resulting in a notable decrease in virus accumulation and symptom development. Similarly, in two separate studies, Aman et al. (2018) employed an RNAtargeting LwaCas13a (formerly known as C2c2) to obstruct the RNA genome of turnip mosaic virus (TuMV), demonstrating virus RNA aiming both *N. benthamiana* and *Arabidopsis*. The remarkable efficiency of Cas13 for targeting and interfering with RNA viruses in plants could potentially be leveraged to confer immunity against economically significant plant-infecting viroids.

5.6.3.2 Targeting Plant Susceptible (S) Genes

CRISPR-mediated targeting of host susceptibility (S) genes overcomes the difficulties connected to direct targeting of the viral genome. In this technique, the guide RNA (sgRNA) targets the plant gene rather than the viral genome, which is susceptible to evasion due to its high copy number and recombination rate. Among the S genes, eIF4E is extensively targeted in CRISPR-mediated virus resistance engineering. Since viruses don't have ribosomes in their virions, viral protein synthesis is strongly reliant on the host translation machinery. In plants, eIF4E and its isoforms are important translation initiation factors that recruit ribosomes to mRNA untranslated regions (UTRs), allowing viral proteins to be translated more efficiently.

In *Arabidopsis* and cucumber, targeting *eIF4E* genes with CRISPR has resulted in effective resistance to numerous viruses, including TuMV, *cucumber vein yellowing virus*, *zucchini yellow mosaic virus*, and *papaya ring spot mosaic virus*-W. Subsequent research targeting eIF4E isoforms in different crop species has resulted in broad-spectrum viral resistance across multiple hosts.

Furthermore, a biomimetic technique was used to generate CRISPR-mediated changes in Arabidopsis eIF4E1, aligning the sequence with the virus resistance allele observed in *Pisum sativum*. This technique effectively provided resistance to *clover yellow vein virus*. Similarly, Macovei et al. (2018) developed resistance in rice-to-rice tungro spherical virus via biomimicking of eIF4G alleles (Table 5.3).

TABLE 5.3
Advances in Genome Editing for Disease Resistance Empowering Plant Immunity

Disease	Pathogen	Host Crop	Target Gene	Endonuclease	Remarks	References
A. Genome Editing for Resistance against Phytobacterial Pathogen						
Rice bacterial blight	*Xanthomonas oryzae* pv. *oryzae* (*Xoo*)	Rice	OsSWEET14/ promoter	TALEN	Reduced virulence of the pathogen	Oliva et al. (2019)
Rice bacterial blight	*Xanthomonas oryzae* pv. *oryzae* (*Xoo*)	Rice	*OsSWEET13/* exon	CRISPR/Cas9	Reduced virulence of the pathogen	Zhou et al. (2015)
Citrus canker	*Xanthomonas citri* subsp. *citri*	Citrus	CsLOB1/exon	CRISPR/Cas9	Enhanced resistance against pathogen	Jia et al. (2017)
Citrus canker	Xanthomonas citri subsp. citri	Citrus	CsLOB1/ promoter	CRISPR/Cas9	Enhanced resistance against pathogen	Peng et al. (2017)
Bacterial leaf spot	*Pseudomonas syringae* pv. *tomato*	Tomato	*SlDMR6-1*	CRISPR/Cas9	Null mutants of the gene showed resistance to the bacteria	Thomazella et al. (2016)

(*Continued*)

TABLE 5.3 (*Continued*)
Advances in Genome Editing for Disease Resistance Empowering Plant Immunity

Disease	Pathogen	Host Crop	Target Gene	Endonuclease	Remarks	References
		B. Genome Editing for Resistance against Phytopathogenic Fungi				
Powdery mildew	*Blumeria graminis* f. sp. *tritici*	Wheat	*TaMLO*/exon	TALEN	Resistance to Powdery mildew	Wang et al. (2014)
Powdery mildew	*Blumeria graminis* f. sp. *tritici*	Wheat	*TaEDR1*/exon	CRISPR/Cas9	Resistance to powdery mildew pathogen	Zhang et al. (2017)
Rice blast	*Magnaporthe oryzar*	Rice	*OsERF922*/ exon	CRISPR/Cas9	Resistant to rice blast disease	Wang et al. (2016)
Powdery mildew of tomato	*Oidium neolycopersici*	Tomato	*SlMlo1*/exon	CRISPR/Cas9	Resistant to powdery mildew	Nekrasov et al. (2017)
		C. Genome Editing for Resistance against Phytopathogenic Virus				
DNA viral disease	BSCTV	*Arabidopsis*	Replication origin	AZP	Suppression of viral infection	Sera (2005)
DNA viral disease	TYLCCNV and TbCSV	*Nicotiana benthamiana*	Rep	ZFN	Suppression of viral replication	Chen et al. (2014)
DNA viral disease	TYLCCNV, TbCSV, and TLCYnV	*Nicotiana benthamiana*	Rep	TALE	Suppression of viral replication	Cheng et al. (2015)
DNA viral disease	TYLCV, BCTV, and MeMV	*Nicotiana benthamiana*	IR, CP, *RCRII*	CRISPR/Cas9	Reduction of virus titer	Ali et al. (2015)
RNA viral disease	CMV	*Nicotiana benthamiana* and *Arabidopsis*	ORF1a, ORFCP, and 3′-UTR	CRISPR/Cas9	Notable reduction in virus accumulation and symptom development	Zhang et al. (2018)
RNA viral disease	TuMV	*Nicotiana benthamiana*	GFP, Hc-Pro, and CP	CRISPR/ Cas13	Reduction of viral load	Aman et al. (2018)
RNA viral disease	CVYV, ZYMV, and PRSV-W	*Cucumber*	*eIF4E*/exon of host plant	CRISPR/Cas9	Broad-spectrum virus resistance across various hosts	Chandrasekaran et al. (2016)
RNA viral disease	TuMV	*Arabidopsis*	*eIF(iso)4E*/ exon of host plant	CRISPR/Cas9	Broad-spectrum virus resistance across various hosts	Gomez et al. (2017)
Rice tungro disease	RTSV	*Oryza sativa L. japonica*	eIF4G of host plant	CRISPR/Cas9	Broad-spectrum virus resistance across various hosts	Macovei et al. (2018)

5.7 CONCLUSION AND THE WAY AHEAD

The employment of ecologically safe and environmentally friendly ways to protect crops from disease is becoming increasingly important. However, diseases continue to damage plants, necessitating ongoing research in this field to discover more effective ways. Among the different strategies for controlling plant diseases, host resistance remains the most effective one. GE is one of the best options available nowadays to induce resistance in our crop of interest. It causes alteration of the genetic composition of plants through *Agrobacterium*-mediated gene transfer (R gene, PR gene, etc.), thereby enhancing resistance. In addition, comparatively, a nascent technique called genome editing is also being used to induce disease resistance in crops. Genome editing brings about modifications in the endogenous plant DNA (susceptibility gene, executer genes, decoy genes) through deletions, insertions, and substitutes of DNA at specific targets. The new genome editing tool, i.e., CRISPR/Cas, enables accurate modification of endogenous genes for disease resistance and will be instrumental in future crop breeding toward improved disease resistance. Advancements in bioinformatics tools and genome sequencing have resulted in the identification of novel R genes/alleles, which will equip us to win the arms race against plant pathogens. However, the incorporation of pathogen resistance with abiotic stress tolerance should also be our priority to get optimal results. Despite the several advantages of the GE approach to thwart disease, almost none of the currently available GM solutions have reached growers. Gaining the confidence of consumers and policymakers regarding the safety of GE crops remains the biggest challenge. Genome editing can solve that bottleneck and will be a ray of hope.

REFERENCES

Ali, Z., Abulfaraj, A., Idris, A., Ali, S., Tashkandi, M., & Mahfouz, M. M. (2015). CRISPR/Cas9-mediated viral interference in plants. *Genome Biology*, *16*, 1–11.

Aman, R., Ali, Z., Butt, H., Mahas, A., Aljedaani, F., Khan, M. Z., & Mahfouz, M. (2018). RNA virus interference via CRISPR/Cas13a system in plants. *Genome Biology*, *19*(1), 1–9.

Andika, I. B., Kondo, H., & Tamada, T. (2005). Evidence that RNA silencing-mediated resistance to Beet necrotic yellow vein virus is less effective in roots than in leaves. *Molecular Plant-Microbe Interactions*, *18*(3), 194–204.

Bebber, D. P., Holmes, T., Smith, D., & Gurr, S. J. (2014). Economic and physical determinants of the global distributions of crop pests and pathogens. *New Phytologist*, *202*(3), 901–910.

Berger, S., El Chazli, Y., Babu, A. F., & Coste, A. T. (2017). Azole resistance in Aspergillus fumigatus: A consequence of antifungal use in agriculture? *Frontiers in Microbiology*, *8*, 1024.

Bradeen, J. M., Iorizzo, M., Mollov, D. S., Raasch, J., Kramer, L. C., Millett, B. P., & Carputo, D. (2009). Higher copy numbers of the potato RB transgene correspond to enhanced transcript and late blight resistance levels. *Molecular Plant-Microbe Interactions*, *22*(4), 437–446.

Brunner, S., Stirnweis, D., Diaz Quijano, C., Buesing, G., Herren, G., Parlange, F., & Keller, B. (2012). Transgenic Pm3 multilines of wheat show increased powdery mildew resistance in the field. *Plant Biotechnology Journal*, *10*(4), 398–409.

Carroll, D. (2011). Genome engineering with zinc-finger nucleases. Genetics, 188(4), 773–782.

Chandrasekaran, J., Brumin, M., Wolf, D., Leibman, D., Klap, C., Pearlsman, M., & Gal-On, A. (2016). Development of broad virus resistance in non-transgenic cucumber using CRISPR/Cas9 technology. *Molecular Plant Pathology*, *17*(7), 1140–1153.

Chen, W., Qian, Y., Wu, X., Sun, Y., Wu, X., & Cheng, X. (2014). Inhibiting replication of begomoviruses using artificial zinc finger nucleases that target viral-conserved nucleotide motif. *Virus Genes*, *48*, 494–501.

Cheng, X., Li, F., Cai, J., Chen, W., Zhao, N., Sun, Y., & Wu, X. (2015). Artificial TALE as a convenient protein platform for engineering broad-spectrum resistance to begomoviruses. *Viruses*, *7*(8), 4772–4782.

Christian, M., Cermak, T., Doyle, E. L., Schmidt, C., Zhang, F., Hummel, A., & Voytas, D. F. (2010). Targeting DNA double-strand breaks with TAL effector nucleases. Genetics, 186(2), 757–761.

Cong, L., Ran, F. A., Cox, D., Lin, S., Barretto, R., Habib, N., & Zhang, F. (2013). Multiplex genome engineering using CRISPR/Cas systems. Science, 339(6121), 819–823.

Dai, L., Wang, D., Xie, X., Zhang, C., Wang, X., Xu, Y., & Zhang, J. (2016). The novel gene VpPR4-1 from Vitis pseudoreticulata increases powdery mildew resistance in transgenic Vitis vinifera L. *Frontiers in Plant Science*, *7*, 695.

Dong, O. X., & Ronald, P. C. (2019). Genetic engineering for disease resistance in plants: Recent progress and future perspectives. *Plant Physiology*, *180*(1), 26–38.

Fuchs, M. (2017). Pyramiding resistance-conferring gene sequences in crops. *Current Opinion in Virology*, *26*, 36–42.

Ghag, S. B., Shekhawat, U. K., & Ganapathi, T. R. (2014). Host-induced post-transcriptional hairpin RNA-mediated gene silencing of vital fungal genes confers efficient resistance against *Fusarium* wilt in banana. *Plant Biotechnology Journal*, *12*(5), 541–553.

Ghislain, M., Byarugaba, A. A., Magembe, E., Njoroge, A., Rivera, C., Román, M. L., & Kiggundu, A. (2019). Stacking three late blight resistance genes from wild species directly into African highland potato varieties confers complete field resistance to local blight races. *Plant Biotechnology Journal*, *17*(6), 1119–1129.

Gomez, M. A., Lin, Z. D., Moll, T., Luebbert, C., Chauhan, R. D., Vijayaraghavan, A., & Bart, R. (2017). Simultaneous CRISPR/Cas9-mediated editing of cassava eIF4E isoforms nCBP-1 and nCBP-2 confers elevated resistance to cassava brown streak disease. *BioRxiv*, *209874*. https://doi.org/10.1101/209874.

Gonsalves, D. (2006). Transgenic papaya: Development, release, impact and challenges. *Advances in Virus Research*, *67*, 317–354.

Govindarajulu, M., Epstein, L., Wroblewski, T., & Michelmore, R. W. (2015). Host-induced gene silencing inhibits the biotrophic pathogen causing downy mildew of lettuce. *Plant Biotechnology Journal*, *13*(7), 875–883.

Haverkort, A. J., Boonekamp, P. M., Hutten, R., Jacobsen, E., Lotz, L. A. P., Kessel, G. J. T., & Visser, R. G. F. (2016). Durable late blight resistance in potato through dynamic varieties obtained by cisgenesis: Scientific and societal advances in the DuRPh project. *Potato Research*, *59*, 35–66.

Hily, J. M., Ravelonandro, M., Damsteegt, V., Bassett, C., Petri, C., Liu, Z., & Scorza, R. (2007). Plum pox virus coat protein gene Intron-hairpin-RNA (ihpRNA) constructs provide resistance to plum pox virus in *Nicotiana benthamiana* and *Prunus domestica*. *Journal of the American society for Horticultural science*, *132*(6), 850–858.

Horvath, D. M., Pauly, M. H., Hutton, S. F., Vallad, G. E., Scott, J. W., Jones, J. B., & Deynze, A. V. (2013, June). The pepper Bs2 gene confers effective field resistance to bacterial leaf spot and yield enhancement in Florida tomatoes. In *IV International Symposium on Tomato Diseases* (Vol. 1069, pp. 47–51). https://doi.org/10.17660/ActaHortic.2015.1069.5

Jia, H., Zhang, Y., Orbović, V., Xu, J., White, F. F., Jones, J. B., & Wang, N. (2017). Genome editing of the disease susceptibility gene Cs LOB 1 in citrus confers resistance to citrus canker. *Plant Biotechnology Journal*, *15*(7), 817–823.

Jo, K. R., Kim, C. J., Kim, S. J., Kim, T. Y., Bergervoet, M., Jongsma, M. A., & Vossen, J. H. (2014). Development of late blight resistant potatoes by cisgene stacking. *BMC Biotechnology*, *14*, 1–10.

Jun, Z. Y., Xia, Y. H., Jun, W. X., Rui, Z. J., Yan, W. H., & Qun, L. D. (2017). Expression analysis and functional characterization of a pathogen-induced thaumatin-like gene in wheat conferring enhanced resistance to *Puccinia triticina*. *Journal of Plant Interactions*, *12*(1), 332–339.

Kaur, B., Bhatia, D., & Mavi, G. S. (2021). Eighty years of gene-for-gene relationship and its applications in identification and utilization of R genes. *Journal of Genetics*, *100*, 1–17.

Kim, H. J., Kim, M. J., Pak, J. H., Im, H. H., Lee, D. H., Kim, K. H., & Chung, Y. S. (2016). RNAi-mediated Soybean mosaic virus (SMV) resistance of a Korean Soybean cultivar. *Plant Biotechnology Reports*, *10*, 257–267.

Klümper, W., & Qaim, M. (2014). A meta-analysis of the impacts of genetically modified crops. *PloS One*, *9*(11), e111629.

Kunwar, S., Iriarte, F., Fan, Q., Evaristo da Silva, E., Ritchie, L., Nguyen, N. S., & Paret, M. L. (2018). Transgenic expression of EFR and Bs2 genes for field management of bacterial wilt and bacterial spot of tomato. *Phytopathology*, *108*(12), 1402–1411.

Li, P., Pei, Y., Sang, X., Ling, Y., Yang, Z., & He, G. (2009). Transgenic indica rice expressing a bitter melon (Momordica charantia) class I chitinase gene (McCHIT1) confers enhanced resistance to Magnaporthe grisea and Rhizoctonia solani. *European Journal of Plant Pathology*, *125*, 533–543.

Liu, Y., Li, R., Tan, S., Wang, S., & Chen, F. (2017). Overexpression of Jatropha curcas defensin (JcDef) enhances sheath blight disease resistance in tobacco. *Journal of Phytopathology*, *165*(1), 15–21.

Macovei, A., Sevilla, N. R., Cantos, C., Jonson, G. B., Slamet-Loedin, I., Čermák, T., ... & Chadha-Mohanty, P. (2018). Novel alleles of rice *eIF4G* generated by CRISPR/Cas9-targeted mutagenesis confer resistance to *Rice tungro spherical virus*. *Plant Biotechnology Journal*, *16*(11), 1918–1927.

Malathi, P., Muzammil, S. A., Krishnaveni, D., Balachandran, S. M., & Mangrauthia, S. K. (2019). Coat protein 3 of Rice tungro spherical virus is the key target gene for development of RNAi mediated tungro disease resistance in rice. *Agri Gene*, *12*, 100084.

McLaughlin, J. E., Darwish, N. I., Garcia-Sanchez, J., Tyagi, N., Trick, H. N., McCormick, S., & Tumer, N. E. (2021). A lipid transfer protein has antifungal and antioxidant activity and suppresses Fusarium head blight disease and DON accumulation in transgenic wheat. *Phytopathology®*, *111*(4), 671–683.

Missiou, A., Kalantidis, K., Boutla, A., Tzortzakaki, S., Tabler, M., & Tsagris, M. (2004). Generation of transgenic potato plants highly resistant to potato virus Y (PVY) through RNA silencing. *Molecular Breeding*, *14*, 185–197.

Moosa, A., Farzand, A., Sahi, S. T., & Khan, S. A. (2018). Transgenic expression of antifungal pathogenesis-related proteins against phytopathogenic fungi–15 years of success. *Israel Journal of Plant Sciences, 65(1–2)*, 38–54.

Mundt, C. C. (2018). Pyramiding for resistance durability: Theory andpractice. *Phytopathology*, *108*(7), 792–802.

Muramoto, N., Tanaka, T., Shimamura, T., Mitsukawa, N., Hori, E., Koda, K., & Imaeda, T. (2012). Transgenic sweet potato expressing thionin from barley gives resistance to black rot disease caused by Ceratocystis fimbriata in leaves and storage roots. *Plant Cell Reports*, *31*, 987–997.

Nandi, A. K. (2016). Application of antimicrobial proteins and peptides in developing disease-resistant plants. *Plant Pathogen Resistance Biotechnology*, *3*, 51–70. https://doi.org/10.1002/9781118867716.ch3

Nekrasov, V., Wang, C., Win, J., Lanz, C., Weigel, D., & Kamoun, S. (2017). Rapid generation of a transgene-free powdery mildew resistant tomato by genome deletion. *Scientific Reports*, *7*(1), 482.

Nowara, D., Gay, A., Lacomme, C., Shaw, J., Ridout, C., Douchkov, D., & Schweizer, P. (2010). HIGS: Host-induced gene silencing in the obligate biotrophic fungal pathogen Blumeria graminis. *The Plant Cell*, *22*(9), 3130–3141.

Oliva, R., Chonghui, J., Genelou, A.-G., José, C., Huguet-Tapia, A., Perez-Quintero, T. Li, Joon-Seob, E. et al. (2019). Broad-spectrum resistance to bacterial blight in rice using genome editing. *Nature Biotechnology*, *37*(11), 1344–1350.

Patil, B. L., Ogwok, E., Wagaba, H., Mohammed, I. U., Yadav, J. S., Bagewadi, B., & Fauquet, C. M. (2011). RNAi-mediated resistance to diverse isolates belonging to two virus species involved in Cassava brown streak disease. *Molecular Plant Pathology*, *12*(1), 31–41.

Paula de Toledo Thomazella, D., Brail, Q., Dahlbeck, D., & Staskawicz, B. (2016). CRISPR-Cas9 mediated mutagenesis of a DMR6 ortholog in tomato confers broad-spectrum disease resistance. *BioRxiv*, 064824. https://doi.org/10.1101/064824

Peng, A., Chen, S., Lei, T., Xu, L., He, Y., Wu, L., & Zou, X. (2017). Engineering canker-resistant plants through CRISPR/Cas9-targeted editing of the susceptibility gene Cs LOB 1 promoter in citrus. *Plant Biotechnology Journal*, *15*(12), 1509–1519.

Pradeep, K., Satya, V. K., Selvapriya, M., Vijayasamundeeswari, A., Ladhalakshmi, D., Paranidharan, V., & Velazhahan, R. (2012). Engineering resistance against Tobacco streak virus (TSV) in sunflower and tobacco using RNA interference. *Biologia Plantarum*, *56*, 735–741.

Que, Q., Chilton, M. D. M., de Fontes, C. M., He, C., Nuccio, M., Zhu, T., & Shi, L. (2010). Trait stacking in transgenic crops: Challenges and opportunities. *GM Crops*, *1*(4), 220–229.

Ramirez, C. L., Foley, J. E., Wright, D. A., Müller-Lerch, F., Rahman, S. H., Cornu, T. I., & Joung, J. K. (2008). Unexpected failure rates for modular assembly of engineered zinc fingers. *Nature methods*, 5(5), 374–375.

Savary, S., Willocquet, L., Pethybridge, S. J., Esker, P., McRoberts, N., & Nelson, A. (2019). The global burden of pathogens and pests on major food crops. *Nature Ecology & Evolution*, *3*(3), 430–439.

Senthilraja, C., Gurivi Reddy, M., Rajeswaran, J., Kokiladevi, E., & Velazhahan, R. (2018). RNA interference-mediated resistance to Tobacco streak virus in transgenic peanut. *Australasian Plant Pathology*, *47*, 227–230.

Sera, T. (2005). Inhibition of virus DNA replication by artificial zinc finger proteins. *Journal of Virology*, *79*(4), 2614–2619.

Singh, D., Haicour, R., Sihachakr, D., & Rajam, M. V. (2015). Expression of rice chitinase gene in transgenic eggplant confers resistance to fungal wilts. *Indian Journal of Biotechnology*, *14*, 233–240

Smyth, S. J., Kerr, W. A., & Phillips, P. W. (2015). Global economic, environmental and health benefits from GM crop adoption. *Global Food Security*, *7*, 24–29.

Sundaresha, S., Manoj Kumar, A., Rohini, S., Math, S. A., Keshamma, E., Chandrashekar, S. C., & Udayakumar, M. (2010). Enhanced protection against two major fungal pathogens of groundnut, Cercospora arachidicola and Aspergillus flavus in transgenic groundnut over-expressing a tobacco β 1–3 glucanase. *European Journal of Plant Pathology*, *126*, 497–508.

Sunpapao, A., & Pornsuriya, C. (2016). Overexpression of β-1, 3-glucanase gene in response to Phytophthora palmivora infection in leaves of Hevea brasiliensis clones. *Walailak Journal of Science and Technology (WJST)*, *13*(1), 35–43.

Van Esse, H. P., Reuber, T. L., & van der Does, D. (2020). Genetic modification to improve disease resistance in crops. *The New phytologist*, *225*(1), 70–86. https://doi.org/10.1111/nph.15967

van Loon, L. C., Rep, M., & Pieterse, C. M. (2006). Significance of inducible defense-related proteins in infected plants. *Annual Review of Phytopathology*, *44*, 135–162.

Wang, F., Wang, C., Liu, P., Lei, C., Hao, W., Gao, Y., & Zhao, K. (2016). Enhanced rice blast resistance by CRISPR/Cas9-targeted mutagenesis of the ERF transcription factor gene OsERF922. *PloS One*, *11*(4), e0154027.

Wang, N., Xiao, B., & Xiong, L. (2011). Identification of a cluster of PR4-like genes involved in stress responses in rice. *Journal of Plant Physiology*, *168*(18), 2212–2224.

Wang, Y., Cheng, X., Shan, Q., Zhang, Y., Liu, J., Gao, C., & Qiu, J. L. (2014). Simultaneous editing of three homoeoalleles in hexaploid bread wheat confers heritable resistance to powdery mildew. *Nature Biotechnology*, *32*(9), 947–951.

Yang, X., Yang, J., Wang, Y., He, H., Niu, L., Guo, D., & Dong, Y. (2019). Enhanced resistance to sclerotinia stem rot in transgenic soybean that overexpresses a wheat oxalate oxidase. *Transgenic Research*, *28*, 103–114.

Yin, K., & Qiu, J. L. (2019). Genome editing for plant disease resistance: Applications and perspectives. *Philosophical Transactions of the Royal Society B*, *374*(1767), 20180322.

Zhang, T., Jin, Y., Zhao, J. H., Gao, F., Zhou, B. J., Fang, Y. Y., & Guo, H. S. (2016). Host-induced gene silencing of the target gene in fungal cells confers effective resistance to the cotton wilt disease pathogen Verticillium dahliae. *Molecular Plant*, *9*(6), 939–942.

Zhang, T., Zheng, Q., Yi, X., An, H., Zhao, Y., Ma, S., & Zhou, G. (2018). Establishing RNA virus resistance in plants by harnessing CRISPR immune system. *Plant Biotechnology Journal*, *16*(8), 1415–1423.

Zhang, Y., Bai, Y., Wu, G., Zou, S., Chen, Y., Gao, C., & Tang, D. (2017). Simultaneous modification of three homoeologs of Ta EDR 1 by genome editing enhances powdery mildew resistance in wheat. *The Plant Journal*, *91*(4), 714–724.

Zhou, J., Peng, Z., Long, J., Sosso, D., Liu, B. O., Eom, J. S., & Yang, B. (2015). Gene targeting by the TAL effector PthXo2 reveals cryptic resistance gene for bacterial blight of rice. *The Plant Journal*, *82*(4), 632–643.

Zhou, Y., Yuan, Y., Yuan, F., Wang, M., Zhong, H., Gu, M., & Liang, G. (2012). RNAi-directed down-regulation of RSV results in increased resistance in rice (*Oryza sativa* L.). *Biotechnology Letters*, *34*, 965–972.

Zhu, S., Li, Y., Vossen, J. H., Visser, R. G., & Jacobsen, E. (2012). Functional stacking of three resistance genes against Phytophthora infestans in potato. *Transgenic Research*, *21*, 89–99.

6 Phytohormones
Crucial Factors for Plant Stress Tolerance

Priyanka Saha, Bishal Debnath, Neha Sharma, Suman Dutta, Anamika Barman, Anurag Bera, and Swagata Mandal

6.1 INTRODUCTION

Plant productivity, yield, and resilience to environmental stresses stand as pivotal focal points in agriculture. The growth and development of plants are intricately shaped by a combination of biotic and abiotic factors. In response to these influences, plants undergo numerous internal changes. Biotic factors encompass interactions between organisms and plants, manifesting both positive and negative effects. Positive effects can contribute to beneficial impacts on plant growth, while negative effects may involve allelopathy, herbivory influences, or pathogen infections in plants (Ciura and Kruk, 2018). Abiotic stress pertains to the detrimental impact of non-living factors on plants within a particular environment, leading to a variety of reactions that affect biological functions like gene expression, cell metabolism, and overall growth and development (Zhang et al., 2023). It encompasses various stressors such as extreme temperatures, drought, flooding, salinity, metal toxicity, and nutrient deficiencies, each eliciting distinct responses (Zhang et al., 2020). Extreme temperatures, drought, and high salinity in soils represent primary environmental challenges that constrain the survival and geographical distribution of plants in natural settings (Zhang et al., 2022). For instance, elevated temperatures and periods of drought commonly coincide. The scientific investigation of how plants perceive, transduce, and respond to stress signals is a topic of considerable interest in biology. To counteract adverse effects, plants employ defense systems featuring various chemical compounds (Li et al., 2019; Riaz et al., 2021). Plant cell walls serve as reservoirs for proteins with antimicrobial properties and secondary metabolites, forming a robust molecular layer of polysaccharides that resist physical penetration and growth of pathogens. Phytohormones, also known as plant growth hormones, are organic, natural, and small lipophilic compounds. They play a crucial role in responding to diverse biotic and abiotic stresses and coordinating various developmental and growth functions in plants (Jiang and Asami, 2018). Additionally, they effectively regulate cellular processes, exhibiting responsiveness to external stimuli and changing environmental conditions even at low concentrations (Nowicka et al., 2018). Phytohormones with low molecular weights frequently serve as plant defense mechanisms, precisely responding to external stimuli against biotic stresses (Cao et al., 2011; Parveen et al., 2020). While jasmonic acid (JA), salicylic acid (SA), ethylene (ET), auxins (indole-3-acetic acid), gibberellins (GAs), abscisic acid (ABA), and cytokinins (CKs) have traditionally been extensively studied by plant biologists based on their physiological functions and chemical structures, ongoing research is exploring other compounds such as brassinosteroids (BRs), JA-derived compounds, CKs (zeatin), SA-derived compounds, strigolactones, and peptides as potential plant hormones (Li et al., 2021).

6.2 PHYTOHORMONES AND THEIR CHARACTERISTICS

Phytohormones, often called plant hormones, are chemical messengers that control a wide range of plant physiological functions. These hormones play a crucial role in coordinating growth, development, and responses to environmental stimuli (Bhatt et al., 2020). A single phytohormone can

DOI: 10.1201/9781003543237-6

regulate many cellular and developmental processes, while multiple hormones often influence a single process. Several different internal and environmental factors might interfere with a plant's growth and development (Wolters and Jürgens, 2009). The major function of phytohormones is to control the development, nutrition source or sink, growth, and maintenance processes in plants so that they can adapt to various environmental conditions. Many factors contribute to a plant's reaction to abiotic stress, but, among these endogenous molecules, phytohormones serve as the most crucial, regulating the molecular, physiological, and biochemical processes that are essential to the plant's survival (Fahad et al., 2015). Plant hormones are chemical substances that occur in very low concentrations in plants. They are derived from indole (auxins), terpenes (GAs), adenine (CKs), carotenoids (ABA), and gases (ET) (Murti & Upreti, 2000; Bajguz & Piotrowska-Niczyporuk, 2023). Major plant hormones are categorized into two categories: plant growth promoters (auxins, GAs, CKs) and plant growth inhibitors (ABAs). ET is one such hormone that operates as both a growth promoter and an inhibitor (Pierik et al., 2006). It influences numerous physiological processes and is one of the most often used hormones in agriculture. This chapter extensively explains plant hormones and their functions in plant development, including current achievements and prospects, as well as their role in growth and drought stress responses. In this chapter, plant hormones and their functions in plant development are thoroughly elucidated, with a particular focus on their crucial role in stress tolerance. A number of phytohormones, notably CKs, auxins, ET, GAs, ABA, jasmonates, BRs, and two recently identified phytohormones, SA and strigalactone, have been found to have a role in drought stress tolerance (Ullah et al., 2018).

Phytohormones are signaling molecules that are produced in one part of the plant and transported to another, where they elicit specific responses. Phytohormones work in both mobile and non-mobile modes; they can be created at the site of action or transferred to where they have the greatest need (Peleg & Blumwald, 2011). Balanced plant development and growth depend on hormone homeostasis, which is controlled at several levels such as synthesis of hormones, catabolism, perception, and transduction. Several pathways are involved in the hormone's internal transport in plants, including cytoplasmic streaming, slow diffusion, and vascular tissue (xylem and phloem) transportation (Figure 6.1). These multiple movement mechanisms enable plants to coordinate their reactions to external stimuli, as well as their growth and development, across different areas of the plant. The combination of short- and long-distance transport facilitates optimal communication and integration of hormonal signals throughout the plant body. Though not all plant cells are hormone-responsive, the ones that have this characteristic are triggered at particular stages of development. At certain points in the cell's lifetime, the impacts are at their peak, and they're significantly reduced at other times. Hormones are essential for plant growth but at precise intervals and in certain places (Wolter & Jürgens, 2009). When hormones are no longer required, plants must likewise deactivate their actions. Hormone synthesis frequently takes place at meristem active growth regions prior to complete cell differentiation. They are either sent to other areas of the plant after production to have an immediate impact, or they are kept in cells to be released at a later time (Sun et al., 2009). To control the amount of chemicals needed to biosynthesize hormones, plants use an array of mechanisms that regulate internal hormone amounts and their effects. Hormones can be stored in cells, rendered inactive, or even absorbed by cells that have previously produced them through the process of conjugation with carbohydrates, amino acids, or peptides. Another way that plants can degrade hormones is by chemically breaking them down. In many cases, one plant hormone will control how much of another plant hormone is present. For example, auxins, a type of plant hormone, can influence the synthesis and distribution of CKs, another type of hormone. Similarly, GAs, another type of plant hormone, can interact with auxins to regulate activities such as cell elongation and germination. The hormone concentrations needed to elicit reactions from plants are extremely low, ranging from 10^{-6} to 10^{-5} mol/L. This poses a significant challenge for researchers studying plant hormones because they need to accurately manipulate and measure these low concentrations to understand their effects on plant physiology and development (Santner et al., 2009). Despite the challenges, advances in analytical techniques and experimental methodologies

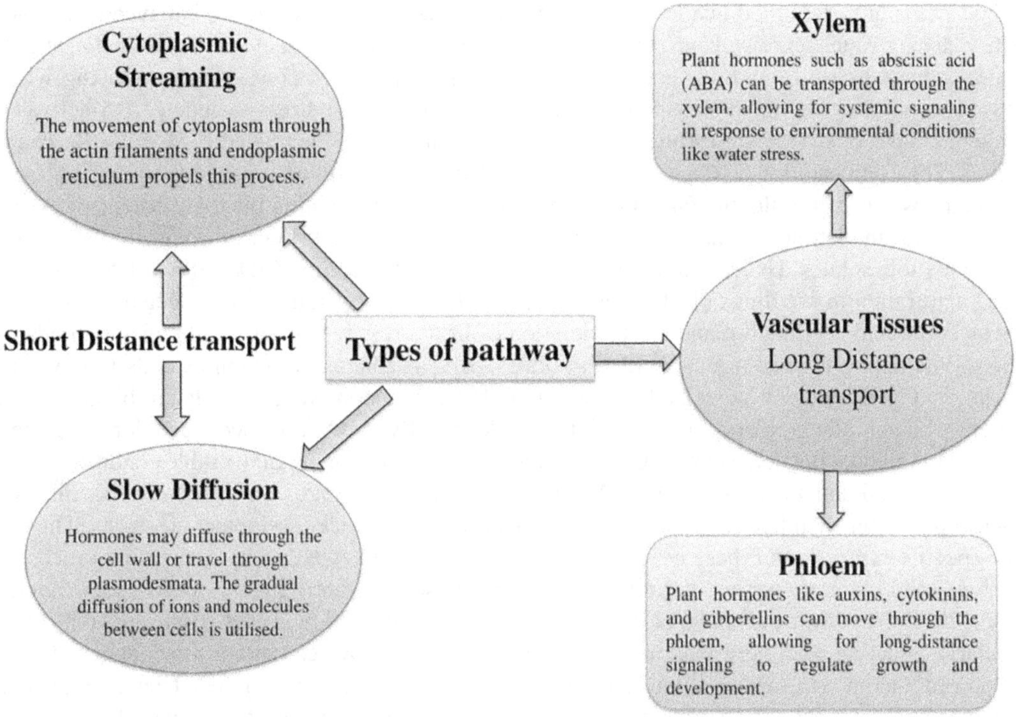

FIGURE 6.1 Pathways that are involved in the internal transport of hormones in plants.

have allowed scientists to delve into the complexities of plant hormone signaling pathways and their regulatory roles in various biological processes. A more fascinating investigation into these plant hormones, as well as the discovery of new ones, is underway globally.

6.3 SEVERAL PHYTOHORMONES AND THEIR REGULATORY FUNCTION IN STRESS TOLERANCE

Phytohormones, also known as plant growth regulators, exert their effects at remarkably low concentrations, overseeing diverse cellular processes and orchestrating plant responses to shifting environmental conditions (Fahad et al., 2015; Wani et al., 2016). Among the extensively studied plant hormones are CKs, auxins (Aux), GAs, ABA, ET, BRs, and jasmonates. Conventionally, it is believed that JA and ET play roles in plant defense, whereas Aux, CK, GA, and BR are linked to plant development. ABA stands out as the pivotal hormone governing plant responses to abiotic stresses (Kazan, 2015).

6.3.1 Abscisic Acid

ABA serves as a central regulator in the intricate mechanisms that plants employ to tolerate environmental stress. Across diverse stressors such as drought, salinity, and extreme temperatures, ABA plays a pivotal role in orchestrating adaptive responses. In instances of drought stress, ABA facilitates stomatal closure, conserving water by signaling guard cells to limit stomatal opening. It has been observed that ABA strongly affects plant leaf senescence. Senescence-associated gene 113 (SAG113) positively regulates ABA-induced leaf senescence by inhibiting stomata closure, which accelerates water loss in senescing leaves (Xue-Xuan et al., 2010) Additionally, ABA influences root system modifications, promoting the growth of deeper roots to enhance soil exploration during water scarcity. Under drought stress, ABA alters ion transport in guard cells, stimulating stomatal closure and preventing stomatal opening, and as a consequence reducing water loss. The pathway that mediates stomatal closure during drought stress includes ABA-activated protein kinase, open stomata 1 kinase (OST1), slow anion channel-associated 1 (SLAC1), K+ channel (KAT1), and catalase 3 (CAT3) (Sah et al., 2016). Under salinity stress, ABA contributes to ion homeostasis by regulating ion uptake and transportation, mitigating the toxic effects of excessive salt accumulation. Furthermore, ABA assists in osmotic adjustment through the synthesis of compatible solutes. Under osmotic stress, LEA, ABI5, PP2C, SnRK2D/E/I, and HSP regulate seed maturation in seedlings (Trebichalský et al., 2017). In temperature stress scenarios, ABA is involved in cold acclimation and the modulation of heat stress responses, aiding plants in adapting to fluctuating temperatures (Yu et al., 2020). ABA's regulatory function extends to the transcriptional level, where it governs the expression of stress-responsive genes related to antioxidant defense, osmoprotection, and stress perception. Additionally, ABA influences seed dormancy and germination, inducing dormancy to prevent premature germination in unfavorable conditions. The exploration and understanding of NCED genes have unveiled their role in regulating specific biological processes by influencing endogenous ABA levels at distinct developmental stages. The tissue-specific expression of these genes has been identified as a key factor in modulating processes such as seed maturation, seed germination, and responses to abiotic stresses (Martínez-Andújar et al., 2011). Notably, triple and quadruple mutants of pyl exhibited modified ABA sensitivity, particularly affecting seed germination and growth. Conversely, overexpression lines demonstrated enhanced tolerance to various abiotic stresses. This research sheds light on the intricate regulatory mechanisms governed by NCED genes and their consequential impact on crucial physiological processes in plants (Santiago et al., 2009). Understanding the multifaceted role of ABA provides valuable insights for developing stress-resilient crops and enhancing agricultural sustainability in the face of evolving environmental challenges.

6.3.2 Ethylene

ET, a gaseous phytohormone, emerges as a pivotal regulator in plants' stress tolerance mechanisms, showcasing its versatility in responding to a spectrum of environmental challenges. In instances of abiotic stress, ET contributes significantly to plant resilience. During drought stress, it plays a role in regulating stomatal closure and promoting adaptive root development, crucial strategies to mitigate water loss through transpiration. In flooded conditions, ET acts as a signaling molecule, fostering the formation of adventitious roots (ARs) and aiding plant survival in oxygen-deprived environments (Khan et al., 2020). It allows gas exchange with the environment by promoting the growth of shoots toward the water's surface, the emergence of ARs, and the maintenance of the fundamental metabolism needed to survive submersion (Fukao and Bailey-Serres, 2008; Fukao et al., 2006). Moreover, under salt stress, ET modulates ion uptake and transport, influencing the expression of genes related to ion homeostasis, thereby aiding plants in coping with saline soils (Riyazuddin et al., 2020). By promoting antioxidant defense, enhancing nitrate and sulfate absorption, and preserving the equilibrium of Na^+/K^+, nutrients, and reactive oxygen species (ROS), ET primarily regulates salinity stress responses. In the realm of biotic stress, ET is a cornerstone of plant defense mechanisms (Zehra et al., 2021). It orchestrates responses against pathogens by regulating the expression of defense-related genes and activating immune responses. Similarly, in defense against herbivores, ET induces the production of volatile organic compounds that attract natural predators of herbivores, creating a dynamic and layered defense strategy. Beyond stress responses, ET plays a vital role in plant developmental processes. It is a key regulator of senescence, influencing the timing and progression of aging in plant tissues. Recent research on flower senescence has demonstrated that ET's role and its interactions with sugars and other growth regulators directly control flower senescence by affecting the genes responsible for ET-sensitive flower senescence (Dar et al., 2021). Additionally, ET governs fruit ripening by promoting the synthesis of enzymes responsible for softening and the production of aroma compounds, impacting the quality and timing of fruit ripening (Gao et al., 2020). The APETALA2/ET response factor (AP2/ERF) superfamily includes the ERFs, which function to control the expression of ET-responsive genes by acting downstream of the ET signaling pathway. ET plays a major role in climacteric fruits as the primary regulator of the ripening process. It has been demonstrated that ERFs, the final elements of the ET signal pathway and the direct mediators controlling ET-responsive genes, are involved in a number of ripening processes, such as ET biosynthesis, color change, fruit softening, and flavor formation in different fleshy fruit species (Gao et al., 2020; Feng et al., 2016). One notable feature of ET's regulatory function is its role in integrating responses to both abiotic and biotic stresses, creating a central hub for crosstalk between various signaling pathways (Ju et al., 2015). This integration allows plants to prioritize and optimize their responses when facing simultaneous environmental challenges. Understanding the multifaceted regulatory role of ET in stress tolerance provides valuable insights for developing strategies to enhance crop resilience and productivity under diverse and changing environmental conditions.

6.3.3 Jasmonic Acid

The name 'jasmonic acid' originates from the jasmine plant due to the significance of its methyl ester in the plant's scent. This compound has been a vital part of the perfume industry. Jasmonates are also linked to tuberonic acid, another hydroxylated substance that triggers potato tuberization, along with its methyl ester and glycosides. JA is synthesized from linolenic acid, and it is presumed that JA is the precursor of tuberonic acid. One of the primary functions of JA in abiotic stress management is its involvement in the regulation of antioxidant defense systems. Abiotic stress often leads to the generation of ROS, causing oxidative damage to plant cells. JA induces the expression of genes involved in antioxidant defense, leading to the synthesis of enzymes and molecules that scavenge ROS, thereby mitigating oxidative stress. Furthermore, JA has been implicated in the modulation of stomatal closure, regulating water loss and improving drought tolerance in plants. The hormone also participates in the maintenance of osmotic balance by influencing the accumulation of osmoprotectants, such as proline and sugars, which contribute to cellular hydration and protect against water deficit. In response to salinity stress, JA interacts with other signaling pathways, such as ABA, to regulate ion homeostasis and osmotic balance. This crosstalk between JA and ABA helps plants adapt to high-salinity conditions by minimizing ion toxicity and osmotic stress. The intricate interplay between JA and other phytohormones, SA and ET, further highlights the complexity of plant responses to abiotic stress. Comprehending these interactions is crucial to unravel the molecular processes that underlie JA-mediated stress responses and devise tactics to augment crop resistance against stress (Table 6.1).

6.3.4 Salicylic Acid

Salicylates have been identified in willow bark long back, yet their potential as regulatory compounds has only been recently acknowledged. SA is synthesized from phenylalanine-amino acid. It plays a crucial role in abiotic stress management in plants, serving as a crucial signaling molecule mediating a plethora of physiological and biochemical responses. A major constraint for crop production is abiotic stresses, for instance, drought, salinity, extreme temperatures, and heavy metal toxicity, can significantly impact plant growth and productivity. In this milieu, understanding the role of SA in mitigating the adverse effects of abiotic stress has become a focal point in plant science research. One of the primary mechanisms through which SA contributes to abiotic stress management is by inducing a cascade of signaling pathways. SA also activates a network of genes involved in stress tolerance, leading to the synthesis of protective proteins, osmoprotectants, and antioxidants. These molecules collectively help plants withstand the detrimental effects

TABLE 6.1
Role of Jasmonic Acid in Abiotic Stresses in Various Crops Focusing on Sustainable Productivity

Drought	*Oryza sativa*	By increasing PSII's (Fv/Fm) photosynthetic activity and photochemical efficiency, priming with methyl jasmonate mitigated the detrimental effects of water stress on rice seedlings.	Sheteiwy et al. (2018)
	Triticum aestivum	In wheat plants under water stress, jasmonic acid seed priming boosted the percentage of germination, proline, soluble carbohydrate accumulation, and shoot growth.	Ilyas et al. (2017)
	Zea mays	Under water stress, maize plants' ion absorption and antioxidant activities were enhanced when seeds were treated with methyl jasmonate pretreatment.	Abdelgawad et al. (2014)
Salinity	*O. sativa*	MeJA treatment (125 μM) can enhance rice plant fluorescence, gas exchange characteristics, and agronomical indices while mitigating the detrimental effects of salinity stress.	Hussain et al. (2022)
	T. aestivum	Plant tolerance to salt stress is enhanced by the upregulation of stress-responsive genes, which is induced by an increase in JA levels.	Zhu et al. (2022)
	Gossypium hirsutum	*GaJAZ1* overexpression significantly increased salt tolerance.	Zhao et al. (2020)
Low temperature	*Solanum lycopersicum*	Accumulation of JA stimulates the C-repeat binding factor (CBF) pathway and positively impacts the tomato's reaction to cold.	Wang et al. (2016)
	O. sativa	Genes linked to JA biosynthesis, viz *OsLOX2*, *OsAOC*, *OsAOS1*, and *OsAOS2* associated with cold exposure upregulation.	Du et al. (2013)
High temperature	*O. sativa*	By increasing antioxidant capacity, JAs mitigate the damage caused by high-temperature stress to the stigma viability of rice lines.	Chen et al. (2021)
	Arabidopsis thaliana	A number of transcriptional reactions specific to high-temperature stress are regulated by JA specifically.	Balfagón et al. (2019)
Heavy metals	*Vicia faba*	Through the inhibition of Cd, H_2O_2, and MDA buildup and the enhancement of osmolyte and antioxidant activities that lessen oxidative stress, JA helps faba bean plants circumvent the detrimental effects of Cd stress.	Ahmad et al. (2017)
	Solanum lycopersicum	By adjusting the activity of their antioxidant defense system, the JA-primed seeds demonstrated decreased Pb uptake and enhanced development in Pb-toxic environments.	Balia et al. (2019)
	Cajanus cajan	Methyl jasmonate seed priming reduced the toxicity of copper and cadmium by activating the antioxidant machinery, increasing proline production, and glutathione and ascorbic acid content.	Kaushik et al. (2022)

of abiotic stressors. Moreover, SA has been implicated in the modulation of stomatal closure, thereby regulating water loss and improving drought tolerance in plants. SA also participates in the regulation of ion homeostasis under conditions of salinity stress, promoting salt tolerance by minimizing ion toxicity and osmotic stress. In addition to its direct effects, SA interacts with other signaling molecules, such as JA and ET, to fine-tune the plant's response to abiotic stress. This crosstalk between different signaling pathways enables plants to integrate and prioritize stress signals effectively. Understanding the intricate mechanisms by which SA operates in abiotic stress management is critical for developing strategies to enhance stress resilience in crops. Harnessing the potential of SA, either through exogenous application or through the manipulation of endogenous SA levels, holds promise for improving crop productivity in the face of increasing environmental challenges (Table 6.2).

TABLE 6.2
Salicylic Acid Function in Abiotic Stressors in a Range of Crops Emphasizing Sustainable Production

Drought	*Carthamus tinctorius* L.	When SA was applied to plants during a drought, the plants' anthocyanin concentrations and phenylalanine lyase enzyme activity increased.	Ahmad et al. (2023)
	Solanum lycopersicum	Foliar SA spray was found to mitigate the pernicious effects of water deficits.	Chakma et al. (2021)
Salinity	*Triticum aestivum*	SA therapy controlled the glyoxalase system and antioxidant apparatus, indicating a potential role in reducing salt stress.	Talaat et al. (2022)
	Oryza sativa	Different genes were expressed in rice seedlings treated with NaCl and control groups after foliar spraying with SA, indicating that SA may be useful in reducing saline stress.	Khan et al. (2019)
Low temperature	*Triticum aestivum*	The pretreatment of wheat plants with SA substantially mitigated the effects of low temperatures on grain yield, plant height, leaf net photosynthetic rate, biomass output, and growth period extension.	Wang et al. (2021)
	Solanum tuberosum	Due to modifications in the activity of antioxidant enzymes, pretreatment with SA increases the cold resistance of potato plants (superoxide dismutase and guaiacol-dependent peroxidase).	Golovatskaya et al. (2023)
High temperature	*Triticum aestivum*	The most effective way to mitigate yield loss and worse photosynthetic activity under high-temperature stress was to spray SA at the anthesis stage.	Fan et al. (2022)
	Solanum tuberosum	SA showed a protective effect regarding potato microplant shoot growth.	Rudić et al. (2022)
Heavy metals	*Phaseolus vulgaris*	Salicylic acid-soaked seed can be sown to counteract the harmful effects of nickel and/or lead on bean plants.	Khalil et al. (2021)
	Brassica campestris	Exogenous administration of SA reduces oxidative damage under Pb stress, enhances growth and yield, and activates the antioxidant system.	Hasanuzzaman et al. (2019)

6.3.5 Brassinosteroids

6.3.5.1 Salt Stress Tolerance

BRs play a crucial role in plants' physiological and biochemical responses, affecting various aspects such as vascular differentiation, stem elongation, and leaf bending. They contribute to adaptive responses to environmental stresses, including salt stress (Ali et al., 2007). The application of BRs has been documented to alleviate the adverse effects of various abiotic stresses, including drought, salinity, and extreme temperatures, thereby enhancing plant stress tolerance. Specifically, under salt stress conditions, BRs contribute to the restoration of chlorophyll levels, augmentation of nitrate reductase enzyme activity, and a positive influence on nucleic acid content and seed yield (Hayat et al., 2006, 2007a, b). Under salt stress, BRs demonstrate protective effects on nuclei and chloroplasts, mitigating damage (Kulaeva et al., 1991). Particularly noteworthy is their enhanced ameliorative impact observed in salt-tolerant wheat cultivars (Shahbaz et al., 2008). From a physiological standpoint, BRs contribute to enhanced stress tolerance in plants, overcoming the impediments posed by salt to seed germination while simultaneously safeguarding the integrity of the photosynthetic apparatus (Kagale et al., 2007). They improve the quantum yield of PSII and mitigate the deleterious effects of salt stress on photosynthetic pigments, while also enhancing RuBisCO activation (Yu et al., 2004; Anuradha and Rao, 2003). BRs also modulate the antioxidant system, reducing H_2O_2 content and lipid peroxidation (Zhang et al., 2007;

Shahbaz et al., 2008), and elevating the activities of antioxidative enzymes. Moreover, BRs play a role in ion homeostasis by improving K^+/Na^+ ratios, bolstering Ca^{2+} and K^+ uptake, and diminishing Na+ uptake, consequently augmenting salt tolerance (Qasim et al., 2006). The documented potential of BRs in agriculture highlights their beneficial effects on diverse crops under varied stress conditions (Khripach et al., 2000).

6.3.5.2 Drought Stress Tolerance

BRs, a category of plant hormones, have emerged as pivotal regulators in alleviating the detrimental effects of drought on plants (Li et al., 2009; Khan et al., 2020). The challenge to plant growth posed by drought stress involves the induction of osmotic stress and elevated Na+ levels within cells (Lei, 2008; Wang et al., 2011; Basu et al., 2010). The exogenous application of BRs enhances drought tolerance by promoting stomatal closure, regulating antioxidant systems, modulating expression of stress-responsive genes, alterations in response to drought, and influence the rate of photosynthesis in water-stressed rice, cauliflower, soybean, and tomato, indicating a potential role in enhancing CO_2 assimilation under water stress conditions (Divi and Krishna, 2009; Xia et al., 2009; Wang et al., 2015; Farooq et al., 2009; Hnilička et al., 2010; Zhang et al., 2008; Yuan et al., 2010). Moreover, the application of 24-epibrassinolide, a type of BR, to tomato plants under drought stress led to elevated levels of non-enzymatic antioxidants, including ascorbate, carotenoids, and proline. This was accompanied by enhanced activity of antioxidant enzymes such as catalase, superoxide dismutase, and ascorbate peroxidase, ultimately resulting in reduced lipid peroxidation and H_2O_2 content (Behnamnia et al., 2009). In sorghum varieties, both drought-sensitive and tolerant, the application of BRs increased superoxide dismutase and glutathione (GSH) reductase activities but decreased activities of indole-3-acetic acid oxidase, polyphenol oxidase, protease, and ribonuclease (Vardhini et al., 2011).

6.3.5.3 Cold Stress Tolerance

BRs play a crucial role in enhancing stress tolerance in plants, particularly under cold stress conditions, which is a major handicap for crop production worldwide, especially for thermophilic plants (Cui et al., 2011; Zhang et al., 2019). Cold stress may cause various disruptions in plants, including modifications in the flexibility of cell membranes, alterations in macromolecular activities, decreased solute potential, and physical limitations (Xiong et al., 2002). Moreover, it impacts plant photosynthesis, resulting in a lowered CO_2 assimilation rate, photoinhibition at PSI and PSII, and diminished enzyme activity within the Benson–Calvin Cycle (Jiang et al., 2013; Zhang et al., 2019). Plants have evolved diverse strategies for scavenging ROS and providing photoprotection to counter excessive ROS buildup and prevent over-reduction of photosystems, effectively alleviating photoinhibition (Fang et al., 2019; Ahammed et al., 2020). BRs play a pivotal role in coordinating these responses, as indicated by the engagement of ROS signals in the modulation of cold stress tolerance through BR regulation (Cui et al., 2011). Furthermore, BRs have been demonstrated to play a role in photoprotection under cold stress conditions. If plants are subjected to chilling temperatures, there is an accumulation of active BRs, triggering the activation of transcription factors like BZR1. This process leads to an increase in transcript levels of RBOH1 and the production of apoplastic H_2O_2, essential for photoprotection mechanisms (Fang et al., 2019). However, BRs also exhibit a dual role in regulating cold stress responses, as they can negatively influence prolonged cold stress responses (Ye et al., 2019). Importantly, the role of BRs extends beyond intact plants to harvested plant products, such as fruits, where BRs can improve post-harvest quality under low-temperature stress (Aghdam and Mohammadkhani, 2014; Wang et al., 2012). In summary, BRs assume a central role as master regulators, fortifying plant resilience against cold stress through their sophisticated engagement in ROS signaling, photoprotection mechanisms, and the subtle regulation of stress-responsive genes.

6.3.6 Polyamines

6.3.6.1 Salt Stress Tolerance

At the core of higher plant physiology, polyamines (PAs) stand as indispensable nitrogen-containing compounds, synthesized through the decarboxylation of amino acids or aldehyde trans-amination, they play a vital role in fundamental biological processes (Smith, 1975). Their classification into primary, secondary, tertiary, and quaternary amines hinges on the number of amine groups affixed to a single carbon atom. Integral to plant growth, development, and diverse physiological processes, PAs play pivotal roles (Medda et al., 1995). Diamine putrescine (Put), triamine spermidine (Spd), and tetraamine spermine (Spm) serve as key constituents of plant PAs, manifesting in both free and conjugated forms (Bouchereaua et al., 2000). Their involvement spans critical biological processes, including DNA replication, transcription, translation, and adaptive responses to environmental stresses (Cohen, 1998; Tiburcio et al., 1990). Salt-tolerant species exhibit an increased (Spd+Spm)/Put ratio, highlighting a distinctive pattern, and the external application of PAs proves beneficial in augmenting plant resilience to salinity stress (Chattopadhayay et al., 2002; Ndayiragije and Lutts, 2006). The decline in photosynthesis induced by salinity can be ascribed to various factors, including alterations in enzymes, chlorophylls, and carotenoids. The extent and duration of the stress significantly impact these parameters (Mishra et al., 1997). In the domain of stress tolerance against high salinity, endogenous PAs emerge as crucial contributors, particularly in preserving the photochemical efficiency of plants. Meanwhile, the external application of PAs is posited to counteract the salinity-induced reduction in photosynthetic efficiency, contingent upon nuanced factors such as PA concentrations, types, and the prevailing stress levels (Duan et al., 2008).

6.3.6.2 Drought Stress Tolerance

Under drought and osmotic stresses, various studies have reported an elevation in PA concentrations in different plant tissues (Harper, 2008; Rodriguez et al., 2010; Smith et al., 2013; Chen et al., 2014; Walker et al., 2015). The findings by Todorova et al. (2008) underscore the multifaceted impact of moderate or severe drought on young wheat seedlings, revealing not only an escalation in lipid peroxidation but also a concurrent rise in both proline and PA levels. The buildup of PAs, predominantly spermidine (Spd) and/or spermine (Spm), is frequently linked with increased drought tolerance, highlighting the crucial involvement of PAs in plant stress responses and tolerance mechanisms (Yu and Zhou, 2010; Montesinos-Pereira et al., 2014; Basu et al., 2010). For instance, in rice and wheat varieties with varying drought tolerance, sensitive cultivars exhibited putrescine (Put) accumulation, while tolerant varieties showed an increase in Spm and Spd (Basu et al., 2010). In their study, Yu and Zhou (2010) showcased that the augmentation of both free and conjugated spermidine (Spd) and spermine (Spm) in vetiver grass leaves played a pivotal role in aiding plants to withstand water stress. Conversely, the deprivation of water resulted in a notable accumulation of putrescine (Put) and cadaverine in the leaves of a drought-sensitive pepper cultivar (Sziderics et al., 2010). Microsome- and thylakoid-associated PAs were found to be reduced in drought-sensitive cultivars of *Phaseolus vulgaris* and *Zea mays* when exposed to equiosmotic concentrations of sorbitol exhibit responses that merit further exploration in understanding their physiological adaptations under osmotic stress, suggesting their potential as markers of plant stress tolerance (Legocka and Sobieszczuk-Nowicka, 2012). Examining the impact of putrescine (Put), spermine (Spm), and spermidine (Spd) on bolstering rice drought tolerance, notable improvements were observed in various aspects, including heightened net photosynthesis, increased water use efficiency, enhanced leaf water status, and elevated production of free proline, anthocyanins, and soluble phenolics in the solution phase (Farooq et al., 2009b).

6.3.6.3 Heavy Metal Stress Tolerance

PAs like spermidine (Spd) and spermine (Spm) alleviate the effects of heavy metal stress in plants (Yu et al., 2016; Choudhary et al., 2012). PAs have been documented to interact with phytohormones, notably BRs and ET, impacting plant responses to heavy metal stress. Additionally, PAs not only improve growth parameters but also diminish oxidative stress markers like malondialdehyde (MDA) and hydrogen peroxide (H_2O_2), while concurrently enhancing the antioxidant system (Rady et al., 2019; Takahashi et al., 2010; Ahsan et al., 2008; Aloui et al., 2009; Choudhary et al., 2012). Plants employ strategies involving PAs to regulate ion uptake, translocation, and distribution in response to heavy metal stress (Aloui et al., 2009). In the context of copper stress, the concurrent application of spermidine (Spd) and BRs not only downregulates the expression of genes responsible for copper (Cu) uptake into cells but also diminishes the expression of genes encoding heavy metal ATPases involved in copper (Cu) transport. This dual action indirectly curtails the detoxification processes, showcasing a sophisticated approach to mitigating metal stress (Aloui et al., 2009). PAs also play a crucial role in the sequestration of heavy metals, serving as a key process in how plants respond to the stress induced by these metals. The addition of PAs results in elevated levels of metal-chelating agents like phytochelatins and GSH, accompanied by an augmentation in photosynthetic pigment levels (Rady et al., 2019; Aloui et al., 2009).

6.3.7 Sugar-Like Phytohormone-Like Plant Growth Regulators

Plants are autotrophic creatures that supply their energy demands and feed to other heterotrophs by converting inorganic carbon into sugars through photosynthesis (Hennion et al., 2019). These sugars come in a variety of forms, such as polyols; sucrose, glucose, and fructose monosaccharides; and starch (Dong and Beckles, 2019). Glucose and sucrose are particularly useful metabolites that are involved in controlling a range of physiological functions in plants, including growth and development, the supply of energy, and acting as building blocks for different metabolic pathways. In many metabolic and developmental processes, glucose, fructose, sucrose, and trehalose 6-phosphate are important signaling molecules. These sugars function as important regulators of gene expression in addition to acting as osmoprotectants under particular abiotic stress situations (Rosa et al., 2009; Das et al., 2022). Three major uses for assimilated sugars exist: they can be used as precursors for different metabolic processes, retained as reserve energy, or sent as energy to sink organs. The processes of assimilation, transport, and distribution of sugars from leaves to heterotrophic organs such as storage organs, roots, flowers, fruits, and seeds are collectively referred to as 'carbohydrate partitioning' (MacNeill et al., 2017). During the day, some of the sugars are moved to different plant areas, while the leaves momentarily store the digested carbon as starch. On the contrary, starch that has been stored during the night breaks down into glucose or maltose, which is transferred from the chloroplast to the cytoplasm. Furthermore, carbohydrates can be temporarily or permanently stored in the vacuolar area of leaf cells (Büttner, 2010). It is essential to comprehend the mechanisms governing the distribution of carbohydrates and their genetic regulation across a range of physiological and stress scenarios. Through a diversity of sugar transporters, which coordinate signal transduction in response to stress and environmental adaptation, plants control the distribution of carbohydrates (Kong et al., 2019). The involvement of various sugar transporters, such as SWEETs, SUTs, and MSTs, underscores the complexity of sugar transport and its integration into plant signaling pathways (Chen et al., 2015).

6.3.7.1 The Variety of Plant Sugars and their Functions in Abiotic Environments

A wide variety of sugars can be metabolized by plants into both structural and non-structural forms. Long-chain molecules that make up structural carbohydrates like hemicellulose and cellulose contribute to the total biomass and structural integrity of the plant (Hartmann & Trumbore, 2016). Non-structural sugars, on the contrary, are also referred to as soluble sugars and include a wide range of substances such as disaccharides (maltose, trehalose, sucrose), oligosaccharides (raffinose, stachyose), and polysaccharides (fructan, starch). Moreover, galactitol, mannitol, sorbitol, and inositol are sugar alcohols, or polyols, that are generated from sugars (Dong and Beckles, 2019). These soluble sugars are a dynamic metabolite complex found in plants that perform several functions, including energy production for growth and development, metabolic pathway precursory roles, short- and long-range signaling molecules, osmoprotective properties in stressful environments, and involvement in ROS foraging systems (Salmon et al., 2020; Maity et al., 2023). Plants may also accumulate soluble sugars like glucose, sucrose, trehalose, and fructose as well as polyols like mannitol, sorbitol, and ononitol when exposed to abiotic stress conditions (Zulfiqar et al., 2020). Through the action of the glucose hexokinase (HXK) sensor, glucose operates as a highly versatile metabolite and signaling molecule that is essential for controlling a broad range of biological activities (Sakr et al., 2018). The HXK enzyme is in charge of phosphorylating hexose sugars for metabolic processes and detecting glucose levels, which in turn regulates cellular activity. Furthermore, through the target of rapamycin kinase signaling cascade, glucose regulates effector genes. This cascade also controls glycolysis and the metabolism of sugars that are sensitive to stress (Ahmad, 2019). Glucose is transformed into glucose 6-phosphate (G6P) by the action of HXK, which is then used to synthesize polyols. Similar to this, sucrose is the main sugar type that is transportable in plants. It not only makes it easier for sugars to flow from source to sink tissues but also regulates physiological and cellular signaling cascades (Sakr et al., 2018). Moreover, Slama et al. (2015) pointed out that sucrose is essential for osmoprotection in halophytes and higher plants. A recent study has emphasized the strong relationship between reactions to environmental stimuli and the accumulation, gene expression, and metabolism of sucrose. Abiotic stress conditions cause the catabolic enzymes that break down sucrose, such as SUS (sucrose synthase) and INV (invertase), to become active. This process produces fructose, glucose, and uridine diphosphate (UDP)-glucose. The disaccharide sugar trehalose functions as an osmolyte that can stabilize membrane lipids (Suprasanna et al., 2016). An intermediate molecule called T6P (trehalose-6-phosphate) controls reactions to environmental cues, sugar metabolism, and photosynthesis. Furthermore, trehalose forms an amorphous glass structure around polar phospholipid groups or forms hydrogen bonds with amino acids to protect protein structures, and scavenge ROS to preserve membrane integrity (Zulfiqar et al., 2020). In summary, sugars like glucose, sucrose, and trehalose play multifaceted roles, serving as energy sources, signaling molecules, and osmoprotectants. They are regulated by various factors including environmental stressors and genetic mechanisms.

6.3.8 Strigolactone

Strigolactones (SLs) are tiny signaling molecules that are categorized as phytohormones and produced from carotenoids. They regulate a variety of developmental processes and respond to a wide range of environmental stimuli. They also help plants adjust to changes in the distribution of resources by deliberately modifying their growth to make them more nutrient-smart (Alder et al., 2012). SLs are synthesized from carotenoids through a series of enzymatic reactions involving dwarf27 (D27) carotenoid isomerase and carotenoid cleavage dioxygenase genes CCD7 and CCD8 (Dutta et al., 2021; Alder et al., 2012). The conversion of precursor carlactone (CL) to carlactonoic acid (CLA) in *Arabidopsis* is catalyzed by the cytochrome P450 monooxygenase encoded by more axillary growth1 (MAX1). Following this, an unidentified methyltransferase methylates CLA to form methyl carlactonoate (MeCLA) (Seto et al., 2014). Subsequently, lateral branching oxidoreductase facilitates the oxidation of MeCLA to a molecule resembling SLs (Brewer et al., 2016). In contrast, Os900 (CYP711A2), a homolog of MAX1 in rice, changes CL into 4-deoxyorobanchol (4DO), which is then broken down into orobanchol by Os1400 (CYP711A3) (Zhang et al., 2014). SLs and other plant phytohormones share signaling pathways that involve hormone-induced transcriptional regulator targeting for degradation. A/b-fold hydrolases like dwarf14 (D14) in *Oryza sativa* and the F-box component (MAX2) in *Arabidopsis* are probably involved in this process (Hamiaux et al., 2012). According to Yao et al. (2016), D14 has a conserved catalytic triad (Ser-His-Asp) with hydrolase activity that is in charge of identifying and deactivating SL. karrikin insensitive2 (KAI2), which shares structural similarities with D14, can detect SL enantiomers that are not formed naturally, such as GR24ent-5DS, and smoke-derived karrikin (KAR) (Waters et al., 2012). D14 changes its conformation when SL binds to it, making it easier for it to connect with F-box proteins like MAX2 (Zhao et al., 2015). The transcriptional repressor D53 (homologous to SMXL6, SMXL7, and

SMXL8) is ubiquitinated by this complex and subsequently degraded by the 26S proteasome, which allows the transcription of SL-responsive genes in Arabidopsis (Yao et al., 2018). In rice, D53 is essential for regulating the development of axillary buds (Fang et al., 2020). In rice, transcriptional stimulation of genes involved in SL control is mediated by Ideal Plant Architecture1, a transcription factor belonging to the SPL family, and D53 (Song et al., 2017).

6.3.8.1 Biological Functions of SLs

As a class of plant hormones, SLs are essential in determining the general structure of plants. They control a number of characteristics, including root shape, secondary growth, and shoot branching. A variety of internal and external cues govern the regulation of axillary bud development, which in turn affects shoot branching. One interesting phenomenon is apical dominance, in which the primary shoot apex inhibits the development of buds. Auxin produced by the juvenile leaves at the apex is partly responsible for this suppression (Domagalska and Leyser, 2011). SL stimulates the TCP transcription factor branched1 (BRC1), which in turn decreases bud activity. This mechanism leads to the expansion of axillary buds, which is aided by SL (Dun et al., 2013). Moreover, the auxin that emerges from these buds influences the development of axillary buds. Auxin efflux from lateral buds is decreased by SL because it prevents the auxin efflux carrier PIN1 from localizing to the plasma membrane and modifies the auxin feedback on PIN1 internalization. Tillering and other aspects of shoot architecture are established in part by these developmental processes. Reaching the ideal tiller count in rice and wheat is important for increasing grain output, and it may be related to SL exudation (Zhao et al., 2019). Following the application of GR24 in growth chambers, biomass was enhanced in *Brassica napus* (Ma et al., 2017). In *Arabidopsis* and pea mutants lacking in SL signaling, stem thickness was decreased (Agusti et al., 2011). In cereal crops, the particular effects of SL on stem and root thickening may be used to decrease lodging susceptibility. Mutants lacking SL signaling in peas and *Arabidopsis* showed more ARs, indicating that SLs inhibit the development of ARs (Rasmussen et al., 2012). In particular, GR24 was able to restore AR formation in the rice mutant D10, which was lacking in SL biosynthesis but not in the mutant D3, which was poor in SL signaling (Sun et al., 2015). There are contradictory findings across several species, making it difficult to pinpoint the exact process by which SLs affect root growth. Future alterations in root architecture are probably influenced by the complex hormonal interactions that occur between SLs and other types of phytohormones. As a result, during growth and the adaptability of environmental conditions, plants might profit from these interactions. The exudation and dispersion of SL in agricultural soils are influenced by the availability of nutrients, particularly phosphorus (P) shortage (Yoneyama et al., 2012). When compared to controlled normal developing settings, the expression of SL biosynthesis genes in rice roots is higher when nitrogen (N) or P is reduced (Sun et al., 2014). Additionally, the transcription level of PDR1 (Pleiotropic Drug Resistance1), an ABC transporter, increases in the roots of N or P-deficient *Lotus japonicus* and *petunia*, facilitating the translocation of synthesized SLs from the root to the shoot (Liu et al., 2015). SLs affect both above-ground and below-ground architecture, which is important for helping plants adjust to N and P deficits. Suboptimal P concentrations in rice inhibited the development of shoots and the generation of tillers; however, this negative effect was not seen in mutants lacking in SL signaling or SL biosynthesis, suggesting that SLs were involved (Luo et al., 2018). Furthermore, during both P and N famine, SLs play a critical role in controlling nitric oxide-mediated seminal root elongation in rice (Sun et al., 2015). The ability of SLs to adapt to nitrogen scarcity offers hope for creating plans that will improve low-fertility soil production and nutrient usage efficiency. In addition, SLs directly contribute to increased plant resistance to abiotic stressors. Certain cis-acting regions that bind to transcription factors responsive to salt and drought were found by analyzing the promoter sequences of the SL biogenesis genes in *Arabidopsis thaliana* (Marzec and Muszynska, 2015). SL analog AB01 increased grain yield and kernel weight in maize and sunflower during drought circumstances (Chesterfield et al., 2020). Arbuscular mycorrhizal (AM) fungus was more prevalent in the rhizosphere of lettuce plants as a result of the SLs that were secreted from the roots by salt

(Aroca et al., 2013). In contrast to wild-type plants, rice mutants lacking in SL biosynthesis showed reduced AM colonization (Kobae et al., 2018). In conclusion, SLs play a pivotal role in shaping plant development and responses to various environmental stimuli. Understanding their intricate signaling mechanisms and their impact on plant architecture and stress tolerance opens avenues for enhancing agricultural productivity and sustainability in diverse growing conditions.

6.4 CONCLUSION

In summary, the deep and complex interplay between phytohormones and stress tolerance highlights plants' remarkable resilience to a variety of environmental conditions. ABA, SA, JA, ET, and other phytohormones work together to form a complex signaling network that controls how plants react to a variety of stresses, including infections, mechanical injury, and dehydration. As messengers, these signaling molecules coordinate a series of physiological, biochemical, and molecular alterations that cumulatively improve a plant's resistance to harsh environments. Plants can fine-tune their stress reactions through the complex interactions among several phytohormones, which guarantees a balanced and situation-specific response to the individual stressor faced. The comprehensive aspect of plant stress tolerance is further demonstrated by the integration of phytohormonal signaling with other stress-responsive pathways, such as osmotic adjustment and ROS scavenging. Phytohormones play a critical role in coordinating an efficient defense mechanism, as demonstrated by their capacity to control stomatal closure, induce the manufacture of stress-related proteins, and modify gene expression. The potential to use our growing understanding of phytohormonal control to boost crop resilience, lessen the effects of climate change, and increase overall agricultural productivity is becoming more and more clear. Researchers can create focused strategies for crop improvement by deciphering the complexities of phytohormonal networks, which will ultimately contribute to global food security in the face of escalating environmental difficulties. All things considered, the story of plant adaptation and resilience that is revealed by the involvement of phytohormones in stress tolerance is compelling and presents exciting prospects for sustainable agriculture in the years to come.

REFERENCES

Abdelgawad, Z. A., Khalafaallah, A. A., & Abdallah, M. M. (2014). Impact of methyl jasmonate on antioxidant activity and some biochemical aspects of maize plant grown under water stress condition. *Agricultural Sciences*, *5*(12), 1077.

Aghdam, M. S., & Mohammadkhani, N. (2014). Enhancement of chilling stress tolerance of tomato fruit by postharvest brassinolide treatment. *Food and Bioprocess Technology*, *7*(3), 909–914.

Agusti, J., Herold, S., Schwarz, M., Sanchez, P., Ljung, K., Dun, E. A. Brewar, P. B., Beveridge C. A., Seiberer, T., Sehr, E M., Greb, T. (2011). Strigolactone signaling is required for auxin-dependent stimulation of secondary growth in plants. *Proceedings of the National Academy of Sciences of the United States of America*, *108*, 20242–20247. doi:10.1073/pnas.1111902108

Ahammed, G. J., Wang, Y., Mao, Q., Wu, M., Yan, Y., Ren, J., Wang, X., Liu, A., & Chen, S. (2020). Dopamine alleviates bisphenol A-induced phytotoxicity by enhancing antioxidant and detoxifcation potential in cucumber. *Environmental Pollution*, *259*, 113957.

Ahmad, I. Z. (2019). Role of sugars in abiotic stress signaling in plants. In M. Iqbal, R. Khan, P. Sudhakar Reddy, A. Ferrante, & N. A. Khan (Eds.), *Plant signaling molecules* (pp. 207–217). Woodhead Publishing

Ahmad, P., Alyemeni, M. N., Wijaya, L., Alam, P., Ahanger, M. A., & Alamri, S. A. (2017). Jasmonic acid alleviates negative impacts of cadmium stress by modifying osmolytes and antioxidants in faba bean (*Vicia faba* L.). *Archives of Agronomy and Soil Science*, *63*, 1889–1899.

Ahmad, R., & Anjum, M. A. (2023). Mineral nutrition management in fruit trees under salt stress: a review. *Erwerbs-obstbau*, *65*(2), 397–405.

Ahsan, N., Lee, D. G., Alam, I., Kim, P. J., Lee, J. J., Ahn, Y. O. et al. (2008). Comparative proteomic study of arsenic-induced differentially expressed proteins in rice roots reveals glutathione plays a central role during As stress. *Proteomics*, *8*, 3561–3576.

Alder, A., Jamil, M., Marzorati, M., Bruno, M., Vermathen, M., Bigler, P. et al. (2012). The path from b-carotene to carlactone, a strigolactone-like plant hormone. *Science, 335*, 1348–1351. doi:10.1126/science.1218094

Ali, B., Hayat, S., & Ahmad, A. (2007). 28-Homobrassinolide ameliorates the salt stress in chickpea (Cicer arietinum L.). *Environmental and Experimental Botany, 59*, 217–223.

Aloui, A., Recorbet, G., Gollotte, A., Robert, F., Valot, B., Gianinazzi-Pearson, V., Aschi-Smiti, S., & Dumas-Gaudot, E. (2009). On the mechanisms of cadmium stress alleviation in *Medicago truncatula* by arbuscular mycorrhizal symbiosis: A root proteomic study. *Proteomics, 9*, 420–433.

Anuradha, S., & Rao, S. S. R. (2003). Application of brassinosteroids to rice seeds (Oryza sativa L.) reduced the impact of salt stress on growth, prevented photosynthetic pigment loss and increased nitrate reductase activity. *Plant Growth Regulation, 40*, 29–32.

Aroca, R., Ruiz-Lozano, J. M., Zamarreño, A. M., Paz, J. A., García-Mina, J. M., Pozo, M. J. et al. (2013). Arbuscular mycorrhizal symbiosis influences strigolactone production under salinity and alleviates salt stress in lettuce plants. *Journal of Plant Physiology, 170*, 47–55. doi:10.1016/j.jplph.2012.08.020

Bajguz, A., & Piotrowska-Niczyporuk, A. (2023). Biosynthetic pathways of hormones in plants. *Metabolites, 13*(8), 884.

Balfagón, D., Sengupta, S., Gómez-Cadenas, A., Fritschi, F. B., Azad, R. K., & Mittler, R. (2019).. Jasmonic acid is required for plant acclimation to a combination of high light and heat stress. *Plant Physiology, 181*(4), 1668–1682.

Bali, S., Jamwal, V. L., Kaur, P., Kohli, S. K., Ohri, P., Gandhi, S. G. et al. (2019). Role of P-type ATPase metal transporters and plant immunity induced by jasmonic acid against Lead (Pb) toxicity in tomato. *Ecotoxicology and Environmental Safety, 174*, 283–294.

Barbier, F. F., Dun, E. A., Kerr, S. C., Chabikwa, T. G., & Beveridge, C. A. (2019). An update on the signals controlling shoot branching. *Trends in Plant Science, 24*, 220–236. doi:10.1016/j.tplants.2018.12.001

Basu, S., Roychoudhury, A., Saha, P., & Sengupta, D. (2010). Comparative analysis of some biochemical responses of three indica rice varieties during polyethylene glycol-mediated water stress exhibits distinct varietal differences. *Acta Physiologiae Plantarum,, 32*, 551–563.

Behnamnia, M., Kalantari, K. M., & Rezanejad, F. (2009a) Exogenous application of brassinosteroid alleviates drought-induced oxidative stress in *Lycopersicon esculentum* L. *General and Applied Plant Physiology, 35*, 22–34.

Bennett, T., Liang, Y., Seale, M., Ward, S., Müller, D., & Leyser, O. (2016). Strigolactone regulates shoot development through a core signalling pathway. *Biology Open, 5*, 1806–1820.

Bhatt, D., Nath, M., Sharma, M., Bhatt, M. D., Bisht, D. S., & Butani, N. V. (2020). Role of growth regulators and phytohormones in overcoming environmental stress. In . A Roychoudhury, & D Kumar Tripathi (Eds.), *Protective chemical agents in the amelioration of plant abiotic stress: Biochemical and molecular perspectives* (pp. 254–279).

Bhattacharya, S., & Kundu, A. (2020). Sugars and sugar polyols in overcoming environmental stresses. In A Roychoudhury & D Kumar Tripathi (Eds.), *Protective chemical agents in the amelioration of plant abiotic stress: Biochemical and molecular perspectives* (Vol. 71). Wiley-Blackwell.

Bouchereaua, A., Gue´notb, P., & Larhera, F. (2000). Analysis of amines in plant materials. *Journal of Chromatography B, 747*, 49–67.

Brewer, P. B., Dun, E. A., Gui, R., Mason, M. G., & Beveridge, C. A. (2015). Strigolactone inhibition of branching independent of polar auxin transport. *Plant Physiology, 168*, 1820–1829. doi:10.1104/pp.15.00014

Büttner, M. (2010). The *Arabidopsis* sugar transporter (AtSTP) family: An update. *Plant Biology, 12*, 35–41

Chakma, R., Biswas, A., Saekong, P., Ullah, H., & Datta, A. (2021). Foliar application and seed priming of salicylic acid affect growth, fruit yield, and quality of grape tomato under drought stress. *Scientia Horticulturae, 280*, 109904.

Chattopadhayay, M. K., Tiwari, B. S., Chattopadhyay, G., Bose, A., Sengupta, D. N., & Ghosh, B. (2002). Protective role of exogenous polyamines on salinity-stressed rice (Oryza sativa) plants. *Physiologia Plantarum, 116*, 192–199.

Chen, J., Miao, W., Fei, K., Shen, H., Zhou, Y., Shen, Y., Li, C., He, J., Zhu, K., & Wang, Z., & Yang, J. (2021). Jasmonates alleviate the harm of high-temperature stress during anthesis to stigma vitality of photothermosensitive genetic male sterile rice lines. *Frontiers in Plant Science, 12*, 634959.

Chen, L. Q., Cheung, L. S., Feng, L., Tanner, W., & Frommer, W. B. (2015). Transport of sugars. *Annual Review of Biochemistry, 84*, 865–894.

Chesterfield, R. J., Vickers, C. E., & Beveridge, C. A. (2020). Translation of strigolactones from plant hormone to agriculture: Achievements, future perspectives, and challenges. *Trends Plant Science, 25*, 1087–1106. doi:10.1016/j.tplants.2020.06.005

Choudhary, S. P., Kanwar, M., Bhardwaj, R., Yu, J.-Q., & Tran, L.-S. .P. (2012). Chromium stress mitigation by polyamine-brassinosteroid application involves phytohormonal and physiological strategies in *Raphanus sativus* L. *PLoS ONE*, *7*, e33210.

Ciura, J., & Kruk, J. (2018). Phytohormones as targets for improving plant productivity and stress tolerance. *Journal of Plant Physiology*, *229*, 32–40.

Cohen, S. S. (1998). *A guide to the polyamines*. Oxford University Press.

Crawford, S., Shinohara, N., Sieberer, T., Williamson, L., George, G., Hepworth, J. et al. (2010). Strigolactones enhance competition between shoot branches by dampening auxin transport. *Development*, *137*, 2905–2913. doi:10.1242/dev.051987

Cui, J.-X., Zhou, Y.-H., Ding, J.-G., Xia, X.-J., Shi, K., Chen, S.-C., Asami, T., Chen, Z., & Yu, J.-Q. (2011). Role of nitric oxide in hydrogen peroxide-dependent induction of abiotic stress tolerance by brassinosteroids in cucumber. *Plant Cell Environment*, *34*(2), 347–358.

Dar, R. A., Nisar, S., & Tahir, I. (2021). Ethylene: A key player in ethylene sensitive flower senescence: A review. *Scientia Horticulturae*, *290*, 110491.

Das, A., Singh, S., Islam, Z., Munshi, A. D., Behera, T. K., Dutta, S., Weng, Y., & Dey, S. S. (2022). Current progress in genetic and genomics-aided breeding for stress resistance in cucumber (Cucumis sativus L.). *Scientia Horticulturae*, *300*, 111059.

Divi, U. K., & Krishna, P. (2009). Brassinosteroid: A biotechnological target for enhancing crop yield and stress tolerance. *Nature Biotechnology*, *26*, 131–136.

Domagalska, M., & Leyser, O. (2011). Signal integration in the control of shoot branching. *Nature Reviews Molecular Cell Biology*, *12*, 211–221. doi:10.1038/nrm3088

Dong, S., & Beckles, D. M. (2019). Dynamic changes in the starch-sugar interconversion within plant source and sink tissues promote a better abiotic stress response. *Journal of Plant Physiology*, *234*, 80–93

Du, H., Liu, H., & Xiong, L. (2013). Endogenous auxin and jasmonic acid levels are differentially modulated by abiotic stresses in rice. *Frontiers in Plant Science*, *4*, 397.

Du, H., Zhou, X., Yang, Q., Liu, H., & Kurtenbach, R. (2015). Changes in H+-ATPase activity and conjugated polyamine contents in plasma membrane purified from developing wheat embryos under short-time drought stress. *Plant Growth Regulator*, *75*(1), 1–10.

Duan, W. J., Ren, H., Fu, S. L., Wang, J., Yang, L., & Zhang, J. P. (2008). Natural recovery of different areas of a deserted quarry in South China. *Journal of Environmental Science*, *20*, 476–481.

Duan, J., Yu, H., Yuan, K., Liao, Z., Meng, X., Jing, Y. et al. (2019). Strigolactone promotes cytokinin degradation through transcriptional activation of cytokinin oxidase/dehydrogenase 9 in rice. *Proceedings of the National Academy of Sciences of the United States of America*, *116*, 14319. doi:10.1073/pnas.1810980116

Dun, E. A., de Saint, Germain, A., Rameau, C., & Beveridge, C. A. (2013). Antagonistic action of strigolactone and cytokinin in bud outgrowth control. *Plant Physiology*, *158*, 487–489. doi:10.1104/pp.111.186783

Dutta, S., Muthusamy, V., Chhabra, R., Baveja, A., Zunjare, R. U., Mondal, T. K., Yadava, D. K., & Hossain, F. (2021). Low expression of carotenoids cleavage dioxygenase 1 (ccd1) gene improves the retention of provitamin-A in maize grains during storage. *Molecular Genetics and Genomics*, *296*, 141–153.

Fahad, S., Hussain, S., Bano, A., Saud, S., Hassan, S., Shan, D.,Khan, D. A., Khan, F., Chen, Y., Wu, C., Tabassam, M. A., Chun, M., Afzal, M., Jan, A., Jan, M., & Huang, J. (2015). Potential role of phytohormones and plant growth-promoting rhizobacteria in abiotic stresses: Consequences for changing environment. *Environmental Science and Pollution Research*, *22*, 4907–4921.

Fan, Y., Lv, Z., Li, Y., Qin, B., Song, Q., Ma, L., … & Huang, Z. (2022). Salicylic acid reduces wheat yield loss caused by high temperature stress by enhancing the photosynthetic performance of the flag leaves. *Agronomy*, *12*(6), 1386.

Fang, P., Yan, M., Chi, C., Wang, M., Zhou, Y. H., Zhou, J., Shi, K., Xia, X., & Foyer, C. H., Yu, J. (2019). Brassinosteroids act as a positive regulator of photoprotection in response to chilling stress. *Plant Physiology*, *180*(4), 2061–2076.

Fang, Z., Ji, Y., Hu, J., Guo, R., Sun, S., & Wang, X. (2020). Strigolactones and brassinosteroids antagonistically regulate the stability of the D53–OsBZR1 complex to determine FC1 expression in rice tillering. *Molecular Plant*, *13*, 586–597. doi:10.1016/j.molp.2019.12.005

Farooq, M., Wahid, A., Basra, S. M. A., & Islam, D. (2009a). Improving water relations and gas exchange with brassinosteroids in rice under drought stress. *Journal of Agronomy and Crop Science*, *195*, 262–269.

Farooq, M., Wahid, A., & Lee, D. J. (2009b). Exogenously applied polyamines increase drought tolerance of rice by improving leaf water status, photosynthesis and membrane properties. *Acta Physiologiae Plantarum*, *31*, 937–945.

Feng, B. H., Han, Y. C., Xiao, Y. Y., Kuang, J. F., Fan, Z. Q., Chen, J. Y., Lu, W. J. (2016). The banana fruit Dof transcription factor MaDof23 acts as a repressor and interacts with MaERF9 in regulating ripening-related genes. *Journal of Experimental Botany*, *67*(8), 2263–2275.

Fukao, T., & Bailey-Serres, J. (2008). Submergence tolerance conferred by Sub1A is mediated by SLR1 and SLRL1 restriction of gibberellin responses in rice. *Proceedings of the National Academy of Sciences*, *105*(43), 16814–16819.

Gao, J., Zhang, Y., Li, Z., & Liu, M. (2020). Role of ethylene response factors (ERFs) in fruit ripening. *Food Quality and Safety*, *4*(1), 15–20.

Golovatskaya, I. F., Kadyrbaev, M. K., Boyko, E. V., Filonova, M. V. (2023). Salicylic acid improves cold resistance of solanum tuberosum regenerants via regulation of the antioxidant system. *Russian Journal of Plant Physiology*, *70*(5), 112.

Hamiaux, C., Drummond, R. S., Janssen, B. J., Ledger, S. E., Cooney, J. M., Newcomb, R. D. et al. (2012). DAD2 is an a/b hydrolase likely to be involved in the perception of the plant branching hormone, strigolactone. *Current Biology*, *22*, 2032–2036. doi:10.1016/j.cub.2012.08.007

Hartmann, H., & Trumbore, S. (2016). Understanding the roles of nonstructural carbohydrates in forest trees–from what we can measure to what we want to know. *New Phytologist*, *211*, 386–403.

Hasanuzzaman, M., Matin, M. A., Fardus, J., Hasanuzzaman, M., Hossain, M. S., & Parvin, K. (2019). Foliar application of salicylic acid improves growth and yield attributes by upregulating the antioxidant defense system in Brassica campestris plants grown in lead-amended soils. *Acta Agrobotanica*, 72(2).

Hayat, S., Ali, B., & Ahmad, A. (2006). Response of *Brassica juncea*, to 28-homobrassinolide, grown from the seeds exposed to salt stress. *Journal of Plant Biology*, *33*, 169–174.

Hayat, S., Ali, B., & Ahmad, A. (2007a) Effect of 28-homobrassinolide on salinity-induced changes in Brassica juncea. *Turkish Journal of Biology*, *31*, 141–146.

Hayat, S., Ali, B., Hasan, S. A., & Ahmad, A. (2007b) Brassinosteroid enhanced the level of antioxidants under cadmium stress in Brassica juncea. *Environmental and Experimental Botany*, *60*, 33–41.

Hennion, N., Durand, M., Vriet, C., Doidy, J., Maurousset, L., Lemoine, R., & Pourtau, N. (2019). Sugars en route to the roots. Transport, metabolism and storage within plant roots and towards microorganisms of the rhizosphere. *Physiologia Plantarum*, *165*, 44–57.

Hnilička, F., Koudela, M., Martinková, J., Hniličková, H., & Hejnák, V. (2010). Effect of water deficit and application of 24-epibrassinolide on gas exchange in cauliflower plants. *Scientia Agriculturae Bohemica*, *41*, 15–20.

Hussain, S., Zhang, R., Liu, S., Li, R., Wang, Y., Chen, Y., Hou, H., & Dai, Q. (2022). Methyl jasmonate alleviates the deleterious effects of salinity stress by augmenting antioxidant enzyme activity and ion homeostasis in rice (*Oryza sativa* L.). *Agronomy*, *12*(10), 2343.

Ilyas, N., Gull, R., Mazhar, R., Saeed, M., Kanwal, S., Shabir, S., & Bibi, F. (2017). Influence of salicylic acid and jasmonic acid on wheat under drought stress. *Communications in Soil Science and Plant Analysis*, *48*(22), 2715–2723.

Fukao, T., Xu, K., Ronald, P. C, & Bailey-Serres, J. (2006). A variable cluster of ethylene response factor–like genes regulates metabolic and developmental acclimation responses to submergence in rice. *The Plant Cell*, *18*(8), 2021–2034.

Jiang, K., & Asami, T. (2018). Chemical regulators of plant hormones and their applications in basic research and agriculture. *Bioscience, Biotechnology, and Biochemistry*, *82*(8), 1265–1300.

Jiang, Y. P., Huang, L. F., Cheng, F., Zhou, Y. H., Xia, X. J., Mao, W. H., Shi, K., & Yu, J. Q. (2013). Brassinosteroids accelerate recovery of photosynthetic apparatus from cold stress by balancing the electron partitioning, carboxylation and redox homeostasis in cucumber. *Physiologia Plantarum*, *148*(1), 133–145.

Ju, C., & Chang, C. (2015). Mechanistic insights in ethylene perception and signal transduction. *Plant Physiology*, *169*(1), 85–95.

Kagale, S., Divi, U. K., Kronchko, J. E., Keller, W. A., & Krishna, P. (2007). Brassinosteroid conifers tolerance in Arabidopsis thaliana and Brassica napus to a range of abiotic stresses. *Planta*, *225*, 353–364.

Kapulnik, Y., & Delaux, P. M., Resnick, N., Mayzlish-Gati, E., Wininger, S., & Bhattacharya, C. et al. (2011). Strigolactones affect lateral root formation and root-hair elongation in Arabidopsis. *Planta*, *233*, 209–216. doi:10.1007/s00425-010-1310-y.

Kaushik, S., Sharma, P., Kaur, G., Singh, A. K., Al-Misned, F. A., Shafik, H. M., & Sirhindi, G. (2022). Seed priming with methyl jasmonate mitigates copper and cadmium toxicity by modifying biochemical attributes and antioxidants in *Cajanus cajan*. *Saudi Journal of Biological Sciences*, *29*, 721–729.

Kazan, K. (2015). Diverse roles of jasmonates and ethylene in abiotic stress tolerance. *Trends in Plant Science*, *20*(4), 219–229.

Khalil, R., Haroun, S., Bassyoini, F., Nagah, A., & Yusuf, M. (2021). Salicylic acid in combination with kinetin or calcium ameliorates heavy metal stress in Phaseolus vulgaris plant. *Journal of Agriculture and Food Research*, *5*, 100182.

Khan, M. I. R., Trivellini, A., Chhillar, H., Chopra, P., Ferrante, A., Khan, N. A., & Ismail, A. M. (2020). The significance and functions of ethylene in flooding stress tolerance in plants. *Environmental and Experimental Botany*, *179*, 104188.

Khan, M. S., Akther, T., Ali, D. M., & Hemalatha, S. (2019). An investigation on the role of salicylic acid alleviate the saline stress in rice crop (*Oryza sativa* (L)) *Biocatalysis and Agricultural Biotechnology*, *18*, 101027.

Khripach, V., Zhabinskii, V., & De Groot, A. (2000). Twenty years of brassinosteroids: Steroidal plant hormones warrant better crops for the XXI century. *Annals of Botany*, *86*, 441–447.

Kobae, Y., Kameoka, H., Sugimura, Y., Saito, K., Ohtomo, R., Fujiwara, T. et al. (2018). Strigolactone biosynthesis genes of rice are required for the punctual entry of arbuscular mycorrhizal fungi into the roots. *Plant Cell Physiology*, *59*, 544–553. doi:10.1093/pcp/pcy001

Kong, W., An, B., Zhang, Y., Yang, J., Li, S., Sun, T., & Li, Y. (2019). Sugar transporter proteins (STPs) in Gramineae crops: Comparative analysis, phylogeny, evolution, and expression profiling. *Cells*, *8*, 560.

Kulaeva, O. N., Burkhanova, E. A., Fedina, A. B., Khokhlova, V. A., Bokebayeva, G. A.,Vorbrodt, & H. M., Adam, G. (1991). Effect of brassinosteroids on protein synthesis and plant cell ultrastructure under stress conditions. In H. G. Cutler, T. Yokota, & G. Adam (Eds.), *Brassinosteroids: Chemistry, bioactivity and applications. ACS symposium series 474* (pp. 141–155). American Chemical Society.

Lee, H. Y., & Yoon, G. M. (2020). Strigolactone elevates ethylene biosynthesis in etiolated Arabidopsis seedlings. *Plant Signal Behaviour*, *15*, 1805232. doi:10.1080/15592324.2020.1805232

Legocka, J., & Sobieszczuk-Nowicka, E. (2012). Sorbitol and NaCl stresses affect free, microsome-associated and thylakoid-associated polyamine content in *Zea mays* and *Phaseolus vulgaris*. *Acta Physiology Plant*, *34*:, 1145–1151.

Lei, Y. (2008). Physiological responses of *Populus przewalskii* to oxidative burst caused by drought stress. *Russian Journal of Plant Physiology*, *55*(6), 857–864.

Li, F., Asami, T., Wu, X., Tsang, E. W., & Cutler, A. J. (2007). A putative hydroxysteroid dehydrogenase involved in regulating plant growth and development. *Plant Physiology*, *145*, 87–97.

Li, L., & Sheen, J. (2016). Dynamic and diverse sugar signaling. *Current Opinion Plant Biology*, *33*, 116–25

Li, G., Hu, S., Zhao, X., Kumar, S., Li, Y., Yang, J., & Hou, H. (2021). Mechanisms of the morphological plasticity induced by phytohormones and the environment in plants. *International Journal of Molecular Sciences*, *22*(2), 765.

Liu, J., He, H., Vitali, M., Visentin, I., Charnikhova, T., Haider, I. et al. (2015). Osmotic stress represses strigolactone biosynthesis in Lotus japonicus roots: Exploring the interaction between strigolactones and ABA under abiotic stress. *Planta*, *241*, 1435–1451. doi:10.1007/s00425-015-2266-8

Loka, D. A., Oosterhuis, D. M., Mattice, J. D., & McMichael, B. L. (2013). Polyamine metabolism of the cotton flower and its subtending leaf under water-deficit stress in the field. *American Journal of Plant Sciences*, *4*, 84–91.

Loreti, E., & Perata, P. (2020). The many facets of hypoxia in plants. *Plants*, *9*, 745.

Luo, L., Wang, H., Liu, X., Hu, J., Zhu, X., Pan, S. et al. (2018). Strigolactones affect the translocation of nitrogen in rice. *Plant Science*, *270*, 190–197. doi:10.1016/j.plantsci.2018.02.020

Ma, Q. J., Sun, M. H., Lu, J., Kang, H., You, C. X., & Hao, Y. J. (2018). An apple sucrose transporter MdSUT2.2 is a phosphorylation target for protein kinase MdCIPK22 in response to drought. *Plant Biotechnology Journal*, *17*, 625–637.

Ma, N., Hu, C., Wan, L., Hu, Q., Xiong, J., & Zhang, C. (2017). Strigolactones improve plant growth, photosynthesis, and alleviate oxidative stress under salinity in rapeseed (Brassica napus L.) by regulating gene expression. *Front Plant Science*,. *8*, 1671. doi:10.3389/fpls.2017.01671

MacNeill, G. J., Mehrpouyan, S., Minow, M. A., Patterson, J. A., Tetlow, I. J., Emes, M. J., & Raines, C. (2017). Starch as a source, starch as a sink: The bifunctional role of starch in carbon allocation. *Journal of Experimental Botany*, *68*, 4433–4453.

Maity, A., Paul, D., Lamichaney, A., Sarkar, A., Babbar, N., Mandal, N., & Dutta, S., Maity, P. P., & Chakrabarty, S. K. (2023). Climate change impacts on seed production and quality: Current knowledge, implications, and mitigation strategies. *Seed Science and Technology*, *51*(1), 7–38.

Martínez-Andújar, C., Ordiz, M. I., Huang, Z., Nonogaki, M., Beachy, R. N., & Nonogaki, H. (2011). Induction of 9-cis-epoxycarotenoid dioxygenase in Arabidopsis thaliana seeds enhances seed dormancy. *Proceedings of the National Academy of Sciences*, *108*(41), 17225–17229.

Marzec, M., & Muszynska, A. (2015). In silico analysis of the genes encoding proteins that are involved in the biosynthesis of the RMS/MAX/D pathway revealed new roles of strigolactones in plants. *International Journal of Molecular Sciences*, *16*, 6757–6782. doi:10.3390/ijms16046757

Medda, R., Padiglia, A., & Floris, G. (1995). Plant copper-amine oxidase. *Phytochemistry*, *39*, 1–9.

Mishra, S. K., Subralimanyam, D., & Singhal, G. S. (1997). Interactionship between salt and light stress on the primary process of photosynthesis. *Journal of Plant Physiology*, *138*, 92–96.

Montesinos,-Pereira, D., Barrameda,-Medina, Y., Romero, L., Ruiz, J. M., & Sánchez-Rodríguez, E. (2014). Genotype differences in the metabolism of proline and polyamines under moderate drought in tomato plants. *Plant Biology*, *16*(6), 1050–1057.

Murti, G. S. R., Upreti, K. K. (2000). Plant hormones. *Advances in Plant Physiology*, *3*, 109–148.

Ndayiragije, A., & Lutts, S. (2006). Exogenous putrescine reduces sodium and chloride accumulation in NaCl-treated calli of the saltsensitive rice cultivar I Kong Pao. *Plant Growth Regul. 48*, 51–63.

Niño-González, M., Novo-Uzal, E., Richardson, D. N., Barros, P. M., & Duque, P. (2019). More transporters, more substrates: The *Arabidopsis* major facilitator superfamily revisited. *Molecular Plant*, *12*, 1182–1202

Orabi, S. A., Dawood, M. G., & Saleem, S. R. (2015). Comparative study between the physiological role of hydrogen peroxide and salicylic acid in alleviating low temperature stress on tomato plants grown under sandponic culture. *Scientia Agricola*, *9*, 49–59.

Peleg, Z., & Blumwald, E. (2011). Hormone balance and abiotic stress tolerance in crop plants. *Current Opinion Plant Biology*, *14*(3), 290–295.

Phukan, U. J., Mishra, S., & Shukla, R. K. (2016). Waterlogging and submergence stress: Affects and acclimation. *Critical Reviews in Biotechnology*, *36*, 956–966.

Pierik, R., Tholen, D., Poorter, H., Visser, E. J., & Voesenek, L. A. (2006). The Janus face of ethylene: Growth inhibition and stimulation. *Trends Plant Science*, *11*(4), 176–183.

Qasim, A., Habib-ur-rehman, A., & Ashraf, M. (2006). Influence of exogenously applied brassinosteroids on the mineral nutrient status of two wheat cultivars grown under saline conditions. *Pakistan Journal of Botany*, *38*(5), 1621–1632.

Rady, M. M., & Ahmed, S. M., El-Yazal, M. A. S., & Taie, H. A. (2019). Alleviation of cadmium stress in wheat by polyamines. In M. Hasanuzzaman, M. Prasad, & K. Nahar (Eds.), *Cadmium tolerance in plants: Agronomic, molecular, signaling, and omic approaches* (pp. 463–496). Elsevier.

Rasmussen, A., Mason, M. G., De, Cuyper, C., Brewer, P. B., Herold, S., Agusti, J. et al. (2012). Strigolactones suppress adventitious rooting in Arabidopsis and pea. *Plant Physiology*, *158*, 1976–1987. doi:10.1104/pp.111.187104

Rodriguez, M., Parola, R., Andreola, S., Pereyra, C., & Martínez-Noël, G. (2019). TOR and SnRK1 signaling pathways in plant response to abiotic stresses: Do they always act according to the "yin-yang" model? *Plant Science*, *288*, 110220.

Rook, F., Hadingham, S. A., Li, Y., & Bevan, M. W. (2006). Sugar and ABA response pathways and the control of gene expression. *Plant Cell Environment*, *29*, 426–34.

Rosa, M., Prado, C., Podazza, G., Interdonato, R., González, J. A., Hilal, M., & Prado, F. E. (2009). Soluble sugars: Metabolism, sensing and abiotic stress: A complex network in the life of plants. *Plant Signal Behaviour*, *4*, 388–393.

Rudić, J., Pantelić, D., Oljača, J., & Momčilović, I. (2022). Effects of elevated temperature and salicylic acid on heat shock response and growth of potato microplants. *Horticulturae*, *8*(5), 372.

Ruiz-Lozano, J. M., Aroca, R., Zamarreño, Á. M., Molina, S., Andreo-Jiménez, B., Porcel, R. et al. (2016). Arbuscular mycorrhizal symbiosis induces strigolactone biosynthesis under drought and improves drought tolerance in lettuce and tomato. *Plant Cell Environment*, *39*, 441–452. doi:10.1111/pce.12631

Saddhe, A. A., Kundan, K., & Padmanabh, D. (2017). Mechanism of ABA signaling in response to abiotic stress in plants. In G. K. Pandey (Ed.), *Mechanism of plant hormone signaling under stress, II* (pp. 173–195). John Wiley.

Sah, S. K., Reddy, K. R., & Li, J. (2016). Abscisic acid and abiotic stress tolerance in crop plants. *Frontiers in Plant Science*, *7*, 571.

Sakr, S., Wang, M., Dédaldéchamp, F., Perez-Garcia, M. D., Ogé, L., & Hamama, L., Atanassova, R. (2018). The sugar-signaling hub: Overview of regulators and interaction with the hormonal and metabolic network. *International Journal of Molecular Sciences*, *19*, 2506.

Salmon, Y., Lintunen, A., Dayet, A., Chan, T., Dewar, R., Vesala, T., & Hölttä, T. (2020). Leaf carbon and water status control stomatal and nonstomatal limitations of photosynthesis in trees. *New Phytologist*, *226*, 690–703

Santiago, J., Dupeux, F., Round, A., Antoni, R., Park, S. Y., Jamin, M., & … & Márquez, J. A. (2009). The abscisic acid receptor PYR1 in complex with abscisic acid. *Nature*, *462*(7273), 665–668.

Riyazuddin, R., Verma, R., Singh, K., Nisha, N., Keisham, M., Bhati, K. K.,… & Gupta, R. (2020). Ethylene: A master regulator of salinity stress tolerance in plants. *Biomolecules*, *10*(6), 959.

Santner, A., & Calderon-Villalobos, L. I. A., & Estelle, M. (2009). Plant hormones are versatile chemical regulators of plant growth. *Nature Chemical Biology*, *5*(5), 301–307.

Sayyari, M., Ghanbari, F., Fatahi, S., & Bavandpour, F. (2013). Chilling tolerance improving of watermelon seedling by salicylic acid seed and foliar application. *Notulae Scientia Biologicae*, *1*, 67–73.

Seto, Y., Sado, A., Asami, K., Hanada, A., Umehara, M., Akiyama, K. et al. (2014). Carlactone is an endogenous biosynthetic precursor for strigolactones. *Proceedings of the National Academy of Sciences of the United States of America*, *111*, 1640–1645. doi:10.1073/pnas.1314805111

Shahbaz, M., Ashraf, M., & Athar, H. (2008). Does exogenous application of 24-epibrassinolide ameliorate salt induced growth inhibition in wheat (*Triticum aestivum* L.). *Plant Growth Regulator*, *55*, 51–64.

Shen, J., Zhang, Y., Ge, D., Wang, Z., Song, W., Gu, R. et al. (2019). CsBRC1 inhibits axillary bud outgrowth by directly repressing the auxin efflux carrier CsPIN3 in cucumber. *Proceedings of the National Academy of Sciences of the United States of America*, *116*, 17105–17114. doi:10.1073/pnas.1907968116

Sheteiwy, M. S., Gong, D., Gao, Y., Pan, R., Hu, J., & Guan, Y. (2018). Priming with methyl jasmonate alleviates polyethylene glycol-induced osmotic stress in rice seeds by regulating the seed metabolic profile. *Environmental and Experimental Botany*, *153*, 236–248.

Slama, I., Abdelly, C., Bouchereau, A., Flowers, T., & Savoure, A. (2015). Diversity, distribution and roles of osmoprotective compounds accumulated in halophytes under abiotic stress. *Annals of Botany*, *115*, 433–447.

Smith, T. (1975). Recent advances in the biochemistry of plant amines. *Phytochemistry*, *14*, 865–890.

Song, X., & Lu, Z., Yu, H., Shao, G., Xiong, J., & Meng, X. (2017). IPA1 functions as a downstream transcription factor repressed by D53 in strigolactone signaling in rice. *Cell Research*, *27*, 1128–1141. doi:10.1038/cr.2017.102

Sun, B., Xu, Y., Ng, K. H., & Ito, T. (2009). A timing mechanism for stem cell maintenance and differentiation in the Arabidopsis floral meristem. *Genes Development*, *23*(15), 1791–1804.

Sun, H., & Tao, J., Hou, M., Huang, S., Chen, S., Liang, Z. et al. (2015). A strigolactone signal is required for adventitious root formation in rice. *Annals of Botany*, *115*, 1155–1162. doi:10.1093/aob/mcv052

Suprasanna, P., Nikalje, G. C., & Rai, A. N. (2016). Osmolyte accumulation and implications in plant abiotic stress tolerance. In N. Iqbal, N. A. Khan, & R. Nazar (Eds.), *Osmolytes and plants acclimation to changing environment: Emerging omics technologies* (pp. 1–12). Springer.

Sziderics, A. H., Oufir, M., Trognitz, F., Kopecky, D., Matušíková, I., H. ausman J-F., & Wilhelm, E. (2010). Organ-specific defence strategies of pepper (*Capsicum annuum* L.) during early phase of water deficit. *Plant Cell Reports*, *29*, 295–305.

Tabor, C. W., & Tabor, H. (1984). Polyamines. *Annual Review of Biochemistry*, *53*, 749–790.Takahashi, Y., Cong, R., Sagor, G. H. .M., Niitsu, M., Berberich, T., & Kusano, T. (2010). Characterization of five polyamine oxidase isoforms in *Arabidopsis thaliana. Plant Cell Rep*. *29*, 955–965.

Talaat, N. B., & Todorova, D. (2022). Antioxidant machinery and glyoxalase system regulation confers salt stress tolerance to wheat (*Triticum aestivum* L.) plants treated with melatonin and salicylic Acid. *Journal of Soil Science and Plant Nutrition*, *22*(3), 3527–3540.

TH, K. (2010). Guard cell signal transduction network: Advances in understanding abscisic acid, CO_2, and Ca_2, signaling. *Annual Review of Plant Biology*, *61*, 561–591.

Tiburcio, A. F., Kaur-Sawhney, R., & Galston, A. W. (1990). Polyamine metabolism. In P. K. Stumpf & E. E. Conn (Eds.), *The biochemistry of plants* (pp. 283–235). Academic Press.

Todorova, D., Moskova, I., Sergiev, I., Alexieva, V., & Mapelli, S. (2008). *Changes in endogenous polyamines and some stress markers content induced by drought, 4PU-30 and abscisic acid in wheat plants. Abiotic Stress and Plant Responses*. New Delhi, India: IK International Publishing House, pp. 205–215.

Trebichalský, P., Tóth, T., Bajčan, D., Harangozo, L., & Vollmannová, A. (2017). Hormonal changes in spring barley after triazine herbicide treatment and its mixtures of regulators of polyamine biosynthesis. *Potravinarstvo Slovak Journal of Food Sciences*, *11*(1), 156–161.

Ullah, A., Manghwar, H., Shaban, M., Khan, A. H., Akbar, A., Ali, U., & … & Fahad, S. (2018). Phytohormones enhanced drought tolerance in plants: A coping strategy. *Environmental Science and Pollution Research*, *25*, 33103–33118.

Vardhini, B. V., Sujatha, E., & Rao, S. S. (2011b) Brassinosteroids: Alleviation of water stress in certain enzymes of sorghum seedlings. *Journal of Phytologist*, *3*, 38–43.

Wang, B.-Q., & Zhang, Q.-F., Liu, J.-H., & Li, G.-H. (2011). Overexpression of *PtADC* confers enhanced dehydration and drought tolerance in transgenic tobacco and tomato: Effect on ROS elimination. *Biochemical and Biophysical Research Communications*, *413*, 10–16.

Wang, Q., Ding, T., Gao, L., Pang, J., & Yang, N. (2012). Efect of brassinolide on chilling injury of green bell pepper in storage. *Scientia Horticulturae*, *144*, 195–200.

Wang, Z., Zheng, P., Meng, J., & Xi, Z. (2015). Effect of exogenous 24- epibrassinolide on chlorophyll fluorescence, leaf surface morphology and cellular ultrastructure of grape seedlings (*Vitis vinifera* L.) under water stress. *Acta Physiol Plant*, 37, 1729. doi:10.1007/s11738-014-1729-z.

Wang, F., Guo, Z., Li, H., Wang, M., Onac, E., Zhou, J., … & Zhou, Y. (2016). Phytochrome A and B function antagonistically to regulate cold tolerance via abscisic acid-dependent jasmonate signaling. *Plant Physiology*, *170*(1), 459–471.

Wang, W., Wang, X., Huang, M., Cai, J., Zhou, Q., Dai, T., & Jiang, D. (2021). Alleviation of field low-temperature stress in winter wheat by exogenous application of salicylic acid. *Journal of Plant Growth Regulation*, *40*, 811–823.

Wani, S. H., Kumar, V., Shriram, V., & Sah, S. K. (2016). Phytohormones and their metabolic engineering for abiotic stress tolerance in crop plants. *The Crop Journal*, *4*(3), 162–176.

Waters, M. T., Brewer, P. B., Bussell, J. D., Smith, S. M., & Beveridge, C. A. (2012). The Arabidopsis ortholog of rice DWARF27 acts upstream of MAX1 in the control of plant development by strigolactones. *Plant Physiology*, *159*, 1073–1085. doi:10.1104/pp.112.196253

Wolters, H., & Jürgens, G. (2009). Survival of the flexible: Hormonal growth control and adaptation in plant development. *Nature Reviews Genetics*, *10*(5), 305–317.

Wurzinger, B., Nukarinen, E., Nägele, T., Weckwerth, W., & Teige, M. (2018). The SnRK1 kinase as central mediator of energy signaling between different organelles. *Plant Physiology*, *176*, 1085–94

Xia, X. J., Wang, Y. J., Zhou, Y. H., Tao, Y., Mao, W. H., & Shi, K. (2009). Reactive oxygen species are involved in brassinosteroid-induced stress tolerance in cucumber. *Plant Physiology*, *150*, 801–814.

Xiong, L., & Zhu, J.-K. (2002). Molecular and genetic aspects of plant responses to osmotic stress. *Plant Cell Environment*, *25*(2), 131–139.

Xu, J., Zha, M., Li, Y., Ding, Y., & Chen, L., Ding, C. et al. (2015). The interaction between nitrogen availability and auxin, cytokinin, and strigolactone in the control of shoot branching in rice (Oryza sativa L.). *Plant Cell Report*, *34*, 1647–1662. doi:10.1007/s00299-015-1815-8

Xue-Xuan, X., Hong-Bo, S., Yuan-Yuan, M., Gang, X., Jun-Na, S., Dong-Gang, G., & Cheng-Jiang, R. (2010). Biotechnological implications from abscisic acid (ABA) roles in cold stress and leaf senescence as an important signal for improving plant sustainable survival under abiotic-stressed conditions. *Critical Reviews in Biotechnology*, *30*(3), 222–230.

Yang, J., Zhang, J., Liu, K., Wang, Z., & Liu, L. (2007). Involvement of polyamines in the drought resistance of rice. *Journal of Experimental Botany*, *58*(6): 1545–1555.

Yang, W., Yin, Y., Li, Y., Cai, T., Ni, Y., Peng, D., & Wang, Z. (2014). Interactions between polyamines and ethylene during grain filling in wheat grown under water deficit conditions. *Plant Growth Regulator*, *72*(2), 189–201.

Yao, R., Ming, Z., Yan, L., Li, S., Wang, F., Ma, S. et al. (2016). DWARF14 is a non-canonical hormone receptor for strigolactone. *Nature*, *536*, 469–473. doi:10.1038/nature19073

Yao, R., & Wang, L., Li, Y., Chen, L., Li, S., & Du, X. et al. (2018). Rice DWARF14 acts as an unconventional hormone receptor for strigolactone. *Journal of Experimental Botany*, *69*, 2355–2365. doi:10.1093/jxb/ery014

Ye, K., Li, H., Ding, Y., Shi, Y., Song, C., Gong, Z., & Yang, S. (2019). Brassinosteroid-insensitive2 negatively regulates the stability of transcription factor ICE1 in response to cold stress in arabidopsis. *Plant Cell*, *31*(11), 2682–2696.

Yoneyama, K., Xie, X., Kim, H. I., Kisugi, T., Nomura, T., Sekimoto, H. et al. (2012). How do nitrogen and phosphorus deficiencies affect strigolactone production and exudation? *Planta*, *235*, 1197–1207. doi:10.1007/s00425-011-1568-8

Yu, J. Q., Huang, L. F., Hu, W. H., Zhou, Y. H., Mao, W. H., Ye, S. F., & Nogues, S. (2004). A role of brassinosteroids in the regulation of photosynthesis in *Cucumis sativus*. *Journal of Experimental Botany*, *55*, 1135–1143.

Yu, J., Cang, J., Lu, Q., Fan, B., Xu, Q., & Li, W., & Wang, X. (2020). ABA enhanced cold tolerance of wheat 'dn1'via increasing ROS scavenging system. *Plant Signaling & Behavior*, *15*(8), 1780403.

Yu, Y., Jin, C., Sun, C., Wang, J., Ye, Y., Zhou, W., Lu, L., & Lin, X. (2016). Inhibition of Ethylene Production by Putrescine Alleviates Aluminium-Induced Root Inhibition in Wheat Plants. *Science Report*, *6*, 18888.

Yuan, G., Jia, C., Li, Z., Sun, B., Zhang, L., Liu, N., & Wang, Q. (2010). Effect of brassinosteroids on drought resistance and abscisic acid concentration in tomato under water stress. *Scientia Horticulturae*, *126*, 103–108.

Zehra, A., Raytekar, N. A., Meena, M., & Swapnil, P. (2021). Efficiency of microbial bio-agents as elicitors in plant defense mechanism under biotic stress: A review. *Current Research in Microbial Sciences*, *2*, 100054.

Zhang, M., Zhai, Z., Tian, X., Duan, L., & Li, Z. (2008). Brassinolide alleviated the adverse effect of water deficits on photosynthesis and the antioxidant of soybean (*Glycine max* L.). *Plant Growth Regulator*, *56*(3), 257–264

Zhang, S., Hu, J., Zhang, Y., Xie, X. J., & Knapp, A. (2007). Seed priming with brassinolide improves lucerne (Medicago sativa L.) seed germination and seedling growth in relation to physiological changes under salinity stress. *Australian Journal of Agricultural Research*, *58*, 811–815.

Zhang, Y., Liang, Y., Zhao, X., Jin, X., Hou, L., Shi, Y., & Ahammed, G. J. (2019). Silicon compensates phosphorus defcit-induced growth inhibition by improving photosynthetic capacity, antioxidant potential, and nutrient homeostasis in tomato. *Agronomy*, *9*(11), 733.

Zhang, Y., Hartinger, C., Wang, X., & Friml, J. (2020). Directional auxin fluxes in plants by intramolecular domain-domain coevolution of PIN auxin transporters. *New Phytologist*, *227*, 1406–1416. doi:10.1111/nph.16629

Zhang, Y., Van, Dijk, A. D., Scaffidi, A., Flematti, G. R., Hofmann, M., Charnikhova, T. et al. (2014). Rice cytochrome P450 MAX1 homologs catalyze distinct steps in strigolactone biosynthesis. *Nature Chemical Biology*, *10*, 1028–1033. doi:10.1038/nchembio.1660

Zhao, B., Wu, T. T., Ma, S. S., Jiang, D. J., Bie, X. M., Sui, N. et al. (2019). TaD27-B gene controls the tiller number in hexaploid wheat. *Plant Biotechnology Journal*,. *18*, 513–525. doi:10.1111/pbi.13220

Zhao, G., Song, Y., Wang, Q., Yao, D., Li, D., Qin, W., & … & Wu, J. (2020). *Gossypium hirsutum* salt tolerance is enhanced by overexpression of G. *arboreum JAZ1*. *Frontiers in Bioengineering and Biotechnology*, *8*, 157.

Zhao, L. H., Zhao, L. H., Zhou, X. E., Yi, W., Wu, Z., Liu, Y. et al. (2015). Destabilization of strigolactone receptor DWARF14 by binding of ligand and E3-ligase signaling effector DWARF3. *Cell Research*, *25*, 1219–1236. doi:10.1038/cr.2015.122

Zhou, Q., & Yu, B. (2010). Changes in content of free, conjugated and bound polyamines and osmotic adjustment in adaptation of vetiver grass to water deficit. *Plant Physiology and Biochemistry*, *48*, 417–425.

Zhu, M., Liu, Y., Cai, P., Duan, X., & Sang, S, Qiu, & Z. (2022). Jasmonic acid pretreatment improves salt tolerance of wheat by regulating hormones biosynthesis and antioxidant capacity. *Frontiers in Plant Science*, *13*, 968477. doi:10.3389/fpls.2022.968477

Zulfiqar, F., Akram, N. A., & Ashraf, M. (2020). Osmoprotection in plants under abiotic stresses: New insights into a classical phenomenon. *Planta*, *251*, 3.

7 Integration of Multi-Omics Approaches and Speed Breeding for Plant Stress Tolerance

Challenges and Opportunities

Asad Azeem, Furqan Tayyab, Ghulam Mustafa, Muhammad Sarwar Khan, and Faiz Ahmad Joyia

7.1 INTRODUCTION

In the face of burgeoning global population and climate uncertainties, the imperative to enhance crop resilience and ensure food security has never been more pressing. Numerous environmental conditions can cause plant stress, which is a serious threat to agricultural productivity (Zhang & Batley, 2020). Innovative approaches at the nexus of genomes, transcriptomics, proteomics, and metabolomics, collectively, multi-omics, are being combined with state-of-the-art technologies like speed breeding to strengthen our agricultural systems against these threats (Ali et al., 2022). This all-encompassing strategy has great potential for streamlining breeding cycles, elucidating the intricacies of plant stress responses, and clearing the path for the creation of crop types that can withstand stress.

The incorporation of multi-omics methodologies provides a thorough comprehension of the complex molecular and metabolic networks that regulate plant stress responses (Roychowdhury et al., 2023). Through the simultaneous analysis of genomic, transcriptomic, proteomic, and metabolomic alterations, scientists are able to obtain unique insights into the dynamic adaptations that plants go through while they are under stress. By using an integrative approach, it is possible to identify important regulatory nodes and biomarkers linked to stress tolerance, which makes it easier to precisely manipulate features in crops to increase their resilience (Ali et al., 2021).

The ground-breaking idea of speed breeding has emerged as a potent tool to speed up the breeding process when combined with multi-omics (Razzaq et al., 2021). Speed breeding shortens the time needed to produce new plant generations by rapidly accelerating the growth and development cycles of plants through manipulation of environmental parameters like temperature, light, and photoperiod. The combination of multi-omics and speed breeding not only accelerates the process of identifying genes and pathways that respond to stress but also optimizes the breeding pipeline to guarantee a rapid turnover of enhanced crop types (Naik et al., 2024). Nevertheless, there are certain difficulties in combining multi-omics with fast breeding. Significant challenges include the requirement for defined procedures, data integration challenges, and technical complexity. In addition, ethical issues and public acceptability of fast-bred or genetically modified crops demand a fair discussion to guarantee the appropriate and sustainable application of these technologies (Azeem et al., 2023).

The integration of multi-omics and speed breeding for plant stress tolerance was investigated in this study, and we looked at the opportunities and problems that exist at the nexus of these

DOI: 10.1201/9781003543237-7

innovative techniques (Singha & Singha, 2024). To meet the pressing need for stress-tolerant crops in the face of a changing climate and rising global food demand, we seek to clarify a path forward for utilizing the full potential of multi-omics and speed breeding by navigating through the challenges of data integration, technological harmonization, and societal implications.

7.2 SPEED BREEDING: ENHANCING PLANT BREEDING EFFICIENCY

To speed up the conventional breeding process, a novel idea in plant breeding called 'speed breeding' involves changing environmental factors to quicken the growth and development cycles of plants (Figure 7.1) (Kashyap et al., 2023). With the potential to greatly increase plant breeding efficiency and allow researchers and breeders to produce numerous generations of crops in a fraction of the time needed for traditional breeding procedures, this novel approach has attracted a lot of attention. Optimization of environmental parameters like light, temperature, and photoperiod to foster fast plant development is the secret to speed breeding (Rodríguez del Río et al., 2023). Through the use of high-intensity lighting systems, like LED arrays, and prolonged exposure to light, speed breeding promotes faster photosynthesis and, in turn, faster plant development. Furthermore, by creating a continuous development cycle by precisely calibrated modification of day length and controlled temperature regimes, researchers are able to overcome the inherent seasonal limitations of conventional breeding operations (Lin et al., 2023). The fact that speed breeding significantly shortens the time needed to produce new plant generations is among its main benefits. A new crop variety may take several years to generate using traditional breeding procedures; however, speed breeding may shorten this period to a few months (Swami et al., 2023). This acceleration is especially helpful in areas where seasonal variations are significant or in reaction to new issues such as changing pest and disease pressures and climate change.

Not only can speed breeding be used to successfully select for specific genetic combinations, but it can also be applied to a wide variety of crops, such as vegetables, cereals, and ornamental plants Not only can speed breeding be used to successfully select for specific genetic combinations, but it can also be applied to a wide variety of crops, such as vegetables, cereals, and ornamental plants (Potts et al., 2023). Because of its adaptability, speed breeding is a valuable tool for breeding programs that aim to address a variety of agricultural and horticultural needs. Additionally, because plant generations are produced quickly, researchers can identify desirable traits and accelerate the development of improved crop varieties. However, despite its potential advantages, speed breeding is not without its challenges (Schilling et al., 2023). Care must be taken when managing stress factors to prevent unintended consequences on plant physiology and performance. Furthermore, ethical considerations, public acceptance, and regulatory frameworks must be taken into account. To sum up, speed breeding is a revolutionary method in plant breeding that has the potential to

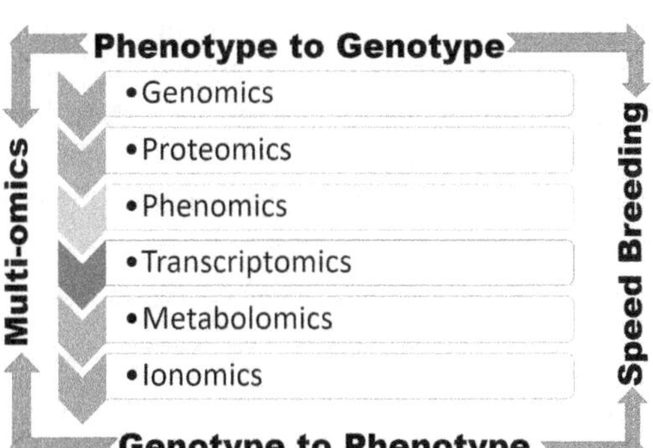

FIGURE 7.1 Integration of multi-omics approach and speed breeding for plant stress.

drastically increase the productivity of creating new crop types. In light of the growing demand for resilient and high-yielding crops worldwide, speed breeding is becoming an increasingly important technique for addressing these issues and advancing secure and sustainable food production in a world that is changing quickly (Hussain et al., 2023).

7.3 MULTI-OMICS: BASIC BIOLOGY TO CROP IMPROVEMENT

Multi-omics is a broad and integrated method that includes proteomics, metabolomics, transcriptomics, and genomes (Figure 7.1). It has become a potent instrument that can be used to revolutionize crop enhancement strategies or unravel fundamental biological processes (Chaturvedi et al., 2024). With the unparalleled degree of detail this comprehensive methodology offers researchers, it becomes possible to comprehend the complex molecular systems that underpin plant biology on a deeper level. Multi-omics has developed into a vital link that spans the fields of fundamental biology and applied agricultural science, linking the former to the latter. Fundamentally, multi-omics entails the methodical examination of all the components of a biological system, including genes (genomics), RNA transcripts (transcriptomics), proteins (proteomics), and metabolites (metabolomics) (Athieniti & Spyrou, 2023). This integration offers a comprehensive understanding of cellular processes and responses to external stimuli, including environmental challenges, by allowing scientists to investigate how information encoded in the plant's genome is transcribed into RNA, translated into proteins, and further metabolically processed.

Multi-omics has been essential in helping us understand the intricacies of plant development, signaling pathways, and stress responses in the context of basic biology (Singhal et al., 2021). By using this method, scientists are able to shed light on the complex relationships that exist between genes, proteins, and metabolites and the regulatory networks that control development, reproduction, and environmental adaptation. Basic research provides the fundamental knowledge for focused studies aimed at improving plant characteristics including yield, nutritional value, and resistance to biotic and abiotic stressors (Gonzalez Guzman et al., 2022). Beyond basic comprehension, multi-omics holds revolutionary potential for crop enhancement. Researchers are able to find important genes, proteins, and metabolic pathways linked to desirable qualities by combining large datasets from genomics, transcriptomics, proteomics, and metabolomics. With the use of this knowledge, precise genome alteration of plants can be achieved through genome editing, genetic engineering, or enhanced breeding methods to produce crops with better qualities (Yang et al., 2021). Using multi-omics data to guide crop improvement efforts, production, nutritional quality, and resistance to pests, diseases, and environmental challenges can all be improved. For the sake of global food security, crop adaptation to survive climatic changes and resource constraints is especially important. The discovery of molecular targets for intervention is made possible by multi-omics-driven techniques, which hastens the manufacturing of crops that can flourish in a variety of tough agroecosystems (Wijayawardene et al., 2023). Although multi-omics has unquestionably transformed our understanding of plant biology and agricultural development, there are still issues that need to be resolved, such as standardizing methodology, integrating massive and complicated information, and ethical concerns around genetic manipulation (Yang et al., 2021). But multi-omics, which offers a revolutionary route from fundamental biology to the creation of robust and high-yielding crops for a changing world, is driving modern biotechnology forward with the promise of realizing the full potential of plant genomes for sustainable agriculture.

7.4 GENOMICS

A key component of contemporary agricultural research is the study of genomes or an organism's whole genetic makeup. This is especially true when trying to design crops that are more resilient to stress (Yang et al., 2021). The demand for robust crops has never been greater as the world's climate continues to change and environmental stressors like salinity, drought, and severe temperatures become more frequent (Aslam et al., 2023). To develop focused crop improvement methods, genomics offers a potent toolkit for comprehending the genetic basis of stress tolerance in plants (Iqbal et al., 2023). The identification and characterization of stress-responsive genes is one of the main ways that genomics has improved crop stress resistance (Azeem, Abbas, et al., 2023). By utilizing methods like transcriptomics, genome sequencing, and comparative genomics, scientists may thoroughly examine the complete genetic composition of plants and identify specific genes that are essential for stress adaptation. Knowing how these genes work sheds light on the molecular processes that underlie stress tolerance and offers useful opportunities for genetic modification. Moreover, the development of molecular markers linked to stress tolerance features is facilitated by genomic information. Within the plant genome, genetic markers function as signposts that let breeders more effectively track and select desired features (Shimizu et al., 2020). Through the use of marker-assisted selection (MAS), which speeds up the conventional breeding process, scientists may now find and grow plants with desirable stress tolerance traits without having to rely entirely on labor- and resource-intensive field trials (Boopathi & Boopathi, 2020). Powerful gene editing tools like CRISPR/Cas9, which allow for precise and targeted alterations of the plant genome, have been made possible by advancements in genomics (Montecillo et al., 2020). With the use of this ground-breaking instrument, scientists can directly alter or add particular genes linked to stress tolerance, giving rise to crops that are more resilient to harsh environmental circumstances. Gene editing's accuracy and effectiveness provide unmatched control over a crop's genetic composition, hastening the creation of stress-tolerant cultivars (Verma et al., 2023).

Deciphering the intricacy of gene networks and regulatory mechanisms involved in stress responses is another important function of genomics. A comprehensive understanding of the plant's adaptive mechanisms can be gained by examining the interactions and coordination of genes in response to stress (Raza et al., 2020). With this information, scientists can create methods for modifying entire genetic networks, providing a more thorough method of crop development for stress tolerance. Even though genomics has a lot of potential to improve crops, there are still obstacles to overcome, such as the requirement for substantial genomic resources, knowledge of bioinformatics, and handling the moral issues raised by genetic modification (Srushti et al., 2022). However, the incorporation of genomics into breeding programs marks a paradigm shift in agriculture that is revolutionary, giving scientists and breeders access to never-before-seen resources to create crops that are resilient to changing environmental conditions. In the end, genomics plays a major role in the effort to maintain sustainable agriculture and global food security in the face of climate change.

7.5 TRANSCRIPTOMICS

A significant tool in crop development, transcriptomics is a field of molecular biology that studies the entire set of RNA transcripts within a cell (Guo et al., 2021). It is especially useful for improving stress tolerance. Transcriptomics offers important insights into the molecular reactions of plants to stress by interpreting the dynamic variations in gene expression under various environmental situations (Ambrosino et al., 2020). This information serves as the foundation for focused crop breeding techniques meant to create cultivars that are resilient to difficult environmental stresses. The identification of genes that respond to stress and the comprehension of their patterns of expression represent one of transcriptomics' main contributions to crop development (Wang et al., 2020). Researchers may profile the whole transcriptome of stressed plants using methods such as RNA sequencing (RNA-seq), which makes the genes that are upregulated or downregulated in response to particular stressors visible. The identification of these differentially expressed genes paves the way for additional research and modification. These genes frequently encode proteins involved in stress tolerance processes (Manna et al., 2021). Deciphering the spatiotemporal patterns of gene expression is essential to understanding the intricacy of stress reactions. Researchers can see how variations in gene expression patterns occur throughout time and across many tissues thanks to transcriptomics. By identifying key points in a plant's life cycle when stress tolerance systems are most active, this information helps to manage crops or breed new plants with focused treatments. Additionally, transcriptomics is essential for identifying the signaling pathways and regulatory networks connected to stress reactions (Shi et al., 2024). Researchers can improve crop stress tolerance by manipulating important transcription factors and signaling molecules involved in stress signaling cascades. With the use of this knowledge, precise breeding techniques to modify the expression of particular genes or pathways linked to stress responses can be developed, including gene editing and molecular breeding. Furthermore, transcriptomics aids in the creation of molecular markers for characteristics related to stress tolerance. Through methods like MAS, specific RNA molecules linked to stress adaptation can be identified and used to create markers for use in breeding programs (Iqbal et al., 2021). This makes it possible to specifically select plants that have the appropriate stress tolerance qualities, which speeds up the breeding process. Transcriptomics has hurdles in its application to crop development, much like any other advanced technology. Careful attention is needed in the fields of data processing, bioinformatics, and the integration of multi-omics information (Ahmar et al., 2020). Furthermore, it continues to be difficult to correlate identified alterations in gene expression with functional characteristics and field performance (Rivera et al., 2021). To sum up, transcriptomics provides a dynamic and comprehensive understanding of how plants react to stress at the molecular level, making it a revolutionary tool for crop improvement. Researchers may deliberately improve crop stress tolerance by understanding the language of gene expression, which will help create resilient cultivars that can flourish in the face of increasingly erratic environmental conditions.

7.6 PROTEOMICS

Large-scale protein analysis has become a potent tool in agricultural research, especially when it comes to crop development for stress tolerance (Pazhamala et al., 2021). Gaining knowledge of the dynamic alterations in the plant proteome during stressful situations is essential for comprehending the molecular processes that control stress reactions (Mergner & Kuster, 2022). Through the process of deciphering the intricate interactions between proteins involved in stress adaptation, proteomics plays a major role in the production of crops that are resilient to environmental stresses. The identification and characterization of stress-responsive proteins are two of the main ways that proteomics has improved crops. Plants undergo complex modifications in protein expression to adapt and survive under stressful situations like drought, salt, or extremely high or low temperatures (Shakespear et al., 2024). Researchers can examine all of the proteins in a plant using proteomic techniques, such as mass spectrometry and two-dimensional gel electrophoresis, to determine which proteins are upregulated or downregulated in response to particular stressors. These proteins that exhibit differential expression frequently have important functions in stress tolerance systems, making them possible targets for crop enhancement techniques.

Understanding the post-translational modifications (PTMs) of proteins – which are crucial for controlling their activity – was made possible by proteomics (Lee et al., 2023). PTMs include ubiquitination, glycosylation, and phosphorylation, which can affect how a protein functions in response to stress. Researchers can target gains in stress tolerance by manipulating stress-responsive proteins by understanding how these adjustments take place (Lohani et al., 2022). Another important application of proteomics is the detection of protein–protein interactions. Through the process of mapping out the intricate networks of interactions between proteins in plants, scientists are able to better comprehend the regulatory processes that underlie stress responses. With this information, therapies to improve stress tolerance can be created either by adjusting the quantity of particular proteins or by changing the way proteins interact with one another. The identification of biomarkers linked to characteristics of stress tolerance was also made possible by proteomics (Lohani et al., 2022). Through the process of discovering proteins that function as markers of a plant's stress tolerance, scientists can create molecular markers that can be applied to breeding initiatives. During the breeding process, these markers facilitate the selection of plants with increased stress tolerance, which speeds up the generation of superior crop varieties. Proteomics has enormous potential for improving crops (Singh et al., 2021), but there are several obstacles in the way, such as the complexity of protein samples, data analysis, and the requirement for specialized equipment. Additionally, interdisciplinary cooperation and integration with other omics methods are necessary to translate proteomic findings into useful applications in crop breeding. To sum up, proteomics is an important factor to consider while searching for crops that can withstand stress. Proteomics gives researchers a full picture of the plant's protein landscape under stress, which helps them uncover important players in stress adaptation. This information then informs focused crop breeding efforts, promoting the emergence of robust cultivars that can flourish in the face of adversity.

7.7 METABOLOMICS

In the field of crop development for stress tolerance, metabolomics – the study of tiny molecules or metabolites inside a biological system – has become a game-changing technology (Raza, 2022). Metabolism offers a comprehensive picture of the physiological reactions and biochemical pathways that control stress adaptation by studying the whole suite of metabolites seen in plants under stress. Having this knowledge is essential for creating crops that can survive in harsh environments. The discovery of metabolites that respond to stress is one of the main ways that metabolomics has improved crops. Plants dynamically modify their metabolic profiles in response to stress to adapt to the unfavorable environment (Singh et al., 2020). Researchers may identify and quantify these changes by using metabolomic techniques such as nuclear magnetic resonance spectroscopy and mass spectrometry, which let them discover compounds that are directly linked to stress tolerance. These metabolites can be targeted for alteration in crop breeding programs and are frequently used as markers for stress response. Researchers have been able to identify the metabolic pathways that become active or inactive in response to stress thanks to metabolomics. Knowing how plants modify their metabolism in response to unfavorable circumstances sheds light on the tactics used for stress tolerance. Through the manipulation of important enzymes or regulatory processes within these metabolic pathways, this knowledge can direct the evolution of crops with increased stress tolerance. The identification of bioactive substances with anti-stress functions is aided by metabolomics. Certain plant metabolites function as signaling molecules, antioxidants, or osmoprotectants and are essential in reducing the negative effects of stress (Nawaz et al., 2022).

Plant stress responses are better understood when metabolomics is combined with other omics techniques including transcriptomics, proteomics, and genomes. By examining the complex relationships between genes, proteins, and metabolites, systems biology provides a comprehensive understanding of the molecular mechanisms controlling stress tolerance. Having this comprehensive knowledge is essential for creating strategies that increase crops' resistance to stress. The identification of metabolic biomarkers linked to features of stress tolerance was also made easier by metabolomics (Hamany Djande et al., 2020). These biomarkers can be utilized for screening and selection in breeding programs, as well as serving as indicators of a plant's resistance to particular stresses. The breeding process is streamlined by the use of metabolic markers, which makes it possible to quickly identify plants that possess the requisite stress tolerance features. Although metabolomics has great potential to improve crops, there are still difficulties, such as the difficulty in identifying metabolites and interpreting data. Furthermore, cooperation between plant biologists, bioinformaticians, and breeders is necessary to translate metabolomic research into useful applications in crop breeding. In summary, metabolomics is an effective technique that offers a dynamic, real-time understanding of plant metabolism in stressful environments. Researchers can learn a great deal about the stress-reduction tactics plants use by deciphering the language of metabolites. This information forms the basis for focused crop breeding interventions, which in turn help create stress-tolerant crops for resilient and sustainable agriculture.

7.8 GENETIC ENGINEERING

Genetic engineering is a targeted strategy to improve features that allow crops to tolerate environmental challenges like drought, salinity, severe temperatures, and pest infestations (Hafeez et al., 2023). It can introduce, change, or mute certain genes in plants. The insertion of genes encoding stress-responsive proteins is one of the main uses of genetic engineering in crop development for stress tolerance. Genes from naturally stress-tolerant organisms – like some bacteria or other plants – are identified by researchers and inserted into agricultural plants (Yadav et al., 2023). These transgenes could encode proteins involved in antioxidant defense, osmotic control, or other defense mechanisms that help plants survive harsh environments. Genetic engineering provides crops with more instruments to lessen the effects of stress by inserting such genes (Lee et al., 2023). Enhancement of current stress tolerance mechanisms in crops is possible through genetic engineering modification of endogenous genes. For instance, scientists can change the expression levels of genes linked to pathways connected to stress, nutrient uptake, and efficient use of water. By adjusting gene expression, a plant can better respond to stress and distribute resources more effectively in demanding environmental circumstances (Kudapa et al., 2023).

The augmentation of crop resilience by the alteration of hormone signaling pathways is another potent use of genetic engineering. Plant hormones that are important in stress reactions, such as jasmonic acid and abscisic acid, can be produced and responded to differently when genetic alterations are made (Singh & Roychoudhury, 2023). Through the manipulation of these pathways, scientists can alter the way plants perceive and respond to stress, thereby improving their capacity for adaptation and survival (Aslam et al., 2023). Crop enhancement also includes the creation of genetically modified crops that are resistant to biotic stressors like pests and diseases (Riaz et al., 2024). Sustainable agricultural practices can be promoted by reducing the need for chemical pesticides by including genes that confer resistance to particular illnesses or pests (Ran et al., 2023). The creation of genetically modified (GM) drought-tolerant crops is a noteworthy example of genetic engineering for stress tolerance. These crops frequently carry genes that control stomatal opening and closure, improve root architecture, or increase water usage efficiency (Vadez et al., 2023). These alterations lessen the detrimental impacts of drought stress by assisting plants in maintaining proper hydration levels during periods of water scarcity. The application of genetic engineering to crop enhancement is not without difficulties and worries, despite the possible advantages. Globally, there are differences in public opinion, ethical issues, and regulatory frameworks regarding GM organisms. For the responsible and sustainable application of this technology, it is imperative to strike a balance between addressing societal issues and maximizing the promise of genetic engineering for stress tolerance. In conclusion, by adding, changing, or controlling particular genes, genetic engineering significantly advances crop development for stress tolerance (Ul-Allah et al., 2023). As scientists work to understand the complexities of plant genomes and stress responses, genetic engineering provides a focused and effective way to create crops that can withstand harsh environmental conditions, promoting sustainable agriculture and global food security.

7.9 GENE EDITING TECHNIQUES

CRISPR/Cas9 gene editing tools have completely changed the field of crop development for stress tolerance (Yadav et al., 2023). Researchers now have unparalleled control over the genetic composition of crops thanks to these focused and accurate technologies that allow them to make specific modifications in plant DNA. In addition to conventional breeding techniques and genetic engineering, gene editing has become a revolutionary tool for creating crops that are more resilient to environmental challenges including salinity, drought, and high temperatures (Yadav et al., 2023). The CRISPR/Cas9 technology, which was modified from the immune system of bacteria, enables researchers to precisely edit or modify particular genes inside the genome of plants (Erdoğan et al., 2023). Using a guide RNA that is designed to match the sequence of the target gene, the Cas9 enzyme is guided to the desired site in the DNA to induce modifications, deletions, or insertions. Crop development and genetic research have advanced more quickly thanks to CRISPR-efficiency Cas9's accuracy and efficiency. Targeting and directly altering genes linked to stress tolerance is one of the main benefits of gene editing. Researchers can utilize CRISPR/Cas9 to optimize or improve the functioning of particular genes that are essential to stress response pathways (Kumar et al., 2023). For instance, it is possible to precisely modify genes related to drought response, osmotic control, or antioxidant defense mechanisms to increase a plant's resistance to environmental stressors like water scarcity.

Crops with better stress-tolerant features, such as more effective water utilization, changed root architecture, or greater nutrient uptake, can be created through gene editing (Kumar et al., 2023). Through precise genetic alterations, researchers can customize these features to meet the particular needs presented by various environmental stressors. The development of non-transgenic crops with decreased vulnerability to pests or illnesses is another noteworthy use of gene editing. Researchers can increase a plant's innate resistance to diseases by altering genes linked to defensive mechanisms. This can lessen the need for chemical pesticides and support sustainable agricultural methods. The quick generation of crop types resistant to stress is also made possible by the effectiveness and speed of gene editing techniques. Gene editing makes it possible to create modified plants more quickly than with traditional breeding approaches, which can take years to obtain desired features. This faster schedule is essential for meeting the pressing demand for crops that can adjust to shifting environmental circumstances (Iqbal et al., 2023). Although gene editing has enormous potential to improve agricultural products, public acceptance, legal frameworks, and ethical issues all have a significant role in how widely this technology is used. Responsible and sustainable agriculture operations require striking a balance between addressing societal concerns and utilizing the promise of gene editing for stress tolerance (Das et al., 2023). To sum up, CRISPR/Cas9 and other gene editing methods have shown to be effective tools in the search for stress-tolerant crops. These methods' accuracy, speed, and adaptability open up new avenues for customizing plant genomes to address the demands of a changing climate, advancing the creation of resilient and sustainable agricultural systems.

7.10 HOW TO INTEGRATE SPEED BREEDING AND OMICS?

Select Target Crops and Stressors:

- Determine which particular crops you want to improve your ability to withstand stress.
- Identify the stresses, such as drought, salt, or extremely high or low temperatures, that are pertinent to your target crops (Zhang et al., 2023).

Optimize Speed Breeding Conditions:

- Optimize the speed breeding procedures to produce the best environment possible for faster plant development (Swami et al., 2023).
- Modify temperature, photoperiod, and light intensity to accelerate the turnover of plant generations and reduce generation times.

Establish Baseline Omics Data:

- Under normal growing conditions, perform baseline Omics analysis (genomics, transcriptomics, proteomics, and metabolomics) on the target crops (Oh et al., 2023).

- Provide extensive datasets that shed light on the crops' genetic, transcriptomic, proteomic, and metabolic characteristics.

Induce Stress Conditions:

- Put the plants through simulated stress situations that are appropriate for the selected stressors (Salse et al., 2023).
- Use stress management techniques throughout particular growth phases to replicate real-world conditions and record dynamic reactions.

Conduct Time-Series Omics Analyses:

- To capture the dynamic changes in protein abundance, metabolite levels, and gene expression during stress, use time-series Omics studies.
- Identify the essential proteins, metabolites, and genes that are sensitive to stress and are involved in the adaptive mechanisms of plants (Chaturvedi et al., 2024).

Integrate Multi-Omics Data:

- Combine data from proteomics, metabolomics, transcriptomics, and genomes to provide a thorough grasp of the molecular reactions to stress.
- Determine important pathways, regulatory nodes, and possible genetic modification targets (Kitashova et al., 2023).

Implement Genetic Editing or Modification:

- Modify particular genes found through the Omics investigations using gene editing methods like CRISPR/Cas9.
- Optimize genetic alterations to improve stress tolerance features, such as increased antioxidant defense systems, better water usage efficiency, or changed root architecture (Akram et al., 2023).

Validate Stress Tolerance Traits:

- Test the stress tolerance qualities by putting modified or altered plants under controlled stress.
- Keep an eye on and evaluate the plants' performance with regard to yield, stress tolerance, and other pertinent characteristics (Wen et al., 2023).

Select and Propagate Improved Varieties:

- To facilitate continued multiplication, choose plants that exhibit the appropriate stress tolerance features.
- To ensure that the desired genetic alterations are inherited and to speed up the breeding process, use MAS (Al-Ashkar et al., 2023).

Iterative Optimization:

- Repeat the procedure, taking into account the input from every iteration to continuously improve the speed breeding and Omics integration.
- Adjust the protocols in light of the crops' observed stress-related performance (Sharma et al., 2023).

Ethical Considerations and Regulatory Compliance:

- Engage with pertinent stakeholders and communicate openly about the techniques used in crop improvement.
- Take ethical issues into account and follow legal requirements pertaining to genetic modification and breeding procedures. This method contributes to resilient and sustainable agriculture by speeding up the breeding cycle, facilitating the quick discovery of stress-responsive characteristics, and offering a methodical way to increase crop stress tolerance. It does this by combining Omics with speed breeding.

7.11 CURRENT APPLICATIONS OF MULTIDISCIPLINARY INTEGRATED APPROACHES FOR PLANT STRESS TOLERANCE

The use of interdisciplinary integrated approaches in plant stress tolerance nowadays reflects a paradigm shift in crop improvement techniques and agricultural research. A comprehensive understanding of plant stress responses is made possible by the integration of multiple disciplines, such as sophisticated phenotyping, proteomics, metabolomics, transcriptomics, and genomes. The following are some noteworthy ways that multidisciplinary integrated techniques might be used to improve plant stress tolerance (Figure 7.2) (Thakur et al., 2023).

Identification of Stress-Responsive Genes and Pathways:

- The identification of important genes and pathways involved in plant stress responses is made possible by the integration of transcriptomics, proteomics, and genomics (Naik et al., 2023).
- Targeted therapies for stress tolerance can be developed with the help of an understanding of the molecular mechanisms.

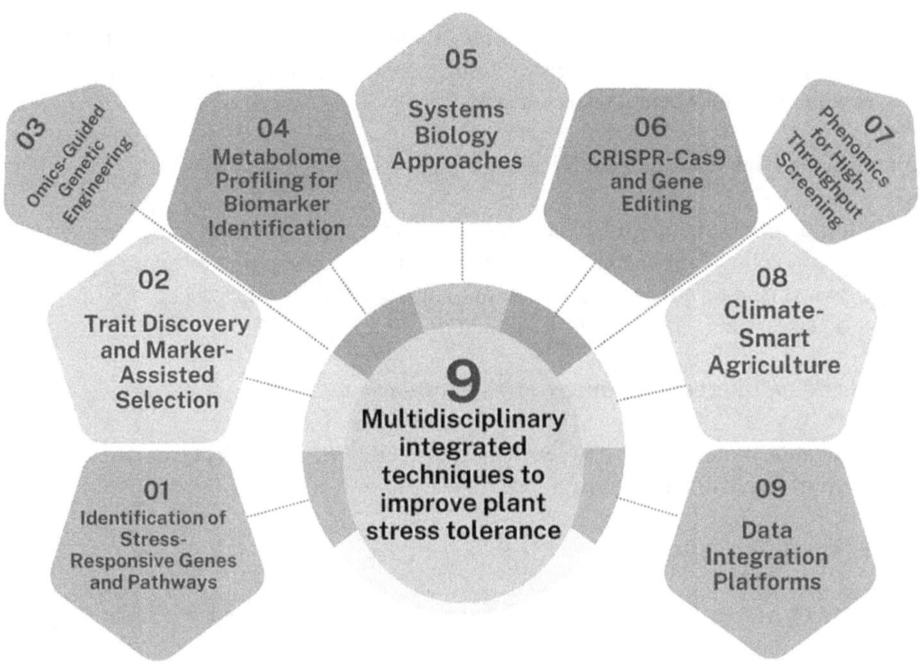

FIGURE 7.2 Multidisciplinary integrated techniques applied to improve plant stress tolerance.

Trait Discovery and MAS:

- Finding features linked to stress in crops is made easier by the combination of modern phenotyping and genomics.
- Breeding is streamlined by the use of genetic markers linked to stress tolerance using MAS (Shende & Shinde, 2023).

Omics-Guided Genetic Engineering:

- Target gene selection for genetic engineering is guided by the integration of diverse omics approaches.
- The goal of genetic changes is to improve stress tolerance features in crops based on extensive omics data (Chao et al., 2023).

Metabolome Profiling for Biomarker Identification:

- Metabolomics is utilized to profile metabolite alterations during stressful situations.
- Crop improvement efforts are guided by the identification of stress-responsive metabolites, which act as biomarkers for stress tolerance (Manickam et al., 2023).

Systems Biology Approaches:

- A thorough understanding of stress responses can be obtained by combining genomes, transcriptomics, proteomics, and metabolomics within a systems biology framework.
- Modeling techniques aid in clarifying the dynamic interactions that occur in stressed plant systems (Afzal et al., 2023).

CRISPR/Cas9 and Gene Editing:

- Transcriptomics and genomics are combined to pinpoint certain genes that CRISPR/Cas9 can be used to modify. Thanks to gene editing technologies, crops can now withstand more stress with precision alterations (Raza et al., 2023).

Phenomics for High-Throughput Screening:

- High-throughput screening of plant stress responses is achieved by integrating sophisticated phenotyping techniques, such as remote sensing and imaging technologies.
- Rapid assessment of stress tolerance in individuals with varying genetic origins is made easier by large-scale data collection (Wen et al., 2023).

Comprehending Interactions among Stress Responses:

- The utilization of several omics methodologies facilitates the deciphering of the complex interplay among distinct stress response routes (Sharma et al., 2024).

Climate-Smart Agriculture:

- The development of climate-smart agricultural practices is aided by multidisciplinary approaches.
- Varieties of crops with improved stress tolerance are engineered to flourish in ever-changing climates.

Data Integration Platforms:

- The creation of data integration platforms makes it easier to combine various omics datasets in a smooth manner and enables researchers to draw conclusions from complicated multidimensional data that have several dimensions.

Open Science Initiatives and Public Databases:

- Collaborative efforts and open-access databases facilitate the integration of diverse data. By integrating multidisciplinary approaches, researchers and breeders gain a comprehensive understanding of plant stress responses, paving the way for the development of stress-tolerant crops crucial for global food security in the face of environmental challenges.
- Shared resources promote transparency, collaboration, and the collective advancement of knowledge in plant stress tolerance.

7.12 IMPLICATIONS AND FUTURE DIRECTIONS

A major turning point in the study of plants was reached with the multi-omics approach and speed breeding for plant stress tolerance, which provided insights into a revolutionary strategy that might completely change crop improvement. The goal of the dynamic synergy of speed breeding and multi-omics (genomics, transcriptomics, proteomics, and metabolomics) is to accelerate the development of stress-tolerant crops (Sharma et al., 2024). Examining the implications and future prospects presented in this paper, we find a landscape characterized by previously unheard-of opportunities as well as obstacles. Wide-ranging effects result from combining multi-omics with speed breeding, especially when it comes to solving the urgent problems brought on by climate change and the rising frequency of environmental stressors. Plant stress responses are intricately understood because of the multi-omics approach, which takes a molecular-level approach (Roychowdhury et al., 2023). Through the simultaneous analysis of genomic, transcriptomic, proteomic, and metabolomic alterations, researchers are able to obtain a comprehensive understanding of the intricate networks that underlie stress adaptation (Zheng et al., 2022). The fast and effective breeding cycles made possible by speed breeding make this abundance of knowledge even more important. The combination not only speeds up the process of identifying genes and pathways that respond to stress but also simplifies the breeding process, making it possible to generate stress-tolerant crops quickly. Nevertheless, there are certain difficulties in combining multi-omics with fast breeding (Yang et al., 2021). Significant obstacles include the complexity of data produced by multi-omics investigations, the requirement for advanced computer tools for data integration, and the moral issues raised by genetic modification. It is crucial to manage the complex dance between technology development and social, ethical, and legal factors to use these innovative methods responsibly.

Looking ahead, the review article's suggested future paths emphasize how important it is to keep coming up with new ideas and working together. Technological developments and tools for data analysis will play a critical role in solving today's problems. Furthermore, the combination of artificial intelligence and machine learning offers great promise for understanding the complex interactions seen in multi-omics datasets and refining speed breeding techniques. The review article also stressed the value of knowledge sharing and interdisciplinary cooperation. Bringing together specialists in data science, plant physiology, genetics, and agricultural sciences will provide a thorough grasp of the subtleties of the integration. Initiatives that encourage cooperation and open data sharing will also advance the field and guarantee that breeders and researchers around the world may take advantage of new technologies. In summary, the combination of multi-omics and fast breeding to increase plant stress tolerance represents a ground-breaking strategy with significant ramifications for worldwide agriculture. Even while there are many obstacles, there are also a lot of previously untapped opportunities to improve our knowledge of plant stress responses and hasten the creation of resilient crops. The merger of multi-omics and speed breeding, with a dedication to responsible research and collaborative efforts, has the potential to significantly impact agriculture in the future and support sustainable food production in the face of climate change.

7.13 CONCLUSION

To sum up, the combination of speed breeding with the multi-omics technique is a novel approach to producing crops that can withstand stress. The synergistic marriage of high-throughput omics technologies with quick breeding cycles holds considerable potential, despite numerous challenges to be addressed. The difficulties, which range from the complexities of deciphering multi-omics data to moral dilemmas surrounding genetic engineering, highlight the necessity of a well-rounded and morally sound strategy. Overcoming these obstacles necessitates teamwork, utilizing knowledge from several fields to maneuver through the intricacies of this combined strategy. In the future, machine learning and data analytics technical developments will be essential for maximizing integrated techniques. These sectors' convergence will boost our ability to glean valuable insights from multi-omics information, enabling more accurate and expedited crop development tactics. The combination of multi-omics and speed breeding, in the ever-changing field of agricultural innovation, not only boosts yields but also builds the foundation for a resilient and sustainable future in the face of changing environmental issues. As we proceed along this revolutionary route, obstacles turn into chances, and by working together with unwavering resolve, we redefine what is possible for agriculture worldwide.

REFERENCES

Afzal, M., Hindawi, S. E. S., Alghamdi, S. S., Migdadi, H. H., Khan, M. A., Hasnain, M. U., Arslan, M., Habib ur Rahman, M., & Sohaib, M. (2023). Potential breeding strategies for improving salt tolerance in crop plants. *Journal of Plant Growth Regulation*, *42*(6), 3365–3387.

Ahmar, S., Gill, R. A., Jung, K.-H., Faheem, A., Qasim, M. U., Mubeen, M., & Zhou, W. (2020). Conventional and molecular techniques from simple breeding to speed breeding in crop plants: Recent advances and future outlook. *International Journal of Molecular Sciences*, *21*(7), 2590.

Akram, F., Sahreen, S., Aamir, F., Haq, I. U., Malik, K., Imtiaz, M., Naseem, W., Nasir, N., & Waheed, H. M. (2023). An insight into modern targeted genome-editing technologies with a special focus on CRISPR/Cas9 and its applications. *Molecular Biotechnology*, *65*(2), 227–242.

Al-Ashkar, I., Al-Doss, A., & Ullah, N. (2023). Accelerating crop improvement through speed breeding. In Hasanuzzaman, M. (Ed.), *Climate-resilient agriculture* (Vol. 1, pp. 821–847). Springer, Cham. https://doi.org/10.1007/978-3-031-37424-1_37.

Ali, A., Altaf, M. T., Nadeem, M. A., Karaköy, T., Shah, A. N., Azeem, H., Baloch, F. S., Baran, N., Hussain, T., & Duangpan, S. (2022). Recent advancement in OMICS approaches to enhance abiotic stress tolerance in legumes. *Frontiers in Plant Science*, *13*, 952759.

Ali, S., Tyagi, A., & Bae, H. (2021). Ionomic approaches for discovery of novel stress-resilient genes in plants. *International Journal of Molecular Sciences*, *22*(13), 7182.

Ambrosino, L., Colantuono, C., Diretto, G., Fiore, A., & Chiusano, M. L. (2020). Bioinformatics resources for plant abiotic stress responses: State of the art and opportunities in the fast evolving-omics era. *Plants*, *9*(5), 591.

Aslam, U., Azeem, A., Mumtaz, A., & Ul-Allah, S. (2023). Crop Responses to Climate Change. In Hasanuzzaman, M. (Ed.), *Climate-resilient agriculture* (Vol. 1, pp. 57–75). Springer, Cham. https://doi.org/10.1007/978-3-031-37424-1_3.

Athieniti, E., & Spyrou, G. M. (2023). A guide to multi-omics data collection and integration for translational medicine. *Computational and Structural Biotechnology Journal*, *21*, 134–149.

Azeem, A., Abbas, N., Azeem, S., Iqbal, Z., & Ul-Allah, S. (2023a). Physiological and molecular mechanism of nanoparticles induced tolerance in plants. In Aftab, T. (Ed.), *Emerging contaminants and plants. Emerging contaminants and associated treatment technologies* (pp. 233–248). Springer, Cham. https://doi.org/10.1007/978-3-031-22269-6_9.

Azeem, A., Ul-Allah, S., Sher, A., Ijaz, M., Sattar, A., Ahmad, W., & Qayyum, A. (2023b). Biosafety and biosecurity in genetically modified crops. In Prakash, C.S., Fiaz, S., Nadeem, M.A., Baloch, F.S., & Qayyum, A. (Eds.), *Sustainable agriculture in the era of the OMICs revolution* (pp. 501–510). Springer, Cham. https://doi.org/10.1007/978-3-031-15568-0_23.

Boopathi, N. M. (2020). Marker-assisted selection. In: Boopathi, N. Manikanda (Eds.), *Genetic mapping and marker assisted selection*. Springer, India. https://doi.org/10.1007/978-81-322-0958-4_8

Chao, H., Zhang, S., Hu, Y., Ni, Q., Xin, S., Zhao, L., Ivanisenko, V. A., Orlov, Y. L., & Chen, M. (2023). Integrating omics databases for enhanced crop breeding. *Journal of Integrative Bioinformatics*, *20*(4), 20230012.

Chaturvedi, P., Pierides, I., Zhang, S., Schwarzerova, J., Ghatak, A., & Weckwerth, W. (2024). Multi-omics for crop improvement. In *Frontier technologies for crop improvement* (pp. 107–141). Springer.

Das, S., Ray, M. K., Panday, D., & Mishra, P. K. (2023). Role of biotechnology in creating sustainable agriculture. *PLOS Sustainability and Transformation*, *2*(7), e0000069.

Erdoğan, İ., Cevher-Keskin, B., Bilir, Ö., Hong, Y., & Tör, M. (2023). Recent developments in CRISPR/Cas9 genome-editing technology related to plant disease resistance and abiotic stress tolerance. *Biology*, *12*(7), 1037.

Gonzalez Guzman, M., Cellini, F., Fotopoulos, V., Balestrini, R., & Arbona, V. (2022). New approaches to improve crop tolerance to biotic and abiotic stresses. *Physiologia Plantarum*, *174*(1), e13547.

Guo, J., Huang, Z., Sun, J., Cui, X., & Liu, Y. (2021). Research progress and future development trends in medicinal plant transcriptomics. *Frontiers in Plant Science*, *12*, 691838.

Hafeez, U., Ali, M., Hassan, S. M., Akram, M. A., & Zafar, A. (2023). Advances in breeding and engineering climate-resilient crops: A comprehensive review. *International Journal of Research and Advances in Agricultural Sciences*, *2*(2), 85–99.

Hamany Djande, C. Y., Pretorius, C., Tugizimana, F., Piater, L. A., & Dubery, I. A. (2020). Metabolomics: A tool for cultivar phenotyping and investigation of grain crops. *Agronomy*, *10*(6), 831.

Hussain, K., Mahrukh, Nisa, R. T., Zaid, A., & Mushtaq, M. (2023). The utilization of speed breeding and genome editing to achieve zero hunger. In Prakash, C.S., Fiaz, S., Nadeem, M.A., Baloch, F.S., & Qayyum, A. (Eds.), *Sustainable agriculture in the era of the OMICs revolution* (pp. 1–15). Springer, Cham. https://doi.org/10.1007/978-3-031-15568-0_1.

Iqbal, Z., Azeem, A., Ul-Allah, S., Sher, A., Ahmad, M. Q., Haider, B., & Asghar, M. (2023). Advances in biotechnological tools and their impact on global climate change and food security. In Hasanuzzaman, M. (Ed.), *Climate-resilient agriculture* (Vol. 2, pp. 667–676). Springer, Cham. https://doi.org/10.1007/978-3-031-37428-9_28.

Iqbal, Z., Iqbal, M. S., Khan, M. I. R., & Ansari, M. I. (2021). Toward integrated multi-omics intervention: Rice trait improvement and stress management. *Frontiers in Plant Science*, *12*, 741419.

Kashyap, A., Tanwar, K., Garg, P., Kumari, S., Thu, H. P. T., Singh, S., & Rao, M. (2023). Speed breeding: Space-inspired crop improvement in the nano-era. In Vishnu D. Rajput, Abhishek Singh, Tatiana M. Minkina, Krishan K. Verma, & Awani Kumar Singh (Eds.), *Nanotechnology for sustainable agriculture* (pp. 205–224). Apple Academic Press.

Kitashova, A., Brodsky, V., Chaturvedi, P., Pierides, I., Ghatak, A., Weckwerth, W., & Nägele, T. (2023). Quantifying the impact of dynamic plant-environment interactions on metabolic regulation. *Journal of Plant Physiology*, *290*, 154116. doi: 10.1016/j.jplph.2023.154116.

Kudapa, H., Barmukh, R., Garg, V., Chitikineni, A., Samineni, S., Agarwal, G., & Varshney, R. K. (2023). Comprehensive transcriptome profiling uncovers molecular mechanisms and potential candidate genes associated with heat stress response in chickpea. *International Journal of Molecular Sciences*, *24*(2), 1369.

Kumar, M., Prusty, M. R., Pandey, M. K., Singh, P. K., Bohra, A., Guo, B., & Varshney, R. K. (2023). Application of CRISPR/Cas9-mediated gene editing for abiotic stress management in crop plants. *Frontiers in Plant Science*, *14*, 1157678.

Lee, J. M., Hammarén, H. M., Savitski, M. M., & Baek, S. H. (2023). Control of protein stability by post-translational modifications. *Nature Communications*, *14*(1), 201.

Lee, T.-M., Lin, J.-Y., Tsai, T.-H., Yang, R.-Y., & Ng, I.-S. (2023). Clustered regularly interspaced short palindromic repeats (CRISPR) technology and genetic engineering strategies for microalgae towards carbon neutrality: A critical review. *Bioresource Technology*, *368*, 128350.

Lin, Y., Li, S., Ye, Y., Li, B., Li, G., Lyv, D., Jin, L., Bian, C., & Liu, J. (2023). Methodological evolution of potato yield prediction: A comprehensive review. *Frontiers in Plant Science*, *14*, 1214006.

Lohani, N., Singh, M. B., & Bhalla, P. L. (2022). Biological parts for engineering abiotic stress tolerance in plants. *BioDesign Research, 2022*, 9819314. doi: 10.34133/2022/9819314.

Manickam, S., Rajagopalan, V. R., Kambale, R., Rajasekaran, R., Kanagarajan, S., & Muthurajan, R. (2023). Plant metabolomics: Current initiatives and future prospects. *Current Issues in Molecular Biology*, *45*(11), 8894–8906.

Manna, M., Thakur, T., Chirom, O., Mandlik, R., Deshmukh, R., & Salvi, P. (2021). Transcription factors as key molecular target to strengthen the drought stress tolerance in plants. *Physiologia Plantarum*, *172*(2), 847–868.

Mergner, J., & Kuster, B. (2022). Plant proteome dynamics. *Annual Review of Plant Biology*, *73*, 67–92.

Montecillo, J. A. V., Chu, L. L., & Bae, H. (2020). CRISPR-Cas9 system for plant genome editing: Current approaches and emerging developments. *Agronomy*, *10*(7), 1033.

Naik, B., Kumar, V., Rizwanuddin, S., Chauhan, M., Choudhary, M., Gupta, A. K., Kumar, P., Kumar, V., Saris, P. E. J., & Rather, M. A. (2023). Genomics, proteomics, and metabolomics approaches to improve abiotic stress tolerance in tomato plant. *International Journal of Molecular Sciences*, *24*(3), 3025.

Naik, Y. D., Zhao, C., Channale, S., Nayak, S. N., Bhutia, K. L., Gautam, A., Kumar, R., Niranjan, V., Shah, T. M., & Mott, R. (2024). Bioinformatics for plant genetics and breeding research. In Pandey, M.K., Bentley, A., Desmae, H., Roorkiwal, M., & Varshney, R.K. (Eds.), *Frontier technologies for crop improvement. Sustainability Sciences in Asia and Africa* (pp. 35–64). Springer, Singapore. https://doi.org/10.1007/978-981-99-4673-0_3.

Nawaz, F., Majeed, S., Farman, M., Sheteiwy, M. S., Al-Mamun, A., Nawaz, M., & Neumann, G. (2022). Biostimulants as regulators of stress metabolites to enhance drought and salinity stress tolerance in plants. In Mirza Hasanuzzaman, Barbara Hawrylak-Nowak, Tofazzal Islam, & Masayuki Fujita (Eds.), *Biostimulants for crop production and sustainable agriculture* (pp. 265–294). CABI GB.

Oh, S. W., Imran, M., Kim, E. H., Park, S. Y., Lee, S. G., Park, H. M., Jung, J. W., & Ryu, T. H. (2023). Approach strategies and application of metabolomics to biotechnology in plants. *Frontiers in Plant Science*, *14*, 1192235. doi: 10.3389/fpls.2023.1192235.

Pazhamala, L. T., Kudapa, H., Weckwerth, W., Millar, A. H., & Varshney, R. K. (2021). Systems biology for crop improvement. *The Plant Genome*, *14*(2), e20098.

Potts, J., Jangra, S., Michael, V. N., & Wu, X. (2023). Speed breeding for crop improvement and food security. *Crops*, *3*(4), 276–291.

Ran, X., Hadiatullah, H., Yuchi, Z., Yang, X., & Zhu, X. (2023). Sustainable use of pesticides. Agriculture, *13*, 1393). https://doi.org/10.3390/agriculture13071393

Raza, A. (2022). Metabolomics: A systems biology approach for enhancing heat stress tolerance in plants. *Plant Cell Reports*, *41*(3), 741–763.

Raza, A., Ashraf, F., Zou, X., Zhang, X., & Tosif, H. (2020). Plant adaptation and tolerance to environmental stresses: Mechanisms and perspectives. In Hasanuzzaman, M. (Ed.), *Plant ecophysiology and adaptation under climate change: Mechanisms and perspectives I* (pp. 117–145). Springer, Singapore. https://doi.org/10.1007/978-981-15-2156-0_5

Raza, A., Tabassum, J., Fakhar, A. Z., Sharif, R., Chen, H., Zhang, C., Ju, L., Fotopoulos, V., Siddique, K. H., & Singh, R. K. (2023). Smart reprograming of plants against salinity stress using modern biotechnological tools. *Critical Reviews in Biotechnology*, *43*(7), 1035–1062.

Razzaq, A., Kaur, P., Akhter, N., Wani, S. H., & Saleem, F. (2021). Next-generation breeding strategies for climate-ready crops. *Frontiers in Plant Science*, *12*, 620420.

Riaz, A., Asad, M., Ali, K., Ullah, S., Ashraf, A., Qadeer, S., Rasheed, M., Abdullah, U., & Zawar, H. (2024). Climate change and household insect pests. In Habib Ali, Youming Hou, & Muhammad Bilal Tahir (Eds.), *Climate change and insect biodiversity* (pp. 196–207). CRC Press.

Rivera, H. E., Aichelman, H. E., Fifer, J. E., Kriefall, N. G., Wuitchik, D. M., Smith, S. J., & Davies, S. W. (2021). A framework for understanding gene expression plasticity and its influence on stress tolerance. *Molecular Ecology*, *30*(6), 1381–1397.

Rodríguez del Río, Á., Monteagudo, A., Contreras-Moreira, B., Kiss, T., Mayer, M., Karsai, I., Igartua, E., & Casas, A. M. (2023). Diversity of gene expression responses to light quality in barley. *Scientific Reports*, *13*(1), 17143.

Roychowdhury, R., Das, S. P., Gupta, A., Parihar, P., Chandrasekhar, K., Sarker, U., Kumar, A., Ramrao, D. P., & Sudhakar, C. (2023). Multi-omics pipeline and omics-integration approach to decipher plant's abiotic stress tolerance responses. *Genes*, *14*(6), 1281.

Salse, J., Barnard, R. L., Veneault-Fourrey, C., & Rouached, H. (2023). Strategies for breeding crops for future environments. *Trends in Plant Science*, *29*(3), 303–318. doi: 10.1016/j.tplants.2023.08.007.

Schilling, S., Melzer, R., Dowling, C. A., Shi, J., Muldoon, S., & McCabe, P. F. (2023). A protocol for rapid generation cycling (speed breeding) of hemp (Cannabis sativa) for research and agriculture. *The Plant Journal*, *113*(3), 437–445.

Shakespear, S., Sivaji, M., Kumar, V., Arumugam Pillai, M., Wani, S. H., Penna, S., & Yasin, J. K. (2024). Navigating through harsh conditions: Coordinated networks of plant adaptation to abiotic stress. *Journal of Plant Growth Regulation*, 1–19. https://doi.org/10.1007/s00344-023-11224-4

Sharma, A., Choudhary, P., Chakdar, H., & Shukla, P. (2024). Molecular insights and omics-based understanding of plant–microbe interactions under drought stress. *World Journal of Microbiology and Biotechnology*, *40*(2), 42.

Sharma, D., Singh, S., Sharma, S. K., & Singh, R. (2023). *Smart Plant Breeding for Field Crops in Post-genomics Era*. Springer, New York.

Shende, R., & Shinde, R. (2023). Marker assisted selection (MAS) for crop improvement. In Khushboo Chandra, S. K. Kaushik, & Devendra Kumar Payasi (Eds.), *Strategy & application of plant breeding*, pp. 263–281.

Shi, X., Zhang, R., Liu, Z., Zhao, G., Guo, J., Mao, X., & Fan, B. (2024). Alternative splicing reveals acute stress response of litopenaeus vannamei at high alkalinity. *Marine Biotechnology*, *26*, 103–115. https://doi.org/10.1007/s10126-023-10281-w

Shimizu, T., Aka Kacar, Y., Cristofani-Yaly, M., Curtolo, M., & Machado, M. A. (2020). Markers, maps, and marker-assisted selection. In Gentile, A., La Malfa, S., & Deng, Z. (Eds.), *The citrus genome. Compendium of plant genomes*. Springer, Cham. https://doi.org/10.1007/978-3-030-15308-3_7.

Singh, A., & Roychoudhury, A. (2023). Abscisic acid in plants under abiotic stress: Crosstalk with major phytohormones. *Plant Cell Reports*, *42*, 961–974. https://doi.org/10.1007/s00299-023-03013-w

Singh, A. K., Dhanapal, S., & Yadav, B. S. (2020). The dynamic responses of plant physiology and metabolism during environmental stress progression. *Molecular Biology Reports*, *47*(2), 1459–1470.

Singh, R. K., Sood, P., Prasad, A., & Prasad, M. (2021). Advances in omics technology for improving crop yield and stress resilience. *Plant Breeding*, *140*(5), 719–731.

Singha, S., & Singha, R. (2024). Crop improvement strategies and principles of selective breeding. In Ahmed Karmaoui (Ed.), *Water-soil-plant-animal nexus in the era of climate change* (pp. 93–113). IGI Global. https://doi.org/10.4018/978-1-6684-9838-5.ch005

Singhal, R. K., Saha, D., Skalicky, M., Mishra, U. N., Chauhan, J., Behera, L. P., Lenka, D., Chand, S., Kumar, V., & Dey, P. (2021). Crucial cell signaling compounds crosstalk and integrative multi-omics techniques for salinity stress tolerance in plants. *Frontiers in Plant Science*, *12*, 670369.

Srushti, S., Biradar, S., Shreedevi, S., Naveen Kumar, M., Mondal, S., & Babu, R. L. (2022). Ethical aspects and public perception on plant genomics. In Singh, R.L., Mondal, S., Parihar, A., & Singh, P.K. (Eds.), *Plant genomics for sustainable agriculture* (pp. 371–399). Springer, Singapore. https://doi.org/10.1007/978-981-16-6974-3_15.

Swami, P., Deswal, K., Rana, V., Yadav, D., & Munjal, R. (2023). Speed breeding: A powerful tool to breed more crops in less time accelerating crop research. In Mohd. Kamran Khan, Anamika Pandey, Mehmet Hamurcu, Om Prakash Gupta, & Sait Gezgin (Eds.), *Abiotic stresses in wheat* (pp. 33–49). Elsevier.

Thakur, N., Nigam, M., Mann, N. A., Gupta, S., Hussain, C. M., Shukla, S. K., Shah, A. A., Casini, R., Elansary, H. O., & Khan, S. A. (2023). Host-mediated gene engineering and microbiome-based technology optimization for sustainable agriculture and environment. *Functional & Integrative Genomics*, *23*(1), 57.

Ul-Allah, S., Azeem, A., Sher, A., Ijaz, M., Sattar, A., Naeem, M., & Ullah, I. (2023). Plant breeding approaches in developing stress tolerance. In Aryadeep Roychoudhury (Ed.), *Biology and biotechnology of environmental stress tolerance in plants* (pp. 167–186). Apple Academic Press.

Vadez, V., Pilloni, R., Grondin, A., Hajjarpoor, A., Belhouchette, H., Brouziyne, Y., Chehbouni, G., Kharrou, M. H., Zitouna-Chebbi, R., & Mekki, I. (2023). Water use efficiency across scales: From genes to landscapes. *Journal of Experimental Botany*, *74*(16), 4770–4788.

Verma, V., Kumar, A., Partap, M., Thakur, M., & Bhargava, B. (2023). CRISPR-Cas: A robust technology for enhancing consumer-preferred commercial traits in crops. *Frontiers in Plant Science*, *14*, 1122940.

Wang, X., Li, N., Li, W., Gao, X., Cha, M., Qin, L., & Liu, L. (2020). Advances in transcriptomics in the response to stress in plants. *Global Medical Genetics*, *7*(02), 030–034.

Wen, T., Li, J. H., Wang, Q., Gao, Y. Y., Hao, G. F., & Song, B. A. (2023). Thermal imaging: The digital eye facilitates high-throughput phenotyping traits of plant growth and stress responses. *Science of The Total Environment*, *899*, 165626. ISSN 0048-9697, https://doi.org/10.1016/j.scitotenv.2023.165626.

Wijayawardene, N. N., Boonyuen, N., Ranaweera, C. B., de Zoysa, H. K., Padmathilake, R. E., Nifla, F., Dai, D.-Q., Liu, Y., Suwannarach, N., & Kumla, J. (2023). OMICS and other advanced technologies in mycological applications. *Journal of Fungi*, *9*(6), 688.

Yadav, R. K., Tripathi, M. K., Tiwari, S., Tripathi, N., Asati, R., Chauhan, S., Tiwari, P. N., & Payasi, D. K. (2023). Genome editing and improvement of abiotic stress tolerance in crop plants. *Life*, *13*(7), 1456.

Yang, Y., Saand, M. A., Huang, L., Abdelaal, W. B., Zhang, J., Wu, Y., Li, J., Sirohi, M. H., & Wang, F. (2021). Applications of multi-omics technologies for crop improvement. *Frontiers in Plant Science*, *12*, 563953.

Zhang, F., & Batley, J. (2020). Exploring the application of wild species for crop improvement in a changing climate. *Current Opinion in Plant Biology*, *56*, 218–222.

Zhang, Y., Xu, J., Li, R., Ge, Y., Li, Y., & Li, R. (2023). Plants' response to abiotic stress: Mechanisms and strategies. *International Journal of Molecular Sciences*, *24*(13), 10915.

Zheng, L., Wu, W., Chen, Q., Zhang, G., Gao, F., & Zhou, Y. (2022). Integrated transcriptomics, proteomics, and metabolomics identified biological processes and metabolic pathways involved in heat stress response in jojoba. *Industrial Crops and Products*, *183*, 114946.

8 Understanding Stress-Related Genes and the Role of Biostimulants in Biotic and Abiotic Stress Response to Improve Plant Breeding and Cultivation Methods

K. Varun and B. S. Lakshmi

LIST OF ABBREVIATIONS

ABA	Abscisic acid
ABC	ATP-bound cassette
ABF	Abscisic acid-responsive transcription factors
ACA	Autoinhibited calcium ion ATPase
ACC	Aminocyclopropane-1-carboxylate
ACL	ATP citrate lyase
ACO	Aminocyclopropane carboxylic acid oxidase
ACS	Acetyl CoA synthetase
ACS	Aminocyclopropane carboxylic acid synthase
AD	Alcohol dehydrogenase
AG	Arabidopsis agamous
AHA	Autoinhibited hydrogen ion ATPase
AIN	Abscisic acid insensitive
AKT	Potassium channel genes
AOS	Allene oxide synthase
AP2	Apetala2
APX	Ascorbate peroxidase
AQP	Aquaporin
AREB	Abscisic acid-responsive element binding
ARG	Arginase
ARO	Argonaute
ASR	Abscisic acid, stress and ripening
ATAF	*Arabidopsis* transcription activator factor
ATG	Autophagy
ATP	Adenosine triphosphate
ATPase	Adenosine triphosphatase
Avr	Avirulent
BADH	Betaine aldehyde dehydrogenase
BAG	Bcl2-associate athanogene

DOI: 10.1201/9781003543237-8

BBX	B box protein
bHLH	Basic helix loop helix
BPH	Brown plant hopper
BSMT	Sabath methyltransferase
BSR	Broad-spectrum resistance
BX	Benzoxazionoids
bZIP	Basic zinc-regulated, iron-regulated transporter protein
BZR	Brassinazole resistant
C4H	Cinnamate-4-hydroxylase
CA	Carbonic anhydrase
CAS	Calcium sensing
CAT	Catalase
CathB2	Cathepsin B2
CAX	Cation exchanger
CBF	C-repeat binding factor
CBL	Calcineurin B-like protein
CCD	Carotenoid cleavage dioxygenase
CCX	Cation calcium exchanger
CDPK	Calcium-dependent protein kinase
CDT1	Chromatin licensing and DNA replication factor 1
CERK	Ceramide kinase
CHS	Chalcone synthase
CIPK	CBL-interacting protein kinase
CLC	Chloride channel
CNR	Cell number regulator
COR	Cold regulated
CRF	Cytokinin response factor
CRT	Calreticulin
CSD	Cold-shock domain
CSL	Cellulase synthase like
CSR	Calcium signaling related
CUC	Cub-shaped cotyledon
CYP	Cytochrome P450
CYS	Cysteine synthase
DHAR	Dehydroascorbate reductase
DHN	Dehydrins
DIR	Dirigen
Dn	Diuraphis noxia
DNA	Deoxyribonucleic acid
DOF	DNA binding with one finger
DR	Drought responsive
DRAP	Drought resistance-associated protein
DREB	Dehydration-responsive element binding
EDR	Erythroid differentiation regulator
ER	Endoplasmic reticulum
ERD	Early response to dehydration
ERF	Ethylene receptor factor
ETI	Effector-triggered immunity
ETR	Ethylene receptor
EXP	Expansin

EXPB	Expansin-like B
FER	Ferritin
Fhb	Fusarium head blight
FMO	Flavin-dependent monooxygenase
FRG	Fusarium resistance gene
GA	Gibberellic acid
GA(n)ox	Gibberellic acid (n) oxidase
GAI	Gibberellic acid insensitive
Gb	Green bug
GDSL	Glycine, aspartic acid, serine, and leucine
GELP	GDSL-type esterase/lipase protein
GH	Glycosyl hydrolysate
GLR	Glutamate receptor
Glu	Glutenin
GOI	Gene of interest
GP	Glutathione peroxidase
GPCR	G-protein-coupled receptor
GPX	Guaiacol peroxidase
GR	Glutathione reductase
GRAS	Gene family of GAI, RGA, and SCR
GRF	Growth regulation factors
GSH	Glutathione
GST	Glutathione S transferase
GSYT	Glycine transferase
$\mathbf{H_2O_2}$	Hydrogen peroxide
HAG	Histone acetyltransferase gene
HAK	High-affinity potassium
HDZ	Home domain leucine zipper
HIPP	Heavy metal-associated isoprenylated plant proteins
HK	House keeping
HKT	High-affinity potassium transporter
HLH	Helix loop helix
HMA	Heavy metal associated
HQT	Hydroxycinnamoyl quinate transferase
HSF	Heat stress transcription factor
HSP	Heat shock protein
Hy	Hypocotyl
IAA	Indole acetic acid
ICS	Isochorismate synthase
JA	Jasmonic acid
JAZ	Jasmonate zinc finger inflorescence meristem
JIG	Jasmonate-induced bHLH gene
JUB	Jungbrunnen
KAT	Potassium channel genes
KIN	Kinase
LAC	Laccase
LAR	Localized acquired resistance
LAX	Like auxin resistance
LEA	Late embryogenesis abundant
LHC	Light-harvesting complex

LOX	Lipoxygenase
LRR	Leucine-rich repeats
LTG	Low-temperature germinability
MADS	Gene family of MCM1, AG, DEF, and SRF
MAPK	Mitogen-activated protein kinase
MCM1	Minichromosome maintenance 1
MDA	Malondialdehyde
MDHAR	Mono dehydroascorbate reductase
MIB	Myoinositol biosynthesis
MIR	miRNA
MM	Mexicana lesion mimic
MPK	Mitogen-activated protein kinase
MT	Metallothionein
MTP	Metal tolerance proteins
MYB	Myeloblastosis
MYC	Myelocytomatosis
NAC	Conserved region of NAM, ATAF, and CUC
NAC	Nascent polypeptide-associated complex
NADH	Nicotinamide adenine dinucleotide hydrogen
NAM	No apical meristem
NB-ARC	Nucleotide-binding adapter APAF-1, R proteins, and CED-4
NBS-LRR	Nucleotide-binding site–leucine-rich repeats
NCED	Nine-cis-epoxy carotenoid dioxygenase
NHX	Sodium hydrogen antiporter
NLR	Nucleotide-binding leucine-rich repeats
NPR	Natriuretic peptide receptor
NRAMP	Natural resistance-associated macrophage proteins
P5CS	Pyrroline-5-carboxylate synthetase
PAL	Phenylalanine ammonia lyase
PAO	Polyamine oxidase
PAT	Phosphinothricin acetyltransferase
PCS	Phytochelatin synthase
PEP	Peptide-expressed pri-miRNA
PETF	Polypeptide ferredoxin
PHT	Phosphate ion transportation
PIF	Phytochrome-interacting factor
PIP	Phosphotidyl inositol 4,5-bisphosphate
PK	Protein kinase
PM	Powdery mildew
PMA	Plasma membrane ATPase
PMTR	Phyto-melatonin receptor
POD	Peroxidase
POR	Protochlorophyllide oxidoreductases
PPase	Phosphate phosphatase
PP2C	Protein phosphatase 2C
PR	Pathogenesis related
PRP	Pathogenesis-related protein
PRT	Proteases
PRX	Peroxiredoxins
PSB	Photo system subunit

PTI	Pattern-triggered immunity
PTP1-I	Potato type 1 protease inhibitor
PYL	Pyrabactin resistance like
QTL	Qualitative trait loci
RAB	Ras-associated binding
RAFS	Raffinose synthase
RD	Dehydration responsive
RFP	Ring finger protein
RGA	Repressor of GAI
RNA	Ribonucleic acid
ROP	Rho-related proteins of plants
RPO	RNA polymerase
RRM	Resistance-related metabolite
SA	Salicylic acid
SALT	Salinity tolerance
SAMDC	S-Adenosyl methionine decarboxylase
SAP	Stress-associated protein
SAPK	Stress-activated protein kinase
SBS	Stillbane synthase
SCR	Scarecrow
SFR	Senstivity to freezing
SIK	Salt inducible kinase
SIN	Salt intolerance
SKC	Shoot potassium content
SNaC	Shoot sodium content
SNF1	Sucrose nonfermenting 1
SnRK	SNF1-related protein kinase
SOD	Superoxide dismutase
SOS	Salt overly sensitive
SP	Spot blotch
SPH	Soybean protein hydrolysate
SR	Stem rust
SRF	Stress response factor
SRK	S locus receptor kinase
ST	Salinity tolerance
STS	Stillbane synthase
SUB	Submergence
TAF	Transcription factor
TPP	Trehalos-6-phosphate phosphatases
TPPs	Thiol peptidase proteins
TPS	Terpene synthase
TPS	Trehalose 6-phosphate synthase
UBC	Ubiquitin conjugate
UDP	Uridine diphosphate
UGT	UDP glyceryl transferase
USP	Universal stress protein
VOZ	Vascular plant one zinc finger
WRKY	Gene family for growth and development
ZIM	Zinc finger inflorescence meristem
ZIP	Zinc-regulated, iron-regulated transporter protein

8.1 INTRODUCTION

Stress is a major factor in plants limiting crop growth and yield and is caused by both abiotic and biotic factors. Abiotic factors are mainly heat, drought, salinity, cold, excess water, and heavy metals, and biotic factors are bacteria, fungi, insects, and viruses (Song et al., 2023). Plants have developed many tolerant mechanisms of stress through evolution and changes in physiological parameters as well as growth characteristics to mitigate the harmful effects of stress (Mishra et al., 2023). Plant breeding programs have been studied in cultivated crop varieties to enhance stress tolerance while maintaining yield potential (Yadav et al., 2023). Breeding programs in the modern age depend upon recognizing and marking desirable traits and correlating them to QTL and gene mapping using marker screening of genomic DNA, which has led to faster plant variety development and also establishment of genome and gene library databases (Sanchez et al., 2023).

Biostimulants in agriculture are used to enhance plant traits and are defined as any microorganism, metabolite, or substance that when applied to plants enhances plant growth, stress tolerance, and yield enhancement (Bashir et al., 2021). Biostimulant development has also helped to mitigate environmental stress without requiring plant variety development to reduce the harmful effects of stress and establish sustainable methods of cultivation and reduction of input cost (Riyazuddin et al., 2023; Ansari et al., 2023; Johnson et al., 2023). The properties of biostimulants have led to the examination of different sources and to determine how they affect plant properties (Cristofano et al., 2021) (Figure 8.1).

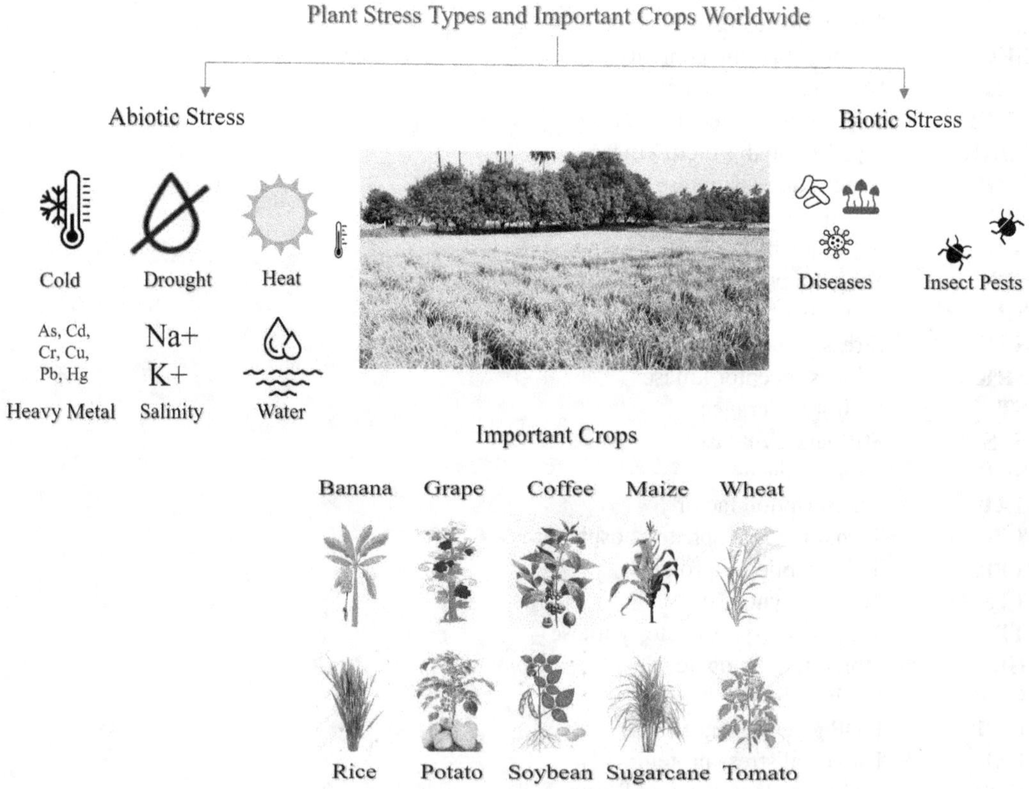

FIGURE 8.1 Types of stress on plants.

The different abiotic and biotic stress factors are as described.

8.1.1 Abiotic Stress

8.1.1.1 Cold

Cold stress in plants is caused due to reduced temperature, leading to reduced plant growth, and different mechanisms for stress alleviation are followed by different plants to deal with cold stress and freezing stress. The mechanisms involved are based on light reception and the triggering of cold stress genes in plants to initiate cold tolerance mechanisms to prevent damage due to water crystals (Ritonga and Chen, 2020).

8.1.1.2 Drought

Drought stress is caused by lack of water in plants, leading to a reduction in plant photosynthesis. Alleviation of drought stress is by preventing water loss in plants, maintaining salt balance, and reducing functions in plants where not required. Different mechanisms exist that help in mitigating drought stress and focus on different plant functions leading to achieving normal plant growth and development (Seleiman et al., 2021; Yang et al., 2021).

8.1.1.3 Heat

Heat stress in plants has a mitigating effect on growth and yield due to reduced photosynthesis in plants. Different mechanisms are followed by plants to reduce the effect of heat stress through changes in membrane proteins and reduction in reactive oxygen species (ROS) such as hydrogen peroxide, inducing heat shock proteins which play a major role in mitigating heat stress and also environmental stresses as well as regulating other plant processes. Osmolytes also help in mitigating heat and other environmental stress conditions in plants (Asthir, 2015)

8.1.1.4 Heavy Metal

Heavy metal stress is due to the presence of high quantities of heavy metals within the soil leading to improper plant growth. It is important to understand how plants store the metal within them as well as prevent oxidative damage caused by the heavy metal ions in plants. Different mechanisms exist to mitigate different heavy metal damage to the plants (Ghori et al., 2019; Kosakivska et al., 2021).

8.1.1.5 Salinity

Salinity stress is caused by the presence of excess minerals in the soil of cultivated crops, which inhibits normal plant growth through the production of hydrogen peroxide and other ROS as well as its effect on the water balance within the plant. Plants mitigate salt stress by improving the number of antioxidant agents within the plant as well as by maintaining salt balance through different mechanisms to maintain the osmotic levels (Khan et al., 2020).

8.1.1.6 Water

Water stress is mainly caused by excess water present at roots and due to flooding. The stress causes a reduction in plant function by damaging the roots, thus affecting plant development and damaging crop plants. Plants change membrane permeability due to high amounts of water and produce antioxidant agents to reduce the amount of ROS such as hydrogen peroxide (Sun et al., 2020; Jia et al., 2021).

8.1.2 Biotic Stress

8.1.2.1 Diseases

Different diseases affect plants differently, and plants follow different methods to mitigate the disease through recognition and secondary metabolite production to reduce the effect of the disease on normal plant growth (Li et al., 2020).

8.1.2.2 Insect Pests

Various insects cause different effects on plants like reducing their growth functions, by feeding on the plants and also inhabiting within the plants. A reduced yield due to such insect infestations is observed. Plant defense mechanisms cause the production of terpenes in plants preventing the insects from feeding or residing within the plant (Boncan et al., 2020). A silicon-based defense mechanism is also triggered in plants to prevent herbivore insects from feeding on plants (Islam et al., 2020).

8.2 GENETIC AND BIOSTIMULANT INFORMATION OF IMPORTANT CROPS WORLDWIDE

The abiotic and biotic stress due to the factors described above causes changes in crops eliciting a tolerant response through gene expression or causing the death of the plants due to lack of tolerance to the stress. Biostimulants enhance tolerance in plants which when combined with plant breeding programs will improve both tolerant and susceptible varieties. This chapter describes the effect of stress on the most valuable crops worldwide and also based on the type of crops cultivated: fruit crops – banana and grape; plantation crop – coffee; cereal crops – maize, rice, and wheat; tuber crop – potato; legume crop – soybean; sugar crop – sugarcane; vegetable crop – tomato (Anders et al., 2021). The value and production volume of the above crops is described (Table 8.1).

8.2.1 Fruit Crop

8.2.1.1 Banana

Banana (*Musa* spp.) is a widely cultivated crop in the regions of southern India, Southeast Asia, and the western part of Australia. It is easily cultivable and has not been fully domesticated by humans as it is easily crossable with wild varieties of bananas.

The genus *Musa* has been divided into four subspecies:

Eumusa covers the primary diversity area except for Melanesia (New Guinea to Fiji islands).
Rhodochlamys covers monsoonal mainland southeast Asia.
Australimusa covers southeastern Indonesia, southern Phillipines to Malenasia.
Callimusa covering Southern Vietnam, Malaysia, Borneo, and Sumatra.

Breeding of banana cultivars is oriented toward biotic stress resistance to pests and diseases as well as abiotic stress tolerance based on the land conditions (de Langhe et al., 2009).

TABLE 8.1
Value and Production Volume of Cultivated Crops Worldwide (Anders et al., 2021)

Crop	Approximate Volume (Million Metric Tons)	Approximate Value (Billion Dollars)
Banana	100	225
Coffee	300	300
Grape	1,100	60
Maize	320	150
Potato	800	75
Rice	300	200
Soybean	1,800	120
Sugarcane	100	70
Tomato	90	150
Wheat	800	125

8.2.1.1.1 Genes for Abiotic and Biotic Stress Tolerance

Autophagy-related gene *ATG8f* is a positive regulator of drought stress, which alleviates lipid peroxidation. ATG8*f* has also been shown to reduce the ROS burst and improve the photosynthetic process by increasing photosynthetic efficiency and carbon assimilation, facilitating nutrient uptake and promotion of carbon transportation under drought stress conditions. The regulation of drought stress response is linked to an increase in endogenous ABA response caused by an overexpression of *ATG8f*, which regulates drought stress response in abscisic acid (ABA) dependent pathway (Li et al., 2019). *ASR1* gene showed an increased expression in banana leaves and roots with roots showing three-fold sensitivity to salinity stress. Further gene expression studies indicated that salt stress tolerance was achieved by regulating the ABA stress response pathway genes, whereas non-ABA pathways were not affected and gene expression was normal. So, *ASR1* is a suitable candidate for salt tolerance in banana breeding (Miao et al., 2014). *Csl* genes are responsible for developing hemicellulose backbones in plants. These genes are shown to be induced due to different stress responses. *CslD4* is one particular gene that shows increased expression from chilling tolerance in banana cultivars when compared to chilling susceptible cultivars. Also related to chilling tolerance are *CslA4/12* and *CslE2*, which also showed increased expression in tolerant varieties when compared to susceptible ones. Based on studies with different plant species the potential of *Csl* genes in abiotic and biotic stress tolerance is promising (Yuan et al., 2021).

DREB1F is a dehydration-responsive element-binding protein gene. These proteins are further classified into six subclasses, with the C-repeat binding factor subclass being responsible for cold acclimation. The *DREB/CBF*-related gene expression has shown an improvement in stress tolerance in plants. Subjecting banana plants to cold, drought, and salt treatment shows an increase in *DREB1F* gene expression, which improved cold and drought tolerance. This gene has been shown to activate or alleviate the repression of genes present in jasmonic acid (JA) and ethylene-related pathways in cold and drought stress conditions. An increase in JA was observed where *DREB1F* activated the promoter of allene oxide cyclase type gene *AOC4* in plants subjected to cold and drought stress and improved cold stress tolerance. Similarly, the promoter of aminocyclopropane carboxylic acid oxidase gene *ACO20* was activated, which led to an increase in ethylene production under cold and drought stress conditions, causing improved plant response to both cold and drought stress. Also, *DREB1F* showed an improvement in the *ACO* gene activity by influencing the transcriptional response of ethylene receptor factors (ERFs). One such receptor factor *ERF9* has been shown to activate the promoter of *ACO1* related to banana fruit ripening (Xu et al., 2023). *Fer1* is a sequence region that enables the induction of ferritin in plant leaves in response to various stress conditions such as cold, drought, and salinity as well as those caused by pathogens. Ferritins are proteins that enable the storage of iron in plants and thus avoid oxidative damage due to high iron content and prevent stress response. Overexpression of *Fer1* showed an increase in increased iron content in the leaves and roots of banana plants as well as tolerance to iron content-induced oxidative stress (Yadav et al., 2017). *MPK5* is a mitogen-activated protein kinase responsible for reducing cold stress in bananas in proline level improvement and decrease in malondialdehyde (MDA) levels in banana plants, thus mitigating the effects of cold stress on plant growth and development by phosphorylation of *NAC042* and *SNAC67* transcription factors, which are important regulators of stress tolerance in banana plants (Tak et al., 2020).

NAC29 is a highly conserved domain responsible for three proteins in plants (*NAM*, *ATAF*, and *CUC)* and is highly expressed in bananas for drought and salinity conditions. To improve tolerance in banana- gallic acid, coumaric acid, naringenin, and chlorogenic acid along with the expression of stress-related genes coding for antioxidants, thiol peptidase proteins, cold-regulated (COR) proteins, late embryogenesis abundant (LEA) proteins, ERFs, basic helix loop helix (bHLH) proteins, jasmonate-zim-domain proteins (JAZ), and WRKY are induced (Negi et al., 2023). The *ROP5g* gene is a part of Rho-related protein of plants gene family, which is an important molecular switch required for environmental stress response. The gene *ROP5g* showed overexpression in response to salt stress conditions in banana plants and also showed an increase in sodium and potassium homeostasis in banana plants in response to salt stress by initiating the sodium and potassium homeostasis pathway proteins such as salt overly sensitive (SOS) and calcium signaling-related pathways as studied in transgenic *Arabidopsis thaliana* seedlings (Miao et al., 2018). *WRKY71* is an important member of the *WRKY* gene family (*WRKYGQK* heptapeptide located at N-terminal domain), which is responsible for regulating banana plant response to biotic and abiotic stress tolerance. The overexpression of the *WRKY71* gene has been shown to improve oxidative and salt stress tolerance in banana plants (Shekhawat and Ganapathi, 2013).

8.2.1.1.2 Biostimulant Application for Abiotic and Biotic Stress Tolerance

Anaerobic fermentation of microbes using plant waste has been used as a biostimulant. An increase in membrane-related genes, photosynthesis-related genes, and osmoregulation-related genes such as nicotinamide adenine dinucleotide hydrogen would improve plant growth and stress tolerance. *rpoC2* gene, which is part of the ribonucleic acid (RNA) polymerase subunit related to chloroplast, is associated with adenosine triphosphate (ATP) synthases in chloroplast, indicating an increase in ATP levels. ATP has shown improved levels due to biostimulant treatment for bioremediation in polluted soils. Thus, improvement in ATP synthases leads to improved abiotic stress tolerance (Chavez-Navarrete et al., 2023). Glycine, betaine and compost combination showed improvement in salinity tolerance of banana plant variety - Grand Nain and Paz, respectively, over a period of six weeks, with Paz showing better salinity tolerance compared to Grand Nain when treated with foliar and soil drenching treatment of biostimulants (Abu Zurayk, 2019). Lipoxygenases (Lox) play an important role in the JA pathway as well as antibacterial and antifungal compound synthesis in plants. *Lox3* gene expression showed overexpression when treated with biostimulants, thus indicating the inducement of biotic stress tolerance. Similarly, lipoxygenases have been shown to improve salinity and drought tolerance due to their overexpression, which when applied to bananas would provide abiotic stress tolerance to the crop (Chavez-Navarrete et al., 2023).

Myc2 transcription factor is a *bHLH* protein that is overexpressed due to mechanical damage and also due to exogenous JA application. Bananas treated with biostimulant showed improvement in *Myc2* expression till 48 hours due to the feedback loop. This indicates that there is enhanced tolerance to biotic as well as abiotic stress (Chavez-Navarrete et al., 2023). Peroxidase proteins, guaiacol peroxidase and β-1,3-glucanase, were induced by biostimulant treatment using tea tree oil in banana plants. Peroxidase proteins are related to the defense mechanism of plants against pathogens indicating that tea tree oil extracts produced a defense response in banana plants. Further mRNA expression of peroxidase-related genes such as *BSMT1*, *NPR1*, and *PR1* in normal bananas that were not infected with Fusarium wilt and in infected plants treated with biostimulants *PR1*, *PR2*, and *PR3* showed increased expression compared to infected untreated plants, indicating an induced biotic stress tolerance through the usage of biostimulants. Also expressed in uninfected treated plants were *ACS*, *ERF1*, and *MYC2*, which help in inducing systemic resistance in banana plants. The peroxidase expression indicated that the biostimulant stress tolerance may be caused by the inducement of the salicylic acid (SA) pathway (Dalio et al., 2020).

8.2.1.1.3 Summary

Csl, *DREB1f*, *Myc2*, and *rpoC2* genes showed expression across multiple plant stress conditions and are vital for tolerance response in banana plants, and the genes *ACS*, *ERF1*, *Myc2*, *Pr*, and *rpoC2* showed increased expression when treated with biostimulants sourced from plant waste, and tea tree oil. Direct application of JA and a combination of glycine betaine and compost would play a key role in improving banana cultivation practices (Table 8.2).

8.2.1.2 Grapes

Grapes (*Vitus vinifera*) belong to the genus *Vitus*. It is a widely cultivated crop from 6,000 to 8,000 years and has been used for wine making. The subspecies *vinifera* for human cultivation originated from the wild grape type *Vitus vinifera* sp sylvestris. The most important cultivars are

TABLE 8.2
Genes for Environmental Stress Tolerance in Banana

Cold stress	*Myc2, Mpk5, Csl, Dreb1f, Rpoc2*
Drought stress	*Lox3, Myc2, ATG8f, Csl, Dreb1f, Rpoc2*
Heat stress	*Myc2, Csl, RpoC2*
Heavy metal stress	*Myc2, Csl, Fer1, RpoC2*
Salinity stress	*Lox3, Myc2, Csl, Asr1, Wrky71, ROP5g, Rpoc2*
Diseases	*Myc2, Csl, Pr, Acs, Erf1*
Insects	*Myc2, Csl*

required to be disease resistant due to the many diseases the grapes crop are prone to and due to the high number of genotypes available (Myles et al., 2011).

8.2.1.2.1 Genes for Abiotic and Biotic Stress Tolerance

AHA8 and *PMA1* genes, responsible for ATP transportation in plants due to oxidative phosphorylation, are highly expressed in cold-tolerant grape varieties. Also, genes related to hormone expression such as *bHLH, NB-ARC, PP2C, PK, WRKY,* and *bZIP*, leading to cold tolerance in grapes, are also reported (Liang et al., 2022). *CAS* calcium signaling gene, which is upregulated by calcium ions in grapes and is responsible for stomatal functions in the leaves, affects transpiration rate and carbon dioxide uptake in leaves. The *CAS* gene also helps in drought stress resistance by maintaining salt and water content in grape plants and decreasing the antioxidant level, thereby reducing cell damage (Cui et al., 2022). *CIPK02* is a gene that is expressed as a stress response is found to be sensitive to salt stress and tolerant to drought stress and also reduces ROS caused by ABA in grape plants. Also, *CIPK02* regulates *ABF3* and *ABF4*, which regulate the ABA signaling pathway to increase ABA (Xu et al., 2020).

DOF gene family, which is DNA binding with one finger, is responsible for providing abiotic stress tolerance in grapes. A particular gene of the family *Dof17d* was found to induce cold tolerance in grapes. Further analysis showed that it was linked to raffinose family oligosaccharides due to the high gene expression of galactinol and raffinose synthase (RAFS) genes (Wang et al., 2021). Similarly, the *Dof8* gene found in the grape cultivar *yeshanensis* was found to be overexpressed in drought stress conditions, leading to a reduction of MDA and H_2O_2 levels as well as an increase in ABA content and causing the expression of different stress response genes (Li et al., 2021). Stress response *GRAS* gene *PAT1* is found to be a positive regulator for cold tolerance in grape plants by inducing the JA pathway, which helps in improved cold tolerance in grapes (Wang et al., 2021). Heat response genes in grapes *HSFA2, HSFA7,* and *PsbP* were overexpressed due to heat stress inducement; *sHSP* and *Petf1* genes were also found to express as a response to ferrodoxin. The ABA pathway was induced in grapes in response to heat stress mitigation (Zha et al., 2020). *HSFA2* gene in *vinifera* and *davidii* varieties of grapes were also compared for heat tolerance, with the *davidii* variety showing higher *HSFA2* gene expression compared to the *vinifera* variety. Similarly, the *davidii* variety also showed higher target gene activation compared to the *vinifera* variety. *HSFA2* also targets the *MBF1c* gene, which is responsible for heat tolerance in grapes (Liu et al., 2023).

Grape variety *amuranis*, which is cold tolerant, has a novel transcription factor *HsfC1*, which is part of *Hsf* regulatory genes that are responsible for abiotic and biotic stress response. *HsfC1* was highly expressed when exposed to cold, heat, and salt tolerance and showed enhanced tolerance to heat and cold stress but was found to be sensitive to salt and oxidative stress in grape cultivars (Jiao et al., 2022). *Hsp* genes were expressed in response to heat stress and drought stress including combined heat and drought stress; *Hsf* genes were expressed in response to only heat and drought stress; *DREB, HLH, MYB,* and *NAC* genes were expressed to combined stress in grape plants (Ju et al., 2021). *LAR2* gene expression during fungal infection causes the overproduction of catechin, which is a secondary metabolite linked to susceptibility in grapes by breaking down antifungal defense compounds in grape plants (Maia et al., 2020). *LOX2* and *LOX3* genes in grapes are responsible for the JA pathway, which produces secondary metabolites to deter herbivore pests. Also upregulated were *BHLH, UBI*, and *ZAT10*, which interact with jasmonates to cause resistance responses to herbivore insects (Jia et al., 2022).

The MTP gene group for grapes was found to cause an abiotic stress tolerance response, with the genes mainly targeting metal toxicity in grape plants and the *MTP8* and *MTP9* genes showing manganese tolerance by storing the manganese in vacuoles of cells. The *MTP9* gene also showed high expression of osmotic stress in grape plants. Overall, the *MTP* gene family has shown heavy metal tolerance response in plants by different expressions of miRNA thus alleviating the stress from plants. Also, drought, salinity, and osmotic stress triggered the expression of *MTP1, MTP3, MTP7*, and *MTP11* gene expression wherein heavy metal toxicity response and salinity stress response are the same (Shirazi et al., 2019). *MYB, NAC, WRKY*, and *bZIP* increased significantly in response to salt stress, and the genes help in proline production in response to salt stress in grapes. Also, the *WRKY* gene was increased two-fold in response to endoplasmic reticulum (ER) and salt stress. Ethylene receptor factor genes were also found to be expressed in grapevine plants in response to salt and ER stress as the genes are expressed in response to abiotic stresses in plants. The salt stress response also led to a reduction in aquaporin genes, which are related to osmotic regulation (Cakir Aydemir et al., 2020). *MYB2* also provided heavy metal tolerance by activating antioxidant and heavy metal transport pathways in grapes (Noor et al., 2022). *MYB108A* was expressed in grape plants treated with melatonin, the gene acting as a mediator during salt tolerance response for grape plants. *MYB108A* is also bound to the promoter *ACS1* to synthesize ethylene, which leads to salt tolerance response pathway expression in grapes (Xu et al., 2019). A novel *NAC* factor *NAC17* was induced by cold, drought, and salinity stress and activated the tolerance pathway of the grape plant, leading to plant survival. The pathway was induced by modulating the ABA pathway (Ju et al., 2020). *ZIP30* gene causes drought tolerance in grapes by inducing the *NAC17* gene, which helps mitigate drought stress in grapes (Tu et al., 2020). *PEP172b* and *PEP3635b* are related to peptide-expressed pri-miRNA in grapes whose function in cold tolerance is to regulate the *MIR* gene, which plays a major role in triggering miRNA cold tolerance gene expression downstream, thus mitigating cold stress in grapes (Chen et al., 2022).

The *PYL9* gene from grapes has shown improved tolerance to drought stress, which was found in *Vitis yeshanenesis* variety. The tolerance was attributed to reduced ROS due to improved antioxidant defense response from the grape plant. The *PYL9* gene showed increased ABA sensitivity, which may have helped in mitigating drought stress effects on grape plants (Liu et al., 2019). *RD22* gene is responsible for drought tolerance in grapes by improving growth and reducing water loss during drought stress by reducing water loss. Other genes that helped drought resistance in grapes are *CBL1, CYP707A2, DREB2A, NAC72*, and *SRK2A* in grapes by stomatal closure and preventing water loss and maintaining turgidity (Cui et al., 2022).

STS29/STS2 is a gene in grapes that causes plant defense response upon infection with fungus. The mechanism involves the *STS* gene expression, which leads to SA signaling, resulting in the expression of *MYB* and *WRKY* transcription factors and other defense genes, thus conferring fungal resistance in grape cultivars (Xu et al., 2019). *WRKY28* gene expression in grapes has a function in cold and salt tolerance. The mechanism involves the expression of the *WRKY28* gene during cold and salt stress, which causes the gene expression of different cold stress-related downstream genes *COR15A, COR47, ERD10, ICS1*, and *PIF4* as well as salt stress-related downstream genes *CAT2, NCED3, P5CS1 SnRK2.4, SOD1*, and *SOS2* in grapes (Liu et al., 2022). *ZIP60* and *HSP83* interact in grape cultivars to mitigate heat stress. The mechanism involved is *ZIP60* gene is expressed in heat stress conditions and interacts with *HSP83*, causing a spliced version of *ZIP60* that mitigates heat tress by renaturation and degradation of denatured proteins, which helps the grape plant to resist heat stress (Zha et al., 2022).

8.2.1.2.2 Biostimulant Activity for Abiotic and Biotic Stress Response

Seaweed extracts used on drought-stressed grapes showed improvement in ABA accumulation, proline, phenols, and antioxidant enzymes; decreased MDA and H_2O_2 levels; and improved the weight of fruit (Irani et al., 2021). Seaweed and glycine betaine biostimulants improve grape quality and improved anthocyanin levels as well as antioxidant properties, which helps alleviate plant stress due to less ROS. Anthocyanin levels were improved through anthocyanin gene expression after biostimulant treatment (Monteiro et al., 2023). Protein hydrolysate-based biostimulant APR showed improved yield potential in plants grown under drought stress by improving the biomass and fruit diameter when applied to the roots of the grape plant (Meggio et al., 2020). Protein hydrolysate biostimulants from soybean and casein helped improve grape yield, total soluble salts, improved anthocyanin content in berries, and improved drought stress tolerance (Boselli et al., 2019). Several defense responses were found to *Plasmopara viticola* infection in grapes, which are related to ethylene, JA, and ABA pathways in grapes. Furthermore, DNA methylation was observed six hours after infection in tolerant grape varieties, indicating a defense response as DNA methylation is indicative of several gene expression pathways related to defense genes such as *ETI* and *PTI* gene pathway, *PR* pathway, and defense-related proteins (Azevedo et al., 2022).

8.2.1.2.3 Summary

ERF and *Hsp* gene expression were observed for multiple stress responses. Silicon treatment led to improved expression of *ERFTF11* and *ERF13* genes. Seaweed-based biostimulants regulated drought stress and reduced ROS as a stress response mechanism. Protein-based biostimulants used on grapes showed alleviation of drought stress as well as drought tolerance; *P. viticola* infection initiated several defense responses in grapes such as *ETI* and *PTI* gene pathways. Based on the gene expression and the biostimulant usage, new methods of grape cultivation can be adopted for improved and sustainable production (Table 8.3).

8.2.2 Plantation Crop

8.2.2.1 Coffee

Coffee belongs to the *Coffea* genus. There are two species that are cultivated for coffee beans *Coffea arabica* originating from Ethiopia and *Coffea canephora*, a native from the central African region mainly Liberia, Tanzania, and Angola. It is a crop that originated from the African continent, but the highest-producing regions are presently in Latin America. Tolerance to diseases and abiotic stress are the main targets of coffee cultivation to enable high coffee production (Teketay, 1998).

8.2.2.1.1 Genes for Abiotic and Biotic Stress Tolerance

Expression levels of *APX1*, *APX5*, *DHN1*, *EDR2*, *ERF017*, *GoIS3*, *NCED*, and *PYL8a* genes have been clearly displayed as a major response to drought stress. *EDR2*, *ERF017*, and *MYB1*

TABLE 8.3
Genes for Environmental Stress Tolerance in Grapes

Cold stress	*Pma1, Aha8, Erf, HsfC1, Gras, Pat1, Dof17d, Pep172, Pep3635b, Wrky28*
Drought stress	*Cas, Rd22, Cipk02, Erf, Hsp, Hsf, Mtp, Nac, Pyl9*
Heat stress	*Erf, Hsfa2, Hsfa7, PsbP, Hsp, Hsf, HsfC1, Zip60, Hsp83*
Heavy metal stress	*Hsp, Petf, Myb, Mtb*
Salinity stress	*Erf, HsfC1, Myb108a, Mtb, Wrky28*
Diseases	*Eti, Lar2, Pti, Sts2, Sts29*
Insects	*Lox2, Lox3*

are overexpressed in drought-tolerant plants. *ERF017* is responsible for various abiotic stress tolerance by expressing several downstream ABA-dependent and independent pathways and is most expressed under low humidity in drought-tolerant plants (de Oliviera Santos et al., 2021). *MYB* is responsible for secondary metabolite production, root growth, and stomatal movement. *EDR2* regulates cell death caused by bacteria, which is mediated by SA, and *NCED* is overexpressed for water-stressed plants. *GOIS3* is highly expressed in drought conditions for coffee plants, and the *DHN1* gene is highly expressed in coffee plants in response to environmental stress (de Oliviera Santos et al., 2021). *DREB1d* is upregulated during drought stress in coffee haploids and is responsible for the heat stress tolerance mechanism by causing the expression of *ERF014* and *ERF053* in coffee and triggering the tolerance mechanism under other abiotic stress conditions also (Alves et al., 2017; Torres et al., 2019). The coffee *DUR3* gene responsible for urea transportation in plants is expressed differently in different parts of the plant, and the highest expression is found in the roots indicating its role in N distribution throughout the plant. The gene has been reported to be also involved in the redistribution of N in plants based on the stress response of the coffee plant to different abiotic stress such as drought and salinity (Dos Santos et al., 2021).

Metallothionein genes are responsible for detoxification to metal toxicity in coffee. *MT3*, *MT4*, *MT8*, and *MT15* are responsible for reducing copper toxicity by activation of *MT* types 1, 2, and 3 in coffee (Bulgarelli et al., 2016). Coffee *NAC* genes have shown upregulation in coffee plants in response to cold stress, and the mechanism showed the involvement of *NAC* genes that make the coffee plants cold tolerant such as *NAC18*, *NAC23*, and *NAC47* (Dong et al., 2019). Several *WRKY* genes were overexpressed in coffee plants grown at low temperatures, which showed a cold tolerance response, with 14 *WRKY*-related genes upregulated as a stress response for coffee plants. *WRKY* activated the pathways leading to cold tolerance such as carbohydrate-related genes, lipid-related genes, and photosynthesis-related genes (Dong et al., 2019)

8.2.2.1.2 Biostimulant Activity for Abiotic and Biotic Stress Response

Biochar incorporation into coffee fields at 8 tons per hectare helped in improving moderate water deficit during testing and also improved photosynthetic activity (Reyes-Herrera et al., 2023). The exogenous application of melatonin helped overcome drought stress in Arabic coffee cultivars by enhancing the photosynthetic activity and antioxidant enzymes and also by delaying leaf senescence. Also, ROS molecules were observed to be reduced to improve upon the drought stress conditions. Chlorophyll-related *PAO* gene, which causes its degradation when overexpressed, was found to be downregulated upon melatonin treatment in coffee. *AREB*, which is responsible for drought stress response by expressing stress-related genes in plant tissues, was found to be downregulated in coffee seedlings that were subjected to melatonin treatment (Cherono et al., 2021). Silicon treatment in coffee reduces insect infestation and disease in coffee plants by improving root and shoot growth, photosynthesis, nutrient absorption, enzyme activation, and the upregulation of *ERFTF11* and *ERF13* (Yang et al., 2022). Selenium application in coffee plants helped alleviate osmotic stress when treated before the onset of the stress effects on the plants (Sousa et al., 2023).

8.2.2.1.3 Summary

Dur3 and *ERF* are expressed due to multiple stresses, where *Dur3* is activated during drought and salinity stress and *ERF* is activated during disease and insect infestation, improving coffee crops. Melatonin application showed that the *AREB1* gene was downregulated, reducing drought stress. Biochar application in coffee cultivation showed that drought tolerance and selenium application improved osmotic stress tolerance, complementing the usage of such genes with biostimulant application, helping improve coffee cultivation practices (Table 8.4).

TABLE 8.4
Genes for Environmental Stress Tolerance in Coffee

Cold stress	*Nac18, Nac23, Nac47, Wrky*
Drought stress	*Areb1, Dreb1d, Erf017, Gois3, Dur3*
Heat stress	*Erf053, Erf014, Dreb1d*
Heavy metal stress	*Mt3, Mt4, Mt8, Mt15*
Salinity stress	*Dur3*
Diseases	*Erftf11, Erf13, edr2*
Insects	*Erftf11, Erf13,*

8.2.3 Cereal Crop

8.2.3.1 Maize

Maize (*Zea mays*) is native to Mexico and Central America and is used throughout the world for human and animal consumption. Many cultivars are available with suitable traits to eliminate heat, water, and salinity stress as well as disease and pest resistance, which will be useful for maize breeding programs (Hossain et al., 2022).

8.2.3.1.1 Genes for Abiotic and Biotic Stress Tolerance

Asr1 gene expression in maize reduced oxidative damage caused by drought stress and allowed maize seedlings to resist drought conditions, which is caused by increased ABA content in maize plants due to exogenous ABA treatment (Yao et al., 2019). *ABC* transporter genes of maize plants are responsible for the transportation of ATP throughout the plant for energy supply. Lead toxicity also induced 15 *ABC* transporter genes, which showed different expression levels due to heavy metal stress and established the role of *ABC* transporter genes in response to heavy metal stress (Guo et al., 2022). *AGO* genes were tested for different stress responses such as heat, cold, drought, and salinity, and it was found that heat and drought stress response from *AGO* was robust but was subdued for salt stress and absent for cold stress, thus helping in identifying the *AGO* gene family expression for stress response (Zhai et al., 2019). *CBL* and *CIPK* play an important role in biotic and abiotic stress tolerance. *CIPK42* was found to improve salt tolerance in different parts of maize plants by interacting with different proteins that provide salt tolerance. Thus, *CIPK42* is a positive regulator of salt stress in maize plants (Chen et al., 2021). *CDPK7* expression in maize causes a tolerant response to heat stress by increasing heat tolerance and improving photosynthesis and antioxidant activity. It regulates the heat tolerance pathway both upstream and downstream to provide resistance to heat stress (Zhao et al., 2021).

Maize *DREB1a* is responsible for plant cold tolerance as it controls raffinose accumulation by directly regulating the *RAFS* gene, which provides for raffinose biosynthesis and improves cold tolerance by improving the photosynthetic product biosynthesis, leading to normal plant growth and development (Han et al., 2020). Three *GA2ox* genes were found to show different expressions under cold and drought stress involved in maize roots and play a major role in gibberellic acid (GA) synthesis and a regulator of stress response. Overall, 13 maize *GA2ox* genes were found to respond to different abiotic stress in maize plants (Li et al., 2021). Two *SNaC* and *SKC* hub genes *GRMZM2G075104* and *GRMZM2G333183*, which caused salt tolerance response in maize plants, were identified (Ma et al., 2021). *HIPP* is a gene associated with lead toxicity in maize that increases lead tolerance and accumulation in cell walls and prevents lead toxicity in maize plants (Ma et al., 2022). *Hsf* and *Hsp70* are major gene families involved with heat tolerance in maize and are upregulated when exposed to high-temperature stress. Both *Hsf* and *hsp70* expression will cause heat stress alleviation in different parts of the maize plant (Jiang et al., 2021).

Ht1, *Ht2*, *Ht3*, and *Htn1* are genes that confer resistance to northern leaf blight in maize as they are alleles of the wall-associated receptor kinase in maize, which affects the cell wall and helps in disease control (Yang et al., 2021). *JAZ* gene is linked to the JA pathway, which is expressed during insect attacks in maize (Han et al., 2021).

The metallothionein gene family is involved in heavy metal tolerance in maize. Nine *MT* genes were identified in maize, which were further analyzed for tolerance to copper, cadmium, and lead metals by expressing in different parts of the maize plant and regulating the uptake of the metals without causing damage to the plants (Gao et al., 2022). The *MM* gene in maize helps in countering diseases such as northern leaf blight, gray leaf spot, and southern corn rust. It helps in disease resistance by downregulating the *MM*-targeting gene *MT3*, which reduces plant immunity and cell death (Wang et al., 2021). The *MYB3R* gene in maize helps confer drought and salt tolerance in maize by regulating stomatal activity and also controlling ABA activity, which helps in promoting antioxidant activity in maize. This indicates that the gene stress response is through an ABA-dependent pathway for stress tolerance and confers higher resistance to drought and salt tolerance (Wu et al., 2019). *PMTR1* is a phyto-melatonin gene receptor that controls phyto-melatonin, which regulates plant growth and stress response. The gene helps promote osmotic and drought stress tolerance in maize (Wang et al., 2022). *PP2C* genes in maize are responsible for abiotic stress response by following the ABA-dependent stress response pathway. The *PP2C* gene family can help identify different stress responses based on their expression during different stress response pathways (He et al., 2019).

8.2.3.1.2 Biostimulant Activity for Abiotic and Biotic Stress Response

Similarly, the ACC deaminase-producing bacteria *Achromobacter* sp. strain A1 was isolated from maize rhizosphere soil and was evaluated for maize cadmium tolerance. The genes responsible for cadmium transportation *HMA3* and *NRAMP5* were downregulated in maize plant roots and shoots. The bacterial inoculant can be used for cadmium resistance (Sun et al., 2022). *Fusarium verticillioides* infection produces a plant defense response in maize, which reduces growth. The phenylpropanoid pathway is triggered, which causes the production of defense compounds. Also, the cytochrome P450 gene pathway is also expressed, leading to further production of defense compounds, followed by glutathione metabolism and glyoxylate and decarboxylate metabolism in maize-tolerant varieties (Yao et al., 2020). A protein hydrolysate-based biostimulant was used to alleviate heat stress in maize, which was tested by checking the gene expression of *DHN*, *Hsp*, and *PRT*. The treatment induced higher transcription of the genes after exposure to heat stress, which led to stress-mitigating effects in maize plants (Vaseva et al., 2022). Seed-priming biostimulant *MWG9* was used to alleviate salt stress in maize seedlings by improving salt stress-effected factors such as water, chlorophyll, proline, and sugar content, and root activities, and also decreasing the levels of MDA in maize (Li et al., 2021).

About 0.5 mM of silymarin was infused into 2% maize grain extract and applied to maize plants through a foliar application, which was exposed to cadmium heavy metal toxicity. Antioxidant genes *APX*, *CAT*, *GR*, *PrxQ*, and *SOD* were expressed in response to cadmium stress, leading to increased antioxidant activity to reduce the ROS in maize plants, resulting in normal growth of the plants (Alharby et al., 2021). Zeatin-type cytokinin biostimulant helped in alleviating the salt stress effects in maize plants by reducing excess H_2O_2 within the plant and improving the plant function, which is reduced during excess salt content in the plant; the treatment also improved photosynthetic activity in the plants (Azzam et al., 2022).

8.2.3.1.3 Summary

GA2ox, *Myb3r*, and *PP2C* gene expression improved tolerance to multiple stresses. Protein hydrolysate-based biostimulant showed improved expression of heat-tolerant genes. Silymarin and maize cereal biostimulant application helped to alleviate cadmium toxicity. *Achromobacter* isolated from maize rhizosphere displayed disease resistance to *Fusarium verticillioides* infection due to defense

TABLE 8.5
Genes for Environmental Stress tolerance in Maize

Cold stress	*Dreb1a, GA2ox, Pp2c*
Drought stress	*Asr1, Ago, GA2ox, Myb3r, Pp2c*
Heat stress	*Ago, Cdpk7, Hsf, Hsp70, Pp2c*
Heavy metal stress	*Abc, Hipp, Mt*
Salinity stress	*Cipk42, Myb3r, Mtr1, Pp2c, SNac, SKc*
Diseases	*Mm, Ht, Htn1*
Insects	*Jaz*

compounds produced. Zeatin cytokinin biostimulant alleviates salt stress and seed treatment using *MWG9* biostimulant in maize. The gene expression providing stress tolerance and the biostimulant application providing tolerance in plants can be combined to improve traditional maize cultivation practices (Table 8.5).

8.2.3.2 Rice

Rice (*Oryza sativa*) belonging to the genus *oryzae* originates from the Southern and East Asian regions. Rice is presently being cultivated throughout the world, with India and China being the largest producers. Even though rice has many cultivars, it still suffers from many stress factors that limit its yield, and hence further breeding and gene identification strategies are required (Fuller and Castillo, 1871; Ali and Wani et al., 2021).

8.2.3.2.1 Genes for Abiotic and Biotic Stress Tolerance

ACA6 is expressed in cold-tolerant rice varieties, which triggers the cold response pathway, causing calcium ions to migrate from apoplast to cytosol. Improved expression of catalase (CAT), superoxide dismutase (SOD), and ascorbate peroxidase (APX) is observed in response to cold stress. *GH3-2* gene causes improvement in cold and osmotic stress tolerance in rice during the vegetative stage. The *SAP1* gene in rice helps stress response by gene regulation or protein interaction. *TPP1* helps in cold tolerance in resistant rice plants. *ZFP245* and *SFR6* genes are highly expressed in response to cold stress wherein *SFR6* triggers the regulation of gene transcription factors and *ZFP245* improves the proline content and antioxidant enzymes (de Freitas et al., 2019). *ASR* genes are triggered by ABA, stress, and ripening and also provides abiotic stress resistance in plants. *ASR2* in rice provides disease resistance against *Xanthomonas oryzae* and *Rhizoctonia solani* as well as drought resistance when overexpressed and regulates the *2H16* gene, which responds to pathogen attacks and osmotic stress tolerance in rice (Li et al., 2018). *ASR6* gene is expressed due to salinity stress as well as H_2O_2 and ABA treatment. The *ASR6* gene showed salt and oxidative stress tolerance, reduced H_2O_2 and MDA levels, and prevented electrolyte leakages. Also, the *NCED1* gene involved in ABA biosynthesis interacts with the ASR6 gene (Zhang et al., 2022).

Bph14 and *Bph15* genes were tested for resistance against *Nilaparvata lugens* rice insect, and it was observed that both genes provide resistance against insects in rice plants (Han et al., 2018). *Bph30* gene confers resistance against brown plant hopper by expressing in sclerenchyma and is part of a gene family which encodes a protein with leucine-rich repeats (LRR) domains (Shi et al., 2021). *Bph33* gene was found to be expressed which was found to be an LRR in tolerant rice varieties which showed both antixenosis and antibiosis against brown plant hopper in rice (Hu et al., 2018). *Bph38* isolated from wild rice variety *Oryza rufipogon* provided high resistance to brown plant hopper and white backed plant hopper in rice cultivars which showed antibiotic and antixenotic effects in both insects upon testing (Yang et al., 2020). Blast resistance *BR* gene in rice encodes an atypical protein that causes broad-spectrum resistance to diseases in rice. The genes encode proteins with nucleotide-binding site (NBS)-LRRs (Zhao et al., 2018). The *BSR2* gene, which provides broad-spectrum resistance to diseases in rice, was found to be encoding cytochrome 450 protein related to the *CYB78A* gene family, which provided resistance to *Rhizoctonia solani* fungus infection. It also caused slow growth in rice and longer grain production in rice crops (Maeda et al., 2019).

DRAP1 gene in rice was found to confer drought tolerance in rice and was also found to cause salt tolerance. The mechanism of salt tolerance is by expression of the *WAK* gene, which is responsible for environmental stress response. *BIERF3* and *ERF105* are also expressed and are characterized to respond to osmotic stress in rice. Also, several *WRKY* genes are expressed, which are involved in biotic stress response. After salt stress, many transcription factor genes such as *ERF*, *MYB*, *NAC*, and *ZIP* were found to be expressed in response to salt stress in rice plants. Also, genes responsible for potassium and sodium salt regulation are expressed to control the ion balance in rice such as *CAX1b* and *NHX2*. Thus, *DRAP1* is responsible to trigger salt-tolerant responses in response to salt stress (Wang et al., 2021). *GSTU30* improved tolerance to chromium toxicity and drought stress (Srivastava et al., 2019). *HIPP* genes have been shown to be responsive to heavy metal toxicity. *HIPP16*, *HIPP34*, and *HIPP60* genes have been transferred to yeast to check sensitivity toward manganese, copper, cadmium, and zinc toxicity. Yeast-transformed cells accumulated more metals in the cells, but their growth status was not affected by the heavy metals (Rono et al., 2019). *LGF1* gene in rice is responsible for flood tolerance by producing hydrophobicity in submerged rice leaves and expressing the C30 alcohol pathway for gas film formation, thus improving the photosynthetic activity of the leaves and wax platelets in rice plants (Kurokawa et al., 2018). *MADS23* is a part of the *MADS* gene family which helps in plants adapting to abiotic stress. It is overexpressed in rice plants to provide drought and salt tolerance by increasing the production of ABA and proline in plants by triggering *NCED2*, *NCED3*, and *NCED4* genes as well as *P5CR* gene, which are key components of ABA and proline biosynthesis. *MADS23* is activated by *SAPK9*-mediated phosphorylation, which is dependent on ABA in rice plants (Li et al., 2021).

MiR396 targets *GRF* genes in rice and inhibits their expression. Studies suggest that increased *GRF* gene expression improves the grain yield of rice and confers resistance to blast fungus *Magnaporthe oryzae*. Thus, improved suppression of the *R396* gene would lead to increased disease resistance and also grain yield in rice based on the differential expression of the *GRF* gene family (Chandran et al., 2019). *NAC006* is a transcription factor in rice that plays an important role in drought and heat tolerance, which is important as its downregulation increases sensitivity to drought and heat stress. It is expressed in different abiotic stress responses such as cold, heat, salt, and drought stress and also modulates H_2O_2, ABA as well as IAA within the rice plants. *NAC* transcription factors play a major role in enhancing growth as well as improving abiotic stress tolerance in rice (Wang et al., 2020). *NAC041* transcription plays a crucial role in the salt tolerance pathway of rice as gene suppression leads to increased salt sensitivity (Bo et al., 2019). Similarly, the *NAC45* gene is expressed in roots when exposed to salt stress conditions, and ABA is responsible for the suppression of the gene expression. The gene is expressed in all the root cells of rice plants and plays a key role in ABA signaling responses. *NAC45* also regulates genes such as *CYP89G1*, *DREB1F*, *ERF104*, *PM1*, *SAMDC2*, and *SIK1* genes in rice plants (Zhang et al., 2020).

PL gene encodes an *MYB* family transcription factor controlling anthocyanin synthesis, heat stress response, and also hormonal signaling. The anthocyanin biosynthesis genes such as *PAL*, *CHS*, *ANS*, and *MYB55* showed anthocyanin regulation, and the *PL* gene downregulated the photosynthetic-related genes in rice. Hormones such as ABA, JA, and IAA were found to be increased in *PL* gene-expressed plants in response to stress (Akhter et al., 2019). Potassium channel genes are responsible for providing abiotic stress tolerance in plants. The genes *AKTs*, *KAT1*, and *KAT2* in rice plants show that gene repression led to salinity tolerance in the roots of rice plants. Also, *AKT* and *KAT* genes have been shown to express and regulate drought, salinity, osmotic, and iron stress tolerance; it was also found to control diseases in rice cultivars (Musavizadeh et al., 2021).

PYL10 receptor in rice is responsible for regulating ABA in rice boot leaf through the expression of *NCED1*, *NCED2*, *NCED3*, *NCED4*, and *ZEP1*. *PYL10* gene expression has been shown to enhance survivability under cold stress. For drought stress, the *PYL10* gene showed the regulation of drought resistance genes such as *ABA45*, *DHN*, *LEA3*, and *RAB16*, which helped in drought tolerance in indica rice varieties (Verma et al., 2019).

QTL of rice *qPSST6* was found to be expressed due to cold stress. Further analysis showed that the gene LOC_Os06g39750 was responsible for the percentage seed set in rice and is established to be a candidate gene in the QTL *qPSST6* of rice responsible for cold tolerance (Sun et al., 2018). *R40c1* is a rice lectin protein-related gene that contributes toward drought tolerance in rice by regulating chromatin proteins *MNB1B* and *H4* as well as the expression of the *SAM2* gene, which is responsible for polyamine production to improve drought tolerance in rice (Sahid et al., 2020). The *RAB7* gene in rice is from a family of proteins that plays a key role in responding to environmental stress. Its overexpression helped in tolerating drought and heat stress wherein the expression of the antioxidant genes such as *APX*, *CAT*, and *SOD* as well as abiotic stress tolerance genes such as *DREB*, *LEA*, *NAC*, *RAB*, and *RD* genes was upregulated as a response to drought and heat stress treatment in rice plants. *Rab7* was also observed to enhance grain yield in rice plants (El-Esawi et al., 2019). *R2R3 MYB* leads to brown plant hopper resistance by improving phenylalanine ammonia lyase (PAL) activity-inducing secondary metabolites, which confers resistance to biotic stress in rice plants. *MYB30* gene upregulates *PAL6* and *PAL8* genes in response to brown plant hopper infestations (He et al., 2020).

SlZ1 gene in rice exhibited high tolerance to salt, drought, and heat stress conditions in rice. This is due to the exclusion of sodium ions from plants when grown in salt-containing soil and the proper storage of potassium ions in rice plants. The gene also produces proline through the activation of the proline synthesis gene *P5CS*, which alleviates damage caused by drought and salt stress in plants (Mishra et al., 2018). The *snRK2* gene family is responsible for abiotic stress tolerance in plants. The Rice *SAPK8* gene was found to improve tolerance to cold, drought, and salt stress in the vegetative stage of rice (Zhong et al., 2020). The analysis of *Sub1*, *Saltol*, *Pl*, and *Xa* gene expression under different stress conditions showed *Sub1* gene provided submergence tolerance and *Saltol* gene provided salt tolerance. *Pl2* and *Pl9* genes were responsible for blast resistance, and *Xa4*, *Xa5*, *Xa13*, and *Xa21* were responsible for bacterial blight resistance in rice plants (Das et al., 2018). *WRKY11* gene in rice has been found to regulate the expression of defense-regulated genes. The *WRKY11* gene bound to the *chitinase 2* gene is responsible for the defense response by the activation of chitinase 2 gene. The gene causes the expression of drought-related genes in response to drought stress. *WRKY11* binds directly to drought stress response gene *RAB21* and activates its transcription. The expression of the *WRKY11* gene is controlled by the ubiquitin-proteosome system in rice plants (Lee et al., 2018).

WRKY67 expression is activated by leaf and panicle blast blight infection and provides resistance to leaf blast, panicle blast, and bacterial blight diseases in rice. *WRKY67* induces defense-related genes related to the SA pathway for enhanced disease resistance in rice and also causes the expression of pathogenesis-related proteins PR1a and PR10 (Liu et al., 2018). *Xa7* is a gene in rice that provides suitable resistance to bacterial leaf blight by causing a hypersensitive response in the cells of rice. It is triggered by the avirulence gene in *X. oryzae avrXa7*, which helps in bypassing the defense response in rice plants (Chen et al., 2021). A resistance gene designated *Xa43* was found in rice variety, which confers resistance to bacterial blight disease (Kim et al., 2019). Similarly, the *xa47* gene in rice also enhances bacterial blight disease tolerance by encoding an LRR protein in rice (Lu et al., 2022). *ZIP73* gene in rice enhances cold tolerance by being expressed in cold stress conditions. The gene reduces ABA levels in anthers and helps in the transportation of soluble sugars from anthers to pollen, thus improving fertility, seed setting, and higher grain yield in rice. The gene *ZIP73* also controls the expression of *LTG3* in rice, which is responsible for cold tolerance in the reproductive stage in rice, and thus helps improve rice yield (Liu et al., 2019)

8.2.3.2.2 Biostimulant Activity for Abiotic and Biotic Stress Response

Ascophyllum nodosum seaweed and Moringa plant extract helped improve the contents of ascorbate and glutathione and also improved the antioxidant enzymes such as AHX, CAT, DHAR, GR, GP, and MDHAR as well as enhancing activity of glyoxalase I and glyoxalase II, which prevented the excess production of methylglyoxal in Cd stressed rice plants (Hasanuzzaman et al., 2023). *Moringa oleifera* plant water extract also plays a major role in improving *SAM2* expression in rice while improving grain yield in the BPT5204 rice variety (Kusampudi et al., 2022). *SAM2* plays a major role in drought tolerance by enabling polyamine production as a response (Sahid et al., 2020). Chitosan helped in alleviating drought stress by decreasing electrolyte leakage, MDA, and H_2O_2; increasing the activities of antioxidant enzymes APX, CAT, GPX, and SOD; and improving starch accumulation under drought stress (Moolphuerk and Pattanagul, 2020). *Microbacterium testaceum*-treated rice variety Pusa Basmati 1, which is susceptible to rice blast disease, showed around 80% reduction in the severity of the blast disease upon foliar application. Gene expression analysis showed induction of *CERK1*, *FMO1*, *NPK1.3*, and *PAD4*, suggestive of improved plant defense against blast disease (Patel et al., 2022). Rice plants treated with red seaweed extract helped mitigate rice blast disease by increasing the levels of defense-related enzymes such as PAL, chitinase, peroxidase, polyphenol oxidase, and phenolic content (Sahana et al., 2022).

Rice variety IR64, which was infected with rice blast fungus *Magnaporthe oryzae* and treated with red seaweed extracts, showed an upregulation of *beta-D-xylosidase* gene, which is responsible for host defense response. Other defense genes were also upregulated in rice plants such as PAL, pathogenesis-related *Bet-v-I* gene family protein, chalcone synthase, chitinases, *AP2/ERF*, *WRKY*, and *MYB* gene family (Banakar et al., 2023). Spermidine and IAA-based biostimulants applied for heat stress response in rice varieties such as Nagina 22, which is heat tolerant, and Manu Ratna, which is heat susceptible, showed improved yield production under heat stress, improved physiological responses such as lower levels of MDA and H_2O_2, as well as improved chlorophyll and proline content in rice plants (Lakshmi et al., 2023). *Trichoderma* species in combination with *koningiopsis* and *asperellum* applied to rice plants showed improvement in salt stress response with the upregulation of stress-related genes such as *APX*, *BADH*, *BZ8*, *GST*, *LYSO*, *MAPK5*, and *NRAMP6*, which reduced the effect on rice plants (Anshu et al., 2022).

8.2.3.2.3 Summary

AKT, *DRAP1*, *KAT*, *PYL10*, *SAPK8*, and *SLZ1* genes showed tolerance to multiple stress conditions. Moringa and fungal-based biostimulants provided tolerance to cadmium toxicity in rice; chitosan treatment provided drought tolerance; red seaweed extract provided rice blast resistance; *Microbacterium testaceum* treatment provided similar rice blast resistance for susceptible rice variety and, combined with *Trichoderma* spp., improved rice salt tolerance. The gene information and biostimulant action helped in improving rice cultivation, making it economical and sustainable (Table 8.6).

TABLE 8.6
Genes for Environmental Stress Tolerance in Rice

Cold stress	*qPSST6, Aca6, Pyl10, Sapk8, Zip73*
Drought stress	*Akt, Drap1, Kat, Nac006, Pyl10, R40c1, Rab7, Sam2, Sapk8, Slz1*
Heat stress	*Nac006, Pl, Slz1*
Heavy metal stress	*Akt, Kat, Gstu30*
Salinity stress	*Adr6, Akt, Drap1, Kat, Nac041, Saltol, Sapk8, Slz1*
Water stress	*Lgf1, Sub1*
Diseases	*Asr2, Br, Bsr2, Grf, Pl, Wrky11, Wrky67, Xa*
Insects	*Bph, Myb30*

8.2.3.3 Wheat

Wheat (*Triticum eastivum*) is a staple crop throughout the world with its origin being southwestern Asian region. Wheat is found to be very adaptable to different climatic regions for cultivation. The top producers are China, India, Russia, and the United States. Many breeding programs are ongoing to produce wheat varieties suitable for different regions of the world resistant to both biotic and abiotic stress resistance (Velimirovic et al., 2021).

8.2.3.3.1 Genes for Abiotic and Biotic Stress Tolerance

AB15 gene in wheat was found to promote cold tolerance by interacting with cold-tolerant *ICE1* transcription factor. The *AB15* gene is linked as a transcription factor for ABA pathways (Liang et al., 2022). *ASR1* gene is responsible for salt tolerance in wheat, which initiates the salt stress response pathway (Hamdi et al., 2020). *Bx6* gene in wheat produces benzoxazionoids to reduce insect infestation of herbivore insects (Shavit et al., 2022). *CNR2* gene in wheat causes heavy metal toxic resistance by inducing the translocation of heavy metals such as cadmium, zinc, and manganese from roots to shoots and also reducing the presence of excess heavy metals in the plant (Qiao et al., 2019). *CYP81D5* gene in wheat provided tolerance to salt stress by reducing ROS, thereby reducing salt stress damage (Wang et al., 2020). *Dn10* provided resistance to Russian wheat aphids in wheat cultivars by reducing the damage caused to leaves by aphid feeding and allowing plants to grow normally, which helps in aphid tolerance (Puterka et al., 2020). *EXPA2* provides improved drought tolerance in wheat by enhancing antioxidant enzyme activity. The *EXPA2* gene is expressed by a *MYB* transcription factor *MPS*, which enables drought tolerance and improved root growth. *MPS* may be a regulator of the *EXPA2* gene conferring drought tolerance in wheat (Yang et al., 2020).

EXPA8 gene was induced by cold and drought stress, and its inducement caused an increase in CAT, peroxidase, and SOD, along with stable MDA and proline content. *EXPA8-A* and *EXPA8-B* were expressed in the roots of wheat (Peng et al., 2019). Wheat species *Thinophyrum elongatum* consists of *Fhb7*, which is resistant to Fusarium head blight disease by *Fusarium* spp. by detoxifying trichothecenes through de-epoxidation. *Fhb7* in wheat confers resistance to Fusarium head blight as well as crown rot in different varieties (Wang et al., 2020). *Gb8* is responsible for greenbug resistance in wheat (Xu et al., 2020). *HAG1* is responsible for salt tolerance in wheat by regulating ROS and also regulating genes responsible for hydrogen peroxide production (Zheng et al., 2021). *HDZ5-6A* is responsible for resisting drought stress in wheat by regulating water loss and improving proline content as well as improving the expression of several drought response genes (Li et al., 2020). *Lr34* gene in wheat is responsible for the ABC pathway, which regulates ABA and provides disease resistance to many fungal pathogens (Krattinger et al., 2019). *NAC4-3A* transcription factor is responsible for increased root growth in salt and osmotic stress and also increased drought resistance by reducing water loss through stomata. The *NAC4-3A* transcription factor leads to the expression of different stress response genes providing drought stress tolerance (Mei et al., 2021).

NAC8-6A gene overexpression in wheat provides increased tolerance to drought by inducing auxin and drought resistance-related pathways and increasing root development (Mao et al., 2020). *NAC48* is part of the *NAC* transcription factor family which regulates plant development and stress responses. *NAC48* helps in drought tolerance by improving proline content and reducing water loss, MDA, and H_2O_2. *AREB3* gene in wheat is responsible for triggering the *NAC48* gene to respond to drought stress (Chen et al., 2021). *NAC071-A* gene in wheat is responsible for drought tolerance by controlling water usage as well as regulating several stress-related genes. *NAC071-A* is controlled by the *MYBL1* gene, which helps improve drought tolerance (Mao et al., 2022). *NHX2* provides salinity tolerance in wheat by maintaining higher salt content in leaves and roots under salt stress; improving protection against cell damage; improving photosynthesis, water content, and proline levels; and reducing MDA, ROS, and H_2O_2 (Mushke et al., 2019). *PIP2;1* aquaporin gene provides salt and drought stress tolerance in wheat by inducing CAT and SOD enzyme activity and also reducing water loss (Ayadi et al., 2019). *Pm68* provides disease resistance to powdery mildew in wheat (He et al., 2021). *PYL1-1B* provides drought tolerance in wheat and is regulated by the *MYB70* transcription factor during drought stress, through which drought tolerance by increased water use efficiency is achieved with an improvement in grain yield (Mao et al., 2022). *SHN1* is a transcription factor that provides cadmium, copper, and zinc tolerance in wheat by reducing the oxidative damage caused to plants through activation of antioxidant enzymes and improved root and biomass accumulation to mitigate the stress effects (Djemal and Khoudi, 2022). *Sb4* confers spot blotch resistance in wheat (Zhang et al., 2020). *Sr26* and *Sr61* gene in wheat imparts disease resistance by activating *NLR* genes to reduce stem rust (Zhang et al., 2021).

TPS11 gene in wheat is responsible for cold tolerance by lowering sucrose content, increasing starch content, and increasing enzyme activity related to sucrose metabolism. The *TPS11* gene also causes the expression of the *SnRK1* gene, which is responsible for catalyzing sucrose in wheat (Liu et al., 2019). *UGT6* gene confers resistance to Fusarium head blight by developing tolerance to the mycotoxin deoxynivalenol in wheat by breaking down the mycotoxin (He et al., 2020). *WRKY30* transcription factor is responsible for providing environmental resistance in wheat. *WRKY30* enabled drought and heat tolerance by improving chlorophyll content, water, and proline content. Improvement in antioxidant enzymes was also observed. Reduction of electrolyte leakage, H_2O_2, and MDA was also observed in *WRKY30* overexpressed plants. The genes regulated by *WRKY30* were *AQP7, DREB1, DREB3, ERF5α, TIP2*, and *WRKY19*, which are stress response genes in wheat (El-Esawi et al., 2019).

8.2.3.3.2 Biostimulant Activity for Abiotic and Biotic Stress Response

A combination of *A. nodosum* and chitosan biostimulant application on wheat resulted in the control of Fusarium head blight with fewer symptoms observed and a reduction in mycotoxins present in seed grains. Defense-related genes *Glu2, PR1, PR2*, and *PR3* were expressed which led to the activation of defense-related enzymes as peroxidases and polyphenol oxidases (Gunupuru et al., 2019). Enzymatic hydrolysate of seaweed *Ulva clathrata* was used as a biostimulant to test efficacy in controlling salt-alkali stress on wheat plants, which showed improved growth during stress conditions by reducing H_2O_2 content in roots and shoots and activation of POD and APX to alleviate oxidative damage. Gene expression of *AQP6, AQP7, AQP8, CCD1, CRT, HKT2:1, NHX2*, and *SOS1* in roots or leaves to reduce ion toxicity and improve water uptake in wheat plants (Li et al., 2022). Humic acid was observed as a biostimulant in wheat varieties, which were tolerant and susceptible to drought stress, drought stress-related H_2O_2, and antioxidant enzymes as well as drought stress-related genes. Analysis of genes such as *AP2, CSD1/CSD2, GRF, LAC*, and plastocyanin showed humic acid treatment exhibited positive regulation of drought stress in all the varieties tested (Arslan et al., 2021). About 2% maize grain extract alleviated combined salinity and drought stress in wheat by reducing sodium and chlorine levels in plants as well as reducing the H_2O_2 and MDA levels in crop plants, leading to better plant growth and yield (Alharby et al., 2020),

Metabolites of soil *Streptomycetes* were used as biostimulant sources to control cyst nematode of wheat, which led to the production of different si and mi RNA in plants, which were complimentary to plant and nematode-produced mRNA, causing the reduction of nematode infestation in wheat crops (Blyuss et al., 2019). *Stipagrostis pennata* rhizosphere isolates were used to test drought tolerance on two wheat cultivars Sirvan (drought resistant) and Sivand (drought susceptible), which showed improvement in shoot and root biomass, improved chlorophyll levels, improved SOD, and improved nutrient uptake (Naderi et al., 2022). Strigolactone was used as a biostimulant on wheat for cadmium stress tolerance, where plants treated with strigolactone showed improved plant growth, reduction in cadmium levels in plants, and activation of oxidative stress response genes *APX, CAT, POD*, and *SOD* in wheat plants, leading to reduced oxidative damage to the plant under cadmium stress (Shah et al., 2023).

8.2.3.3.3 Summary

Genes *EXPA8, PIP1*, and *WRKY30* showed improved expression and tolerance across multiple stress conditions. Biostimulant usage of maize extract provided drought tolerance, seaweed and chitosan-based biostimulants provided Fusarium head blight resistance and reduced mycotoxins

TABLE 8.7
Genes for Environmental Stress Tolerance in Wheat

Cold stress	*Ab15, Expa8, Tps11*
Drought stress	*Expa2, Expa8, Hdz5-6A, Pip1, Nac, Pyl1-1b, Wrky30*
Heat stress	*Wrky30*
Heavy metal stress	*Cnr2, Shn1*
Salinity stress	*Asr1, Cyp81d5, Hag1, Pip1, Nhx2, Nac4-3a*
Diseases	*Fhb7, Lr34, Pm68, Sr26, Sr61, Ugt6*
Insects	*Bx6, Dn10, Gb8*

in seed grain, Strigolactone biostimulant application provided cadmium tolerance, *Ulva clathora* seaweed-based biostimulant treatment provided salinity tolerance, humic acid treatment provided drought tolerance, *Stipagrostis pennata* rhizosphere isolate provided tolerance to drought conditions, and *Streptomycetes* metabolites provided tolerance to cyst nematode upon treatment. The usage of the genes responsible for stress tolerance and analyzing their expression on biostimulant treatment would help in improving wheat cultivation in adverse conditions (Table 8.7).

8.2.4 Tuber Crop

8.2.4.1 Potato

Potato (*Solanum tuberosum*) is a native of the western South American region and is a crop cultivated throughout the world. The major stress factors such as heat, salt, diseases, and drought stress mitigated by different cultivars have been grown by many farmers in the Andean region of South America (de Haan et al., 2016).

8.2.4.1.1 Genes for Abiotic and Biotic Stress Tolerance

AP2/ERF genes in potatoes were found to enhance cadmium tolerance by cadmium accumulation in plants as well as detoxification of cadmium from potatoes (Tian et al., 2022). Also, *MTP9* gene is involved in the control of cadmium stress response and enables the plant to become cadmium tolerant by reducing cadmium content in plants slowly (Li et al., 2021). *CAX* gene is involved in calcium, magnesium, and manganese transport and in maintaining ion homeostasis in plants. During cadmium stress, *CAX1/4* gene was expressed and interacted with many proteins and is regulated by transcription factors such as *WRKY6. CAX* provides high cadmium tolerance in potato (Liu et al., 2023).

CBF1 was observed to exhibit increased cold, salt, and drought tolerance in potato *commersonii* variety in comparison to *tuberosum* variety of potato. The *commersonii CBF* gene was found to upregulate fatty acid desaturase, Calvin cycle, and antioxidant enzyme genes. The *CBF1* enzyme was able to maintain photosynthetic activity under cold stress in *commersonii* variety and enhanced tolerance to drought and salt stress in *commersonii* and *tuberosum* varieties (Li et al., 2022).

CDPK3 and *CDPK23* were induced during drought stress in potatoes, thereby providing drought tolerance. *CDPK* encodes calcium-dependent protein kinase, which helps in plant development and stress response (Bi et al., 2021). Copper ions help induce a robust disease response against *Phytophthora infestans* as ABA synthesis negatively regulates disease resistance to *P. infestans* (Liu et al., 2020). The *CYS1* gene in potatoes was able to enhance salt tolerance by improving proline and chlorophyll content in plants. Also, H_2O_2 scavenging activity and improved cell membrane stability were observed due to *CYS1* gene expression (Liu et al., 2020). Expansin-like B genes *EXPB1 to 6* in potatoes were found to be expressed in drought and heat stress conditions and provided tolerance to heat stress and also improved tuber development (Chen et al., 2019). *Hsp26* and H*sp70* were expressed after prolonged exposure to heat stress in potatoes for three days, indicating their significant role in conferring plant tolerance to heat stress (Tang et al., 2020). Six *NBS-LRR* genes were found to be expressed in potato *pinnatisectum* variety, which showed tolerance to *P. infestans* in comparison to the susceptible *cardophyllum* variety of potatoes (Gu et al., 2020). *NRAMP1, 4,* and *5* were shown to be induced by copper, cadmium, and zinc toxicity in the soil of grown potatoes and showed enhanced tolerance to the heavy metal stress by enabling the absorption of the metal within the plants, thus alleviating heavy metal toxicity in potato plants (Tian et al., 2021).

Ring finger protein *RFP2* is involved in drought tolerance in potatoes by showing increased expression in drought conditions. The gene has also been expressed due to osmotic stress with increased proline content in potatoes. Similarly, under drought stress, overexpression of *RFP2* caused a reduction in MDA levels and improved water retention capacity in potato plants (Qi et al., 2020). *WRKY6* is a transcription factor of potatoes that regulates the *SRO* gene family, which is responsible for improved plant antioxidant activity and enhancing drought and high-salinity stress. The *SRO6* transcription factor was responsible for achieving cadmium tolerance in potato (He et al., 2022). *ZIP65* transcription factor enhances salt tolerance in potatoes, which was induced by exogenous methyl jasmonate application, although the mechanism needs to be further elucidated (Zhao et al., 2020). *C4H* and *HQT* gene expression improved biosynthetic pathways in potatoes, producing higher chlorogenic and caffeic acid in the skin of tubers, leading to reduced feeding by potato tuber moth in tubers (Pacifico et al., 2019).

8.2.4.1.2 Biostimulant Activity for Abiotic and Biotic Stress Response

A low level of arbuscular mycorrhiza fungi in the potato root system helped trigger plant defense response to aphid infestation by expressing ethylene gene *Etr1* during initial infestation and then triggering defense genes *AOC* and *PTP1-I*, leading to less aphid damage in potato (Rizzo et al., 2020). Phosphite acted as a biostimulant and helped improve plant growth and development-related genes as well as improve the expression of secondary metabolite-producing genes, which helped alleviate plant pathogenic stress in potato plants (Han et al., 2021).

8.2.4.1.3 Summary

CBF and *EXPB 1 to 6* expression provided tolerance to multiple stresses and showed improved potato growth. Phosphite-based biostimulants showed improvement in potato growth and plant disease tolerance expression through secondary metabolite pathway activation. The usage of diluted mycorrhiza fungi in potato roots also induced a defense response to aphid infestation. The usage of biostimulant treatment to enhance gene expression would be advantageous to help new cultivation practices to improve potato production (Table 8.8).

8.2.5 Legume Crop

8.2.5.1 Soybean

Soybean (*Glycine max*) originated from China 3,000 years ago and is presently being cultivated worldwide due to its high protein value and has several cultivars for different regions. The highest

TABLE 8.8
Genes for Environmental Stress Tolerance in Potato

Cold stress	*Cbf*
Drought stress	*Cdpk, Cbf, Expb (1–6), Rfp2*
Heat stress	*Hsp26, Hsp70, Expb (1–6)*
Heavy metal stress	*Cax1, Wrky6, Ap2/Erf, Nramp*
Salinity stress	*Cbf, Cys1, Zip65*
Diseases	*Nbs-Lrr, Aba*
Insects	*Aoc, C4h, Hqt, Etr1, P1-1*

producer of soybeans is Brazil due to its climate. Presently soybean is affected by many diseases and abiotic stress problems due to its specific cultivation requirements, and many breeding programs aim to make it more easily cultivable (de Maria et al., 2020).

8.2.5.1.1 Genes for Abiotic and Biotic Stress Tolerance

Drl protein expressed in soybean plants showed a broad immune response to the fungal pathogen *Fusarium virguliforme* as well as insects infesting soybean – spider mites, soybean aphids, and soybean cyst nematode. The mechanism for immunity would involve the induction of JA and SA pathways in soybean plants (Ngaki et al., 2021). The *GELP28* gene of soybean was found to enhance drought and salt tolerance in soybean by increasing adaptability to stress through its overexpression. The *GELP* gene family, which are *GDSL*-type esterase/lipase proteins, plays a major role in mitigating abiotic stress and defense response in plants (Su et al., 2020).

Genes involved in ABA degradation, gibberellin and SA biosynthesis, as well as H_2O_2 production and cadmium transport were upregulated, whereas zeatin ribose, IAA, and methyl jasmonate biosynthesis were downregulated in cadmium and drought stress response of soybean. In tolerant cultivars genes responsible for gibberellins, IAA, and methyl jasmonate were upregulated in soybean (Bashir et al., 2019). *GRAS37* is responsible for drought and salt tolerance in soybeans by being overexpressed in drought and salt stress conditions. The roots where *GRAS37* was overexpressed showed increased upregulation of several salt and drought stress-related genes in soybean such as *DREB1*, *NCED3*, *CLC1*, *SOS1*, *SOD1*, and *SOD2* (Wang et al., 2020). Interleukin 1 receptor nucleotide-binding site–leucine-rich repeat protein type resistance gene in soybean *KR3* was found to enhance resistance to several viral infections. The gene *KR3* played a role in ABA signaling, which enabled it to cause a viral resistance response in soybean (Xun et al., 2019). *J* gene in soybean was found to express salt stress-resistant genes such as *WRKY12*, *WRKY27*, *WRKY54*, *NAC*, and *SIN1* and help improve stress tolerance (Cheng et al., 2020). *MYB12* in soybean helps in the accumulation of flavonoids and abiotic stress resistance in plants by maintaining osmotic balance, membrane integrity, and ROS levels normal in soybean (Wang et al., 2019). *NAC06* in soybean causes tolerance to salt stress by causing the accumulation of proline in plants to mitigate the negative effects of ROS, which may damage the plant and maintain normal plant function and is expressed in the roots of soybean (Li et al., 2020). *NAC109* caused drought and salt stress tolerance by the expression of tolerance genes such as *AREB1*, *AREB2*, *COR15A*, *DREB2A*, and *RD29A* in soybean.

NAC109 increased the expression of ABA-insensitive genes AIN1 and AIN5 without affecting the *NCED3* gene, which is responsible for ABA biosynthesis. *NAC109* also increased auxin expression through the *AIR3* gene, which produces more hairy roots in soybean (Yang et al., 2019). Nucleotide-binding receptor *rsc 4-3* is responsible for resistance to soybean mosaic virus by conferring defense response by recognizing cylindrical inclusion proteins, thus preventing infection in soybean (Yin et al., 2021). *PRX9* gene isolated from soybean soja wild variety was found to increase salt tolerance in plants by increasing the antioxidant activity and maintaining proper water and ion balance in shoot and root through its expression in soybean (Jin et al., 2019).

Photosynthesis-related gene *PsaC* and the ATP synthase gene *ATPsyn* α were found to be upregulated as a defense response to the soybean mosaic virus in plants. Several hormone synthesis-related genes were found to be overexpressed by *ATPsyn* α, whereas *PsaC* was found to upregulate JA and SA pathway-related genes (Bwalya et al., 2022). *SAP16* improved drought and salt tolerance in soybean and increased ABA sensitivity by delayed seed germination and stomatal closure and also showed increased proline content in plants. The overexpression of *SAP16* showed increased expression of stress genes such as *DREB1B*, *NCED3 RD22*, *DREB2*, *NHX1*, and *SOS1* (Zhang et al., 2019). The *SnRK1* gene helps regulate the *ERF7* gene in soybean wild variety to improve adaptability to various environmental stresses by controlling the salt and alkali tolerance in plants (Feng et al., 2020). *TCF1α* was found to be expressed in soybean due to cold stress by increasing plant survival and reducing electrolyte leakage. Cold response gene *COR15a* was observed to exhibit increased expression where *TCF1α* was overexpressed in soybean (Dong et al., 2021). The *TGA13* gene in soybean was highly expressed in salt stress in the roots. It showed salt resistance through the expression of salt resistance genes in soybean (Ke et al., 2022).

UGT gene expression in soybean showed resistance to leaf-chewing insects *Helicoverpa armigera* and *Spodoptera litura* by improving the production of flavonoids in soybean plants, thus reducing insect feeding (Zhang et al., 2022). *VOZ1G* showed overexpression in hairy roots, providing greater tolerance to drought and salt stress in soybean, where reduced MDA was observed with an increase in POD and SOD with the overexpression of hairy roots (Li et al., 2020). *WRKY40* was found to show increased expression due to *Phytopthera sojae* infection in the roots of soybean plants. The resistance to the disease was through interaction with *JAZ* proteins without interaction with *WRKY* or *ZIP* domains in soybean and also through the modulation of H_2O_2, which leads to improved plant disease response in soybean (Cui et al., 2019). *WRKY142* gene helps in increased cadmium tolerance for soybean by inducing two cadmium stress-related genes *CDT1-1* and *CDT1-2*. The *CDT1-1* and *CDT1-2* decrease the cadmium uptake into the plant and improve cadmium tolerance (Cai et al., 2020). *WRKY172* was highly expressed in the leaves and flowers of soybean plants in response to cadmium stress. It caused enhanced cadmium tolerance and reduced cadmium content in shoots, roots, and seeds. There was less accumulation of H_2O_2 and MDA in soybean plants with increased flavonoid and lignin content as well as peroxidase activity. *WRKY172* was also regulating several stress-related pathways in soybean (Xian et al., 2023). Soybean *YC4-2* gene expression provided a broad defense response to bacterial, viral, and fungal diseases in soybean and was observed to improve protein levels and help in the maturity of soybean plant (O' Conner et al., 2021). *ZFP03* in soybean was found to enhance disease resistance to *P. sojae*, which causes root and stem rot disease in soybean roots. *ZFP03* causes the improved gene expression of the *SOD* gene in soybean, leading to improved disease resistance to *P. sojae*. The *ZFP03* gene provides improved disease resistance by targeting SOD promotors in soybean (Li et al., 2023). Soybean *ZIP* transcription factors are responsible for stress response and plant growth in plants. *ZIP2* in soybean is expressed due to multiple stresses affecting soybean and helps in expressing different stress response genes such as *MYB48*, *WD40*, *DHN15*, *GST1*, and *LEA*, thus providing resistance to different abiotic stresses (Yang et al., 2020).

8.2.5.1.2 Biostimulant Activity for Abiotic and Biotic Stress Response

Amino acid-based biostimulants have been used to treat drought stress of soybean with 0.5 kg/ha treatment, showing improved photosynthetic activity by increasing stomatal conductance and improved antioxidant enzymes APX, CAT, and SOD as well as proline and biomass content (Repke et al., 2022). *Arthrobacter woluwensis* strain AK1 was used as a biostimulant to mitigate salt stress in soybean where significantly less salt uptake was observed. Four genes were observed to be upregulated, an auxin-related gene (*LAX1*), a gene responsible for potassium channel (*AKT2*), and two salt-tolerant genes *ST1* and *SALT3*. Two genes related to sodium (*NHX1*) and chlorine (*CLC1*) were downregulated (Khan et al., 2020). *A. nodosum* seaweed and fulvic acid-based biostimulants have been tested for drought stress response in soybean, wherein treated plants showed improved stress response by improving water potential, osmotic adjustments, stomatal conductance and photosynthetic activity, as well as activity of antioxidant enzymes (do Rosario Rosa et al., 2021).

Biostimulant treatment using Elicel6Indures showed an increase in secondary metabolite-producing genes related to JA and SA as well as isoflavonoid and phytoalexin pathways. *MAPK* cascade genes and oxidative stress-related genes were also expressed, leading to improved plant immune response (Decsi et al., 2023). Seaweed-based biostimulant from *A. nodosum* showed improved heat tolerance in soybean by showing an increased rate of carbon dioxide assimilation, stomatal conductance, transpiration rate, and carboxylation efficiency and reduced leaf temperature. Further, increased activity of APX, CAT, and SOD antioxidant enzymes was observed along with maintaining proline levels in the plant (Repke et al., 2022). Trichoderma isolates *harzianum* and *asperellum* treatment showed a defense response in soybean roots against *F. virguliforme* through increased expression of jasmonic and ethylene response genes such as *PR3*, *PR12*, and *AOS* as well as SA genes such as *PR2*, *CHS*, and *IPER* (Pimentel et al., 2020).

TABLE 8.9
Genes for Environmental Stress Tolerance in Soybean

Cold stress	*Myb12, Tcf1a, Zip2*
Drought stress	*Gelp28, Grap37, Myb12, Nac109, Sap16, Voz1g, Zip2*
Heat stress	*Myb12, Zip2*
Heavy metal stress	*Wrky142, Wrky172*
Salinity stress	*Gelp28, Grap37, J, Nac109, Prx9, Sap16, Snrk1, Thai3, Voz1g, Zip2*
Diseases	*Dr1, Kr3, PsaC, Rsc4-3, Wrky40, Yc4-2, Zep03*
Insects	*Dr1, Ugt*

8.2.5.1.3 Summary

GELP, GRAP, MYB12, SAP16, VOZ1G, and *ZIP2* gene expression induced multiple plant stress tolerance responses in soybean. Biostimulant application of *A. woluwensis* strain provided salt tolerance, infusing amino acids into the soil and improved drought tolerance. Seaweed-based biostimulant provided heat tolerance and a combination of seaweed and fulvic acid-induced drought tolerance upon treatment and Elice16indures. Biostimulant product induced secondary metabolite production upon treatment and improved plant defense response. Analyzing gene expression upon biostimulant treatments would thus provide a better understanding to improve soybean cultivation practices (Table 8.9).

8.2.6 Sugar Crop

8.2.6.1 Sugarcane

Sugarcane (*Saccharum officinarum*) is a widely cultivated sugar crop throughout the world. It is believed to have originated from the Indian subcontinent based on documents dating 3,400 years ago. Its wild species is *Saccharum spontaneum*. It then spread from the subcontinent to the east and west due to trade and spread worldwide through Portugal and Spanish trade to the Americas. Many desirable traits required for stress tolerance are available for further improvement of sugarcane crops (Brumbley et al., 2009).

8.2.6.1.1 Genes for Abiotic and Biotic Stress Tolerance

The *ACLA1* gene in sugarcane is responsible for the regulation of several drought-related genes triggering a reduction in ROS as well as altering MDA, POD, and SOD in sugarcane, thus mitigating drought stress (Zhu et al., 2021). The *CBL* gene family in sugarcane was found to be upregulated in response to salt, copper, and disease-related stress and plays an important role in triggering tolerance mechanisms to the stresses within sugarcane (Su et al., 2020). *DHN, DREB2, ERD4*, and *SNAC1* genes were found to be expressed during drought stress and gradually reduced in sugarcane. Invertase and calmodulin functional proteins were expressed throughout the stress and recovery phase of sugarcane drought stress (Devi et al., 2019). *DIR* genes are responsible for stress resistance in sugarcane. The *DIR7* gene was found to promote drought tolerance through analysis of CAT, H_2O_2, MDA, and SOD whose levels indicate its resistance potential (Li et al., 2022). *ERF52* gene expression indicated tolerance to salt stress in sugarcane, whereas four different Soloist gene expressions indicated water stress tolerance in sugarcane (Li et al., 2020). Genes expressed due to insect infestation of fall armyworm were peroxidase for phenylpropanoid synthesis,

PR1 gene for insect defense helped in resisting the insect herbivore. Sugarcane response to fall armyworm was to trigger the gene expression of *ERF, MYB*, and *WRKY* as well as JA and SA pathway-related genes to provide tolerance to the insects during sugarcane growth (Li et al., 2022). The *GPCR1* gene in sugarcane helps in tolerating drought, salinity, and cold stress by expressing genes such as *AD3, CRF2, DHN, DR4, ERF3*, and *SOS1* which are responsible for stress tolerance pathways (Ramasamy et al., 2021). Peroxidase class genes in sugarcane were expressed during virus, cadmium, and salt stress induction, leading to tolerance response pathways being elucidated in sugarcane, causing resistance to sugarcane mosaic disease and salt and cadmium heavy metal presence in soil (Shang et al., 2023). *PHT* genes were involved in phosphate ion transportation and stress response in sugarcane. Salinity stress in roots caused the upregulation of *PHT1-1*, *PHT1-2*, and *PHT1-3* (Murugan et al., 2022). Phytepsin, glutamate synthetase, and stress protein *ddr48* gene expression were observed in sugarcane due to heat stress, leading to the reduction of oxidative stress, heat shock protein activation, and activation of the glutamate pathway (Raju et al., 2020). *PIP2-1* gene in sugarcane, responsible for osmotic stress response, is also expressed under salt stress condition to initiate tolerance response by improving water content and root length, preventing excess ion leakage, lower MDA,, and higher POD, and SOD. Further, salt response genes like *DREB2*, *HKT1*, *KIN2*, *NHX1*, *P5CS1*, *P5CS2*, *RD29A*, and *SOS1* genes have been observed in sugarcane (Tang et al., 2021).

Salt stress response in sugarcane showed the expression of *NHX1*, *Ppase*, and *ATPase* genes, which help in the storage of sodium ions into vacuoles of sugarcane root cells. Also, total soluble sugars were increased in roots and increased proline levels in leaves of sugarcane, in response to salt stress treatment. This helped in increasing photosynthesis, stomatal conductance, and better water usage efficiency in sugarcane (Theerawitaya et al., 2020). *SEC14p* in sugarcane helps in reducing drought and salt stress by increasing SOD but reduced MDA. The response is due to the upregulation of several salt-related stress genes and genes related to the phosphatidylinositol pathway (Ren et al., 2021). The *SOS* gene family in sugarcane mostly maintains salt balance in stems and leaves. *SOS1*, *SOS2*, and *SOS3* gene expression showed improved salt tolerance by reducing/maintaining sodium ion levels in the stem of sugarcane (Brindha et al., 2021). *Sporisorium scitamineum* disease in sugarcane is reduced by the *CAT1* gene from *spontaneum* variety, which could cause disease resistance by the upregulation of genes related to hypersensitive response, ROS, and SA pathways (Wu et al., 2023). Sugarcane mosaic virus infection caused the expression of antioxidant genes such as glutathione transferase 1 in sugarcane-resistant varieties, indicating the mode of resistance, which reduced ROS and is indicative of plant defense response. The *GST1* gene triggered transcription factors such as *AP2*, *HLH*, *NAC*, *WRKY*, and *ZIP*, which help in reducing sugarcane mosaic virus infection in sugarcane (Akbar et al., 2020).

WRKY5 gene was found to be expressed during disease response in sugarcane. The gene expression was increased by SA and ABA and was initiated when infected by *S. scitamineum* smut disease in sugarcane-tolerant varieties. *WRKY5* also enhanced resistance to *Ralstonia solanacearum*. *WRKY5* was also expressed when induced with drought and salt stress (Wang et al., 2020). *WRKY13-2*, *39-1*, *49-3*, and *125-3* were found to show increased expression when exposed to *Xanthomonas* causing leaf scald disease in sugarcane and may play an important role in defense-related pathways of sugarcane. The *WRKY* genes are also a negative regulator of nitrogen and play a major role in different growth pathways (Javed et al., 2022),

8.2.6.1.2 Biostimulant Activity for Abiotic and Biotic Stress Response

Bacillus xiamenensis strain pm14 was isolated from the sugarcane rhizosphere and applied as a biocontrol agent in sugarcane red rot disease, where disease symptoms were suppressed and improved plant growth was observed along with improvement in antioxidant enzymes and proline levels (Xia et al., 2020). *Metarhizium anisopliae* inoculum helped in mitigating stalk- and root-based pests of sugarcane through the application in the field and improved stalk length and sugarcane yield (Siqueira, 2020). *Pantoea dispersa*-AA7 and *Enterobacter asburiae*-BY4 strains from sugarcane rhizosphere microbe isolates showed improvement in sugarcane growth and nitrogen fixation as well as showed resistance to osmotic stress and improved stress-related gene expression of CAT, chitinase, PAL, and SOD as well as induction of defense mechanism in sugarcane (Singh et al., 2021). Seaweed-based biostimulants applied to late-harvest sugarcane in drought conditions showed improved stalk production as well as improved antioxidant enzyme levels and MDA levels

TABLE 8.10
Genes for Environmental Stress Tolerance in Sugarcane

Cold stress	*Gpcr1*
Drought stress	*Acla1, Dir7, Dreb2, Dhn, Erd4, Gpcr1, Sec14p, Snac1*
Heat stress	*Ddr48*
Heavy metal stress	*Cbl1, Prx*
Salinity stress	*Cbl1, Erf52, Gpcr1, Nhx1, Pht, Pip2-1, Prx, Sec14p, Sos*
Diseases	*Cat1, Cbl1, Gst1, Pr1, Prx, Wrky*
Insects	*Erf, Myb, Wrky, Ja, Sa*

in sugarcane plants, leading to improved drought tolerance by showing improved biomass and sugar production (Jacomassi et al., 2022).

8.2.6.1.3 Summary

CBL1, *GPCR1*, and *SEC14p* gene expression provided stress tolerance across multiple stress conditions. Biostimulant treatment of *M. anisopliae* provided tolerance to stalk and root-based insect pests, *B. xiamenensis* strain from sugarcane rhizosphere provided tolerance to red rot disease on application to sugarcane, and seaweed-based biostimulants provided drought tolerance in sugarcane. Gene expression profiling and biostimulant usage optimization would provide for better sugarcane cultivation method optimization (Table 8.10).

8.2.7 Vegetable Crop

8.2.7.1 Tomato

Tomato (*Solanum lycopersicum*) is a widely cultivated crop throughout the world and is used in almost every household. It originated from Central America and presently the highest producers are China, India, and the United States. It is widely disease prone, and resistance cultivars are available for breeding, which provide disease and abiotic stress resistance (Kumar et al., 2011).

8.2.7.1.1 Genes for Abiotic and Biotic Stress Tolerance

BBX17 is a positive regulator of heat stress in tomatoes but causes a reduction in plant growth as a response to heat stress. It reduces heat stress by improving membrane stability and improves antioxidant enzymes, which reduce the ROS. The overexpression of *BBX17* affected genes related to GA biosynthesis, photosynthesis, heat stress, ROS, and other cellular processes (Xu et al., 2022). Brassinosteroids play a major role in plant growth and abiotic stress response in plants. The transcription factor *BZR1* is part of a brassinozole-resistant factor family and was found to regulate tolerance to salt stress in tomato and was expressed in flowers and fruits. The gene *BZR1* interacted with *MYB30*, *MYBL2*, *PIF4*, *HAT1*, WS1, and REF6 (Jia et al., 2021). *CathB2*, *GST*, *UBC5*, and *ARG1* genes were shown to provide heat tolerance in tomato. *CathB2* reduced the ROS in the plant; *GST* plays a major role by expressing as an abiotic stress response; *UBC5* gene is responsible for different abiotic stress responses such as drought, salt, and osmotic stress; *ARG1* controls the plant response to abiotic stress, thus enabling normal plant functioning (Wen et al., 2019). *COR413PM2* gene is responsible for cold tolerance in tomato. The overexpression of *COR413PM2* caused a reduction in cell membrane damage and ROS; it also increased antioxidant enzyme activity and osmotic regulators to enable tolerance to cold stress (Zhang et al., 2021). *DREBA4* gene was upregulated in response to heat, drought, cold, and salt tolerance. The gene *DREBA4* also regulates the downstream expression of several heat shock proteins and calcium-binding proteins. Ethylene, JA, and SA pathway genes were also induced by *DREBA4* and the heat shock protein genes (Mao et al., 2020). *FRG4* and *AGO4a* genes in tomato provide resistance to *Fusarium oxysporum* wilt disease in tomato (Ji et al., 2021).

GATA genes are related to plant growth functions. Recent studies show the genes being expressed in response to drought stress in tomato. The *GATA17* gene was found to overexpress in drought stress conditions through activation of the phenylpropanoid biosynthesis pathway (Zhao et al., 2021). *GRAS4* gene is responsible for drought tolerance in tomatoes by modulating ROS levels in plants. *GRAS4* directly binds to *SnRK2*, which plays a crucial role in ABA signaling and interacting with downstream *AREB1* and *AREB2* genes, which are also important for ABA signaling (Liu et al., 2021). *HK2* downregulation in tomato led to increased tolerance to drought, heat, and combined stress. This was due to improved membrane stability, improved osmotic tolerance, as well as improvement in antioxidant enzyme levels in tomato through the cytokinin signaling pathway (Mushtaq et al., 2022). The *HSP17.7* gene was found to provide cold tolerance in tomato by inducing intracellular sucrose and reducing ROS. High *HSP17.7* expression and low *CCX1* gene levels indicate that *HSP17.7* could be a co-factor of the *CCX1* gene in tomato, which leads to the maintenance of calcium ion homeostasis and cold tolerance response (Zhang et al., 2020). *The hy3* gene in tomato led to the control of the *MIB3* gene, which is responsible for myoinositol levels, leading to increased cold tolerance. The mechanism is through light sensitivity and myoinositol levels, which help in cold tolerance response (Wang et al., 2022). *Hy5* gene plays an important role in regulating *MYB-CBF* cascading pathway to reduce cold stress in tomato. *Hy5* interacts with *MYB15* transcription factor to control the *CBF* gene pathway, which enables cold tolerance in tomato plants as a response to cold stress by expressing *CBF1*, *CBF2*, and *CBF3* genes (Zhang et al., 2020). *JUB1* is a *NAC* transcription factor that responds to abiotic stress in tomato. *JUB1* improved salinity tolerance in plants by improving proline levels as well as improved the water content of the plant (Alshareef et al., 2019). Reduced expression of *MAPK3* enabled higher heat tolerance in tomato by causing less wilting, less membrane damage, less ROS, and increased antioxidant enzyme activity in tomato as well as increased expression of *Hsf* and *Hsp* genes to provide heat stress tolerance (Yu et al., 2019). *MT* and *GR1* were expressed in response to zinc heavy metal toxicity. *MT* is responsible for metal binding and *GR1* causes the reduction of ROS by increasing the amount of antioxidant enzymes in tomato (Bolukbasi and Karakas, 2019).

MTP1, *3*, *4*, *8*, *10*, and *11* showed the highest expression as a response to the heavy metal presence of cadmium, cobalt, manganese, and zinc. The *MTP* genes are part of the metal transportation family that enables the storage and transportation of different heavy metals in the plant, thereby reducing heavy metal toxicity and allowing tomato plants to grow (El-Sappah et al., 2021). *MYB102* is a transcription factor in tomato, which regulates environmental stress response. *MYB102* expression helps in enhancing salt tolerance in the plants by maintaining salt balance and reducing ROS and electrolyte leakage. The genes expressed due to *MYB102* were *SOS1*, *SOS2*, *NHX3*, *NHX4*, *HAK5*, *CPK1*, and *CPK3*, which are responsible for salt tolerance response in tomato (Zhang et al., 2020). *MYC1* gene activates the JA pathway, leading to terpene production in tomato trichomes, thus providing resistance to spider mite infestations in tomato (Hua et al., 2021). *MYC2* is an important gene that is responsible for the JA pathway defense response in tomato against microbial and insect infestation. *MYC2* strongly induces the JIG gene to induce a defense response by activating the terpene synthesis gene *TPS*, which provides a biotic defense response from plants (Cao et al., 2022).

NAM3 transcription factor showed improved photosynthesis, less electrolyte leakages, and less wilting in tomato plants. The cold tolerance was achieved by activating the ethylene-producing genes such as *ACS1a*, *ACS1B*, *ACO1*, and *ACO4* by *NAM3* expression (Dong et al., 2022). *NHX4* transcription factor in tomato was found to improve salinity tolerance by improving proline content, osmotic balance in plants, and sugar content of the plants. It was observed that *NHX4* showed an increase in fruit size and yield of tomato (Maach et al., 2020). Phytochrome gene *PIF4* in tomato is induced by far-red light in phytochrome A and provides cold tolerance in plants. *PIF4* directly binds to the *CBF* gene and regulates plant hormone biosynthesis of ABA, JA, and GA in plants. *PIF4* directly activates the *GAI4* gene, which is responsible for positive regulation of cold tolerance. The light and temperature signals together play a major role to trigger cold tolerance response in tomato

(Wang et al., 2020). *Pti4*, *Pti5*, and *Pti6* genes in tomato were found to express various pathogen resistance genes and prevent *Pseudomonas* infection. The genes also improved fruit ripening and were influenced by ubiquitin-related gene pathways (Wang et al., 2021). E3 Ubiquitin Ligase *RING1* gene in tomato reduced cadmium stress by the expression of cadmium tolerance genes such as *CA1*, *MDHAR*, *GSH1*, and *PCS*. The *RING1* overexpression enhanced chlorophyll content and photosynthetic rates. Also, the overexpression of *RING1* led to reduced cadmium levels in the roots and shoots of tomato plants and also increased antioxidant activity to reduce oxidative damage to the plant (Ahammed et al., 2021).

SnRK1α gene helped in increasing salt tolerance in tomato by leaf damage, increased proline content, and reduced MDA and ROS species by activating antioxidant enzyme genes. The expression of *SnRK1α* increased the expression of several salt-related genes in tomato such as *PP2C*, *PYL4*, *PYL8*, *NAC22*, *NAC42*, and *SnRK2*, which indicates the role of *SnRK1α* in the ABA signaling pathway regulation to provide salt stress tolerance (Wang et al., 2020). The *TAF1* gene in tomato was responsible for salt tolerance by controlling the sodium content in the plant as well as improved stomatal conductance and less loss of water from leaves. *TAF1* also targets stress-related transcription factors, ABA metabolism, and defense-related genes as direct targets (Devkar et al., 2020). Transcription factor *WRKY33* induces cold tolerance in tomato by targeting and inducing many cold stress response genes such as *CDPK11*, *MYBS3*, and *BAG6*, which enhance cold tolerance (Guo et al., 2022).

8.2.7.1.2 Biostimulant Activity for Abiotic and Biotic Stress Response

A. nodosum-based plant biostimulants when applied to tomato plants and subjected to heat stress showed improved fruit yield and heat tolerance as well as improvement in soluble sugars with improved gene expression of heat shock proteins during the flowering stage, leading to mitigation of heat stress effects (Carmody et al., 2020). *Aspergillus flavus* treatment on tomato seedlings exposed to cadmium and chromium stress showed resistance to heavy metal toxicity by the expression of the specific metal-chelating genes such as *GSH1* and *PCS1*, leading to less cadmium and chromium presence in tomato seedlings treated with the fungus (Aziz et al., 2021). *Chlorella* microalgae-based biostimulant showed the upregulation of LEA proteins, oleosins, ethylene, and abscisic-related genes and also nascent polypeptide-associated complex (*NAC*) genes. Genes related to ABA, ethylene, and JA pathways showed upregulation, which is essential for biotic and abiotic stress tolerance (Gitau et al., 2023). A combination of protein hydrolysates and *A. nodosum*-based biostimulant-treated tomato plants showed salinity stress tolerance by maintaining salt ratio balance with the expression of transport genes such as *NHX4* and *HKT1;2* as well as aquaporin genes related to maintaining homeostasis *PIP1.2* and *TIP1.2* and improvement of photosynthetic efficiency and chlorophyll biosynthesis genes *LHC* and *PORC* (Ikuyinminu et al., 2023). Cycloflow biostimulant-treated tomatoes showed higher resistance to elevated temperatures with improved plant height and yield, ascorbic acid levels, and antioxidant activity, leading to less heat stress effect on tomato plants (Francesca et al., 2020). *ERF1* expression was enhanced through the application of *Euphorbia hirta* methanol extract on tomatoes to provide tolerance to *R. solanacearum* and *Xanthomonas axonopodis* pv. vesicatoria infection as well as enhance fruit production and quality (Kusampudi et al., 2023).

Eranthis biostimulant made from seaweed *A. nodosum* and *Laminaria digitata* used to alleviate drought stress in tomato showed lower ABA, MDA, and proline levels, leading to less ROS present in the stressed plants, indicating stress alleviation (Campobenedetto et al., 2021). Humic acid and plant growth-promoting bacteria-based biostimulant combination showed improved disease resistance to bacterial spot disease caused by *Xanthomonas* species in tomato plants. The treatment induced improved shoot and root weight in treated plants and also improved citric acid content in the plant. It induced plant secondary metabolite production from JA, ethylene, SA pathways andPAL enzyme pathway alongwith improvement in antioxidant activity due to increase in CAT and SOD enzymes. Overall, the biostimulant caused an induced defense response upon application (da Silva et al., 2021). JA, when applied to tomato plants grown in soil containing excess lead content, showed improved levels of secondary metabolites (total phenols, polyphenols, flavonoids, and anthocyanin), due to enhanced expression of chalcone synthase and PAL as well as elevated levels of organic acids and metal-chelating compounds. JA upregulated the expression of succinate dehydrogenase and fumarate hydratase, thus alleviating lead toxicity in tomato plants (Bali et al., 2019). L-Glutamate treatment in tomato helped reduce gray mold effect caused by *Botrytis cinerea* in tomato fruit. Gene expression of glutamate receptors (*GLR1.1, GLR1.2, GLR2.1, GLR2.4, GLR2.5, GLR2.6, GLR3.1, GLR3.2*, and *GLR3.3*), as well as pathogen-related gene (*PR1a, PR2a, PR3a*, and *PR3b*) and amino acid-related genes (*GABA, methionine, Lysine*, and *Arginine*), showed marked induction, leading to disease alleviation (Sun et al., 2019). Microalgae-cyanobacteria-based biostimulants developed from *Dunaliella salina, Chlorella ellipsoidea, Aphanothece* sp., and *Arthrospira maxima*, when applied to tomato plants and exposed to salinity stress, showed improved photosynthetic activity and proline level accumulation as well as improved nutrient uptake in plants. Antioxidant activity due to CAT and SOD caused reduced leaf peroxidation leading to improved plant tolerance to salt stress conditions (Mutale-Joan et al., 2021).

Peptone when applied to tomato plants exposed to drought stress showed improved antioxidant activity and improved the levels of JA, IAA, and zeatin and also improved the chlorophyll content (Casadesus et al., 2019). Plasticine biostimulant showed disease and pest resistance to powdery mildew and thrips feeding by upregulating genes *WRKY, ERF, MYB, NAC, bZIP, PRPs*, and *RRMs* during mild infestations in tomato (Rakoczy-Lelek et al., 2023). Protein hydrolysate from sea algae *Chondrus crispus*-treated tomato plants showed higher growth in drought stress conditions by altering ABA and proline levels in plants and causing the expression of drought marker gene Solyc02g084840 (Domingo et al., 2023). Soybean protein hydrolysate SPH18 showed increased expression of PR1 and PR8 pathogen-related genes in tomato Micro Tom variety, leading to enhanced resistance to Pseudomonas syringae DC 3000 bacteria, but no significant effect on *Botrytis cinerea* fungus was observed (Barrada et al., 2022). Tannin-based biostimulant Vivema Twin was tested on tomato plants exposed to salt stress, which showed the upregulation of *WRKY* transcription factor 61, *GST*, and *GSYT*, indicating improvement in root growth, abiotic stress tolerance, and improved nutrient uptake by tomato plants, leading to better growth in salty soil conditions (Campobenedetto et al., 2021).

8.2.7.1.3 Summary

DREBA4, HK2, and *MYC2* showed gene expression and provided tolerance to multiple abiotic and biotic stress conditions. Many biostimulants tested on tomato showed drought tolerance, salinity tolerance, heat tolerance, cadmium chromium and lead toxicity resistance, secondary metabolite production as a defense response, resistance to gray mold and activation of pathogen-related genes, resistance to *P. syringae* D3000, resistance to powdery mildew and thrips feeding, and tolerance to bacterial spot disease, as well as improved secondary metabolite production. The study of gene expression upon biostimulant application provides an opportunity to improve crop growth in adverse conditions and thus improve crop production of tomato (Table 8.11).

8.3 CONCLUSION

Extensive plant breeding programs have been conducted toward understanding plant genomic regions that contain gene factors responsible for reducing the effect of stress on plants. The balancing of allele factors in the breeding program has helped among only limited plant varieties that have high environmental stress tolerance levels with a reduction in yield traits. The other traits have focused on improving yield and food characteristics, leading to a reduction in tolerance traits in the crop plant. Recently, biostimulant usage has been shown to improve the stress tolerance factor of high-yield-producing varieties and to improve the yield of highly environmental stress-tolerant varieties based on the type of biostimulant that has been used. Based on the genes that have been described with their environmental stress alleviation function and the description of different biostimulants that have been used, there is a high chance of developing newer strategies of analyzing gene expressions as markers to evaluate the usage of biostimulants and the effective enhancement of the desirable alleles using molecular biology analysis techniques. This method has the potential to enhance breeding programs by analyzing biostimulant usage on existing crop varieties and to reduce the load of developing new varieties using plant breeding programs (Figure 8.2).

TABLE 8.11
Genes for Environmental Stress Tolerance in Tomato

Cold stress	*Cor413pm2, Dreba4, Hsp17.7, Hy3. Hy5, Nam3, Pif4, Wrky33*
Drought stress	*Dreba4, Gatal, Gras4, Hk2*
Heat stress	*Bbx17, Dreba4, Hk2, Mapk3, Cathb2, Gst, Ubc5, Argl*
Heavy metal stress	*Grl, Mtp, Mt, Ringl*
Salinity stress	*Bzrl, Dreba4, Jubl, Myb102, Nhx4, SnRKla, Sitafl*
Diseases	*Arog4a, Erfl, Frg4, Glr, Myc2, Pr, Pti*
Insects	*Erfl, Mycl, Myc2*

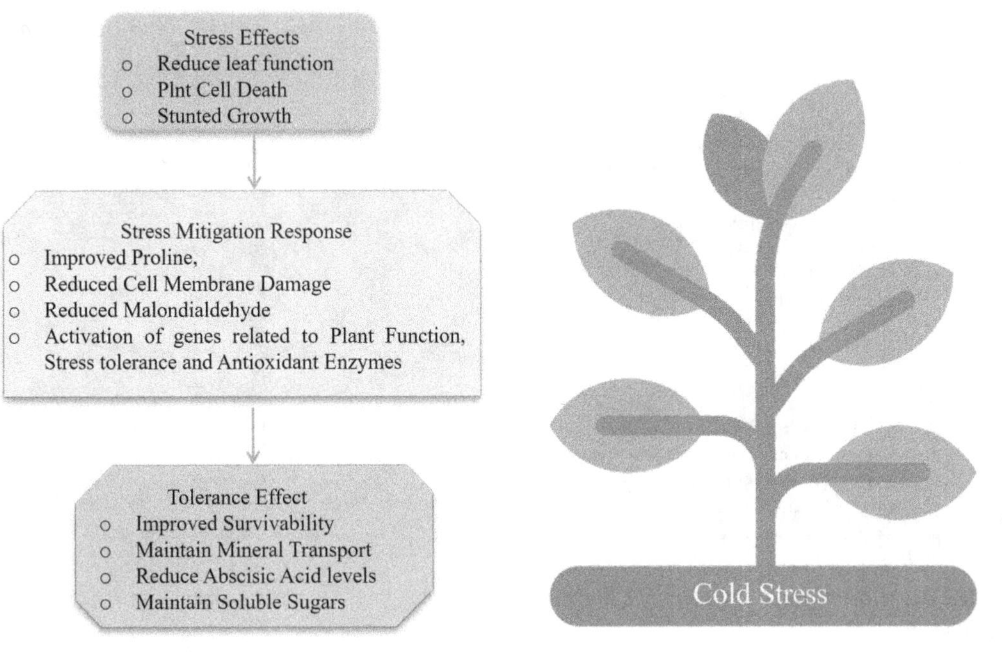

FIGURE 8.2 Environmental stress response, breeding, and biostimulant usage strategies for crop improvement.

(Continued)

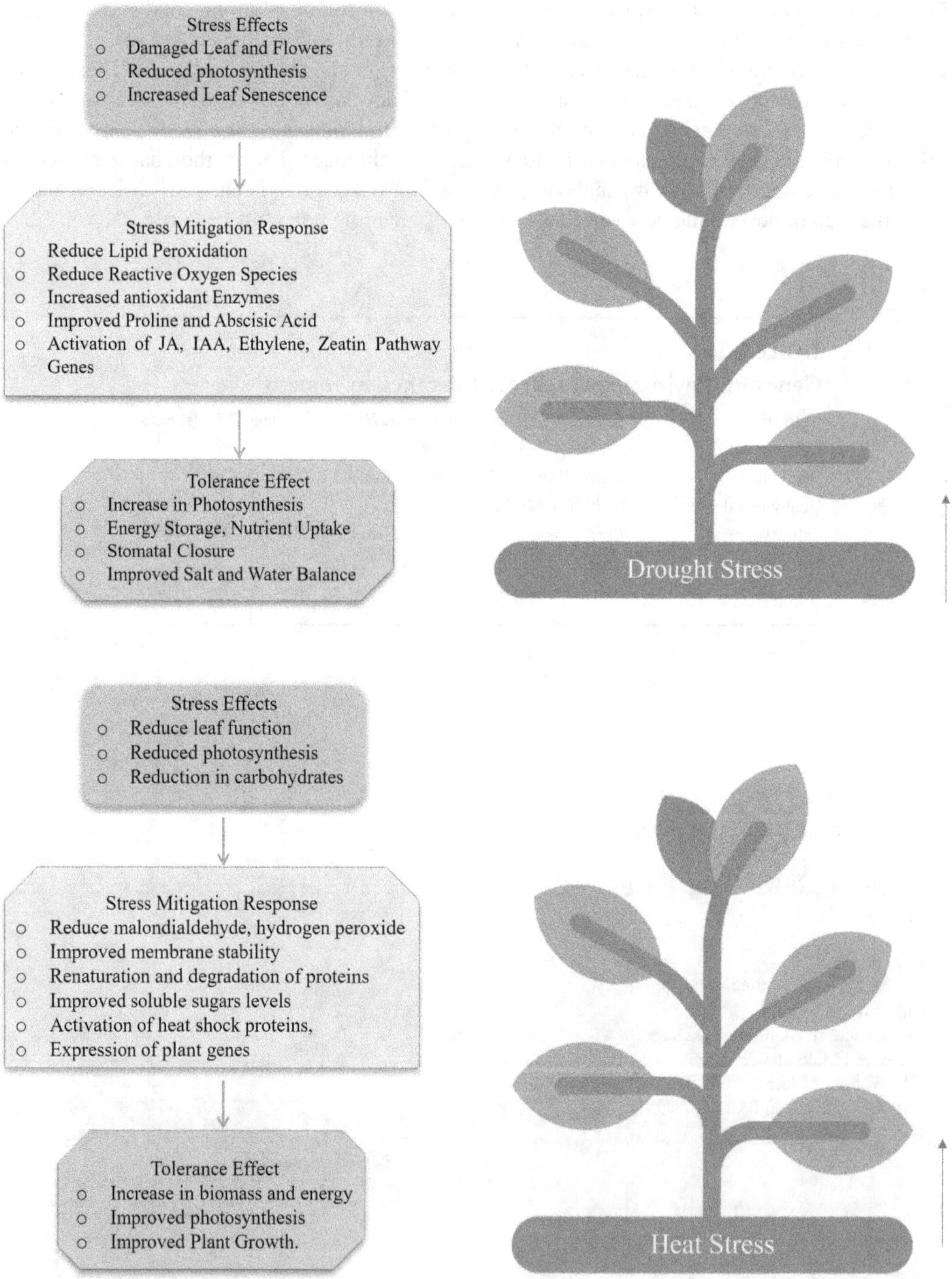

FIGURE 8.2 (*Continued*) Environmental stress response, breeding, and biostimulant usage strategies for crop improvement.

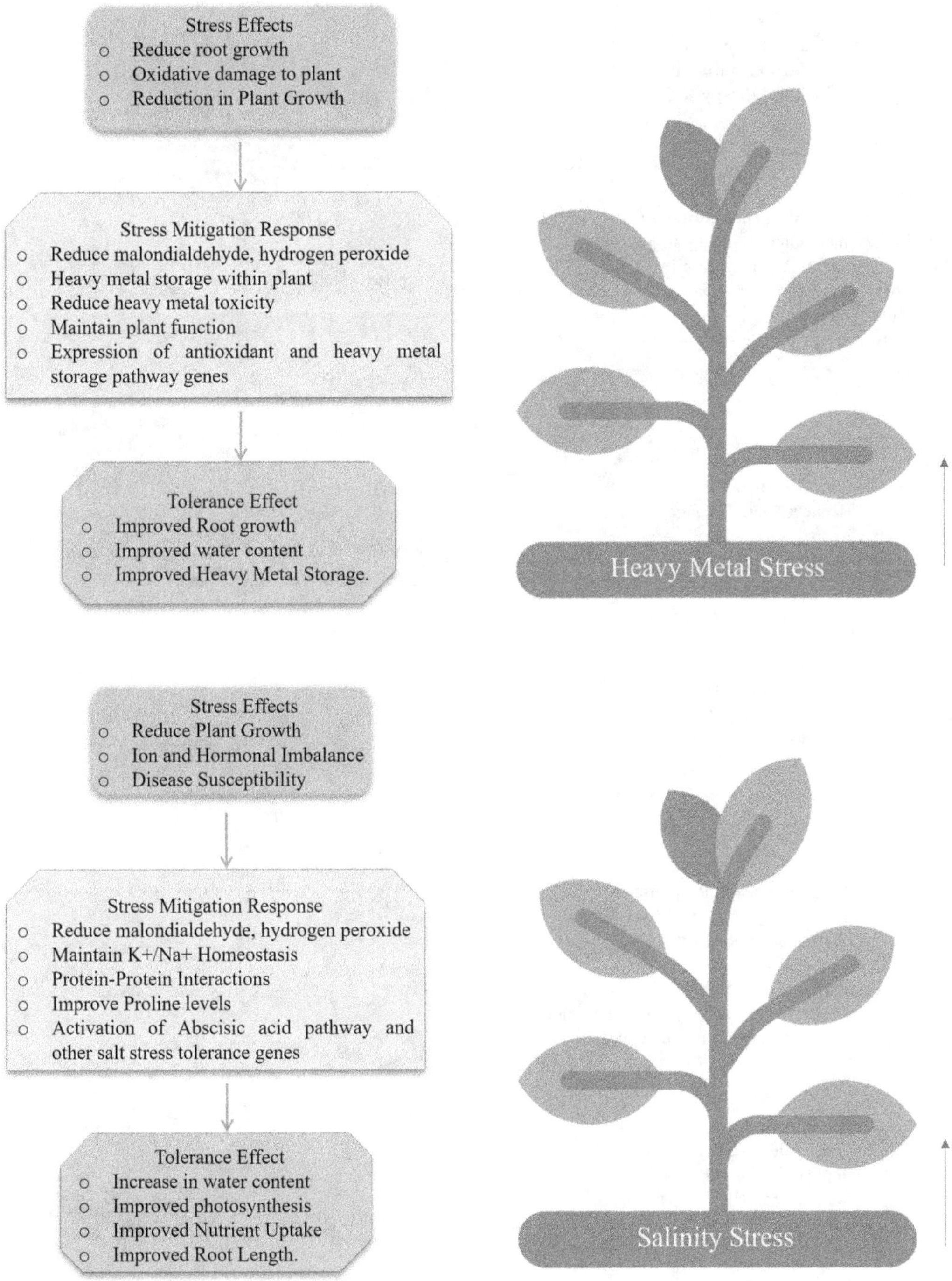

FIGURE 8.2 (*Continued*) Environmental stress response, breeding, and biostimulant usage strategies for crop improvement.

(*Continued*)

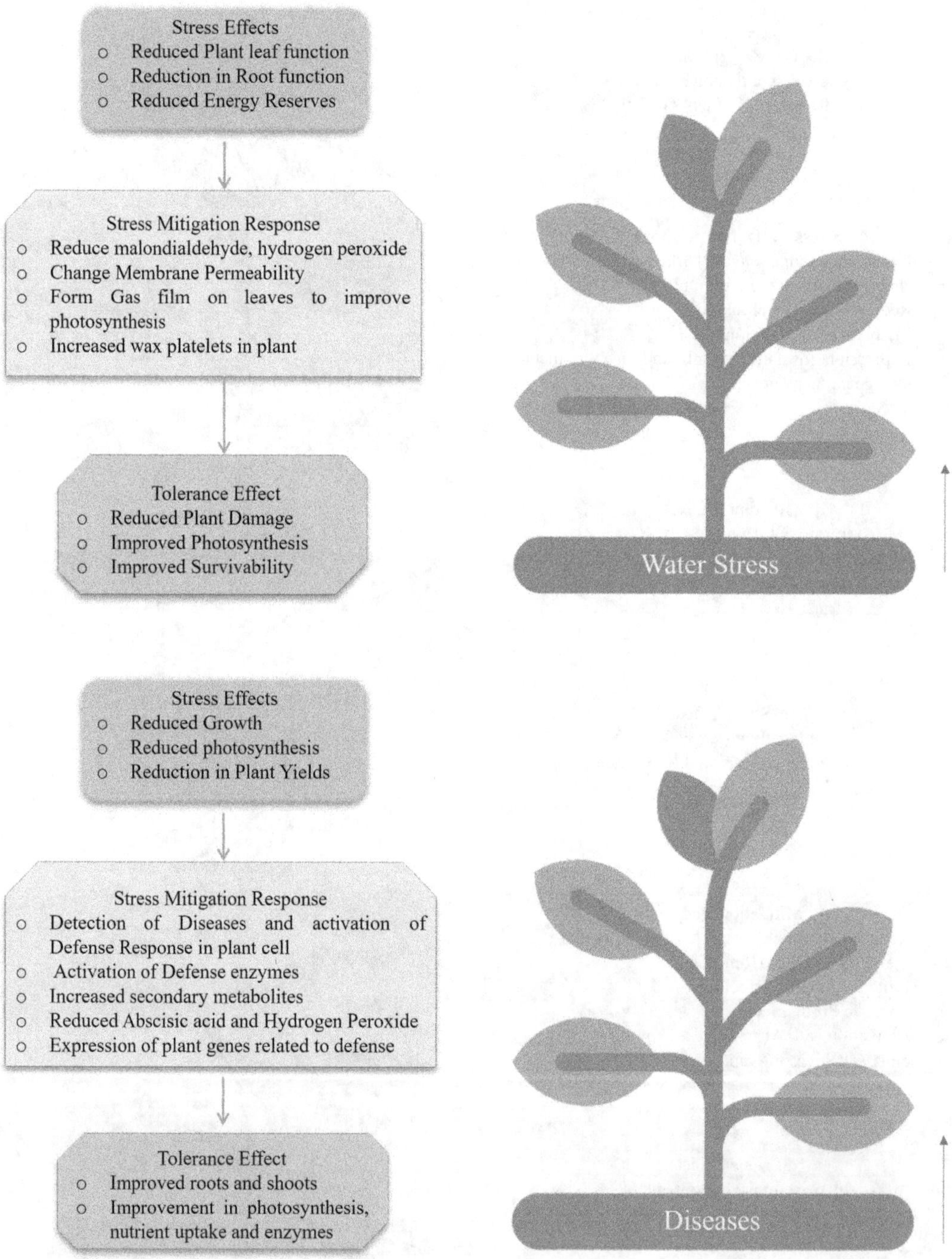

FIGURE 8.2 (*Continued*) Environmental stress response, breeding, and biostimulant usage strategies for crop improvement.

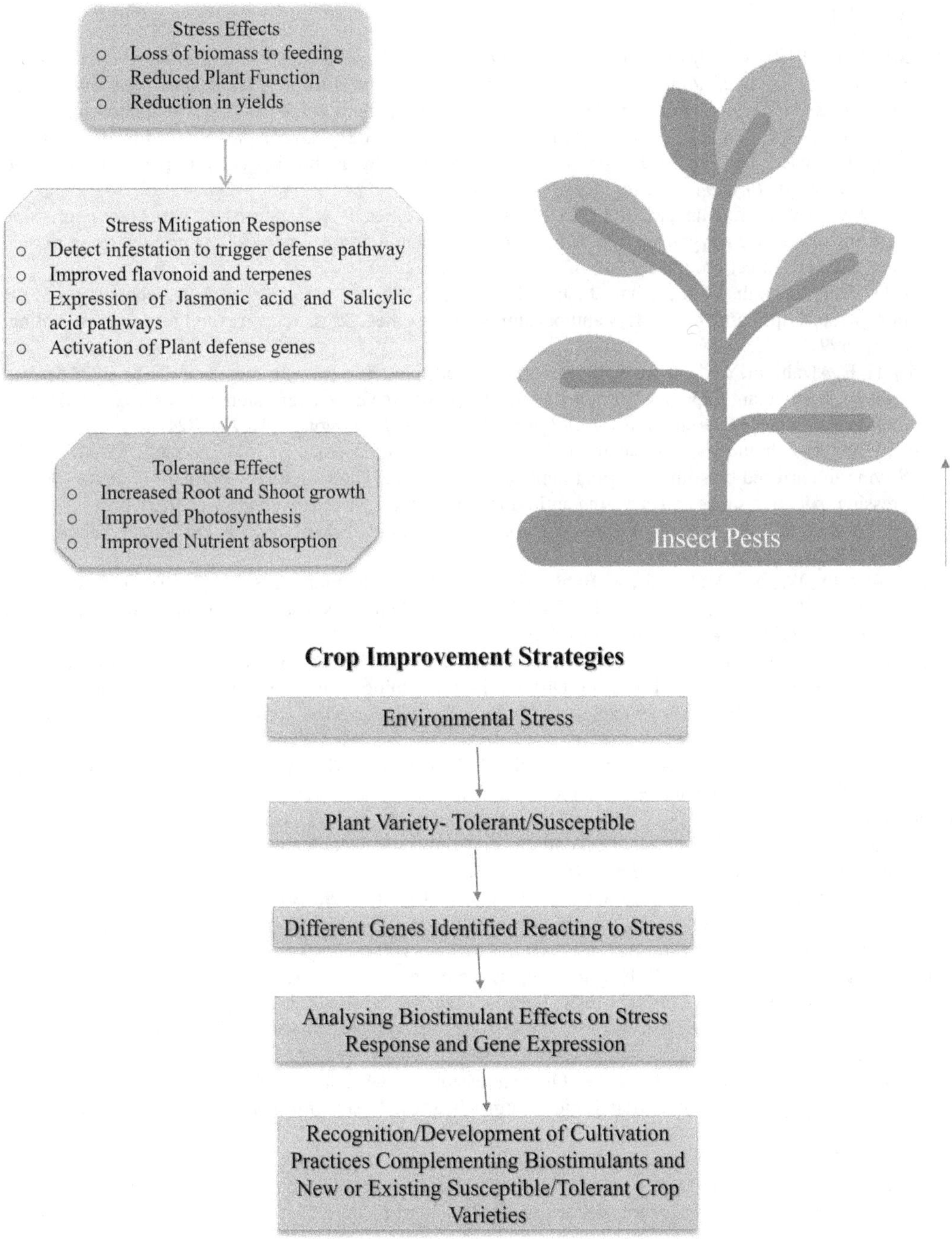

FIGURE 8.2 (*Continued*) Environmental stress response, breeding, and biostimulant usage strategies for crop improvement.

REFERENCES

Abu Zurayk, I. (2019). *A study of the impact of salinity on growth and development of important tomato and banana cultivars of the Jordan Valley with an assessment of the efficacy of biostimulants as ameliorants for increased tolerance to salinity* [Doctoral dissertation, Stellenbosch University, Stellenbosch].

Ahammed, G. J., Li, C. X., Li, X., Liu, A., Chen, S., & Zhou, J. (2021). Overexpression of tomato *RING E3* ubiquitin ligase gene *SlRING1* confers cadmium tolerance by attenuating cadmium accumulation and oxidative stress. *Physiologia plantarum*, *173*(1), 449–459.

Akbar, S., Wei, Y., Yuan, Y., Khan, M. T., Qin, L., Powell, C. A., Chen, B. & Zhang, M. (2020). Gene expression profiling of reactive oxygen species (ROS) and antioxidant defense system following Sugarcane mosaic virus (SCMV) infection. *BMC Plant Biology*, *20*(1), 1–12.

Akhter, D., Qin, R., Nath, U. K., Eshag, J., Jin, X., & Shi, C. (2019). A rice gene, OsPL, encoding a MYB family transcription factor confers anthocyanin synthesis, heat stress response and hormonal signaling. *Gene*, *699*, 62–72.

Alharby, H. F., Alzahrani, Y. M., & Rady, M. M. (2020). Seeds pretreatment with zeatins or maize grain-derived organic biostimulant improved hormonal contents, polyamine gene expression, and salinity and drought tolerance of wheat. *International Journal of Agriculture and Biology*, *24*(4), 714–724.

Alharby, H. F., Al-Zahrani, H. S., Hakeem, K. R., Alsamadany, H., Desoky, E. S. M., & Rady, M. M. (2021). Silymarin-enriched biostimulant foliar application minimizes the toxicity of cadmium in maize by suppressing oxidative stress and elevating antioxidant gene expression. *Biomolecules*, *11*(3), 465.

Ali, J., & Wani, S. H. (2021). *Rice improvement: Physiological, molecular breeding and genetic perspectives* (p. 498). Springer.

Alshareef, N. O., Wang, J. Y., Ali, S., Al-Babili, S., Tester, M., & Schmöckel, S. M. (2019). Overexpression of the NAC transcription factor JUNGBRUNNEN1 (JUB1) increases salinity tolerance in tomato. *Plant Physiology and Biochemistry*, *140*, 113–121.

Alves, G. S. C., Torres, L. F., Déchamp, E., Breitler, J. C., Joët, T., Gatineau, F., Andrade, A. C., Bertrand, B., Marraccini, P., & Etienne, H. (2017). Differential fine-tuning of gene expression regulation in coffee leaves by *CcDREB1D* promoter haplotypes under water deficit. *Journal of Experimental Botany*, *68*(11), 3017–3031.

Anders, S., Cowling, W., Pareek, A., Gupta, K. J., Singla-Pareek, S. L., & Foyer, C. H. (2021). Gaining acceptance of novel plant breeding technologies. *Trends in Plant Science*, *26*(6), 575–587.

Ansari, M., Devi, B. M., Sarkar, A., Chattopadhyay, A., Satnami, L., Balu, P., Choudhary, M., Shahid, M. A. & Jailani, A. A. K. (2023). Microbial exudates as biostimulants: Role in plant growth promotion and stress mitigation. *Journal of Xenobiotics*, *13*(4), 572–603.

Anshu, A., Agarwal, P., Mishra, K., Yadav, U., Verma, I., Chauhan, S., Srivastava, P. K., & Singh, P. C. (2022). Synergistic action of *Trichoderma koningiopsis* and *T. asperellum* mitigates salt stress in paddy. *Physiology and Molecular Biology of Plants*, *28*(5), 987–1004.

Arslan, E., Agar, G., & Aydin, M. (2021). Humic acid as a biostimulant in improving drought tolerance in wheat: The expression patterns of drought-related genes. *Plant Molecular Biology Reporter*, *39*(3), 508–519.

Asthir, B. (2015). Protective mechanisms of heat tolerance in crop plants. *Journal of Plant Interactions*, *10*(1), 202–210.

Ayadi, M., Brini, F., & Masmoudi, K. (2019). Overexpression of a wheat aquaporin gene, *Td PIP2;1*, enhances salt and drought tolerance in transgenic durum wheat cv. Maali. *International Journal of Molecular Sciences*, *20*(10), 2389.

Azevedo, V., Daddiego, L., Cardone, M. F., Perrella, G., Sousa, L., Santos, R. B., Malhó, R., Bergamini, C., Marsico, A. D., Figueiredo, A. & Alagna, F. (2022). Transcriptomic and methylation analysis of susceptible and tolerant grapevine genotypes following *Plasmopara viticola* infection. *Physiologia Plantarum*, *174*(5), e13771.

Aziz, L., Hamayun, M., Rauf, M., Iqbal, A., Husssin, A., Khan, S. A., Irshad, M., & Lee, I. J. (2021). Aspergillus Flavus reprogrammed morphological and chemical attributes of *Solanum lycopersicum* through *SlGSH1* and *SlPCS1* genes modulation under heavy metal stress. *Journal of Plant Interactions*, *16*(1), 104–115.

Azzam, C. R., Zaki, S. N. S., Bamagoos, A. A., Rady, M. M., & Alharby, H. F. (2022). Soaking maize seeds in zeatin-type cytokinin biostimulators improves salt tolerance by enhancing the antioxidant system and photosynthetic efficiency. *Plants*, *11*(8), 1004.

Bali, S., Jamwal, V. L., Kohli, S. K., Kaur, P., Tejpal, R., Bhalla, V., Ohri, P., Gandhi, S. G., Bhardwaj, R., Al-Huqail, A. A. & Ahmad, P. (2019). Jasmonic acid application triggers detoxification of lead (Pb) toxicity in tomato through the modifications of secondary metabolites and gene expression. *Chemosphere*, *235*, 734–748.

Banakar, S. N., PrasannaKumar, M. K., Parivallal, P. B., Pramesh, D., Mahesh, H. B., Sarangi, A. N., Puneeth, M. E., & Patil, S. S. (2023). Rice-Magnaporthe transcriptomics reveals host defense activation induced by red seaweed biostimulant in rice plants. *Frontiers in Genetics*, *14*, 1132561.

Barrada, A., Delisle-Houde, M., Nguyen, T. T. A., Tweddell, R. J., & Dorais, M. (2022). Drench application of soy protein hydrolysates increases tomato plant fitness, fruit yield, and resistance to a hemibiotrophic pathogen. *Agronomy*, *12*(8), 1761.

Bashir, M. A., Rehim, A., Raza, H. M. A., Zhai, L., Liu, H., & Wang, H. (2021). Biostimulants as plant growth stimulators in modernized agriculture and environmental sustainability. In F. Ahmad & M. Sultan (Eds.), *Technology in agriculture* (pp. 311–322). IntechOpen.

Bashir, W., Anwar, S., Zhao, Q., Hussain, I., & Xie, F. (2019). Interactive effect of drought and cadmium stress on soybean root morphology and gene expression. *Ecotoxicology and Environmental Safety*, *175*, 90–101.

Bi, Z., Wang, Y., Li, P., Sun, C., Qin, T., & Bai, J. (2021). Evolution and expression analysis of *CDPK* genes under drought stress in two varieties of potato. *Biotechnology Letters*, *43*, 511–521.

Blyuss, K. B., Fatehi, F., Tsygankova, V. A., Biliavska, L. O., Iutynska, G. O., Yemets, A. I., & Blume, Y. B. (2019). RNAi-based biocontrol of wheat nematodes using natural poly-component biostimulants. *Frontiers in Plant Science*, *10*, 483.

Bo, W., Zhaohui, Z., Huanhuan, Z., Xia, W., Binglin, L., Lijia, Y., Xiangyan, H., Deshui, Y., Xuelian, Z., Chunguo, W., Wenqin, S., Chengbin C., & Yong, Z. (2019). Targeted mutagenesis of NAC transcription factor gene, *OsNAC041*, leading to salt sensitivity in rice. *Rice Science*, *26*(2), 98–108.

Bolukbasi, E., & Karakas, M. (2019). Effect of zinc heavy metal on stress-related genes in tomato (*Solanum lycopersicum* L.) plants. *International Journal of Agriculture Environment and Biotechnology (IJAEB)*, *4*, 1211–1216.

Boncan, D. A. T., Tsang, S. S., Li, C., Lee, I. H., Lam, H. M., Chan, T. F., & Hui, J. H. (2020). Terpenes and terpenoids in plants: Interactions with environment and insects. *International Journal of Molecular Sciences*, *21*(19), 7382.

Boselli, M., Bahouaoui, M. A., Lachhab, N., Sanzani, S. M., Ferrara, G., & Ippolito, A. (2019). Protein hydrolysates effects on grapevine (*Vitis vinifera* L., cv. Corvina) performance and water stress tolerance. *Scientia Horticulturae*, *258*, 108784.

Brindha, C., Vasantha, S., Raja, A. K., & Tayade, A. S. (2021). Characterization of the salt overly sensitive pathway genes in sugarcane under salinity stress. *Physiologia Plantarum*, *171*(4), 677–687.

Brumbley, S. M., Snyman, S. J., Gnanasambandam, A., Joyce, P., Hermann, S. R., da Silva, J. A., McQualter, R. B., Wang, M. L., Egan, B. T., Paterson, A. H. & Moore, P. H. (2009). Sugarcane. In C. Kole, & T. C. Hall (Eds.), *Compendium of transgenic crop plants* (pp. 1–58). Blackwell Publishing.

Bulgarelli, R. G., Araujo, P., Tezotto, T., Mazzafera, P., & Andrade, S. A. L. (2016). Expression of metallothionein genes in coffee leaves in response to the absence or excess of Cu and Zn. *Theoretical and Experimental Plant Physiology*, *28*, 371–383.

Bwalya, J., Alazem, M., & Kim, K. H. (2022). Photosynthesis-related genes induce resistance against soybean mosaic virus: Evidence for involvement of the RNA silencing pathway. *Molecular Plant Pathology*, *23*(4), 543–560.

Cai, Z., Xian, P., Wang, H., Lin, R., Lian, T., Cheng, Y., Ma, Q., & Nian, H. (2020). Transcription factor *GmWRKY142* confers cadmium resistance by up-regulating the cadmium tolerance 1-like genes. *Frontiers in Plant Science*, *11*, 724.

Çakır Aydemir, B., Yüksel Özmen, C., Kibar, U., Mutaf, F., Büyük, P. B., Bakır, M., & Ergül, A. (2020). Salt stress induces endoplasmic reticulum stress-responsive genes in a grapevine rootstock. *PLoS One*, *15*(7), e0236424.

Campobenedetto, C., Agliassa, C., Mannino, G., Vigliante, I., Contartese, V., Secchi, F., & Bertea, C. M. (2021). A biostimulant based on seaweed (*Ascophyllum nodosum* and *Laminaria digitata*) and yeast extracts mitigates water stress effects on tomato (*Solanum lycopersicum* L.). *Agriculture*, *11*(6), 557.

Campobenedetto, C., Mannino, G., Beekwilder, J., Contartese, V., Karlova, R., & Bertea, C. M. (2021). The application of a biostimulant based on tannins affects root architecture and improves tolerance to salinity in tomato plants. *Scientific Reports*, *11*(1), 354.

Carmody, N., Goñi, O., Łangowski, Ł., & O'Connell, S. (2020). Ascophyllum nodosum extract biostimulant processing and its impact on enhancing heat stress tolerance during tomato fruit set. *Frontiers in Plant Science*, *11*, 807.

Casadesús, A., Polo, J., & Munné-Bosch, S. (2019). Hormonal effects of an enzymatically hydrolyzed animal protein-based biostimulant (Pepton) in water-stressed tomato plants. *Frontiers in Plant Science*, *10*, 758.

Cao, Y., Liu, L., Ma, K., Wang, W., Lv, H., Gao, M., .Wang, X., Zhang, X., Ren, S., & Guo, Y. D. (2022). The jasmonate-induced *bHLH* gene *SlJIG* functions in terpene biosynthesis and resistance to insects and fungus. *Journal of Integrative Plant Biology*, *64*(5), 1102–1115.

Chandran, V., Wang, H., Gao, F., Cao, X. L., Chen, Y. P., Li, G. B., Zhu, Y., Yang, X. M., Zhang, L. L., Zhao, Z. X. & Wang, W. M. (2019). miR396-OsGRFs module balances growth and rice blast disease-resistance. *Frontiers in Plant Science*, *9*, 1999.

Chavez-Navarrete, T., Sanchez-Timm, L., Pacheco-Coello, R., Baisakh, N., & Santos-Ordóñez, E. (2023). Identification of differential-expressed genes in banana-biostimulant interaction using suppression subtractive hybridization. *Agronomy*, *13*(2), 415.

Chen, J., Gong, Y., Gao, Y., Zhou, Y., Chen, M., Xu, Z., Guo, C., & Ma, Y. (2021). *TaNAC48* positively regulates drought tolerance and ABA responses in wheat (*Triticum aestivum* L.). *The Crop Journal*, *9*(4), 785–793.

Chen, Q. J., Zhang, L. P., Song, S. R., Wang, L., Xu, W. P., Zhang, C. X., Wang, S. P., Liu, H. F. & Ma, C. (2022). *vvi-miPEP172b* and *vvi-miPEP3635b* increase cold tolerance of grapevine by regulating the corresponding MIRNA genes. *Plant Science*, *325*, 111450.

Chen, X., Chen, G., Li, J., Hao, X., Tuerxun, Z., Chang, X., Gao, S., & Huang, Q. (2021). A maize calcineurin B-like interacting protein kinase ZmCIPK42 confers salt stress tolerance. *Physiologia Plantarum*, *171*(1), 161–172.

Chen, X., Liu, P., Mei, L., He, X., Chen, L., Liu, H., Shen, S., Ji, Z., Zheng, X., Zhang, Y., Gao, Z., Zeng, D., Qian, Q., & Ma, B. (2021). Xa7, a new executor R gene that confers durable and broad-spectrum resistance to bacterial blight disease in rice. *Plant Communications*, *2*, 3.

Chen, Y., Zhang, B., Li, C., Lei, C., Kong, C., Yang, Y., & Gong, M. (2019). A comprehensive expression analysis of the expansin gene family in potato (Solanum tuberosum) discloses stress-responsive expansin-like B genes for drought and heat tolerances. *PloS One*, *14*(7), e0219837.

Cheng, Q., Gan, Z., Wang, Y., Lu, S., Hou, Z., Li, H., Xiang, H., Liu, B., Kong, F., & Dong, L. (2020). The soybean gene *J* contributes to salt stress tolerance by up-regulating salt-responsive genes. *Frontiers in Plant Science*, *11*, 272.

Cherono, S., Ntini, C., Wassie, M., Mollah, M. D., Belal, M. A., Ogutu, C., & Han, Y. (2021). Exogenous application of melatonin improves drought tolerance in coffee by regulating photosynthetic efficiency and oxidative damage. *Journal of the American Society for Horticultural Science*, *146*(1), 24–32.

Cristofano, F., El-Nakhel, C., & Rouphael, Y. (2021). Biostimulant substances for sustainable agriculture: Origin, operating mechanisms and effects on cucurbits, leafy greens, and nightshade vegetables species. *Biomolecules*, *11*(8), 1103.

Cui, X., Yan, Q., Gan, S., Xue, D., Wang, H., Xing, H., Zhao, J., & Guo, N. (2019). *GmWRKY40*, a member of the WRKY transcription factor genes identified from *Glycine max* L., enhanced the resistance to *Phytophthora sojae*. *BMC Plant Biology*, *19*(1), 1–15.

Cui, X., Zhang, P., Chen, C., & Zhang, J. (2022). *VyCAS*, a calcium sensing receptor from the Chinese wild *Vitis yeshanensis*, confers drought tolerance in transgenic V. *vinifera*. *Scientia Horticulturae*, *305*, 111382.

da Silva, A. P. S., Olivares, F. L., Sudré, C. P., Peres, L. E. P., Canellas, N. A., da Silva, R. M., Cozzolino, V., & Canellas, L. P. (2021). Attenuations of bacterial spot disease *Xanthomonas euvesicatoria* on tomato plants treated with biostimulants. *Chemical and Biological Technologies in Agriculture*, *8*, 1–9.

Dalio, R. J., Maximo, H. J., Roma-Almeida, R., Barretta, J. N., José, E. M., Vitti, A. J., Blachinsky, D., Reuveni, M., & Pascholati, S. F. (2020). Tea tree oil induces systemic resistance against Fusarium wilt in banana and Xanthomonas infection in tomato plants. *Plants*, *9*(9), 1137.

Das, G., Rao, G. J., Varier, M., Prakash, A., & Prasad, D. (2018). Improved Tapaswini having four BB resistance genes pyramided with six genes/QTLs, resistance/tolerance to biotic and abiotic stresses in rice. *Scientific Reports*, *8*(1), 2413.

de Freitas, G. M., Thomas, J., Liyanage, R., Lay, J. O., Basu, S., Ramegowda, V., do Amaral, M. N., Benitez, L. C., Bolacel Braga, E. J., & Pereira, A. (2019). Cold tolerance response mechanisms revealed through comparative analysis of gene and protein expression in multiple rice genotypes. *PLoS One*, *14*(6), e0218019.

de Haan, S., & Rodriguez, F. (2016). Potato origin and production. In J. Singh, & L. Kaur (Eds.), *Advances in potato chemistry and technology* (pp. 1–32). Academic Press.

de Langhe, E., Vrydaghs, L., De Maret, P., Perrier, X., & Denham, T. (2009). Why bananas matter: An introduction to the history of banana domestication. *Ethnobotany Research and Applications*, *7*, 165–177.

de Maria, M., Robinson, E. J., Kangile, J. R., Kadigi, R., Dreoni, I., Couto, M., Howai., N. Peci, J., Fiennes, S. & Peci, J. (2020). Global soybean trade-the geopolitics of a bean. UK Research and Innovation Global Challenges Research Fund (UKRI GCRF) Trade, Development and the Environment Hub, pp 1–55.

de Oliveira Santos, M., Coelho, L. S., Carvalho, G. R., Botelho, C. E., Torres, L. F., Vilela, D. J. M., Andrade, A. C., & Silva, V. A. (2021). Photochemical efficiency correlated with candidate gene expression promote coffee drought tolerance. *Scientific Reports*, *11*(1), 7436.

Decsi, K., Kutasy, B., Hegedűs, G., Alföldi, Z. P., Kálmán, N., Nagy, Á., & Virág, E. (2023). Natural immunit stimulation using ELICE16INDURES® plant conditioner in field culture of soybean. *Heliyon*, *9*, 1

Devi, K., Prathima, P. T., Gomathi, R., Manimekalai, R., Lakshmi, K., & Selvi, A. (2019). Gene expression profiling in sugarcane genotypes during drought stress and rehydration. *Sugar Technology*, *21*(5), 717–733.

Devkar, V., Thirumalaikumar, V. P., Xue, G. P., Vallarino, J. G., Turečková, V., Strnad, M., Alisdair, R. F., Rainer, H., Bernd, M., & Balazadeh, S. (2020). Multifaceted regulatory function of tomato *SlTAF1* in the response to salinity stress. *New Phytologist*, *225*(4), 1681–1698.

Djemal, R., & Khoudi, H. (2022). The ethylene-responsive transcription factor of durum wheat, *TdSHN1*, confers cadmium, copper, and zinc tolerance to yeast and transgenic tobacco plants. *Protoplasma*, *259*(1), 19–31.

do Rosário Rosa, V., Dos Santos, A. L. F., da Silva, A. A., Sab, M. P. V., Germino, G. H., Cardoso, F. B., & de Almeida Silva, M. (2021). Increased soybean tolerance to water deficiency through biostimulant based on fulvic acids and *Ascophyllum nodosum* (L.) seaweed extract. *Plant Physiology and Biochemistry*, *158*, 228–243.

Domingo, G., Marsoni, M., Álvarez-Viñas, M., Torres, M. D., Domínguez, H., & Vannini, C. (2023). The role of protein-rich extracts from *Chondrus crispus* as biostimulant and in enhancing tolerance to drought stress in tomato plants. *Plants*, *12*(4), 845.

Dong, X., Jiang, Y., Yang, Y., Xiao, Z., Bai, X., Gao, J., Tan, S., Hur, Y., Hao, S., & He, F. (2019). Identification and expression analysis of the NAC gene family in *Coffea canephora*. *Agronomy*, *9*(11), 670.

Dong, X., Yang, Y., Zhang, Z., Xiao, Z., Bai, X., Gao, J., Hur, Y., Hao, S., & He, F. (2019). Genome-wide identification of WRKY genes and their response to cold stress in *Coffea canephora*. *Forests*, *10*(4), 335.

Dong, Y., Tang, M., Huang, Z., Song, J., Xu, J., Ahammed, G. J., Yu, J., & Zhou, Y. (2022). The miR164a-NAM3 module confers cold tolerance by inducing ethylene production in tomato. *The Plant Journal*, *111*(2), 440–456.

Dong, Z., Wang, H., Li, X., & Ji, H. (2021). Enhancement of plant cold tolerance by soybean RCC1 family gene GmTCF1a. *BMC Plant Biology*, *21*(1), 1–16.

Dos Santos, T. B., Baba, V. Y., Vieira, L. G. E., Pereira, L. F. P., & Domingues, D. S. (2021). The urea transporter DUR3 is differentially regulated by abiotic and biotic stresses in coffee plants. *Physiology and Molecular Biology of Plants*, *27*(2), 203–212.

El-Esawi, M. A., & Alayafi, A. A. (2019). Overexpression of rice Rab7 gene improves drought and heat tolerance and increases grain yield in rice (*Oryza sativa* L.). *Genes*, *10*(1), 56.

El-Esawi, M. A., Al-Ghamdi, A. A., Ali, H. M., & Ahmad, M. (2019). Overexpression of *AtWRKY30* transcription factor enhances heat and drought stress tolerance in wheat (*Triticum aestivum* L.). *Genes*, *10*(2), 163.

El-Sappah, A. H., Elrys, A. S., Desoky, E. S. M., Zhao, X., Bingwen, W., El-Sappah, H. H., Zhu, Y., Zhou, W., Zhao, X., & Li, J. (2021). Comprehensive genome wide identification and expression analysis of MTP gene family in tomato (*Solanum lycopersicum*) under multiple heavy metal stress. *Saudi Journal of Biological Sciences*, *28*(12), 6946–6956.

Feng, X., Feng, P., Yu, H., Yu, X., Sun, Q., Liu, S., Minh, T. N., Chen, J., Wang, D., Zhang, Q. and Cao, L., Zhou, C., Li, Q., Xiao, J., Zhong, S., Wang, A., Wang, L., Pan, H., & Ding, X. (2020). *GsSnRK1* interplays with transcription factor *GsERF7* from wild soybean to regulate soybean stress resistance. *Plant, Cell & Environment*, *43*(5), 1192–1211.

Francesca, S., Arena, C., Hay Mele, B., Schettini, C., Ambrosino, P., Barone, A., & Rigano, M. M. (2020). The use of a plant-based biostimulant improves plant performances and fruit quality in tomato plants grown at elevated temperatures. *Agronomy*, *10*(3), 363.

Fuller, D. Q., & Castillo, C. (1871). Rice: Origins and development. *Bulletin de la Société Géologique de France*, *24*, 692–717.

Gao, C., Gao, K., Yang, H., Ju, T., Zhu, J., Tang, Z., Zhao, L., & Chen, Q. (2022). Genome-wide analysis of metallothionein gene family in maize to reveal its role in development and stress resistance to heavy metal. *Biological Research*, *55*(1), 1–13.

Ghori, N. H., Ghori, T., Hayat, M. Q., Imadi, S. R., Gul, A., Altay, V., & Ozturk, M. (2019). Heavy metal stress and responses in plants. *International Journal of Environmental Science and Technology*, *16*, 1807–1828.

Gitau, M. M., Shetty, P., & Maróti, G. (2023). Transcriptional analysis reveals induction of systemic resistance in tomato treated with *Chlorella* microalgae. *Algal Research*, *72*, 103106.

Gu, B., Cao, X., Zhou, X., Chen, Z., Wang, Q., Liu, W., Chen, Q., & Zhao, H. (2020). The histological, effectoromic, and transcriptomic analyses of *Solanum pinnatisectum* reveal an upregulation of multiple NBS-LRR genes suppressing *Phytophthora infestans* infection. *International Journal of Molecular Sciences, 21*(9), 3211.

Gunupuru, L. R., Patel, J. S., Sumarah, M. W., Renaud, J. B., Mantin, E. G., & Prithiviraj, B. (2019). A plant biostimulant made from the marine brown algae Ascophyllum nodosum and chitosan reduce Fusarium head blight and mycotoxin contamination in wheat. *PLoS One, 14*(9), e0220562.

Guo, M., Yang, F., Liu, C., Zou, J., Qi, Z., Fotopoulos, V., Lu, G., Yu, J., & Zhou, J. (2022). A single-nucleotide polymorphism in *WRKY33* promoter is associated with the cold sensitivity in cultivated tomato. *New Phytologist, 236*(3), 989–1005.

Guo, Z., Yuan, X., Li, L., Zeng, M., Yang, J., Tang, H., & Duan, C. (2022). Genome-wide analysis of the ATP-binding cassette (ABC) transporter family in *Zea mays* L. and its response to heavy metal stresses. *International Journal of Molecular Sciences, 23*(4), 2109.

Hamdi, K., Brini, F., Kharrat, N., Masmoudi, K., & Yakoubi, I. (2020). Abscisic acid, stress, and ripening (TtASR1) gene as a functional marker for salt tolerance in durum wheat. *BioMed Research International, 1*, 7876357.

Han, Q., Qi, J., Hao, G., Zhang, C., Wang, C., Dirk, L. M., ...Downie, A. B., & Zhao, T. (2020). *ZmDREB1A* regulates Raffinose Synthase controlling raffinose accumulation and plant chilling stress tolerance in maize. *Plant and Cell Physiology, 61*(2), 331–341.

Han, X., Xi, Y., Zhang, Z., Mohammadi, M. A., Joshi, J., Borza, T., & Wang-Pruski, G. (2021). Effects of phosphite as a plant biostimulant on metabolism and stress response for better plant performance in *Solanum tuberosum*. *Ecotoxicology and Environmental Safety, 210*, 111873.

Han, Y., & Luthe, D. (2021). Identification and evolution analysis of the JAZ gene family in maize. *BMC Genomics, 22*, 1–21.

Han, Y., Wu, C., Yang, L., Zhang, D., & Xiao, Y. (2018). Resistance to Nilaparvata lugens in rice lines introgressed with the resistance genes Bph14 and Bph15 and related resistance types. *PLoS One, 13*(6), e0198630.

Hasanuzzaman, M., Raihan, M. R. H., Nowroz, F., & Nahar, K. (2023). Insight into the physiological and biochemical mechanisms of biostimulating effect of *Ascophyllum nodosum* and *Moringa oleifera* extracts to minimize cadmium-induced oxidative stress in rice. *Environmental Science and Pollution Research, 30*(19), 55298–55313.

He, H., Liu, R., Ma, P., Du, H., Zhang, H., Wu, Q., Yang, L., Gong, S., Liu, T., Huo, N., Gu, Y. Q., & Zhu, S. (2021). Characterization of Pm68, a new powdery mildew resistance gene on chromosome 2BS of Greek durum wheat TRI 1796. *Theoretical and Applied Genetics, 134*, 53–62.

He, J., Liu, Y., Yuan, D., Duan, M., Liu, Y., Shen, Z., Yang, C., Qiu, Z., Liu, D., Wen, P., & Wan, J. (2020). An R2R3 MYB transcription factor confers brown planthopper resistance by regulating the phenylalanine ammonia-lyase pathway in rice. *Proceedings of the National Academy of Sciences, 117*(1), 271–277.

He, Y., He, G., Lou, F., Zhou, Z., Liu, Y., Zhang, Y., & He, T. (2022). Identification of the Major Effector *StSROs* in Potato: A Potential StWRKY-SRO6 Regulatory Pathway Enhances Plant Tolerance to Cadmium Stress. *International Journal of Molecular Sciences, 23*(22), 14318.

He, Y., Wu, L., Liu, X., Jiang, P., Yu, L., Qiu, J., Wang, G., Zhang, X., & Ma, H. (2020). *TaUGT6*, a novel UDP-Glycosyltransferase gene enhances the resistance to FHB and DON accumulation in wheat. *Frontiers in Plant Science, 11*, 574775.

He, Z., Wu, J., Sun, X., & Dai, M. (2019). The maize clade A PP2C phosphatases play critical roles in multiple abiotic stress responses. *International Journal of Molecular Sciences, 20*(14), 3573.

Hossain, F., Muthusamy, V., Bhat, J. S., Zunjare, R. U., Kumar, S., Prakash, N. R., & Mehta, B. K. (2022). Maize breeding. In D. K. Yadava, H. K. Dikshit, G. P. Mishra, & S. Tripathi (Eds.), *Fundamentals of field crop breeding* (pp. 221–258). Springer Nature.

Hu, J., Chang, X., Zou, L., Tang, W., & Wu, W. (2018). Identification and fine mapping of Bph33, a new brown planthopper resistance gene in rice (Oryza sativa L.). *Rice, 11*, 1–12.

Hua, B., Chang, J., Wu, M., Xu, Z., Zhang, F., Yang, M., Xu, H., Wang, L. J., Chen, X., & Wu, S. (2021). Mediation of JA signalling in glandular trichomes by the woolly/*SlMYC1* regulatory module improves pest resistance in tomato. *Plant Biotechnology Journal, 19*(2), 375–393.

Ikuyinminu, E., Goñi, O., Łangowski, Ł., & O'Connell, S. (2023). Transcriptome, Biochemical and Phenotypic Analysis of the Effects of a Precision Engineered Biostimulant for Inducing Salinity Stress Tolerance in Tomato. *International Journal of Molecular Sciences, 24*(8), 6988.

Irani, H., ValizadehKaji, B., & Naeini, M. R. (2021). Biostimulant-induced drought tolerance in grapevine is associated with physiological and biochemical changes. *Chemical and Biological Technologies in Agriculture*, *8*(1), 1–13.

Islam, W., Tayyab, M., Khalil, F., Hua, Z., Huang, Z., & Chen, H. Y. (2020). Silicon-mediated plant defense against pathogens and insect pests. *Pesticide Biochemistry and Physiology*, *168*, 104641.

Javed, T., Zhou, J. R., Li, J., Hu, Z. T., Wang, Q. N., & Gao, S. J. (2022). Identification and expression profiling of WRKY family genes in sugarcane in response to bacterial pathogen infection and nitrogen implantation dosage. *Frontiers in Plant Science*, *13*, 917953.

Jacomassi, L. M., Viveiros, J. D. O., Oliveira, M. P., Momesso, L., de Siqueira, G. F., & Crusciol, C. A. C. (2022). A seaweed extract-based biostimulant mitigates drought stress in sugarcane. *Frontiers in Plant Science*, *13*, 865291.

Ji, H. M., Mao, H. Y., Li, S. J., Feng, T., Zhang, Z. Y., Cheng, L., Luo, S. J., Borkovich, K. A., & Ouyang, S. Q. (2021). *Fol-milR1*, a pathogenicity factor of *Fusarium oxysporum*, confers tomato wilt disease resistance by impairing host immune responses. *New Phytologist*, *232*(2), 705–718.

Jia, C., Zhao, S., Bao, T., Zhao, P., Peng, K., Guo, Q., Gao, X., & Qin, J. (2021). Tomato BZR/BES transcription factor *SlBZR1* positively regulates BR signaling and salt stress tolerance in tomato and Arabidopsis. *Plant Science*, *302*, 110719.

Jia, H., Li, T., Haider, M. S., Sadeghnezhad, E., Pang, Q., Han, J., Zhang, P., Su, L., Jia, H., & Fang, J. (2022). Comparative Transcriptomic and Metabolomic Profiling of Grapevine Leaves (cv. Kyoho) upon Infestation of Grasshopper and *Botrytis cinerea*. *Plant Molecular Biology Reporter*, *40*(3), 539–555.

Jia, W., Ma, M., Chen, J., & Wu, S. (2021). Plant morphological, physiological and anatomical adaption to flooding stress and the underlying molecular mechanisms. *International Journal of Molecular Sciences*, *22*(3), 1088.

Jiang, L., Hu, W., Qian, Y., Ren, Q., & Zhang, J. (2021). Genome-wide identification, classification and expression analysis of the *Hsf* and *Hsp70* gene families in maize. *Gene*, *770*, 145348.

Jiao, S. Z., Guo, C., Yao, W. K., Zhang, N. B., Zhang, J. Y., & Xu, W. R. (2022). An Amur grape *VaHsfC1* is involved in multiple abiotic stresses. *Scientia Horticulturae*, *295*, 110785.

Jin, T., Sun, Y., Zhao, R., Shan, Z., Gai, J., & Li, Y. (2019). Overexpression of peroxidase gene *GsPRX9* confers salt tolerance in soybean. *International Journal of Molecular Sciences*, *20*(15), 3745.

Johnson, R., Joel, J. M., & Puthur, J. T. (2023). Biostimulants: The futuristic sustainable approach for alleviating crop productivity and abiotic stress tolerance. *Journal of Plant Growth Regulation*, *43(3)*, 659–674.

Ju, Y. L., Min, Z., Zhang, Y., Zhang, K. K., Liu, M., & Fang, Y. L. (2021). Transcriptome profiling provide new insights into the molecular mechanism of grapevine response to heat, drought, and combined stress. *Scientia Horticulturae*, *286*, 110076.

Ju, Y. L., Yue, X. F., Min, Z., Wang, X. H., Fang, Y. L., & Zhang, J. X. (2020). *VvNAC17*, a novel stress-responsive grapevine (*Vitis vinifera* L.) NAC transcription factor, increases sensitivity to abscisic acid and enhances salinity, freezing, and drought tolerance in transgenic Arabidopsis. *Plant Physiology and Biochemistry*, *146*, 98–111.

Ke, D., He, Y., Fan, L., Niu, R., Cheng, L., Wang, L., & Zhang, Z. (2022). The soybean TGA transcription factor *GmTGA13* plays important roles in the response to salinity stress. *Plant Biology*, *24*(2), 313–322.

Khan, M. A., Asaf, S., Khan, A. L., Jan, R., Kang, S. M., Kim, K. M., & Lee, I. J. (2020). Rhizobacteria AK1 remediates the toxic effects of salinity stress via regulation of endogenous phytohormones and gene expression in soybean. *Biochemical Journal*, *476*(16), 2393–2409.

Kim, S. M., & Reinke, R. F. (2019). A novel resistance gene for bacterial blight in rice, Xa43 (t) identified by GWAS, confirmed by QTL mapping using a bi-parental population. *PloS One*, *14*(2), e0211775.

Kosakivska, I. V., Babenko, L. M., Romanenko, K. O., Korotka, I. Y., & Potters, G. (2021). Molecular mechanisms of plant adaptive responses to heavy metals stress. *Cell Biology International*, *45*(2), 258–272.

Krattinger, S. G., Kang, J., Bräunlich, S., Boni, R., Chauhan, H., Selter, L. L., Robinson, M. D., Schmid, M. W., Wiederhold, E., Hensel, G., & Keller, B. (2019). Abscisic acid is a substrate of the ABC transporter encoded by the durable wheat disease resistance gene Lr34. *New Phytologist*, *223*(2), 853–866.

Kumar, S., Rai, M & Singh, B. (2011). Tomato. In H.P. Singh, V. A. Parthasarathy, & K. N. Babu (Eds.) *Advances in horticulture biotechnology regeneration systems: Vegetables, ornamentals and tuber crops* (Vol. 2, pp. 1–12). Westville Publishing House.

Kurokawa, Y., Nagai, K., Huan, P. D., Shimazaki, K., Qu, H., Mori, Y., Toda, Y., Kuroha, T., Hayashi, N., Aiga, S., & Ashikari, M. (2018). Rice leaf hydrophobicity and gas films are conferred by a wax synthesis gene (LGF 1) and contribute to flood tolerance. *New Phytologist*, *218*(4), 1558–1569.

Kusampudi, V., Kusampudi, S., & Lakshmi, B. S. (2023). Effect of foliar-applied Euphorbia hirta towards controlling bacterial diseases in tomato crops and enhancing fruit yield and shelf life. *Plant Production Science*, *26*(1), 1–16.

Kusampudi, V., Kusampudi, S., & Lakshmi, B. S. (2022). Seed priming using Moringa oleifera leaf decoction increases BPT 5204 rice yield. *Israel Journal of Plant Sciences*, *70*(1–2), 65–75.

Lakshmi, G., Beena, R., Soni, K. B., Viji, M. M., & Jha, U. C. (2023). Exogenously applied plant growth regulator protects rice from heat-induced damage by modulating plant defense mechanism. *Journal of Crop Science and Biotechnology*, *26*(1), 63–75.

Lee, H., Cha, J., Choi, C., Choi, N., Ji, H. S., Park, S. R., Lee, S., & Hwang, D. J. (2018). Rice WRKY11 plays a role in pathogen defense and drought tolerance. *Rice*, *11*(1), 1–12.

Li, A. M., Wang, M., Chen, Z. L., Qin, C. X., Liao, F., Wu, Z., He, W. Z., Lakshmanan, P., Pan, Y. Q., & Huang, D. L. (2022). Integrated transcriptome and metabolome analysis to identify sugarcane gene defense against fall armyworm (*Spodoptera frugiperda*) herbivory. *International Journal of Molecular Sciences*, *23*(22), 13712.

Li, B., Liu, G., Wang, Y., Wei, Y., & Shi, H. (2019). Overexpression of banana ATG8f modulates drought stress resistance in Arabidopsis. *Biomolecules*, *9*(12), 814.

Li, B., Zheng, J. C., Wang, T. T., Min, D. H., Wei, W. L., Chen, J., Zhou, Y. B., Chen, M., Xu, Z. S., & Ma, Y. Z. (2020). Expression analyses of soybean VOZ transcription factors and the role of *GmVOZ1G* in drought and salt stress tolerance. *International Journal of Molecular Sciences*, *21*(6), 2177.

Li, C., Sun, Y., Li, J., Zhang, T., Zhou, F., Song, Q., Liu, Y., Brestic, M., Chen, T. H., & Yang, X. (2022). ScCBF1 plays a stronger role in cold, salt and drought tolerance than *StCBF1* in potato (*Solanum tuberosum*). *Journal of Plant Physiology*, *278*, 153806.

Li, D., He, G., Tian, W., Saleem, M., Huang, Y., Meng, L., Wu, D., & He, T. (2021). Comparative and systematic omics revealed low cd accumulation of potato *StMTP9* in yeast: Suggesting a new mechanism for heavy metal detoxification. *International Journal of Molecular Sciences*, *22*(19), 10478.

Li, F., Wu, Z., Zuo, S., Fan, L., Wei, Z., Ma, L., Sun, W., Du, C., Liu, R., & Wang, P. (2022). Application of enzymatic hydrolysate of *Ulva clathrata* as biostimulant improved physiological and metabolic adaptation to salt-alkaline stress in wheat. *Journal of Applied Phycology*, *34*(3), 1779–1789.

Li, G., Xu, W., Jing, P., Hou, X., & Fan, X. (2021). Overexpression of *VyDOF8*, a Chinese wild grapevine transcription factor gene, enhances drought tolerance in transgenic tobacco. *Environmental and Experimental Botany*, *190*, 104592.

Li, H., Yue, H., Li, L., Liu, Y., Zhang, H., Wang, J., & Jiang, X. (2021). Seed biostimulant Bacillus sp. MGW9 improves the salt tolerance of maize during seed germination. *AMB Express*, *11*(1), 1–15.

Li, M., Chen, R., Jiang, Q., Sun, X., Zhang, H., & Hu, Z. (2021). *GmNAC06*, a NAC domain transcription factor enhances salt stress tolerance in soybean. *Plant Molecular Biology*, *105*, 333–345.

Li, N., Wei, S., Chen, J., Yang, F., Kong, L., Chen, C., Ding, Z., & Chu, Z. (2018). *OsASR2* regulates the expression of a defence-related gene, *Os2H16*, by targeting the GT-1 cis-element. *Plant Biotechnology Journal*, *16*(3), 771–783.

Li, P., Chai, Z., Lin, P., Huang, C., Huang, G., Xu, L., Deng, Z., Zhang, M., Zhang, Y., & Zhao, X. (2020). Genome-wide identification and expression analysis of *AP2/ERF* transcrip.,tion factors in sugarcane (*Saccharum spontaneum* L.). *BMC Genomics*, *21*(1), 1–17.

Li, S., Chen, N., Li, F., Mei, F., Wang, Z., Cheng, X., Kang, Z., & Mao, H. (2020). Characterization of wheat homeodomain-leucine zipper family genes and functional analysis of TaHDZ5-6A in drought tolerance in transgenic Arabidopsis. *BMC Plant Biology*, *20*(1), 1–23.

Li, W., Deng, Y., Ning, Y., He, Z., & Wang, G. L. (2020). Exploiting broad-spectrum disease resistance in crops: From molecular dissection to breeding. *Annual Review of Plant Biology*, *71*, 575–603.

Li, W., Zheng, X., Cheng, R., Zhong, C., Zhao, J., Liu, T. H., Yi, T., Zhu, Z., Xu, J., Meksem, K., Dai, L., & Liu, S. (2023). Soybean ZINC FINGER PROTEIN03 targets two SUPEROXIDE DISMUTASE1s and confers resistance to Phytophthora sojae. *Plant Physiology*, *192*(1), 633–647.

Li, X., Liu, Z., Zhao, H., Deng, X., Su, Y., Li, R., & Chen, B. (2022). Overexpression of sugarcane *ScDIR* genes enhances drought tolerance in *Nicotiana benthamiana*. *International Journal of Molecular Sciences*, *23*(10), 5340.

Li, X., Yu, B., Wu, Q., Min, Q., Zeng, R., Xie, Z., & Huang, J. (2021). OsMADS23 phosphorylated by SAPK9 confers drought and salt tolerance by regulating ABA biosynthesis in rice. *PLoS Genetics*, *17*(8), e1009699.

Li, Y., Shan, X., Jiang, Z., Zhao, L., & Jin, F. (2021). Genome-wide identification and expression analysis of the GA2ox gene family in maize (*Zea mays* L.) under various abiotic stress conditions. *Plant Physiology and Biochemistry*, *166*, 621–633.

Liang, G., Ma, Z., Lu, S., Ma, W., Feng, L., Mao, J., & Chen, B. (2022). Temperature-phase transcriptomics reveals that hormones and sugars in the phloem of grape participate in tolerance during cold acclimation. *Plant Cell Reports*, *41*(6), 1357–1373.

Liang, Y., Xia, J., Jiang, Y., Bao, Y., Chen, H., Wang, D., Zhang, D., Yu, J., & Cang, J. (2022). Genome-wide identification and analysis of bZIP gene family and resistance of *TaABI5* (*TabZIP96*) under freezing stress in wheat (*Triticum aestivum*). *International Journal of Molecular Sciences*, *23*(4), 2351.

Liu, C., Schläppi, M. R., Mao, B., Wang, W., Wang, A., & Chu, C. (2019). The bZIP73 transcription factor controls rice cold tolerance at the reproductive stage. *Plant Biotechnology Journal*, *17*(9), 1834–1849.

Liu, H. F., Xue, X. J., Yu, Y., Xu, M. M., Lu, C. C., Meng, X. L., Zhang, B. G., Ding, X. H., & Chu, Z. H. (2020). Copper ions suppress abscisic acid biosynthesis to enhance defence against Phytophthora infestans in potato. *Molecular Plant Pathology*, *21*(5), 636–651.

Liu, J., Zhao, F. L., Guo, Y., Fan, X. C., Wang, Y. J., & Wen, Y. Q. (2019). The ABA receptor-like gene *VyPYL9* from drought-resistance wild grapevine confers drought tolerance and ABA hypersensitivity in Arabidopsis. *Plant Cell, Tissue and Organ Culture (PCTOC)*, *138*, 543–558.

Liu, M. M., Li, Y. L., Li, G. C., Dong, T. T., Liu, S. Y., Pei, L., & Wang, Q. G. (2020). Overexpression of *StCYS1* gene enhances tolerance to salt stress in the transgenic potato (*Solanum tuberosum* L.) plant. *Journal of Integrative Agriculture*, *19*(9), 2239–2246.

Liu, Q., Li, X., Yan, S., Yu, T., Yang, J., Dong, J., Zhang, S., Zhao, J., Yang, T., Mao, X., Zhu, X., & Liu, B. (2018). OsWRKY67 positively regulates blast and bacteria blight resistance by direct activation of PR genes in rice. *BMC Plant Biology*, *18*(1), 1–13.

Liu, W., Liang, X., Cai, W., Wang, H., Liu, X., Cheng, L., Song, P., Luo, G., & Han, D. (2022). Isolation and functional analysis of *VvWRKY28*, a Vitis Vinifera WRKY transcription factor gene, with functions in tolerance to cold and salt stress in transgenic Arabidopsis thaliana. *International Journal of Molecular Sciences*, *23*(21), 13418.

Liu, X., Chen, H., Li, S., Lecourieux, D., Duan, W., Fan, P., Liang, Z., & Wang, L. (2023). Natural variations of HSFA2 enhance thermotolerance in grapevine. *Horticulture Research*, *10*(1), uhac250.

Liu, X., Fu, L., Qin, P., Sun, Y., Liu, J., & Wang, X. (2019). Overexpression of the wheat trehalose 6-phosphate synthase 11 gene enhances cold tolerance in *Arabidopsis thaliana*. *Gene*, *710*, 210–217.

Liu, Y., He, G., He, Y., Tang, Y., Zhao, F., & He, T. (2023). Discovery of cadmium-tolerant biomacromolecule (*StCAX1/4* transport proteins) in potato and its potential regulatory relationship with WRKY transcription factors. *International Journal of Biological Macromolecules*, *228*, 385–399.

Liu, Y., Wen, L., Shi, Y., Su, D., Lu, W., Cheng, Y., & Li, Z. (2021). Stress-responsive tomato gene *SlGRAS4* function in drought stress and abscisic acid signalling. *Plant Science*, *304*, 110804.

Lu, Y., Zhong, Q., Xiao, S., Wang, B., Ke, X., Zhang, Y., Yin, F., Zhang, D., Jiang, C., Liu, L., Li, J., Yu, T., Wang, L., Cheng, Z., & Chen, L. (2022). A new NLR disease resistance gene Xa47 confers durable and broad-spectrum resistance to bacterial blight in rice. *Frontiers in Plant Science*, *13*, 1037901.

Ma, L., An, R., Jiang, L., Zhang, C., Li, Z., Zou, C., Yang, C., Pan, G., Lübberstedt, T., & Shen, Y. (2022). Effects of *ZmHIPP* on lead tolerance in maize seedlings: Novel ideas for soil bioremediation. *Journal of Hazardous Materials*, *430*, 128457.

Ma, L., Zhang, M., Chen, J., Qing, C., He, S., Zou, C., Yuan, G., Yang, C., Peng, H., Pan, G., Lubberstedt, T., & Shen, Y. (2021). GWAS and WGCNA uncover hub genes controlling salt tolerance in maize (*Zea mays* L.) seedlings. *Theoretical and Applied Genetics*, *134*, 3305–3318.

Maach, M., Baghour, M., Akodad, M., Gálvez, F. J., Sánchez, M. E., Aranda, M. N., Venema, K., & Rodríguez-Rosales, M. P. (2020). Overexpression of *LeNHX4* improved yield, fruit quality and salt tolerance in tomato plants (*Solanum lycopersicum* L.). *Molecular Biology Reports*, *47*(6), 4145–4153.

Maeda, S., Dubouzet, J. G., Kondou, Y., Jikumaru, Y., Seo, S., Oda, K., Matsui, M., Hirochika, H., & Mori, M. (2019). The rice CYP78A gene BSR2 confers resistance to Rhizoctonia solani and affects seed size and growth in Arabidopsis and rice. *Scientific Reports*, *9*(1), 587.

Maia, M., Ferreira, A. E. N., Nascimento, R., Monteiro, F., Traquete, F., Marques, A. P., Cunha, J., Eiras-Dias, J.E., Cordeiro, C., Figueiredo, A., & Sousa Silva, M. (2020). Integrating metabolomics and targeted gene expression to uncover potential biomarkers of fungal/oomycetes-associated disease susceptibility in grapevine. *Science Report*, *10*, 15688.

Mao, H., Li, S., Chen, B., Jian, C., Mei, F., Zhang, Y., Li, F., Chen, N., Li, T., Du, L., & Kang, Z. (2022). Variation in cis-regulation of a NAC transcription factor contributes to drought tolerance in wheat. *Molecular Plant*, *15*(2), 276–292.

Mao, H., Li, S., Wang, Z., Cheng, X., Li, F., Mei, F., Mei, F., Chen, N., & Kang, Z. (2020). Regulatory changes in *TaSNAC8-6A* are associated with drought tolerance in wheat seedlings. *Plant Biotechnology Journal*, *18*(4), 1078–1092.

Mao, H., Jian, C., Cheng, X., Chen, B., Mei, F., Li, F., Zhang, Y., Li, S., Du, L., Li, T., Hao, C., Wang, X., & Kang, Z. (2022). The wheat ABA receptor gene *TaPYL1-1B* contributes to drought tolerance and grain yield by increasing water-use efficiency. *Plant Biotechnology Journal*, *20*(5), 846–861.

Mao, L., Deng, M., Jiang, S., Zhu, H., Yang, Z., Yue, Y., & Zhao, K. (2020). Characterization of the DREBA4-type transcription factor (*SlDREBA4*), which contributes to heat tolerance in tomatoes. *Frontiers in Plant Science*, *11*, 554520.

Meggio, F., Trevisan, S., Manoli, A., Ruperti, B., & Quaggiotti, S. (2020). Systematic investigation of the effects of a novel protein hydrolysate on the growth, physiological parameters, fruit development and yield of grapevine (*Vitis vinifera* L., cv Sauvignon Blanc) under water stress conditions. *Agronomy*, *10*(11), 1785.

Mei, F., Chen, B., Li, F., Zhang, Y., Kang, Z., Wang, X., & Mao, H. (2021). Overexpression of the wheat NAC transcription factor *TaSNAC4-3A* gene confers drought tolerance in transgenic Arabidopsis. *Plant Physiology and Biochemistry*, *160*, 37–50.

Miao, H., Sun, P., Liu, J., Wang, J., Xu, B., & Jin, Z. (2018). Overexpression of a novel ROP gene from the banana (*MaROP5g*) confers increased salt stress tolerance. *International Journal of Molecular Sciences*, *19*(10), 3108.

Miao, H., Wang, Y., Liu, J., Jia, C., Hu, W., Sun, P., Jin, Z & Xu, B. (2014). Molecular cloning and expression analysis of the *MaASR1* gene in banana and functional characterization under salt stress. *Electronic Journal of Biotechnology*, *17*(6), 287–295.

Mishra, N., Jiang, C., Chen, L., Paul, A., Chatterjee, A., & Shen, G. (2023). Achieving abiotic stress tolerance in plants through antioxidative defense mechanisms. *Frontiers in Plant Science*, *14*, 1110622.

Mishra, N., Srivastava, A. P., Esmaeili, N., Hu, W., & Shen, G. (2018). Overexpression of the rice gene OsSIZ1 in Arabidopsis improves drought-, heat-, and salt-tolerance simultaneously. *PLoS One*, *13*(8), e0201716.

Monteiro, E., De Lorenzis, G., Ricciardi, V., Baltazar, M., Pereira, S., Correia, S., Ferreira, H., Alves, F., Cortez, I., Gonalves, B. & Castro, I. (2023). Exploring seaweed and glycine betaine biostimulants for enhanced phenolic content, antioxidant properties and gene expression of Vitis vinifera cv. 'Touriga Franca' berries. *International Journal of Molecular Sciences*, *25*(10), 5335.

Moolphuerk, N., & Pattanagul, W. (2020). Pretreatment with different molecular weight chitosans encourages drought tolerance in rice (*Oryza sativa* L.) seedling. *Notulae Botanicae Horti Agrobotanici Cluj-Napoca*, *48*(4), 2072–2084.

Murugan, N., Palanisamy, V., Channappa, M., Ramanathan, V., Ramaswamy, M., Govindakurup, H., & Chinnaswamy, A. (2022). Genome-wide in silico identification, structural analysis, promoter analysis, and expression profiling of PHT gene family in sugarcane root under salinity stress. *Sustainability*, *14*(23), 15893.

Musavizadeh, Z., Najafi-Zarrini, H., Kazemitabar, S. K., Hashemi, S. H., Faraji, S., Barcaccia, G., & Heidari, P. (2021). Genome-wide analysis of potassium channel genes in rice: Expression of the *OsAKT* and *OsKAT* genes under salt stress. *Genes*, *12*(5), 784.

Mushke, R., Yarra, R., & Kirti, P. B. (2019). Improved salinity tolerance and growth performance in transgenic sunflower plants via ectopic expression of a wheat antiporter gene (*TaNHX2*). *Molecular Biology Reports*, *46*(6), 5941–5953.

Mushtaq, N., Wang, Y., Fan, J., Li, Y., & Ding, J. (2022). Down-regulation of cytokinin receptor gene *SlHK2* improves plant tolerance to drought, heat, and combined stresses in tomato. *Plants*, *11*(2), 154.

Mutale-Joan, C., Rachidi, F., Mohamed, H. A., Mernissi, N. E., Aasfar, A., Barakate, M., Mohammed, D., Sbabou, L., & Arroussi, H. E. (2021). Microalgae-cyanobacteria–based biostimulant effect on salinity tolerance mechanisms, nutrient uptake, and tomato plant growth under salt stress. *Journal of Applied Phycology*, *33*, 3779–3795.

Myles, S., Boyko, A. R., Owens, C. L., Brown, P. J., Grassi, F., Aradhya, M. K., Prins, B., Reynolds, A., Chia, J., Ware, D., Bustamante, C.D., & Buckler, E. S. (2011). Genetic structure and domestication history of the grape. *Proceedings of the National Academy of Sciences*, *108*(9), 3530–3535.

Naderi, K., Etesami, H., Alikhani, H. A., & Arani, A. M. (2022). Potential use of endophytic and rhizosheath bacteria from the desert plant *Stipagrostis pennata* as biostimulant against drought in wheat cultivars. *Rhizosphere*, *24*, 100617.

Negi, S., Bhakta, S., Ganapathi, T. R., & Tak, H. (2023). *MusaNAC29*-like transcription factor improves stress tolerance through modulation of phytohormone content and expression of stress responsive genes. *Plant Science*, *326*, 111507.

Ngaki, M. N., Sahoo, D. K., Wang, B., & Bhattacharyya, M. K. (2021). Overexpression of a plasma membrane protein generated broad-spectrum immunity in soybean. *Plant Biotechnology Journal*, *19*(3), 502–516.

Noor, I., Sohail, H., Sun, J., Nawaz, M. A., Li, G., Hasanuzzaman, M., & Liu, J. (2022). Heavy metal and metalloid toxicity in horticultural plants: Tolerance mechanism and remediation strategies. *Chemosphere*, *303*, 135196.

O'Conner, S., Zheng, W., Qi, M., Kandel, Y., Fuller, R., Whitham, S. A., & Li, L. (2021). *GmNF-YC4-2* increases protein, exhibits broad disease resistance and expedites maturity in soybean. *International Journal of Molecular Sciences*, *22*(7), 3586.

Pacifico, D., Musmeci, S., del Pulgar, J. S., Onofri, C., Parisi, B., Sasso, R., Mandolino, G., & Lombardi-Boccia, G. (2019). Caffeic acid and α-chaconine influence the resistance of potato tuber to *Phthorimaea operculella* (Lepidoptera: Gelechiidae). *American Journal of Potato Research*, *96*, 403–413.

Patel, A., Sahu, K. P., Mehta, S., Balamurugan, A., Kumar, M., Sheoran, N., Kumar, S., Krishnappa, C., Ashajyothi, M., Kundu, A., Goyal, T., Narayanasamy, P., & Kumar, A. (2022). Rice leaf endophytic *Microbacterium testaceum*: Antifungal actinobacterium confers immunocompetence against rice blast disease. *Frontiers in Microbiology*, *13*, 1035602.

Peng, L. N., Xu, Y. Q., Wang, X., Feng, X., Zhao, Q. Q., Feng, S. S., Zhao, Z. Y., Hu, B. Z., & Li, F. L. (2019). Overexpression of paralogues of the wheat expansin gene *TaEXPA8* improves low-temperature tolerance in Arabidopsis. *Plant Biology*, *21*(6), 1119–1131.

Pimentel, M. F., Arnão, E., Warner, A. J., Subedi, A., Rocha, L. F., Srour, A., Bond, J. P., & Fakhoury, A. M. (2020). Trichoderma isolates inhibit *Fusarium virguliforme* growth, reduce root rot, and induce defense-related genes on soybean seedlings. *Plant Disease*, *104*(7), 1949–1959.

Puterka, G. J., Xu, X., Li, G., Carver, B. F., & Guo, P. (2020). Mechanisms of resistance of new wheat gene Dn10 in comparison with other Dn genes resistant to Russian wheat aphid. *Crop Science*, *60*(4), 1782–1788.

Qi, X., Tang, X., Liu, W., Fu, X., Luo, H., Ghimire, S., Zhang, N., & Si, H. (2020). A potato RING-finger protein gene *StRFP2* is involved in drought tolerance. *Plant Physiology and Biochemistry*, *146*, 438–446.

Qiao, K., Wang, F., Liang, S., Wang, H., Hu, Z., & Chai, T. (2019). Improved Cd, Zn and Mn tolerance and reduced Cd accumulation in grains with wheat-based cell number regulator TaCNR2. *Scientific Reports*, *9*(1), 870.

Raju, G., Shanmugam, K., & Kasirajan, L. (2020). High-throughput sequencing reveals genes associated with high-temperature stress tolerance in sugarcane. *3 Biotech*, *10*(5), 198.

Rakoczy-Lelek, R., Czernicka, M., Ptaszek, M., Jarecka-Boncela, A., Furmanczyk, E. M., Kęska-Izworska, K., Grzanka, M., Skoczylas, E., Kuznik, N., Smolen, S., Macko-Podgarni, A., Gaska, K., Chalanska, A., Ambroziak, K., & Kardasz, H. (2023). Transcriptome dynamics underlying planticine®-induced defense responses of tomato (*Solanum lycopersicum* L.) to biotic stresses. *International Journal of Molecular Sciences*, *24*(7), 6494.

Ramasamy, M., Damaj, M. B., Vargas-Bautista, C., Mora, V., Liu, J., Padilla, C. S., Irigoyen, S., Saini, T., Sahoo, N., DaSilva, J.A., & Mandadi, K. K. (2021). A sugarcane G-protein-coupled receptor, *ShGPCR1*, confers tolerance to multiple abiotic stresses. *Frontiers in Plant Science*, *12*, 745891.

Ren, Y., Mao, H., Su, W., Su, Y., Zhang, C., Shi, M., & Que, Y. (2021). Improved drought and salt tolerance in transgenic Nicotiana benthamiana by overexpressing sugarcane ScSEC14p gene. *Sugar Technology*, *23*, 326–335.

Repke, R. A., Silva, D. M. R., dos Santos, J. C. C., & de Almeida Silva, M. (2022). Increased soybean tolerance to high-temperature through biostimulant based on *Ascophyllum nodosum* (L.) seaweed extract. *Journal of Applied Phycology*, *34*(6), 3205–3218.

Repke, R. A., Silva, D. M. R., dos Santos, J. C. C., & de Almeida Silva, M. (2022). Alleviation of Drought Stress in Soybean by Applying a Biostimulant Based on Amino Acids and Macro-and Micronutrients. *Agronomy*, *12*(10), 2244.

Reyes-Herrera, D. F., Sanchez-Reisono, A. D., Lombardini, L., & Restrepo-Diaz, H. (2023). Physiological responses of coffee (*Coffea arabica* L.) plants to biochar application under water deficit conditions. *Notulae Botanicae Horti Agrobotanici Cluj-Napoca*, *51*(3), 12873–12873.

Ritonga, F. N., & Chen, S. (2020). Physiological and molecular mechanism involved in cold stress tolerance in plants. *Plants*, *9*(5), 560.

Riyazuddin, R., Singh, K., Iqbal, N., Nisha, N., Rani, A., Kumar, M., Khatri, N., Siddiqui, M. H., Yasheshwar., Kim, S. T., Attila, F.,.. & Gupta, R. (2023). Iodine: An emerging biostimulant of growth and stress responses in plants. *Plant and Soil*, *486*(1), 119–133.

Rizzo, E., Sherman, T., Manosalva, P., & Gomez, S. K. (2020). Assessment of local and systemic changes in plant gene expression and aphid responses during potato interactions with arbuscular mycorrhizal fungi and potato aphids. *Plants*, *9*(1), 82.

Rono, J. K., Zhang, B. Q., Liu, X. S., Wang, M. Q., Wang, L. L., Wu, X. C., Chen, X., Cao, H. W., & Yang, Z. M. (2019). Identification of novel rice (*Oryza sativa*) HPP and HIPP genes tolerant to heavy metal toxicity. *Ecotoxicology and Environmental Safety*, *175*, 8–18.

Sahana, B. N., PrasannaKumar, M. K., Mahesh, H. B., Buela Parivallal, P., Puneeth, M. E., Gautam, C., Girish, T. R., Nori, S., & Suryanarayan, S. (2022). Biostimulants derived from red seaweed stimulate the plant defence mechanism in rice against *Magnaporthe oryzae*. *Journal of Applied Phycology*, *34*, 659–665.

Sahid, S., Roy, C., Paul, S., & Datta, R. (2020). Rice lectin protein r40c1 imparts drought tolerance by modulating S-adenosylmethionine synthase 2, stress-associated protein 8 and chromatin-associated proteins. *Journal of Experimental Botany*, *71*(22), 7331–7346.

Sanchez, D., Sadoun, S. B., Mary-Huard, T., Allier, A., Moreau, L., & Charcosset, A. (2023). Improving the use of plant genetic resources to sustain breeding programs' efficiency. *Proceedings of the National Academy of Sciences*, *120*(14), e2205780119.

Seleiman, M. F., Al-Suhaibani, N., Ali, N., Akmal, M., Alotaibi, M., Refay, Y., Dindaroglu, T., Abdul-Wajid, H. H., & Battaglia, M. L. (2021). Drought stress impacts on plants and different approaches to alleviate its adverse effects. *Plants*, *10*(2), 259.

Shah, T., Khan, Z., Asad, M., Imran, A., Niazi, M. B. K., & Alsahli, A. A. (2023). Alleviation of cadmium toxicity in wheat by strigolactone: Regulating cadmium uptake, nitric oxide signalling, and genes encoding antioxidant defense system. *Plant Physiology and Biochemistry*, *202*, *107916*.

Shang, H., Fang, L., Qin, L., Jiang, H., Duan, Z., Zhang, H., Yang, Z., Cheng, C., Bao, Y., Xu, J., Yao, W., & Zhang, M. (2023). Genome-wide identification of the class III peroxidase gene family of sugarcane and its expression profiles under stresses. *Frontiers in Plant Science*, *14*, 1101665.

Shavit, R., Batyrshina, Z. S., Yaakov, B., Florean, M., Köllner, T. G., & Tzin, V. (2022). The wheat dioxygenase BX6 is involved in the formation of benzoxazinoids in planta and contributes to plant defense against insect herbivores. *Plant Science*, *316*, 111171.

Shekhawat, U. K., & Ganapathi, T. R. (2013). *MusaWRKY71* overexpression in banana plants leads to altered abiotic and biotic stress responses. *PLoS One*, *8*(10), e75506.

Shi, S., Wang, H., Nie, L., Tan, D. I., Zhou, C., Zhang, Q., Li, Y., Du, B., Guo, J., Huang, J., Wu, D., Zheng, X., Guan, W., Shan, J., Zhu, L., Chen, R., Xue, L., Walling, L.L., & He, G. (2021). Bph30 confers resistance to brown planthopper by fortifying sclerenchyma in rice leaf sheaths. *Molecular Plant*, *14*(10), 1714–1732.

Shirazi, Z., Abedi, A., Kordrostami, M., Burritt, D. J., & Hossain, M. A. (2019). Genome-wide identification and characterization of the metal tolerance protein (MTP) family in grape (*Vitis vinifera* L.). *3 Biotech*, *9*, 1–17.

Singh, P., Singh, R. K., Li, H. B., Guo, D. J., Sharma, A., Lakshmanan, P., .Malviya, M. K., Song, X. P., Solanki, M. K., Verma, K. K., & Li, Y. R. (2021). Diazotrophic bacteria *Pantoea dispersa* and *Enterobacter asburiae* promote sugarcane growth by inducing nitrogen uptake and defense-related gene expression. *Frontiers in Microbiology*, *11*, 600417.

Siqueira, A. C. O. (2020). *Development of a Metarhizium-based biostimulant for the sugarcane crop: Effects on pests and crop yield* [Doctoral dissertation, Universidade de São Paulo].

Srivastava, D., Verma, G., Chauhan, A. S., Pande, V., & Chakrabarty, D. (2019). Rice (*Oryza sativa* L.) tau class glutathione S-transferase (*OsGSTU30*) overexpression in *Arabidopsis thaliana* modulates a regulatory network leading to heavy metal and drought stress tolerance. *Metallomics*, *11*(2), 375–389.

Song, W., Shao, H., Zheng, A., Zhao, L., & Xu, Y. (2023). Advances in roles of salicylic acid in plant tolerance responses to biotic and abiotic stresses. *Plants*, *12*(19), 3475.

Sousa, G. F. D., Silva, M. A., Carvalho, M. R. D., Morais, E. G. D., Benevenute, P. A. N., Van Opbergen, G. A. Z., Van Opbergen, G. G. A. Z., & Guilherme, L. R. G. (2023). Foliar selenium application to reduce the induced-drought stress effects in coffee seedlings: Induced priming or alleviation effect? *Plants*, *12*(17), 3026.

Su, H. G., Zhang, X. H., Wang, T. T., Wei, W. L., Wang, Y. X., Chen, J., Zhou, Y. B., Chen, M., Ma, Y. Z., Xu, Z. S., & Min, D. H. (2020). Genome-wide identification, evolution, and expression of GDSL-type esterase/lipase gene family in soybean. *Frontiers in Plant Science*, *11*, 726.

Su, W., Huang, L., Ling, H., Mao, H., Huang, N., Su, Y., Ren, Y., Wang, D., Xu, L., Muhammad, K., & Que, Y. (2020). Sugarcane calcineurin B-like (CBL) genes play important but versatile roles in regulation of responses to biotic and abiotic stresses. *Scientific Reports*, *10*(1), 167.

Sun, C., Jin, L., Cai, Y., Huang, Y., Zheng, X., & Yu, T. (2019). L-Glutamate treatment enhances disease resistance of tomato fruit by inducing the expression of glutamate receptors and the accumulation of amino acids. *Food Chemistry*, *293*, 263–270.

Sun, J., Yang, L., Wang, J., Liu, H., Zheng, H., Xie, D., Zhang, M., Feng, M., Jia, Y., Zhao, H., & Zou, D. (2018). Identification of a cold-tolerant locus in rice (*Oryza sativa* L.) using bulked segregant analysis with a next-generation sequencing strategy. *Rice*, *11*, 1–12.

Sun, L., Zhang, X., Ouyang, W., Yang, E., Cao, Y., & Sun, R. (2022). Lowered Cd toxicity, uptake and expression of metal transporter genes in maize plant by ACC deaminase-producing bacteria Achromobacter sp. *Journal of Hazardous Materials*, *423*, 127036.

Sun, Y., Wang, C., Chen, H. Y., & Ruan, H. (2020). Response of plants to water stress: A meta-analysis. *Frontiers in Plant Science*, *11*, 978.

Tak, H., Negi, S., Rajpurohit, Y. S., Misra, H. S., & Ganapathi, T. R. (2020). *MusaMPK5*, a mitogen activated protein kinase is involved in regulation of cold tolerance in banana. *Plant Physiology and Biochemistry*, *146*, 112–123.

Tang, H., Yu, Q., Li, Z., Liu, F., Su, W., Zhang, C., Ling, H., Luo, J., Su, Y., & Que, Y. (2021). A PIP-mediated osmotic stress signalling cascade plays a positive role in the salt tolerance of sugarcane. *BMC Plant Biology*, *21*, 1–17.

Tang, R., Gupta, S. K., Niu, S., Li, X. Q., Yang, Q., Chen, G., .Zhu, W., & Haroon, M. (2020). Transcriptome analysis of heat stress response genes in potato leaves. *Molecular Biology Reports*, *47*, 4311–4321.

Teketay, D. (1998). History, botany and ecological requirements of coffee. *Walia*, *1998*(20), 28–50.

Theerawitaya, C., Tisarum, R., Samphumphuang, T., Singh, H. P., Takabe, T., & Cha-Um, S. (2020). Expression levels of vacuolar ion homeostasis-related genes, Na+ enrichment, and their physiological responses to salt stress in sugarcane genotypes. *Protoplasma*, *257*, 525–536.

Tian, W., He, G., Qin, L., Li, D., Meng, L., Huang, Y., & He, T. (2021). Genome-wide analysis of the NRAMP gene family in potato (Solanum tuberosum): Identification, expression analysis and response to five heavy metals stress. *Ecotoxicology and Environmental Safety*, *208*, 111661.

Tian, W., Huang, Y., Li, D., Meng, L., He, T., & He, G. (2022). Identification of *StAP2/ERF* genes of potato (*Solanum tuberosum*) and their multiple functions in detoxification and accumulation of cadmium in yest: Implication for Genetic-based phytoremediation. *Science of the total Environment*, *810*, 152322.

Torres, L. F., Reichel, T., Déchamp, E., de Aquino, S. O., Duarte, K. E., Alves, G. S. C., Silva, A. T., Cotta, M. G., Costa, T. S., Diniz, L. E. C., & Marraccini, P. (2019). Expression of DREB-like genes in Coffea canephora and C. arabica subjected to various types of abiotic stress. *Tropical Plant Biology*, *12*, 98–116.

Tu, M., Wang, X., Yin, W., Wang, Y., Li, Y., Zhang, G., Li, Z., Song, J., & Wang, X. (2020). Grapevine VlbZIP30 improves drought resistance by directly activating VvNAC17 and promoting lignin biosynthesis through the regulation of three peroxidase genes. *Horticulture Research*, *7*, 150–165.

Vaseva, I. I., Simova-Stoilova, L., Kostadinova, A., Yuperlieva-Mateeva, B., Karakicheva, T., & Vassileva, V. (2022). Heat-Stress-Mitigating effects of a protein-Hydrolysate-Based biostimulant are linked to changes in protease, DHN, and HSP gene expression in maize. *Agronomy*, *12*(5), 1127.

Verma, R. K., Santosh Kumar, V. V., Yadav, S. K., Pushkar, S., Rao, M. V., & Chinnusamy, V. (2019). Overexpression of ABA receptor PYL10 gene confers drought and cold tolerance to indica rice. *Frontiers in Plant Science*, *10*, 1488.

Velimirović, A., Jovović, Z., & Pržulj, N. (2021). From neolithic to late modern period: Brief history of wheat. *Genetika*, *53*(1), 407–417.

Wang, B., Zhong, Z., Wang, X., Han, X., Yu, D., Wang, C., Song, W., Zheng, X., Chen, C., & Zhang, Y. (2020). Knockout of the *OsNAC006* transcription factor causes drought and heat sensitivity in rice. *International Journal of Molecular Sciences*, *21*(7), 2288.

Wang, D., Wang, L., Su, W., Ren, Y., You, C., Zhang, C., Que, Y., & Su, Y. (2020). A class III WRKY transcription factor in sugarcane was involved in biotic and abiotic stress responses. *Scientific Reports*, *10*(1), 20964.

Wang, F., Chen, X., Dong, S., Jiang, X., Wang, L., Yu, J., & Zhou, Y. (2020). Crosstalk of PIF4 and DELLA modulates CBF transcript and hormone homeostasis in cold response in tomato. *Plant Biotechnology Journal*, *18*(4), 1041–1055.

Wang, F., Ren, X., Zhang, F., Qi, M., Zhao, H., Chen, X., Ye, Y., Yang, J., Li, S., Zhang, Y., Niu, Y., & Zhou, Q. (2019). A R2R3-type MYB transcription factor gene from soybean, *GmMYB12*, is involved in flavonoids accumulation and abiotic stress tolerance in transgenic Arabidopsis. *Plant Biotechnology Reports*, *13*, 219–233.

Wang, F., Wang, X., Zhang, Y., Yan, J., Ahammed, G. J., Bu, X., Sun, X., Liu, Y., Xu, T., Qi, H., Qi, M.,& Li, T. (2022). *SlFHY3* and *SlHY5* act compliantly to enhance cold tolerance through the integration of myo-inositol and light signalling in tomato. *New Phytologist*, *233*(5), 2127–2143.

Wang, H., Hou, J., Ye, P., Hu, L., Huang, J., Dai, Z., Zhang, B., Dai, S., Que, J., Min, H., Chen, G., Wang, Y., Jiang, M., Liang, Y., Li, L., Zhang, X.,& Lai, Z. (2021). A teosinte-derived allele of a MYB transcription repressor confers multiple disease resistance in maize. *Molecular Plant, 14*(11), 1846–1863.

Wang, H., Sun, S., Ge, W., Zhao, L., Hou, B., Wang, K., Lyu, Z., Chen, L., Xu, S., Guo, J., Li, M., Su, P., Li, X., Wang, G., Bo, C., Fang, X., Zhuang, W., Cheng, X., Wu, J., Dong, L., Chen, W., Li, W., Xiao, G., Zhao, J., Hao, Y., Xu, Y., Gao, Y., Liu, W., Liu, Y., Tin, H., Li, J., Li, X., Zhao, Y., Wang, X., Ni, F., Ma, X., Li, A., Xu, S.S., Bai, G., Nevo, E., Gao, C., Ohm, H., & Kong, L. (2020). Horizontal gene transfer of Fhb7 from fungus underlies Fusarium head blight resistance in wheat. *Science, 368*, 5435.

Wang, L. F., Lu, K. K., Li, T. T., Zhang, Y., Guo, J. X., Song, R. F., & Liu, W. C. (2022). Maize Phytomelatonin Receptor 1 functions in plant tolerance to osmotic and drought stress. *Journal of Experimental Botany, 73*(17), 5961–5973.

Wang, M., Yuan, J., Qin, L., Shi, W., Xia, G., & Liu, S. (2020). Ta CYP 81D5, one member in a wheat cytochrome P450 gene cluster, confers salinity tolerance via reactive oxygen species scavenging. *Plant Biotechnology Journal, 18*(3), 791–804.

Wang, T. T., Yu, T. F., Fu, J. D., Su, H. G., Chen, J., Zhou, Y. B., Chen, M., Guo, J., Ma, Y.Z., Wei, W.L., & Xu, Z. S. (2020). Genome-wide analysis of the GRAS gene family and functional identification of *GmGRAS37* in drought and salt tolerance. *Frontiers in Plant Science, 11*, 604690.

Wang, W. R., Liang, J. H., Wang, G. F., Sun, M. X., Peng, F. T., & Xiao, Y. S. (2020). Overexpression of PpSnRK1α in tomato enhanced salt tolerance by regulating ABA signaling pathway and reactive oxygen metabolism. *BMC Plant Biology, 20*, 1–15.

Wang, Y., Feng, G., Zhang, Z., Liu, Y., Ma, Y., Wang, Y., Ma, F., Zhou, Y., Gross, R., Xu, H., Wang, R., Xiao, F., Liu, Y.,& Niu, X. (2021). Overexpression of Pti4, Pti5, and Pti6 in tomato promote plant defense and fruit ripening. *Plant Science, 302*, 110702.

Wang, Y., Huang, L., Du, F., Wang, J., Zhao, X., Li, Z., Wang, W., Xu, J., & Fu, B. (2021). Comparative transcriptome and metabolome profiling reveal molecular mechanisms underlying OsDRAP1-mediated salt tolerance in rice. *Scientific Reports, 11*(1), 5166.

Wang, Z., Wong, D. C. J., Wang, Y., Xu, G., Ren, C., Liu, Y., Kuang, Y., Fan, P., Li, S., Xin, H., & Liang, Z. (2021). GRAS-domain transcription factor PAT1 regulates Jasmonic acid biosynthesis in grape cold stress response. *Plant Physiology, 186*(3), 1660–1678.

Wang, Z., Wang, Y., Tong, Q., Xu, G., Xu, M., Li, H., Fan, P., Li, S., & Liang, Z. (2021). Transcriptomic analysis of grapevine Dof transcription factor gene family in response to cold stress and functional analyses of the *VaDof17d* gene. *Planta, 253*, 1–14.

Wen, J., Jiang, F., Weng, Y., Sun, M., Shi, X., Zhou, Y., .Yu, L.,& Wu, Z. (2019). Identification of heat-tolerance QTLs and high-temperature stress-responsive genes through conventional QTL mapping, QTL-seq and RNA-seq in tomato. *BMC Plant Biology, 19*, 1–17.

Wu, J., Jiang, Y., Liang, Y., Chen, L., Chen, W., & Cheng, B. (2019). Expression of the maize MYB transcription factor *ZmMYB3R* enhances drought and salt stress tolerance in transgenic plants. *Plant Physiology and Biochemistry, 137*, 179–188.

Wu, Q., Chen, Y., Zou, W., Pan, Y. B., Lin, P., Xu, L., Grisham, M.P., Ding, Q., Su, Y., & Que, Y. (2023). Genome-wide characterization of sugarcane catalase gene family identifies a *ScCAT1* gene associated disease resistance. *International Journal of Biological Macromolecules, 232*, 123398.

Xia, Y., Farooq, M. A., Javed, M. T., Kamran, M. A., Mukhtar, T., Ali, J., Tabassum, T., ur Rehman, S., Munis, M.F.H., Sultan, T., & Chaudhary, H. J. (2020). Multi-stress tolerant PGPR Bacillus xiamenensis PM14 activating sugarcane (Saccharum officinarum L.) red rot disease resistance. *Plant Physiology and Biochemistry, 151*, 640–649.

Xian, P., Yang, Y., Xiong, C., Guo, Z., Alam, I., He, Z., Zhang, Y., Cai, Z., & Nian, H. (2023). Overexpression of *GmWRKY172* enhances cadmium tolerance in plants and reduces cadmium accumulation in soybean seeds. *Frontiers in Plant Science, 14*, 1133892.

Xu, L., Xiang, G., Sun, Q., Ni, Y., Jin, Z., Gao, S., & Yao, Y. (2019). Melatonin enhances salt tolerance by promoting MYB108A mediated ethylene biosynthesis in grapevines. *Horticulture Research, 6*, 114–128.

Xu, W., Ma, F., Li, R., Zhou, Q., Yao, W., Jiao, Y., Zhang, C., Zhang, J., Wang, X., Xu, Y., & Wang, Y. (2019). *VpSTS29/STS2* enhances fungal tolerance in grapevine through a positive feedback loop. *Plant, Cell & Environment, 42*(11), 2979–2998.

Xu, W., Shen, W., Ma, J., Ya, R., Zheng, Q., Wu, N., Yu, Q., Yao, W., Zhang, N., & Zhang, J. (2020). Role of an Amur grape CBL-interacting protein kinase *VaCIPK02* in drought tolerance by modulating ABA signalling and ROS production. *Environmental and Experimental Botany, 172*, 103999.

Xu, X., Li, G., Carver, B. F., & Armstrong, J. S. (2020). Gb8, a new gene conferring resistance to economically important greenbug biotypes in wheat. *Theoretical and Applied Genetics, 133*, 615–622.

Xu, X., Wang, Q., Li, W., Hu, T., Wang, Q., Yin, Y., Liu, X., He, S., Zhang, M., Liang, Y., Zhu, J.,& Zhan, X. (2022). Overexpression of *SlBBX17* affects plant growth and enhances heat tolerance in tomato. *International Journal of Biological Macromolecules*, *206*, 799–811.

Xu, Y., Hu, W., Song, S., Ye, X., Ding, Z., Liu, J., Wang, Z., Li, J., Hou, X., Xu, B., & Jin, Z. (2023). *MaDREB1F* confers cold and drought stress resistance through common regulation of hormone synthesis and protectant metabolite contents in banana. *Horticulture Research*, *10*(2), uhac275.

Xun, H., Yang, X., He, H., Wang, M., Guo, P., Wang, Y., Pang, J., Dong, Y., Feng, X., Wang, S., & Liu, B. (2019). Over-expression of GmKR3, a TIR–NBS–LRR type R gene, confers resistance to multiple viruses in soybean. *Plant Molecular Biology*, *99*, 95–111.

Yadav, K., Patel, P., Srivastava, A. K., & Ganapathi, T. R. (2017). Overexpression of native ferritin gene MusaFer1 enhances iron content and oxidative stress tolerance in transgenic banana plants. *PLoS One*, *12*(11), e0188933.

Yadav, R. K., Tripathi, M. K., Tiwari, S., Tripathi, N., Asati, R., Chauhan, S., Tiwari, P. N., & Payasi, D. K. (2023). Genome editing and improvement of abiotic stress tolerance in crop plants. *Life*, *13*(7), 1456.

Yang, J., Song, J., & Jeong, B. R. (2022). Drenched silicon suppresses disease and insect pests in coffee plant grown in controlled environment by improving physiology and upregulating defense genes. *International Journal of Molecular Sciences*, *23*(7), 3543.

Yang, J., Zhang, G., An, J., Li, Q., Chen, Y., Zhao, X., Wu, J., Wang, Y., Hao, Q., Wang, W., & Wang, W. (2020). Expansin gene *TaEXPA2* positively regulates drought tolerance in transgenic wheat (Triticum aestivum L.). *Plant Science*, *298*, 110596.

Yang, M., Lin, J., Cheng, L., Zhou, H., Chen, S., Liu, F., Li, R., & Qiu, Y. (2020). Identification of a novel planthopper resistance gene from wild rice (*Oryza rufipogon* Griff.). *The Crop Journal*, *8*(6), 1057–1070.

Yang, P., Scheuermann, D., Kessel, B., Koller, T., Greenwood, J. R., Hurni, S., Herren, G., Zhou, S., Marande, W., Wicker, T., Krattinger, S. G., Ouzunovo, M., & Keller, B. (2021). Alleles of a wall-associated kinase gene account for three of the major northern corn leaf blight resistance loci in maize. *The Plant Journal*, *106*(2), 526–535.

Yang, X., Kim, M. Y., Ha, J., & Lee, S. H. (2019). Overexpression of the soybean NAC gene GmNAC109 increases lateral root formation and abiotic stress tolerance in transgenic Arabidopsis plants. *Frontiers in Plant Science*, *10*, 1036.

Yang, X., Lu, M., Wang, Y., Wang, Y., Liu, Z., & Chen, S. (2021). Response mechanism of plants to drought stress. *Horticulturae*, *7*(3), 50.

Yang, Y., Yu, T. F., Ma, J., Chen, J., Zhou, Y. B., Chen, M., Ma, Y. Z., Wei, W. L., & Xu, Z. S. (2020). The soybean bZIP transcription factor gene *GmbZIP2* confers drought and salt resistances in transgenic plants. *International Journal of Molecular Sciences*, *21*(2), 670.

Yao, C., Zhang, F., Sun, X., Shang, D., He, F., Li, X., Zhang, J., & Jiang, X. (2019). Effects of S-abscisic acid (S-ABA) on seed germination, seedling growth, and ASR1 gene expression under drought stress in maize. *Journal of Plant Growth Regulation*, *38*, 1300–1313.

Yao, L., Li, Y., Ma, C., Tong, L., Du, F., & Xu, M. (2020). Combined genome-wide association study and transcriptome analysis reveal candidate genes for resistance to Fusarium ear rot in maize. *Journal of Integrative Plant Biology*, *62*(10), 1535–1551.

Yin, J., Wang, L., Jin, T., Nie, Y., Liu, H., Qiu, Y., Tang, Y., Li, B., Zhang, J., Wang, D., Li, K., Xu. K., & Zhi, H. (2021). A cell wall-localized NLR confers resistance to Soybean mosaic virus by recognizing viral-encoded cylindrical inclusion protein. *Molecular Plant*, *14*(11), 1881–1900.

Yu, W., Wang, L., Zhao, R., Sheng, J., Zhang, S., Li, R., & Shen, L. (2019). Knockout of SlMAPK3 enhances tolerance to heat stress involving ROS homeostasis in tomato plants. *BMC Plant Biology*, *19*(1), 1–13.

Yuan, W., Liu, J., Takáč, T., Chen, H., Li, X., Meng, J., Tan, Y., Ning, T., He, Z., Yi, G., & Xu, C. (2021). Genome-wide identification of banana csl gene family and their different responses to low temperature between chilling-sensitive and tolerant cultivars. *Plants*, *10*(1), 122.

Zha, Q., Xi, X., He, Y., & Jiang, A. (2020). Transcriptomic analysis of the leaves of two grapevine cultivars under high-temperature stress. *Scientia Horticulturae*, *265*, 109265.

Zha, Q., Xi, X., He, Y., Yin, X., & Jiang, A. (2022). Interaction of *VvbZIP60s* and *VvHSP83* in response to high-temperature stress in grapes. *Gene*, *810*, 146053.

Zhai, L., Teng, F., Zheng, K., Xiao, J., Deng, W., & Sun, W. (2019). Expression analysis of Argonaute genes in maize (*Zea mays* L.) in response to abiotic stress. *Hereditas*, *156*, 1–10.

Zhang, J., Hewitt, T. C., Boshoff, W. H., Dundas, I., Upadhyaya, N., Li, J., Li, J., Patpour, M., Chandramohan, S., Pretorius, Z. A., Hovmeller, M., Schnippenkoetter, W., Park, R. F., Mago, R., Perivannan, S., Bhatt, D., Hoxha, S., Chakraborty, S., Luo, M., Dodds, P., Steuernagel, B., Wulff, B. B. H., Avliffe, M., Mcintosh, R. A., Zhang, P., & Lagudah, E. S. (2021). A recombined Sr26 and Sr61 disease resistance gene stack in wheat encodes unrelated NLR genes. *Nature Communications*, *12*(1), 3378.

Zhang, L., Jiang, X., Liu, Q., Ahammed, G. J., Lin, R., Wang, L., Shao, S., Yu, J., & Zhou, Y. (2020). The HY5 and MYB15 transcription factors positively regulate cold tolerance in tomato via the CBF pathway. *Plant, Cell & Environment*, *43*(11), 2712–2726.

Zhang, L., Guo, X., Zhang, Z., Wang, A., & Zhu, J. (2021). Cold-regulated gene *LeCOR413PM2* confers cold stress tolerance in tomato plants. *Gene*, *764*, 145097.

Zhang, N., Zhao, H., Shi, J., Wu, Y., & Jiang, J. (2020). Functional characterization of class I *SlHSP17.7* gene responsible for tomato cold-stress tolerance. *Plant Science*, *298*, 110568.

Zhang, P., Guo, G., Wu, Q., Chen, Y., Xie, J., Lu, P., Li, B., Dong, L., Li, M., Wang, R., & Liu, Z. (2020). Identification and fine mapping of spot blotch (*Bipolaris sorokiniana*) resistance gene Sb4 in wheat. *Theoretical and Applied Genetics*, *133*, 2451–2459.

Zhang, Q., Liu, Y., Jiang, Y., Li, A., Cheng, B., & Wu, J. (2022). *OsASR6* enhances salt stress tolerance in rice. *International Journal of Molecular Sciences*, *23*(16), 9340.

Zhang, X., Chen, L., Shi, Q., & Ren, Z. (2020). *SlMYB102*, an R2R3-type MYB gene, confers salt tolerance in transgenic tomato. *Plant Science*, *291*, 110356.

Zhang, X. Z., Zheng, W. J., Cao, X. Y., Cui, X. Y., Zhao, S. P., Yu, T. F., Chen, J., Zhou, Y. B., Chen, M., Chal, S. C., Xu, Z. S., & Ma, Y. Z. (2019). Genomic analysis of stress associated proteins in soybean and the role of GmSAP16 in abiotic stress responses in Arabidopsis and soybean. *Frontiers in Plant Science*, *10*, 1453.

Zhang, X., Long, Y., Huang, J., & Xia, J. (2020). OsNAC45 is involved in ABA response and salt tolerance in rice. *Rice*, *13*(1), 1–13.

Zhang, Y., Guo, W., Chen, L., Shen, X., Yang, H., Fang, Y., Ouyang, W., Mai, S., Chen, H., Chen, S., Hao, Q., Yuan, S., Zhang, C., Huang, Y., Shah, Z., Yang, Z., Qiu, D., Zhou, X., Cao, D., Li, X., & Jiao, Y. (2022). CRISPR/Cas9-mediated targeted mutagenesis of *GmUGT* enhanced soybean resistance against leaf-chewing insects through flavonoids biosynthesis. *Frontiers in Plant Science*, *13*, 802716.

Zhao, H., Wang, X., Jia, Y., Minkenberg, B., Wheatley, M., Fan, J., Jia, M. H., Famoso, A., Edwards, J. D., Wamishe, Y., Valent, B., Wang, G. L., & Yang, Y. (2018). The rice blast resistance gene Ptr encodes an atypical protein required for broad-spectrum disease resistance. *Nature Communications*, *9*(1), 2039.

Zhao, P., Ye, M., Wang, R., Wang, D., & Chen, Q. (2020). Systematic identification and functional analysis of potato (*Solanum tuberosum* L.) bZIP transcription factors and overexpression of potato bZIP transcription factor *StbZIP-65* enhances salt tolerance. *International Journal of Biological Macromolecules*, *161*, 155–167.

Zhao, T., Wu, T., Pei, T., Wang, Z., Yang, H., Jiang, J., Zhang, H., Chen, X., Li, J., & Xu, X. (2021). Overexpression of *SlGATA17* promotes drought tolerance in transgenic tomato plants by enhancing activation of the phenylpropanoid biosynthetic pathway. *Frontiers in Plant Science*, *12*, 634888.

Zhao, Y., Du, H., Wang, Y., Wang, H., Yang, S., Li, C., Chen, N., Yang, H., Zhang, Y., Zhu, Y., Yang, l., & Hu, X. (2021). The calcium-dependent protein kinase *ZmCDPK7* functions in heat-stress tolerance in maize. *Journal of Integrative Plant Biology*, *63*(3), 510–527.

Zheng, M., Lin, J., Liu, X., Chu, W., Li, J., Gao, Y., An, K., Song, W., Xin, M., Yao, Y., Peng, H., Ni, Z., Sun, Q., & Hu, Z. (2021). Histone acetyltransferase TaHAG1 acts as a crucial regulator to strengthen salt tolerance of hexaploid wheat. *Plant Physiology*, *186*(4), 1951–1969.

Zhong, R., Wang, Y., Gai, R., Xi, D., Mao, C., & Ming, F. (2020). Rice SnRK protein kinase *OsSAPK8* acts as a positive regulator in abiotic stress responses. *Plant Science*, *292*, 110373.

Zhu, K., Huang, C., Phan, T. T., Yang, L. T., Zhang, B. Q., Xing, Y. X., & Li, Y. R. (2021). Overexpression of SoACLA-1 gene confers drought tolerance improvement in sugarcane. *Plant Molecular Biology Reporter*, *39*, 489–500.

9 ABC Proteins as Regulators of Plant Tolerance to Biotic and Abiotic Stresses

Gohar Sahakyan and Naira Sahakyan

9.1 INTRODUCTION

Plants are subject to biotic and abiotic stress factors, which can significantly influence their physiological and metabolic processes. Artificial factors are particularly harmful and can pose a significant threat to agricultural plants and the biodiversity of ecosystems. The severity and duration of the stressor influence can inhibit plant growth and development, decreasing yield and biodiversity (Lin et al., 2020; Sinha R., 2021; Nawaz M. et al., 2023). Biotic stressors, such as insect pests, bacterial and viral diseases, and competition with other plants, can also have adverse effects (Zhang et al., 2023). Abiotic stress is caused by extreme temperatures, mineral toxicity, drought, nutrient deficiencies, and harmful components of soil, such as heavy metals. This type of stress can negatively impact plant growth (Boulc'h PN, 2020). To develop a more resilient ecosystem with stress-resistant plants, it's crucial to gain a better understanding of the effects of these environmental hazards on them.

Plants have evolved various structural adaptations, such as the development of specialized tissues like trichomes and stomata, which help them regulate water loss through transpiration and protect against herbivores and pathogens. Furthermore, some plants have evolved specific mechanisms to tolerate extreme environmental conditions, such as drought, salinity, and high temperatures. For example, succulent plants like cacti have developed water-storing tissues and mechanisms to minimize water loss, while halophytes can tolerate high levels of salt in their environment through mechanisms such as salt exclusion or salt secretion. In essence, plants have evolved a multitude of strategies to cope with environmental stresses, ranging from physiological and biochemical mechanisms to symbiotic relationships and structural adaptations, all of which contribute to their survival and success in diverse ecosystems.

There is plenty of literature data regarding the plants' adaptation to distinct stressors (Zhao et al., 2017; Hasanuzzaman et al., 2013; Trchounian et al., 2016; Barreca, 2020; Waadt et al., 2022; Balasubramaniam et al., 2023). Indeed, numerous studies have contributed to our understanding of the molecular and physiological mechanisms that enable plants to respond to stress. These investigations have revealed intricate signaling pathways, regulatory networks, and biochemical processes that plants employ to perceive, transduce, and mitigate stress conditions. However, there remains a need for further understanding of these response mechanisms. Specific genes activated under stress conditions and proteins/enzymes contributing to stress tolerance have been pinpointed. Exploring the involvement of plant hormones like abscisic acid (ABA), researchers have discovered how they mediate plant responses to various stressors. ABA is a crucial phytohormone involved in regulating plant growth, development, and stress responses, particularly in response to abiotic stresses such as drought, salinity, and cold.

ATP-binding cassette (ABC) transporters, which are specialized membrane proteins, play a crucial role in the establishment of growth and development of plants through sophisticated mechanisms for nutrient acquisition and allocation (Kretzschmar et al., 2011; Do et al., 2018;

DOI: 10.1201/9781003543237-9

Dahuja et al., 2020). ABC transporters play diverse and crucial roles in plant biology, including detoxification of xenobiotics, sequestration of heavy metals, transport of secondary metabolites, and protection against herbivores and pathogens. By mediating the transport and compartmentalization of harmful substances, ABC transporters contribute to plant adaptation and survival in challenging environmental conditions (An et al., 2019; Dahuja et al., 2020; Banasiak et al., 2022). They mediate the transport of iron across the membranes of mitochondria and chloroplast, crucial for iron metabolism (Zuo et al., 2019; Kim et al., 2021). Tonoplast-situated ABC transporters protect against DNA impairment and maintain cell growth due to the internal detoxification of metal ions (Huang et al., 2012; Yang et al., 2021). ABC transporters play a crucial role in relocating the secondary metabolites inside the cells and suppressing the seed's embryonic growth in unfavorable conditions (Dahuja et al., 2020; Nogia et al., 2021).

This review seeks to offer an understanding of plant ABC proteins as transporters, drawing primarily from research conducted in *Arabidopsis thaliana*, a well-established model plant. Additionally, it incorporates findings from a select number of studies conducted in other plant species.

9.2 STRUCTURE AND FUNCTIONS OF ABC TRANSPORTERS

Plants harbor an array of specific natural compounds that are synthesized and accumulated meticulously, serving as pivotal defense mechanisms against herbivores, pathogens, and facilitators of symbiotic interactions or pollination. The synthesis of these compounds typically entails coordination among multiple organelles within individual cells and across diverse cell types within an organism. Crucially, transporting both precursors and final products across membranes is indispensable for ensuring these compounds' successful production and functional deployment, thereby underpinning essential aspects of plant survival, adaptation, and ecological interactions. ABC proteins represent the largest superfamily of transporters found in cells across all species (Thomas et al., 2018; Xiong et al., 2015). These transporters serve as pivotal gatekeepers, orchestrating the translocation of a diverse array of substrates across membranes through the dynamic processes of ATP binding and hydrolysis. Operating as importers or exporters, their intricate mechanisms dictate the directionality of substrate movement, thus regulating vital cellular processes essential for the maintenance of cellular homeostasis and the adaptation of organisms to their ever-changing environments (Lewinson et al., 2017; Choi et al., 2021).

ABC transporters have two main components: transmembrane domains (TMDs) and cytoplasmic nucleotide-binding domains (NBDs). The TMDs are responsible for substrate recognition and specificity and are characterized by multiple membrane-spanning α-helices. On the contrary, NBDs comprise highly conserved motifs, including Walker A, Q-loop, Walker B, D-loop, switch H-loop, and signature motifs. The D-loop facilitates dimer connection, the switch H-loop mediates interaction with TMDs, and the Walker A and B motifs form the ATP-binding P-loop. The Q-loop and H-loop interact with ATP's γ-phosphate ATPases (Zolnerciks et al., 2011; Wilkens, 2015). The signature motif is exclusively found in ABC proteins, distinguishing them from others (Nikaido, 2002).

Plant ABC transporters are grouped into eight distinct subfamilies (ABCA-ABCG and ABCI) based on thorough examinations of sequence homology in their NBDs and TMDs, as well as their gene structure and arrangement of functional domains (Navarro-Quiller et al., 2018; Lefèvre & Boutry, 2018) (Figure 9.1).

On the contrary, the ABCH subfamily stands out as an exclusive feature of the animal genome, distinguishing it from plant ABC transporters (Thi et al., 2021). ABC proteins are stratified into three types according to their structural domains: full-length transporters, characterized by two TMDs and two NBDs; half-length transporters, possessing one TMD and one NBD; and incomplete or divergent transporters, featuring two NBD domains without TMDs (Thomas et al., 2020).

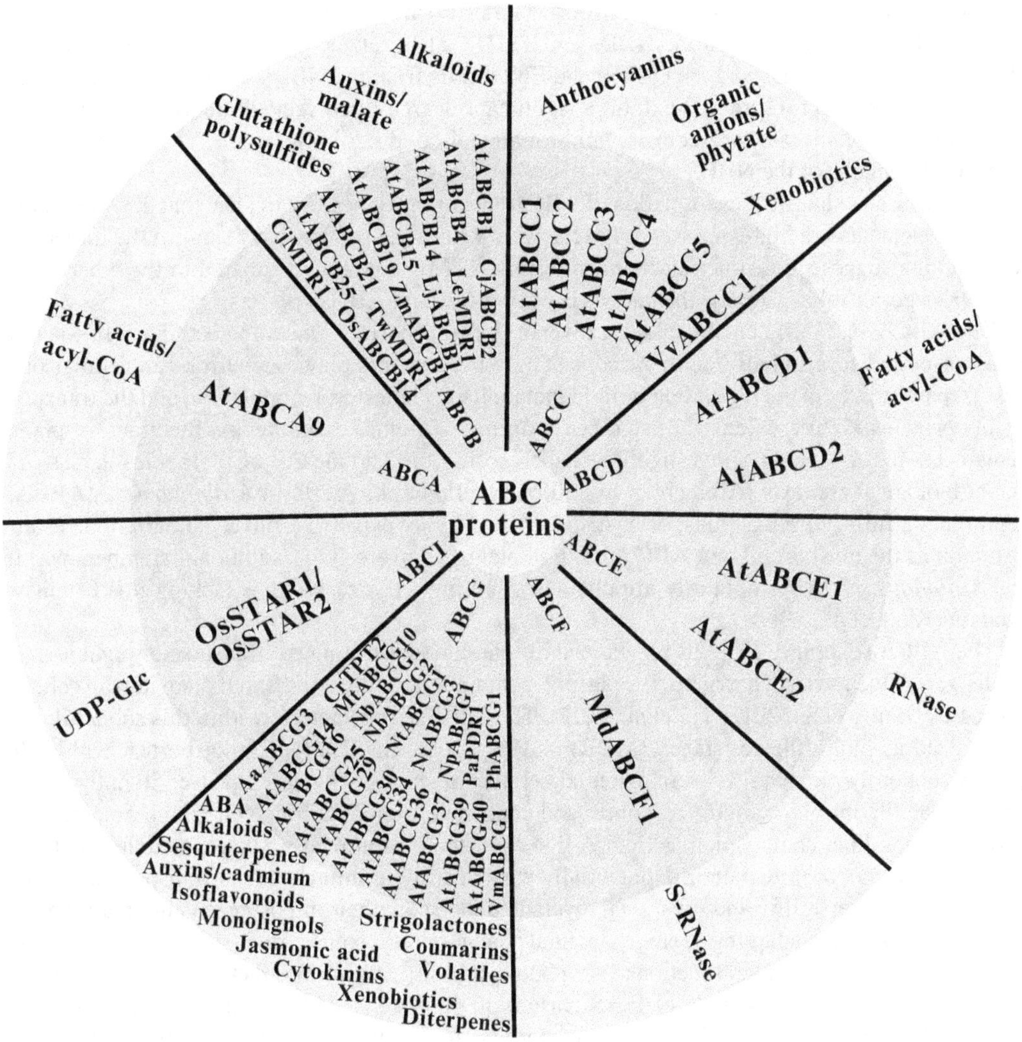

FIGURE 9.1 Diversity of sub-types and substrates of plant ABC transporters. See text for details and references.

The translocation of substances across the cell membrane predominantly occurs within the TMDs, which act as conduits regulating substrate movement and controlling its accessibility between the intra- and extracellular environments. However, the energy required for this process is provided by conformational changes in the NBDs induced by ATP binding and subsequent hydrolysis.

The coordination of structural alterations within the NBDs must be intricately connected with changes occurring in the TMDs. Extensive research on the maltose transporter of *Escherichia coli* has elucidated the coupling mechanism responsible for linking NBDs with TMDs. Findings have highlighted the significance of coupling helices as the principal structural element facilitating NBD–TMD coupling (Wen et al., 2011).

Despite being covalently linked to the remainder of the TMDs, the coupling helices can undergo conformational separation in the absence of NBDs. The robust coupling of these helices to the TMD core and their structural integrity is contingent upon the interactions facilitated by the presence of NBDs.

The functionality of the maltose transporter relies upon a complex interplay of structural elements at the interface between the NBDs and TMDs. These elements encompass the EAA loops (L-loops) from the TMDs and the Q-loop and ENI motifs from the NBDs. In smaller ABC importers, these three structural motifs exhibit a significant degree of correlated motion and synergistically operate to facilitate the requisite conformational coupling between the TMD core and the helical subdomains of the NBD.

Following an exhaustive examination of ABC transporters, it became evident that the three-motif coupling elements are highly conserved exclusively within the subfamily of small ABC importers. This finding suggests a distinct mechanism of NBD–TMD coupling compared to the other major ABC transporter folds, namely, the large ABC importers and ABC exporters.

The ABCA subfamily encompasses a diversity of transporters, including a solitary full-length transporter and several half-length variants. These transporters play crucial roles in cellular processes, particularly in the regulation of lipid metabolism, cholesterol modulation, and the transport of lipoproteins. While extensively studied in animal systems, their precise functions in plants remain an area of ongoing investigation, underscoring the complexity of their roles across different biological contexts (Albrecht et al., 2007; Pasello et al., 2020). In *Arabidopsis*, AtABCA1 represents a full-size ABC transporter distinguished by its expansive linker domain, thus establishing it as the most prominent ABC protein in plants. Conversely, 11 additional members within the *Arabidopsis* ABCA subfamily are characterized by half-size proteins lacking this extensive domain (Kang et al., 2011).

The ABCB subfamily proteins play a pivotal role in orchestrating the transport of organic molecules across the intricate barrier of the plasma membrane, serving as essential mediators in cellular processes (Bailly et al., 2012; Thi et al., 2021). The full-size transporters within this subfamily are recognized as multidrug resistance (MDR) or P-glycoproteins. Half-size transporters within the ABCB subfamily play vital roles in essential cellular functions. They are involved in antigen processing, facilitating the immune response, and contribute to maintaining cell membrane integrity and regulating lipid composition as lipid A-like exporters. Additionally, some function as putative ABC transporters of mitochondria, potentially participating in mitochondrial transport processes. Their structural versatility and functional diversity underscore their importance in maintaining cellular homeostasis and adapting to environmental changes. For instance, iron–sulfur (Fe–S) proteins are crucial for vital cellular functions like photosynthesis, respiration, metabolism, and genome integrity, requiring the assembly of Fe–S clusters in plant cells' mitochondria, chloroplasts, and cytosol (Pearson et al., 2021). Half-size transporters from the ABCB subfamily, like ABC transporter of mitochondrion 3 (ATM3; ABCB25), are found in mitochondria and play a role in cytosolic Fe–S cluster biosynthesis, vital for heavy metal tolerance in eukaryotes (Kruse et al., 2017).

Within *Arabidopsis*, the ABCB subgroup encompasses a comprehensive array of proteins, comprising 21 full-size and 7 half-size variants, as delineated by Deslauriers et al. (2021) and Mellor et al. (2022). Notably, the localization pattern of these proteins reveals a predominant presence of full-size ABCB proteins at the plasma membrane, implying their pivotal role in cellular transport processes and signaling cascades (Lefèvre et al., 2008). Schaedle et al. (2014) identified three half-size ABCB transporters within the mitochondria. Furthermore, AtABCB26 is detected within the chloroplast, and AtABCB27 is localized in the membrane of the central vacuole (Ferro et al., 2010; Kang et al., 2011).

In *Coptis japonica*, a medicinal plant, the presence of two full-size ABCB proteins, ABCB1 and ABCB2, has been confirmed. These proteins play a crucial role in the transportation of the alkaloid berberine. Berberine, synthesized in the plant roots, undergoes translocation to the rhizomes, where it eventually accumulates within the central vacuoles. It is hypothesized that CjABCB1 and CjABCB2 facilitate the uptake of berberine into the cytosol of rhizome cells, thereby aiding in the translocation of berberine from the roots to the rhizomes (Shitan et al., 2013).

The ABCC subfamily contains predominantly half-size MDR-associated proteins (MRP), which realize the vacuolar sequestration of glutathione (GSH) and glucuronide conjugates (Klein et al., 2006; Lefèvre et al., 2018). This subfamily comprises a group of ABC transporters that play vital roles in several fundamental physiological processes in plants. These include the regulation of stomatal aperture, the transport of phytate, folate, and chlorophyll catabolites into vacuoles and the defense against heavy metal and cadmium stress. Through their involvement in these diverse functions, ABCC transporters contribute significantly to plant growth, development, and adaptation to environmental challenges.

In *Arabidopsis*, the ABCC protein subfamily exclusively comprises full-size ABC proteins characterized by the presence of an additional N-terminal TMD known as TMD0, which is unique to this subfamily (Klein et al., 2006; Tan et al., 2019). The precise function of the TMD0 domain in plants remains uncertain, yet in human and yeast ABCC proteins, this domain is recognized for its involvement in protein targeting processes (Tusnády et al.,2006; Bickers et al., 2021).

ABCC proteins in *Arabidopsis* exhibit a distribution pattern spanning the vacuolar membrane and tonoplasts within the plant cell. Such widespread localization underscores the diverse roles ABCC transporters play in mediating various physiological processes within *Arabidopsis* cells, including but not limited to ion transport, detoxification, and homeostasis regulation (Kang et al., 2011; Behrens et al.; 2019). Specifically, under metal stress conditions, AtABCC1 and AtABCC2 ABC transporters are activated, playing crucial roles as critical molecules in facilitating vacuolar phytochelatin transport within *Arabidopsis*, as outlined by Behrens et al. (2019).

The transporters of the ABCD subfamily are mainly TMD–NBD-oriented half-size peroxisomal membrane proteins associated with the peroxisomal transport of fatty acids across the plant membrane. Within the *Arabidopsis* genome, the ABCD subfamily is characterized by the presence of two distinct members, encompassing both half-size and full-size variants. Notably, AtABCD1 stands out as the full-size representative, predominantly localized within the peroxisomes. This subcellular localization hints at its pivotal role in peroxisomal functions, such as lipid metabolism, beta-oxidation of fatty acids, and the synthesis of various metabolites crucial for plant growth and development (Baker et al., 2015).

The ABCE, ABCF, and ABCI subfamilies are comprised of soluble proteins featuring two NBDs without TMD. Among their essential functions are facilitating RNA interference mechanisms, orchestrating the regulation of gene expression, and participating in the intricate synthesis of plastid lipids (Lane et al., 2016; Navarro-Quiles et al., 2018).

Within *Arabidopsis*, the ABCE and ABCF subfamilies are represented by three and five members, respectively. Notably, members of the ABCI subfamily are proposed to target the chloroplast or mitochondria, with two encoded by the mitochondrial genome. This dual localization suggests a crucial role for ABCI subfamily members in facilitating essential processes within these organelles. Such intricate spatial distribution underscores the complexity and significance of their functions in coordinating cellular activities (Mõttus et al., 2021; Lane et al., 2016; Kim et al., 2020).

The largest subfamily of ABC transporters is the ABCG type performed by reverse domain-organized (NBD–TMD) proteins. Consisting of both full-sized pleiotropic drug resistance (PDR) transporters and half-sized white-brown complex (WBC) transporters, this group predominantly localizes within the plasma membrane. Their pivotal function lies in coordinating the transport of phytohormones and reinforcing plant defense mechanisms against pathogens (Lane et al., 2016; Dhara et al., 2021).

Numerous studies underscore the indispensable role of ABC proteins in orchestrating the transport of toxic metal ions and shielding plants from their detrimental effects (Dahuja et al., 2020; Won-Yong et al., 2014; Wang et al., 2019; Naaz et al., 2023). It has been demonstrated that a sophisticated defense mechanism mediated by ABC transporters facilitates plants' ability to resist heavy metal toxicity. This mechanism encompasses various processes, such as the uptake of heavy metals, their binding to metallothioneins and phytochelatins, the sequestration of compound conjugates within the central vacuole, and both internal and external excretion pathways (Ai et al., 2022; Seregin et al., 2022; Gautam et al., 2023; Naaz et al., 2023).

In *Arabidopsis*, the ABCG subfamily is distinguished by its composition of 28 half-size WBCs and 15 full-size PDRs. A notable characteristic of these transporters is their predominant localization within the plasma membrane, with the exception of AtABCG 19 known as WBC homolog protein 19 (Dahuja et al., 2020; Amrhein et al., 2021; Cho et al., 2021).

ABC transporters contribute to plants' diverse physiological and metabolic processes, allowing them to adapt to challenging environmental conditions and deal with biotic and abiotic stresses (Kretzschmar et al., 2011; Nawaz et al., 2023; Boulc'h et al., 2020; Dahuja et al., 2021). Additionally, these proteins play a fundamental role in mediating the transport of secondary metabolites, which are vital for bolstering the plant's defense mechanisms against a spectrum of abiotic stressors (Hwang et al., 2016). Heavy metal pollution, notably cadmium and mercury, poses a significant environmental health concern for plant life. Several studies have underscored the indispensable roles of ABC transporters in enabling tolerance mechanisms against these metals. Specifically, ABCC-type transporters have been identified as pivotal players in facilitating the translocation of cadmium into the central vacuole, where detoxification processes are initiated (Lin et al., 2016; Yang et al., 2021; Ai et al., 2022). The involvement of the plasma membrane-localized full-size transporter protein AtABCG36/AtPDR8 is shown in the processes of the resistance and intoxication of plants due to the influence of pathogens and toxic ions (Kim et al., 2010; Dhara et al., 2021). Elevated resistance to Pb and Cd stress, achieved through augmented translocation to the shoots, is ascribed to the overexpression of AtATM3. Identified as the ABC transporter of mitochondria in *A. thaliana*, AtATM3 assumes a pivotal role in orchestrating this protective response against heavy metal toxicity (Thi Do et al., 2021).

9.3 THE ROLE OF ABC TRANSPORTERS IN PHYTOHORMONE TRANSPORT

Phytohormones are the cornerstone in orchestrating plant responses to a spectrum of biotic and abiotic environmental stimuli. As master regulators, these specialized signaling molecules govern plant growth, development, and adaptation to fluctuating conditions (Petrosyan et al., 2015; Poghosyan et al., 2021). Their intricate signaling cascades empower plants to mount dynamic responses to biotic challenges, such as pathogen invasions, and navigate through adverse abiotic stressors like drought, salinity, and temperature fluctuations. The classical plant hormones auxins, ABA, cytokinins (CKs), brassinosteroids, salicylic acid, jasmonate, and ethylene recognize and transfer environmental stimuli and endogenous signals to regulate plant survival mechanisms to various abiotic stresses (Sytar et al., 2019).

In plants, hormones are synthesized at distinct sites from where they exert their actions. Consequently, the export of hormones from producing cells, their subsequent distribution within the plant body, and their targeted uptake into responsive cells are critical for optimal plant functioning. ABC transporters, alongside other transporter proteins, play a pivotal role in facilitating the transport of essential hormones. Notably, they are involved in translocating auxin, a primary developmental regulator, and ABA, a stress hormone derived from carotenoids. ABA, particularly, is implicated in orchestrating responses to both biotic and abiotic stresses, including pathogen attacks, salinity, extreme temperatures, mechanical injury, and drought (Borghi et al., 2015; Sytar et al., 2019).

Auxin serves as a vital plant growth regulator, mediating diverse physiological functions through cell-to-cell transport mechanisms. Beyond its regulatory role in growth, auxins dynamically contribute to enhancing plant tolerance to nutrient deficiencies. However, under conditions of drought and other stressors, intricate adjustments in auxin synthesis, metabolism, transport, and activity take place to enable plants to adapt and cope with the environmental challenges they face (Korver et al., 2018; Feng et al., 2022). ABCB transporters play a pivotal role in facilitating the polar distribution of auxin within plants. Among the 22 full-size ABCB isoforms identified in *Arabidopsis*, a specific subset, including AtABCB1, 4, 6, 14, 15, 19, 20, and 21, has been implicated in mediating polar auxin transport. Notably, the activities of AtABCB1, 4, 6, 19, 20, and 21 are recognized as crucial for this transport process. Among these, *Arabidopsis* AtABCB1, 6, 19, and 20 and tomato ABCB4 are identified as specific auxin exporters (Ofori et al., 2018; Hao et al., 2021). The *Arabidopsis* transporters AtABCB4 and AtABCB21, as well as the rice transporter OsABCB14, are purported to serve as versatile regulators capable of both importing and exporting indole-3-acetic acid, the primary natural auxin in higher plants (Xu et al., 2014; Hao et al., 2021). An in-depth investigation into the membrane-intrinsic transporters AtABCB1 and AtABCB19 has elucidated the pivotal role of ABC proteins in intercellular auxin transport. Single-knockout studies of Atabcb1 and Atabcb19 revealed a modest retardation in growth, whereas simultaneous double knockout resulted in a profound growth impairment resembling an auxin-deficient phenotype. Notably, the double knockout mutant exhibited a significant reduction in auxin transport, exceeding 70% (Cecchetti et al., 2015).

A noteworthy observation arises from the intricate relationship between the two auxin efflux systems. While the PIN-FORMED (PIN) proteins represent plants' most extensively studied family of auxin transporters, recent findings suggest their interaction with ABC transporters, forming a complex regulatory network that modulates auxin transport activity (Mellor et al., 2022). Additionally, research has revealed that the immunophilin-like protein TWD1 interacts with ABCB transporters, leading to a substantial enhancement of auxin transport (Geisler et al., 2004).

ABA plays a crucial role as a signaling mediator in plants, facilitating responses to diverse environmental stresses. This multifaceted hormone initiates and regulates numerous physiological processes essential for plant survival and adaptation. For instance, ABA is pivotal in regulating seed germination, ensuring proper timing and conditions for seedling emergence. It modulates bud dormancy, controlling the timing of bud break and ensuring optimal growth conditions (Wang et al., 2020). ABA is instrumental in regulating stomatal closure, a critical mechanism for minimizing water loss during drought stress and maintaining plant water balance (Chen et al., 2020). This phytohormone exerts precise control over stress response genes through both transcriptional and post-transcriptional regulatory mechanisms (Wilkinson et al., 2002; Yang et al., 2022). ABA is pivotal in bolstering plant stress tolerance by regulating gene expression, particularly in enhancing antioxidant activities crucial for scavenging reactive oxygen species (ROS). Through control over transcriptional processes, ABA orchestrates the upregulation of genes encoding key antioxidant enzymes and proteins. Thus, ABA contributes significantly to fortifying plant defenses against oxidative damage, thus enhancing their resilience to adverse conditions (Nguyen et al., 2023).

The G-subfamily ABC transporters, ABCG25 and ABCG40, play a pivotal role in the transmembrane transport of ABA. This process facilitates the movement of ABA from vascular cells to the xylem and subsequently into guard cells. Consequently, it initiates stomatal closure, particularly in response to drought stress. The half-size transporter, AtABCG25, synthesized in the vascular parenchyma of both root and shoot tissues, functions as an ABA exporter, demonstrating a notable affinity for ABA in heterologous transport systems (Kuromori et al., 2010). The full-size transporter, AtABCG40, which was initially speculated to participate in heavy metal resistance mechanisms, has been identified as a pivotal player in facilitating the import of ABA across the stomatal plasma membrane (Hu et al., 2020; Bano et al, 2023).

Several full-size ABC transporters in plants have been proposed to function as ABA importers. Among them, AtABCB14, akin to AtABCG40, is localized in the stomatal plasma membrane and proposed to regulate the import of apoplastic malate, which accumulates during photosynthesis and acts as a signaling molecule influencing the stomatal aperture. As carbon dioxide concentrations rise due to heightened photosynthetic activity, malate is exported from guard cells to the apoplast. AtABCB14 then facilitates the import of malate back into guard cells, instigating physiological responses that lead to stomatal closure, ultimately contributing to plant adaptation to changing atmospheric conditions (Lee et al., 2008). An exemplary case is observed in AtABCB4, demonstrating a nuanced role as an ABA importer or exporter contingent upon auxin concentration in the cytosol or external medium. This dynamic functionality underscores the intricate interplay between hormonal signaling pathways within plant cells. In elevated cytosolic auxin levels, AtABCB4 predominantly functions as an ABA exporter, facilitating the efflux of ABA from the cell. Conversely, when auxin concentration is low, AtABCB4 transitions to act as an ABA importer, facilitating the uptake of ABA into the cell. This sophisticated regulatory mechanism allows AtABCB4 to finely tune intracellular ABA levels in response to fluctuations in auxin signaling, thereby integrating the signaling pathways of these pivotal hormones to orchestrate plant growth and stress responses (Hellsberg et al., 2015).

Furthermore, ABCG25 and ABCG40, alongside AtABCG30 and AtABCG31, are reported as key players in the transportation of ABA from the endosperm of seed coats to the embryo, thereby contributing to the maintenance of seed dormancy. Studies have demonstrated that ABCG25 and ABCG31 facilitate the export of ABA from the endosperm, whereas ABCG40 and ABCG30 are essential for the uptake of ABA into the embryo. This coordinated transport mechanism is critical for regulating seed dormancy and ensuring proper seed germination under favorable conditions (Kang et al., 2015; Ma et al., 2018). Notably, the abcg25 and abcg31 endosperm double loss-of-function plant maintained at least 60% of the transport activity (Park et al., 2017). The half-size ABCG22 transporter is involved in drought stress tolerance. Still, the activity of ABCG22 in independent systems needs to be clarified entirely, leaving this ABC protein's identity as an additional ABA transporter open (Qin et al., 2021).

As a result of the integration of internal and external signals, phytohormones play crucial roles in plant morphogenesis, development, and interactions with microbes. ABC proteins belonging to the ABCG and ABCB subfamilies are critical in translocating specific hormones within plants, facilitating their movement from the sites of synthesis or accumulation to target organs and external plant surfaces (Borghi et al., 2019). This mechanism is indispensable for orchestrating the multifaceted aspects of plant growth and their reactions to environmental signals. Moreover, it assumes a critical function in mediating interactions with microbial organisms.

Legume species such as beans, soybeans, chickpeas, peanuts, lentils, and lupins can interact symbiotically with rhizobial nitrogen-fixing bacteria. This interaction requires the specific organs, the root nodules. Phytohormones, particularly CKs, govern the initiation of nodule formation, regulating cellular processes within the root cortex. By modulating specific genes and signaling pathways, CKs promote the emergence of nodule primordia from pericycle or inner cortical cells, laying the groundwork for developing mature nodules. In essence, CKs serve as master regulators, steering the early stages of nodule formation and facilitating the establishment of symbiotic interactions between plants and rhizobia (Schiessl et al., 2019; Suzaki et al., 2013).

Half-sized ABCG transporters play a role in distributing CKs in *Arabidopsis* (Ko et al., 2014). In *Medicago truncatula* various full-sized ABCG transporters exhibit specific expression patterns in roots and respond to signaling molecules produced by rhizobia (Jarzyniak et al., 2021). For example, MtABCG56, identified as a CK transporter, shows increased epidermal expression following infection with *Ensifer meliloti* or treatment with NOD factor. According to Jarzyniak et al., this transporter likely facilitates the efflux of CKs from the epidermis upon infection. Additionally, Shen et al. discovered that LjABCB1 expression is exclusive to developing nodules in Fabaceae, particularly in cells adjacent to the infection zone (Shen et al., 2015). These cells are believed to support nitrogen fixation in infected cells and counteract oxygen, thereby maintaining nitrogenase activity (Takanashi et al., 2012).

9.4 THE ROLE OF ABC TRANSPORTERS IN CELLULAR DETOXIFICATION

In plant systems, potentially harmful compounds conjugate with the tripeptide glutathione, forming glutathione conjugates, which are sequestered into vacuoles independent of vacuolar proton pump activity (Hasanuzzaman et al., 2017). Analogous to animal systems, detoxification mechanisms in plants involve cytochrome P450 proteins and a variety of transferases, including glutathione S-transferases, glucuronosyltransferases, and glycosyltransferases (Brazier-Hicks et al., 2017). However, unlike animals, which typically metabolize and eliminate toxins, plants sequester toxins within central vacuoles.

The initial insights into ABC-mediated transport mechanisms in plants emerged from studies conducted by Martinoia et al., focusing on the detoxification of xenobiotics (Martinoia et al., 1993).

Follow-up investigations uncovered two specialized transporters, AtABCC1 and AtABCC2, located within the vacuoles of *Arabidopsis* plants (Raichaudhuri, 2016; Behrens et al., 2019). These transporters are pivotal in transporting many substrates, including glutathione conjugates, glucuronates, and chlorophyll catabolites. Significantly, AtABCC1 exhibits an additional function as an effective transporter for folate, a crucial molecule intricately involved in methylation processes within cellular metabolism (Ofori et al., 2018).

Recent studies with *Arabidopsis* suggest a potential role for AtABCG19 in mediating the plant's defense against kanamycin-induced stress (Dhara et al., 2021). This transporter has a high specificity to kanamycin transport and no connection with the resistance to other clinically significant aminoglycoside antibiotics, and it is suggested that AtABCG19 can serve as a resistance marker for creating transgenic plants.

Heavy metal pollution is a pressing environmental issue with detrimental impacts on ecosystems. Cadmium (Cd), mercury (Hg), and lead (Pb) are among the most common heavy metal pollutants, posing severe threats to plants. Even at low concentrations, these metals, arsenic (As), chromium (Cr), and others, can induce profound toxicity in plant systems. The adverse effects manifest in various ways, including reduced biomass accumulation, water uptake, and distribution disturbances leading to water imbalance, chlorosis, diminished photosynthetic capacity, and plant mortality (Singh et al., 2016). Penetrating the cells, heavy metal ions are found to activate ROS production in plants, causing oxidative stress. The presence of heavy metals in plants can lead to changes in gene expression, disrupting cellular metabolism and impacting plant growth (Angulo-Bejarano et al., 2021). Additionally, these metals can alter the structure of macromolecules and replace vital metal ions in biomolecules, which serve as essential cofactors for enzymes (Schutzendubel & Polle, 2002).

Numerous studies investigating ABC transporters have highlighted their crucial involvement in imparting plant resistance to heavy metals like Cd, Pb, and Hg. The defense against heavy metal toxicity relies on ABC transporter-dependent mechanisms, including the uptake of heavy metals, binding to phytochelatins and metallothioneins, separating and storing compound conjugates within the large central vacuole, and internal/external excretion. In *Arabidopsis*, the AtABCC1 and AtABCC2 transporters have been identified as participating in the transport of phytochelatins to the vacuole during metal stress (Ofori et al., 2018; Raichaudhuri, 2016; Behrens et al., 2019). Additionally, recent investigations show that plasma membrane-intrinsic transporter AtABCG36/AtPDR8 appears to mediate the plant's defense mechanisms against pathogenic invaders and regulate the homeostasis of toxic ions within the cell membrane (Kim et al., 2010; Dhara et al., 2021).

Remarkably, heightened resistance to lead (Pb) and cadmium (Cd) stress is achieved through the overexpression of the mitochondrial ABC transporter AtATM3, facilitating their pronounced translocation to the aerial parts of the plant. This phenomenon underscores the pivotal role of AtATM3 in mediating the plant's adaptive response to heavy metal exposure, particularly by promoting the efficient translocation of Pb and Cd from the roots to the shoots (Thi Do et al., 2021).

As mentioned, ABCC-type transporters relocate the phytochelatins and complexes formed between phytochelatins and heavy metals to the central vacuole for detoxification. At the same time, plants survive by secreting toxic compounds from the above-ground and below-ground cells into the apoplast, extracellular space outside the plasma membrane, the cuticle, the protective outer layer, and the rhizosphere, the soil environment surrounding the roots. The protein AtABCG36/AtPDR8, known for its involvement in pathogen defense in *Arabidopsis*, has also been implicated in detoxification processes. Thus, Kim et al. revealed the crucial role of ABC transporter AtABCG36 in cadmium transport in *Arabidopsis* (Kim et al., 2010). Terpenoid derivates and cyanogenic glycosides are antimicrobial plant secondary metabolites contributing to defense against pathogens, inhibiting the fungal and bacterial proliferation on plant surfaces around local infection sites (Mahizan et al., 2019; Avetisyan et al., 2017; Sahakyan, 2023; Ginovyan et al., 2023). In Solanaceae plants, full-size ABCG transporters have been recognized for their role in producing and releasing specialized terpenoids that contribute to defense mechanisms (Shibata et al., 2016; Banasiak et al., 2022).

For instance, in Tex-Mex tobacco plants, NpABCG1 transports an antifungal diterpene sclareol secreted in the leaves (Stukkens et al., 2005).

The NtABCG1 protein in cultivated tobacco (*Nicotiana tabacum*) is responsible for transporting two types of diterpenes: sclareol, which is a bicyclic diterpene alcohol, and cembrene, a monocyclic diterpene (Crouzet et al., 2013). Research indicates that capsidiol can influence the ATPase activity of the NtABCG1 protein (Pierman et al., 2017). In *Nicotiana benthamiana* plants with mutations in both NbABCG1 and NbABCG2 genes, there was a reduction in capsidiol secretion, which is crucial for defense against *Phytophthora* after invasion, leading to increased susceptibility to *Phytophthora* infection. Moreover, these transporters are closely associated with the diterpenes necessary for pre-invasion resistance mechanisms, as indicated by studies conducted by Shibata et al. (2016) and Rin et al. (2017).

Wang et al. (2019) identified an ABC transporter gene, PtoABCG36, from *Populus tomentosa*. They observed that the PtoABCG36 transcript was present in leaves, stems, and roots, with higher expression noted in the roots, particularly under Cd stress conditions. The researchers found that the PtoABCG36 protein localized to the plasma membrane. By overexpressing PtoABCG36 in yeast and *Arabidopsis*, they observed reduced Cd accumulation and increased Cd tolerance in the transgenic lines. Notably, Cd uptake was decreased in the roots of transgenic *Arabidopsis* plants. The authors concluded that PtoABCG36 serves as a cadmium extrusion pump, thereby enhancing Cd tolerance by reducing Cd content in plants.

The AtABCG40 transporter, identified as an ABA importer, has been implicated as a key factor in lead (Pb) resistance in *Arabidopsis* plants (Lee et al., 2005). Knockout mutants lacking the AtABCG40 gene were found to exhibit greater susceptibility to Pb compared to wild-type plants. Conversely, plants overexpressing the AtABCG40 gene showed reduced lead accumulation compared to wild-type plants. The authors proposed that the AtABCG40 transporter may function either as an efflux pump, facilitating the export of lead from plants, or as an inhibitor of lead uptake into root cells by releasing chelating agents (Lee et al., 2005). Another ABC transporter in *Arabidopsis*, AtABCB25/AtATM3, is involved in cadmium (Cd) resistance by transporting glutamine synthetase-conjugated Cd across the mitochondrial membrane (Wang et al., 2019).

ABCC subfamily transporters are also implicated in heavy metal resistance. For instance, both full-size AtABCC1 and AtABCC2 proteins function as transporters of vacuolar phytochelatins, contributing to resistance against arsenic (As), mercury (Hg), and cadmium (Cd). Upon exposure to Cd treatment, the expression level of AtABCC6/AtMRP6 is upregulated in *Arabidopsis* plants (Park et al., 2011; Gaillard et al., 2008). Additionally, AtABCC3 has been shown to induce the Cd-sensitive phenotype in the ycf1 mutant of *Saccharomyces cerevisiae* (Brunetti et al., 2015).

Rice ABCC transporter OsABCC1 contributes to arsenic detoxification and decreases the accumulation of metal ions in the rice grains, and its homolog PtABCC1 increases tolerance to Cd in poplar (Song et al., 2014; Hayashi et al., 2017; Sun et al., 2018).

The upregulation of AtABCB1 expression in *Arabidopsis* confers heightened resistance to a range of herbicides, encompassing dicamba, pendimethalin, oryzalin, monosodium acid, and methane arsenate. This resistance is attributed to either decreased xenobiotic retention or increased efflux from cells (Remy et al., 2014).

9.5 THE ROLE OF ABC TRANSPORTERS IN PATHOGEN DEFENSE

Plants have developed intricate defense mechanisms to safeguard themselves against detrimental pathogens. These mechanisms involve complex signaling pathways, molecular interactions, and physiological responses to detect, recognize, and neutralize pathogenic invaders. Terpenoid derivatives and cyanogenic glycosides are powerful antimicrobial plant secondary metabolites that contribute to this defense. These compounds demonstrate remarkable efficacy in suppressing the proliferation of fungi and bacteria on plant surfaces surrounding infection sites, thereby playing a pivotal role in containing the spread of disease and preserving plant health (Mahizan et al., 2019).

Recent research has revealed that full-sized ABCGs from Solanaceae are particularly effective in specialized terpenoid secretion related to defense mechanisms (Shibata et al., 2016; Banasiak et al., 2022). In Tex-Mex tobacco plants, for example, NpABCG1 is responsible for transporting the antifungal diterpene sclareol that is secreted in the leaves (Stukkens et al., 2005).

Gaining insights into the mechanisms through which plants defend themselves against pathogens holds paramount importance for agriculture and human health. By leveraging these innate defense mechanisms, there exists the potential to develop novel and more potent strategies to safeguard crops and curb the transmission of diseases.

The NtABCG1 protein in *Nicotiana tabacum*, a species of cultivated tobacco, facilitates the transport of two diterpenes: sclareol and cembrene (Lefèvre et al., 2018). Notably, the sesquiterpene capsidiol has been found to stimulate the ATPase activity of this protein (Pierman et al., 2017). In double mutant *Nicotiana benthamiana* plants lacking both NbABCG1 and NbABCG2, the secretion of capsidiol, crucial for post-invasion defense against *Phytophthora*, is diminished. This reduction in capsidiol secretion renders *Nicotiana benthamiana* plants more susceptible to *Phytophthora* infection. Additionally, these transporters are closely linked to the presence of diterpenes, which play a crucial role in pre-invasion resistance mechanisms (Shibata et al., 2016; Rin et al., 2017).

Petunia (*Petunia hybrida*) produces two terpenoids, petuniasterone and petuniolide, which act as a defense mechanism against herbivores. These terpenoids are transported by the PhABCG2 protein, as reported by Sasse et al. in 2016. Additionally, the AaABCG3 transporter from *Artemisia annua* has been shown to export β-caryophyllene. This sesquiterpenoid mediates antagonistic interactions with bacterial pathogens, according to a study by Fu et al. in 2017.

Studies conducted in rice have revealed that approximately 50% of ABCG transporters are upregulated in response to jasmonic acid and salicylic acid, two phytohormones critical for combating biotic stresses. Moons et al. (2017) identified three full-length ABCG transporters in rice leaves during infection by the biotrophic fungus *Magnaporthe grisea*. Conversely, in *Arabidopsis*, the disruption of the AtABCG30 function leads to significant alterations in the composition of root exudates, resulting in marked changes in the soil microflora within the rhizosphere (Badri et al., 2012).

In leguminous plants, isoflavonoids and flavonoids serve as essential specialized metabolites that play roles in defending against pathogens, attracting beneficial microorganisms, and facilitating the recognition of symbiotic partners during the establishment of legume-rhizobia symbiosis (Liu et al., 2016).

ABCG proteins have been implicated as putative transporters of genistein, an isoflavone with multifaceted roles including UV filtration, plant pigmentation, and facilitation of symbiotic nitrogen fixation. This compound is secreted by the roots of Glycine max soybeans (Sugiyama et al., 2007).

In *Medicago truncatula*, the MtABCG10 transporter is responsible for transporting precursors of the phenylpropanoid (PP) biosynthetic pathway, such as 4-coumarate and liquiritigenin (Biala et al., 2017). Liquiritigenin is the substrate for legume-specific 5-deoxyisoflavonoid synthesis, associated with defense responses and the production of phytoalexin medicarpin (Naoumkina et al., 2007). Biala et al. demonstrated that silencing MtABCG10 results in decreased concentrations of medicarpin and its precursors in roots and root exudates, consequently increasing the plants' susceptibility to fungal infections (Biala et al., 2017).

In *Arabidopsis*, the PEN3/AtABCG36 transporter facilitates the secretion of tryptophan-derived indole glucosinolates upon bacterial, fungal, or oomycete infection, potentially limiting the proliferation of microbial pathogens and promoting the accumulation of callose induced by flg22. Additionally, PEN3, in conjunction with AtABCG40, is involved in the secretion of camalexin, another tryptophan-derived metabolite associated with plant defense. The full-size AtABCG34 transporter is believed to be responsible for camalexin transport (Lefèvre et al., 2018; Matern et al., 2019; Aryal et al., 2019; Khare et al., 2017).

9.6 THE ROLE OF ABC TRANSPORTERS IN THE CUTICLE FORMATION AND FUNCTIONING

The plant cuticle comprises cuticular waxes embedded within a cutin matrix, providing a protective layer that envelops the epidermis of above-ground plant organs. This layer regulates interactions between plant surfaces and their environment. Moreover, the cuticle is essential in organ development, preventing fusion and influencing organ morphology. These insights are supported by research conducted by Ingram et al. (2017) and Berhin et al. (2018).

The cuticle on the leaf surface is continuous and relatively impermeable, helping to reduce excessive water loss from the surface of the leaves. Gas exchange predominantly takes place through stomata, which are capable of adjusting their aperture in response to various environmental and physiological cues. During water stress, the stomata closes, allowing the leaf to conserve water that is retained due to the presence of the cuticle. This was demonstrated in a study by Tominaga et al. in 2015.

Plant cuticles are a protective layer that limits the harmful effects of UV radiation, as noted in a study by Solovchenko et al. in 2003. Changes in the structure and composition of cuticles can significantly affect how plants interact with pathogens and beneficial microorganisms. The properties of the cuticle also play a vital role in plant reproduction, affecting processes such as the fusion of reproductive organs and the germination of pollen grains on female floral organs. Ingram et al. (2017) documented this role in their study. Finally, cuticles also play a role in attracting pollinators, as noted in a study by Moyroud et al. in 2017.

The importance of ABCG proteins in transporting cutin monomers across various angiosperm species cannot be overstated. The ABCG protein cluster G32, which includes well-known transporters like AtABCG32/AtPDR4/permeable cuticle 1 (Pecl) from *Arabidopsis*, SlABCG36 and SlABCG42 from tomato *Solanum lycopersicum*, and OsABCG31 and HvABCG31/EIBI1 from monocot rice and barley *Hordeum vulgare*, respectively, is widely distributed. These transporters play a crucial role in cuticle formation in young leaves, flowers, and fruits (Chen et al., 2011; Banasiak et al., 2022; Elejalde-Palmett et al., 2021).

Recent transport experiments validate that AtABCG32 and SlABCG42 indeed export sn-2-monoacylglycerols (sn-2 MAGs), such as mono-10,16-diOH C16:0-2-glycerol, essential building blocks for cutin biosynthesis (Elejalde-Palmett et al., 2021). Moreover, besides full-size ABCG proteins, half-size ABCG transporters such as DSO/WBC11/AtABCG11, CER5/AtABCG12, and AtABCG13 also contribute to cuticle formation (Banasiak et al., 2022; Elejalde-Palmett et al., 2021; Wang et al., 2023).

Genes encoding AtABCG11/12/13 homologous proteins are present in the liverwort Marchantia polymorpha and moss *Physcomitrella patens* genomes, highlighting the importance of these transporters across different species (Banasiak et al., 2022; Zhang et al., 2023).

9.7 CONCLUSION

In conclusion, the intricate roles of ABC proteins in regulating plant tolerance to biotic and abiotic stresses underscore their significance as key players in plant defense and adaptation mechanisms. These versatile transporters exhibit a broad substrate specificity, facilitating the transport of a diverse array of molecules across cellular membranes. In the context of biotic stress, ABC proteins are pivotal components of plant immunity, orchestrating the transport of antimicrobial compounds, defense-related metabolites, and signaling molecules crucial for mounting effective defense responses against pathogens and pests.

Moreover, ABC proteins play vital roles in mitigating the adverse effects of abiotic stresses by mediating the uptake, sequestration, and detoxification of various stress-inducing agents, including heavy metals, salt ions, and ROS. ABC transporters help maintain cellular homeostasis and protect plants from environmental damage by regulating these harmful substances' intracellular and extracellular distribution.

Furthermore, emerging evidence suggests that ABC proteins are intricately involved in hormonal signaling pathways, including those associated with ABA, auxins, and strigolactones, exerting additional influence on plant stress responses and developmental processes.

As research advances, further elucidation of ABC proteins' diverse functions and regulatory mechanisms promises to provide valuable insights into their roles in plant stress tolerance. These insights will deepen our understanding of plant defense strategies and pave the way for the development of innovative approaches to enhance crop resilience, improve yield stability, and ensure food security in the face of escalating environmental challenges. By harnessing the potential of ABC transporters, scientists can unlock new avenues for sustainable agriculture and contribute to the global effort to address the pressing challenges posed by biotic and abiotic stresses on crop production.

REFERENCES

Ai, H., Wu, D., Li, C., & Hou, M. (2022). Advances in molecular mechanisms underlying cadmium uptake and translocation in rice. *Frontiers in Plant Science*, *13*, 1003953. https://doi.org/10.3389/fpls.2022.1003953

Albrecht, C., & Viturro, E. (2007). The ABCA subfamily-gene and protein structures, functions and associated hereditary diseases. *Pflügers ArchivEuropean Journal of Physiology*, *453*(5), 581–589. https://doi.org/10.1007/s00424-006-0047-8.

Amrhein, N., & Martinoia, E. (2021). An ABC transporter of the ABCC subfamily localized at the plasma membrane confers glyphosate resistance. *Proceedings of the National Academy of Sciences of the United States of America*, *118*(18). https://doi.org/10.1073/pnas.2104746118

An, L., Ma, Q., Du, J., Yu, M., Li, F., Luan, J., Li H. (2019). Preliminary classification of the ABC transporter family in Betula halophila and expression patterns in response to exogenous phytohormones and abiotic stresses. *Forests*, *10*, 722.

Angulo-Bejarano, P. I., Puente,-Rivera, J., & Cruz,-Ortega, R. (2021). Metal and metalloid toxicity in plants: an overview on molecular aspects. *Plants*, *10*(4). https://doi.org/10.3390/plants10040635

Aryal, B., Huynh, J., Schneuwly, J., Siffert, A., Liu, J., Alejandro, S., Ludwig,-Müller J., Martinoia, E., & Geisler, M. (2019). ABCG36/PEN3/PDR8 is an exporter of the auxin precursor, indole-*3*-butyric acid, and involved in auxin-controlled development. *Frontiers in Plant Science*, *10*, 899. https://doi.org/10.3389/fpls.2019.00899

Avetisyan, A., Markosian, A., Petrosyan, M. et al. (2017). Chemical composition and some biological activities of the essential oils from basil *Ocimum* different cultivars. *BMC Complementary Medicine and Therapies*, *17*, 60. https://doi.org/10.1186/s12906-017-1587-5

Badri, D. V., Chaparro, J. M., Manter, D. K., Martinoia, E., & Vivanco, J. M. (2012). Influence of atp-binding cassette transporters in root exudation of phytoalexins, signals, and in disease resistance. *Frontiers in Plant Science*, *3*. https://doi.org/10.3389/fpls.2012.00149

Bailly, A., Yang, H., Martinoia, E., Geisler, M., & Murphy, A. S. (2012). Plant lessons: exploring ABCB functionality through structural modeling. *Frontiers in Plant Science*, *2*, 108. https://doi.org/10.3389/fpls.2011.00108

Baker, A., Carrier, D. J., Schaedler, T., Waterham, H. R., van Roermund, C. W., & Theodoulou, F. L. (2015). Peroxisomal ABC transporters: functions and mechanism. *Biochemical Society Transactions*, *43*(5), 959–965. https://doi.org/10.1042/BST20150127.

Balasubramaniam, T., Shen, G., Esmaeili, N., & Zhang, H. (2023). Plants' response mechanisms to salinity stress. *Plants*, *12*, 2253. https://doi.org/10.3390/plants12122253

Banasiak, J., & Jasiński M. (2022). ATP-binding cassette transporters in nonmodel plants. *New Phytologist*, *233*, 1597–1612. https://doi.org/10.1111/nph.17779

Bano, A., Singh, K., Singh, S. P., & Sharma, P. (2023). Abscisic acid: metabolism, signaling, and crosstalk with other phytohormones under heavy metal stress. *Stresses*, *3*(4), 665–686. https://doi.org/10.3390/stresses3040046

Barreca, D. (2020). Mechanisms of plant antioxidants action. *Plants (Basel)*, *10*(1), 35. https://doi.org/10.3390/plants10010035.

Behrens, C. E., Smith, K. E., Iancu, C. V., Choe, J. Y., & Dean, J. V. (2019). Transport of anthocyanins and other flavonoids by the arabidopsis ATP-binding cassette transporter AtABCC2. *Sci Report*, *9*(1), 437. https://doi.org/10.1038/s41598-018-37504-8.

Behrens, C. E., Smith, K. E., Iancu, C. V., Choe, J., & Dean, J. V. (2019). Transport of anthocyanins and other flavonoids by the arabidopsis atp-binding cassette transporter AtABCC2. *Scientific Reports*, *9*(1), 1–15. https://doi.org/10.1038/s41598-018-37504-8

Berhin, A., De, Bellis, D., Franke, R. B., Buono, R. A., Nowack, M. K., & Nawrath, C. (2019). The root cap cuticle: a cell wall structure for seedling establishment and lateral root formation. *Cell*, *176*(6), 1367–1378.e8. https://doi.org/10.1016/j.cell.2019.01.005

Biala, W., Banasiak, J., Jarzyniak, K., Pawela, A., & Jasinski, M. (2017). Medicago truncatula ABCG10 is a transporter of 4-coumarate and liquiritigenin in the medicarpin biosynthetic pathway. *Journal of Experimental Botany*, *68*(12), 3231–3241. https://doi.org/10.1093/jxb/erx059.

Bickers, S. C., Benlekbir, S., Rubinstein, J. L., & Kanelis, V. (2021). Structure of Ycf1p reveals the transmembrane domain TMD0 and the regulatory region of ABCC transporters. *Proceedings of the National Academy of Sciences of the United States of America*, *118*(21), e2025853118. https://doi.org/10.1073/pnas.2025853118.

Borghi, L., Kang, J., Ko, D., Lee, Y., & Martinoia, E. (2015). The role of ABCG-type ABC transporters in phytohormone transport. *Biochemical Society Transactions*, *43*(5), 924–30. https://doi.org/10.1042/BST20150106.

Boulc'h, P. N., Caullireau, E., Faucher, E., Gouerou, M., Guérin, A., Miray, R., & Couée, I. (2020)Abiotic stress signalling in extremophile land plants. *Journal of Experimental Botany*, *71*(19), 5771–5785. https://doi.org/10.1093/jxb/eraa336.

Brazier,-Hicks, M., Gershater, M., Dixon, D., Edwards, R. (2017). Substrate specificity and safener inducibility of the plant UDP-glucose-dependent family 1 glycosyltransferase super-family. *Plant Biotechnology Journal*, *16*(1), 337–348. https://doi.org/10.1111/pbi.12775

Brunetti, P., Zanella, L., De, Paolis, A., Di, Litta, D., Cecchetti, V., Falasca, G., Barbieri, M., Altamura, M. M., Costantino, P., & Cardarelli, M. (2015)Cadmium-inducible expression of the ABC-type transporter AtABCC3 increases phytochelatin-mediated cadmium tolerance in Arabidopsis. *Journal of Experimental Botany*, *66*(13), 3815–29. https://doi.org/10.1093/jxb/erv185.

Cecchetti, V., Brunetti, P., Napoli, N., Fattorini, L., Altamura, M. M., Costantino, P., Cardarelli, M. (2015). ABCB1 and ABCB19 auxin transporters have synergistic effects on early and late Arabidopsis anther development. *Journal of Integrative Plant Biology*, *57*(12), 1089–1098. https://doi.org/10.1111/jipb.12332

Chen, K., Li, G. J., Bressan, R. A., Song, C. P., Zhu, J. K., & Zhao, Y. (2020)Abscisic acid dynamics, signaling, and functions in plants. *Journal of Integrative Plant Biology*, *62*(1), 25–54. https://doi.org/10.1111/jipb.12899.

Chen, G., Komatsuda, T., Ma, J. F., Nawrath, C., Pourkheirandish, M., Tagiri, A., Hu, G., Sameri, M., Li, X., Zhao, X., Liu, Y., Li, C., Ma, X., Wang, A., Nair, S., Wang, N., Miyao, A., Sakuma, S., Yamaji, N., & Nevo, E. (2011). An ATP-binding cassette subfamily G full transporter is essential for the retention of leaf water in both wild barley and rice. *Proceedings of the National Academy of Sciences of the United States of America*, *108*(30), 12354–12359. https://doi.org/10.1073/pnas.1108444108

Cho, C. H., Jang, S., Choi, B. Y., Hong, D., Choi, D. S., Choi, S., Kim, H., Han, S. K., Kim, S., Kim, S., Palmgren, M., Sohn, K. H., Yoon, H. S., & Lee, Y. (2021). Phylogenetic analysis of ABCG subfamily proteins in plants: Functional clustering and coevolution with ABCGs of pathogens. *Physiologia Plantarum*, *172*(3), 1422–1438. https://doi.org/10.1111/ppl.13052

Choi, C. C., & Ford, R. C. (2021). ATP binding cassette importers in eukaryotic organisms. *Biological Reviews*, *96*(4), 1318–1330. https://doi.org/10.1111/brv.12702

Crouzet, J., Roland, J., Peeters, E., Trombik, T., Ducos, E., Nader, J., Boutry, M. (2013). NtPDR1, a plasma membrane ABC transporter from Nicotiana tabacum, is involved in diterpene transport. *Plant Molecular Biology*, *82*, 181–192

Dahuja, A., Kumar, R. R., Sakhare, A., Watts, A., Singh, B., Goswami, S., Sachdev, A., & Praveen, S. (2021) Role of ATP-binding cassette transporters in maintaining plant homeostasis under abiotic and biotic stresses. *Physiologia Plantarum*, *171*(4), 785–801. https://doi.org/10.1111/ppl.13302.

Demurtas, O. C., Diretto, G., Ferrante, P., Frusciante, S., Pietrella, M., Aprea, G., Borghi, L., Feeney, M., Frigerio, L., Coricello, A., Costa, G., Alcaro, S., Martinoia, E., & Giuliano, G. (2019). ABCC transporters mediate the vacuolar accumulation of crocins in saffron stigmas. *The Plant Cell*, *31*(11), 2789–2804. https://doi.org/10.1105/tpc.19.00193

Deslauriers, S. D., & Spalding, E. P. (2021). Electrophysiological study of Arabidopsis ABCB4 and PIN2 auxin transporters: Evidence of auxin activation and interaction enhancing auxin selectivity. *Plant Direct*, *5*(11). https://doi.org/10.1002/pld3.361

Dhara, A., & Raichaudhuri, A. (2021). ABCG transporter proteins with beneficial activity on plants. *Phytochemistry*, *184*, 112663. https://doi.org/10.1016/j.phytochem.2021.112663

DoH, T., Martinoia, E., & Lee, Y. (2018). Functions of ABC transporters in plant growth and development. *Current Opinion in Plant Biology*, *41*, 32–38. https://doi.org/10.1016/j.pbi.2017.08.003

Elejalde,-Palmett, C., Martinez, San, Segundo, I., Garroum, I., Charrier, L., De, Bellis, D., Mucciolo, A., Guerault, A., Liu, J., Zeisler,-Diehl, V., Aharoni, A., Schreiber, L., Bakan, B., Clausen, M. H., Geisler, M., & Nawrath, C. (2021). ABCG transporters export cutin precursors for the formation of the plant cuticle. *Curr Biology*, *31*(10), 2111–2123.e9. https://doi.org/10.1016/j.cub.2021.02.056

Feng, Y., Bayaer, E., & Qi, Y. (2022). Advances in the biological functions of auxin transporters in rice. *Agriculture*, *12*(7), 989. https://doi.org/10.3390/agriculture12070989

Ferro, M., Brugière S., Salvi, D., Seigneurin,-Berny, D., Court, M., Moyet, L., Ramus, C., Miras, S., Mellal, M., Le, Gall, S., Kieffer,-Jaquinod, S., Bruley, C., Garin, J., Joyard, J., Masselon, C., & Rolland, N. (2010). AT_CHLORO, a comprehensive chloroplast proteome database with subplastidial localization and curated information on envelope proteins. *Mol Cell Proteomics*. *9*(6), 1063–84. https://doi.org/10.1074/mcp.M900325-MCP200

Fu, X., Shi, P., He, Q., Shen, Q., Tang, Y., Pan, Q., Ma, Y., Yan, T., Chen, M., Hao, X., Liu, P., Li, L., Wang, Y., Sun, X., & Tang, K. (2017). AaPDR3, a PDR transporter 3, is involved in sesquiterpene β-caryophyllene transport in artemisia annua. *Frontiers in Plant Science*, *8*, 723. https://doi.org/10.3389/fpls.2017.00723

Gaillard, S., Jacquet, H., Vavasseur, A., Leonhardt, N., & Forestier, C. (2008). AtMRP6/AtABCC6, an ATP-Binding Cassette transporter gene expressed during early steps of seedling development and up-regulated by cadmium in *Arabidopsis thaliana*. *BMC Plant Biology*, *8*, 22. https://doi.org/10.1186/1471-2229-8-22

Gautam, N., Tiwari, M., Kidwai, M., Dutta, P., & Chakrabarty, D. (2023). Functional characterization of rice metallothionein OsMT-I-Id: Insights into metal binding and heavy metal tolerance mechanisms. *Journal of Hazardous Materials*, *458*, 131815. https://doi.org/10.1016/j.jhazmat.2023.131815

Geisler, M., Girin, M., Brandt, S., Vincenzetti, V., Plaza, S., Paris, N., Kobae, Y., Maeshima, M., Billion, K., Kolukisaoglu, Ü. H., Schulz, B., & Martinoia, E. (2004). Arabidopsis immunophilin-like TWD1 functionally interacts with vacuolar ABC transporters. *Molecular Biology of the Cell*, *15*(7), 3393–3405. https://doi.org/10.1091/mbc.E03-11-0831

Ginovyan, M., Babayan, A., Shirvanyan, A., Minasyan, A., Qocharyan, M., Kusznierewicz, B., Koss,-Mikołajczyk I., Avtandilyan, N., Vejux, A., Bartoszek, A., & Sahakyan, N. (2023)The action mechanisms, anti-cancer and antibiotic-modulation potential of Vaccinium myrtillus L. extract. *Discovery Medicine*, *35*, 590–611. https://doi.org/10.24976/Discov.Med.202335177.59

Hao, P., Xia, J., Liu, J., Donato, M. D., Pakula, K., Bailly, A., Jasinski, M., & Geisler, M. (2020). Auxin-transporting ABC transporters are defined by a conserved D/E-P motif regulated by a prolylisomerase. *The Journal of Biological Chemistry*, *295*(37), 13094–13105. https://doi.org/10.1074/jbc.RA120.014104

Hasanuzzaman, M., Nahar, K., Alam, M. M., Roychowdhury, R., & Fujita, M. (2013). Physiological, biochemical, and molecular mechanisms of heat stress tolerance in plants. *International Journal of Molecular Sciences*, *14*(5), 9643–84. https://doi.org/10.3390/ijms14059643.

Hasanuzzaman, M., Nahar, K., Anee, T. I., & Fujita, M. (2017). Glutathione in plants: Biosynthesis and physiological role in environmental stress tolerance. *Physiology and Molecular Biology of Plants*, *23*(2), 249–268. https://doi.org/10.1007/s12298-017-0422-2

Hayashi, S., Kuramata, M., Abe, T., Takagi, H., Ozawa, K., & Ishikawa, S. (2017). Phytochelatin synthase OsPCS1 plays a crucial role in reducing arsenic levels in rice grains. *Plant Journal*, *91*(5), 840–848. https://doi.org/10.1111/tpj.13612.

Hellsberg, E., Montanari, F., & Ecker, G. F. (2015). The ABC of phytohormone translocation. *Planta Medica*, *81*(6), 474. https://doi.org/10.1055/s-0035-1545880

Hu, B., Deng, F., Chen, G., Chen, X., Gao, W., Long, L., Xia, J., & Chen, H. (2020). Evolution of Abscisic Acid Signaling for Stress Responses to Toxic Metals and Metalloids. *Frontiers in Plant Science*, *11*. https://doi.org/10.3389/fpls.2020.00909

Huang, C. F., Yamaji, N., Chen, Z., & Ma, J. F. (2012). A tonoplast-localized half-size ABC transporter is required for internal detoxification of aluminum in rice. *Plant Journal*, *69*(5), 857–67. https://doi.org/10.1111/j.1365-313X.2011.04837.x

Hwang, J., Song, W., Hong, D., Ko, D., Yamaoka, Y., Jang, S., Yim, S., Lee, E., Khare, D., Kim, K., Palmgren, M., Yoon, H. S., Martinoia, E., & Lee, Y. (2016). Plant ABC transporters enable many unique aspects of a terrestrial plant's lifestyle. *Molecular Plant*, *9*(3), 338–355. https://doi.org/10.1016/j.molp.2016.02.003

Ingram, G., & Nawrath, C. (2017). The roles of the cuticle in plant development: Organ adhesions and beyond. *Journal of Experimental Botany*, *68*(19), 5307–5321. https://doi.org/10.1093/jxb/erx313

Jarzyniak, K., Banasiak, J., Jamruszka, T., Pawela, A., Di, Donato, M., Novák O., Geisler, M., & Jasiński, M. (2021). Early stages of legume–rhizobia symbiosis are controlled by ABCG-mediated transport of active cytokinins. *Nature Plants*, *7*(4), 428–436. https://doi.org/10.1038/s41477-021-00873-6

Kang, J., Yim, S., Choi, H., Kim, A., Lee, K. P., Lopez,-Molina, L., Martinoia, E., & Lee, Y. (2015). Abscisic acid transporters cooperate to control seed germination. *Nat Communication*. *6*, 8113. https://doi.org/10.1038/ncomms9113.

Kang, J., Park, J., Choi, H., Burla, B., Kretzschmar, T., Lee, Y., & Martinoia, E. (2011). Plant ABC transporters: The Arabidopsis book. *American Society of Plant Biologists*, 9. https://doi.org/10.1199/tab.0153

Khare, D., Choi, H., Huh, S. U., Bassin, B., Kim, J., Martinoia, E., Sohn, K. H., Paek, K., & Lee, Y. (2017). Arabidopsis ABCG34 contributes to defense against necrotrophic pathogens by mediating the secretion of camalexin. *Proceedings of the National Academy of Sciences*, *114*(28), E5712–E5720. https://doi.org/10.1073/pnas.1702259114

Kim, A., Chen, J., Khare, D., Jin, Y., Maeshima, M., Zhao, Y., Martinoia, E., Hwang, U., & Lee, Y. (2020). Non-intrinsic ATP binding cassette proteins ABCI19, ABCI20 and ABCI21 modulate cytokinin response at endoplasmic reticulum in Arabidopsis thaliana. *Plant Cell Reports*, *39*(4), 473. https://doi.org/10.1007/s00299-019-02503-0

Kim, L. J., Tsuyuki, K. M., Hu, F., Park, E. Y., Zhang, J., Iraheta, J. G., Chia, C., Huang, R., Tucker, A. E., Clyne, M., Castellano, C., Kim, A., Chung, D. D., DaVeiga, C. T., Parsons, E. M., Vatamaniuk, O. K., & Jeong, J. (2021). Ferroportin 3 is a dual-targeted mitochondrial/chloroplast iron exporter necessary for iron homeostasis in Arabidopsis. *The Plant Journal*, *107*(1), 215–236. https://doi.org/10.1111/tpj.15286

Kim, Y., Jin, Y., Alejandro, S., Martinoia, E., & Lee, Y. (2010). Overexpression of AtABCG36 improves drought and salt stress resistance in Arabidopsis. *Physiologia Plantarum*, *139*(2), 170–180. https://doi.org/10.1111/j.1399-3054.2010.01353.x

Klein, M., Burla, B., & Martinoia, E. (2006). The multidrug resistance-associated protein (MRP/ABCC) subfamily of ATP-binding cassette transporters in plants. *FEBS Letters*, *580*(4), 1112–1122. https://doi.org/10.1016/j.febslet.2005.11.056

Ko, D., Kang, J., Kiba, T., Park, J., Kojima, M., Do, J., Kim, K. Y., Kwon, M., Endler, A., Song, W., Martinoia, E., Sakakibara, H., & Lee, Y. (2014). Arabidopsis ABCG14 is essential for the root-to-shoot translocation of cytokinin. *Proceedings of the National Academy of Sciences*, *111*(19), 7150–7155. https://doi.org/10.1073/pnas.1321519111

Korver, R. A., Koevoets, I. T., & Testerink, C. (2018). Out of shape during stress: a key role for auxin. *Trends in Plant Science*, *23*(9), 783–793. https://doi.org/10.1016/j.tplants.2018.05.011

Kretzschmar, T., Burla, B., Lee, Y., Martinoia, E., & Nagy, R. (2011). Functions of ABC transporters in plants. *Essays Biochemistry*, *50*(1), 145–60. https://doi.org/10.1042/bse0500145.

Kruse, I., Maclean, A. E., Hill, L., & Balk, J. (2017). Genetic dissection of cyclic pyranopterin monophosphate biosynthesis in plant mitochondria. *Biochemical Journal*, *475*(2), 495–509. https://doi.org/10.1042/BCJ20170559

Kuromori, T., Miyaji, T., Yabuuchi, H., Shimizu, H., Sugimoto, E., Kamiya, A., Moriyama, Y., & Shinozaki, K. (2010). ABC transporter AtABCG25 is involved in abscisic acid transport and responses. *Proceedings of the National Academy of Sciences of the United States of America*, *107*(5), 2361–6. https://doi.org/10.1073/pnas.0912516107.

Lane, T. .S., Rempe, C. .S., Davitt J. et al. (2016). Diversity of ABC transporter genes across the plant kingdom and their potential utility in biotechnology. *BMC Biotechnology*, *16*, 47 . https://doi.org/10.1186/s12896-016-0277-6

Lee, M., Choi, Y., Burla, B., Kim, Y. Y., Jeon, B., Maeshima, M., Yoo, J. Y., Martinoia, E., & Lee, Y. (2008). The ABC transporter AtABCB14 is a malate importer and modulates stomatal response to CO_2. *Nat Cell Biology*, *10*(10), 1217–23. https://doi.org/10.1038/ncb1782.

Lee, M,, Lee, K., Lee, J., Noh, E. W., & Lee, Y. (2005)AtPDR12 contributes to lead resistance in Arabidopsis. *Plant Physiology*, *138*, 827–836.

Lefèvre, F., & Boutry, M. (2018). Towards identification of the substrates of ATP-binding cassette transporters. *Plant Physiology*, *178*(1), 18–39. https://doi.org/10.1104/pp.18.00325

Lewinson, O., & Livnat,-Levanon, N. (2017). Mechanism of action of ABC importers: conservation, divergence, and physiological adaptations. *Journal of Molecular Biology*, *429*(5), 606–619. https://doi.org/10.1016/j.jmb.2017.01.010

Lin, T, Chen, J, Zhou, Sh,, Wenhui, Yu,, Chen, G, Chen, L, Wang, X, Shi, H., Han, Sh,,& Zhang, F. (2020), Testing the elemental defense hypothesis with a woody plant species: Cadmium accumulation protects Populus yunnanensis from leaf herbivory and pathogen infection, *Chemosphere*, *247*, 125851. https://doi.org/10.1016/j.chemosphere.2020.125851.

Liu, C. W., & Murray, J. D. (2016). The role of flavonoids in nodulation host-range specificity: an update. *Plants (Basel)*, *5*(3), 33. https://doi.org/10.3390/plants5030033.

Ma, Y., Cao, J., He, J., Chen, Q., Li, X., & Yang, Y. (2018). Molecular mechanism for the regulation of aba homeostasis during plant development and stress responses. *International Journal of Molecular Sciences*, *19*(11). https://doi.org/10.3390/ijms19113643

Mahizan, N. A., Yang, K., Moo, L., Ai,-Lian, Song, A., Chong, M., Chong, W., Abushelaibi, A., Lim, H. E., & Lai, S. (2019). Terpene derivatives as a potential agent against antimicrobial resistance (AMR) pathogens. *Molecules*, *24*(14). https://doi.org/10.3390/molecules24142631

Martinoia, E., Grill, E., Tommasini, R., Kreuz, K., & Amrhein, N. (1993). ATP-dependent glutathione S-conjugate 'export' pump in the vacuolar membrane of plants. *Nature*, *364*(6434), 247–249. https://doi.org/10.1038/364247a0

Matern, A., Böttcher C., Eschen,-Lippold, L., Westermann, B., Smolka, U., Döll S., Trempel, F., Aryal, B., Scheel, D., Geisler, M., & Rosahl, S. (2019). A substrate of the ABC transporter PEN3 stimulates bacterial flagellin (flg22)-induced callose deposition in Arabidopsis thaliana. *Journal of Biological Chemistry*, *294*(17), 6857–6870. https://doi.org/10.1074/jbc.RA119.007676

Mellor, N. L., Voß, U., Ware, A., Janes, G., Barrack, D., Bishopp, A., Bennett, M. J., Geisler, M., Wells, D. M., Band, L. R. (2022). Systems approaches reveal that ABCB and PIN proteins mediate co-dependent auxin efflux. *Plant Cell*. *34*(6), 2309–2327. https://doi.org/10.1093/plcell/koac086.

Mellor, N. L., Voß, U., Ware, A., Janes, G., Barrack, D., Bishopp, A., Bennett, M. J., Geisler, M., Wells, D. M., & Band, L. R. (2022). Systems approaches reveal that ABCB and PIN proteins mediate co-dependent auxin efflux. *The Plant Cell*, *34*(6), 2309–2327. https://doi.org/10.1093/plcell/koac086

Moon, S., Chandran, A. K. N., Gho, Y. S., Park, S. A., Kim, S. R., Yoo, Y. H., & Jung, K. H. (2018). Integrated omics analysis of root-preferred genes across diverse rice varieties including Japonica and indica cultivars. *J Plant Physiology*, *220*, 11–23. https://doi.org/10.1016/j.jplph.2017.10.003.

Mõttus J., Maiste, S., Eek, P., Truve, E., & Sarmiento, C. (2020). Mutational analysis of Arabidopsis thaliana ABCE2 identifies important motifs for its RNA silencing suppressor function. *Plant Biology (Stuttgart, Germany)*, *23*(1), 21–31. https://doi.org/10.1111/plb.13193

Moyroud, E., & Glover, B. J. (2017). The physics of pollinator attraction. *New Phytologist*, *216*(2), 350–354. https://doi.org/10.1111/nph.14312

Naoumkina, M., Farag, M. A., Sumner, L. W., Tang, Y., Liu, C., & Dixon, R. A. (2007). Different mechanisms for phytoalexin induction by pathogen and wound signals in Medicago truncatula. *Proceedings of the National Academy of Sciences*, *104*(46), 17909–17915. https://doi.org/10.1073/pnas.0708697104

Navarro,-Quiles, C., Mateo,-Bonmatí, E., & Micol, J. L. (2018). ABCE proteins: from molecules to development. *Frontiers in Plant Science*, *9*. https://doi.org/10.3389/fpls.2018.01125

Nawaz, M., Sun, J., Shabbir, S., Khattak, W. A,, Ren, G., Nie, X., Bo, Y., Javed, Q., Du, D., & Sonne, Ch. (2023). A review of plants strategies to resist biotic and abiotic environmental stressors, *Science of The Total Environment*, *900*, 165832, https://doi.org/10.1016/j.scitotenv.2023.165832.

Nguyen, C. H., Yan, D., & Nambara, E. (2023). Persistence of abscisic acid analogs in plants: chemical control of plant growth and physiology. *Genes (Basel)*, *14*(5), 1078. https://doi.org/10.3390/genes14051078.

Nikaido, H. (2002). How are the ABC transporters energized? *Proceedings of the National Academy of Sciences of the United States of America*, *99*(15), 9609–9610. https://doi.org/10.1073/pnas.162375699

Nogia, P, & Pati, P. K. (2021). Plant secondary metabolite transporters: diversity, functionality, and their modulation. *Frontiers in Plant Science*, *12*. https://doi.org/10.3389/fpls.2021.758202

Ofori, P. A., Mizuno, A., Suzuki, M., Martinoia, E., Reuscher, S., & Aoki, K. (2018). Genome-wide analysis of ATP-binding cassette (ABC) transporters in tomato. *PLoS One*, *13*(7), e0200854. https://doi.org/10.1371/journal.pone.0200854

Ofori, P. A., Geisler, M., Di, Donato, M., Pengchao, H., Otagaki, S., Matsumoto, S., & Shiratake, K. (2018). Tomato ATP-binding cassette transporter slabcb4 is involved in auxin transport in the developing fruit. *Plants*, *7*(3), 65. https://doi.org/10.3390/plants7030065

Park, J., Song, W. Y., Ko, D., Eom, Y., Hansen, T. H., Schiller, M., Lee, T. G., Martinoia, E., & Lee, Y. (2012). The phytochelatin transporters AtABCC1 and AtABCC2 mediate tolerance to cadmium and mercury. *Plant Journal*, *69*(2), 278–88. https://doi.org/10.1111/j.1365-313X.2011.04789.x.

Park, J., Lee, Y., Martinoia, E, & Geisler, M. Plant hormone transporters: what we know and what we would like to know. *BMC Biology*, *15*, 93 (2017). https://doi.org/10.1186/s12915-017-0443-x

Pasello, M., Giudice, A. M., & Scotlandi, K. (2020). The ABC subfamily A transporters: Multifaceted players with incipient potentialities in cancer. *Seminars in Cancer Biology*, *60*, 57–71. https://doi.org/10.1016/j.semcancer.2019.10.004

Pearson, S. A., & Cowan, J. (2021). Evolution of the human mitochondrial ABCB7 [2Fe–2S](GS)4 cluster exporter and the molecular mechanism of an E433K disease-causing mutation. *Archives of Biochemistry and Biophysics*, *697*, 108661. https://doi.org/10.1016/j.abb.2020.108661

Petrosyan, M., Shcherbakova, Y., Sahakyan, N., Vardanyan Z., Poladyan A., Popov Yu., Trchounian A. (2015). *Alkanna orientalis* (L.) Boiss. plant isolated cultures and antimicrobial activity of their extracts: phenomenon, dependence on different factors and effects on some membrane-associated properties of bacteria. *Plant Cell Tissue Organic Culture*, *122*, 727–738. https://doi.org/10.1007/s11240-015-0806-3.

Pierman, B., Toussaint, F., Bertin, A., Lévy D., Smargiasso, N., De, Pauw, E., & Boutry, M. (2017). Activity of the purified plant ABC transporter NtPDR1 is stimulated by diterpenes and sesquiterpenes involved in constitutive and induced defenses. *Journal of Biological Chemistry*, *292*(47), 19491–19502. https://doi.org/10.1074/jbc.M117.811935.

Poghosyan, E., Sahakyan, N., Petrosyan, M., Batlutskaya, I., & Trchounian, K. (2021). Isolated culture of *Areptance* L. its' morphological and growth features. *BIO Web Conference*, *40*, 01001.

Qin, M., Li, X., Tang, S., Huang, Y., Li, L., & Hu, B. (2021). Expression of AhATL1, an ABA transport factor gene from peanut, is affected by altered memory gene expression patterns and increased tolerance to drought stress in arabidopsis. *International Journal of Molecular Sciences*, *22*(7). https://doi.org/10.3390/ijms22073398

Raichaudhuri, A. (2016). Arabidopsis thaliana MRP1 (AtABCC1) nucleotide binding domain contributes to arsenic stress tolerance with serine triad phosphorylation. *Plant Physiology and Biochemistry*, *108*, 109–120. https://doi.org/10.1016/j.plaphy.2016.07.005

Remy, E., & Duque, P. (2014). Beyond cellular detoxification: A plethora of physiological roles for MDR transporter homologs in plants. *Frontiers in Physiology*, *5*. https://doi.org/10.3389/fphys.2014.00201

Rin, S., Mizuno, Y., Shibata, Y., Fushimi, M., Katou, S., Sato, I., Chiba, S., Kawakita, K., & Takemoto, D. (2017). EIN2-mediated signaling is involved in pre-invasion defense in Nicotiana benthamiana against potato late blight pathogen, Phytophthora infestans. *Plant Signaling & Behavior*, *12*(4). https://doi.org/10.1080/15592324.2017.1300733

Sahakyan, N. Z. Lamiaceae family plants: one of the potentially richest sources of antimicrobials. *Pharmaceutical Chemistry Journal*, *57*, 565–572 (2023). https://doi.org/10.1007/s11094-023-02921-1

Sasse, J., Schlegel, M., Borghi, L., Ullrich, F., Lee, M., Liu, G. W., Giner, J. L., Kayser, O., Bigler, L., Martinoia, E., Kretzschmar T. (2016). Petunia hybrida PDR2 is involved in herbivore defense by controlling steroidal contents in trichomes. *Plant Cell Environment*, *39*, 2725–2739.

Schaedler, T. A., Thornton, J. D., Kruse, I., Schwarzländer, M., Meyer, A. J., Van, Veen, H. W., & Balk, J. (2014). A conserved mitochondrial atp-binding cassette transporter exports glutathione polysulfide for cytosolic metal cofactor assembly. *Journal of Biological Chemistry*, *289*(34), 23264–23274. https://doi.org/10.1074/jbc.M114.553438

Schiessl, K., Lilley, L. S., Lee, T., Tamvakis, I., Kohlen, W., Bailey, P. C., Thomas, A., Luptak, J., Ramakrishnan, K., Carpenter, M. D., Mysore, K. S., Wen, J., Ahnert, S., Grieneisen, V. A., & Oldroyd, E. D. (2019). Nodule inception recruits the lateral root developmental program for symbiotic nodule organogenesis in *Medicago truncatula*. *Current Biology*, *29*(21), 3657–3668. https://doi.org/10.1016/j.cub.2019.09.005

Schützendübel, A., & Polle, A. (2002). Plant responses to abiotic stresses: Heavy metal-induced oxidative stress and protection by mycorrhization. *Journal of Experimental Botany*, *53*(372), 1351–1365. https://doi.org/10.1093/jexbot/53.372.1351

Seregin, I. V., & Kozhevnikova, A. D. (2022). Phytochelatins: sulfur-containing metal(loid)-chelating ligands in plants. *International Journal of Molecular Sciences*, *24*(3), 2430. https://doi.org/10.3390/ijms24032430

Shen, C. J., Yue, R. Q., Bai, Y. H., Feng, R., Sun, T., Wang, X. F., Yang, Y. J., Tie, S. G., & Wang, H. Z. (2015). Identification and analysis of Medicago truncatula auxin transporter gene families uncover their roles in responses to Sinorhizobium meliloti Infection. *Plant and Cell Physiology*, *56*, 1930–1943.

Shibata, Y., Ojika, M., Sugiyama, A., Yazaki, K., Jones, D. A., Kawakita, K., & Takemoto, D. (2016). The full-size ABCG transporters Nb-ABCG1 and Nb-ABCG2 function in pre- and postinvasion defense against phytophthora infestans in nicotiana benthamiana. *The Plant Cell*, *28*(5), 1163–1181. https://doi.org/10.1105/tpc.15.00721

Shitan, N., Dalmas, F., Dan, K., Kato, N., Ueda, K., Sato, F., Forestier, C., Yazaki, K. (2013). Characterization of Coptis japonica CjABCB2, an ATP-binding cassette protein involved in alkaloid transport. *Phytochemistry*, *91*, 109–116.

Singh, S., Parihar, P., Singh, R., Singh, V. P., & Prasad, S. M. (2016). Heavy metal tolerance in plants: role of transcriptomics, proteomics, metabolomics, and ionomics. *Frontiers in Plant Science*, *6*, 1143. https://doi.org/10.3389/fpls.2015.01143

Sinha, R., Fritschi, F., Zandalinas, S., & Mittler, R. (2021). The impact of stress combination on reproductive processes in crops, *Plant Science*, *311*, 111007. https://doi.org/10.1016/j.plantsci.2021.111007.

Solovchenko, A., & Merzlyak, M. (2003). Optical properties and contribution of cuticle to UV protection in plants: experiments with apple fruit. *Photochemical & Photobiological Sciences*, *2*(8), 861–6. https://doi.org/10.1039/b302478d.

Song, W. Y., Yamaki, T., Yamaji, N., Ko, D., Jung, K. H., Fujii,-Kashino, M., An, G., Martinoia, E., Lee, Y., & Ma, J. F. (2014). A rice ABC transporter, OsABCC1, reduces arsenic accumulation in the grain. *Proceedings of the National Academy of Sciences of the United States of America*, *111*(44), 15699–704. https://doi.org/10.1073/pnas.1414968111.

Song,, W.-Y., Park, J., Eisenach, C., Maeshima,, M., Lee, Y., Martinoia,, E. (2014). ABC transporters and heavy metals. In M. Geisler (Ed.), *Plant ABC transporters* (pp. 1–17). Springer International Publishing Switzerland.

Stukkens, Y., Bultreys, A., Grec, S., Trombik, T., Vanham, D., & Boutry, M. (2005). NpPDR1, a pleiotropic drug resistance-type ATP-binding cassette transporter from nicotiana plumbaginifolia, plays a major role in plant pathogen defense. *Plant Physiology*, *139*(1), 341–352. https://doi.org/10.1104/pp.105.062372

Sugiyama, A., Shitan, N., & Yazaki, K. (2007). Involvement of a soybean ATP-binding cassette-type transporter in the secretion of genistein, a signal flavonoid in legume-Rhizobium symbiosis. *Plant Physiology*, *144*(4), 2000–8. https://doi.org/10.1104/pp.107.096727.

Sun, L., Ma, Y., Wang, H., Huang, W., Wang, X., Han, L., Sun, W., Han, E., & Wang, B. (2018). Overexpression of PtABCC1 contributes to mercury tolerance and accumulation in Arabidopsis and poplar. *Biochemical and Biophysical Research Communications*, *497*(4), 997–1002. https://doi.org/10.1016/j.bbrc.2018.02.133

Suzaki, T., Ito, M., & Kawaguchi, M. (2013). Genetic basis of cytokinin and auxin functions during root nodule development. *Frontiers in Plant Science*, *4*, 42. https://doi.org/10.3389/fpls.2013.00042

Sytar, O., Kumari, P., Yadav, S., Brestic, M., & Rastogi, A. (2019). Phytohormone priming: regulator for heavy metal stress in plants. *Journal of Plant Growth Regulator*, *38*, 739–752. https://doi.org/10.1007/s00344-018-9886–8

Takanashi, K., Takahashi, H., Sakurai, N., Sugiyama, A., Suzuki, H., Shibata, D., Nakazono, M., & Yazaki, K. (2012). Tissue-specific transcriptome analysis in nodules of Lotus japonicus. *Molecular Plant-Microbe Interactions*, *25*, 869–876. https://doi.org/10.1094/MPMI-01-12-0011-R

Tan, X., Li, K., Wang, Z., Zhu, K., Tan, X. & Cao, J. (2019). A review of plant vacuoles: formation, located proteins, and functions. *Plants*, *8*, 327.

Thi, Do, T. H., Martinoia, E., Lee, Y., & Hwang, U. (2021). 2021 update on ATP-binding cassette (ABC) transporters: How they meet the needs of plants. *Plant Physiology*, *187*(4), 1876–1892. https://doi.org/10.1093/plphys/kiab193

Thomas, C., Aller, S. G., Beis, K., Carpenter, E. P., Chang, G., Chen, L., Dassa, E., Dean, M., Duong, Van, Hoa, F., Ekiert, D., Ford, R., Gaudet, R., Gong, X., Holland, I. B., Huang, Y., Kahne, D. K., Kato, H., Koronakis, V., Koth, C. M., Lee, Y., Lewinson, O., Lill, R., Martinoia, E., Murakami, S., Pinkett, H. W., Poolman, B., Rosenbaum, D., Sarkadi, B., Schmitt, L., Schneider, E., Shi, Y., Shyng, S. L., Slotboom, D. J., Tajkhorshid, E., Tieleman, D. P., Ueda, K., Váradi A., Wen, P. C., Yan, N., Zhang, P., Zheng, H., Zimmer, J., & Tampé, R. (2020). Structural and functional diversity calls for a new classification of ABC transporters. *FEBS Letter*, *594*(23), 3767–3775. https://doi.org/10.1002/1873-3468.13935

Thomas, C., & Tampé, R. (2018). Multifaceted structures and mechanisms of ABC transport systems in health and disease. *Current Opinion in Structural Biology*, *51*, 116–128. https://doi.org/10.1016/j.sbi.2018.03.016

Tominaga, J., & Kawamitsu, Y. (2015). Cuticle affects calculations of internal co 2 in leaves closing their stomata. *Plant and Cell Physiology*, *56*(10), 1900–1908. https://doi.org/10.1093/pcp/pcv109

Trchounian, A., Petrosyan, M., Sahakyan, N. (2016). Plant cell redox homeostasis and reactive oxygen species. In D. Gupta, J. Palma, & F. Corpas (Eds.), *Redox state as a central regulator of plant-cell stress responses*. Springer. https://doi.org/10.1007/978-3-319-44081-1_2

Tusnády, G. E., Sarkadi, B., Simon, I., & Váradi A. (2006). Membrane topology of human ABC proteins. *FEBS Letters*, *580*(4), 1017–1022. https://doi.org/10.1016/j.febslet.2005.11.040

Vráblová, M., Marková, D., Vrábl, D., Koutník, I., Sokolová, B., & Hronková, M. (2021). Surface plasmon resonance: An innovative method for studying water permeability of plant cuticles. *Plant Science*, *310*, 110978. https://doi.org/10.1016/j.plantsci.2021.110978

Waadt, R., Seller, C. A., Hsu, P. K., Takahashi, Y., Munemasa, S., & Schroeder, J. I. (2022). Plant hormone regulation of abiotic stress responses. *Nature Reviews Molecular Cell Biology*, *23*(10), 680–694. https://doi.org/10.1038/s41580-022-00479-6. Erratum in: Nat Rev Mol Cell Biol.

Wang, L., Yao, W., Zhang, X., Tang, Y., Van, Nocker, S., Wang, Y., & Zhang, C. (2023). The putative ABCG transporter VviABCG20 from grapevine (Vitis vinifera) is strongly expressed in the seed coat of developing seeds and may participate in suberin biosynthesis. Physiol Mol Biol Plants. *29*(1), 23–34. https://doi.org/10.1007/s12298-022-01276-3.

Wang, H., Liu, Y., Peng, Z., Li, J., Huang, W., Liu, Y., Wang, X., Xie, S., Sun, L., Han, E., Wu, N., Luo, K., & Wang, B. (2019). Ectopic Expression of Poplar ABC Transporter PtoABCG36 Confers Cd Tolerance in Arabidopsis thaliana. *International Journal of Molecular Sciences*, *20*(13). https://doi.org/10.3390/ijms20133293

Wang, Z., Ren, Z., Cheng, C., Wang, T., Ji, H., Zhao, Y., Deng, Z., Zhi, L., Lu, J., Wu, X., Xu, S., Cao, M., Zhao, H., Liu, L., Zhu, J., & Li, X. (2020). Counteraction of ABA-mediated inhibition of seed germination and seedling establishment by ABA signaling terminator in arabidopsis. *Molecular Plant*, *13*(9), 1284–1297. https://doi.org/10.1016/j.molp.2020.06.011

Wen, P. C., & Tajkhorshid, E. (2011). Conformational coupling of the nucleotide-binding and the transmembrane domains in ABC transporters. *Biophys Journal*, *101*(3), 680–90. https://doi.org/10.1016/j.bpj.2011.06.031.

Wilkens, S. (2015). Structure and mechanism of ABC transporters. *F1000Prime Reports*, *7*. https://doi.org/10.12703/P7-14

Wilkinson, S, & Davies, W. J. (2002). ABA-based chemical signalling: the co-ordination of responses to stress in plants. *Plant, Cell and Environment*, *25*, 195–210

Xiong, J., Feng, J., Yuan, D., Zhou, J., Miao, W. (2015). Tracing the structural evolution of eukaryotic ATP binding cassette transporter superfamily. *Scientific Reports*, *5*(1), 1–15. https://doi.org/10.1038/srep16724

Xu, Y., Zhang, S., Guo, H., Wang, S., Xu, L., Li, C., Qian, Q., Chen, F., Geisler, M., Qi, Y., & de Jiang, A. (2014). OsABCB14 functions in auxin transport and iron homeostasis in rice (Oryza sativa L.). *Plant Journal*, *79*(1), 106–17. https://doi.org/10.1111/tpj.12544.

Yang, G., Fu, S., Huang, J., Li, L., Long, Y., Wei, Q., Wang, Z., Chen, Z., & Xia, J. (2021). The tonoplast-localized transporter OsABCC9 is involved in cadmium tolerance and accumulation in rice. *Plant Science*, *307*, 110894. https://doi.org/10.1016/j.plantsci.2021.110894

Yang, X., Jia, Z., Pu, Q., Tian, Y., Zhu, F., & Liu, Y. (2022). ABA mediates plant development and abiotic stress via alternative splicing. *International Journal of Molecular Sciences*, *23*(7). https://doi.org/10.3390/ijms23073796

Zhang, L., Sasaki-Sekimoto, Y., Kosetsu, K., Aoyama, T., Murata, T., Kabeya, Y., Sato, Y., Koshimizu, S., Shimojima, M., Ohta, H., Hasebe, M., & Ishikawa, M. (2023). An ABCB transporter regulates anisotropic cell expansion via cuticle deposition in the moss Physcomitrium patens. *New Phytologist*, *241*(2), 665–675. https://doi.org/10.1111/nph.19337

Zhang, Y., Xu, J., Li, R., Ge, Y., Li, Y., & Li, R. (2023). Plants' Response to Abiotic Stress: Mechanisms and Strategies. *International Journal of Molecular Sciences*, *24*(13), 10915. https://doi.org/10.3390/ijms241310915.

Zhao, Y., Gao, J., Im, Kim, J., Chen, K., Bressan, R. A., & Zhu, J. K. (2017). Control of plant water use by ABA induction of senescence and dormancy: an overlooked lesson from evolution. *Plant Cell Physiology*, *58*(8), 1319–1327. https://doi.org/10.1093/pcp/pcx086.

Zolnerciks, J. K., Andress, E. J., Nicolaou, M., Linton, & K. J. (2011). Structure of ABC transporters. *Essays Biochemistry*, *50*(1), 43–61. https://doi.org/10.1042/bse0500043.

Zuo, J., Wu, Z., Li, Y., Shen, Z., Feng, X., Zhang, M., & Ye, H. (2017). Mitochondrial ABC transporter ATM3 is essential for cytosolic iron-sulfur cluster assembly. *Plant Physiology*, *173*(4), 2096–2109. https://doi.org/10.1104/pp.16.01760.

10 Plant Genome Editing for Pest Management
Current Achievements and Future Directions

Prodipto Bishnu Angon, Md. Shafiul Islam, Sharah Jabeen Mou, Arpita Rani Roy, Md. Inzamam-Ul-Haque, Akivha Akter Sweety, and Monisha Mondol

10.1 INTRODUCTION

The survival and advancement of society are based on agriculture. The goal of agriculture in the twenty-first century is sustainable agriculture (Hafeez et al., 2023), which aims to provide food security while reducing environmental impact. A major issue that needs to be addressed in the context of increasing population growth and climatic variability is increasing crop productivity (Kogo et al., 2021). Several biotic and abiotic factors are a constant source of difficulty for agriculture in the twenty-first century. The primary source of crop loss among biotic causes is insects (Pratap et al., 2020). Insects are the reason for the reduced production and the poorer productivity. Around the world, despite management efforts, insects cause significant yield losses in commercially significant crops. However, farmers have resorted to chemical pesticides, whose widespread use may be hazardous to both humans and the environment (Zaller and Zaller, 2020), in an attempt to control pests. These problems have inspired researchers all across the world to develop innovative, environmentally responsible pest management strategies. Thus, the primary objective of modern agriculture has been to increase yields on the land and resources currently at hand to guarantee agricultural sustainability and global food security (Tyagi et al., 2020; Bisht et al., 2019). The primary objective of breeders is to create pest-resistant cultivars since insect pests, in addition to other biotic stresses like weeds, are thought to diminish agricultural productivity globally by 28 percent annually (Bhat and Kumar, 2022).

A genetic modification technique known as 'gene editing', or 'genome editing', entails adding, removing, labeling, altering, or replacing DNA inside a living organism's genome to produce a desired trait (Maeder et al., 2016). The effective and ecologically friendly technique of developing pest-resistant crop varieties might lead to sustainable agriculture. Despite its success in recent decades, conventional resistance breeding is not without its limits. Initially, it can only be done between plants that are able to reproduce. Second, it is heavily reliant on populations of plants with adequate genetic diversity. Third, in addition to the resistance trait of interest, it frequently introduces a variety of characteristics, some of which have unfavorable consequences for yield. Fourth, segregant progeny selection and genetic crossings often require a lot of work and time. Thus, it is difficult for traditional breeding to keep up with the ever-changing weeds and insects as well as the rising need for food, especially in light of the current global climate change.

The recent developments in genetic engineering and plant biotechnology have allowed pest management approaches to advance and accomplish several firsts (Giudice et al., 2021). Numerous molecular methods for genetic pathways have been documented, and these investigations have

DOI: 10.1201/9781003543237-10

been replicated to carry out in vitro tests. With the use of several instruments, such as vectors for transporting DNA into cells, scientists have been able to develop novel methods for altering DNA through extensive research into the molecular genetics and biochemistry of organisms like bacteria and viruses. These methods enable the genetic alteration of higher species, such as plants, in addition to transgenic bacteria.

Oligonucleotide-directed mutagenesis (ODM), zinc finger nucleases (ZFNs), transcription activator-like effector nucleases (TALENs), and clustered regularly interspaced short palindromic repeats (CRISPR)/CRISPR-associated proteins (Cas) are the four primary categories of sequence-specific nucleases used in gene editing to date (Moon et al., 2022). The most recent and advanced method of editing DNA is through the use of CRISPR/Cas RNA-guided nucleases, which were adapted from type II CRISPR/Cas9 mechanisms and bacterial innate immune mechanisms (Doudna and Charpentier, 2014). To achieve Sustainable Development Goal 2 (zero hunger) and provide a food supply for the world's growing population, modern agricultural techniques that make use of genetically modified (GE) and stress-resistant crops are prioritized. Using synthetic chemical pesticides to manage pests during agricultural production is costly and environmentally and humanly dangerous.

In this chapter, we outlined the specifics of the most modern genetic tools for controlling pests, such as weeds and insects. We also discussed the benefits and drawbacks of using those tools for pest control. From this chapter, academic students and researchers may quickly obtain knowledge about the existing achievements of such tools and future directions.

10.2 GENOME EDITING TOOLS FOR PEST MANAGEMENT

10.2.1 Zinc Finger Nucleases

ZFNs revolutionized the field of genome manipulation research by being the first genuinely targeted protein reagents (Ahmar et al., 2020). Zinc finger proteins offer a variety of methods for using most of this information and applying it for the development of new traits and the enhancement of plants (Novak, 2019). ZFNs are fusions of the *FokI* endonuclease's cleavage domain and zinc finger protein's DNA recognition domain, as outlined in 1996 (Hamdan et al., 2022). The non-specific type II restriction enzyme *FokI* makes up the cleavage domain, and the zinc finger domain (ZFN) and *FokI* are located at the C and N-terminal ends of polar ZFNs, respectively (Vyshnavi et al., 2022). The zinc fingers mainly come into contact with 3 bp of DNA, while a group of three to six fingers can identify 9–18 bp (Sushmita et al., 2021). Since *FokI* must dimerize to become active, ZFNs should be created in pairs to make sure that *FokI* dimerizes in the right orientation and at the right distance apart (Hamdan et al., 2022). With their use, the target genome can be transformed to induce a variety of genetic modifications, such as deletions, insertions, inversions, translocations, and point mutations (Yadav et al., 2023).

10.2.1.1 Use of ZFNs in Pest Management

One application of ZFNs in crop breeding affected the inositol phosphate profile of developing maize seeds and enhanced herbicide resistance through the disruption of the endogenous ZmIPK1 gene in maize through the insertion of PAT gene cassettes (Yadav et al., 2023). ZFNs are utilized to direct the aad1 herbicide tolerance gene to a locus that already has an integrated TLP and the PAT-herbicide tolerance gene (Davies et al., 2017). To identify the desired events in these situations among the background of randomly inserted events, many plants must be screened (Davies et al. 2017). ZFNs were utilized to target the *Caenorhabditis briggsae* homolog of the *Caenorhabditis elegans* gene sdc-2, which controls the Dosage compensation condensin (DCC) to coalesce onto X and trigger the sexual differentiation of hermaphrodites (Wood et al., 2011). ZFNs were used to target two DICER-like genes in soybeans, DCL4a and DCL4b, under the control of an estrogen-inducible promoter (Table 10.1). The duplicate genes were successfully altered (Satheesh et al., 2019).

TABLE 10.1
Details on ZFNs, based on Genome Editing in Insects (Hillary et al., 2020)

Tool	Insects	Species	Gene Targeted	Mutation
ZFNs	Fruitflies	*Drosophila melangonstor*	Coil/psk	Knockout
			y/ry/bw	
		D. melangonstor		
	Mosquitoes	*Aedes aegypti*	Orco	Knockout
		Ae. aegypti	npylr1	
		Ae. aegypti	AaegGr3	
	Butterflies	*Monarch*	cry2	Knockout
	Silkworm	*Bombyx mori*	Bmfib-H	Knockout

TABLE 10.2
List of Some Practical Achievements of ZFNs (Kanda et al., 2022)

Plant	DNA Repair Mechanism	Modified Gene
Arabidopsis	NHEJ	ABI4-ABA and sugar insensitivity ADH1 TT4
Maize	HDR/TSI	AAD1-herbicide resistance
	HR HDR/TSI HR	IPK1: phylate accumulation in grain
		PAT-herbicide resistance ZmTLP-trait stacking
Tobacco	NHEJ	ALS, SURA, herbicide resistance SURB
Soybean	NHEJ	DCL4a, DCL4b, RDR, HEN
Tomato		Mitochondrial malate dehydrogenase
Rice		OsQQR-trait stacking

10.2.1.2 Advantages of ZFNs

1. Compared to previous methods, ZFNs are more efficient, have greater specificity, and have fewer off-target effects (Kanda et al., 2022).
2. ZFNs are simple in design, as they may be created by combinatorially assembling pre-existing zinc fingers with recognized patterns (Table 10.2) (Khalil et al., 2020).

10.2.1.3 Disadvantages of ZFNs

The ZFN's 'context-dependent specificity', or how well they cleave target sequences, is one of their main drawbacks (Khalil et al., 2020). The use of ZFNs in genome editing has always been difficult and has not been universally accepted by the research community, despite the existence of multiple successful experiments (Satheesh et al., 2019).

ZFNs were not used largely in the field of pest management. They have the potential to be an effective tool in research, provided certain challenges are addressed, such as improving the ZFNs' target region and sequence specificity (Hillary et al., 2020). Globally, scientists are utilizing gene editing instruments to explore the unexplored possibilities of these techniques, which have demonstrated great promise in the realm of agricultural research in particular. CRISPR is seen as a very potent tool to address a variety of issues in the near future (Saurabh et al., 2021).

10.2.2 Transcription Activator-Like Effector Nucleases

TALENs, which stands for transcription activator-like effector nucleases, is a genome editing tool that is responsible for precise changes in DNA.

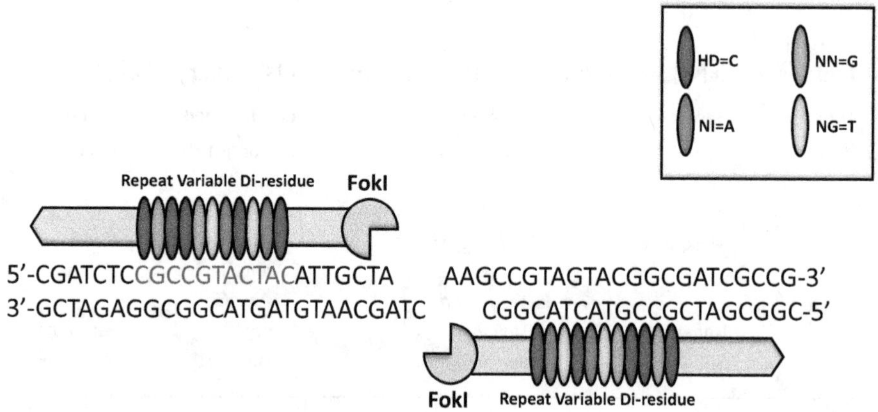

FIGURE 10.1 Structure of transcription activator-like effector nucleases (TALENs).

Natural biologist: *Xanthomonas* spp., plant pathogenic bacteria, have a pathogenic strategy where TALE plays a crucial role for this strategy. This bacterium, which is called nature's molecular biologist because it has a sequence-specific transcription factor, uses its weapon for disease susceptibility and resistance genes in plants (White, 2016).

TALE: It is a base-specific protein or amino acid sequence that is particular to an individual nitrogenous base of the DNA. This protein manipulates the transcription of the host plant. These changes instigate the virulence and pathogenicity of the respective microbes (Figure 10.1).

Gene function of TALE: *Xanthomonas* spp. attack the transcription machinery of the host plant by targeting the promoter region of the respective genes. The idea that the TALE impacted or engaged with the host transcription machinery in some manner was based on their characteristics (Becker and Boch, 2021b). This DNA-binding protein acts as an activator or repressor (Becker and Boch, 2021a). Activation of the gene, by imitating the activator and recruiting the co-activator, leads to upregulation and tends to benefit the bacteria for the acquisition of nutrients from the host. On the contrary, when it acts like a repressor, it inhibits the transcription of the host plant, and hence the gene expression gets called off, which is responsible for the defense.

This mechanism was engineered to use TALENs as a tool for genome editing. In TALENs, there are two types of domain work. Unlike the science of *Xanthomonas*, the extra DNA-cleavage domain in TALEN helps ease the task of editing the genome.

Working principle of TALENs: Like the mechanism of the *Xanthomonas* TALE protein, the DNA-binding domain attaches to the specific targeted DNA sequence of the organism because this repeated unit of amino acids can recognize the base. In this repeat, each of the two amino acids plays a crucial role in binding themselves to DNA bases; they are called RVD, or repeat variable di-residue. This RVD can be customized, and the customized sequence can be used as a weapon to target and manipulate the genome of interest.

Nuclease domain: The cleaving procedure is carried out by this domain. Often, the *FokI* restriction endonuclease is used for it (Becker and Boch, 2021b). There is a dimerization process that takes place in this work. The actual work of cleaving occurs only when the dimerization process comes to be executed, which means the *FokI* can only be functional when the two parts of *FokI* come from opposite positions to each other (Table 10.3).

DNA repairing process: Two types of mechanisms can occur: non-homologous end joining (NHEJ), and another one is homology-directed repair (HDR). However, in these DNA repair processes, mutations can happen and are considered to be error-prone (Shamshirgaran et al., 2022). Between these two, the HDR is considered to be the better one, as there is less chance of mutation here and it gets the guiding template to be fixed. However, in the case of NHEJ, the insertion or deletion of a gene can take place and lead to a mutation (Figure 10.2) (Becker and Boch, 2021b).

TABLE 10.3
An Overview of Nuclease Architecture Featuring Fok1

Restriction Enzyme	Fusion Zone	Linker	Functional Domain	References
FokI	C-terminal	No	*FokI* (198-aa)	Mussolino et al. (2011)
FokI	C-terminal	GGSGGSGS	*FokI* (198-aa)	Miller et al. (2011)
FokI	C-terminal	GS	*FokI* (198-aa)	Miller et al., 2011)

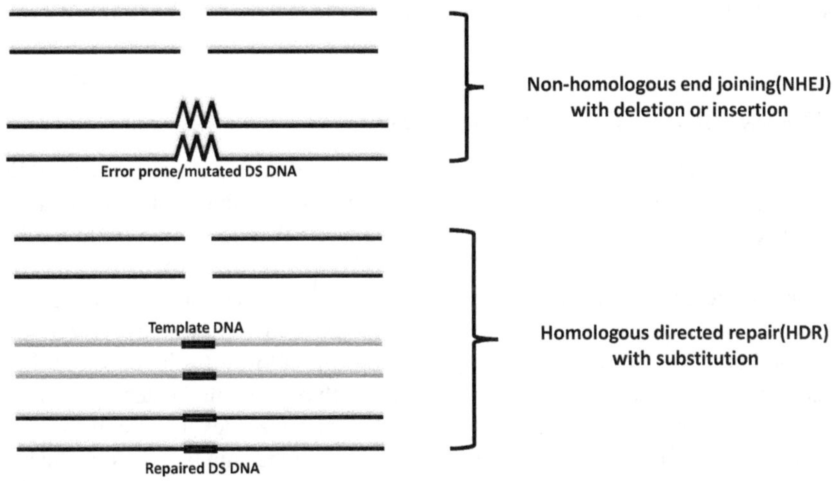

FIGURE 10.2 DNA repair technique.

10.2.2.1 Uses of TALENs in Pest Management

Gene knockout or knock-in: These techniques can be the strategy for controlling the pest, and the TALENs can be applied for this purpose (Xu et al., 2019). In the repair steps, the small insertion or deletion of genes can be beneficial for humankind, as these can prohibit the pathogenicity of the pests. Genome modification utilizing different genome editing tools – TALENs is one of them – is being extended, which is leading a path toward pest control strategy (Suresh et al., 2021)

The male sterility technique for controlling insect pests is considered the best one as it is environment friendly, and here genome editing can assist in carrying out the technique (Kandul et al., 2019). When it comes to enhancing resistance, different biotic and abiotic stress-resistant crops' development has become easier using genome editing tools like TALENs, ZFN, and CRISPR/Cas9 (Razzaq et al., 2019). Genome editing tools have created a way to control different parasitic weeds and insect vectors (Pixley et al., 2019).

Genes can be manipulated through this technique to produce herbicide-resistant plant species (Hussain et al., 2021). In recent times, many genome editing tools have made it easy to work on sustainable agriculture by improving crops, and TALENs is one of them (Gosavi, G. et al., 2022). Gene driving can create resilient plant species, which will be the components of smart agriculture (Barrett et al., 2019).

The application of TALENs in reversing the resistance of pesticides and herbicides could be promising in the upcoming days (Miglani, 2017). Herbicide recurrence can be reduced by using TALENs to damage the main genes of weeds, therefore reversing their resistance to such compounds (Kumam et al., 2023).

10.2.2.2 Limitation of TALENs

Some limitations are faced while working with TALENs, such as the off-target effects, which can produce some mutations that are not of interest. Further, in DNA repair, competition between the HDR and NHEJ can make the process difficult to have a good result. The deletion or insertion of genes interferes with the process. Sometimes customized TALENs cannot accomplish the goal, as the interaction between the DNA-binding domain and DS DNA is specific (Miglani, 2017).

In the case of smart agriculture, the negative effects of using pesticides should always be considered. To mitigate the bad effects of these, gene editing techniques can assist by providing a field where natural resources can be engineered and used in pest management. Environment smart pest control strategy can also be possible by using this technology (Zhang et al., 2018). It is a promising technique to work in different fields associated with genome editing, and it has shown its achievements.

10.2.3 Oligonucleotide-Directed Mutagenesis

ODM is based on the site-specific correction or direct mutation (base substitution, addition, deletion) of the target gene by introducing a chemically synthesized complementary chimeric DNA and/or RNA oligonucleotide designed to have one or more bases that do not pair with the endogenous gene sequence.

ODM is one of the unique gene editing tools to generate targeted mutagenesis. ODM uses specific synthetic oligonucleotides (10–100 bps long) that share homology with a target sequence in the genome, except that it contains a single base pair modification toward achieving site-directed editing of the gene/sequence of interest. Oligonucleotides with a single mismatch bind to homologous sequences in the genome. This is recognized by DNA repair enzymes in the cell that introduce an altered nucleotide using their repair mechanism. Then it generates the target single nucleotide base editing in the plant genome for a novel trait. This mechanism has been demonstrated in *Arabidopsis thaliana*, *Nicotiana tabacum*, *Oryza sativa*, *Zea mays*, and *Brassica napus* for herbicide resistance (Tyagi et al., 2020).

In contrast to ZFN, TALENs, and CRISPR/Cas9, the ODM technique does not require a DNA DSB or nuclease. The mechanism of ODMs is based on the use of chemically synthesized oligonucleotides for the induction of site-specific mutations in the genome.

One to four adjacent nucleotides are affected by the alteration, resulting in point mutations (Modrzejewski, 2020). In the beginning, ODM was applied to maize and tobacco, and it was applied later to *Brassica napus*. The tobacco acetolactate synthase (ALS) gene and proline-196 were converted into threonine by ODM. Then it was found that cultured cells have chlorosulfuran resistance. Similarly, the acetohydroxyacid synthase gene was mutated by ODM, in which serine-621-asparagine and proline-165-alanine with a single construct were used to induce resistance against both sulfonylurea and imidazoline herbicides (Rehman and Yousaf, 2021).

An agritech company, Cibus LLC, recently developed and trademarked an example of ODM, the Rapid Trait Development System. This company developed a sulfonylurea herbicide tolerance GE canola, which is commercially available in the United States and Canada. A mutated form of the ALS enzyme is involved in the biosynthesis of essential amino acids such as valine, leucine, and isoleucine. This amino acid enables this plant to survive the application of the specific herbicide, inhibiting the activity of the enzyme in its original form and causing the death of target weeds (Lombardo et al., 2016).

Significant and practical gene editing frequency has been demonstrated with ODM techniques in many plant species nowadays. However, correction rates have been relatively low, and editing has depended on large amounts of the introduced oligonucleotide. For enhancing conversion efficiency, several treatments have been explored (Sauer et al., 2016).

10.2.4 CRISPR

10.2.4.1 Utilizing CRISPR/Cas to Manage Weed in Crops

One of the most effective methods for editing genomes is the use of CRISPR/CRISPR-associated (Cas) systems, which are CRISPRs. CRISPR/Cas has surfaced as a viable and effective technology to expedite crop breeding (Table 10.1). With the ability to reverse pesticide and herbicide resistance in weeds and insects, CRISPR/Cas-produced gene drives may be able to stop the spread of disease, manage harmful invasive species, and assist agriculture. Although there is great potential for genome editing with both classes of CRISPR systems, class 2 systems' effector complexes are more straightforward, making them the preferred option for genetic engineering. Three stages are involved in the CRISPR class 2 immunity process; they are spacer acquisition, CRISPR RNA (crRNA) biogenesis, and interference. The process of identifying, preparing, and integrating foreign DNA into the CRISPR locus as a new spacer is called spacer acquisition or adaptation. CRISPR locus transcription, which is involved in biogenesis, frequently starts as a single precrRNA and is processed into mature crRNAs later on. The crRNA recognizes and eliminates any phage or plasmid-bearing sequence that is complementary to the spacer sequence of the crRNA during the interface stage, which is triggered by an effector complex. Diverse strategies have been employed to broaden the scope of the CRISPR/Cas system (Miglani, 2017)

Herbicides are frequently used in fields to eradicate undesirable vegetation without endangering the intended crops. However, because desired crops have minimal resistance to these chemicals, herbicides can occasionally cause harm to them. Undesired weeds frequently grow, which lowers crop yield and has a significant global effect on the health of people. Herbicides utilized by the industry, like chlorsulfuron, primarily target the ALS genes (ALS1 and ALS2), which can undergo a mutation that enhances resistance to these herbicides (Sun et al., 2016). The procedure known as knockout is used to swap out the gene that is wanted for the unwanted gene. Rice (*Oryza sativa*), wheat (*Triticum aestivum*), *Nicotiana benthamiana*, and *Arabidopsis thaliana* were the first plants whose genomes were edited using the CRISPR–Cas system (Table 10.4) (Chen, K. 2019). A CRISPR/Cas9 mutation that targets ALS1 and ALS2 elevated maize's (Svitashev et al., 2015) and soybean's (Li et al., 2015) resistance to herbicides but didn't show success in rice. Finally, to give rice herbicide resistance, CBE has been applied (Shimatani, Z. et al., 2017). Similarly, the same approach was applied to watermelon (Tian, S. et al., 2018) and *Arabidopsis* (Chen, Y. et al., 2017). The thermosensitive male-sterile TMS5 lines in rice (Zhou, H. et al., 2016) and maize (Svitashev et al., 2016), Ms45 in wheat (Singh, M. et al., 2018), and other crops have been created using CRISPR/Cas. Bispyribac sodium resistance in rice was developed by CRISPR/Cas point mutation into specific OsALS regions (Sun et al., 2016), and modified EPSPS, which has two amino acids, T102I and P106S, was created by CRISPR/Cas utilizing the second exon of the rice endogenous gene, 5-enolpyruvylshikimate-3-phosphate synthase (EPSPS). This enables rice plants to withstand glyphosate (Li et al., 2016). Moreover, single-stranded oligonucleotides (ssODN) and CRISPR/Cas9 together allowed for the discovery of the herbicide resistance trait in flax (Sauer et al., 2016).

Using newly discovered TadA variants called TadA9, Yan et al. (2021) concurrently generated novel SNPs in four endogenous herbicide target genes (OsALS1, OsGS1, OsTubA2, OsACC) in rice. In addition, rice was engineered to acquire APP herbicide resistance through the production of a novel W2125S mutation using the design of sgRNAs in the carboxyltransferase domain of the ACCase gene (Liu et al., 2020).

It is necessary to modify the plant genome to improve crops resistant to herbicides, and CRISPR/Cas technology offers great promise as an accurate and simple genome editing tool (Xie et al., 1975–1983). Even though this area of genome editing has seen a number of successes, there are still several issues that need to be resolved, and more work is needed to do so. On the one hand, selective herbicides are the main target of the researchers to achieve; on the other hand, one study

TABLE 10.4
Identifying Genes Involved in Resistance to Herbicide through Genome Editing with CRISPR/Cas

Editing	Species	Target Gene	References
Replacement and insertion of gene	Rice	ALS	Chen et al. (2019)
		EPSPS	Li et al. (2016)
		BEL	Xu et al. (2014)
		ACCase	Lei et al. (2014)
		OsACC-T1	Li et al. (2018)
		OsTubA2	Liu et al. (2021)
		SF3B1	Zhang et al. (2019); Kuang et al. (2020)
Base modification		ALS, C287	Shimatani et al. (2017)
Editing of multiplex genome		C287T, ALS	Shimatani et al. (2017)
		OsALS1, OsGS1, OsTubA2, and OsACC	Yan et al. (2021)
Replacement and insertion of gene	Wheat	ACCase	Xie et al. (2017)
		TaDEP1, TaGW2, and TaEPSPS	Li et al. (2018)
	Potato	ALS	Choudhury et al. (2016)
		StALS, StGBSS	Zong et al. (2018)
		SLALS1	Veillet et al. (2019)
	Flax	EPSPS	Sauer et al. (2016)
	Cassava	EPSPS	Hummel et al. (2018)
	Tomato	SLALS1	Veillet et al. (2019)
	Rapeseed	RGA, IAA7, and ALS	Cheng et al. (2021)
		EPSPS	Sauer et al. (2016)
	lettuce	LsGGP	Yu et al. (2015)
	Maize	ALS	Li et al.,2020)
	Arabidopsis	ALS, ACCase	Ali et al. (2020)
	Soybean	ALS	Cai et al. (2015)
Producing targeted mutagenesis and integrating genes		DD20 and DD43	Li et al. (2015)
Base modification	Watermelon	ALS	Tian et al. (2018)
	Tobacco	ALS	Veillet et al. (2019)

has examined non-selective herbicides with broad-spectrum herbicidal qualities that increase crop resistance (Dong et al., 2021). One important component that determines CRISPR/Cas specificity, which is a significant challenge, is the distinct PAM site that SpCas9 requires (Hsu et al., 2014). This challenge is minimized by discovering a novel SpCas9 variant (xCas9) that is capable of recognizing a wide variety of PAM sequences, such as NG, GAA, and GAT (Hu et al., 2018). The recently identified Casl4a system, which can only target ssDNA and does not require PAM, can also be used to get around the restriction caused by PAM specificity (Harrington et al., 2018). Researchers can also use this system to introduce new herbicide-resistant plants. Another challenge is that the genome editing research that has been documented thus far only demonstrates the concept in small settings (Mishra et al., 2018). Cultivating modified crops in the field is necessary for genuinely successful genome editing for agricultural improvement. Regulating and biosafety concerns have prevented the majority of genome-edited crops that have been reported to date from being planted. The idea that these plants could be marketed as non-GMO GE plants is supported, even though genome-edited plants are debatable (Zafar et al., 2020).

10.2.4.2 Utilizing CRISPR/Cas to Manage Insects in Crops

Insect pests are one of the biggest biotic and abiotic strains that agriculture faces in the twenty-first century. So, farmers become independent of insecticides for the management of insects which is harmful to the environment and human health. Additionally, it has a detrimental effect on non-target insects like pollinators and biocontrol agents, and, as insecticide resistance develops, it gradually loses its effectiveness. To solve this issue, researchers developed various strategies to control insects rather than using insecticides. Genome editing is one of the most famous and latest strategies in all to control insects, and CRISPR/Cas is the most promising tool to develop a sustainable and globally food-secured situation.

The process of editing an insect's genome to control it through gene drive propagation involves eliminating specific sites that could be fatal, affect the sex ratio, or reduce the fecundity of the affected insects (Deepak et al., 2019). By following the strategy, a systematic decline in the number of offspring can be anticipated by using the technique, which may lessen or shut off female receptivity (Rato et al., 2021). Numerous instances exist where the CRISPR/Cas9 genome editing tool has been utilized to modify insect genomes to control pests and mitigate insecticide resistance (Table 10.5). The white gene and the sex-lethal gene, which cause the development of white eyes or the recovery of females with changed genitalia and ovaries, have been targeted by injecting plasmid DNA encoding Cas9 and gRNAs into embryos (Li and Scott, 2016). *Drosophila* sp. is the primary cause of soft-skinned fruits. By disrupting the white (we) and paired (Ccprd; necessary for segmentation of the developing embryo) genes, *Ceratitis capitata* also proved suitable for CRISPR/Cas9 actions. Embryos with white eyes (we) or abnormalities in the development of the embryo were produced by injecting Cas9 and gRNA RNP complexes into the embryos (Meccariello et al., 2017). Furthermore, focusing on the worm *Spodoptera litura*, which causes significant losses globally, Abdominal-A (Abd-A) is responsible for proper segment development during embryogenesis in *Drosophila melanogaster*. By microinjecting Abd-A-targeting mRNA and gRNA into embryos, aberrantly segmented larvae were recovered (Bi et al., 2016). Using gRNAs targeted at two genes involved in eye pigmentation – cinnabar and white – a dangerous rice pest and viral disease vector known as *Nilaparvata lugens* demonstrated heritably altered eyes after embryo injection with RNPs (Xue et al., 2018). A group developed male offspring that were 100% sterile by using *D. melanogaster* by using a method called 'precision-guided sterile insect technique', in which two lines express gRNAs and Cas9, respectively, to interfere with genes necessary for male fertility (b-Tubulin 85D, fuzzy onions, protease A, spermatocyte arrest) and female survival (sex lethal, transformer, double sex) (Kandul et al., 2019). CRISPR/Cas was used to identify microorganisms named *Beauveria bassiana* genes involved in virulence, and an in-cis genetic complementation system was used to complement a *B. bassiana* mutant for the polyketide synthesis 15 (Pks15) gene, which is essential for insect virulence (Toopang et al., 2017). Researchers discovered a mutation in the CHS1 gene of BPU-resistant *P. xylostella* that can replace isoleucine (I) with methionine (M) (I1042M), and then they introduced this mutation into *Drosophila* to create homozygous lines that were extremely resistant to BPUs, etoxazole, and buprofezin (I1056M/F) (Douris et al., 2016).

The release of CRISPR/Cas9-edited insects carrying gene drives into the environment raises several biosafety concerns because the modified insects have the chance to alter the entire population. Moreover, there is a chance that any gene flow between the target species and its non-target relatives could exacerbate the situation. However, modifying the effector-binding element instead of the gene itself offers a workable workaround for obtaining disease resistance without compromising the plant's ability to function (Li et al., 2012). Consequently, scientists must concentrate on identifying sources of resistance that could provide the basis for insect control. To accomplish this, it is required to use multi-omics techniques to identify desired genes and assess the available germplasm for pest response, including wild relatives of specific crops. Genes that are identified so far should be utilized to achieve the desired goal in the pest management of agriculture. By using these methods, it is possible to increase the cultivated gene pool's resistance to insect pests.

TABLE 10.5
Identifying Genes Involved in Resistance to Insects through Genome Editing with CRISPR/Cas

Host Plant Species	Insect Species	Target Gene	Improved Traits	References
Rice		CYP71A1	Increased ability to withstand stripped stem borers and hoppers	Lu et al. (2018)
	Nilaparvata lugens	White and cinnabar genes	Changes in the pigmentation of the eyes	Xue et al. (2018)
Cherry, raspberry, blackberry, blueberry, strawberry, peach, plums, grapes, and other soft fruits	*Drosophila suzukii*	Sex-lethal and white genes	Females with changed ovaries and genitalia or white eyes	Li and Scott (2016)
Cassava, cotton, sweet potatoes, tobacco, tomato, pepper, cucumber, aubergine	*Bemisia tabaci*	White	Modified pigmentation of the eyes	Heu et al. (2020)
Avocado, banana, bitter melon, carambola, coffee, guava, mango, papaya, peppers, and persimmon	*Ceratitis capitata*	Paired and white	Developing abnormalities in embryos or white eyes	Meccariello et al. (2017)
Bell pepper, sunflower, cabbage, cotton, and pigweed	*Spodoptera exigua*	Ryanodine receptor		Zuo et al. (2017)
Tobacco, cotton, soybean, beet, cabbage, and chickpeas	*Spodoptera litura*	Abdominal-A	Unexplained larval segmentation	Bi et al. (2016)
Sweet potato, tobacco Chinese cabbage, cowpeas, and sugarcane	*Spodoptera littoralis*	Olfactory receptor co-receptor (Orco) gene		Koutroumpa et al. (2016)
Sugarbeet		Cry1Ab	An increase in mortality and the duration of development and powerful insect feeding inhibitory effect	Sedighi et al. (2011)
Maize, rice, wheat, sorghum, alfalfa, barley, clover, cabbage, lettuce, cucumber, and cowpea	*Spodoptera frugiperda*	Cry1Fa		Huanga et al. (2020), Flagel (2018), Jin et al. (2020)
		SfABCC2		Abdelgaffar et al., (2020), Jin et al. (2018)
		ABC transporters	The potency of two cry1 toxins produced by *Bacillus thuringiensis* against pests and ability to withstand both cry1fa and cry1ab toxins	Jin et al. (2020)
		Cry1Ac	Focuses on various SfCad gene exons	Zhang et al. (2020)
		Cry1Ab Cry1C		Zhang et al. (2020), Kimoto et al. (2001), Hernández-Rodríguez et al. (2013), Jin et al. (2020)

(Continued)

TABLE 10.5 (*Continued*)
Identifying Genes Involved in Resistance to Insects through Genome Editing with CRISPR/Cas

Host Plant Species	Insect Species	Target Gene	Improved Traits	References
Brussels broccoli, mustard, radish, turnip, sprouts, cabbage, kale, kohlrabi, watercress Chinese cabbage, cauliflower, collard	*Plutella xylostella*	Pxabd-A		Huang et al. (2016)
		PxCHS1		Douris et al. (2016)
Maize, lentil, rice, beans, peas, groundnut, oats, almond, rye, sorghum, wheat Brazil nut, barley, and walnuts	*Tribolium castaneum*	E-cadherin gene, EGFP	Resistance to insects	Gilles et al. (2015)
Sorghum, barley, maize, rice, grasses, bamboo leaves, and other Gramineae plants	*Locusta migratoria*	Odorant receptor co-receptor (Orco) gene		Li et al. (2016)
Vegetables and ripened fruits (especially bananas)	*Drosophila melanogaster*	Chitin synthase 1		Douris et al. (2016)
		Nicotinic acetylcholine receptor α6		Zimmer et al. (2016)
Maize	*Bicyclus anynana*	Yellow-h3	Resistance to insects	Zhang et al. (2017)
Strawberry, eggplant, cucumber, orange, and beans	*Tetranychus urticae*	PSST		Bajda et al. (2017)
Field crops	*Helicoverpa armigera*	Cry1Ac	Strong level of resistance to insects	Stewart et al. (2001), Wang et al. (2017), Wang et al. (2005), Wang et al. (2016), Karim et al. (2000), Chen, L. et al. (2018)
		Cry2Ab		Chen et al. (2018), Wang et al. (2017)
Soyabean		GmUGT (QTL-M)	Resistance to insects	Zhang et al. (2022)
Pigeon pea		Cry2Aa	Strong level of resistance to insects	Singh et al. (2018), Wang et al. (2017)
Sugar beet	*Tetanops myopaeformis*	Delta-endotoxin genes	Resistance to insects	Wilhite et al. (2000), Smigocki. et al. (2003)
Mango, cucumber, pumpkin, guava, Lichi, orange, chili, tomato, potato	All species of diptera	Cry2	Resistance to insects	Yuan et al. (2007)
Several crops	All species of coleopteran	Cry3	Possibility of targeting this gene with CRISPR/Cas 9 to produce insecticidal Bt toxins	Pauchet et al. (2016)

10.3 FUTURE PERSPECTIVES

It is imperative to develop novel pest management techniques that effectively suppress insect pests while having no adverse effects on the environment, people, or natural enemies (Akhter et al., 2023). The CRISPR/Cas9 system has become a sophisticated and reasonably priced genetic technology during the last decade, and it is anticipated that soon it will be widely used in pest control applications for crop enhancement (Singh et al., 2022). Current research in genomics aims to understand the genetic basis of plant disease resistance. With genome editing, it is possible to modify endogenous genes in crops without introducing transgenes across species boundaries. On the contrary, a variety of applications in crop improvement involve base editing and CRISPR/Cas9-mediated gene knockout (Namo and Belachew, 2021). CRISPR/Cas and other gene editing technologies are being used to create crops with greater resistance to a range of diseases. Although using the CRISPR/Cas9 system to regulate transcription may be a successful strategy, the resulting genetically edited creature will not be 'transgene-free' because it will have a transgene for both the gRNA expression cassette and dead Cas9 (dCas9). Ovarian targeting peptide ligand, or BAPC, facilitates the transport of the injected RNP complex in flies, both nymphs and adults. This is a recent example of another innovative strategy that has been investigated (Singh et al., 2022). This technique is more convenient than microinjecting CRISPR components into eggs or embryos because it allows the components to be administered close to the ovary of an adult female fly. There are many applications for this strategy, and a similar idea can be used to manage pests with different insect orders. Its wide range of applications and study has been made possible by the most recent discoveries about the genomes of important insects and the understanding of gene activities in the physiological, behavioral, and reproductive sciences of this diverse group of insects. The development of gene editing technologies and their use to confer beneficial features on insects during containment studies demonstrated increased potential and promise. A precise and environmentally friendly method of pest management is the purposeful introduction of genetic components into wild insect species through the use of CRISPR/Cas technology. This strategy involves introducing deadly mutations or changing the sex ratio of the population to lower the number of pests. Accordingly, there is cause for concern over the prospect of insect resistance developing as a result of a CRISPR-mediated gene drive, with possible consequences for theory and experimentation (Komal et al., 2023). Concerns about the possible establishment of so-called 'better adapted' strains arise when GE crops are cultivated extensively and/or for an extended period of time, particularly in monocultures. The goal of genetic modification is to enhance beneficial traits like insect resistance and higher crop yields. However, monocultures' intrinsic homogeneity creates an environment that makes certain adaptations easier to evolve. The long-term cultivation of GE plants has the capacity to exert selective pressure on natural pest or pathogen populations, which may lead to the emergence of more robust and adaptive strains that are able to overcome the engineered resistance mechanisms. The unintended consequence raises concerns about the long-term sustainability and effectiveness of GE crops and emphasizes the need for careful monitoring and sustainable farming practices to address and reduce potential risks associated with the widespread production of GE organisms.

10.4 CONCLUSION

The development of genome editing tools has revolutionized both fundamental and applied plant research in the modern period. The functional genomics of insects have already been altered by genome editing methods based on CRISPR/Cas, despite their recent development. We can now easily add, delete, or change DNA practically anywhere in any crop or insect species using CRISPR/Cas to make plants resistant to insect pests. Thus, to develop agricultural plants that are resistant to insect pests, this technique has to be improved. It is imperative that scientists worldwide focus on identifying sources of resistance that might serve as a foundation for managing weeds and insects. The applicability of several novel targets to lessen off-target impacts on target insects is being considered. A crucial clue that requires ground-breaking attention is the extensive phenotyping and enrichment of the resistance gene pool. To identify the genes that respond to stress, multi-omics techniques must be used to identify the accessible germplasm, which includes wild relatives of certain crops. The details of employing genome editing to manage weeds and insects are found in the multiplex editing of such newly discovered resistance genes utilizing high-throughput transformation methods. In addition to promoting scientific advancement, these factors and regulatory concerns about GE crops may make technology successful in gaining cultural acceptability.

REFERENCES

Abdelgaffar, H., Perera, O. P., & Jurat-Fuentes, J. L. (2020). ABC transporter mutations in Cry1F-resistant fall armyworm (Spodoptera frugiperda) do not result in altered susceptibility to selected small molecule pesticides. *Pest Management Science*, *77*(2), 949–955. https://doi.org/10.1002/ps.6106

Ahmar, S., Saeed, S., Khan, M. H. U., Ullah Khan, S., Mora-Poblete, F., Kamran, M., ... & Jung, K. H. (2020). A revolution toward gene-editing technology and its application to crop improvement. *International Journal of Molecular Sciences*, *21*(16), 5665.

Akhter, A., Iqbal, I., & Jan, U. (2023). Review on: CRISPR-Cas 9: A novel genome editing tool for insect pest management. *Biological Forum – An International Journal*, *15*(1), 550–553. https://www.researchgate.net/profile/Amreena-Akhter/publication/371165464_Review_on_CRISPR-Cas_9_A_Novel_Genome_Editing_Tool_for_Insect_Pest_Management/links/6476daf96a3c4c6efbf44103/Review-on-CRISPR-Cas-9-A-Novel-Genome-Editing-Tool-for-Insect-Pest-Management.pdf

Ali, Z., Shami, A., Sedeek, K., Kamel, R., Alhabsi, A., Tehseen, M., Hassan, N., Butt, H., Kababji, A., Hamdan, S. M., & Magdy, M. M. (2020). Fusion of the Cas9 endonuclease and the VirD2 relaxase facilitates homology-directed repair for precise genome engineering in rice. *Communications Biology*, *3*, 44.

Bajda, S., Dermauw, W., Panteleri, R., Sugimoto, N., Douris, V., Tirry, L., Osakabe, M., Vontas, J., & Leeuwen, T. V. (2017). A mutation in the PSST homologue of complex I (NADH:ubiquinone oxidoreductase) from Tetranychus urticae is associated with resistance to METI acaricides. *Insect Biochemistry and Molecular Biology*, *80*, 79–90. https://doi.org/10.1016/j.ibmb.2016.11.010

Bhat, S., & Kumar, S. (2022). Improvement of resistance in plants against insect-pests using genome editing tools. In Shabir Hussain Wani, & Goetz Hensel (Eds.), *Genome editing: Current technology advances and applications for crop improvement* (pp. 237–249). Springer.

Bi, H.-L., Xu, J., Tan, A.-J., & Huang, Y.-P. (2016). CRISPR/Cas9-mediated targeted gene mutagenesis in Spodoptera litura. *Insect Science*, *23*, 469–477. https://doi.org/10.1111/1744-7917.12341

Bisht, D. S., Bhatia, V., & Bhattacharya, R. (2019, December). Improving plant-resistance to insect-pests and pathogens: The new opportunities through targeted genome editing. In *Seminars in cell & developmental biology* (Vol. 96, pp. 65–76).

Cai, Y., Chen, L., Liu, X., Sun, S., Wu, C., Jiang, B., Han, T., & Hou, W. (2015). CRISPR/Cas9-mediated genome editing in soybean hairy roots. *PLoS One*, *10*, e0136064.

Chen, K., Wang, Y., Zhang, R., Zhang, H., & Gao, C. (2019). CRISPR/Cas genome editing and precision plant breeding in agriculture. *Annual Review of Plant Biology*, *70*, 667–697.

Chen, L., Wei, J., Liu, C., Zhang, W., Wang, B., Niu, L. et al. (2018). Specific binding protein ABCC1 is associated with Cry2Ab toxicity in Helicoverpa armigera. *Frontiers in Physiology*, *19* (9), 745. https://doi.org/10.3389/fphys.2018.00745

Chen, Y., Wang, Z., Ni, H., Xu, Y., Chen, Q., & Jiang, L. (2017). CRISPR/Cas9-mediated base-editing system efficiently generates gain-offunction mutations in arabidopsis. *Science China Life Sciences*, *60*, 520–523.

Cheng, H., Hao, M., Ding, B., Mei, D., Wang, W., Wang, H., Zhou, R., Liu, J., Li, C., & Hu, Q. (2021). Base editing with high efciency in allotetraploid oilseed rape by A3A-PBE system. *Plant Biotechnology Journal*, *19*(1), 87–97. https://doi.org/10.1111/pbi.13444

Choudhury, M. D., Das, S., & Tarafdar, S. (2016). Effect of loading history on visco-elastic potato starch Gel. *Colloids and Surfaces A: Physicochemical and Engineering Aspects*, *492*, 47–53.

Davies, J. P., Kumar, S., & Sastry-Dent, L. (2017). Use of zinc-finger nucleases for crop improvement. *Progress in Molecular Biology and Translational Science*, *149*, 47–63.

Deepak Singh, B., Varnika, B., & Ramcharan, B. Seminars in cell and developmental biology. https://doi.org/10.1016/j.semcdb.2019.04.008

Dong, H., Huang, Y., & Wang, K. (2021). The development of herbicide resistance crop plants using CRISPR/Cas9-mediated gene editing. *Genes*, *12*, 912. https://doi.org/10.3390/genes12060912

Doudna, J. A., & Charpentier, E. (2014). The new frontier of genome engineering with CRISPR-Cas9. *Science, 346*(6213), 1258096.

Douris, V., Steinbach, D., Panteleri, R., Livadaras, I., Pickett, J. A., Van Leeuwen, T., Nauen, R., & Vontas, J. (2016). Resistance mutation conserved between insects and mites unravels the benzoylurea insecticide mode of action on chitin biosynthesis. *Proceedings of the National Academy of Sciences of the United States of America, 113*, 14692–14697, https://doi.org/10.1073/pnas

Flagel, L., Lee, Y. W., Wanjugi, H., Swarup, S., Brown, A., Wang, J. et al. (2018). Mutational disruption of the ABCC2 gene in fall armyworm Spodoptera frugiperda confers resistance to the Cry1Fa and Cry1A.105 insecticidal proteins. *Science Report, 8*, 7255. https://doi.org/10.1038/s41598-018-25491-9

Gilles, A. F., Schinko, J. B., & Averof, M. (2015). Efficient CRISPR-mediated gene targeting and transgene replacement in the beetle Tribolium castaneum. *Development, 142*, 2832–2839, https://doi.org/10.1242/dev.125054.

Giudice, G., Moffa, L., Varotto, S., Cardone, M. F., Bergamini, C., De Lorenzis, G., Velasco, R., Nerva, L., Chitarra, W., & Chitarra, W. (2021). Novel and emerging biotechnological crop protection approaches. *Plant Biotechnology Journal, 19*(8), 1495–1510.

Gurbachan, S. M. (2017). Genome editing in crop improvement: Present scenario and future prospects, *Journal of Crop Improvement, 31*(4), 453–559, https://doi.org/10.1080/15427528.2017.1333192

Hafeez, A., Ali, B., Javed, M. A., Saleem, A., Fatima, M., Fathi, A., Afridi, M. S., Aydin, V., Oral, M. A., & Soudy, F. A. (2023). Plant breeding for harmony between sustainable agriculture, the environment, and global food security: An era of genomics-assisted breeding. *Planta, 258*(5), 97.

Hamdan, M. F., Karlson, C. K. S., Teoh, E. Y., Lau, S. E., & Tan, B. C. (2022). Genome editing for sustainable crop improvement and mitigation of biotic and abiotic stresses. *Plants, 11*(19), 2625.

Harrington, L. B., Burstein, D., Chen, J. S., Paez-Espino, D., Ma, E., Witte, I. P., Cofsky, J. C., Kyrpides, N. C., Banfield, J. F., & Doudna, J. A., (2018). Programmed DNA destruction by miniature CRISPR-Cas14 enzymes. *Science, 362*, 839–842. https://doi.org/10.1126/science.aav4294

Hernández-Rodríguez, C. S., Hernández-Martinez, P., Rie, J. V., Escriche, B., & Ferre, J. (2013). Shared midgut binding sites for Cry1A.105, Cry1Aa, Cry1Ab, Cry1Ac and Cry1Fa proteins from Bacillus thurigenesis in two important corn pests, Ostrinia nubilais and Spodoptera frugiperda. *PloS One, 8*(7), e68164. https://doi.org/10.1371/journal.pone.0068164

Heu, C. C., McCullough, F. M., Luan, J., & Rasgon, J. L. (2020). CRISPRCas9-based genome editing in the silverleaf W\whitefly (Bemisia tabaci). *Cris Journal, 3*, 89–96. https://doi.org/10.1089/crispr.2019.0067

Hillary, V. E., Ceasar, S. A., & Ignacimuthu, S. (2020). Genome engineering in insects: Focus on the CRISPR/Cas9 system. In Vijai Singh & Pawan K. Dhar (Eds.), *Genome engineering via CRISPR-Cas9 system* (pp. 219–249). Academic Press.

Hsu, P. D., Lander, E. S., & Zhang, F. (2014). Development and applications of CRISPR-Cas9 for genome engineering. *Cell, 157*, 1262–1278. https://doi.org/10.1093/hmg/ddy120

Hu, X., Meng, X., Liu, Q., Li, J., & Wang, K. (2018). Increasing the efficiency of CRISPR-Cas9-VQR precise genome editing in rice. *Plant Biotechnology Journal, 16*, 292–297. https://doi.org/10.1111/pbi.12771

Huang, Y., Chen, Y., Zeng, B., Wang, Y., James, A. A., Gurr, G. M., Yang, G., Lin, X., Huang, Y. & You, M. (2016). CRISPR/Cas9 mediated knockout of the abdominal-A homeotic gene in the global pest, diamondback moth (Plutella xylostella), *Insect Biochemistry and Molecular Biology, 75*, 98–106. https://doi.org/10.1016/j.ibmb.2016.06.004.

Huanga, J., Xua, Y., Zuoa, Y., Yanga, Y., Tabashnikb, B. E., & Wua, Y. (2020). Evaluation of five candidate receptors for three Bt toxins in the beet armyworm using CRISPR-mediated gene knockouts. *Insect Biochemistry and Molecular Biology, 121*, 103361. https://doi.org/10.1016/j.ibmb.2020.103361

Hummel, A. W., Chauhan, R. D., Cermak, T., Mutka, A. M., Vijayaraghavan, A., Boyher, A., Starker, C. G., Bart, R., Voytas, D. F., & Taylor, N. J. (2018). Allele exchange at the EPSPS locus confers glyphosate tolerance in cassava. *Plant Biotechnology Journal, 16*, 1275–1282.

Jin, L., Wang, J., Guan, F., Zhang, J., Yu, S., Liu, S. et al. (2018). Dominant point mutation in a tetraspanin gene associated with field-evolved resistance of cotton bollworm to transgenic Bt cotton. *Proceedings of the National Academy of Sciences (PNAS), 115*, 11760–11765. https://doi.org/10.1073/pnas.1812138115

Jin, M., Yang, Y., Shan, Y., Chakrabarty, S., Cheng, Y., Soberon, M. et al. (2020). Two ABC transporters are differentially involved in the toxicity of two Bacillus thuringiensis Cry1 toxins to the invasive crop-pest Spodoptera frugiperda (*J. E. Smith). Pest Management Science, 77*, 1492–1501. https://doi.org/10.1002/ps.6170

Kanda, T., & Srivastava, R., & Singh, N., & Yadav, S., & Prajapati, R., & Yadav, S., & Atri, N. (2022). Advances in agriculture and allied technologies. https://doi.org/10.22271/ed.book.1919.

Kandul, N. P., Liu, J., Sanchez, C. H. M. et al (2019). Transforming insect population control with precision guided sterile males with demonstration in flies. *Nature Communication*, *10*, 84. https://doi.org/10.1038/s41467-018-07964-7

Karim, S., Riazuddin, S., Gould, F., & Dean, D. H. (2000). Determination of receptor binding properties of Bacillus thuringiensis delta-endotoxins to cotton bollworm (Helicoverpa zea) and pink bollworm (Pectinophora gossypiella) midgut brush border membrane vesicles. *Pesticide Biochemistry and Physiology*, *67*, 198–216. https://doi.org/10.1006/pest.2000.2491

Khalil, A. M. (2020). The genome editing revolution. *Journal of Genetic Engineering and Biotechnology*, *18*(1), 1–16.

Kimoto, Y., & Shimamoto, Y. (2001). Difference in toxicity to larvae of cabbage armyworm between transgenic sugar beet lines with Cry1Ab and Cry1AC. *Journal of the American Society of Sugar Beet Technologists*, *43*, 20–23.

Kogo, B. K., Kumar, L., & Koech, R. (2021). Climate change and variability in Kenya: A review of impacts on agriculture and food security. *Environment, Development and Sustainability*, *23*, 23–43.

Komal, J., Desai, H. R., Samal, I., Mastinu, A., Patel, R. D., Kumar, P. D., ... & Bhoi, T. K. (2023). Unveiling the genetic symphony: Harnessing CRISPR-cas genome editing for effective insect pest management. *Plants*, *12*(23), 3961.

Koutroumpa, F. A., Monsempes, C., François, M.-C., de Cian, A., Royer, C., Concordet, J.- P., Jacquin-Joly, E. (2016). Heritable genome editing with CRISPR/Cas9 induces anosmia in a crop pest moth, *Science Report*, *6*, 29620, https://doi.org/10.1038/srep29620

Kuang, Y., Li, S., Ren, B., Yan, F., Spetz, C., Li, X., Zhou, X., & Zhou, H. Base-editing-mediated artificial evolution of OsALS1 in planta to develop novel herbicide-tolerant rice germplasms. *Molecular Plant*, 2020, *13*, 565–572.

Lei, Y., Li, L., Liu, H.-Y., Li, S., Xing, F., & Chen, L.-L. (2014). CRISPR-P: A web tool for synthetic single-guide RNA design of CRISPR-system in plants. *Molecular Plant*, *7*(9), 1494–1496.

Li, C., Li, W., Zhou, Z., Chen, H., Xie, C., & Lin, Y. (2020a) A new rice breeding method: CRISPR/Cas9 system editing of the Xa13 promoter to cultivate transgene-free bacterial blight-resistant rice. *Plant Biotechnology Journal*, *18*(2), 313–315. https://doi.org/10.1111/pbi.13217

Li, C., Zhang, R., Meng, X., Chen, S., Zong, Y., Lu, C., Qiu, J. L., Chen, Y. H., Li, J., & Gao, C. (2020b) Targeted, random mutagenesis of plant genes with dual cytosine and adenine base editors. *Nature Biotechnology*, *38*(7), 875–882. https://doi.org/10.1038/s41587-019-0393-7

Li, C., Zong, Y., Wang, Y., Jin, S., Zhang, D., Song, Q., Zhang, R., & Gao, C. (2018a) Expanded base editing in rice and wheat using a Cas9-adenosine deaminase fusion. *Genome Biology*, *19*(1), 1–9. https://doi.org/10.1186/s13059-018-1443-z

Li, F., & Scott, M. J. (2016). CRISPR/Cas9-mediated mutagenesis of the white and Sex lethal loci in the invasive pest, Drosophila suzukii. *Biochemical and Biophysical Research Communications*, *469*, 911–916. https://doi.org/10.1016/j.bbrc.2015.12.081

Li, H. Y., Li, J. Y., Chen, J. L., Yan, L., & Xia, L. (2020c) Precise modifcations of both exogenous and endogenous genes in rice by prime editing. *Molecular Plant*, *13*(5), 671–674. https://doi.org/10.1111/pbi.13395

Li, J., Meng, X., Zong, Y., Chen, K., Zhang, H., Liu, J., Li, J., & Gao, C. (2016). Gene replacements and insertions in rice by intron targeting using CRISPR?Cas9. *Nature Plants*, *2*, 16139. https://doi.org/10.1038/nplants.2016.139

Li, T., Liu, B., Spalding, M. H., Weeks, D. P., & Yang, B. (2012). High-efficiency TALEN-based gene editing produces disease-resistant rice, *Nature Biotechnology*, *30*, 390–392, https://doi.org/10.1038/nbt.2199

Li, T., Yang, X., Yu, Y., Si, X., Zhai, X., Zhang, H., Dong, W., Gao, C., & Xu, C. (2018b) Domestication of wild tomato is accelerated by genome editing. *Nature Biotechnology*, *36*(12), 1160–1163. https://doi.org/10.1038/nbt.4273

Li, Y., Lin, Z., Huang, C., Zhang, Y., Wang, Z., Tang, Y.-J., Chen, T., Zhao, X. (2015). Metabolic engineering of Escherichia coli using CRISPR-Cas9 meditated genome editing. *Metabolic Engineering*, *31*, 13–21.

Li, Y., Zhang, J., Chen, D., Yang, P., Jiang, F., Wang, X., & Kang, L. (2016). CRISPR/Cas9 in locusts: Successful establishment of an olfactory deficiency line by targeting the mutagenesis of an odorant receptor co-receptor (Orco). *Insect Biochemistry and Molecular Biology*, *79*, 27–35. https://doi.org/10.1016/J.IBMB.2016.10.003

Li, Y., Zhu, J., Wu, H., Liu, C., Huang, C., Lan, J., Zhao, Y., & Xie, C. (2020d) Precise base editing of non-allelic acetolactate synthase genes confers sulfonylurea herbicide resistance in maize. *Crop Journal*, *8*(3), 449–456. https://doi.org/10.1016/j.cj.2019.10.001

Li, Z., Liu, Z.-B., Xing, A., Moon, B. P., Koellhoffer, J. P., Huang, L., Ward, R. T., Clifton, E., Falco, S. C., Cigan, A. M. (2015). Cas9-guide RNA directed genome editing in soybean. *Plant Physiology*, *169*, 960–970.

Liu, L., Kuang, Y., Yan, F., Li, S., Ren, B., Gosavi, G., Spetz, C., Li, X., Wang, X., Zhou, X., & Zhou, H. (2021). Developing a novel artifcial rice germplasm for dinitroaniline herbicide resistance by base editing of OsTubA2. *Plant Biotechnology Journal*, *19*(1), 5–7. https://doi.org/10.1111/pbi.13430

Liu, X., Qin, R., Li, J., Liao, S., Shan, T., Xu, R., Wu, D., & Wei, P. (2020). A CRISPR-Cas9-mediated domain-specific base-editing screen enables functional assessment of ACCase variants in rice. *Plant Biotechnology Journal*, *18*, 1845–1847.

Lombardo, L., Coppola, G., & Zelasco, S. (2016). New technologies for insect-resistant and herbicide-tolerant plants. *Trends in Biotechnology*, *34*(1), 49–57.

Lu, H., Luo, T., Fu, H. et al. (2018). Resistance of rice to insect pests mediated by suppression of serotonin biosynthesis. *Nat Plants*, *4*, 338–344. https://doi.org/10.1038/s41477-018-0152-7

Maeder, M. L., & Gersbach, C. A. (2016). Genome-editing technologies for gene and cell therapy. *Molecular Therapy*, *24*(3), 430–446.

Meccariello, A., Monti, S. M., Romanelli, A. et al (2017). Highly efficient DNA-free gene disruption in the agricultural pest Ceratitis capitata by CRISPR-Cas9 ribonucleoprotein complexes. *Science Report*, *7*, 10061. https://doi.org/10.1038/s41598-017-10347-5

Mishra, R., Joshi, R. K., & Zhao, K. (2018). Genome editing in rice: Recent advances, challenges, and future implications. *Frontiers in Plant Science*, *9*, 1361. https://doi.org/10.3389/fpls.2018.01361

Modrzejewski, D. (2020). Evidence synthesis on the impact of genome editing on plant breeding (Doctoral dissertation, Dissertation, Göttingen, Georg-August Universität, 2020).

Moon, T. T., Maliha, I. J., Khan, A. A. M., Chakraborty, M., Uddin, M. S., Amin, M. R., & Islam, T. (2022). CRISPR-Cas Genome Editing for Insect Pest Stress Management in Crop Plants. *Stresses*, *2*(4), 493–514.

Namo, F. M., & Belachew, G. T. (2021). Genome editing technologies for crop improvement: Current status and future prospective. *Plant Cell Biotechnology and Molecular Biology*, *22*, 1–19.

Novak, S. (2019). Plant biotechnology applications of zinc finger technology. In Sandeep Kumar, Pierluigi Barone, & Michelle Smith (Eds.), *Transgenic plants: Methods and protocols* (pp. 295–310). https://link.springer.com/protocol/10.1007/978-1-4939-8778-8_20

Pauchet, Y., Bretschneider, A., Augustin, S., & Heckel, D. G. (2016). A P-glycoprotein is linked to resistance to the Bacillus thuringiensis Cry3Aa toxin in a leaf beetle. *Toxins (Basel)*, *8*, 362. https://doi.org/10.3390/toxins8120362

Pratap, A., Douglas, C., Prajapati, U., Kumari, G., War, A. R., Tomar, R., Pandey, A. K. & Dubey, S. (2020). Breeding progress and future challenges: Biotic stresses. *The Mungbean Genome*, 55–80.

Rato, C., Carvalho, M. F., Azevedo, C. et al. (2021). Genome editing for resistance against plant pests and pathogens. *Transgenic Research*, ***30***, 427–459. https://doi.org/10.1007/s11248-021-00262-x

Rehman, T., & Yousaf, S. (2021). Site-directed mutagenesis in plants. In Tariq Ahmad Bhat (Ed.), *Mutagenesis, cytotoxicity and crop improvement: Revolutionizing food science* (p. 413).

Satheesh, V., Zhang, H., Wang, X., & Lei, M. (2019, December). Precise editing of plant genomes–prospects and challenges. *Seminars in Cell & Developmental Biology, 96*, 115–123.

Sauer, N. J., Mozoruk, J., Miller, R. B., Warburg, Z. J., Walker, K. A., Beetham, P. R., ... & Gocal, G. F. (2016). Oligonucleotide-directed mutagenesis for precision gene editing. *Plant Biotechnology Journal*, *14*(2), 496–502.

Sauer, N. J., Narváez-Vásquez, J., Mozoruk, J., Miller, R. B., Warburg, Z. J., Woodward, M. J., Mihiret, Y. A., Lincoln, T. A., Segami, R. E., Sanders, S. L. et al. (2016). Oligonucleotide-mediated genome editing provides precision and function to engineered nucleases and antibiotics in plants. *Plant Physiology*, *170*, 1917–1928.

Saurabh, S. (2021). Genome editing: Revolutionizing the crop improvement. *Plant Molecular Biology Reporter*, *39*(4), 752–772.

Sedighi, L., Rezapanah, M., & Aghdam, H. R. (2011). Efficacy of Bt transgenic sugar beet lines expressing cry1Ab gene against Spodoptera littoralis Boisd (Lepidoptera:] *Noctuidae*, *Noctuidae). Journal of the Entomological Research Society*, *13* (1), 61–69

Shimatani, Z., Kashojiya, S., Takayama, M., Terada, R., Arazoe, T., Ishii, H. et al. (2017). Targeted base editing in rice and tomato using a CRISPR-Cas9 cytidine deaminase fusion. *Nature Biotechnology*, *35*, 441–443. https://doi.org/10.1038/nbt.3833

Shimatani, Z., Kashojiya, S., Takayama, M., Terada, R., Arazoe, T., Ishii, H., Teramura, H., Yamamoto, T., Komatsu, H., Miura, K. et al. (2017). Targeted base editing in rice and tomato using a CRISPR-Cas9 cytidine deaminase fusion. *Nature Biotechnology*, *35*, 441–443.

Shimatani, Z., Kashojiya, S., Takayama, M., Terada, R., Arazoe, T., Ishii, H., Teramura, H., Yamamoto, T., Komatsu, H., Miura, K. et al. (2017). Targeted base editing in rice and tomato using a CRISPR-Cas9 cytidine deaminase fusion. *Nature Biotechnology*, *35*, 441–443.

Singh, M., Kumar, M., Albertsen, M. C., Young, J. K., & Cigan, A. M. (2018). Concurrent modifications in the three homeologs of Ms45 gene with CRISPR-Cas9 lead to rapid generation of male sterile bread wheat (Triticum Aaestivum, *L.). Plant Molecular Biology*, *97*, 371–383.

Singh, S., Kumar, N. R., Maniraj, R., Lakshmikanth, R., Rao, K. Y. S., Muralimohan, N. et al. (2018). Expression of Cry2Aa, a Bacillus thuringiensis insecticidal protein in transgenic pigeon pea confers resistance to Gram pod borer, Helicoverpa armigera. *Science Report*, *8*, 8820. https://doi.org/10.1038/s41598-018-26358-9

Singh, S., Rahangdale, S., Pandita, S., Saxena, G., Upadhyay, S. K., Mishra, G., & Verma, P. C. (2022). CRISPR/Cas9 for insect pests management: A comprehensive review of advances and applications. *Agriculture*, *12*(11), 1896.

Smigocki, A. C., Ivic, S. D., Wilson, D., Wozniak, C. A., Campbell, L., Dregseth, R. et al. (2003). Molecular approaches for control of the sugar beet root maggot. In *1st joint IIRB-ASSBT Congress*, 26th Feb–1st March 2003, San Antonio, USA, pp. 416–428.

Stewart, S. D., Adamczyk, J. J., Kinghten, K. S., & Davis, F. M. (2001). Impact of Bt cotton expressing one or two insecticidal proteins of Bacillus thuringiensis Berliner on growth and survival of Noctuidae (Lepidoptera) larvae. *Journal of Economic Entomology*, *94*(3), 752–760. https://doi.org/10.1603/0022–0493–94.3.752

Sun, Y., Zhang, X., Wu, C., He, Y., Ma, Y., Hou, H., Guo, X., Du, W., Zhao, Y., Xia, L. (2016). Engineering herbicide-resistant rice plants through CRISPR/Cas9-mediated homologous recombination of acetolactate synthase. *Molecular Plant*, *9*, 628–63

Sushmita, Kaur, G., Upadhyay, S. K., & Verma, P. C. (2021). *An overview of genome-engineering methods. Genome Engineering for Crop Improvement*, John Wiley & Sons Ltd. pp. 1–21.

Svitashev, S., Schwartz, C., Lenderts, B., Young, J. K., & Mark Cigan, A. (2016). Genome editing in maize directed by CRISPR–Cas9 ribonucleoprotein complexes. *Nature Communication*, *7*, 13274.

Svitashev, S., Young, J. K., Schwartz, C., Gao, H., Falco, S. C., & Cigan, A. M. (2015). Targeted mutagenesis, precise gene editing, and site-specific gene insertion in maize using Cas9 and guide RNA. *Plant Physiology*, *169*, 931–945.

Tian, S., Jiang, L., Cui, X., Zhang, J., Guo, S., Li, M., Zhang, H., Ren, Y., Gong, G., Zong, M. et al. (2018). Engineering herbicide-resistant watermelon variety through CRISPR/Cas9-mediated base-editing. *Plant Cell Reports*, *37*, 1353–1356.

Toopaang, W., Phonghanpot, S., Punya, J. et al. (2017). Targeted disruption of the polyketide synthase gene pks15 affects virulence against insects and phagocytic survival in the fungus Beauveria bassiana. *Fungal Biology*, *121*, 664–675. https://doi.org/10.1016/j.funbio.2017.04.007

Tyagi, S., Kesiraju, K., Saakre, M., Rathinam, M., Raman, V., Pattanayak, D., & Sreevathsa, R. (2020). Genome editing for resistance to insect pests: An emerging tool for crop improvement. *ACS Omega*, *5*(33), 20674–20683.

Veillet, F., Perrot, L., Chauvin, L., Kermarrec, M. P., Guyon-Debast, A., Chauvin, J. E., Nogué, F., & Mazier, M. (2019). Transgene-free genome editing in tomato and potato plants using agrobacterium-mediated delivery of a CRISPR/Cas9 cytidine base editor. *International Journal of Molecular Sciences*, *20*(2), 402. https://doi.org/10.3390/ijms20020402

Vyshnavi, R. G., & Dawar, K. B. P. (2022). Genome Editing: The Next Generation Breeding Strategy. *Technologies*, *91*, 91–101.

Wang, G., Liang, G., Wu, K., & Guo, Y. (2005). Gene cloning and sequencing of aminopeptidase N3, a putative receptor for Bacillus thuringiensis insecticidal Cry1Ac toxin in Helicoverpa armigera (Lepidoptera: noctuidae). *European Journal of Entomology*, *102*, 13–19. https://doi.org/10.14411/eje.2005.002

Wang, J., Wang, H., Liu, S., Liu, L., Tay, W. T., Walsh, T. K. et al. (2017a). CRISPR/Cas9 mediated genome editing of Helicoverpa armigera with mutations of an ABC transporter gene HaABCA2 confers resistance to Bacillus thuringiensis Cry2A toxins. *Insect Biochemistry and Molecular Biology*, *87*, 147–153. https://doi.org/10.1016/j.ibmb.2017.07.002

Wang, J., Zhang, H., Wang, H., Zhao, S., Zuo, Y., Yang, Y. et al. (2016). Functional validation of cadherin as a receptor of Bt toxin Cry1Acin Helicoverpa armigera utilizing the CRISPR/Cas9 system. *Insect Biochemistry and Molecular Biology*, *76*, 11–17. https://doi.org/10.1016/j.ibmb.2016.06.008

Wilhite, S. E., Elden, T. C., Puizdar, V., Armstrong, S., & Smigocki, A. C. (2000). Inhibition of aspartyl and serine proteinases in the midgut of sugarbeet root maggot with proteinase inhibitors. *Entomologia Experimentalis et Applicata*, *97*, 229–233. https://doi.org/10.1046/j.1570–7458.2000.00734.x

Wood, A. J., Lo, T. W., Zeitler, B., Pickle, C. S., Ralston, E. J., Lee, A. H., ... & Meyer, B. J. (2011). Targeted genome editing across species using ZFNs and TALENs. *Science, 333*(6040), 307–307.

Xie, K., Yang, & Y. (2013). RNA-guided genome editing in plants using a CRISPR-Cas system. *Molecular Plant, 6*, 1975–1983.

Xie, X., Ma, X., Zhu, Q., Zeng, D., Li, G., & Liu, Y.-G. (2017). CRISPR-GE: A convenient software toolkit for CRISPRbased genome editing. *Molecular Plant, 10*(9), 1246–1249.

Xu, R., Li, H., Qin, R., Wang, L., Li, L., Wei, P. et al. (2014). Gene targeting using the Agrobacterium tumefaciens mediated CRISPR-Cas system in rice. *Rice, 7*, 5. https://doi.org/10.1186/s12284-014-0005-6

Xue, W.-H., Xu, N., Yuan, X.-B. et al. (2018). CRISPR/Cas9-mediated knockout of two eye pigmentation genes in the brown planthopper, Nilaparvata lugens (Hemiptera: Delphacidae). *Insect Biochemistry and Molecular Biology, 93*, 19–26. https://doi.org/10.1016/j.ibmb.2017.12.003

Yadav, R. K., Tripathi, M. K., Tiwari, S., Tripathi, N., Asati, R., Chauhan, S., ... & Payasi, D. K. (2023). Genome editing and improvement of abiotic stress tolerance in crop plants. *Life, 13*(7), 1456.

Yan, D., Ren, B., Liu, L., Yan, F., Li, S., Wang, G., Sun, W., Zhou, X., & Zhou, H. (2021). High-efficiency and multiplex adenine base editing in plants using new TadA variants. *Molecular Plant, 14*, 722–731.

Yu, Q., Jalaludin, A., Han, H., Chen, M., Sammons, R. D., & Powers, S. B. (2015). Evolution of a double amino acid substitution in the 5-enolpyruvylshikimate-3-phosphate synthase in Eleusine indica conferring high-level glyphosate resistance. *Plant Physiology, 167*, 1440–1447.

Yuan, Q., Metterville, D., Briscoe, A. D., & Reppert, S. M. (2007). Insect cryptochromes: Gene duplication and loss define diverse ways to construct insect circadian clocks. *Molecular Biology and Evolution, 24* (4), 948–955. https://doi.org/10.1093/molbev/msm011

Zafar, K., Sedeek, K. E. M., Rao, G. S., Khan, M. Z., Amin, I., Kamel, R., Mukhtar, Z., Zafar. M., Mansoor, S., & Mahfouz, M. M. (2020). Genome editing technologies for rice improvement: Progress, prospects, and safety concerns. *Frontiers in Genome Editing, 2*, 5. https://doi.org/10.3389/fgeed.2020.00005

Zaller, J. G., & Zaller, J. G. (2020). Pesticide impacts on the environment and humans. In Johann G. Zaller (Ed.), *Daily Poison: Pesticides-an Underestimated Danger* (pp. 127–221). https://link.springer.com/chapter/10.1007/978-3-030-50530-1_2

Zhang, J., Jin, M., Yang Yliu, L., Yang, Y., Gómez, I., Bravo, A. et al. (2020a). The cadherin protein is not involved in susceptibility to Bacillus thuringiensis Cry1Ab or Cry1Fa toxins in Spodoptera frugiperda. *Toxins (Basel), 12* (6), 375. https://doi.org/10.3390/toxins12060375

Zhang, Y., & Qi, Y. (2019). CRISPR enables directed evolution in plants. *Genome Biology, 20*, 83.

Zhang, Y., Guo, W., Chen, L., Shen, X., Yang, H., Fang, Y. et al. (2022). CRISPR/Cas9- mediated targeted mutagenesis of GmUGT enhanced soybean resistance against leaf chewing insects through flavonoids biosynthesis. *Frontiers in Plant Science, 13*, 802716. https://doi.org/10.3389/fpls.2022.802716

Zhang, Z., Liu, X., Shiotsuki, T., Wang, Z., Xu, X., Huang, Y. et al. (2017). Depletion of juvenile hormone esterase extends larval growth in Bombyx mori. *Insect Biochemistry and Molecular Biology, 81*, 72–79. https://doi.org/10.1016/j.ibmb.2017.01.001

Zhou, H., He, M., Li, J., Chen, L., Huang, Z., Zheng, S., Zhu, L., Ni, E., Jiang, D., Zhao, B. et al. (2016). Development of commercial thermo-sensitive genic male sterile rice accelerates hybrid rice breeding using the CRISPR/Cas9-Mediated TMS5 editing system. *Science Report, 6*, 37395.

Zimmer, C. T., Garrood, W. T., Puinean, A. M., Eckel-Zimmer, M., Williamson, M. S., Davies, T. G. E., & Bass, C. (2016). A CRISPR/Cas9 mediated point mutation in the alpha 6 subunit of the nicotinic acetylcholine receptor confers resistance to spinosad in Drosophila melanogaster, *Insect Biochemistry and Molecular Biology, 73*, 62–69, https://doi.org/10.1016/j.ibmb.2016.04.007

Zong, Y., Song, Q., Li, C., Jin, S., Zhang, D., Wang, Y., Qiu, J. L., & Gao, C. (2018). Efcient C-to-T base editing in plants using a fusion of nCas9 and human APOBEC3A. *Nature Biotechnology, 36*(10), 950–953. https://doi.org/10.1038/nbt.4261

Zuo, Y., Wang, H., Xu, Y., Huang, J., Wu, S., Wu, Y., & Yang, Y. (2017). CRISPR/Cas9 mediated G4946E substitution in the ryanodine receptor of Spodoptera exigua confers high levels of resistance to diamide insecticides. *Insect Biochemistry and Molecular Biology, 89*, 79–85, https://doi.org/10.1016/J.IBMB.2017.09.005.

11 Plant Genome Editing for Viral Disease Management

Current Achievements and Future Directions

Shitosri Mondal, Prodipto Bishnu Angon, Md. Shakil Uddin, Arpan Das, Monisha Mondol, and Md. Shafiul Islam

11.1 INTRODUCTION

A fundamental and significant obstacle to agricultural progress and global food security is disease. Diseases impact the quality of agricultural products in the field and during storage, which lowers crop yields by negatively influencing plant growth and development (Yin and Qiu, 2019). Pests and illnesses are further barriers to the expansion of global food production, in addition to the limited resources of arable land and water. Pests pose a threat to the yield of cultivated plants, resulting in yield losses of 20–40%; bacterial and fungal infections cause crop yield reductions of approximately 15%; viruses produce yield reductions of 3–7% (Zhao et al., 2019). Incomplete statistical data indicates that illnesses can cause a 20–40% reduction in crop output worldwide (Yin and Qiu, 2019).

Global warming has accelerated the rate at which viral infection in crops has worsened, encouraging the growth of viruses and the movement of insect vectors that help spread viruses, posing a major threat to plant ecology (Mushtaq et al., 2020). Diseases have a significant impact on crop quality and productivity, endangering the world's food security. Editing speeds up resistance breeding and offers new strategies for genetically enhancing plant disease resistance (Yin and Qiu, 2019). According to predictions, plants that have undergone genetic transformation and expressed viral sequences may develop resistance to viral infection (Baulcombe et al., 2007). While the losses from viruses are not as large as those from pests or other diseases, the outbreak of viral infections can provide significant challenges. Viral illnesses can cost the economy between $60 and $80 billion a year. This is because, while chemical pesticides are effective against insect pests and bacterial or fungal diseases, there isn't yet a conventional chemical pesticide that specifically targets viral diseases (Zhao et al., 2019). Integrants come in two varieties: those that can cause episomal viral infections and those that do not. When hybrid plant hosts are under stress, integrants of three pararetroviruses – banana streak virus (BSV), tobacco vein clearing virus (TVCV), and petunia vein clearing virus (PVCV) – can cause episomal infections. For BSV and PVCV, there are fewer integrant loci than for TVCV (Hull et al., 2000). Controlling viral infections is essential for sustainable agriculture and for preserving the quantity and quality of food produced (Abdelkhalek and Hafez, 2019).

A cell's or organism's genomic DNA can be precisely altered using a process called genome editing (GE). Utilizing sequence-specific nucleases (SSNs) to identify particular DNA sequences and create double-strand breaks (DSBs) in the DNA at predetermined locations forms the basis of GE technology (Yin and Qiu, 2019). Foods and goods developed by GE improve food supply and security worldwide with speed, efficacy, and precision that are unmatched by conventional plant breeding technologies (Van et al., 2022). The non-homologous end joining pathway and the homologous

DOI: 10.1201/9781003543237-11

recombination pathway are the two primary processes through which DSBs are repaired (Yin and Qiu, 2019). Traditional crop breeding techniques have several drawbacks, including high time and resource requirements and biosafety issues. The recently developed GE technologies, which use SSNs to precisely alter DNA sequences at the genomic level, may be able to circumvent these restrictions (Mishra and Zhao, 2018).

The development of four main types of SSNs that can precisely bind to a user-selected genomic area and cause alterations in the target gene is the result of decades of advancements in GE: transcription activator-like effector nucleases (TALENs), zinc finger nucleases (ZFNs), meganucleases, and clustered, regularly interspaced, short palindromic repeats (CRISPR)–CRISPR-associated (Cas) systems (Uranga and Daròs, 2022). SSNs are among the most advanced gene editing techniques currently available. ZFNs were the first nucleases to be used for GE in 2002. The first genuinely target-specific protein reagents, or ZFNs, completely changed the genome modification field. DNA-binding domains known as ZFNs can identify three base pairs at the target region with particularity (Vats et al., 2019). Engineered meganucleases, of which yeast I-*Sce*I meganucleases are the most known, are naturally occurring mediators of DNA targeting and were the first SSNs to be employed for plant GE. Acting normally as a dimer of two identical subunits, these proteins can detect long DNA sequences (20–40 bp). Since it is difficult to build such a lengthy recognition sequence by protein redesign, meganucleases are presently utilized much less frequently than other SSNs, despite their tiny size (about 165 amino acids for a monomer) (Khandagale and Nadaf, 2016; Uranga and Daròs, 2022). A particular zinc finger protein domain that attaches to DNA and is connected to the catalytic domain of the *Fok*I endonuclease (restriction enzyme class IIS, for which the DNA cut points are not inside the regions recognized by the enzyme) makes up ZFNs, which are synthetic bipartite enzymes with a molecular weight of approximately 310 amino acids (Uranga and Daròs, 2022). TALEs, or transcription activator-like effectors, are made by plant pathogenic bacteria, primarily *Xanthomonas* species. These bacteria are the source of the DNA recognition domain of TALENs. A specific DNA sequence in a target gene of the host plant can be recognized and bound to by each TALE. The tandemly arranged 34-amino acid sequence known as the TALE repeat, which is highly conserved except for locations 12 and 13, is present in 13–28 copies inside the DNA-binding region (Weeks et al., 2015; Uranga and Daròs, 2022). *Xanthomonas oryzae pv. oryzae (Xoo)* is the source of the rice vascular bundle disease known as rice bacterial blight, which is one of the main diseases affecting rice and has a high potential for spreading epidemics. It results in 10–20% yield losses; however, in pathogen-friendly conditions (high humidity), this loss can increase to 50% or possibly cause a total loss of production. TALE proteins are secreted into host cells by *Xoo* via the type III secretion system. Many TALE proteins target S genes and trigger the expression of those genes to facilitate infection success (Yin and Qiu, 2019).

A great substitute for the first generation of site-directed nucleases is the recently developed CRISPR/Cas technique, which is based on the type II prokaryotic adaptive immune system and aids bacteria or archaea against invading phages. Gene editing has been accomplished with TALENs and ZFNs, but CRISPR/Cas offers a number of benefits over other techniques in terms of design, specificity, multiplexing, affordability, and flexibility (Vats et al., 2019; Singh, 2020). GE techniques based on CRISPR/Cas have transformed plant research and made it possible to create germplasm with advantageous features (Zhang et al., 2022; Weeks et al., 2015). By focusing on distinct viral genome sequences in infected *Nicotiana benthamiana* or *Arabidopsis* plants, the GE technique known as the CRISPR/Cas9 system has recently been successfully used to engineer resistance to DNA geminiviruses (family: Geminiviridae) (Hadidi et al., 2016).

This chapter describes several GE tools and strategies related to viral disease management and plant responses to viral infections. The application of GE methods for plant viral diseases, a comparison of various technology platforms, possible roadblocks, and prospects are the next areas of emphasis.

11.2 PLANT RESPONSES AGAINST VIRAL DISEASES

Plant viruses have very simple structures but present a significant threat to crops as they undergo complex mutations, exhibit a broad host range, and result in substantial economic losses (He and Krainer, 2020; Chung et al., 2013; Jones, 2021). They are like tiny freeloaders that rely on their host plants for everything – from replication to moving around. They're intracellular moochers with small genomes (Nelson and Citovsky, 2005). Even though plant viruses don't have a lot of proteins, the way they interact with the host plant's stuff has confused scientists for a long time. It's only recently that scientists have started to figure out how complicated these interactions can be (Liu et al., 2017). Following plant infestation, viruses interact with host factors, inducing alterations in the plant's physiological and biochemical features. This includes the production of volatile substances and secondary metabolites and leads to changes in metabolic, signaling, and nutrient pathways. These changes manifest as symptoms like localized greening and yellowing, ultimately impacting the overall plant growth and development (Casteel et al., 2015; Mochizuki et al., 2014).

The study of plant responses to infection has proved very important in the development of plant virology. For studying virus structure and replication, the early focus was on host responses, but the field shifted toward molecular biological techniques after 1945. In recent years, a renewed interest in host responses has increased due to economic losses and biological interest. Most plant–virus interactions result in minimal or symptomless infections in natural plant communities, but cultivated plants often exhibit observable diseases. Plant virologists primarily study these plant responses, which vary widely and lead to macroscopic symptoms dependent on cellular and organelle changes triggered by virus infection. Despite the continuous growth of host plants and their complex structures, progress has been made through molecular studies, electron microscopy, and insights into plant biochemistry, physiology, and development (Matthews, 1980).

Plants' immune defense system is almost similar to that of higher animals and has a very important regulatory role in plant defense against different pests. In 2006, the plant immune system concept was first introduced by some scientists in America in the *Nature* Journal (Jone and Dangl, 2006). Multiple defense mechanisms are employed by plants to restrict viral replication, for example, gene silencing, hormone-mediated defense, immune receptor signaling, degradation of proteins, and metabolism regulation (Incarbone and Dunoyer, 2013). This immune or defense response of plants can mitigate or delay the development and occurrence of plant disease and reduce synthetic fertilizer and pesticide use (De-wen, 2014). Some of this response results from plant innate immunity, like disease resistance genes, cell surface receptors, and systemic acquired resistance. There are also some plant immunity agents such as ningnamycin, dufulin, vanisulfane, methiadinil, cytoinpeptidemycin, and oligosaccharins, and these can enhance the immune system by activating plants to resist the harm caused by pathogenic microorganisms (Huang et al., 2023).

Animal viruses utilize host surface receptors and endocytic mechanisms for invading host cells. In contrast, plant viruses are introduced into cells through either insect vectors or mechanical injuries. The viral particles, which are typically composed of nucleic acids enclosed by a coat protein or capsid, undergo disassembly upon entry into the cell. This disassembly releases the viral genome, initiating the infectious cycle. Key processes involve the expression and replication of the viral genome, movement of viral particles or genome between cells, long-distance movement within the plant, and transmission to new hosts facilitated by vectors (Calil and Fontes, 2017).

A virus particle invades a leaf cell by mechanical inoculation, multiplies, and spreads to neighboring cells, potentially causing macroscopically visible local lesions such as dead or chlorophyll-depleted cell patches. The virus can then move systemically through the plant, typically in the phloem. Systemic infection results in diverse symptoms, as outlined in Table 11.1. These symptoms often manifest in combination with specific diseases, and the disease progression for a given host-virus pairing commonly involves the sequential emergence of various symptom types (Matthews, 1980).

TABLE 11.1
Plant Responses to Virus Infection/Diseases

Plant Part	Response/Symptom
Whole plant	Size reduction
	Necrosis
	Wilting
Leaves	Vein clearing
	Mosaic pattern
	Vein banding
	Yellowing of veins
	Ringspot
	Blistering
	Lamina size reduction
	Enations
	Leaf abcission
Flowers	Malformation
	Necrosis
	Variegation of petal pigmentation
Fruits	Mottling patterns
	Ring and line patterns
	Size reduction
	Malformation
	Necrosis
Stems, roots, leaf veins, and petioles	Tumorlike growths

In response to virus stress, plants deploy defense mechanisms to restrict virus infestation, replication, and dispersal (Huang et al., 2023). Immune receptor signaling, particularly through disease resistance genes (R genes) for specific recognition and cell surface receptors for broad-spectrum recognition, represents key mechanisms in the plant's innate immunity against pathogenic microorganisms (Balmer et al., 2013; Elmore et al., 2011).

Plants possess innate immune responses against viruses, but the constant evolution of viruses can undermine these host resistance mechanisms. To address this challenge, GE technologies are being explored to bolster antiviral defenses. This involves using viral diagnosis to detect viral genetic information and then employing GE techniques to cut viral genomes or modify plant genomes, enhancing the innate immunity of plants. However, the practical application of gene editing technologies is currently limited and has not been extensively implemented across various plant species and large-scale field applications (Chavhan et al., 2018; Hinge et al., 2021).

11.3 GE APPROACHES FOR VIRAL DISEASE RESILIENCE

Attacks of viral diseases have become a major threat to plants in recent times. Scientists are coming up with many possible solutions, and genome alteration is one of them. GE can be done with conventional breeding or mutant breeding. This approach has eminent possibilities for creating viral disease resilience in plants. However using conventional breeding to produce resilience is a time-consuming and quite complex process (Hadidi et al., 2016). In the previous century, recombinant DNA technology was discovered and it opened the door to transgenic breeding which can produce precise alteration. Sometimes the outcome of these genomic changes can be predicted to some extent (Uranga and Daròs, 2023). Programmable endonucleases used in creating transgenic technology can potentially break down the viral genome and thus obliterate or control the spread of viruses (De Silva Feelixge et al., 2018). RNA silencing, ZFN, TALEN, and CRISPR/Cas are the modern technologies that are being used to engineer the genome of a target plant to create resistance to a specific virus or a group of viruses.

11.3.1 RNA Silencing

When a virus attacks a living cell, it injects its viral dsRNA (double-stranded RNA) inside the cell. Then the dsRNA of the virus produces positive and negative-stranded RNA. Using these RNA strands, the virus then produces a positive-stranded mRNA, which later translates structural proteins of the viral body called capsid proteins. This positive-stranded mRNA produces a negative-stranded mRNA as well, which will bind with a positive-stranded mRNA or create one (Kovalev et al.,2014). To multiply the genetic material (dsRNA) the process goes on, and, after some time, many copies of viral dsRNA are generated. Finally, these copies of dsRNA, along with capsid proteins, give birth to many viruses (Figure 11.1) (Unchwaniwala et al., 2021).

Even if a plant is attacked with a single virus, the infection spreads very rapidly by this method. To prevent the regeneration of the viral dsRNA, a new technology called RNA silencing or RNA interference (RNAi) is proven to be effective. RNA silencing is a relatively new technology that was discovered in 1990 (Napoli et al., 1990). It silences a gene by using ncRNA (non-coding RNA) like siRNA (small interfering RNA) and miRNA (micro-RNA).

In the RNA interference (RNAi) process, when a viral dsRNA is injected into a living cell, it's cut down into pieces by DICER, producing a number of siRNAs. Here, DICER is the DCL protein that cleaves dsRNA into smaller pieces. These fragments create miRNA, which acts as a precursor to siRNA. These siRNAs get incorporated with Argonautes or AGO proteins and create a complex called RNA-induced silencing complex (RISC), which binds with complementary strands of mRNA. This coupling of RISC and mRNA cleaves the mRNA or prevents it from translation. The virus needs to multiply the genomic components and produce capsid proteins to regenerate, but the

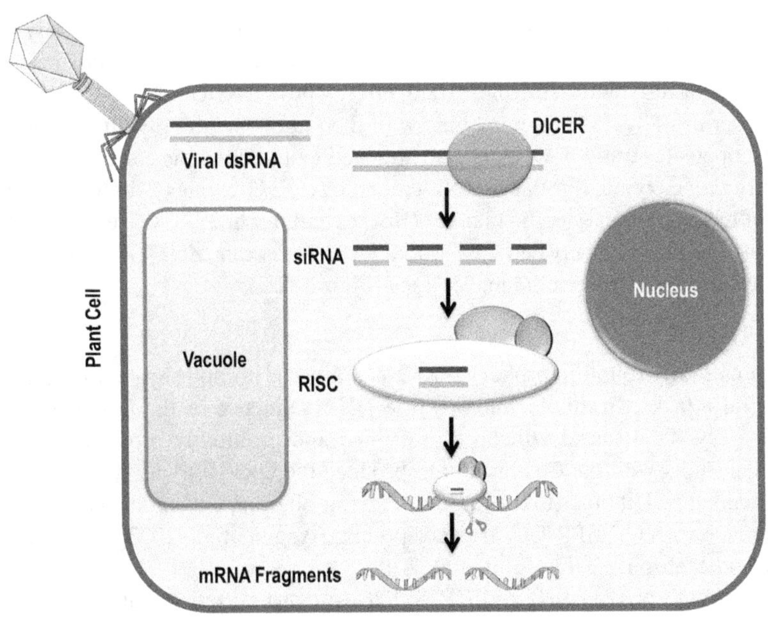

FIGURE 11.1 RNA silencing. (a) The virus inserts its dsRNA into the plant cell. (b) DICER binds with viral dsRNA. (c) DICER cuts the dsRNA and produces siRNA (small interfering RNA). (d) siRNA interacts with AGO proteins and produces RNA-induced silencing complex (RISC). (e) RISC binds with target mRNA. (f) RISC cleaves the mRNA and stops viral regeneration.

whole process gets interrupted by these ncRNAs. As a result, viruses can't reproduce, which results in the prevention of viral diseases (Jiang et al., 2022; Zhao et al., 2020; Flores et al., 2020).

11.3.2 Genome Editing

Viruses require cellular machinery to replicate their genetic materials and multiply their number. With this mission, viruses attack plants, but not every plant is infected by a particular virus. In simple words, every virus has a specific host because, to cause an infection, a host must have all the host factors of that virus. The proviral host factor comprises the things that supply the virus with a suitable environment and components for multiplication. Certain genes of host plants are responsible for being susceptible to viruses (Garcia-Ruiz, 2018). If that particular gene or group of genes is removed from the plant genome or silenced then susceptibility to viral diseases will be lost. To remove a gene from a plant genome we can use ZFNs, TALENs, or CRISPR/Cas technologies.

11.3.2.1 ZFN

Zinc finger nucleases (ZNFs) are a group of endonucleases that can cleave the DNA at the targeted site. These ZNFs are programmable; i.e., one can design them according to the base of the target site. Along with zinc fingers (proteins) that bind with DNA nucleases, a restriction enzyme called *Fok*1 is attached, which cuts. To make a double-stranded cut, two ZNFs need to be designed.

When designed ZNFs are transfected into plant cells, and they bind with DNA. Each zinc finger protein binds to three nucleotides of one strand of DNA and at one point of the ZFN half of the *Fok*1 restriction enzyme is attached. Another ZFN is attached to the opposite strand's nucleotides of the target site and the other half of the restriction enzyme is attached to this. Then these two halves of the *Fok*1 enzyme form a dimer bond and join together to create a functional endonuclease. With this bond being formed a double-stranded cut is made into the DNA, and this cut is non-specific (Carroll, 2011; Cathomen and Joung, 2008; Urnov et al., 2010).

11.3.2.2 TALENs

TALENs are a bunch of restriction enzymes that can be designed according to any DNA sequence to be cut. TALEN has TAL proteins that are found in nature in *Xanthomonas* bacteria (Boch and Bonas, 2010) and a restriction enzyme called *Fok*1. These TALE proteins can bind to a specific nucleotide. A series of TALE proteins binds with the target site along with half of the *Fok*1 endonuclease. The opposite strand of DNA is also attached to another series of TALE proteins with half the *Fok*1 restriction enzyme. Similar to ZNF technology, *Fok*1 creates a dimer bond and cleaves the two strands of DNA in a non-specific manner (Becker and Boch, 2021; Gaj et al., 2013; Wright et al., 2014). The basic difference between ZNFs and TALENs is that ZNFs bind to a codon (3 bases), while TALENs bind only to specific nucleotides (Figure 11.2).

11.3.2.3 CRISPR/Cas

CRISPRs are naturally found in prokaryotes. It is a kind of immune system for bacteria to protect itself from viral attacks. In this technology, the DNA sequence of the target site is inserted into the CRISPR array. This strand will be transcripted and eventually produce single-guided RNA (sgRNA), which links with the Cas protein to produce the CRISPR/Cas system. When CRISPR is inserted into a plant cell, it binds to the target site. After binding, Cas endonuclease cleaves the DNA strand. Cuts made with CRISPR/Cas are very precise (Angon et al., 2023; De Silva Feelixge et al., 2018; Mahas and Mahfouz, 2018; Zaidi et al.,2018).

All these tools (ZFN, TALEN, and CRISPR/Cas) can be used to cut DNA strands, and they provide a chance to eliminate or mutate the susceptibility gene in plants. As the DNA breaks, the cell's natural response is to repair it. During the repair process, the target gene has a high possibility of mutation, or we can insert the gene of our choice to create resilience to viral diseases. Although cuts can be made with restriction enzymes like *EcoRI*, which have very specific recognition sites,

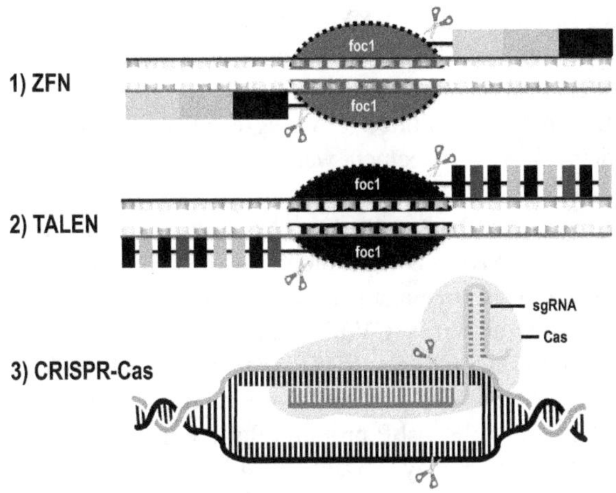

FIGURE 11.2 Genome editing tools. (1) zinc finger nuclease (ZFN), zinc finger proteins bind with codons, and *Fok*1 endonuclease cleaves DNA by producing a dimer bond. (2) transcription activator-like effector nuclease (TALEN), TALE proteins bind with each nucleotide, and *Fok*1 creates a dimmer bond and cuts DNA. (3) CRISPR/Cas, sgRNA binds with the target site, and Cas endonuclease makes a precise cut in the target site.

this will only make cuts at sites that have specific nucleotides. Sometimes the target site may not involve the specific nucleotide of the restriction enzymes. In these types of cases, these tools can be engineered according to the base pairs of the target site DNA sequence. Thus, any region of DNA can be cut and the plant genome can be altered to produce resilience to viral diseases.

11.4 CURRENT ACHIEVEMENTS OF THE RELATED FIELD

Previously, traditional breeding and crossing techniques served as the primary methods for developing resistance to viral diseases. However, these methods had limitations and could not completely control viral diseases in plants. Modern GE tools such as CRISPR/Cas9 and RNAi have revolutionized the development of resistance to viral diseases in plants, leading to disease-free crops and improved food security.

Techniques such as ZFN, TALEN, and Cas9 endonucleases have provided novel approaches to combating viruses by manipulating host susceptibility genes or improving the expression of virus resistance genes. Particularly, CRISPR/Cas9 has gained widespread acceptance due to its versatility and proven effective in a wide range of plants to develop resistance to viral diseases by selectively targeting nucleotide sites (Nisa et al., 2022). CRISPR/Cas9 technology has already been utilized successfully to induce virus resistance in important crops, namely *Arabidopsis thaliana* and *Nicotiana benthamiana* (Borrelli et al., 2018), as well as in wheat through the incorporation of geminivirus-based vectors in plant genome engineering (Ramesh et al., 2017). Multiplex CRISPR/Cas-mediated GE-generated resistance to the cotton leaf curl virus, which enhanced long-term cotton productivity in Pakistan (Binyameen et al., 2021).

Rice ranks second in the world in area and production and third in global production after wheat and corn. Viruses negatively impact rice yields globally, demanding immediate efforts to develop virus-resistant rice varieties. Rice (*Oryza sativa*) has developed resistance to rice stripe virus (RSV) by targeted deletion of the OsSERK2 gene, which produces a receptor-like kinase that serves as the primary entry point for RSV. Applying CRISPR/Cas9 to modify the eukaryotic elongation factor eIF4G in rice yielded plants entirely immune to the rice tungro bacilliform virus (RTBV). Upon RTBV infection, genome-edited plants surpassed their wild-type counterparts in yield and lacked viral protein presence (Macovei et al., 2018). Rice yellow mottle virus (RYMV) poses a significant threat to production, which could lead to a devastating loss of nearly 80–100% in crop yield (Iannella et al., 2021). To introduce RYMV resistance, the genomes of the two rice nucleoporin paralogs, OsCPR5.1 and OsCPR5.2, were edited on a cultivar of japonica named Kitaake. Only OsCPR5.1 loss-of-function mutants were found to be completely resistant to RYMV, whereas OsCPR5.2 loss-of-function mutants exhibited some degree of susceptibility. OsCPR5.1's N-terminal domain short in-frame deletions or replacement were observed to increase the susceptibility of engineered lines to RYMV (Arra et al., 2023).

CRISPR/Cas9 technology has also been used successfully to develop virus resistance in horticultural crops. By employing CRISPR/Cas9 technology for the deletion of the eIF4e gene in cucumber plants (*Cucumis sativus*), resistance against papaya ringspot virus-W, zucchini yellow mosaic virus, and cucumber vein yellowing virus (CVYV) has been induced (Chandrasekaran et al., 2016). Precise silencing of the tomato susceptible 1 (SUS1) gene using CRISPR/Cas9 technology can confer resistance to potato virus X (PVX) and tobacco mild green mosaic virus (TMGMV). Targeted degradation of the ectodomain of the PsbP protein allows citrus species to develop resistance to the citrus tristeza virus, which effectively prevents viral infections and blocks the binding of viral receptors. Cauliflower mosaic virus has undergone GE to produce robust viral resistance using Cas9-mediated multiplex targeting of the viral coat protein sequence (Liu et al., 2018).

In addition to the CRISPR/Cas9 technique, RNA interference-based techniques have enabled precise targeting of viral transcripts and manipulation of viral genomes and host factors (Viswanath et al., 2023). RNA silencing in soybeans has been used to study gene function and develop tolerance against viral infections (Kasai and Kanazawa, 2012). RNAi has also been reported to control cassava brown streak disease significantly, with 85% of transgenic plant lines becoming resistant to upland species (Patil et al., 2011). Australian papaya cultivars have similarly gained immunity to papaya ringspot virus through post-transcriptional gene silencing (Lines et al., 2002). Likewise, sugarcane resistance to begomoviruses has risen due to multiplex gRNA-based GE (Surya Krishna et al., 2023). Resistance to the papaya ringspot virus coat protein gene was introduced into transgenic papaya plants using RNAi expression (Halder et al., 2022), which was also shown in apples treated with apple stem groove virus (ASGV) (Rajput et al., 2021). Barley stripe mosaic virus-mediated gene silencing hindered turnip yellow virus propagation in rapeseed. Suppressed PLRV CP gene expression via RNAi interrupts potato genome packaging, creating PLRV-resistant potato plants (Kumar et al., 2021).

The utilization of viral vectors to deliver GE elements to plant cells has further improved the efficiency of GE and facilitated the treatment of plant viral diseases. For example, turnip mosaic virus can be controlled with GE techniques by using *Agrobacterium tumefaciens* to introduce infectious viral DNA into plants (Grimsley et al., 1986). GE techniques have been employed to treat the Tobacco Etch Virus after identifying the capsid protein gene and surrounding genomic regions. Success in GE for combating viral attacks has been achieved in sugarcane plants as the first transgenic event for virus resistance came from inserting the sugarcane mosaic virus coat protein gene through particle bombardment. The modified PNRSV viral vector successfully silences host factors in Prunus fruit trees, which might help manage virus infections and aid in functional genomic research (Cui and Wang, 2017).

In summary, GE techniques such as CRISPR/Cas9, RNAi, and viral vectors have shown significant potential for combating viral plant diseases, increasing crop yields, and ensuring food security worldwide (Table 11.2). The future of plant genetics and agriculture holds great potential with the use of advanced technologies.

11.5 KNOWLEDGE GAPS AND FUTURE DIRECTIONS

In particular, for commercially significant crops, CRISPR/Cas systems provide an abundance of novel concepts and tools for producing loss- and gain-of-function mutations in genes in plants. The interactions and conflicts between host plants and viruses are endless. Plant virologists have several difficulties in their efforts to exploit the CRISPR/Cas9 system to stop the spread of plant DNA viruses. While plant viruses can evolve due to sequence differences induced by GE methods,

TABLE 11.2
Current Achievements of Plant Genome Editing Technologies

S. No.	Achievements	Crops	Genome Editing Techniques Used	References
1	CRISPR/Cas9-generated resistance to PRSV-W, ZVMV, and CVYV in cucumber plants	Cucumber (*Cucumis sativus)*	CRISPR/Cas9	Chandrasekaran et al. (2016)
2	Rice plants become entirely immune to RTBV through CRISPR/Cas9 edits to elF4G	Rice (*Oryza sativa*)	CRISPR/Cas9	Macovei et al. (2018)
3	Silencing of tomato susceptible I (SUS1) gene by CRISPR/Cas9 leads to PVX and TMGMV resistance	Tomato (*Solanum lycopersicum*)	CRISPR/Cas9	Viswanath et al. (2023)
4	Begomoviruse resistance in sugarcane via multiplex gRNA-based genome editing	Sugarcane (*Saccharum officinarum*)	CRISPR/ Cas9+multiplex gRNA	Surya Krishna et al. (2023)
5	Wheat acquires geminivirus resistance via genome editing and geminivirus-based vectors	Wheat (*Triticum aestivum*)	Geminivirus-based vectors	Ramesh et al. (2017)
6	Post-transcriptional gene silencing in papaya cultivars to gain immunity from PRSV	Papaya (*Carica papaya*)	Post-transcriptional gene silencing	Lines et al. (2002)
7	Cassava brown streak disease controlled by RNAi	Cassava (*Manihot esculenta*)	RNAi	Patil et al. (2011)
8	PLRV-resistant potato plants generated via RNAi suppression of CP gene expression	Potato (*Solanum tuberosum*)	RNAi	Kumar et al. (2021)
9	Resistance to Apple stem grooving virus (ASGV) in apple trees through RNAi	Apple (*Malus domestica*)	RNAi	Rajput et al. (2021)
10	Introduction of RYMV resistance to rice via OsCPR5.1 gene editing	Rice (*Oryza sativa*)	Knockout of genes	Arra et al. (2023)
11	Gene function study and development of tolerance against viral infections	Soybean (*Glycine max* L.)	RNA silencing	Kasai and Kanazawa (2012)
12	Generated resistance to cotton leaf curl virus	Cotton (*Gossypium herbaceum*)	CRISPR/Cas9	Binyameen et al. (2021)
13	Developed virus resistance to cauliflower mosaic virus	Cauliflower (*Brassica oleracea* var. *botrytis*)	CRISPR/Cas9	Liu et al. (2018)

endogenous genes can go dormant through mutation or recombination. Recombination and mutation are the driving forces behind plant RNA/DNA viruses (Cao et al., 2020). By employing CRISPR/Cas9 to target viral open reading frames, more virus variants may emerge. These variations may then cause different degrees of virus escape events, raising the possibility that this system might produce and disseminate mutant viruses from plants even in the presence of numerous gRNAs (Wani et al., 2023.) When two single gRNAs are combined, especially if they are far apart, resistance breakdown is much delayed as compared to a single sgRNA (Wani et al., 2023). The off-target consequences of the CRISPR/Cas9 system in *A. thaliana* plants suggest that there may be unforeseen risks associated with this technique (Zhang et al., 2020). Additionally, it should be taken into consideration when using CRISPR/Cas9 in the future since the presence of folded dsRNA domains in sgRNAs may cause the formation of small interfering RNAs and a decrease in the abundance of sgRNAs. Consequently, viruses may be able to evade the CRISPR/Cas-mediated antiviral defense mechanism in plants, indicating that the application of this technique in antiviral engineering may have drawbacks. This approach presents the crucial concern of novel viral strains or species formed as byproducts of GE, which may accelerate viral evolution or cause transgenic plants to lose resistance to their particular virus, even though it has the ability to alter or destroy the viral genome.

In plant biology research, GE may be able to provide new avenues for investigation. Gene editing has been made easier by CRISPR/Cas technology, which has shown promise in enhancing a number of plant characteristics, including resistance to viruses, fungi, bacteria, and insects, as well as increased yield (Varanda et al., 2021). Consequently, the application of CRISPR/Cas9 technology may soon offer a successful and widely accepted technique for crop enhancement against viruses. While multiple research studies have shown that genome engineering methods are effective in developing plant viral resistance, it is yet unclear if these methods work in open-field trials in natural settings. We have not yet evaluated the designed resistance's durability. The lack of a host component required by the virus results in recessive resistance, which is expected to be more permanent than dominant R genes because there are fewer selection pressures on the virus to create counterdefense measures. Testing for resistance to recessive mutants created by CRISPR/Cas9 is still ongoing.

Worldwide food production is at risk due to the significant decrease in crop production caused by DNA/RNA viruses. The lack of naturally occurring germplasm resources containing viral resistance within the crop gene pool is the main obstacle to the development of virus-resistant crops. Before making any attempts to modify a gene, it is crucial to comprehend its entire function. Due to intellectual property rights, CRISPR/Cas-mediated GE is an expensive technique; as a result, farmers with limited resources may not be able to afford CRISPR/Cas-modified crops. A major obstacle to CRISPR/Cas-mediated GE is the Cas9 enzyme's reduced selectivity, which can result in off-target DSBs. While the genome of rice is relatively tiny, the genomes of maize and wheat are comparatively large and repetitive/duplicated, which can increase the off-target effect (Raghurami Reddy et al., 2022). However, by carefully crafting the gRNAs, off-target alterations can be reduced to a minimum. When designing guide RNAs with little to no off-target action using bioinformatic analysis, the target cultivar's gene/genome sequence availability might be useful. To reduce the consequences of off-target mutations, genome-edited plants can be backcrossed with their wild-type parent, although this might delay the distribution of the finished variety to consumers. While next-generation sequencing cannot correct errors caused by off-target mutations, recent advancements in the technology provide potential ways to swiftly detect lines with such mutations. To develop the desired line with only the intended mutation, it can be paired with backcrossing, where the off-target mutations can be segregated out.

Faster breeding for novel pest- and disease-resistant cultivars has been made possible by CRISPR/Cas-mediated GE and the recent development of phenomics technology. It may also be used to add desired traits through cis-genics within the constraints of the regulatory framework. By appropriately modifying the plant epigenome, the present CRISPR/Cas-mediated GE tool for plants may be expanded to protect pathogen development and colonization. Even if there are still obstacles to overcome, CRISPR/Cas-mediated GE has opened up a new path for faster breeding that can be used to create crop varieties resistant to biotic and abiotic stressors.

11.6 CONCLUSION

Recent developments in plant GE have improved production, quality, and resilience to biotic and abiotic stress in important crops, including wheat, maize, and rice. The development of new crop varieties with broad-spectrum disease resistance and the best genotypic combinations should now get top priority. Growing advances in plant genetics and biotechnology offer promising avenues for environmentally benign viral disease management. In addition to providing higher resistance to several hazardous species, stacking genes increases their longevity and lowers the likelihood that new herbivore resistance may evolve. The application of modern genomic breeding techniques holds considerable potential for producing disease-resistant crops in the future. Genetically modified crops are becoming accepted by many countries as well, although more attention has to be paid to food safety and resistance management. It is important to consider local systems, their constraints, and socio-economic effects before introducing such genetically engineered materials.

Importantly, there is a growing need to look for management strategies that consider viral biology, plant–virus interactions, and their effects on natural enemies to increase the usefulness of resistant crops.

REFERENCES

Abdelkhalek, A., & Hafez, E. (2019, November 28). Plant viral diseases in Egypt and their control. In *Cottage industry of biocontrol agents and their applications* (pp. 403–421). https://doi.org/10.1007/978-3-030-33161-0_13

Angon, P. B., Mondal, S., Akter, S., Sakil, M. A., & Jalil, M. A. (2023). Roles of CRISPR to mitigate drought and salinity stresses on plants. *Plant Stress*, 8, 100169.

Arra, Y., Auguy, F., Stiebner, M., Chéron, S., Wudick, M. M., Miras, M., Schepler-Luu, V., Köhler, S., Cunnac, S., Frommer, W. B., & Albar, L. (2023). Rice Yellow Mottle Virus resistance by genome editing of the Oryza sativa L. ssp. *japonica* nucleoporin gene *OsCPR5.1* but not *OsCPR5.2*. *Plant Biotechnology Journal*. https://doi.org/10.1111/pbi.14266

Balmer, D., Planchamp, C., & Mauch-Mani, B. (2013). On the move: Induced resistance in monocots. *Journal of Experimental Botany*, *64*(5), 1249–1261. https://doi.org/10.1093/jxb/ers248

Baulcombe, D. C., Hamilton, W. D. O., Mayo, M. A., & Harrison, B. D. (2007, September 28). Resistance to viral disease through expression of viral genetic material from the plant genome. In *CIBA Foundation Symposium 133: Plant Resistance to Virus* (pp. 170–184). https://doi.org/10.1002/9780470513569.ch12

Becker, S., & Boch, J. (2021). TALE and TALEN genome editing technologies. *Gene and Genome Editing*, *2*, 100007.

Binyameen, B., Khan, Z., Khan, S. H., Ahmad, A., Munawar, N., Mubarik, M. S., Riaz, H., Ali, Z., Khan, A. A., Qusmani, A. T., Abd-Elsalam, K. A., & Qari, S. H. (2021). Using multiplexed CRISPR/Cas9 for suppression of cotton leaf curl virus. *International Journal of Molecular Sciences*, *22*, 12543. https://doi.org/10.3390/ijms222212543

Boch, J., & Bonas, U. (2010). Xanthomonas AvrBs3 family-type III effectors: Discovery and function. *Annual review of Phytopathology*, *48*, 419–436.

Borrelli, V. M. G., Brambilla, V., Rogowsky, P., Marocco, A., & Lanubile, A. (2018). The enhancement of plant disease resistance using CRISPR/Cas9 TECHNOLOGY. *Frontiers in Plant Science*, *9*, 1–15.

Calil, I. P., & Fontes, E. P. (2017). Plant immunity against viruses: Antiviral immune receptors in focus. *Annals of Botany*, *119*(5), 711–723. https://doi.org/10.1093/aob/mcw200

Cao, Y., Zhou, H., Zhou, X., & Li, F. (2020). Control of plant viruses by CRISPR/Cas system-mediated adaptive immunity. *Frontiers in Microbiology*, *11*, 593700.

Carroll, D. (2011). Genome engineering with zinc-finger nucleases. *Genetics*, *188*(4), 773–782.

Casteel, C. L., De Alwis, M., Bak, A., Dong, H., Whitham, S. A., & Jander, G. (2015). Disruption of ethylene responses by Turnip mosaic virus mediates suppression of plant defense against the green peach aphid vector. *Plant physiology*, *169*(1), 209–218. https://doi.org/10.1104/pp.15.00332

Cathomen, T., & Joung, J. K. (2008). Zinc-finger nucleases: The next generation emerges. *Molecular Therapy*, *16*(7), 1200–1207.

Chandrasekaran, J., Brumin, M., Wolf, D., Leibman, D., Klap, C., Pearlsman, M., Sherman, A., Arazi, T., & Gal-On, A. (2016). Development of broad virus resistance in non-transgenic cucumber using CRISPR/Cas9 technology. *Molecular Plant Pathology*, *17*, 1140–1153. https://doi.org/10.1111/mpp.12375

Chavhan, R. L., Verma, M. K., Hinge, V. R., Kadam, U. S., Kokane, S. B., & Chakrabarty, P. K. (2018). Sequence analysis of coat protein and molecular profiling of sunflower necrosis virus (SNV) strains from Indian subcontinent. *Journal of Plant Biochemistry and Biotechnology*, *27*, 28–35. https://doi.org/10.1007/s13562-017-0412-z

Chung, B., Nam, N., Ju-Yeon, Y., & Peter, P. (2013). Engineered resistance in potato against potato leafroll virus, potato virus A and potato virus Y. *Virus Genes*, *47*, 86–92. https://doi.org/10.1007/s11262-013-0904-4

Cui, H., & Wang, A. (2017). An efficient viral vector for functional genomic studies of *Prunus* fruit trees and its induced resistance to *Plum pox virus* via silencing of a host factor gene. *Plant Biotechnology Journal*, *15*, 344–356. https://doi.org/10.1111/pbi.12629

De Silva Feelixge, H. S., Stone, D., Roychoudhury, P., Aubert, M., & Jerome, K. R. (2018). CRISPR/Cas9 and genome editing for viral disease: Is resistance futile? *ACS Infectious Diseases*, *4*(6), 871–880.

De-wen, Q. I. U. (2014). Progress and prospect of plant immunity inducer. *Journal of Agricultural Science & Technology*, *16*(1) 1008–0864.

Elmore, J. M., Lin, Z. J. D., & Coaker, G. (2011). Plant NB-LRR signaling: Upstreams and downstreams. *Current Opinion in Plant Biology*, *14*(4), 365–371. https://doi.org/10.1016/j.pbi.2011.03.011

Flores, R., Navarro, B., Delgado, S., Serra, P., & Di Serio, F. (2020). Viroid pathogenesis: A critical appraisal of the role of RNA silencing in triggering the initial molecular lesion. *FEMS Microbiology Reviews*, *44*(3), 386–398.

Gaj, T., Gersbach, C. A., & Barbas, C. F. (2013). ZFN, TALEN, and CRISPR/Cas-based methods for genome engineering. *Trends in Biotechnology*, *31*(7), 397–405.

Garcia-Ruiz, H. (2018). Susceptibility genes to plant viruses. *Viruses*, *10*(9), 484.

Grimsley, N., Hohn, B., Hohn, T., & Walden, R. (1986). "Agroinfection," an alternative route for viral infection of plants by using the Ti plasmid. *Proceedings of the National Academy of Sciences*, *83*, 3282–3286. https://doi.org/10.1073/pnas.83.10.3282

Hadidi, A., Flores, R., Candresse, T., & Barba, M. (2016). Next-generation sequencing and genome editing in plant virology. *Frontiers in Microbiology*, *7*, 1325.

Halder, K., Chaudhuri, A., Abdin, M. Z., Majee, M., & Datta, A. (2022). RNA interference for improving disease resistance in plants and its relevance in this clustered regularly interspaced short palindromic repeats-dominated era in terms of dsRNA-based biopesticides. *Frontiers in Plant Science*, *13*, 885128. https://doi.org/10.3389/fpls.2022.885128

He, S., & Krainer, K. M. C. (2020). Pandemics of people and plants: Which is the greater threat to food security?. *Molecular plant*, *13*(7), 933–934. https://doi.org/10.1016/j.molp.2020.06.007

Hinge, V. R., Chavhan, R. L., Kale, S. P., Suprasanna, P., & Kadam, U. S. (2021). Engineering resistance against viruses in field crops using CRISPR-Cas9. *Current Genomics*, *22*(3), 214. https://doi.org/10.2174/1389202922666210412102214

Huang, M., Wu, Z., Li, J., Ding, Y., Chen, S., & Li, X. (2023). Plant protection against viruses: An integrated review of plant immunity agents. *International Journal of Molecular Sciences*, *24*(5), 4453. https://doi.org/10.3390/ijms24054453

Hull, R., Harper, G., & Lockhart, B. (2000, September). Viral sequences integrated into plant genomes. *Trends in Plant Science*, *5*(9), 362–365. https://doi.org/10.1016/s1360-1385(00)01723-4

Iannella, M., De Simone, W., D"Alessandro, P., & Biondi, M. (2021). Climate change favours connectivity between virus-bearing pest and rice cultivations in sub-Saharan Africa, depressing local economies. *Peer Journal*, *9*, e12387. https://doi.org/10.7717/peerj.12387

Incarbone, M., & Dunoyer, P. (2013). RNA silencing and its suppression: Novel insights from in planta analyses. *Trends in Plant Science*, *18*(7), 382–392. https://doi.org/10.1016/j.tplants.2013.04.001

Jiang, L., Du, Z., Zhang, G., Wang, T., & Jin, G. (2022). Advances in RNA-silencing-related resistance against viruses in potato. *Genes*, *13*(5), 731.

Jones, J. D., & Dangl, J. L. (2006). The plant immune system. *Nature*, *444*(7117), 323–329. https://doi.org/10.1038/nature05286

Jones, R. A. (2021). Global plant virus disease pandemics and epidemics. *Plants*, *10*(2), 233. https://doi.org/10.3390/plants10020233

Kasai, M., & Kanazawa, A. (2012). RNA silencing as a tool to uncover gene function and engineer novel traits in soybean. *Breeding Science*, *61*, 468–479. https://doi.org/10.1270/jsbbs.61.468

Khandagale, K., & Nadaf, A. (2016, November). Genome editing for targeted improvement of plants. *Plant Biotechnology Reports*, *10*(6), 327–343. https://doi.org/10.1007/s11816-016-0417-4

Kovalev, N., Pogany, J., & Nagy, P. D. (2014). Template role of double-stranded RNA in tombusvirus replication. *Journal of Virology*, *88*(10), 5638–5651.

Kumar, J., Kumar, R. R., Das, D. K., Mohanty, A., Rajani, K., Kumari, N., Kumar, V., Kumar, S., Kumbhar, B. V., & Ranjan, T., (2022). Knockdown of capsid protein encoding novel ATPase domain inhibits genome packaging in potato leafroll virus. *3 Biotech*, *12*(3), 66.

Lines, R. E., Persley, D., Dale, J. L., Drew, R., & Bateson, M. F. (2002). Genetically engineered immunity to Papaya ringspot virus in Australian papaya cultivars. *Molecular Breeding*, *10*, 119–129. https://doi.org/10.1023/A:1020381110181

Liu, H., Soyars, C. L., Li, J., Fei, Q., He, G., Peterson, B. A., Meyers, B. C., Nimchuk, Z. L., & Wang, X. (2018). CRISPR /Cas9-mediated resistance to cauliflower mosaic virus. *Plant Direct*, *2*, e00047. https://doi.org/10.1002/pld3.47

Liu, J. Z., Li, F., & Liu, Y. (2017). Plant immunity against viruses. *Frontiers in Microbiology*, *8*, 520. https://doi.org/10.3389/978-2-88945-269-9

Macovei, A., Sevilla, N. R., Cantos, C., Jonson, G. B., Slamet-Loedin, I., Čermák, T., Voytas, D. F., Choi, I., & Chadha-Mohanty, P. (2018). Novel alleles of rice *eIF4G* generated by CRISPR/Cas9-targeted mutagenesis confer resistance to *Rice tungro spherical virus*. *Plant Biotechnology Journal*, *16*, 1918–1927. https://doi.org/10.1111/pbi.12927

Mahas, A., & Mahfouz, M. (2018). Engineering virus resistance via CRISPR–Cas systems. *Current Opinion in Virology*, *32*, 1–8.

Matthews, R. E. F. (1980). Host plant responses to virus infection. In *Comprehensive virology: Virus-host interactions: Viral invasion, persistence, and disease* (Vol. 16, pp. 297–359). Springer. https://doi.org/10.1007/978-1-4613-3129-2_7

Mishra, R., & Zhao, K. (2018, January 23). Genome editing technologies and their applications in crop improvement. *Plant Biotechnology Reports*, *12*(2), 57–68. https://doi.org/10.1007/s11816-018-0472-0

Mochizuki, T., Ogata, Y., Hirata, Y., & Ohki, S. T. (2014). Quantitative transcriptional changes associated with chlorosis severity in mosaic leaves of tobacco plants infected with C ucumber mosaic virus. *Molecular Plant Pathology*, *15*(3), 242–254. https://doi.org/10.1111/mpp.12081

Mushtaq, M., Mukhtar, S., Sakina, A., Dar, A. A., Bhat, R., Deshmukh, R., Molla, K., Kundoo, A. A., & Dar, M. S. (2020, February). Tweaking genome-editing approaches for virus interference in crop plants. *Plant Physiology and Biochemistry*, *147*, 242–250. https://doi.org/10.1016/j.plaphy.2019.12.022

Napoli, C., Lemieux, C., & Jorgensen, R. (1990). Introduction of a chimeric chalcone synthase gene into petunia results in reversible co-suppression of homologous genes in trans. *Plant Cell*, *2*, 279–289.

Nelson, R. S., & Citovsky, V. (2005). Plant viruses. Invaders of cells and pirates of cellular pathways. *Plant Physiology*, *138*(4), 1809–1814. https://doi.org/10.1104/pp.104.900167

Nisa, R. T., Jan, S. K., Bhat, F. A., Rather, T. R., Nabi, A., Wani, A. A., Akbar, R., Anjum, N., Bashir, M., Shah, I. M., & Shabir, Z. (2022). Review on "crispr-cas9-a genome" editing tool for plant disease management. *Plant Cell Biotechnology and Molecular Biology*, 1–14. https://doi.org/10.56557/pcbmb/2022/v23i15-167619

Patil, B. L., Ogwok, E., Wagaba, H., Mohammed, I. U., Yadav, J. S., Bagewadi, B., Taylor, N. J., Kreuze, J. F., Maruthi, M. N., Alicai, T., & Fauquet, C. M. (2011). RNAi-mediated resistance to diverse isolates belonging to two virus species involved in Cassava brown streak disease. *Molecular Plant Pathology*, *12*, 31–41. https://doi.org/10.1111/j.1364-3703.2010.00650.x

Podhajska, A. J., & Szybalski, W. (1985, January). Conversion of the FokI endonuclease to a universal restriction enzyme: Cleavage of phage M13mp7 DNA at predetermined sites. *Gene*, *40(2–3)*, 175–182. https://doi.org/10.1016/0378-1119(85)90040-x

Raghurami Reddy, M., Acaso, J. T., Alakonya, A. E., Mangrauthia, S. K., Sundaram, R. M., Balachandran, S. M., & Biswal, A. K. (2022). Accelerating cereal breeding for disease resistance through genome editing. In Kaijun Zhao, Rukmini Mishra, & Raj Kumar Joshi (Eds.), *Genome editing technologies for crop improvement* (pp. 323–347). Springer.

Rajput, M., Choudhary, K., Kumar, M., Vivekanand, V., Chawade, A., Ortiz, R., & Pareek, N. (2021). RNA interference and CRISPR/cas gene editing for crop improvement: Paradigm shift towards sustainable agriculture. *Plants*, *10*, 1914. https://doi.org/10.3390/plants10091914

Ramesh, S., Sahu, P., Prasad, M., Praveen, S., & Pappu, H. (2017). Geminiviruses and plant hosts: A closer examination of the molecular arms race. *Viruses*, *9*, 256. https://doi.org/10.3390/v9090256

Singh, R. (2020, December 10). *Genome editing for plant disease resistance*. Springer. https://doi.org/10.1007/978-981-15-6275-4_25

Surya Krishna, S., Viswanathan, R., Valarmathi, R., Lakshmi, K., & Appunu, C. (2023). CRISPR/Cas-mediated genome editing approach for improving virus resistance in sugarcane. *Sugar Technology*, *25*, 735–750. https://doi.org/10.1007/s12355-023-01252-5

Unchwaniwala, N., Zhan, H., den Boon, J. A., & Ahlquist, P. (2021). Cryo-electron microscopy of nodavirus RNA replication organelles illuminates positive-strand RNA virus genome replication. *Current Opinion in Virology*, *51*, 74–79.

Uranga, M., & Daròs, J. (2022, June 14). Tools and targets: The dual role of plant viruses in CRISPR–Cas genome editing. *The Plant Genome*, *16*(2). https://doi.org/10.1002/tpg2.20220

Uranga, M., & Daròs, J. A. (2023). Tools and targets: The dual role of plant viruses in CRISPR–Cas genome editing. *The Plant Genome*, *16*(2), e20220.

Urnov, F. D., Rebar, E. J., Holmes, M. C., Zhang, H. S., & Gregory, P. D. (2010). Genome editing with engineered zinc finger nucleases. *Nature Reviews Genetics*, *11*(9), 636–646.

Van Vu, T., Das, S., Hensel, G., & Kim, J. Y. (2022, May 19). Genome editing and beyond: What does it mean for the future of plant breeding? *Planta*, *255*(6). https://doi.org/10.1007/s00425-022-03906-2

Varanda, C. M., Félix, M. D. R., Campos, M. D., Patanita, M., & Materatski, P. (2021). Plant viruses: From targets to tools for CRISPR. *Viruses*, *13*(1), 141.

Vats, S., Kumawat, S., Kumar, V., Patil, G. B., Joshi, T., Sonah, H., Sharma, T. R., & Deshmukh, R. (2019, November 4). Genome editing in plants: Exploration of technological advancements and challenges. *Cells*, *8*(11), 1386. https://doi.org/10.3390/cells8111386

Viswanath, K. K., Hamid, A., Ateka, E., & Pappu, H. R. (2023). CRISPR/Cas, multiomics, and RNA interference in virus disease management. *Phytopathology*. https://doi.org/10.1094/PHYTO-01-23-0002-V

Wani, F., Rashid, S., Wani, S., Saleem Bhat, S., Bhat, S., Tufekci, E. D., El Sabagh, A., Wani, S. H. & Hamid, A. (2023). Applications of genome editing in plant virus disease management: CRISPR/Cas9 plays a central role. *Canadian Journal of Plant Pathology*.

Weeks, D. P., Spalding, M. H., & Yang, B. (2015, August 11). Use of designer nucleases for targeted gene and genome editing in plants. *Plant Biotechnology Journal*, *14*(2), 483–495. https://doi.org/10.1111/pbi.12448

Wright, D. A., Li, T., Yang, B., & Spalding, M. H. (2014). TALEN-mediated genome editing: Prospects and perspectives. *Biochemical Journal*, *462*(1), 15–24.

Yin, K., & Qiu, J. L. (2019, January 14). Genome editing for plant disease resistance: Applications and perspectives. *Philosophical Transactions of the Royal Society B: Biological Sciences*, *374*(1767), 20180322. https://doi.org/10.1098/rstb.2018.0322

Zaidi, S. S. E. A., Mukhtar, M. S., & Mansoor, S. (2018). Genome editing: Targeting susceptibility genes for plant disease resistance. *Trends in Biotechnology*, *36*(9), 898–906.

Zhang, C., Liu, S., Li, X., Zhang, R., & Li, J. (2022, September 6). Virus-Induced Gene Editing and Its Applications in Plants. *International Journal of Molecular Sciences*, *23*(18), 10202. https://doi.org/10.3390/ijms231810202

Zhang, P., Du, H., Wang, J., Pu, Y., Yang, C., Yan, R., Yang, H., Cheng, H. & Yu, D. (2020). Multiplex CRISPR/Cas9-mediated metabolic engineering increases soya bean isoflavone content and resistance to soya bean mosaic virus. *Plant Biotechnology Journal*, *18*(6), 1384–1395.

Zhao, Y., Yang, X., Zhou, G., & Zhang, T. (2019, November 6). Engineering plant virus resistance: From RNA silencing to genome editing strategies. *Plant Biotechnology Journal*, *18*(2), 328–336. https://doi.org/10.1111/pbi.13278

12 CRISPR Plants against Fungal Diseases

Methods and Applications

Lara Jadhav, Richa Das, Shreni Agrawal, Indrani Bhattacharya, Pradeep Kumar, Amit Kumar Singh, Vishnu D. Rajput, Tatiana Minkina, Kavindra Nath Tiwari, and Sunil Kumar Mishra

12.1 INTRODUCTION

Due to the increasing population in the world and rapid urbanization, food demand is continuously increasing. In such a scenario, crop damage caused by pests and phytopathogenic microorganisms such as viruses, bacteria, and fungi is a major challenge against crop productivity and food supply worldwide. Among all pathogens, fungi are the most common cause of plant diseases and are severely infecting crops, thereby incurring heavy economic loss. There are 100,000 species of fungi known, out of which 10,000 can cause plant diseases [1]. Frequent outbreaks of fungal diseases in crops such as wheat rust caused by *Puccinia graminis*, potato late blight caused by *Phytophthora infestans*, rice blast caused by *Magnaporthe oryzae*, and soybean rust caused by *Phakopsora pachyrhizi* impose a growing threat to food security [2]. According to Fisher et al. [3] and Savary et al. [4], fungal infections were estimated to destroy at least 125 million tons of agricultural produce of crops such as wheat, rice, maize, and potatoes every year. According to Simion et al. [5], fungi contaminating animal feed can reduce milk production by up to 5–10% in cattle, which is a serious hurdle in fulfilling global nutritional needs. The crop damages caused by such fungal infection cost approximately $60 billion every year in agriculture globally (source: *Agriculture Today: The National Agriculture Magazine* 2012) [1]. In addition to this, 10% of the crops are spoiled post-harvest [6]. If it is possible to save some crops from damage by fungal infection, more than three million people can be fed [1]. Phytopathogenic fungi have become the reason for the untold misery of the farmers.

To control fungal infections in crops, several approaches have been used like chemical pesticides and fungicides, but these approaches have been proven harmful to the environment and public health because the residues of fungicides accumulated in crops and the environment. These accumulations can cause several environmental risks and threats to public health. A detailed description of different conventional methods used for controlling fungal diseases in plants is given in the upcoming sub-headings of this chapter.

To achieve fungal disease resistance in plants and to overcome the limitations of traditional methods, it is becoming necessary to reassess our strategies by adapting approaches such as developing mutation in plants that favor disease resistance against fungi. Making site-specific mutations in the bacterial and fungal genome has been readily possible since the mid-1980s when the process of homologous recombination could be used to insert a particular DNA sequence or a set of mutations (generated in vitro) into the genome of any organism at the locus of interest. However, creating site-specific mutations in the genes of plants and animals is very laborious in contrast to bacteria and fungi [7]. However, the advent of the clustered regularly interspaced palindromic

DOI: 10.1201/9781003543237-12

repeat (CRISPR)/CRISPR-associated (Cas9) gene editing tool allowed scientists to change the plant genomic sequences in a more error-free, simple, and precise way [8]. Different variants of CRISPR are extensively used in crop improvement applications like prolonging shelf-life, abiotic and biotic stress tolerance, improving flavor, grain quality, and nutritional enhancement because of its versatility and precision in gene editing [9]. Currently, researchers are extensively working toward the development of CRISPR/Cas9-mediated immunity in crops against bacterial, viral, and fungal pathogens [10]. This chapter is mainly focused on the application of CRISPR technology in crop improvement strategies and its use in making crops resistant to fungal diseases.

12.2 CONVENTIONAL METHODS OF FUNGAL DISEASE CONTROL

Conventional methods focus on improving plant growth by providing sufficient moisture and nutrients and by developing an environment in agricultural fields that is unfavorable for pathogenic microorganisms [11]. In 1997, Ogle and Brown stated that, when there is proper spacing and good ventilation provision between plants in the greenhouse, humidity is reduced to a great extent, which stops the growth of pathogens and leads to the inhibition of diseases caused by *Botrytis* spp. [12]. Crop rotation is an effective pathogen control strategy against various pathogenic fungi such as *Colletotrichum*, *Pyrenophora tritici-repentis*, *Phoma* spp., and *Gaeumannomyces graminis*. Root-rot-causing fungi such as *Aphanomyces*, *Pythium*, and *Fusarium* spp. and vascular wilt-causing fungi such as *Macrophomina phaseoli* and *Sclerotinia* spp. can be controlled by crop rotation. Crop rotation is one of the most ancient agricultural practices which facilitates weed and fungal control. Certain pathogens survive only in the presence of specific host plants, which means crop rotation practice can highly influence the soil microflora, thereby reducing the occurrence of pathogenic fungi and breaking the disease cycle [11]. Crop rotation is very important because continuous cropping can lead to an imbalance in soil microflora that leads to a decrease in the number of beneficial microbes and an increase in pathogenic bacteria [13–15]. However, crop rotation requires a specific type of machinery and equipment, which means higher costs for farmers, and also they should have proper knowledge of cropping patterns for different crops otherwise their agricultural practice will be a failure.

Another conventional method is sanitation practices, which involves the elimination of the source of inoculum by removing the infected plant parts in case of silver leaf of plums caused by *Chondrostereum purpureum*, apple canker caused by *Nectria galligena*, and powdery mildew of apples caused by *Podosphaera leucotricha* [11].

Tillage is another traditional agricultural practice that suppresses the spread of pathogens indirectly. This practice involves transferring the pathogens from the topsoil to the bottom layers of the soil, such that they are buried deep inside the soil. In addition to this, tillage practices create physiological changes such as changes in temperature, bulk density, aeration, and moisture that influence disease incidence in agricultural fields. However, there is a limitation associated with tillage machinery as it can become a source of infection, and minimum tillage practices can increase the growth of those pathogens that grow on plant residues left on the surface of the soil [11]. As mentioned before, conventional tillage practices have several effects on soil, such as physical, biological, or chemical properties that alter soil productivity and increase the risk of soil erosion. However, conservation tillage is a sustainable agricultural technique that can overcome the limitations of traditional tillage practices. Conservation tillage improves the crop water use efficiency, thereby improving the crop yield [16,17].

Burning is another technique that involves the elimination of infected plant parts, and this method kills the plant pathogens immediately [18]. Temperature of burning is an important factor in pathogen mortality, which was demonstrated in a study conducted in New South Wales, Australia. In this study, rice stubble was burned to eradicate *Rhizoctonia oryzae* and *R. oryzae-sativae*, which causes sheath diseases in rice. At that time, the amount of inoculum was reduced, but the duration of heat exposure and variable temperature in the rice crops may not have been enough to kill all the sclerotia of *R. oryzae*. Later on, another experiment showed that a temperature of 107°C or above for at least 90 seconds was needed to destroy 100% sclerotia [19]. Most often, burning of crop residues has been an efficient way of controlling pathogenic fungi when performed in combination with other techniques such as crop rotation or chemical tillage. Such combinations have been used to control Karnal bunt or partial bunt in wheat caused by *Tilletia indica* [20], ascochyta blight in chickpea caused by *Ascochyta rabiei* [21], and fusarium disease of sorghum and wheat [22]. In China, Fusarium wilt of banana caused by *Fusarium oxysporum f. sp. cubense* was controlled by removing and burning infected plant parts followed by fungicide treatment with triadimefon three times at the interval of 25 days [23]. Nevertheless, there are many examples where burning has reduced the incidence of disease but has not completely eradicated them. Burning has been used in many successful and unsuccessful pathogen control programs, but neither the tests for viable fungal pathogens in leftover debris and ash after burning nor the duration and temperature of burning have been recorded [24].

For a long time, chemical fungicides have been used to control diseases in crops. Chemical fungicides are highly effective in killing fungal pathogens, but their frequent use has led to several environmental issues, soil pollution, residual toxicity, and the development of resistance in pathogenic fungi against these fungicides [25]. Application of chemical fungicides to soil alters the microbial equilibrium. As fungicides are prepared to kill fungi, it is obvious to assume that all the fungal species, beneficial or harmful, will be susceptible. So, there is a high chance that beneficial fungi can be destroyed whose physiological actions contribute to higher agricultural yield. A fungicide named difenoconazole when applied at 500 mg/kg DW led to an overall decrease in microbial activity and activity of soil enzymes [26]. Many microorganisms are involved in the fixation of atmospheric nitrogen to nitrates. The process of nitrification and denitrification are disrupted because of the application of fungicides called dinitrophenyl and chlorothalonil [27]. A fungicide called ethylenebisdithiocarbamate is reported to be toxic to humans, and its implications include allergic skin reactions and inflammation in the upper respiratory tract and eyes [28–30]. Chlorothalonil and captafol are tested among the cancer-causing pesticides by the National Research Council [31]. Thiram reported acute oral toxicity in mammals that reduced egg-laying capacity in hens when elemental sulfur was sprayed against powdery mildew disease in grapes [32]. Records of groundwater pollution due to the accumulation of synthetic fungicides are also available [31]. The accumulation of copper-based fungicides used in banana farming caused the soil in Costa Rica to become toxic, making it unusable for producing any kind of crop [33]. As we have seen synthetic fungicides can impose negative implications on the environment, public health, and microbial and animal biodiversity, and such implications have compelled us to go for environment-friendly biopesticides and botanical fungicides. There are numerous plant metabolites reported to have antifungal properties. Some botanical fungicides are even commercialized, yet their market share is very small because this topic is a matter of intensive research [25].

12.3 MECHANISM OF CRISPR

CRISPR/Cas9 Technology is based on the type II acquired immune system, which is observed in bacteria. This system helps the bacteria protect against infectious pathogens and archaea. Initially, this system was observed in *Streptococcus pyogenes* [34]. It consists of two components: sgRNA (single-guide RNA) and Cas9. Cas9 is a DNA endonuclease that cleaves complementary and non-complementary DNA strands. The cleavage by Cas9 leads to blunt end formation in target DNA, which subsequently disrupts base pairs by the formation of frameshift mutation [35,36]. The second component is called sgRNA (single-guide RNA) or gRNA (guide RNA). sgRNA is formed by the fusion of crRNA and tracrRNA [37]. The role of sgRNA is to instruct and guide Cas9 endonuclease toward the targeted DNA so that Cas9 endonuclease can cleave the target DNA by generating double-strand breaks (DSBs) [38,39]. The sgRNA is engineered to possess two components as demonstrated in Figure 12.1.

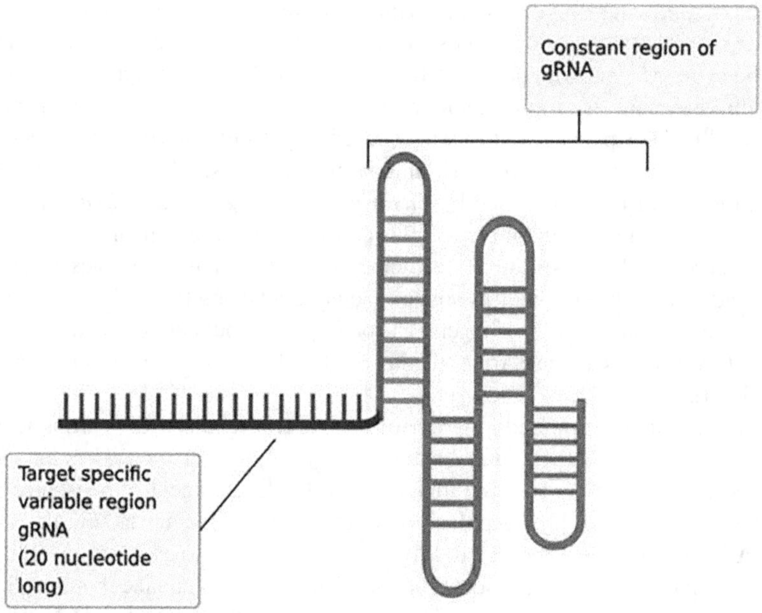

FIGURE 12.1 A diagrammatic representation of two components of sgRNA. The region in red color depicts the constant region responsible for binding between Cas9 protein and sgRNA. The region with green shows a 20-nucleotide-long target-specific variable region complementary to the target sequence on the host. sgRNA is engineered in such a way that it is complementary to the host sequence, and the variable region is edited according to the host sequences.

The first component of sgRNA is called the constant region, which helps in the scaffolding of Cas9 protein binding by forming various stem-loop structures. The second component of sgRNA is called the variable region, which is 20 bp long and can be altered according to the target sequence [40]. sgRNA and Cas9 protein combine to form a complex that recognizes the target DNA site containing PAM (protospacer adjacent motif). The homologous target site in a host which is recognized by this complex has two regions. The first region is complementary to gRNA, and the second region is known as PAM which is bound to Cas9. In other words, PAM (located in the host) serves Cas9 endonuclease to mark the target site in the host [40]. Cas9 proteins in *Staphylococcus aureus* (SaCas9) and *S. pyogenes* (SpCas9) prefer to bind with PAM sites rich in guanidine. Example of such PAM sites are the sequences such as 5'-NNGRRT-3' and 5'-NGG-3', respectively, in host homologous sequences. "NNGRRT" and "NGG" are the standard IUPAC notations where "NN" means any nucleotide can occupy this position, "G" stands for guanine, "RR" stands for purine (either adenine or guanine- this position is flexible for both purines) and "T" is for thymine [38,39]. Once the PAM sites are recognized, Cas9 binds with PAM sequences to produce DSBs [41,40]. The formation of DSBs activates DNA reparation pathways like non-homologous end joining (NHEJ) or homology-directed repair (HDR) [41,34]. A step-by-step mechanism of how CRISPR works is given in Figure 12.2.

The ability of CRISPR/Cas9 technology to form DSBs in the genes makes it an amazing tool for editing the genome of fungal pathogens and host plants severely affected by fungal infections.

Mildew-resistance LOCUS (MLO), which is a crucial gene in powdery mildew disease, was edited with the help of the CRISPR/Cas9 tool, which led to the resistance against *Blumeria graminis f. sp. hordei* (*Bgh*) in wheat. Mutations were induced in three homoeoalleles, namely, TaMLO-A1, TaMLO-B1, and TaMLO-D1, which encode for MLO loci in wheat, and these mutations conferred resistance against powdery mildew in wheat [42]. Two 'S' genes, DMR and MLO-6, were edited

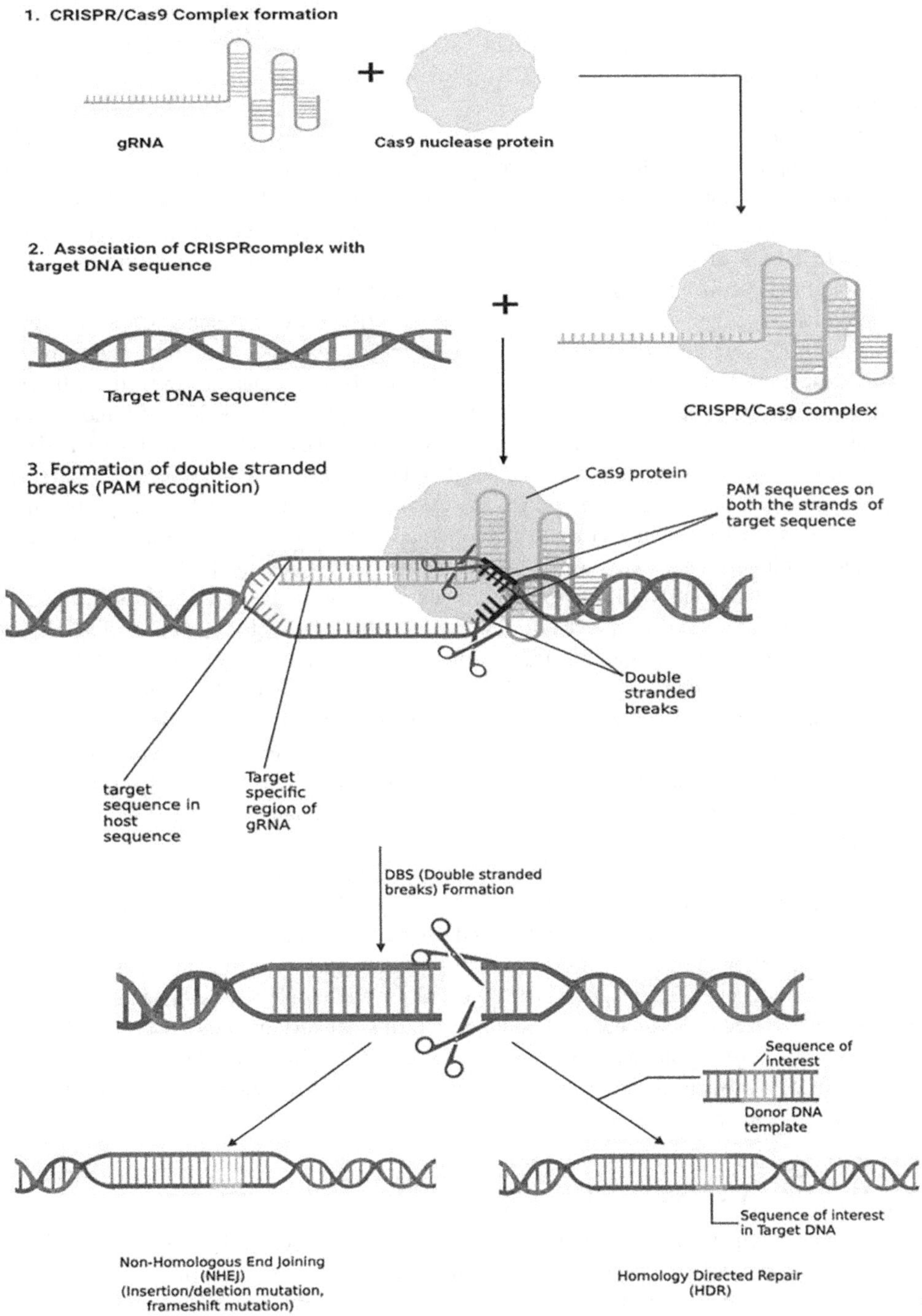

FIGURE 12.2 Step-by-step working of CRISPR/Cas9 tool in gene editing. There are mainly three steps involved. The first step is the formation of the CRISPR/Cas9 complex that is formed by the assembly of Cas9 nuclease protein and sgRNA. The second step involves the association of the CRISPR/Cas9 complex with the target DNA sequence that is supposed to be edited. In the third step, PAM sequences located on either strand of the target sequence are recognized and double-strand breaks (DSB) are formed. If the donor template is available, the target DNA sequence will be repaired by homology-directed repair (HDR); otherwise, the target DNA will be repaired by non-homologous end joining (NHEJ). And, finally, we achieve our transgenic DNA sequence.

with the help of the CRISPR/Cas9 tool that triggered resistance against downy mildew and powdery mildew (caused by (*Erysiphe necator*) in grapes [43]. StDMR6-1 and StDMR6-2 are the two genes that were simultaneously edited using the CRISPR/Cas9 tool, and this led to resistance against late blight disease in potatoes [44].

Something that plays an important role in the efficiency of gene editing in plants is the delivery of CRISPR components into the higher plant systems. Under the regulation of suitable promoters, the Cas9 and gRNA molecules are cloned in appropriate vectors, and then these vectors containing our CRISPR components are delivered into the plant systems with the help of different methods such as particle bombardment/biolistic transformation of plant, *Agrobacterium*-mediated T-DNA delivery, and PEG-mediated transformation of protoplast [45].

In the biolistic mediated transformation method, the CRISPR components (Cas9 nuclease proteins and sgRNA) were coated with tungsten, silver, or gold particles, and, with the help of high pressure, these coated particles were transferred into the plant systems; however, the efficiency of gene editing might be reduced because of the high pressure [46]. This method demands optimized conditions such as target distance, helium pressure, type of plant explant used, and particle size [46]. The main advantage of this delivery technique is that there is no need for any type of binary vector. However, some limitations of this method are lesser editing efficiency, arbitrary integration into the genome, being expensive in comparison to other methods, and also the bombardment into sites such as the nucleus, mitochondria, cytoplasm, or plastids cannot be controlled [47].

Agrobacterium-mediated CRISPR/Cas9 complex delivery is the most widely used delivery method in a wide range of plant species. In this method, a binary vector containing CRISPR components and their expression cassette is transformed into the strain of *Agrobacterium*, followed by the genetic transformation of the CRISPR/Cas9 construct into the favorable explant such as a leaf, callus, or floral parts [46]. The major advantage of this method is its higher efficiency in gene editing in comparison to other methods, easy availability, cost-effectiveness, and broad applicability. Stable transgene integration with single-copy integration is possible because of this method [48]. The only disadvantage of this method is that it introduces an alien gene in the plant body and it needs a binary vector [46].

In the PEG-mediated transformation of the protoplast, the plasmid containing CRISPR components is incubated with the protoplast in the presence of polyethylene glycol (PEG) [46]. The plasmid DNA is first sterilized by dissolving in water and filtering, and then it is mixed with a suspension of protoplast. After some time, PEC in the required concentrations is added to the protoplast. Then the protoplasts are regenerated in appropriate regeneration medium [49]. The main advantage of this method is that it is vector-less, and plants edited with this delivery method will probably have no regulatory or ethical issues and will be more acceptable because this method gives rise to DNA-free edited mutants of plants. However, protoplast isolation and establishment of suspension cells are the weaknesses of this method [46] (Table 12.1).

12.4 MULTIPLEX GENE EDITING

Multiplexing or multiple gene editing (MGE) is a versatile tool based on CRISPR/Cas9 technology that allows the generation of direct mutations simultaneously at multiple loci in a gene or multiple genes with high precision, and this method is known for its ability to introduce mutations at multiple nucleotide level in our target genome [56]. This tool uses Cas9 nucleases to generate two or more DSBs precisely and simultaneously in one round at specific loci inside the genome [57]. For example, 14 distinct genes were mutated simultaneously by a single construct and this is demonstrated in *Arabidopsis* [58].

Furthermore, MGE can help incorporate several desirable crop improvements such as yielding seedless crop species and better resistance against diseases [59,60]. MGE was used to develop nutritionally enhanced and climate-resilient crops, safeguard food security, and attain zero hunger goals [61]. MGE was used to increase disease resistance in *Brassica napus*, against *Sclerotinia*

TABLE 12.1
Fungal Resistance in Different Crops Achieved through CRISPR/Cas9 Technology

Plant Name	Target Gene	Type of Delivery Method	Trait Improvement	References
Solanum tuberosum L	StDMR6-1 and StDMR6-2	Agrobacterium C58	Editing confers increased late blight resistance	[44]
Vitis vinifera	VvWRKY52	Agrobacterium EHA105	Resistance to graymold (Botrytis cinerea)	[50]
Vitis vinifera	MLO-7	CRISPR-RNPs/ PEG-protoplast transformation	Resistance against grape wine powdery mildew (Erysiphe necator)	[51]
Triticum aestivum	TaMLO-A1	Biolistic bombardment	Heritable broad-spectrum resistance to Blumeria graminis f.sp. tritici (powdery mildew)	[52]
Cotton	Gh14-3-3d	Agrobacterium VIGS-mediated GV3101	Resistance to Verticillium dahlia infestation	[53]
Papaya	alEPIC8/cds region	Agrobacterium EHA105	Increased resistance against Phytophthora palmivora	[54]
Theobroma cacao	TcNPR3, a suppressor of the defense response	Agrobacterium infiltration	Increased resistance to infection with the cacao pathogen Phytophthora tropicalis	[55]

sclerotiorum (a pathogenic fungus) by editing two copies of the *BnWRKY11* gene and four copies of the *BnWRKY70* gene [62]. A number of multiplex genome editing strategies were revealed in crops, and all of them were based on a common approach, which was bringing multiple gRNAs together under the control of U3 or U6 promoter into a single construct [63–66]. The ISU maize CRISPR platform allows the cloning of up to four gRNAs in maize for gene targeting [67]. Furthermore, recent multiplex gene editing systems use the self-cleavage ability of RNA molecules that contain tRNA sequences, and this strategy is employed in polycistronic tRNA–gRNA (PTG) systems. PTG system-based multiplexing was performed to cause hereditable mutations in *TaLpx-1* and *TaMLO* genes of hexaploid wheat. Gene constructs with alternating sequences of sgRNA and tRNA under the control of either U3 or U6 promoter help in reducing the construct size and in limiting the risk of silencing due to repetitions of promoter sequences [68]. The PTG system-based multiplexing and gene targeting is shown in Figure 12.3.

So, we can say that, in multiplex gene editing, a number of gRNAs are assembled to form a gene construct, where each gRNA carries its promoter and generates edits on targeted regions of the desired gene [69]. The advantage of delivering several gRNA (multiplexing) is that the remodeling of different intended and unintended genomic targets can be addressed properly at the same time with a single CRISPR system transported in a plant system [70]. Contrary to the CRISPR/Cas9 approach that requires numerous components including gRNA, polymerase III promoter like U3 or U6, and terminator, the multiplexing approach is a streamlined approach of gene targeting and editing that speeds up the construction of these gene editing components mentioned above [71]. Multiplexing is also used for molecular farming, controlling regulatory pathways, metabolic engineering, and stacking multiple traits [72].

12.5 APPLICATIONS OF CRISPR IN CROP IMPROVEMENT

CRISPR/Cas9 tool is a very important tool for efficient, fruitful, eco-friendly, and climate-resilient agricultural practices. This gene editing tool provides a huge potential to speed up the pace of research areas in fungi such as medical mycology, plant physiology, and biotechnology by reducing the time

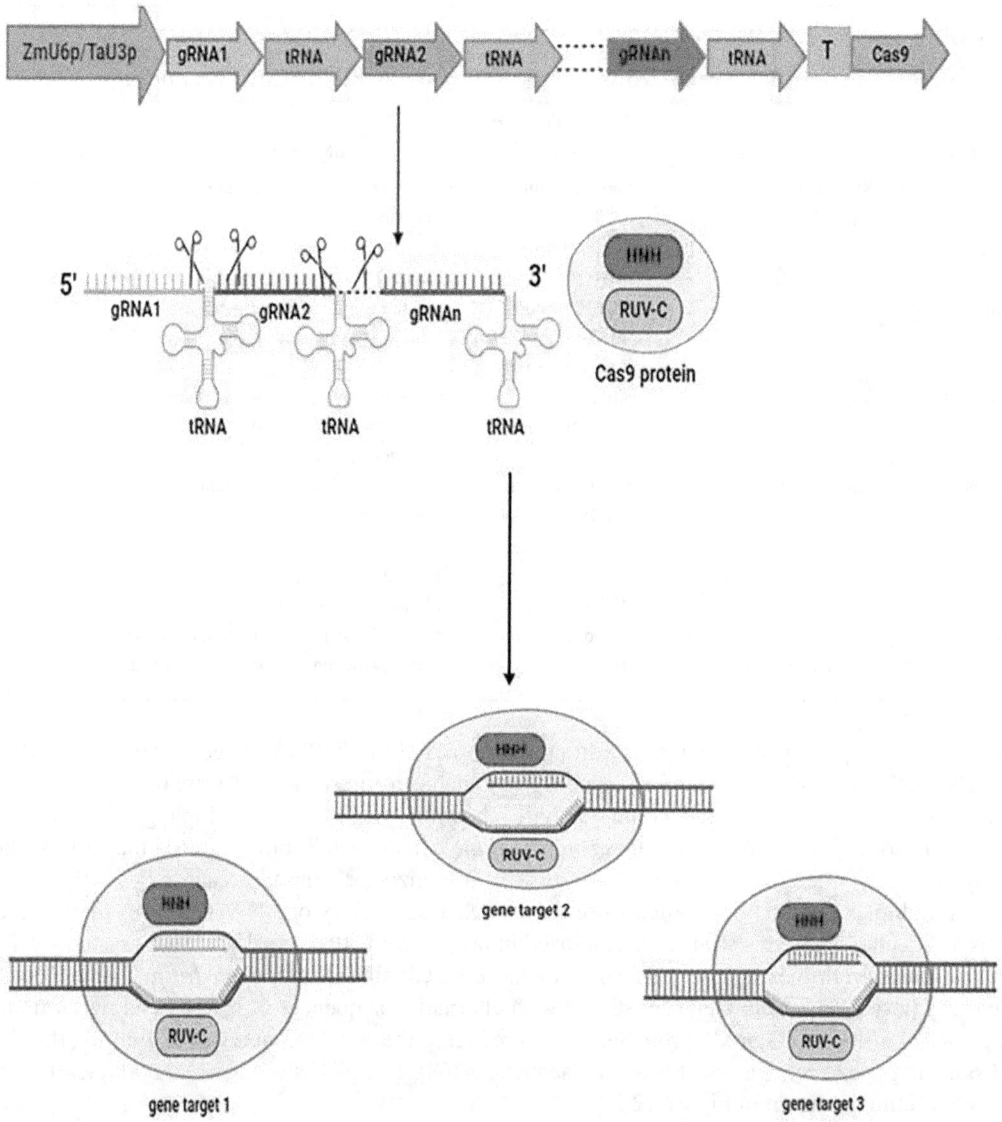

FIGURE 12.3 Diagram of polycistronic tRNA–gRNA (PTG) gene construct and Cas9 gene targeting activity. PTG is upregulated by the ZmU6 promoter or TaU3 promoter, and it is made up of alternating sequences of sgRNA and tRNA. HNH and RUV-C are the catalytic domains for Cas9 and Cpf1 (alternative to SpCas9), respectively. Here the endogenous endonuclease is cutting the tRNA ends, and corresponding gene targeting is carried out. Different gRNA is demonstrated with different colors (i.e., yellow, violet, and red).

needed to achieve several objectives like deletion of targeted genes, single nucleotide changes, or overexpression [73]. This technology also offers researchers the ability to target gene families and make multiple mutations [74] as well as make mutants in polyploidy and dikaryotic fungi [75].

Rice blast fungus *M. oryzae* uses a CRISPR/Cas9 genome editing system to target single genes in vivo, targeting Cas9 and CRISPR components [76]. However, due to labor-intensive cloning procedures, this method of producing mutants is rarely extensively used. Therefore, Foster et al. developed a method called ribonucleoprotein-CRISPR/Cas9 (RNP-CRISPR/Cas9) system for genome editing in *M. oryzae* using purified Cas9 and in vitro generated sgRNA, to extend CRISPR-based gene editing applications. Based on previous observations, they found that stable expression of Cas9 in *M. oryzae* was toxic. So, Foster et al. wanted to see if CRISPR/Cas9 genome editing could be accomplished by transiently introducing purified Cas9 protein and in vitro synthesized sgRNAs into *M. oryzae* protoplasts. They obtained nuclear-localized Cas9 (Cas9-NLS) from the commercial suppliers and complexed it with sgRNAs capable of directing Cas9 to the ALB1 gene. They also tried a second RNP that targeted the RSY1 gene. They inserted Cas9-NLS-sgRNA ribonucleoprotein complexes into the wild-type *M. oryzae* strain Guy11, together with donor DNAs bearing an insertion harboring a selectable marker. The researchers demonstrated highly efficient targeting of both genes, with RSY1 mutation being almost 100% successful in several studies. Targeting ALB1 was also more effective than 50% of the time, with 70–80% of albino transformants produced. The findings show that RNP-CRISPR/Cas9 produced DSBs that induce HR repair, enabling efficient gene manipulation in *M. oryzae* [73]. The transient nature of CRISPR/Cas9-ribonucleoprotein production provides an efficient method of avoiding Cas9 toxicity and rapidly creating gene-edited mutants in *M. oryzae*. The main benefits of RNP-mediated CRISPR/Cas9 editing are its efficiency, precision, and, most importantly, its speed. Employing RNP-CRISPR/Cas9 will potentially help in manipulating the gene in *M. oryzae* and other fungi quickly and precisely, thereby speeding up research on this economically significant plant disease [77].

Cotton cancer, also called cotton verticillium wilt, caused by *Verticillium dahliae* causes a severe economic loss of US$250–310 million in China annually [78]. The development of disease-resistant cultivars is the best measure to avoid economic loss caused by *V. dahliae*. In *Gossypium hirsutum* (a commercial species of cotton plant), few resistant genes against *Verticillium dahliae* are found naturally. Thus, it is important to develop defense mutants using the CRISPR/Cas9 tool targeting the negative regulators of disease resistance in cotton such as cotton *14-3-3c/d*, *CYP82D*, and *NINJA* [79–82]. Silencing of these two genes called *14-3-3c* and *14-3-3d* through virus-induced gene silencing (VIGS) can immensely increase the resistance against *V. dahliae* in cotton [79]. Zhang et al. created a CRISPR/Cas9-mediated genome editing system in plants with simple and practical target sequences. Using this system, they edited the cotton target gene, Gh14-3-3d, to produce indels at predicted target locations. The CRISPR/Cas9 genome editing system produced many nucleotide insertions and deletions in T0 plants at the predicted locations of the Gh14-3-3d target genes. They screened 16 T1 lines and obtained two transgene-clean editing plants (designated ce1 and ce2 lines in T2) with homozygous indels in the tetraploid cotton At and Dt subgenomes. These plants have high resistance to *V. dahliae* compared to the wild type and can be directly used as a germplasm to breed resistant cultivars [82].

F. oxysporum, which is caused by *F. oxysporum f.sp. niveum* (FON), is a devastating infection of watermelon worldwide. FON spreads in xylem vessels and results in hollow dried-out stems [83]. Therefore, Zhang et al. [84] employed CRISPR/Cas9 for mutagenesis of a gene called Clpsk1 gene to generate mutant lines of watermelon which showed enhanced resistance to FON disease. Clpsk1 encodes for phytosulfokine precursors, which is a plant hormone, and it is known to play a role in plant immunity [85]. They found that the knockout of Clpsk1 rendered watermelon seedlings more resistant to FON infection. The study highlights the effectiveness of CRISPR/Cas9-mediated gene modification for improving disease resistance in watermelons.

CRISPR/Cas9 technology was also applied in the simultaneous editing of three homoeoalleles in hexaploid bread wheat using TALENs and CRISPR/Cas9 technologies. Wang et al. conducted a study to confer heritable resistance to powdery mildew. Specifically, the study shows that it is possible to introduce heritable resistance to powdery mildew, a destructive plant disease caused *B. graminis f.* sp. *tritici*, by mutating all three TaMLO homoeologs in bread wheat. This broad-spectrum resistance to powdery mildew can be highly desirable in improving the productivity and sustainability of polyploid crops. Additionally, the study highlights the feasibility of targeted DNA insertion in bread wheat through NHEJ, which opens up possibilities for introducing other beneficial genetic traits in polyploid crops. Overall, this research offers a promising approach to enhancing the genetic potential and resilience of polyploid crops through precise genome editing techniques like CRISPR/Cas9 [42].

12.6 CHALLENGES AND FUTURE ASPECTS

Gene editing technologies are a game changer in basic and applied plant research. CRISPR-based accurate gene editing has resulted in crops with high nutritional content and improved agronomic traits. Apart from this CRISPR has helped improve plant architecture and increase photosynthetic efficiency to overcome problems of malnutrition and to achieve zero hunger goals [61].

In addition to the genes that are found in the nucleus, there are several hundred genes that negatively regulate photosynthesis and respiration present in the genome of organelles like chloroplast and mitochondria. These genomes have immense potential for engineering more efficient crops by CRISPR-based manipulation that would result in higher photosynthetic efficiency and biomass production. Conversion of C3 crops like rice, barley, and wheat into C4 crops (that have higher photosynthetic efficiency and higher water use efficiency) can be achieved in the near future through CRISPR/Cas9 [61].

To reach zero hunger objectives, the CRISPR/Cas technology can improve the efficiency and accuracy of crop breeding. It is possible to reduce time and boost output by generating two-line hybrids. The high expenses of raising a segregating F2 population, on the contrary, can be mitigated by retaining hybrid vigor by CRISPR. This has the potential to alter hybrid development and breeding programs, allowing for the production of smart and efficient crops such as climate-resilient crops, self-compatibility in economically vital food crops, and material preparation for cross-species hybrids [61].

A very important technology that enables gene editing is plant transformation. Development in the methods of delivering CRISPR/Cas9 system into plant genomes have been made, but researchers need to develop even better delivery methods as the transformation ability is still very low in plants with complex genomes such as cotton or wheat. Also, CRISPR/Cas9 technology requires tissue culture, which is laborious and time-consuming. It takes 12 months to regenerate the mutant plants of rice, tomato, cotton, and wheat [61]. The success of every delivery method is determined by the tissue type and subsequent regeneration into entire plants. The characteristics of plant species, tissue type, and culture method all influence regeneration problems. Regeneration concerns vary depending on the kind of plant species, tissue, and culture technique. As a result, a study conducted by Sandhy et al. highlighted the necessity of creating novel delivery systems for CRISPR/Cas9 components, such as pollen magnetofection and nanoparticle-mediated delivery. With these two viable delivery strategies, researchers may avoid the tedious and time-consuming tissue culture process [46]. Also, a number of plant species such as chili pepper (*Capsicum frutescens*) are recalcitrant or non-compatible with tissue-culturing. As a result, there is a need for tissue culture-free genome editing systems such as using RNA-virus-based systems for delivering gRNAs [86].

Natural selection in crops is a process that takes many years to acquire desirable traits, and this is a limitation that we are facing in achieving the zero hunger goal. This limitation can be overcome by CRISPR/Cas-mediated directed evolution of target genes, which can be helpful to produce highly functional genes within a short period [87]. This technology can help evolve essential plant genes to increase plant yield and also bypass natural selection [88].

Conventional crop domestication, a process to make a crop suitable for cultivation, is an extremely long process, and this can be sped up by introducing CRISPR-based MGE systems [89]. Despite of many advantages of CRISPR, certain factors limit the utility of this technology such as the designing of sgRNA, specificity of PAM, types of promoters, and non-target or off-target editing. Off-target genome editing and continuous evolution of the pathogen genome can lead to additional abnormalities and mutations [34]. Several reports have stated different types of 'off-target' effects by CRISPR systems, which include toxic effects on the product and raise ethical and safety implications [90]. Therefore, it is important to develop more well-organized and precise strategies such as unbiased and biased off-target detection methods, improved Cas variants, gRNA engineering and modification, development of anti-CRISPR proteins, efficient CRISPR system delivery methods, and effective prime-editing and base-editing systems [91].

Plants possess a large number of genes that control numerous traits. Therefore, the development of CRISPR-based systems to insert large parts of multiple genes or chromosomes (pyramiding and gene stacking, respectively) can be extremely advantageous for crop improvement [61].

12.7 CONCLUSION

As mentioned in the above section, the increasing global population and urbanization have led to a growing threat to food security due to crop damage caused by pests and phytopathogenic microorganisms. Fungi, the most common cause of plant diseases, are responsible for destroying at least 125 million tons of agricultural produce annually. Traditional methods like chemical pesticides and fungicides have been proven harmful to the environment and public health. In this chapter, we have also discussed conventional agricultural approaches which try to increase plant development by supplying moisture and nutrients while also creating an unfavorable environment for harmful microbes. Crop rotation is a successful disease management approach for fungi such as *Colletotrichum*, *P. tritici-repentis*, *Phoma* spp., and *G. graminis*. Crop rotation, on the contrary, necessitates the use of specialized gear and equipment, which can be costly for farmers. Tillage indirectly reduces disease transmission by shifting pathogens from topsoil to soil, whereas sanitation methods entail the removal of sick plant components. Conservation tillage is a long-term approach for increasing crop water usage efficiency and production. Another strategy that eliminates plant diseases instantly is burning, although its efficiency has been restricted. Chemical fungicides have been employed to manage agricultural diseases, but their widespread usage has resulted in environmental difficulties. To combat this, researchers are developing CRISPR/Cas9 gene editing tools to create site-specific mutations in plants that favor disease resistance against fungi. The CRISPR/Cas9 technology, which disrupts base pairs and creates DSBs, is based on the type II acquired immune system found in bacteria. It consists of two components: sgRNA and Cas9. Because CRISPR/Cas9 technology can create DSBs in genes, it is an effective tool for modifying the genomes of fungal pathogens and host plants that are badly damaged by fungal infections. Table 12.1 shows a list of different crops that have achieved resistance against fungal diseases through CRISPR/Cas technology. Section 12.3 elaborates different delivery systems used to deliver CRISPR components into the plant body.

We have focused on multiplex gene editing separately as it is an extremely wondrous tool based on CRISPR/Cas9 technology. We have mentioned how CRISPR/Cas9 has been applied in various crop plants to achieve resistance against fungal pathogens like *Magnaporthe oryzae*, *Verticillium dahlia*, *F. oxysporum f.sp. niveum*, and *B. graminis f. sp. Tritici*. There are numerous such fascinating studies conducted by researcher enthusiasts all across the world on different crops, which is beyond the scope of this chapter.

We have also discussed the prospects of this technology, and we mentioned how CRISPR-based manipulation of organelle genes can enhance crop efficiency by converting C3 crops like rice, barley, and wheat into C4 crops with higher photosynthetic and water use efficiency; how CRISPR/Cas technology can enhance crop breeding efficiency, reduce time, and produce two-line hybrids, potentially reducing costs and enabling smarter, climate-resilient, and economically vital crops; how CRISPR/Cas technology can bypass natural selection in crops by rapidly evolving essential genes, increasing yield and bypassing the long-term process of acquiring desirable traits. Apart from this, we discussed different challenges and limitations of the CRISPR/Cas9 tool that need to be addressed through new scientific studies.

Plant biotechnology has undergone a revolution thanks to CRISPR/Cas since it has made it possible to modify specific genomes, diversify breeding methods, and create hundreds of crop varieties with exceptional agronomic performance. Researchers may now include desired mutations in the genome thanks to the development of innovative techniques including transgene-free plant genome editing, prime editing, and base editing during the past ten years. Numerous plant biotechnologies have been created or enhanced, such as multiplexed gene editing techniques for functional genomics research, morphogenetic regulators, transgene-free plant genome editing, and targeted gene expression regulation. However, before CRISPR/Cas may be widely used in plants, further research is required as these methods have not yet satisfied all requirements for altering plant genomes. The CRISPR technique has the potential to transform agriculture by increasing crop species' resilience to pests, diseases, and abiotic stress, enabling sustainable crop development and a green revolution, and potentially becoming a breeding tool for next-generation agriculture.

REFERENCES

1. Mahmood, S., Lakra, N., Marwal, A., Sudheep, N. Á., & Anwar, K. (2017). Crop genetic engineering: An approach to improve fungal resistance in plant systems. In D. P. Singh, H. B. Singh, & R. Prabha (Eds.), *Plant-microbe interactions in agro-ecological perspectives: Microbial interactions and agro-ecological impacts* (Vol. 2, pp. 581–591). Springer, Singapore.
2. Pennisi, E. (2010). Armed and dangerous. *Science (New York, NY), 327*(5967), 804–805.
3. Fisher, M. C., Henk, D. A., Briggs, C. J., Brownstein, J. S., Madoff, L. C., McCraw, S. L., & Gurr, S. J. (2012). Emerging fungal threats to animal, plant and ecosystem health. *Nature, 484*(7393), 186–194.
4. Savary, S., Ficke, A., Aubertot, J. N., & Hollier, C. (2012). Crop losses due to diseases and their implications for global food production losses and food security. *Food Security, 4*(4), 519–537.
5. Simion, V. E. (2018). Dairy cows health risk: Mycotoxins. In M. Abubakar (Ed.), *Ruminants: The husbandry, economic and health aspects* (pp. 85–104). IntechOpen.
6. Fisher, M. C., Hawkins, N. J., Sanglard, D., & Gurr, S. J. (2018). Worldwide emergence of resistance to antifungal drugs challenges human health and food security. *Science, 360*(6390), 739–742.
7. Newman, M., & Ausubel, F. M. (2016). Introduction to gene editing and manipulation using CRISPR/Cas9 technology. *Current Protocols in Molecular Biology, 115*(1), 31–4.
8. Knott, G. J., & Doudna, J. A. (2018). CRISPR-Cas guides the future of genetic engineering. *Science, 361*(6405), 866–869.
9. Liu, Q., Yang, F., Zhang, J., Liu, H., Rahman, S., Islam, S., ... & She, M. (2021). Application of CRISPR/Cas9 in crop quality improvement. *International Journal of Molecular Sciences, 22*(8), 4206.
10. Zaidi, S. S. E. A., Mahas, A., Vanderschuren, H., & Mahfouz, M. M. (2020). Engineering crops of the future: CRISPR approaches to develop climate-resilient and disease-resistant plants. *Genome Biology, 21*(1), 1–19.
11. Kannan, K., Nivas, D., Kannan, V. R., & Bastas, K. K. (2015). 5 agro-traditional practices of plant pathogens control. *Sustainable Approaches to Controlling Plant Pathogenic Bacteria*, 111–121.
12. Brown, J. F., & Ogle, H. J. (1997). *Plant pathogens and plant diseases.* Rockvale Publications for the Division of Botany, University of New England.
13. Lu, L., Yin, S., Liu, X., Zhang, W., Gu, T., Shen, Q., & Qiu, H. (2013). Fungal networks in yield-invigorating and-debilitating soils induced by prolonged potato monoculture. *Soil Biology and Biochemistry, 65*, 186–194.
14. Li, J. G., Ren, G. D., Jia, Z. J., & Dong, Y. H. (2014). Composition and activity of rhizosphere microbial communities associated with healthy and diseased greenhouse tomatoes. *Plant and Soil, 380*, 337–347.
15. Fu, H., Zhang, G., Zhang, F., Sun, Z., Geng, G., & Li, T. (2017). Effects of continuous tomato monoculture on soil microbial properties and enzyme activities in a solar greenhouse. *Sustainability, 9*(2), 317.
16. Hobbs, P. R., Sayre, K., & Gupta, R. (2008). The role of conservation agriculture in sustainable agriculture. *Philosophical Transactions of the Royal Society B: Biological Sciences, 363*(1491), 543–555.
17. Zhang, X., Li, Q., Zhu, A., Liang, W., Zhang, J., & Steinberger, Y. (2012). Effects of tillage and residue management on soil nematode communities in North China. *Ecological Indicators, 13*(1), 75–81.
18. Ebbels, D. L. (2003). *Principles of plant health and quarantine.* CABI Publishing.
19. Lanoiselet, V. M., Cottier, E. J., Ash, G. J., Hind-Lanoiselef, T. L., Murray, G. M., & Harper, J. D. I. (2005). Prevalence and survival, with emphasis on stubble burning, of Rhizoctonia spp., causal agents of sheath diseases of rice in Australia. *Australasian Plant Pathology, 34*, 135–142.
20. Singh, B. B., Aujla, S. S., & Sharma, I. (1993). Integrated management of wheat Karnal bunt. *International Journal of Pest Management, 39*(4), 431–434.
21. Gan, Y. T., Siddique, K. H. M., MacLeod, W. J., & Jayakumar, P. (2006). Management options for minimizing the damage by ascochyta blight (Ascochyta rabiei) in chickpea (Cicer arietinum L.). *Field Crops Research, 97(2–3)*, 121–134.
22. Burgess, L. W., Backhouse, D., Swan, L. J., & Esdaile, R. J. (1996). Control of Fusarium crown rot of wheat by late stubble burning and rotation with sorghum. *Australasian Plant Pathology, 25*, 229–233.

23. Lin, X. (2004). Investigation of the occurrence of banana wilting disease and its control. *South China Fruits, 33*, 64–65.
24. Sosnowski, M. R., Fletcher, J. D., Daly, A. M., Rodoni, B. C., & Viljanen-Rollinson, S. L. H. (2009). Techniques for the treatment, removal and disposal of host material during programmes for plant pathogen eradication. *Plant Pathology, 58*(4), 621–635.
25. Yoon, M. Y., Cha, B., & Kim, J. C. (2013). Recent trends in studies on botanical fungicides in agriculture. *The Plant Pathology Journal, 29*(1), 1.
26. Muñoz-Leoz, B., Garbisu, C., Charcosset, J. Y., Sánchez-Pérez, J. M., Antigüedad, I., & Ruiz-Romera, E. (2013). Non-target effects of three formulated pesticides on microbiallymediated processes in a clay-loam soil. *Science Total Environment, 449*, 345–354.
27. Man, L. A. N. G., & Zucong, C. A. I. (2009). Effects of chlorothalonil and carbendazim on nitrification and denitrification in soils. *Journal of Environmental Sciences, 21*(4), 458–467.
28. Laborie, F., Laborie, R., & Dedieu, E. H. (1964). Allergy to fungicides of the Maneb and Zineb type prophylaxis. *Archives des Maladies Professionnelles, 25*, 419–424.
29. Wolf, F., & Jung, M. D. (1970). Akute kontaktdermatiden nachumgang nut Nemtin (Vapam). *Z ges Hyg Grenzgeb, 16*, 423.
30. Zorin, P. M. (1970). Ahergicheskiy dermatit voznikshiy v.rezultate kontakta tsinebom. *Vestnik dermatologii i venerologii, 44*(2), 65.
31. Sanjay Kumar, G., Vineeta, S., Hillol, C., & Prassan, C. (2018). Harmful effects of fungicides-current status. *Plant Pathology, 11,* 1011–1019.
32. Waibel, P. E., Pomeroy, B. S., & Johnson, E. L. (1955). Effect of Arasan-treated corn on laying hens. *Science, 80*(121), 401–402.
33. Thrupp, L. A. (1991). Long-term losses from accumulation of pesticide residues: A case of persistent copper toxicity in soils of Costa Rica. *Geoforum, 22*, 1–15.
34. Talakayala, A., Ankanagari, S., & Garladinne, M. (2022). CRISPR-Cas genome editing system: A versatile tool for developing disease resistant crops. *Plant Stress, 3*, 100056.
35. Doudna, J. A., & Charpentier, E. (2014). The new frontier of genome engineering with CRISPR-Cas9. *Science, 346*(6213), 1258096.
36. Makarova, K. S., Wolf, Y. I., Alkhnbashi, O. S., Costa, F., Shah, S. A., Saunders, S. J., … & Koonin, E. V. (2015). An updated evolutionary classification of CRISPR: Cas systems. *Nature Reviews Microbiology, 13*(11), 722–736.
37. Jinek, M., Chylinski, K., Fonfara, I., Hauer, M., Doudna, J. A., & Charpentier, E. (2012). A programmable dual-RNA–guided DNA endonuclease in adaptive bacterial immunity. *Science, 337*(6096), 816–821.
38. Ran, F. A., Cong, L., Yan, W. X., Scott, D. A., Gootenberg, J. S., Kriz, A. J., … & Zhang, F. (2015). In vivo genome editing using *Staphylococcus aureus* Cas9. *Nature, 520*(7546), 186–191.
39. Kaya, H., Mikami, M., Endo, A., Endo, M., & Toki, S. (2016). Highly specific targeted mutagenesis in plants using *Staphylococcus aureus* Cas9. *Scientific Reports, 6*(1), 26871.
40. Zhu, L. J. (2015). Overview of guide RNA design tools for CRISPR-Cas9 genome editing technology. *Frontiers in Biology, 10*(4), 289–296.
41. Muñoz, I. V., Sarrocco, S., Malfatti, L., Baroncelli, R., & Vannacci, G. (2019). CRISPR-Cas for fungal genome editing: A new tool for the management of plant diseases. *Frontiers in Plant Science, 10*, 135.
42. Wang, Y., Cheng, X., Shan, Q., Zhang, Y., Liu, J., Gao, C., & Qiu, J. L. (2014). Simultaneous editing of three homoeoalleles in hexaploid bread wheat confers heritable resistance to powdery mildew. *Nature Biotechnology, 32*(9), 947–951.
43. Giacomelli, L., Zeilmaker, T., Malnoy, M., Rouppe van der Voort, J., & Moser, C. (2018, July). Generation of mildew-resistant grapevine clones via genome editing. In *XII International Conference on Grapevine Breeding and Genetics* (Vol. 1248, pp. 195–200). Bordeaux, France
44. Kieu, N. P., Lenman, M., Wang, E. S., Petersen, B. L., & Andreasson, E. (2021). Mutations introduced in susceptibility genes through CRISPR/Cas9 genome editing confer increased late blight resistance in potatoes. *Scientific Reports, 11*(1), 4487.
45. Kalinina, N. O., Khromov, A., Love, A. J., & Taliansky, M. E. (2020). CRISPR applications in plant virology: Virus resistance and beyond. *Phytopathology, 110*(1), 18–28.
46. Sandhya, D., Jogam, P., Allini, V. R., Abbagani, S., & Alok, A. (2020). The present and potential future methods for delivering CRISPR/Cas9 components in plants. *Journal of Genetic Engineering and Biotechnology, 18*, 1–11.
47. Baltes, N. J., Gil-Humanes, J., & Voytas, D. F. (2017). Chapter one: Genome engineering and agriculture: Opportunities and challenges. In: D. P. Weeks & B. Yang (Eds.), *Progress in molecular biology and translational science* (vol, 149, pp. 1–26). Academic Press.

48. Travella, S., Ross, S. M., Harden, J., Everett, C., Snape, J. W., & Harwood, W. A. (2005). A comparison of transgenic barley lines produced by particle bombardment and Agrobacterium-mediated techniques. *Plant Cell Report*, *23*(12), 780–789.
49. Lowder, L. G., Zhang, D., Baltes, N. J., Paul, J. W., Tang, X., Zheng, X., Voytas, D. F., Hsieh, T.-F., Zhang, Y., & Qi, Y. (2015). A CRISPR/Cas9 toolbox for multiplexed plant genome editing and transcriptional regulation. Plant Physiology, *169*(2), 971–985.00636.
50. Leisen, T., Bietz, F., Werner, J., Wegner, A., Schaffrath, U., Scheuring, D., … & Hahn, M. (2020). CRISPR/Cas with ribonucleoprotein complexes and transiently selected telomere vectors allows highly efficient marker-free and multiple genome editing in Botrytis cinerea. *PLoS Pathogens*, *16*(8), e1008326.
51. Malnoy, M., Viola, R., Jung, M. H., Koo, O. J., Kim, S., Kim, J. S., … & Nagamangala Kanchiswamy, C. (2016). DNA-free genetically edited grapevine and apple protoplast using CRISPR/Cas9 ribonucleoproteins. *Frontiers in Plant Science*, *7*, 1904.
52. Wang, Y., Cheng, X., Shan, Q., Zhang, Y., Liu, J., Gao, C., & Qiu, J. L. (2014). Simultaneous editing of three homoeoalleles in hexaploid bread wheat confers heritable resistance to powdery mildew. *Nature Biotechnology*, *32*(9), 947–951.
53. Zhang, T., Zheng, Q., Yi, X., An, H., Zhao, Y., Ma, S., & Zhou, G. (2018). Establishing RNA virus resistance in plants by harnessing CRISPR immune system. *Plant Biotechnology Journal*, *16*(8), 1415–1423.
54. Gumtow, R., Wu, D., Uchida, J., & Tian, M. (2018). A Phytophthora palmivora extracellular cystatin-like protease inhibitor targets papain to contribute to virulence on papaya. *Molecular Plant-Microbe Interactions*, *31*(3), 363–373.
55. Fister, A. S., Landherr, L., Maximova, S. N., & Guiltinan, M. J. (2018). Transient expression of CRISPR/Cas9 machinery targeting TcNPR3 enhances defense response in Theobroma cacao. *Frontiers in Plant Science*, *9*, 268.
56. Modamed, A., Zheng, W., & Kaijun, Z. (2021). Multiplex genome editing technologies for revolutionizing plant biology and crop improvement. *Frontiers in Plant Science*, *12*, 1–15.
57. Cong, L., Ran, F. A., Cox, D., Lin, S., Barretto, R., Habib, N., … & Zhang, F. (2013). Multiplex genome engineering using CRISPR/Cas systems. *Science*, *339*(6121), 819–823.
58. Peterson, B. A., Haak, D. C., Nishimura, M. T., Teixeira, P. J., James, S. R., Dangl, J. L., & Nimchuk, Z. L. (2016). Genome-wide assessment of efficiency and specificity in CRISPR/Cas9 mediated multiple site targeting in Arabidopsis. *PloS One*, *11*(9), e0162169.
59. Naim, F., Dugdale, B., Kleidon, J., Brinin, A., Shand, K., Waterhouse, P., & Dale, J. (2018). Gene editing the phytoene desaturase alleles of Cavendish banana using CRISPR/Cas9. *Transgenic Research*, *27*, 451–460.
60. Tripathi, J. N., Ntui, V. O., Ron, M., Muiruri, S. K., Britt, A., & Tripathi, L. (2019). CRISPR/Cas9 editing of endogenous banana streak virus in the B genome of Musa spp. overcomes a major challenge in banana breeding. *Communication Biology*, *2*, 46.
61. Ahmad, S., Tang, L., Shahzad, R., Mawia, A. M., Rao, G. S., Jamil, S., … & Tang, S. (2021). CRISPR-based crop improvements: A way forward to achieve zero hunger. *Journal of Agricultural and Food Chemistry*, *69*(30), 8307–8323.
62. Sun, Q.; Lin, L.; Liu, D.; Wu, D.; Fang, Y.; Wu, J.; Wang, Y. (2018). CRISPR/Cas9-mediated multiplex genome editing of the BnWRKY11 and BnWRKY70 genes in *Brassica napus L. International Journal of Molecular Sciences*, *19*(9), 2716.
63. Ma, Y., Zhang, L., & Huang, X. (2014). Genome modification by CRISPR/Cas9. *The FEBS Journal*, *281*(23), 5186–5193.
64. Xing, H. L., Dong, L., Wang, Z. P., Zhang, H. Y., Han, C. Y., Liu, B., … & Chen, Q. J. (2014). A CRISPR/Cas9 toolkit for multiplex genome editing in plants. *BMC Plant Biology*, *14*(1), 1–12.
65. Zhou, H., Liu, B., Weeks, D. P., Spalding, M. H., & Yang, B. (2014). Large chromosomal deletions and heritable small genetic changes induced by CRISPR/Cas9 in rice. *Nucleic Acids Research*, *42*(17), 10903–10914.
66. Xu, R., Yang, Y., Qin, R., Li, H., Qiu, C., Li, L., … & Yang, J. (2016). Rapid improvement of grain weight via highly efficient CRISPR/Cas9-mediated multiplex genome editing in rice. *Journal of Genetics and Genomics*, *43*(8), 529–532.
67. Char, S. N., Neelakandan, A. K., Nahampun, H., Frame, B., Main, M., Spalding, M. H., ... & Yang, B. (2017). An Agrobacterium-delivered CRISPR/Cas9 system for high-frequency targeted mutagenesis in maize. *Plant Biotechnology Journal*, *15*(2), 257–268.
68. Wang, W., Pan, Q., He, F., Akhunova, A., Chao, S., Trick, H., & Akhunov, E. (2018). Transgenerational CRISPR-Cas9 activity facilitates multiplex gene editing in allopolyploid wheat. *The CRISPR Journal*, *1*(1), 65–74.
69. Borrelli, V. M., Brambilla, V., Rogowsky, P., Marocco, A., & Lanubile, A. (2018). The enhancement of plant disease resistance using CRISPR/Cas9 technology. *Frontiers in Plant Science*, *9*, 1245.

70. Das, D., Singha, D. L., Paswan, R. R., Chowdhury, N., Sharma, M., Reddy, P. S., & Chikkaputtaiah, C. (2022). Recent advancements in CRISPR/Cas technology for accelerated crop improvement. *Planta, 255*(5), 109.
71. Xie, K., Minkenberg, B., & Yang, Y. (2015). Boosting CRISPR/Cas9 multiplex editing capability with the endogenous tRNA-processing system. *Proceedings of the National Academy of Sciences of the United States of America, 112*(11), 3570–3575.
72. Najera, V. A., Twyman, R. M., Christou, P., & Zhu, C. (2019). Applications of multiplex genome editing in higher plants. *Current Opinion in Biotechnology, 59*, 93–102.
73. Shi, T. Q., Liu, G. N., Ji, R. Y., Shi, K., Song, P., Ren, L. J., … & Ji, X. J. (2017). CRISPR/Cas9-based genome editing of the filamentous fungi: The state of the art. *Applied Microbiology and Biotechnology, 101*, 7435–7443.
74. Fan, Y., & Lin, X. (2018). Multiple applications of a transient CRISPR-Cas9 coupled with electroporation (TRACE) system in the *Cryptococcus neoformans* species complex. *Genetics, 208*(4), 1357–1372.
75. Zhang, G. C., Kong, I. I., Kim, H., Liu, J. J., Cate, J. H., & Jin, Y. S. (2014). Construction of a quadruple auxotrophic mutant of an industrial polyploid Saccharomyces cerevisiae strain by using RNA-guided Cas9 nuclease. *Applied and Environmental Microbiology, 80*(24), 7694–7701.
76. Arazoe, T., Miyoshi, K., Yamato, T., Ogawa, T., Ohsato, S., Arie, T., & Kuwata, S. (2015). Tailor-made CRISPR/Cas system for highly efficient targeted gene replacement in the rice blast fungus. *Biotechnology and Bioengineering, 112*(12), 2543–2549.
77. Foster, A. J., Martin-Urdiroz, M., Yan, X., Wright, H. S., Soanes, D. M., & Talbot, N. J. (2018). CRISPR-Cas9 ribonucleoprotein-mediated co-editing and counterselection in the rice blast fungus. *Scientific Reports, 8*(1), 14355.
78. Wang, F., Wang, C., Liu, P., Lei, C., Hao, W., Gao, Y., … & Zhao, K. (2016). Enhanced rice blast resistance by CRISPR/Cas9-targeted mutagenesis of the ERF transcription factor gene OsERF922. *PloS One, 11*(4), e0154027.
79. Gao, W., Long, L., Zhu, L. F., Xu, L., Gao, W. H., Sun, L. Q., ... & Zhang, X. L. (2013). Proteomic and virus-induced gene silencing (VIGS) analyses reveal that gossypol, brassinosteroids, and jasmonic acid contribute to the resistance of cotton to Verticillium dahliae. *Molecular & Cellular Proteomics, 12*(12), 3690–3703.
80. Sun, L., Zhu, L., Xu, L., Yuan, D., Min, L., & Zhang, X. (2014). Cotton cytochrome P450 CYP82D regulates systemic cell death by modulating the octadecanoid pathway. *Nature Communications, 5*(1), 5372.
81. Wang, L., Wu, S. M., Zhu, Y., Fan, Q., Zhang, Z. N., Hu, G., … & Wu, J. H. (2017). Functional characterization of a novel jasmonate ZIM-domain interactor (NINJA) from upland cotton (Gossypium hirsutum). *Plant Physiology and Biochemistry, 112*, 152–160.
82. Zhang, Z., Ge, X., Luo, X., Wang, P., Fan, Q., Hu, G., ... & Wu, J. (2018). Simultaneous editing of two copies of Gh14-3-3d confers enhanced transgene-clean plant defense against Verticillium dahliae in allotetraploid upland cotton. *Frontiers in Plant Science, 9*, 842.
83. Zhang, M., Xu, J. H., Liu, G., Yao, X. F., Li, P. F., & Yang, X. P. (2015). Characterization of the watermelon seedling infection process by *Fusarium oxysporum* f. sp. niveum. *Plant Pathology, 64*(5), 1076–1084.
84. Zhang, M., Liu, Q., Yang, X., Xu, J., Liu, G., Yao, X., ... & Lou, L. (2020). CRISPR/Cas9-mediated mutagenesis of Clpsk1 in watermelon to confer resistance to Fusarium oxysporum f. sp. niveum. *Plant Cell Reports, 39*, 589–595.
85. Hammes, U. Z. (2015). Novel roles for phytosulfokine signalling in plant–pathogen interactions: Commentary on Rodiuc et al. "Evolutionarily distant pathogens require the Arabidopsis phytosulfokine signalling pathway to establish disease". *Plant, Cell & Environment, 39*(7), 1393–1395.
86. Kim, S. G. (2020). The way to true plant genome editing. *Nature Plants, 6*(7), 736–737.
87. Butt, H., Eid, A., Momin, A. A., Bazin, J., Crespi, M., Arold, S. T., & Mahfouz, M. M. (2019). CRISPR directed evolution of the spliceosome for resistance to splicing inhibitors. *Genome Biology, 20*(1), 1–9.
88. Butt, H., Zaidi, S. S. E. A., Hassan, N., & Mahfouz, M. (2020). CRISPR-based directed evolution for crop improvement. *Trends in Biotechnology, 38*(3), 236–240.
89. Lacchini, E., Kiegle, E., Castellani, M., Adam, H., Jouannic, S., Gregis, V., & Kater, M. M. (2020). CRISPR-mediated accelerated domestication of African rice landraces. *PLoS One, 15*(3), e0229782.
90. Ahmad, N., Rahman, M. U., Mukhtar, Z., Zafar, Y., & Zhang, B. (2020). A critical look on CRISPR-based genome editing in plants. *Journal of Cellular Physiology, 235*(2), 666–682.
91. Naeem, M., Majeed, S., Hoque, M. Z., & Ahmad, I. (2020). Latest developed strategies to minimize the off-target effects in CRISPR-Cas-mediated genome editing. *Cells, 9*(7), 1608.

13 Plant Bacterial Disease Tolerance

Molecular Mechanisms and Breeding Strategies

Abdul Razak Ahmed, Kübra Budak Tek, and Mümin Ibrahim Tek

13.1 INTRODUCTION

Bacteria, microscopic single-celled organisms typically measuring 1–2 μm in size, play a dual role in association with plants, acting as either beneficial symbionts or harmful pathogens. Plant surfaces host a diverse array of microbes, termed epiphytes, while others, known as endophytes, reside within plant tissues. These microorganisms, including bacteria, undergo successive colonization of plants as they mature. While individual bacterial cells remain invisible without microscopic aid, they coagulate into visible populations, forming biofilms on plant surfaces, viscous suspensions obstructing plant vessels, or colonies on laboratory Petri dishes. To exert their beneficial or pathogenic effects, bacterial populations typically need to reach higher densities, enabling them to utilize as effective control agents or induce infectious diseases.

Plant pathogenic bacteria are responsible for numerous serious diseases affecting plants globally (Vidhyasekaran, 2008). However, their prevalence is lower compared to fungi or viruses, resulting in relatively less damage and economic cost (Kennedy & Alcorn, 1980). Many plants, including both cultivated and wild varieties, possess innate immunity or resistance mechanisms against various pathogens. Nonetheless, numerous plants can host plant pathogens without exhibiting visible symptoms, a condition referred to as asymptomatic infection.

Bacteria typically undergo reproduction via binary fission, a process wherein one cell divides into two; however, this mechanism is intricate. Extrachromosomal DNA elements, known as plasmids, are typically replicated concurrently with the bacterial chromosome but can undergo natural loss or elimination through chemical manipulation, a process referred to as 'curing', under specific conditions. Numerous plant pathogens harbor plasmids: for instance, strains of *Pantoea* (formerly labeled as *Erwinia*) *stewartii* subsp. *stewartii*, responsible for Stewart's wilt of corn disease, may carry up to 13 plasmids, the functions of which remain undefined (Coplin et al., 1981). Genetic diversity within natural environments, such as agricultural fields, is likely underestimated due to insufficient sampling and characterization. For example, a single field may host at least seven pathogenic variants of *Clavibacter michiganensis* subsp. *nebraskensis*, the causative agent of Goss's wilt and blight in corn (Smidt & Vidaver, 1987).

Phytopathogenic bacteria are particularly prolific in warm climates, presenting a formidable challenge to global agricultural productivity and food security. Among these, gram-negative bacteria such as *Pseudomonas* spp., *Erwinia* sp., and *Ralstonia* sp. stand out as highly destructive, while gram-positive strains like *Clavibacter michiganensis* spp. and *Streptomyces* spp. also contribute significantly to economic losses. These pathogens infiltrate plant systems through various entry points, including natural openings such as hydathodes, stomata, lenticels, stigma, and nectarthodes, as well as through wounds inflicted by mechanical injury or insect vectors. As established

 DOI: 10.1201/9781003543237-13

in previous studies (Burkholder, 1948b; Harveson et al., 2015; Soto et al., 2009), once the bacteria are established within the plant's apoplast, they incite a spectrum of diseases marked by leaf and fruit spots, blights, chlorosis, necrosis, cankers, vascular wilts, and rots, as documented previously. Furthermore, the presence of seed-borne bacteria like *Xanthomonas euvesicatoria* pv. *euvesicatoria*, *X. euvesicatoria* pv. *perforans*, *Xanthomonas hortorum* pv. *gardneri*, and others *Xanthomonas vesicatoria* adds to the complexity of control efforts, as highlighted previously (Osdaghi et al., 2021). Given their widespread distribution and diverse manifestations, bacterial infections can result in yield losses of up to 44%, imposing severe economic repercussions on farmers globally.

Specifically, gram-negative bacteria belonging to the Pseudomonadaceae and Enterobacteriaceae are more effective in the colonization of the apoplast. Relatively, bacteria that colonize these portions of the plants trigger symptoms such as rots, spots, wilts, cankers, and blights in the host plants as clearly explained by Alfano and Collmer (1996). By extension, most of their life is spent in the intercellular spaces of the plant organs. The rate at which they eliminate their host plants is determined by the level of enzymes they secrete.

The general acceptance of bacteria as disease-causing agents in plants was much slower with regards to its acceptance in the research community in comparison to its other pathogen counterpart (fungi). Thus, general sensitization was much clearer in the early 1900s. During that period, disease-causing bacteria in plants are generally classified into two major genera-motile with peritrichous flagella *Erwinia* and polar flagellate *Phytomonas* (Burkholder, 1948a).

It is long overdue to develop effective strategies for bacterial disease resistance in plants. The global scientific community recommends the utilization of antibiotics, gene editing (GE), and cultural methods to combat phytopathogenic diseases. In a review paper authored by Verhaegen et al. (2023), antibiotics were proposed for the treatment of crucial bacterial plant pathogens, such as the use of streptomycin to control *Erwinia amylovora*. However, caution was advised regarding the potential development of antibiotic resistance by these pathogens, as observed in cases like *Xanthomonas oryzae* pv. *oryzae*'s resistance to gentamicin aminoglycoside, which could pose future risks. Fortunately, advanced breeding and biotechnology offer cutting-edge tools to aid in developing plants with enhanced resilience against disease. Genomic selection (GS) analyses genomic data to predict breeding values (Jannink et al., 2010). CRISPR/Cas9 enables targeted GE for precise modifications (Ahmar et al., 2020). Marker-assisted selection (MAS) identifies genetic markers linked to desired traits (Francia et al., 2005). Together, these techniques allow for accelerated breeding of robust plants with heightened bacterial disease tolerance, promising major strides in crop protection.

This chapter will elucidate the intricate interplay between plants and bacterial pathogens, shedding light on the mechanisms through which plants recognize and respond to microbial invaders. By understanding the molecular pathways involved in plant defense against bacteria, researchers can identify potential targets for enhancing disease tolerance. Additionally, the chapter will investigate how environmental factors influence the expression of plant resistance genes and explore novel approaches for manipulating these factors to bolster crop immunity. Through a comprehensive analysis of cutting-edge research, this chapter seeks to provide valuable insights into the development of sustainable agricultural practices aimed at mitigating the impact of bacterial diseases on crop yields and global food security.

13.2 UNVEILING NATURE'S ARSENAL: MOLECULAR WARFARE AGAINST PHYTOPATHOGENIC INVADERS

Plants endure a constant barrage of biotic and abiotic stressors throughout their lifecycle. In response, they have evolved intricate defense mechanisms, particularly against microbial pathogens. Plants rely on innate immunity as their primary weapon. While nematodes and aphids directly pierce host cells with stylets and fungi infiltrate through hyphal growth, bacteria often exploit natural openings or wounds for entry. Remarkably, even bacterial flagella, utilized for motility, can trigger host plant responses. These pathogen structures involved in host penetration are collectively termed effectors. Notably, bacterial flagellar responsiveness plays a crucial role in pathogenicity (Zipfel & Felix, 2005). A classic example is *flg22*, the flagellin subunit, which binds to the FLS2 receptor in *Arabidopsis thaliana*, initiating pathogen-associated molecular pattern (PAMP) immunity. In addition to flagella, bacteria employ an array of sophisticated mechanisms to subvert plant defenses and establish infections. These include the secretion of effector proteins directly into host cells, manipulation of host cellular processes, and evasion of plant immune surveillance (Chinchilla et al., 2006).

PAMPs are encoded by pathogens (Ma et al., 2013). Their recognition by pattern-recognition receptors, often encoded by receptor-like proteins, launches a cascade of cellular defenses (Wu et al., 2014). These include bursts of nitric oxide and reactive oxygen species, increased calcium ion (Ca^{2+}) influx, and the fortification of cell walls with callose to hinder pathogen penetration. Bacterial virulence further depends on type III secretion system effectors and, at times, host-selective toxins. To counter this, plants may trigger programmed cell death (PCD) to confine the spread of infection. As PCD is crucial to limiting phytobacteria, these pathogens have devised strategies to suppress it. For instance, *P. syringae* pv. *phaseolica* inhibits defense genes crucial for PCD in *Phaseolus vulgaris* (Jakobek et al., 1993). Curiously, *AvrPtoB* expression prevents PCD in *Nicotiana tabacum* (tobacco) but not in tomato, despite triggering PCD responses in the latter (Kim et al., 2002). The result of plant–bacteria interaction can be estimated in the early stages of inoculation. Nevertheless, successful pathogens that subdue or escape host defense responses can encourage the activities of non-pathogens (Jakobek et al., 1993). Such interactions offer valuable clues for breeding resistant cultivars. When resistance (*R*) genes recognize an invader, they induce a defensive response.

The genesis of the infection cycle of bacteria is decided by various factors. The foremost is the triggering of hypersensitive reactions (HR) in the host that do not allow for full colonization by the pathogen and, as such, may try to contain the pathogen within a defined location in its system. Necrotrophic bacteria have diverse and distinct pathogenic capabilities. There is a growing body of evidence that indicates that hypersensitive response and pathogenicity (*hrp*) genes are prevalent in these groups of bacteria. In addition, there is an array of variations in the aggressiveness in the necrotic-causing abilities as the employment of excessive impact may lead to the destruction of parenchyma tissues during the process of pathogenesis as opposed to another group (biotrophs) who are relatively slow in the cause of necrosis since they tend to adequately multiply in the host cell before they do so (Collmer & Bauer, 1994).

HR caused by biotrophs tends to be the result of incomplete interaction, especially with crops that are not the main hosts of the pathogen. *Xanthomonas campestris* strains and *P. syringae* are grouped into pathovars based on host specificity and accompanying phenotypic changes and sometimes between pathovars based on interactions with different cultivars of the host. *X. campestris* pv. *campestris* and *P. syringae* pv. *glycinea* trigger a hypersensitive response in tobacco, but the former causes different diseases in other crops (black rot in crucifers) and the latter causes bacterial blight in *Glycine max*. Also, the transfer of avrP-phB to *P. syringae* pv. *pisi* demonstrated avirulence toward all distinct pea cultivars. Conversely, avrPpiA in *P. syringae* pv. *phaseolicola* exhibited a distinctive pattern of interaction with bean cultivars, differing from any known race of the corresponding pathogen. The *P. syringae* pv. tomato strain DC3000, containing avrPphB, was reported to elicit a response in *A. thaliana* accessions bearing the RPS5 gene (Fillingham et al., 1992). The observed variations in specificity levels may stem from the accessibility of routine screening systems tailored for specific host–pathogen interactions, rather than originating from fundamental disparities in the genetic nature of the involved genes.

Several of these genes (type III secretion system) exhibit significant conservation not only among plant pathogens but also across various animal pathogens, spanning from *Yersinia* to *Shigella*. Consequently, they have been reclassified as hrc (H conserved) genes, which encode a protein secretion system known as the type III system. The type III system operates like a needle and plunger to transport pathogen-produced proteins that affect disease or trigger defense (Ponciano et al., 2003). These mechanisms have sometimes shown surprising and unexpected similarities to those found in animal and human pathogens (Cao et al., 2001). There are even a few strains of bacteria that cross-infect the human and animal kingdoms. The genetic basis for such novelty is of immense interest and significance regarding the basis of infectious disease.

Scientists have identified numerous R genes effective against phytobacteria (Ballvora et al., 2002; Brandwagt et al., 2000; Brueggeman et al., 2002) and have successfully utilized them to engineer transgenic plants demonstrating disease resistance (Hammond-Kosack & Parker, 2003). Resistance genes frequently encompass intricate loci, often involving multiple gene copies (Jones, 1996). However, research findings have shown that, in the case of the resistance gene RPM1, *Arabidopsis* possesses a single copy of the gene, while *Brassica napus* harbors two distinct copies (Vivian & Arnold, 2000). Additionally, recessive alleles are characterized by the deletions of the gene from the respective chromosome. Understanding these molecular battles paves the way for identifying key immunity genes and pathways, thereby accelerating the development of crops with fortified natural defenses. This deeper understanding enables researchers to strategically target specific components of the plant immune system, leading to more effective and sustainable approaches for managing bacterial diseases in agriculture. Through continued research and innovation, the deployment of R genes and the enhancement of plant immunity hold great promise for ensuring global food security in the face of evolving bacterial pathogens.

13.2.1 Plant Tolerance Mechanism to Bacteria

Ideally, the two major ways by which plant hosts can contain plant pathogens are resistance and tolerance. Resistance restricts the multiplication of the pathogen as a tolerance mechanism to mitigate the negative impact of the pathogen infection on the plant. Tolerance does not necessarily limit the replication of the pathogen, but resistance does so. As such, the oscillation of epidemics is the rebate of the pathogen's pathogenicity that may eventually lead to the breakdown of the resistance against the pathogen. On the contrary, the advantage of tolerance as it does not exert selection pressure on pathogens, and therefore it is a more reliable and sustainable protection alternative. More often than not, tolerance of plant hosts against bacteria and other pathogens as a whole is less of research interest to scientists. Also, the indiscriminate usage of the term to signify the resolution of the plant host to endure bacterial infection as to partial or complete susceptibility of the host serves as a source of concern (Pagan & Garcia-Arenal, 2020).

13.2.2 Avr Proteins

Avr protein from *P. syringae* pv. *glycinea* was the first to be isolated and recorded as avrA. The avr genes of a couple of phytobacteria (*P. syringae* and *X. campestris*) regulate the type of host they infect. The differences in hrp and avr genes are evident in the widespread strains of plant bacteria pathogens. Dissimilar to hairpins, avr possesses no distinct phenotypic features in addition to the non-triggering reaction from host plants when they are introduced into them.

Phytobacteria differ in their ability and effectiveness to cause disease in the host crop. The degree of successful interaction between R genes in the hosts with the avr of the pathogen usually outlines the basis for describing the race of the pathogen. Mostly, in these pathogens, the signaling and interaction with the host system are initiated by the type III hrp protein secretion system, together with a distinct group of hrp genes whose major roles are to signal and enter interaction between the avr/R genes. Indications exist suggesting the potential transfer of Avr gene products into plant cells via the hrp system, alongside other proteins integral to pathogenicity. Within this context, the term 'effector protein' has been posited to encapsulate all proteins traversing the hrp pathway into the plant milieu, regardless of their specific functionalities as elicitors or virulence factors (van Dijk et al., 1999). Various effector genes and their resultant products have been demonstrated to rely on a functional hrp pathway for their effects within the plant host. These encompass the products of avr-vir (virulence) genes and hop (hrp-dependent outer protein) genes

(Vivian & Arnold, 2000). Several avr genes were identified; however, the confirmation of the genetic basis of resistance occurred with the demonstration of co-segregation of resistance among an F2 population of pea to *P. syringae* pv. *pisi* race 2 and a race 1 strain containing the cloned avrPpiAl gene. This marked the initial validation of the genetic underpinnings of resistance. The gene avrRxv, originating from *X. campestris* pv. *vesicatoria*, a pathogen affecting tomato (*Lycopersicon esculentum*), when transferred to *X. campestris* pv. *phaseoli*, induced distinct responses in bean cultivars, consequently facilitating the identification of a partially dominant resistance gene in beans, named Rxv. In a similar vein to tomato, a homolog of the gene was discovered and reported by Whalen et al. (1993).

In contrast, bacteria are haploid, meaning they either carry a functional avr gene or do not. As regularly explained in a quadratic check, specificity is determined by the compatibility of the avr gene with its corresponding R gene, with any other interaction leading to disease development. Pathovars do not consistently align with singularly defined genomic groups, implying that phylogenetically distinct ancestors have evolved concurrently to induce phenotypically similar diseases within the same host.

The rationale behind why bacteria would benefit from carrying avr genes, which empower plants to impede their access to nutritional resources, has long remained perplexing. Nonetheless, for a limited subset of avr genes, it has been demonstrated that they provide a growth advantage to their bacterial hosts within the plant host environment.

The mutation of the avrBs2 gene in *X. campestris* pv. *vesicatoria* leads to the loss of HR induction in peppers with the Bs2 resistance gene and decreased growth in compatible pepper lines. This gene is conserved across various *X. campestris* pathovars, indicating its crucial role in pathogen fitness. AvrBs2 protein shares similarities with genes involved in glucan modification, suggesting its potential role in plant signaling. Notably, the Bs2 resistance gene was isolated from pepper, conferring resistance to avrBs2-carrying strains in tomato. Mutation of avrE in strain PT23 of *P. syringae* pv. tomato led to significant reductions in bacterial virulence and growth in tomato plants. However, mutating a homolog in the related strain DC3000 did not affect virulence toward tomato, highlighting the influence of bacterial genetic background on the phenotype conferred by an avr gene (Lorang et al., 1994). A similar scenario may account for the disparate observations concerning the gene avrRpm1, which shares considerable similarities with *avrPpiA* (Dangl et al., 1992). The disruption of *avrRpml* in *P. syringae* pv. *maculicola* resulted in a notable reduction in bacterial growth in compatible *Arabidopsis* accessions at low inoculum levels (Ritter & Dangl, 1995). Conversely, the disruption of the homologous *avrPpiA* gene in *P. syringae* pv. *pisi* exhibited no discernible impact on bacterial growth or pathogenicity toward compatible pea cultivars (Gibbon et al., 1997). This underscores the significance of bacterial genetic background in determining the phenotypic outcomes conferred by avr genes.

13.3 BEYOND TRADITION: MASTERING THE ART OF CROP EVOLUTION WITH CUTTING-EDGE BREEDING TECHNIQUES

Plant breeding presents a multitude of methods for enhancing crop traits, including bacterial resistance in plants. While traditional approaches rely on controlled crosses and natural variation, recent advances such as GS, genome-wide association studies (GWAS), MAS, and CRISPR enable precision and accelerated development. These innovative tools not only enhance yields but also bolster disease and pest resistance, improve nutritional value, and enhance climate resilience – all crucial components for ensuring global food security and fostering sustainable agriculture. By harnessing the power of these cutting-edge breeding techniques, researchers can accelerate the pace of crop improvement and cultivate resilient plant varieties capable of thriving in diverse environmental conditions. This transformation in breeding strategies represents a paradigm shift toward more efficient, precise, and sustainable agricultural practices, paving the way for a brighter future in food production.

13.3.1 Genome-Wide Association Study

GWAS provides a potent method for establishing links between specific genetic variations and desirable traits. Originating in human health research (MacArthur et al., 2017), GWAS has now become a transformative tool in plant breeding, facilitating the identification of key genes influencing traits in various crops such as maize, wheat, soybean, and more (Tibbs Cortes et al., 2021). GWAS analyses diverse populations using extensive datasets of markers, including single-nucleotide polymorphisms (SNPs), to map associations between these markers and the genes they influence. Notably, GWAS has been instrumental in pinpointing genes enhancing northern corn leaf blight resistance in maize (Kump et al., 2011), providing robust targets for breeding programs aiming to manage this widespread disease. Moreover, GWASs have successfully mapped bacterial stem blight tolerance caused by *P. syringae* pv. *syringae* in *Medicago sativa* (Desbiez et al., 2022) and have been employed for the fine-mapping of disease-resistance genes in *A. thaliana* interacting with bacterial pathogens (Bartoli & Roux, 2017). In a GWAS conducted on the same plant, the function of the SOLO DANCERS gene, which contains a cyclin-like protein, was confirmed as being a susceptibility gene to the pathogen, *Rastolnia solanacearum* during the early stages of disease infection (Aoun et al., 2020). When coupled with next-generation sequencing technologies, GWAS offers highly precise mapping, furnishing invaluable insights into the genetic architecture of disease resistance.

Close to two decades ago, the possibility of GWA mapping in classifying genes was reported. The study was exemplified by the cloning of genes, which signifies the presence or absence of hypersensitive response genes in the different accessions of the model plant, A. thaliana (Aranzana et al., 2005). Further on, that motivated researchers' interest in the field. Reportedly, there is a relative disproportionate number of research conducted on the mapping of pathogens in wild and domesticated plant species. GWAS is commonly used in studies of bacterial pathogen resistance in wild plant species, comprising about 55% of such research. Studies on oomycetes and fungi resistance follow at around 19%. In domesticated plant species, fungal pathogen resistance is the most studied (56%), while viruses (19%) and oomycetes (17%) also receive considerable focus (Demirjian et al., 2023). This disproportionate distribution may be explained by the devastating effects of fungi on cultivated crops in terms of yield, post-harvest losses, and loss of market value due to reduced physical appeal of crop products. To add to this, the fungi, *Puccinia striiformis* f. sp. *Tritici*, which causes stripe rust in wheat, cause a serious economic impact (Chen, 2013). This was aggregated by the utilization of natural reservoirs (wild type of *A. thaliana*) for bacteria and other pathogens. Wild and cultivated *A. thaliana* are favored hosts for several bacteria pathogens (*P. syringae*, *X. campestris* pv. *campestris* (Xcc), and *R. solanacearum*). Although some bacteria (*Agrobacterium tumefaciens* and *E. amylovora*) are of high importance to plant pathology research and crop cultivation, no significant studies on GWAS have been reported for these pathogens (Mansfield et al., 2012). Similar instances were also observed for the rust pathogens (*Puccinia* spp.), which only accrued less than 5% of the total research of GWAS on plant pathogens despite their economic significance in cereal production.

More recently, the use of co-GWAS on the pathogen–host interaction between *A. thaliana–X. arboricola* indicated the availability of QTLs that conferred resistance to specific strains of the pathogen (Wang et al., 2018). In Germany, the sequencing of a couple of *P. syringae* strains from half a dozen populations of the model crop, *A. thaliana*, showed variations. Also, the co-GWAS in *X. arboricola* showed a clearer mapping of genes responsible for the pathogenicity of the pathogen of interest.

13.3.2 Marker-Assisted Selection

MAS employs molecular markers like SNPs, AFLPs, RAPDs, and others to target genes of economic importance. This technique allows breeders to precisely determine resistance genes (including their susceptible counterparts), accelerating the development of new, resilient cultivars. Examples include using CAPS and SCAR markers to pinpoint sugarcane mosaic virus resistance genes (Dussle et al.,

2002) and deploying marker combinations to map a bacterial wilt resistance locus in *Solanum commersonii* (Siri et al., 2009). MAS, often combined with gene pyramiding strategies, plays a crucial role in addressing diseases like bacterial leaf blight in rice caused by *X. oryzae* pv. *oryzae* (Chukwu et al., 2019). Notably, it has successfully been utilized to map important resistance genes against this disease (Huang et al., 1997; Jiang et al., 2004; Joseph et al., 2004). Through the same technology, the phenotypic and genotypic qualities of other several crops have been improved against bacterial blight disease. For example, the incorporation of xa13 and xa21 in Pusa Basmati, xa5, xa13, and xa21 in Samba Mahsuri, respectively, conferred resistance against the disease (Kottapalli et al., 2010). Still on rice, the incorporation of Xa4, xa5, and Xa21, the Mangeumbye genotype of rice was reportedly successful in conferring resistance to the aforementioned pathogen (Dokku et al., 2013). Aside from these, other successes through the use of MAS to confer resistance to bacteria and other pathogens have been reported in several rice genotypes (Shanti et al., 2010; Suh et al., 2013; Swamy et al., 2006). Moreover, MAS has aided in mapping resistance genes against *Ascochyta* blight and pea-seedborne mosaic virus in pea crops (Carpenter et al., 2018; Gao et al., 2004). These successes exemplify the power of MAS to rapidly create more disease-resistant varieties. To further optimize outcomes, MAS can be integrated into breeding programs. For instance, In-Del and SNP markers analyzed with platforms like kompetitive allele-specific PCR (KASP) can expedite the selection of key quantitative trait loci (QTLs) associated with disease resistance. The markers associated with the disease resistance can then be used in a breeding program via MAS. While effective for major qualitative traits, MAS has limitations with complex traits influenced by multiple genes. The markers associated with the disease resistance can then be used in a breeding program via MAS and MABC. For instance, the use of In-Del markers to select for the QTL region and a panel of SNP markers assayed on the KASP platform. Moreover, combined with MAS, speed breeding (SB) allows breeders to rapidly screen generations of plants for desired traits marked by genetic markers. This powerful synergy has facilitated breeding for tolerance to diseases such as leaf rust and stem rot in bread wheat (Alahmad et al., 2018; Watson et al., 2018).

13.3.3 Genomic Selection

GS offers a potent, genome-wide approach to accelerating breeding programs. By harnessing whole-genome marker data, GS can predict breeding values, even for traits controlled by complex networks of small-effect genes that may be missed by phenotypic observation alone (Duangjit et al., 2016). When integrated with techniques like SB, GS dramatically expedites crop development by enabling the analysis of multiple generations annually. The combination of cost-effective genotyping with high-density SNP markers allows for precise genomic predictions to guide selection throughout the breeding process. This strategy has yielded promising results in enhancing desired traits, such as improved fruit quality in tomatoes (Duangjit et al., 2016). Specifically for bacterial disease resistance, GS has shown superiority over conventional phenotypic selection, particularly for traits like resistance to X. euvesicatoria (Liabeuf, 2016). By leveraging the power of genomic information, GS not only accelerates the breeding process but also enhances the efficiency and effectiveness of selecting for disease resistance, ultimately contributing to the development of more resilient crop varieties crucial for global food security.

13.3.4 CRISPR in Plant Bacterial Disease Resistance

CRISPR-based GE technologies have sparked a revolution in plant science, offering the capability for targeted and precise genome modifications. Unlike earlier breeding methods, CRISPR employs guide RNA (gRNA) and a palindromic adjacent motif to induce site-specific double-strand DNA breaks. These breaks, repaired through homologous directed repair or non-homologous end joining, enable specific gene insertions or deletions (Doudna & Charpentier, 2014; Langner et al., 2018). This tool complements already-existing GE tools such as TALENS and zinc finger nucleases. CRISPR presents numerous advantages over traditional breeding, including speed, accuracy, and, critically, often bypassing transgene regulatory frameworks due to its non-transgenic nature (Waltz, 2018). CRISPR technology holds immense potential for engineering bacterial disease resistance in plants. Studies have showcased its efficacy in editing susceptibility genes involved in disease pathways. Notable examples of susceptible genes that have been identified and reported are the Sugars Will Eventually Be Exported Transporters (SWEET) genes. The genes are responsible for the regulation of the sugar groups (glucose and sucrose) within plant systems. In *A. thaliana* and rice, the *AtSWEET* and *osSWEET* genes, respectively, were reported to enhance the colonization of host plants by biotrophic bacteria and fungi by virtue of the rich contents of these genes (Chen et al., 2019). During the infection of this rice with *X. oryzae* pv. *oryzae*, a strong positive regulation of *osSWEET11*, *Xa13*, and /Os8N3 was observed (Yang et al., 2006). There is a higher expression of the former gene during the invasion through the effector of PXO99A (Xoo strain)-PthXo1. Consequently, the importance of these genes in the successful infection establishment by these genes cannot be underestimated. As such, several breeding tools have been used in the editing of these genes with successes recorded. The use of RNA interference (RNAi) to induce mutations in the *osSWEET11* locus was successful in eliminating susceptibility to Xoo in mutant lines (Yang et al., 2006). Also, natural mutations triggering changes in the amino acid sequences of the *OsSWEET13* and *OsSWEET14* genes led to the development of resistant cultivars in a study conducted by Zaka et al. (2018). Complimentary to these studies, the use of the current CRISPR/Cas tool to edit the *osSWEET11*, *osSWEET11*, and *osSWEET14* led to the generation of resistant cultivars against the rice blight pathogen (Wei et al., 2021). Reports from the study showed that there was the generation of broad-spectrum resistance as mutant lines were resistant to both the African strain of Xoo (AXO1947) and Asian strain (PXO86) and the Asian strain of the pathogen. The blockage of the pathogen's effector access to these sugar transporters may explain the reason for their failed invasion. Further on, it was proposed that the effect of editing this locus could have some effect on plant height, and, as such, further studies need to be conducted to understand the relation between the locus and the trait.

Another notable locus responsible for the successful invasion of bacteria (*P. syringae*) in tomato hosts is the *SlJAZ2* gene in tomato against P. syringae (Ortigosa et al., 2019). The jasmonate-zim-domain (JAZ) gene is responsible for the regulation of the jasmonic acid pathway and, in addition, blocks the interaction of transcription factors with the mediator complex through the use of MED25 (Cevik et al., 2012). An ortholog of this gene *AtJAZ2* was discovered to be one of the receptors that regulates the opening and closing of the stomata during bacterial infection. As such, this gene is hijacked by the bacteria during invasion and deregulates the opening and closure of the stomata, which further makes it easier for the colonization of the *P. syringae* (Gimenez-Ibanez et al., 2017). As such, the heterozygous (JAZ2Δjas) and homozygous (jaz2Δjas) forms of mutation did yield resistant cultivars as the bacteria were not able to establish successful penetration and, hence, no disease occurrence. In tomatoes, CRISPR cas9 generated homozygous mutants (Sljaz2Δjas) were more effective in the control of *P. syringae* penetration and no disease infection in addition to no phenotypic or physiologic penalties (Ortigosa et al., 2019).

In citrus, the *Citrus sinensis* lateral organ boundary 1 (*CsLOB1*), *CsLOB2*, and *CsLOB3* genes have been previously reported to be a susceptibility gene for *Xanthomonas citri* subsp. *citri* (Xcc), which causes the citrus bacterial canker disease in citruses (Peng et al., 2017; Zhang et al., 2017). Orthologs of these genes have been discovered in *A. thaliana* (*AtAS2/AtLBD12*) and apples (*MdLBD11*), but their role in the initiation of bacterial infection in their respective hosts was not elucidated (Wang et al., 2013). The mutation of the CsLOB1 using the CRISPR ca9 tool in a study conducted by Peng et al. (2017) to generate homozygous mutants that had the desired mutations was confirmed to be resistant to the disease. Data from the study mutants harboring mutations with whole deletion of the EBE_{PthA4} allele had the highest resistance to the pathogen. Complementary to this study, it was further reported that the use of the CRISPR tool to mutant the same locus for Hamlin sweet orange and Carrizo citrange resulted in biallelic mutants. Interestingly, these mutants were resistant to Xcc (Huang et al., 2022).

Aside from research that focuses on the use of the CRISPR tool to generate resistant plant varieties against plant bacteria, a study reported on the action of phages' CRISPR system within the plant pathogen, *X. oryzae*, and it was discovered that there is a palindromic motif adjoining the pathogen's phage (Xop411) protospacers (Semenova et al., 2009).

CRISPR has played a pivotal role in banana disease research (Tripathi et al., 2021). Following the targeting of these genes, CRISPR-mediated editing leads to alterations with insertion or deletion in their sequences, thereby affecting the corresponding proteins' structure or function. This can result in enhanced resistance to bacterial pathogens by disrupting key interactions.

Extracellular proteins of plant and animal bacteria are important in virulence. Many of these are secreted through the type I sec-independent and the type II sec-dependent pathways. Recently, a third distinct pathway, involved in the secretion of effectors, has been discovered in *Yersinia*. This pathway has homology with pathways in plant pathogenic bacteria that are putatively involved in the secretion of proteins active on plant cells, such as *harpin* and possibly with *avr* gene (Van Gijsegem et al., 1993).

Furthermore, CRISPR technology allows for the simultaneous editing of multiple genes involved in disease-resistance pathways, offering the possibility of developing plants with enhanced resilience to a range of bacterial pathogens. This multiplex editing capability significantly expands the scope of breeding programs aimed at bolstering crop protection and improving global food security in the face of evolving microbial threats. Compared to traditional breeding, CRISPR offers targeted alterations at precise loci with the potential to modify complex polyploid crops. While techniques like GWAS and MAS identify regions or genes of interest, CRISPR allows direct modification of these targets, thereby accelerating targeted improvements such as enhanced bacterial resistance. This integration of CRISPR technology into breeding programs holds promise for revolutionizing crop resilience to bacterial diseases, thus contributing to sustainable agriculture and global food security.

13.4 CHALLENGES IN BACTERIAL DISEASE RESISTANCE AND FUTURE DIRECTIONS AND OPPORTUNITIES

Despite major advances in breeding methods, several obstacles hinder the development of durable bacterial disease resistance in crops. A primary challenge lies in the ability of phytopathogens to evolve rapidly, potentially creating new strains that overcome previously deployed resistance mechanisms. Moreover, many bacterial infections go undetected in their early stages due to asymptomatic presentation, leading to rapid spread and limited treatment options as conventional bactericides can prove ineffective. One specific method by which bacteria thwart plant defenses is through suppressing pathogen-triggered immunity. Their type III secretion systems (T3SS) inject effector proteins directly into plant cells, bypassing typical recognition by the plant's immune receptors. Pathogen virulence can be compromised by mutations within the *hrp* genes encoding elements of this T3SS (Van Gijsegem et al., 1993). Studies comparing hrp gene sequences reveal conserved regions across diverse pathogens like *R. solanacearum*, *X. compestris hrp. vesicatoria*, and *P. syringae* (Fenselau et al., 1992). While the introduction of specific R genes can offer valuable resistance in the short term, this success is often fleeting. Pathogen evolution, especially within those effector proteins targeted by *R* genes, quickly creates new pathogen races capable of breaking through this defense. This continuous arms race underscores the ongoing need for novel breeding strategies.

The future of bacterial disease-resistance research and breeding holds great potential for developing more resistant crop varieties. This ongoing research is vital to safeguard global food supplies, promote sustainable agriculture, and protect ecosystems. Bacterial diseases remain a major threat to crop health, causing significant yield losses and economic damage. Cutting-edge breeding techniques like GWAS, GS, and CRISPR offer precise tools for creating bacterial-resistant varieties and deepening our understanding of plant–bacteria interactions. Further improvements in GE, such as prime editing and base editing, promise even greater potential, although their flexibility may be somewhat limited. While creating crops with broad-spectrum resistance against diverse bacterial pathogens remains a challenge, these advanced technologies could lead to the development of new cultivars with more durable and effective resistance, minimizing crop losses and advancing sustainable agricultural practices.

REFERENCES

Ahmar, S., Saeed, S., Khan, M. H. U., Khan, S. U., Mora-Poblete, F., Kamran, M., Faheem, A., Maqsood, A., Rauf, M., Saleem, S., Hong, W. J., & Jung, K. H. (2020). A revolution toward gene-editing technology and its application to crop improvement. *International Journal of Molecular Sciences*, *21*(16), 1–28. https://doi.org/10.3390/ijms21165665

Alahmad, S., Dinglasan, E., Leung, K. M., Riaz, A., Derbal, N., Voss-Fels, K. P., Able, J. A., Bassi, F. M., Christopher, J., & Hickey, L. T. (2018). Speed breeding for multiple quantitative traits in durum wheat. *Plant Methods*, *14*(1), 1–15. https://doi.org/10.1186/s13007-018-0302-y

Alfano, J. R., & Collmer, A. (1996). Bacterial pathogens in plants: Life up against the wall. *The Plant Cell*, *8*(10), 1683.

Aoun, N., Desaint, H., Boyrie, L., Bonhomme, M., Deslandes, L., Berthomé, R., & Roux, F. (2020). A complex network of additive and epistatic quantitative trait loci underlies natural variation of Arabidopsis thaliana quantitative disease resistance to Ralstonia solanacearum under heat stress. *Molecular Plant Pathology*, *21*(11), 1405–1420. https://doi.org/10.1111/mpp.12964

Aranzana, M. J., Kim, S., Zhao, K., Bakker, E., Horton, M., Jakob, K., Lister, C., Molitor, J., Shindo, C., Tang, C., Toomajian, C., Traw, B., Zheng, H., Bergelson, J., Dean, C., Marjoram, P., & Nordborg, M. (2005). Genome-wide association mapping in Arabidopsis identifies previously known flowering time and pathogen resistance genes. *PLoS Genetics*, *1*(5). https://doi.org/10.1371/journal.pgen.0010060

Ballvora, A., Ercolano, M. R., Weiß, J., Meksem, K., Bormann, C. A., Oberhagemann, P., Salamini, F., & Gebhardt, C. (2002). The R1 gene for potato resistance to late blight (Phytophthora infestans) belongs to the leucine zipper/NBS/LRR class of plant resistance genes. *Plant Journal*, *30*(3), 361–371. https://doi.org/10.1046/j.1365-313X.2001.01292.x

Bartoli, C., & Roux, F. (2017). Genome-wide association studies in plant pathosystems: Toward an ecological genomics approach. *Frontiers in Plant Science*, *8*(May). https://doi.org/10.3389/fpls.2017.00763

Brandwagt, B. F., Mesbah, L. A., Takken, F. L. W., Laurent, P. L., Kneppers, T. J. A., Hille, J., & Nijkamp, H. J. J. (2000). A longevity assurance gene homolog of tomato mediates resistance to *Altemaria alternata f.* sp. lycopersici toxins and fumonisin B1. Proceedings of the National Academy of Sciences of the United States of America, *97*(9), 4961–4966. https://doi.org/10.1073/pnas.97.9.4961

Brueggeman, R., Rostoks, N., Kudrna, D., Kilian, A., Han, F., Chen, J., Druka, A., Steffenson, B., & Kleinhofs, A. (2002). The barley stem rust-resistance gene Rpg1 is a novel disease-resistance gene with homology to receptor kinases. *Proceedings of the National Academy of Sciences of the United States of America*, *99*(14), 9328–9333. https://doi.org/10.1073/pnas.142284999

Burkholder, W. H. (1948a). Bacteria as plant pathogens. *Annual Review of Microbiology*, *2*(1), 389–412.

Cao, H., Baldini, R. L., &Rahme, L.. (2001). Common mechanisms for pathogens of plants and animals. *Annual Review of Phytopathology*, *39*, 259–284.

Carpenter, M. A., Goulden, D. S., Woods, C. J., Thomson, S. J., Kenel, F., Frew, T. J., Cooper, R. D., & Timmerman-Vaughan, G. M. (2018). Genomic selection for ascochyta blight resistance in pea. *Frontiers in Plant Science*, *871*(December), 1–13. https://doi.org/10.3389/fpls.2018.01878

Cevik, V., Kidd, B. N., Zhang, P., Hill, C., Kiddle, S., Denby, K. J., Holub, E. B., & Al., E. (2012). MEDIATOR25 acts as an integrative hub for the regulation of jasmonate-responsive gene expression in arabidopsis. *Plant Physiology*, *160*, 541–555.

Chen, H. Q., Hou, B. H., S., L., H., T., M.L., H., X.Q., Q., W.J., G., Kim, J. G., & Underwood W., & Chaudhuri, B. (2019). Sugar transporters for intercellular exchange and nutrition of pathogens. *Nature*, *468*, 527–532.

Chen, X. (2013). High-temperature adult-plant resistance, key for sustainable control of stripe rust. *American Journal of Plant Sciences*, *4*, 608–627.

Chinchilla, D., Bauer, Z., Regenass, M., Boller, T., & Felix, G. (2006). The Arabidopsis receptor kinase FLS2 binds flg22 and determines the specificity of flagellin perception. *Plant Cell*, *18*(2), 465–476. https://doi.org/10.1105/tpc.105.036574

Chukwu, S. C., Rafii, M. Y., Ramlee, S. I., Ismail, S. I., Hasan, M. M., Oladosu, Y. A., Magaji, U. G., Akos, I., & Olalekan, K. K. (2019). Bacterial leaf blight resistance in rice: A review of conventional breeding to molecular approach. *Molecular Biology Reports*, *46*(1), 1519–1532. https://doi.org/10.1007/s11033-019-04584-2

Collmer, A., & Bauer, D. W. (1994). Erwinia chrysanthemi and *Pseudomonas syringae*: Plant pathogens trafficking in extracellular virulence proteins. In Jeffery L. Dangl (Ed.), *Bacterial pathogenesis of plants and animals: Molecular and cellular mechanisms* (pp. 43–78). Springer Berlin, Heidelberg.

Coplin, D. L., Rowan, R. G., & Chisholm, D. A., & Whitmoyer, R. E. (1981). Characterization of plasmids in Erwinia stewartii. *Applied and Environmental Microbiology*, *42*(4), 599–604.

Demirjian, C., Vailleau, F., Berthomé, R., & Roux, F. (2023). Genome-wide association studies in plant pathosystems: Success or failure? *Trends in Plant Science*, *28*(4), 471–485. https://doi.org/10.1016/j.tplants.2022.11.006

Desbiez, C., Domingo-Calap, M. L., Pitrat, M., Wipf-Scheibel, C., Girardot, G., Ferriol, I., Lopez-Moya, J. J., & Lecoq, H. (2022). Specificity of resistance and tolerance to cucumber vein yellowing virus in melon accessions and resistance breaking with a single mutation in VPg. *Phytopathology*, *112*(5), 1185–1191. https://doi.org/10.1094/PHYTO-06-21-0263-R

Dokku, P., Das, K. M., & Rao, G. J. N. (2013). Genetic enhancement of host plant-resistance of the Lalat cultivar of rice against bacterial blight employing marker-assisted selection. *Biotechnology Letters*, *35*, 1339–1348.

Doudna, J. A, & Charpentier, E. (2014). The new frontier of genome engineering with CRISPR-Cas9. *Science*, *346*(6213), 1258096. https://doi.org/https://doi.org/10.1126/science.1258096

Duangjit, J., Causse, M., & Sauvage, C. (2016). Efficiency of genomic selection for tomato fruit quality. *Molecular Breeding*, *36*(3), 1–16. https://doi.org/10.1007/s11032-016-0453-3

Dussle, C. M., Quint, M., Xu, M. L., Melchinger A.E., & Lübberstedt, T. (2002). Conversion of AFLP fragments tightly linked to SCMV resistance genes Scmv1 and Scmv2 into simple PCR based markers. *Theoritical Applied Genetics*, *105*(8), 1190–1195. https://doi.org/https://doi.org/10.1007/s00122-002-0964-7

Fenselau, S., Balbo, I., & Bonas, U. (1992). Determinants of pathogenicity in *Xanthomonas campestris* pv. *vesicatoria* are related to proteins involved in secretion in bacterial pathogens of animals. *Molecular Plant Microbe Interactions*, *5*, 390.

Fillingham, A. J., Wood, J., Bevan, J. R., Crute, I. R., Mansfield, J. W., Taylor, J. D, & Vivian, A. (1992). Avirulence genes from Pseudomonas syringae pathovars phaseolicola and pisi con- fer specificity towards both host and non-host species. *Physiological and Molecular Plant Pathology*, *40*, 1–15.

Francia, E., Tacconi, G., Crosatti, C., Barabaschi, D., Bulgarelli, D., Dall'Aglio, E., & Valè, G. (2005). Marker assisted selection in crop plants. *Plant Cell, Tissue and Organ Culture*, *82*(3), 317–342. https://doi.org/10.1007/s11240-005-2387-z

Gao, Z., Eyers, S., Thomas, C., Ellis, N., & Maule, A. (2004). Identification of markers tightly linked to sbm recessive genes for resistance to Pea seed-borne mosaic virus. *Theoretical and Applied Genetics*, *109*(3), 488–494. https://doi.org/10.1007/s00122-004-1652-6

Gibbon, M. J., Jenner, C., Mur, L. A. J., Pur, I. N., Mansfield, J. W., Taylor, J. D, & Vivian, A. (1997). Avirulence gene avrPpiA from Pseudomonas syringe pv. pisi is not required for full virulence on pea. *Physiological and Molecular Plant Pathology*, *50*, 219–236.

Gimenez-Ibanez, S., Boter, M., Ortigosa, A., García-Casado, G., Chini, A., Lewsey, M. G., Ecker, J. R., Ntoukakis, V., & Solano, R. (2017). JAZ2 controls stomata dynamics during bacterial invasion. *New Phytologist*, *213*, 1378–1392.

Hammond-Kosack, K. E., & Parker, J. E. (2003). Deciphering plant-pathogen communication: Fresh perspectives for molecular resistance breeding. *Current Opinion in Biotechnology*, *14*(2), 177–193. https://doi.org/10.1016/S0958-1669(03)00035-1

Harveson, R. M., Schwartz, H. F., Urrea, C. A, & Yonts, C. D. (2015). Bacterial wilt of dry-edible beans in the central high plains of the US: Past, present, and future. *Plant Disease*, *99*, 1665–1677.

Huang, N., Angeles, E. R., Domingo, J., Magpantay, G., Singh, S., Zhang, G., Kumaravadivel, N., Bennett, J., & Khush, G. S. (1997). Pyramiding of bacterial blight resistance genes in rice: Marker-assisted selection using RFLP and PCR. *Theoretical and Applied Genetics*, *95*(3), 313–320. https://doi.org/10.1007/s001220050565

Huang, X., Wang, Y., & Wang, N. (2022). Highly efficient generation of canker-resistant sweet orange enabled by an improved CRISPR/Cas9 system. *Frontiers in Plant Science*, *12*(January), 1–12. https://doi.org/10.3389/fpls.2021.769907

Jakobek, J. L., Smith, J. A., & Lindgren, P. B. (1993). Suppression of bean defense responses by Pseudomonas syringae. *Plant Cell*, *5*(1), 57–63. https://doi.org/10.2307/3869428

Jannink, J. L., Lorenz, A. J., & Iwata, H. (2010). Genomic selection in plant breeding: From theory to practice. *Briefings in Functional Genomics and Proteomics*, *9*(2), 166–177. https://doi.org/10.1093/bfgp/elq001

Jiang, G.., Xu, C.., Tu, J.., Li, X.., He, Y.Q, & Zhang, Q. F. (2004). Pyramiding of insect- and disease-resistance genes into an elite indica, cytoplasm male sterile restorer line of rice, 'Minghui 63.' *Plant Breeding*, *123*, 112–116. https://doi.org/10.1046/j.1439-0523.2003.00917.x

Joseph, M., Gopalakrishnan, S., Sharma, R. K., Singh, V. P., Singh, A. K., K. Singh, N., & Mohapatra, T. (2004). Combining bacterial blight resistance and Basmati quality characteristics by phenotypic and molecular marker-assisted selection in rice. *Molecular Breeding*, *13*(4), 377–387. https://doi.org/10.1023/B:MOLB.0000034093.63593.4c

Kennedy, B. W., & Alcorn, S. M. (1980). Estimates of U.S. crop losses to prokaryote plant pathogens. *Plant Disease*, *64*, 674–676.

Kim, Y. J., Lin, N.-C., & Martin, G. B. (2002). Two distinct pseudomonas effector proteins interact with the pto kinase and activate plant immunity of host disease resistance (R) genes. A variety of defense responses are then activated including the hypersensi-tive response (HR), a localized form of pr. *Cell*, *109*, 589–598. https://www.cell.com/cgi/contenct/full/109/5/589/DC1

Kottapalli, K., Lakshmi, N. M., & Jena, K. (2010). Effective strategy for pyramiding three bacterial blight resistance genes into fine grain rice cultivar, Samba mahsuri, using sequence tagged site markers. *Biotechnology Letters*, *32*, 989–996. https://doi.org/doi:10.1007/s10529-010-0249-1

Kump, K. L., Bradbury, P. J., Wisser, R. J., Buckler, E. S., Belcher, A. R., Oropeza-Rosas, M. A., Zwonitzer, J. C., Kresovich, S., McMullen, M. D., Ware, D., Balint-Kurti, P. J., & Holland, J. B. (2011). Genome-wide association study of quantitative resistance to southern leaf blight in the maize nested association mapping population. *Nature Genetics*, *43*(2), 163–168. https://doi.org/10.1038/ng.747

Langner, T., Kamoun, S., & Belhaj, K. (2018). CRISPR Crops: Plant genome editing toward disease resistance. *Annual Review of Phytopathology*, *56*, 479–512. https://doi.org/10.1146/annurev-phyto-080417-050158

Liabeuf, D. (2016). *Development of processing tomato lines resistant to xanthomonas gardneri: From screening to breeding*. Ohio State University. https://rave.ohiolink.edu/etdc/view?acc_num=osu1480522067298667

Lorang, J. M., Shen, H., Kobayashi, D., Cooksey, D., & Keen, N. T. (1994). avrA and avrE in Pseudomonas syringae pv. tomato PT23 play a role in virulence on tomato plants. *Molecular Plant Microbe Interactions*, *7*(4), 508–515.

Ma, Y., Zhao, Y., Walker, R. K., & Berkowitz, G. A. (2013). Molecular steps in the immune signaling pathway evoked by plant elicitor peptides: Ca2+-dependent protein kinases, nitric oxide, and reactive oxygen species are downstream from the early Ca2+ signal. *Plant Physiology*, *163*(3), 1459–1471. https://doi.org/10.1104/pp.113.226068

MacArthur, J., Bowler, E., Cerezo, M., Gil, L., Hall, P., Hastings, E., Junkins, H., McMahon, A., Milano, A., Morales, J., MayPendlington, Z., Welter, D., Burdett, T., Hindorff, L., Flicek, P., Cunningham, F., & Parkinson, H. (2017). The new NHGRI-EBI Catalog of published genome-wide association studies (GWAS Catalog). *Nucleic Acids Research*, *45*(D1), D896–D901. https://doi.org/10.1093/nar/gkw1133

Mansfield, J., Genin, S., Magori, S., Citovsky, V., Sriariyanum, M., Ronald, P., Dow, M., Verdier, V., Beer, S. V., Machado, M. A., Toth, I., Salmond, G., & Foster, G. D. (2012). Top 10 plant pathogenic bacteria in molecular plant pathology. *Molecular Plant Pathology*, *13*(6), 614–629. https://doi.org/10.1111/j.1364-3703.2012.00804.x

Ortigosa, A., Gimenez-Ibanez, S., Leonhardt, N., & Solano, R. (2019). Design of a bacterial speck resistant tomato by CRISPR/Cas9-mediated editing of SlJAZ2. *Plant Biotechnology Journal*, *17*(3), 665–673. https://doi.org/10.1111/pbi.13006

Osdaghi, E., Jones, J. B., Sharma, A., Goss, E. M., Abrahamian, P., Newberry, E. A., Potnis, N., Carvalho, R., Choudhary, M., Paret, M. L., Timilsina, S., & Vallad, G. E. (2021). A centenary for bacterial spot of tomato and pepper. *Molecular Plant Pathology*, *22*(12), 1500–1519. https://doi.org/10.1111/mpp.13125

Pagan, I., & Garcia-Arenal, F. (2020). Tolerance of plants to pathogens: A unifying view. *Annual Review of Phytopathology*, *58*, 77–96. https://doi.org/10.1146/annurev-phyto-010820-012749

Peng, A., Chen, S., Lei, T., Xu, L., He, Y., Wu, L., Yao, L., & Zou, X. (2017). Engineering canker-resistant plants through CRISPR/Cas9-targeted editing of the susceptibility gene CsLOB1 promoter in citrus. *Plant Biotechnology Journal*, *15*(12), 1509–1519. https://doi.org/10.1111/pbi.12733

Ponciano, G., Ishihara, H., Tsuyumu, S, & Leach, J. E. (2003). Bacterial effectors in plant disease and defense: Keys to durable resistance. *Plant Disease*, *87*, 1271–1282.

Ritter, C, & Dangl, J. L. (1995). The avrRpml gene of Pseudomonas syringae pv. maculicolais required for virulence on Arabidopsis. *Molecular Plant-Microbe Interactions*, *8*, 444–453.

Semenova, E., Nagornykh, M., Pyatnitskiy, M., Artamonova, I. I., & Severinov, K. (2009). Analysis of CRISPR system function in plant pathogen Xanthomonas oryzae. *FEMS Microbiology Letters*, *296*(1), 110–116. https://doi.org/10.1111/j.1574-6968.2009.01626.x

Shanti, M. L., Shenoy, V. V., Devi, G. L., Kumar, V. M., Premalatha, P., Kumar, G. N., Shashidhar, H.E., Zehr, U.B., & Freeman W.H. (2010). Marker-assisted breeding for resistance to bacterial leaf blight in popular cultivar and parental lines of hybrid rice. *Journal of Plant Pathology*, *92*, 495–501. https://doi.org/doi:10.4454/jpp.v92i2.194

Siri, M. I., Galván, G. A., Quirici, L., Silvera, E., Villanueva, P., Ferreira, F., Franco Fraguas, L., & Pianzzola, M. J. (2009). Molecular marker diversity and bacterial wilt resistance in wild *Solanum commersonii* accessions from Uruguay. *Euphytica, 165*(2), 371–382. https://doi.org/10.1007/s10681-008-9800-8

Smidt, M. L., & Vidaver, A. K. (1987). Variation among strains of Clavibacter michiganense subsp. nebraskense isolated from a single popcorn field. *Phytopathology, 77*(2), 388–392.

Soto, M. J., Domínguez-Ferreras, A., Pérez-Mendoza, D., Sanjuán, J., & Olivares, J. (2009). Mutualism versus pathogenesis: The give-and-take in plant-bacteria interactions. *Cellular Microbiology, 11*(3), 381–388. https://doi.org/10.1111/j.1462-5822.2009.01282.x

Suh, J. P., Jeung, J. U., Noh, T. H., Cho, Y. C., Park, S. H., Park, H. S., Shin, M. S., Kim, C. K., & Jena, K. K.(2013). Development of breeding lines with three pyramided resistance genes that confer broad-spectrum bacterial blight resistance and their molecular analysis in rice. *Rice, 6*, 1–11. https://doi.org/doi:10.1186/1939-8433-6-5

Swamy, P., Panchbhai, A. N., Priti, D., Vaishali, N., Panchbhai, S. D., & Zehr, U., Azhakanandam, K., Char, B. (2006). Evaluation of bacterial blight resistance in rice lines carrying multiple resistance genes and Xa21 transgenic lines. *Current Science, 90*, 818–824.

Tibbs Cortes, L., Zhang, Z., & Yu, J. (2021). Status and prospects of genome-wide association studies in plants. *Plant Genome, 14*(1), 1–17. https://doi.org/10.1002/tpg2.20077

Tripathi, J. N., Ntui, V. O., Shah, T., & Tripathi, L. (2021). CRISPR/Cas9-mediated editing of DMR6 orthologue in banana (Musa spp.) confers enhanced resistance to bacterial disease. *Plant Biotechnology Journal, 19*(7), 1291–1293. https://doi.org/10.1111/pbi.13614

van Dijk, K., Fouts, D. E., Rehm, A. H., Hill, A. R., Collmer, A., & Alfano, J. R. (1999). The Avr (effector) proteins HrmA (HopPsyA) and AvrPto are secreted in culture from Pseudomonas syringae pathovars via the Hrp (type III) protein secretion system in a temperature- and pH-sensitive manner. *Journal of Bacteriology, 181*, 4790–4797.

Van Gijsegem, F., Genin, S., & Boucher, C. (1993). Conservation of secretion pathways for pathogenicity determinants of plant and animal bacteria. *Trends in Microbiology, 1*(5), 175–180. https://doi.org/10.1016/0966-842X(93)90087-8

Verhaegen, M., Bergot, T., Liebana, E., Stancanelli, G., Streissl, F., Mingeot-Leclercq, M. P., Mahillon, J., & Bragard, C. (2023). On the use of antibiotics to control plant pathogenic bacteria: A genetic and genomic perspective. *Frontiers in Microbiology, 14*(June). https://doi.org/10.3389/fmicb.2023.1221478

Vidhyasekaran, P. (2008). *Bacterial disease resistance in plants: Molecular biology and biotechnological applications*. Routledge Taylor & Francis.

Vivian, A.., & Arnold, D. L. (2000). Bacterial effector genes and their role in host-pathogen interactions. *Journal of Plant Pathology, 82*(3), 163–178.

Waltz, E. (2018). With a free pass, CRISPR-edited plants reach market in record time. *Nature Biotechnology, 36*(1), 6–8.

Wang, M., Roux, F., Bartoli, C., Huard-Chauveau, C., Meyer, C., Lee, H., & Bergelson, J. (2018). Two-way mixed-effects methods for joint association analysis using both host and pathogen genomes. *Proceedings of the National Academy of Sciences, 115*(24), 5440–5449.

Wang, X., Zhang, S., Su, L., Liu, X. and, & Hao, Y. (2013). A genome-wide analysis of the LBD (lateral organ boundaries domain) gene family in Malus domestica with a functional characterization of MdLBD11. *PLoS One, 8*. https://doi.org/doi:10.1371/journal.pone.0057044

Watson, A., Ghosh, S., Williams, M. J., Cuddy, W. S., Simmonds, J., Rey, M. D., Asyraf Md Hatta, M., Hinchliffe, A., Steed, A., Reynolds, D., Adamski, N. M., Breakspear, A., Korolev, A., Rayner, T., Dixon, L. E., Riaz, A., Martin, W., Ryan, M., Edwards, D., Batley, J., Raman, H., Carter, J., Rogers, J., Domoney, C., Moore, G., Harwood, W., Nicholson, P., Dieters, M. J., DeLacy, I. H., Zhou, J., Uauy, C., Boden, S. A., Park, R. F, Wulff B. B. H., & Hickey L. T. (2018). Speed breeding is a powerful tool to accelerate crop research and breeding. *Nature Plants, 4*(1), 23–29. https://doi.org/10.1038/s41477-017-0083-8

Wei, Z., Abdelrahman, M., Gao, Y., Ji, Z., Mishra, R., Sun, H., Sui, Y., Wu, C., Wang, C., & Zhao, K. (2021). Engineering broad-spectrum resistance to bacterial blight by CRISPR-Cas9-mediated precise homology directed repair in rice. *Molecular Plant, 14*(8), 1215–1218. https://doi.org/10.1016/j.molp.2021.05.012

Whalen, M. C., Wang, J. F., Carland, F. M., Heiskell, M. E., Dahlbeck, D., Minsavage, G. V., Jones, J. B., Scott, J. W., Stall, R.E., & Staskawicz, B. J. (1993). Avirulence gene avrRxv from Xanthomonas campestris pv. vesicatoria specifies resistance on tomato line Hawaii. *Molecular Plant-Microbe Interactionsteractions, 6*, 616–627.

Wu, S., Shan, L., & He, P. (2014). Microbial signature-triggered plant defense responses and early signaling mechanisms. *Plant Science, 228*, 118–126. https://doi.org/10.1016/j.plantsci.2014.03.001

Yang, B., Sugio, A., & White, F. F. (2006). Os8N3 is a host disease-susceptibility gene for bacterial blight of rice. *Proceedings of the National Academy of Sciences*, *103*(27), 10503–10508.

Zaka, A., Grande, G., Coronejo, T., Quibod, I. L., Chen, C. W., Chang, S. J., Szurek, B., Arif, M., Cruz, C. V., & Oliva, R. (2018). Natural variations in the promoter of OsSWEET13 and OsSWEET14 expand the range of resistance against Xanthomonas oryzae pv. Oryzae. *PLoS One*, *13*(9), 1–15. https://doi.org/10.1371/journal.pone.0203711

Zhang, J., Huguet-Tapia, J. C., Hu, Y., Jones, J., Wang, N., Liu, S., & White, F. F. (2017). Homologues of CsLOB1 in citrus function as disease susceptibility genes in citrus canker. *Molecular Plant Pathology*, *18*(6), 798–810. https://doi.org/10.1111/mpp.12441

Zipfel, C., & Felix, G. (2005). Plants and animals: A different taste for microbes? *Current Opinion in Plant Biology*, *8*(4), 353–360. https://doi.org/10.1016/j.pbi.2005.05.004

14 Alternative Strategies to Control Bacterial Diseases in Plants

Khadija Benamar, Ikram Leghrifi, Mohammed Radi, Kaoutar Bourak, Salma Benchlih, Tourya Sagouti, and Rachid Lahlali

14.1 INTRODUCTION

Agriculture plays a vital role in the economies of both developing and developed nations. However, the industrial practices involved, such as the inherent vulnerabilities linked with domestication, lack of genetic diversity, and monoculture, make the crops susceptible to a variety of abiotic and biotic stresses (Begna, 2020; Gupta et al., 2020). Plant diseases, contributing to around 10–13% of food production losses, vary based on climatic conditions, crops, and production systems (Ezrari et al., 2023). More than 200 species associated with 25 bacterial genera pose a threat to various plant species, leading to significant crop losses. Examples of diseases caused by bacterial agents include bacterial leaf blight impacting rice (*Xanthomonas oryzae* pv. *oryzae*) (Zhang & Wang, 2013), fire blight affecting pear and apple (*Erwinia amylovora*) (Billing, 2011), bacterial wilt affecting potato, tomato, and other crops (*Ralstonia solanacearum* and other *Ralstonia* species) (Caldwell et al., 2017), and bacterial canker affecting kiwifruit (*Pseudomonas syringae* pv. *actinidiae*) (Vanneste, 2017). Additionally, fastidious bacteria like *Xylella fastidiosa* causing diseases such as olive decline and pierce's disease pose significant challenges globally (Almeida & Nunney, 2015).

Recent advances in high-throughput DNA sequencing and metagenomic analysis have revolutionized the ability to rapidly and accurately identify pathogenic bacteria (Adams et al., 2018; Piombo et al., 2021; Studholme, 2011). These technological advances allow a comprehensive exploration of the diversity and dynamics of bacterial populations in plant environments (Nizamani et al., 2023). By uncovering the genomic subtleties of pathogenic bacteria, researchers can determine new intervention strategies and develop an in-depth understanding of the factors that influence disease prevalence (Eichmeier et al., 2019). This fundamental knowledge forms the basis for developing precise and effective control strategies.

In the pursuit of sustainable alternatives, biological control agents are emerging as promising candidates (Poveda et al., 2021). Beneficial microorganisms, such as mycorrhizal fungi and plant growth-promoting rhizobacteria (PGPR) (Bouizgarne, 2013; Desai et al., 2016), have demonstrated their ability to suppress pathogenic bacteria through mechanisms such as competition for resources and the induction of systemic resistance in plants (Annapurna et al., 2013). In addition, bacteriophages, viruses that infect and destroy bacteria, represent another avenue for biological control, as they enable pathogenic strains to be specifically targeted while leaving non-target organisms unharmed (Jamal et al., 2018; Vu & Oh, 2020). This biological approach is in line with the growing importance attached to sustainable agriculture and the preservation of natural ecosystems.

The deployment of resistant plant varieties, identified through meticulous breeding programs, offers another sustainable and cost-effective approach to enhancing resilience against specific pathogens (Orgwsundin et al. et al., 2016). Moreover, the introduction of gene editing techniques,

DOI: 10.1201/9781003543237-14

prominently exemplified by CRISPR/Cas9, allows for precise manipulation of plant genomes to confer resistance to bacterial infections (Doerflinger et al., 2016).

The intersection of nanotechnology and agriculture has given rise to innovative strategies for controlling bacterial diseases with unprecedented precision (Gao et al., 2014). Nanoparticles, with their unique physicochemical properties, act as vectors for the targeted release of antimicrobial compounds at the site of infection (Singh et al., 2020). This approach minimizes collateral damage to non-target organisms and the environment, making a compelling case for sustainable disease management (Al-Samarrai., 2012). The controlled release of antimicrobial agents via nanocarriers improves the effectiveness of the intervention while mitigating the problems associated with overuse and persistence in the environment (Khan, 2014).

In this chapter, we will provide a comprehensive overview of advanced detection techniques, biological control methods, genetic strategies, and nanobiotechnology-based approaches to control bacterial diseases in plants.

14.1.1 High-Throughput Sequencing

14.1.1.1 High-Throughput DNA Sequencing

High-throughput sequencing (HTS) refers to novel sequencing technologies recently developed. They are also known as third, fourth, or next-generation sequencing methods. In fact, HTS offers distinct advantages in terms of cost-effectiveness and rapid processing (Pradhan et al., 2019). HTS includes genomics-focused DNA sequencing, which is achieved through whole-genome and exome sequencing techniques (Lyra et al., 2020). Using this approach, millions to billions of DNA molecules can be sequenced rapidly (Ghosh et al., 2018).

14.1.1.2 Role of HTS in Plant Pathology Management

As shown in Figure 14.1, HTS plays an important role in plant pathology management. Indeed, it allows rapid and precise pathogen identification, surpassing conventionnel methods such as serological tests and microscopy (Adams et al., 2018). Unlike these methods, which suffer from limitations in specificity and sensitivity, HTS offers a wide range of advantages, including the rapid rate of sequencing, cost-effectiveness in mitigating losses, and increasing the effectiveness of plant disease management strategies. Its high accuracy ensures reliable pathogen identification, thereby facilitating the development of appropriate preventive and curative control measures (Piombo et al., 2021; Tedersoo et al., 2021). Moreover, HTS allows the detection of emerging and novel pathogens, through the genetic analysis of pathogens in infected samples. For instance, using HTS, Morrow et al. (2020) have characterized bacterial communities transmitted by psyllids insects to Rutaceae in

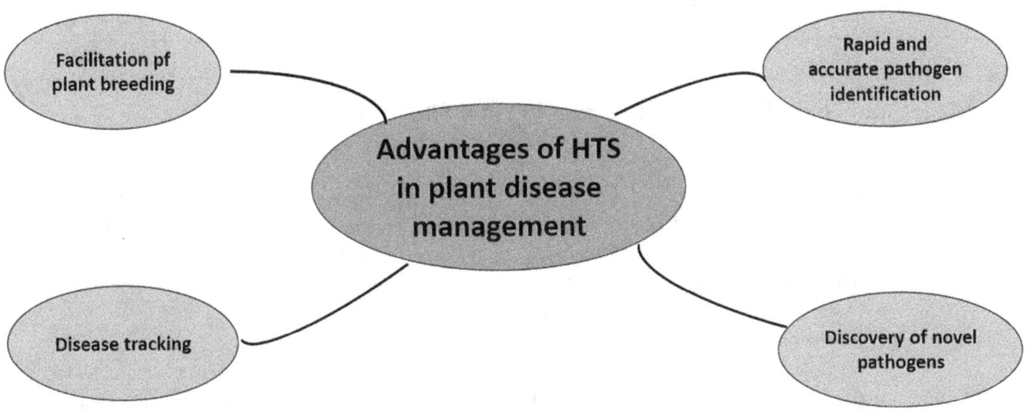

FIGURE 14.1 Advantages of high-throughput sequencing (HTS) in plant disease management.

Bhutan, identifying several bacteria including *Candidatus carsonella ruddii*, *Candidatus profftella armatura*, *Candidatus liberibacter*, and *Wolbachia* strains.

In addition, HTS aids in disease tracking by providing deep genetic data on pathogens, which increase significantly the ability to detect disease spread, identify new strains and variants, and oversee pathogen evolution (Suminda et al., 2022). Furthermore, it facilitates plant breeding, enhancing disease resistance by implementing disease-resistant genes and, therefore, obtaining robust crop varieties with a high defense system against several pathogens. These advantages allow to achieve food security and promote sustainable agriculture. In this context, HTS has been employed to reveal tomato pathogen interactions toward sustainable plant breeding. Four distinct bacterial species of *Xanthomonas* (*Xanthomonas vesicatoria*, *Xanthomonas euvesicatoria*, *Xanthomonas gardneri*, and *Xanthomonas perforans*) were found to cause bacterial spots. Hence, resistant varieties were developed characterized by the upregulation of stress signaling genes such as osmotin-like proteins and pathogenesis-related proteins genes, along with the induction of defense response pathways (Campos et al., 2021).

14.1.2 Metagenomic Analysis

14.1.2.1 Metagenomic-Based Approaches

In contrast to targeted molecular methods (polymerase chain reaction (PCR)-based techniques, real-time quantitative PCR (qPCR), and loop-mediated isothermal amplification), which are designed to identify one or a few specific target pathogens in a single reaction mixture, metagenomic-based approaches, including metabarcoding and shotgun metagenomic, represent untargeted method. Indeed, these metagenomic approaches permit the detection of all microorganisms in a single analysis (Piombo et al., 2021). Thus, metagenomic sequencing possesses the ability to detect and identify multiple, if not all, plant pathogens in a single comprehensive analysis.

14.1.2.1.1 Metabarcoding

Metabarcoding refers to sequencing a barcode region of a group of organisms present in a sample, typically using ITS regions for fungi and 16S for bacteria. As shown in Table 14.1, metabarcoding offers several advantages, including its low cost, ease of bioinformatic analysis, ability to identify all microorganisms present in the sample (including rare and low abundant taxa) (Porter & Hajibabaei, 2018), and the potential of quantifying detected organisms through the addition of control DNA quantities to samples before DNA extraction (Tkacz et al., 2018). However, metabarcoding also has limitations, primarily related to its lack of specificity. Because full barcode sequencing is challenging, scientists often focus on sub-regions like ITS1 and ITS2 for fungi, and hypervariable regions from V1 to V9 for bacteria (Abdelfattah et al., 2018). Yet, these short markers may not provide

TABLE 14.1
Advantages and Limitations of Metabarcoding and Shotgun Metagenomic

	Metabarcoding	Shotgun Metagenomic
Advantages	• Low cost. • Easiness to manage bioinformatic obtained data. • Ability to detect and identification of all microorganisms present in the sample. • Possibility of quantifying detected organisms.	• Suitable for general surveillance. • Detection of all microorganisms included in a sample without PCR-associated biases. • Ability to identify dominant microorganisms at the species level.
Limitations	• Problem of specificity. • Necessity of selecting primers. • Increased contamination risk when working with low-biomass samples.	• Complex operations for data analysis. • Difficult taxonomic assessment of all species in the presence of complex communities.

enough variability to distinguish closely related taxa, making it difficult to precisely identify microbial species. Another limitation is the need to select primers carefully. Microbial genomes can vary in the primer-amplified regions, leading to the exclusive or preferential detection of certain taxa based on primer sets with variation in one or a few bases (Legeay et al., 2019). Hence, it's important to select carefully appropriate primers for the desired application, to ensure the accurate identification and detection of the target pathogens. In addition, metabarcoding presents the risk of sample contamination, which can be more important when working with low-biomass samples (Piombo et al., 2021).

14.1.2.1.2 Shotgun Metagenomic

Shotgun metagenomic represents the untargeted sequencing of the entire DNA fragments present in a sample. Similar to metabarcoding, shotgun metagenomics offers several advantages and some limitations (Table 14.1). Among its advantages, it serves as an effective tool for general surveillance and enables the detection of all bacteria and fungi present in a sample. Additionally, it has the capacity to achieve strain-level resolution for certain abundant species. However, when sequencing complex microbial communities, marker genes may be recovered only at low frequencies, making taxonomic determination of all species impossible or challenging. This limitation requires complex operations for data analysis (Knight et al., 2018; Piombo et al., 2021).

14.1.2.2 Examples of Metagenomic Application in the Detection of Bacterial Plant Pathogens

The metagenomic approach has been successfully applied to the identification of pathogenic bacteria in plants. For instance, Kim et al. (2018) employed metabarcoding to identify pathogenic bacteria present in the phyllosphere of *Brassica rapa* subsp. *Pekinensis* (Chinese cabbage). They sequenced the V5 and V6 regions 16S rRNA gene and found dominant bacteria such as *Pseudomonas*, *Sphingomonas*, *Chryseobacterium*, and *Aurantimonadacea*. Other bacterial pathogens, including *Clostridium*, *Salmonella*, *Staphylococcus*, *Bacillus*, *Yersinia*, *Klebsiella*, *Erwinia*, and *Pantoea*, were also identified. In another study (McKernan et al., 2016), metagenomic analysis has allowed to identify several toxigenic bacteria infecting medicinal Cannabis, including *Stenotrophomonas maltophilia*, *Salmonella enterica*, *Ralstonia pickettii*, *Pseudomonas aeruginosa*, *Acinetobacter baumannii*, and *Escherichia coli*. More interestingly, Román-Reyna et al. (2021) used metagenomic sequencing to detect and identify subspecies and strains of the plant pathogenic bacterium *X. fastidiosa* transmitted by insects. This approach has allowed them to rapidly identify different subspecies and strains of single and mixed plant samples infected by the bacterium.

Overall, metagenomic analysis holds promise as a diagnostic tool for plant diseases, through the identification of multiple pathogenic microorganisms, not only at the genus level but also at subspecies and strain levels.

14.2 BIOLOGICAL CONTROL

14.2.1 Beneficial Microorganisms

Recently, plant disease control using beneficial microbes has gained significant importance due to the global demand for environmentally sustainable alternatives to chemical pesticides and fertilizers (Fira et al., 2018; Syed Ab Rahman et al., 2018). Biocontrol, a relatively new approach in agricultural practices, involves harnessing microbial benefits as a sustainable strategy for controlling crop diseases (Gao et al., 2016). Various bacterial genera have shown great potential as biocontrol agents for diverse plant diseases, while fungi also play a significant role in the successful control of major crop diseases (Zubair et al., 2021).

Understanding the disease cycle is crucial for successful biological control of bacterial diseases, as these diseases can rapidly escalate under favorable environmental conditions. Methods for delivering biological disease control agents directly to plants include treating seeds, covering crops with biocontrol agents, or dipping seedlings in biocontrol agent suspensions (Triwidodo, 2014). Notably, bacterial plant diseases like crown gall (*Agrobacterium tumefaciens*) and fire blight (*E. amylovora*) are known to be resistant to chemical control methods (Hassan Abd-El-Khair, 2019).

Given the urgent need for new methods to control various plant diseases and pests, the application of novel biopesticides holds great promise. Implementing these new approaches can significantly enhance environmental sustainability and economic stability in current food production systems while also contributing to global food security (Dimkić et al., 2022).

14.2.1.1 Bacteria

Using beneficial bacteria to combat various bacterial, fungal, and nematode diseases is environmentally efficient and improves plant growth (Ayaz et al., 2023).

In addition, such bacteria are capable of using various mechanisms for limiting bacterial development in plant pathogens. This includes the colonization and competitive removal of the pathogen from sites of infection, the secretion of high-activity antimicrobials like antibiotics or lytic cell wall enzymes, and the promotion of plant resistance (Berendsen et al., 2012).

A wide range of bacterial genera, particularly *Bacillus*, *Pseudomonas*, and *Streptomyces*, have been identified for their ability to protect plants against bacterial and fungal pathogens (Bonaterra et al., 2022).

14.2.1.1.1 Bacillus Spp.

The extensively studied and widely utilized *Bacillus* genus holds immense potential for agricultural applications (Hamdache et al., 2013). Depending on morphological characteristics, *Bacillus* spp. are described as being rod-shaped, gram-positive, facultative aerobic or anaerobic, and positive catalases. They exhibit resilience to harsh environmental conditions due to their diverse physiological capacity and the production of endospores, making them ubiquitous in various habitats, particularly soil. In fact, *Bacillus* species often dominate soil and rhizosphere bacterial communities, constituting up to 95% of gram-positive bacteria (Prashar et al., 2014). Moreover, they are frequently found as endophytic bacteria within plant tissues (De Silva et al., 2019). *Bacillus* spp. have a high capacity to control disease by inhibiting the pathogen's growth through the synthesis of various low-molecular-weight metabolites, enzymes, and compounds. These include ribosomal peptides, polyketides, non-ribosomal peptides, or volatiles, which can impede pathogen resistance mechanisms (Dimkić et al., 2022). Various studies have revealed the ability of *Bacillus* spp. to biologically control various pathogenic bacteria.

For instance, Mirik et al. (2008) employed *Bacillus* strains isolated from the rhizosphere of greenhouse and field peppers to control populations of *Xanthomonas axonopodis* pv. *vesicatoria*, effectively reducing disease development in both greenhouse and field conditions. Application of these treatments resulted in improved vegetative growth parameters including root elongation, stem diameter, and root and shoot dry weight, contributing to a better yield response in field experiments, indicating successful biological control of bacterial spot.

Similarly, endophyte bacterial species, notably *Bacillus amyloliquefaciens* (strains A1, A3 and A13), *Bacillus subtilis* (strain A15), *Bacillus methylotrophicus* (strain A2), and two rhizosphere bacilli (strains D29 and H8), demonstrated significant inhibitory effects when tested against *X. oryzae* pv. *oryzae* in vitro. In the greenhouse experiments, the inhibition rate of four bacterial strains was significantly improved over that of the pathogen, accompanied by an improvement in the fresh and dry plant weight of treated rice plants. These strains showed their ability to synthesize siderophores and indole-3-acetic acid and to solubilize phosphate (El-shakh et al., 2015).

Furthermore, the antibiosis and hemolysis exhibited by four *Bacillus* isolates positively affected all *X. campestris pv. campestris* strains, the causative agents of black rot in crucifers. Antimicrobial and hemolytic activities were found to be correlated, suggesting that lipopeptides produced by *Bacillus* isolates could trigger antibiosis mechanisms. These findings underscore the ability of

Bacillus species to produce bioactive compounds or surfactants, particularly during the late growth phase (Monteiro et al., 2005).

14.2.1.1.2 Pseudomonas Spp

Biological control of plant diseases using *Pseudomonas* species has emerged as an environmentally friendly and effective approach for plant disease management (Mohammed et al., 2020). *Pseudomonas* spp. are gram-negative, rod-shaped, anaerobic, non-endosporic bacteria with one or more flagella, belonging to the Proteobacteria group (Yadav et al., 2014). They are inherently saprophytic, and are found in water, soil, and wetlands (Singh et al., 2021). *Pseudomonas* strains possess the capacity to produce a diverse array of antibiotics with suppressive activity against plant pathogens. These compounds include phenazine-1-carboxylic acid (PCA), pyoluteorin (Plt), pyrrolnitrin (Prn), 2,4-diacetylphloroglucinol (DAPG), hydrogen cyanide (HCN), and bacteriocins (Voisard et al., 1989).

Numerous *Pseudomonades* species are extensively utilized as bioagents against various bacterial diseases. For example, *Pseudomonas fluorescens* (strain F113), which produces DAPG, has been revealed to inhibit *Erwinia carotovora* subsp. *atroseptica* both in vitro and in vivo. This strain effectively reduced the softening of potato tuber wounds caused by the pathogen and decreased the population of *E. carotovora* subsp. *atroseptica* in soil and on potato tubers. This suggests the potential of *P. fluorescens* strain F113 as a biocontrol agent against *E. carotovora* subsp. *atroseptica* (Cronin et al., 1997).

Similarly, *Pseudomonas* sp. (*strain AM12*), *P. aeruginosa* (*strain T1*), *Pseudomonas* sp. (*strain BH25*), *Pseudomonas* sp. (*strain AM13*), and *P. putida* (*strain R6*) showed an antagonistic action against *R. solanacearum* (*strain Tom5*) obtained directly from a tomato wilted plant. Among these strains, BH25 significantly reduced pathogenic activity against Tom5 and also improved the fresh and dry weights as well as the vigor index of seedlings in biological assays (Maji & Chakrabartty, 2014).

14.2.1.1.3 Streptomyces spp.

Bacterial antagonists, particularly *Streptomyces* spp., have gained significant attention for their potential in bio-controlling plant diseases (Gebily et al., 2021). *Streptomyces* spp. belong to the actinobacteria family and represent the largest group within this family, with over 500 characterized species. They are filamentous, gram-positive bacteria and are commonly found in habitats similar to fungi (Gowdar et al., 2018).

One of the remarkable features of *Streptomycess* spp. is its ability to produce several secondary metabolites, including antibiotics, volatile organic compounds, and enzymes. These compounds not only have direct antimicrobial properties but also promote plant resistance to pathogens (Pacios-Michelena et al., 2021)

A number of studies have investigated the efficacy of *Streptomyces* species against many bacterial plant diseases. Doolotkeldieva et al. (2016) have shown that *Streptomyces diastatochromogenes sk-6* reduces soft rot disease in stored potatoes, by pre-treating potato tubers and protecting them from phytopathogenic bacteria during the early reproductive stages. In another study by Van Hop et al. (2014), significant reductions in yield loss due to *X. oryzae* pv. *oryzae* were observed in infected rice varieties in field experiments, compared to healthy rice varieties. This indicates the potential of *Streptomyces* spp. in mitigating the impact of bacterial diseases on crop yield (Van Hop et al., 2014). Additionally, *Streptomyces* sp. (strain C1–4) was found to alleviate symptoms of fire blight in leaf tissue from apple and pear shoots (Doolotkeldieva & Bobusheva, 2016).

14.2.1.2 Fungal

The use of fungal biocontrol agents, particularly those belonging to the *Trichoderma* genus, is gaining momentum in agriculture due to their high efficiency against plant pathogens and their ability to generate a variety of secondary metabolites (Kariuki et al., 2020; Konappa et al., 2020). *Trichoderma* spp. exhibit target specificity and rapid reproduction and can shift from parasitism to saprotrophism in the absence of a host, allowing them to persist in the environment sustainably (Thambugala et al., 2020).

Trichoderma strains have been extensively studied for their beneficial properties in crop protection, bio-stimulation, and bio-fertilization (Ghorbanpour et al., 2018). Their efficacy in agriculture depends on their metabolic potential and their interactions with plants and other microorganisms. *Trichoderma* spp. efficiently colonize various plant surfaces and roots, generating a number of metabolites presenting antimicrobial compounds such as cell wall-degrading enzymes, antibiotics, and volatile/non-volatile compounds, as well as biostimulants like phytohormones and phytoregulators (Jaroszuk-ściseł et al., 2019).

Li et al. (2014) demonstrated that *Trichoderma pseudokoningii* (SMF2 strain) elicited systemic resistance against the soft rot of Chinese cabbage by producing trichokonin, which activates the salicylic acid signaling pathway. Another study by Idowu et al. (2016) showed that *Trichoderma asperellum* significantly reduced the adverse effect of *E. carotovora* on orka seedlings.

Guo et al. (2021) examined the fungal regulators produced by three *Trichoderma* spp. (*Trichoderma harzianum*, *Trichoderma virens*, and *Trichoderma koningii*) and tested them on three different growth media (STP, MOF, and SuM) to assess their antibacterial properties in vitro as well as in vivo in controlling the bacterial wilt disease of tomato plants caused by *R. solanacearum*. The most effective fungal metabolites of *T. harzianum* against *R. solanacearum* were obtained on an STP medium. *T. harzianum* metabolites exhibited the most significant reduction in the number of soil bacterial communities, reduced the severity of the disease, and enhanced plant growth. These findings suggest that *T. harzianum* metabolites have the potential to serve as ecological, economical, and successful tools for controlling *R. solanacearum* in tomato plants and eventually other crops.

14.2.2 Induced Systemic Resistance

Plants have an innate ability to detect invading microorganisms and activate defense responses through microbial or pathogen-associated molecular patterns, like fungal chitin and bacterial flagellin, by pattern-recognition receptors (PRRs) located in the transmembrane (Pandey et al., 2017). This process stimulates the first layer of immune defense, known as pattern-triggered immunity (PTI). However, pathogenic microorganisms have evolved mechanisms to suppress PTI signaling or prevent recognition by PRRs through the secretion of virulence molecules called effectors. This evasion prompts the activation of the second layer of plant immunity, named effector-triggered immunity (ETI), which has evolved to recognize pathogen-secreted effectors (Yu et al., 2022).

The key elements of PTI and ETI can be stimulated and induced in plants either by beneficial microorganisms or by plant secondary metabolites, leading to what is known as induced systemic resistance (ISR) in plants. In contrast to systemic acquired resistance, which is induced by pathogenic microorganisms, ISR is specifically induced by beneficial microbes (Xie et al., 2018). However, it's worth noting that ISR can also be activated by certain plant secondary metabolites, suggesting a complex interplay between microbial and plant-derived signals in regulating plant immunity (Figure 14.2).

14.2.2.1 Plant Growth-Promoting Microorganisms

Plant growth-promoting microorganisms play an important role in inducing systemic resistance in plants. The mechanism of their action is not yet well known. However, several studies have shown the activation of ISR in plants by beneficial microbes through the production of different elicitors and by regulating small RNAs.

14.2.2.1.1 Small RNAs Regulation

Small RNAs are non-coding RNAs having an approximate length ranging from 20 to 30 nucleotides. These RNAs play a crucial role in the regulation of biological processes, including reproduction, development, and stress responses, and enhance plant immunity systems against pathogenic

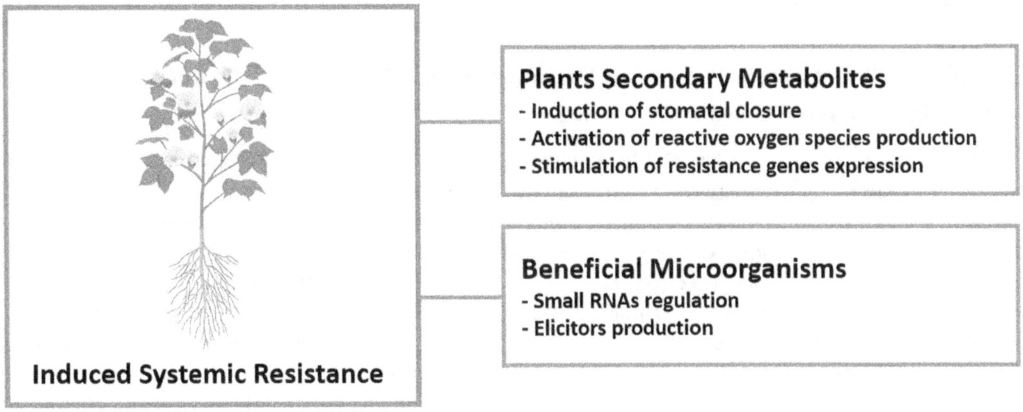

FIGURE 14.2 Involved factors in the stimulation of induced systemic resistance in plants.

viruses, fungi, and bacteria (Yu et al., 2022). Beneficial microorganisms such as PGPR have a potential effect in regulating small RNAs in plants. Niu et al. (2016) have found that the PGPR *Bacillus cereus* AR156 has induced systemic resistance in *Arabidopsis* against *P. syringae* pv. *tomato* DC 3000. Indeed, the plant treatment using these beneficial bacteria has resulted in the downregulation of two small RNAs: micro-RNA 825 (miR825) and miR825*, which targeted two resistance genes related to ubiquitin-protein ligase, toll-interleukin-like receptor nucleotide-binding site, and leucine-rich repeat associated genes. Hence, the expression of resistance genes was negatively correlated with the expression of the small RNAs miR825 and miR825*. Furthermore, transgenic plants having low expression of miR825 and miR825* showed a higher resistance to bacterial infection with *P. syringae* DC3000 in comparison with transgenic plants with an overexpression of these two small RNAs. Similarly, Jiang et al. (2020) found that *Bacillus cereus* AR156 ISR in *Arabidopsis* against *P. syringae* pv. *tomato* DC 3000 by downregulating the plant small RNAs mi R472. The expression of these RNAs was revealed to be negatively correlated with coiled-coil nucleotide-binding site and leucine-rich repeat resistance genes. In another study, Xie et al. (2018) demonstrated that the PGPR *B. amyloliquefaciens* FZB42 ISR in *Arabidopsis* against *P. syringae* pv. *tomato* DC 3000, through the suppression of the small RNAs miR846 expression and the upregulation of the expression of jasmonic acid-related genes.

14.2.2.1.2 Bacterial Elicitors Production

Originally, elicitors referred to compounds that are able to induce the production and accumulation of phytoalexins in plant cells. However, currently, this term refers also to biotic and abiotic molecules that stimulate and support any mechanism of plant defense, such as changes in the activity of catalase, peroxidase, and superoxide dismutase, and changes in phenolic compounds synthesis (Luiz et al., 2017). Microbial elicitors constitute an important class of elicitors. They are composed of a wide range of secondary metabolites produced by microorganisms and induce systemic resistance in plants. These substances induce the activation of jasmonic acid, ethylene, and salicylic acid signaling pathways. Wu et al. (2018) studied the role of nine mutants of the PGPR *B. amyloliquefaciens* SQR9 deficient in the production of extracellular compounds (including surfactin, fengycin, bacillomycin D, macrolactin, 2,3-butandiol, difficidin, bacilysin, and exopolysaccharide) in inducing systemic resistance against *P. syringae* pv. *tomato* DC3000 in *Arabidopsis*. Their results showed that the deficiency in the production of any of the mentioned compounds resulted in a decrease in plant resistance against the studied bacterial pathogen, reflecting therefore the role of these elicitor substances in the induction of plant resistance. Furthermore, volatile compounds produced by beneficial microorganisms were also found to be effective elicitors in inducing resistance in plants. For instance, *B. subtilis GB*03 produced organic volatile compounds triggering the activation of salicylic acid

TABLE 14.2
Examples of Plant Growth-Promoting Microorganisms (PGPM) Recently Reported to Induce Plant Bacterial Disease Resistance

PGPM	Mechanism of Action	Host Plant	Bio-Aggressor	References
Bacillus cereus AR156	Regulation of plant small RNAs miR825 and miR825 expression	*Arabidopsis*	*Pseudomonas syringae* pv. tomato DC3000	Niu et al. (2016)
B. cereus AR156	Regulation of plant small RNAs miR472 expression	*Arabidopsis*	*P. syringae* pv. tomato DC3000	Jiang et al. (2020)
Bacillus amyloliquefaciens FZB42	Regulation of plant small RNA expression	*Arabidopsis*	*P. syringae* pv. tomato DC3000	Xie et al. (2018)
B. amyloliquefaciens SQR9	Production of elicitors	*Arabidopsis*	*P. syringae* pv. tomato DC3000	Wu et al. (2018)
Bacillus artrophaeus LSSC22 and *B. amyloliquefaciens* FZB42	Production of elicitors	Tobacco	*Ralstonia solanacearum* TBBS1	Tahir et al. (2017)

and jasmonic acid pathways (Sharifi & Ryu, 2016). Furthermore, it was demonstrated that *Bacillus artrophaeus* LSSC22 and *B. amyloliquefaciens* FZB42 exhibited an important effect in inducing systemic resistance in tobacco against bacterial wilt caused by *R. solanacearum* TBBS1. Indeed, the volatile compounds including benzaldehyde, 1,3-butadiene, and 1,2-benzisothiazol-3(2H) regulated significantly the expression of genes responsible for colonizing pathogen resistance (Tahir et al., 2017) (Table 14.2).

14.2.2.2 Plants Secondary Metabolites

Plants produce different secondary metabolites that act as elicitors to induce systemic resistance against pathogens. Volatile compounds released by plants play a crucial role in inducing systemic resistance. For instance, indole, which is a volatile compound commonly produced by grass species in response to herbivore damage, it has induced ISR against pathogens by stimulating the production of reactive oxygen species and enhancing the expression of defense-related genes (Shen et al., 2018). Similarly, fumigation of *Arabidopsis* with the two terpenes, α and β-pinene, increases the expression of resistance genes against biotrophic pathogens, with reactive oxygen species playing a key role in the signaling cascade (Riedlmeier et al., 2017). Esters of (Z)-3-hexanal have also been identified as elicitors of plant resistance against bacterial infection by stimulating stomata closure. In tomato plants, the emission of these volatiles was found to be essential for effective defense against bacterial infection by *P. syringae* pv. *tomato* (López-Gresa et al., 2018). In addition, the two volatile plant defense hormones, methyl jasmonate and methyl salicylate, provide plants with the ability to systemically induce resistance responses in distant plant parts from the first infection site without the need of having transit of the signals through the vascular system. For instance, repeated applications of methylsalicylate to *Nicotiana benthamiana* seedlings infected by *P. syringae* pv *tabaci* and *Pectobacterium carotovorum* subsp. *carotovorum* resulted in greater protection against these two bacterial pathogens compared to untreated control plants (Song & Ryu, 2018). Similar defense patterns inductions were observed for plants after the applications of methyl jasmonate (Lundborg et al., 2019).

Plants polysaccharides and essential oils have been shown to effectively induce systemic resistance in plants. In this respect, tea tree or palmarosa essential oils mixed with an aqueous suspension of *Aloe* polysaccharides and tea tree essential oil emulsion or palmarosa essential oil micro-emulsion were applied on two strawberry plants cultivars (San Andreas and Albion). The results showed that these preparations reduced significantly bacterial disease severity caused by *Xanthomonas fragariae*. Indeed, after treatment of plants with the emulsions, increased activities of superoxide dismutase, catalase, peroxidase, and dismutase were observed in the more susceptible cultivar (Albion), whereas a higher total phenolic content was revealed in the less susceptible cultivar (San Andreas). Overall, these emulsions were demonstrated effective in inducing systemic resistance in both strawberry cultivars against the bacterial angular leafspot disease resulting from the infection by *X. fragariae* (Luiz et al., 2017).

14.2.3 Use of Bacteriophages

Bacteriophages are viruses capable of infecting and killing targeted bacteria, making them promising candidates for managing bacterial infections in plants (Sanz-Gaitero et al., 2021).

Bacteriophages exhibit two main life cycles: lytic and lysogenic. These cycles determine their use either for plant bacterial infection control or for plant bacterial disease diagnosis. For both cycles, the first step of bacterial infection by bacteriophages starts with the absorption of these entities with specific receptors localized in the bacterial surface and then the injection of bacteriophage genetic material inside the bacterial host cells. In the lytic cycle, after gene injection, bacteriophages use bacterial ribosomes for their protein synthesis. At the end stage of the lytic cycle, bacteriophages encode two enzymes known as endolysins, namely holins and lysins. These enzymes allow to lyse the bacteria and release the bacteriophage progenies (Dy et al., 2018). In the lysogenic cycle, the injected bacteriophage genes are integrated with bacterial chromosomes, and then they are replicated and transferred to daughter bacterial cells (Dennehy & Abedon, 2020). Hence, lytic bacteriophages are generally used for plant bacterial disease management, because they are able to kill directly bacterial pathogens, whereas lysogenic bacteriophages, due to their gene's integration to bacterial chromosomes, can be used for plant bacterial pathogens diagnosis.

Several bacteriophages have been demonstrated effective in controlling bacterial pathogens, either when applied individually, or in mixtures as shown in Table 14.3. The application of

TABLE 14.3
Recent Studies Demonstrating Bacteriophages Used for the Protection of Plants from Bacterial Diseases

Bacteriophages Application	Bacteriophages	Targeted Bacteria	Host Plants	References
Single	RsPod1EGY	*Ralstonia solanacearum*	Tomato	Elhalag et al. (2018)
	XcpSFC211	*Xanthomonas campestris pv. campestris*	Broccoli	Nagai et al. (2017)
	KΦ1	*Xanthomonas euvesicatoria*	Pepper	Gašić et al. (2018)
	ϕ6	*Pseudomonas syringae pv. actinidiae*	Kiwi	Pinheiro et al. (2020)
	phiEaP-8	*Erwinia amylovora* and *Erwinia pyrifoliae*	Apple and pear	Park et al. (2018)
Cocktail	NJ-P3, NB-P21; NC-P34; NN-P42	*R. solanacearum*	Tomato	X. Wang et al. (2019)
	M5, M8	*R. solanacearum*	Banana	Ramírez et al. (2020)
	φMA11; φMA12; φMA13; φMA14	*Pectobacterium carotovorum* subsp. *carotovorum*	Onion	Zaczek-Moczydłowska et al. (2020a)
	φMA1; φMA1A; φMA2; φMA5 ; φMA6; φMA7	*Pectobacterium atrosepticum* and *P. carotovorum* subsp. *carotovorum*	Potato	Zaczek-Moczydłowska et al. (2020b)

(*Continued*)

TABLE 14.3 (*Continued*)
Recent Studies Demonstrating Bacteriophages Used for the Protection of Plants from Bacterial Diseases

Bacteriophages Application	Bacteriophages	Targeted Bacteria	Host Plants	References
	vB_PatP_CB1 ; vB_PatP_CB3 ; vB_PatP_CB4	*P. atrosepticum*	Potato	Buttimer et al. (2018)
	P-PSG-1; P-PSG-2; P-PSG-3; P-PSG-7; P-PSG-8; P-PSG-9	*R. solanacearum*	Potato	Wei et al. (2017)
	vB_PsyM_KIL1; vB_PsyM_KIL2; vB_PsyM_KIL3; vB_PsyM_KIL4; vB_PsyM_KIL5; vB_PsyM_KIL3b	*P. syringae pv. porri*	Leek	Rombouts et al. (2016)

individual bacteriophages has been revealed effective against a wide range of bacteria including *R. solanacearum, Xanthomonas campestris pv. campestris, Xanthomonas euvesicatoria, P. syringae* pv. *actinidiae, E. amylovora*, and *Erwinia pyrifoliae*, which cause diseases in crops like tomato, broccoli, pepper, kiwi, apple, and pear, respectively (Elhalag et al., 2018; Gašić et al., 2018; Nagai et al., 2017; Park et al., 2018; Pinheiro et al., 2020). However, the use of individual bacteriophages may lead to the development of bacterial resistance (Vu & Oh, 2020). To address this, bacteriophage cocktails containing multiple phages targeting the same or different bacterial pathogens have been developed (Buttimer et al., 2018; Ramírez et al., 2020; Rombouts et al., 2016; Wei et al., 2017). These cocktails have shown efficacy against a broader range of bacterial diseases and can help overcome host-range restrictions of individual phages (Vu & Oh, 2020).

14.3 GENETIC STRATEGIES

The use of genetic tools also constitutes a good strategy for controlling bacterial diseases in plants. Hence, it's important to give an insight into plant breeding and genetic engineering tools, which are greatly important methods for harnessing plant resistance and improving the genetics of different crop plants, to overcome disease threats caused by pathogens.

14.3.1 Resistant Varieties

Plants are characterized by two types of resistance: qualitative and quantitative resistance. The understanding of these resistance types is crucial for designing effective breeding strategies.

14.3.1.1 Qualitative Resistance

Qualitative resistance involves a hypersensitive response to pathogens, often leading to cell death (Kushalappa et al., 2016). Resistance genes responsible for qualitative resistance, also known as major genes, typically confer complete resistance to specific pathogens (Nelson et al., 2018).

14.3.1.2 Quantitative Resistance

The reduced or the remaining susceptibility of plants to pathogens is considered to be the quantitative resistance (Kushalappa et al., 2016). This kind of resistance is controlled by several genes of a small effect known as minor genes. Recently, these genes have been identified in different crops. For instance, in rice, WRKY genes have been implicated in providing quantitative resistance against bacterial blight and blast diseases. Additionally, CCCH genes have been found to contribute to quantitative resistance against rice bacterial blight (Nelson et al., 2018).

14.3.1.3 Resistance Breeding

Breeding programs aimed at developing varieties with durable and robust disease resistance must consider both qualitative and quantitative resistance. While single resistance genes may provide immediate protection, they are often non-durable. Combining these genes with quantitative disease resistance can lead to more effective crop protection. Although creating complex resistant forms is challenging, developing polygenic resistance through genotypic and phenotypic selection is essential for long-term disease management (Mundt, 2014).

Breeding programs are based essentially on the development of elite lines which are characterized by a good adaptation to environmental conditions and have similar resistance alleles. However, this focus on elite germplasm has resulted in reduced genetic diversity and erosion of resistance (Nelson et al., 2018). Hence, as new disease problems steel arising, there is a huge need for novel resistance. Non-elite germplasm may harbor multiple disease-resistance traits, but it can also carry undesirable alleles, presenting challenges for breeding and resistance identification (Wiesner-Hanks & Nelson, 2016).

Several genetic mapping designs, including genomic sequencing technologies, are currently available and facilitate rapid identification of quantitative resistance locus, which can be exploited for marker-assisted selection. Developing new varieties with resistance genes can be obtained through backcross breeding, often supported by genetic markers. However, this process may require several years and multiple generations to introduce quantitative resistance loci from non-elite germplasm into elite germplasm without undesirable traits (Nelson et al., 2018). Therefore, preemptive approaches for predicting and identifying dominant pathogens and corresponding quantitative resistance loci are crucial before epidemics occur.

Furthermore, varieties characterized by complex resistance can be obtained through recurrent selection. This breeding method is based on repeated cycles of recombination and selection using diverse sets of parents having complementary sets of genes (Wang et al., 2017). Hence, using this approach, favorable alleles are accumulated.

Genomic selection constitutes another breeding strategy where HTS is used to assign selection value to alleles at thousands of loci throughout the genome. This technique predicts individuals with the best allelic combinations for desired traits. Genomic selection has been successfully employed for disease-resistance selection in various crops, including maize and cassava (Poland & Rutkoski, 2016).

14.3.2 Genetic Engineering

Despite the contribution of plant breeding in crop genetic improvement, it can be associated with some limitations such as the transfer of other undesirable traits, along with desired ones. Genetic engineering also helps strongly in producing different desired allelic combinations and is not concerned by such limitations. Genetic engineering refers to methods that target the introduction of modifications to crop genes; the case widely applied lies in the insertion of foreign genes or genes to the original gene of the crop. This process is named genetic transformation. Recently, genomic editing technologies have also emerged, allowing genetic material to be alternated, added, or removed from certain locations of the genome.

14.3.2.1 Genetic Transformation

Genetic transformation includes four steps, namely, DNA extraction, gene cloning, gene design, and transformation (Figure 14.3).

14.3.2.1.1 DNA Extraction

This step consists of extracting the genetic material from an organism having the desired traits. The targeted DNA should be separated from the other parts of its containing cells.

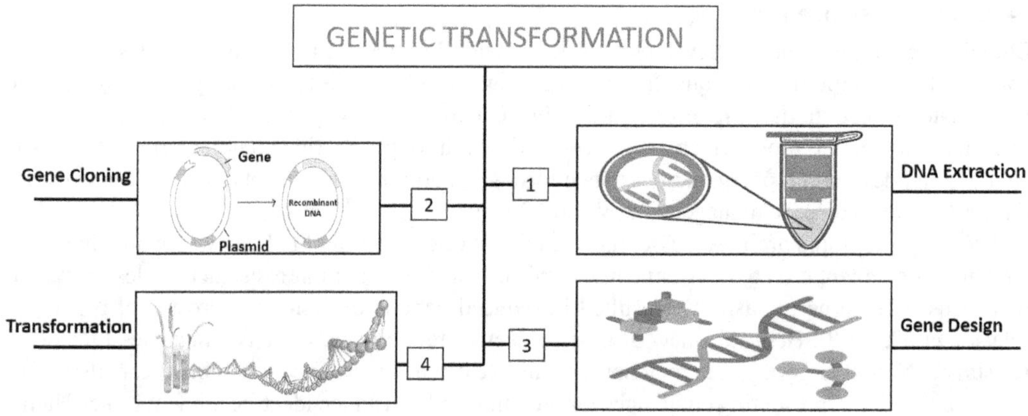

FIGURE 14.3 Steps of plant genetic transformation.

14.3.2.1.2 Gene Cloning

After extraction, the DNA obtained contained multiple genes including those responsible for the desired characteristics and other genes. The gene cloning step relies on separating the gene of interest from the other DNA extracted and its multiplication to obtain several copies from it.

For this purpose, several steps are needed to perform, namely the insertion of a DNA fragment carrying the gene to be cloned into a vector that is composed of circular DNA molecules. Hence, the obtained genetic material is called recombinant DNA. The vector allows the transport of the gene in a host cell, usually a bacterium, to produce several copies of the gene. After cell host division, the recombinant DNA molecules are transferred to the progeny, and further vector replication occurs. After several cell divisions, a clone of host cells is obtained, containing copies of the recombinant DNA molecule (Brown, 2020).

14.3.2.1.3 Gene Design

After gene cloning, gene design takes place. This step is based on cutting the gene of interest using restriction enzymes and induction of gene modifications to express the desired features, through the alteration of some regions, such as replacing the existing promoter with the new promoter (Suza & Lee, 2021).

14.3.2.1.4 Transformation

This process includes the insertion of the desired gene into cells of the plant species. Indeed, cells from the targeted cultivar are isolated to submit transformation, and then a transformation method is applied to ensure the transfer of the gene to plant cells. The two transformation methods commonly used are the gene gun method also known as biolistic transformation and *A. tumefaciens*-mediated transformation. The first method uses high pressure to introduce tungsten or gold beads containing the gene into plant cells, whereas the second method relies on the use of the bacteria *A. tumefaciens*, which ensures the transfer of the desired gene to plant cells (Suza & Lee, 2021).

14.3.2.2 Genomic Editing

Genomic editing technologies have recently emerged providing precise, effective, and rapid engineering of plant genes. These technologies allow not only the enhancement of plant resistance to biotic and abiotic stress but also grain yield increase (Wada et al., 2020).

The well-known editing technology is clustered regularly interspaced short palindromic repeats (CRISPR)-CRISPR-associated Cas (CRISPR/Cas). This editing platform includes specific endonucleases like *Streptococcus pyogenes* Cas9 (SpCas9) and RNA molecules that determine the nuclease specified. SpCas9 can recognize and cut DNA fragments, which are complementary to the variable region of the guide RNA. Indeed, this RNA region can be programmed to direct the Cas9 enzyme to generate a double-stranded break at certain targets of the genome. The subsequent process of DNA repair often includes modifications to DNA at the breakpoints (Dong & Ronald, 2019). Repair using non-homologous end joining often leads to insertions or deletions usually responsible for the creation of knocking out alleles, whereas homology-directed repair allows precise gene insertion at breakpoints (Rinaldo & Ayliffe, 2015).

In addition to SpCas9, other types of Cas nucleases were identified (Wu et al., 2018). Some of these enzymes recognize RNA as substrates (Abudayyeh et al., 2017; Cox et al., 2017). The diversity of Cas nucleases expands therefore the range of targets that can be edited using CRISPR/Cas technology.

Despite the different benefits of CRISPR/Cas technology, untended mutations that can occur constitute the major limitation of this genetic engineering tool. Indeed, Kosicki et al. (2018) have noticed several unexpected deletions in mammalian cells after the use of CRISPR/Cas9 genome editing. Such unexpected deletions have not yet been reported in plants. However, their occurrence could be possible; therefore, it's necessary to take these risks into account.

CRISPR/Cas9 technology has been applied to different plant species against bacterial pathogens (Jaganathan et al., 2018). For instance, CRISPR/Cas9 has been successfully used to design resistant tomato to bacterial speck disease caused by *P. syringae* pv. *tomato* DC3000 (Ortigosa et al., 2019).

14.4 NANOBIOTECHNOLOGY-BASED APPROACHES

Nanotechnology indeed presents a promising avenue for combating plant pathogens and enhancing crop quality. The ultra-small size of nanoparticles enables their effective application in managing plant diseases (Abd-Elsalam & Prasad, 2018).

The mechanism of antimicrobial action of nanoparticles has been reported to be related to the loss of cell membrane integrity; production of reactive oxygen species; and lipid, DNA, and protein damages. These mechanisms can occur in association or individually, depending on the composition, size, and shape of nanoparticles (Hernández-Díaz et al., 2021).

Several nanoparticles have been demonstrated efficient against different bacterial plants and pathogens. For instance, Imada et al. (2016) have tested magnesium oxide nanoparticles against *R. solanacearum* infecting tomato cultivars. Additionally, studies have explored the use of plant-based nanoparticles or 'green nanoparticles', which have also exhibited antimicrobial activity. In this context, Vanti et al. (2019) assessed the activity of silver (Ag) nanoparticles obtained using an extract from the stems of *Gossypium hirsutum*, against the phytopathogenic bacteria *Xanthomonas campestris* pv. *Campestris* and *X. axonopodis* pv. *Malvacearum*. An effective impact on the bacteria growth was obtained. Furthermore, cadmium oxide nanoparticles (Cdo) obtained using the leaf extracts of *Leucaena leucocephala* were revealed effective against *P. aeruginosa* (Savale et al., 2017). This bacterium was previously demonstrated to be phytopathogenic to *Arabidopsis* and sweet basil plants (Walker et al., 2004). In addition, copper nanoparticles (Cu) synthesized in combination with *Ocimum sanctum* leaf extracts were also found efficient against several phytopathogenic bacteria, such as *Xanthomonas axonopodix* pv. *Punicae*, *Rhizoctonia solani*, and *X. axonopodix* pv. Citri (Shende et al., 2016). Similarly, Chen et al. (2019) have shown the potential effect of copper oxide (CuO) nanoparticles, synthesized with papaya leaf extract, against the phytopathogenic bacteria *R. solanacearum* responsible for bacterial wilt disease. In addition, recently, Bibi et al. (2023) have demonstrated the potential antibacterial effect of silver nanoparticles obtained with the extract of *Hedera nepalensis* against two bacterial phytopathogens, namely, *Erwinia carotovora* and *R. solanacearum*.

While these studies underscore the potential of nanoparticles and nanoparticle–plant extract combinations in controlling phytopathogenic bacteria, further field studies are needed to fully realize the benefits of nanobiotechnology on a larger scale.

14.5 CONCLUSIONS AND PROSPECTS

Various crop varieties are suffering from the impact of phytopathogenic bacteria, leading to substantial economic losses in agriculture. The harmful effects of these pathogens have a negative influence on the global economy and pose a threat to world food production. Therefore, the multifaceted nature of bacterial diseases in plants necessitates a comprehensive approach for effective control. The advancements in detection techniques, particularly high-throughput DNA sequencing and metagenomic analysis, offer unique insights into pathogen identification and dynamics. Biological control strategies, encompassing the utilization of beneficial microorganisms such as PGPR and Arbuscular Mycorrhizal Fungi (AMF), along with ISR and the application of bacteriophages, present promising avenues for mitigating the impact of bacterial infections. Genetic strategies, focusing on the development of resistant plant varieties and employing genetic engineering techniques, provide further tools to enhance plant defenses. Additionally, emerging nanobiotechnology-based approaches showcase the potential for precision-targeted interventions. Integrating these diverse strategies into a holistic management framework holds the key to sustainable and resilient plant health in the face of bacterial diseases. Further, interdisciplinary research and collaborative efforts are imperative to refine and optimize these approaches, ensuring their applicability across diverse agricultural settings. The synthesis of these alternative strategies marks a pivotal step toward fostering agricultural sustainability and global food security.

ACKNOWLEDGMENTS

This study received financial support from the Phytopathology Unit within the Department of Plant Protection and Environment at the Ecole Nationale d'Agriculture de Meknes, Morocco.

REFERENCES

Abdelfattah, A., Malacrinò, A., Wisniewski, M., Cacciola, S. O., & Schena, L. (2018). *Metabarcoding: A powerful tool to investigate microbial communities and shape future plant protection strategies*. Elsevier. https://www.sciencedirect.com/science/article/pii/S1049964417301500

Abd-Elsalam, K. A., & Prasad, R. (2018). *Nanobiotechnology applications in plant protection*. Springer. https://www.springer.com/series/15921

Abudayyeh, O. O., Gootenberg, J. S., Essletzbichler, P., Han, S., Joung, J., Belanto, J. J., Verdine, V., Cox, D. B. T., Kellner, M. J., Regev, A., Lander, E. S., Voytas, D. F., Ting, A. Y., & Zhang, F. (2017). RNA targeting with CRISPR-Cas13. *Nature*, *550*(7675), 280–284. https://doi.org/10.1038/nature24049

Adams, I. P., Fox, A., Boonham, N., Massart, S., & De Jonghe, K. (2018). The impact of high throughput sequencing on plant health diagnostics. *European Journal of Plant Pathology*, *152*(4), 909–919. https://doi.org/10.1007/s10658-018-1570-0

Almeida, R. P. P., & Nunney, L. (2015). How do plant diseases caused by xylella fastidiosa emerge? *Plant Disease*, *99*(11), 1457–1467. https://doi.org/10.1094/PDIS-02-15-0159-FE

Al-Samarrai, A. (2012). Nanoparticles as alternative to pesticides in management plant diseases-a review. *International Journal of Scientific and Research Publications*. https://www.academia.edu/download/33984664/pap_IJ07_-_Perceptual_study_of_behavioral_implication_of_usage_of_ICT_for_sustainable_e-Governance_in_Rural_India__(april-2012-59).pdf#page=262

Annapurna, K., & Kumar, A. (2013). PGPR-induced systemic resistance (ISR) in plant disease management. In K. Annapurna, A. Kumar, L. V. Kumar, V. Govindasamy, P. Bose, & D. Ramadoss (Eds.), *Bacteria in agrobiology: Disease management* (pp. 405–425). Springer. https://doi.org/10.1007/978-3-642-33639-3_15

Ayaz, M., Li, C. H., Ali, Q., Zhao, W., Chi, Y. K., Shafiq, M., Ali, F., Yu, X. Y., Yu, Q., Zhao, J. T., Yu, J. W., Qi, R. De, & Huang, W. K. (2023). Bacterial and fungal biocontrol agents for plant disease protection: Journey from lab to field, current status, challenges, and global perspectives. *Molecules*, *28*(18). https://doi.org/10.3390/molecules28186735

Begna, T. (2020). Major challenging constraints to crop production farming system and possible breeding to overcome the constraints. *International Journal of Research Studies in Agricultural Sciences*, *6*(7). https://doi.org/10.20431/2454-6224.0607005

Berendsen, R. L., Pieterse, C. M. J., & Bakker, P. A. H. M. (2012). The rhizosphere microbiome and plant health. *Trends in Plant Science*, *17*(8), 478–486. https://doi.org/10.1016/j.tplants.2012.04.001

Bibi, Z., Ali, M., Abohashrh, M., Ahmad, I., Khan, H., Ali, M., Akbar, F., Ahmad, N., Iqbal, A., Ullah, F., Ullah, Z., & Muhmood, T. (2023). Biologically synthesized silver nanoparticles efficiently control plant pathogenic bacteria-erwinia carotovora and ralstonia solanacearum. *Inorganics*, *11*(7). https://doi.org/10.3390/inorganics11070309

Billing, E. (2011). Fire blight. Why do views on host invasion by Erwinia amylovora differ? *Plant Pathology*, *60*(2), 178–189. https://doi.org/10.1111/j.1365-3059.2010.02382.x

Bonaterra, A., Badosa, E., Daranas, N., Francés, J., Roselló, G., & Montesinos, E. (2022). Bacteria as biological control agents of plant diseases. *Microorganisms*, *10*(9). https://doi.org/10.3390/microorganisms10091759

Bouizgarne, B. (2013). Bacteria for plant growth promotion and disease management. In *Bacteria in agrobiology: Disease management* (pp. 15–47). Springer. https://doi.org/10.1007/978-3-642-33639-3_2

Brown, T. A. (2020). *Gene cloning and DNA analysis: An introduction*. Wiley Blackwell.

Buttimer, C., Hendrix, H., Lucid, A., Neve, H., Noben, J. P., Franz, C., O'Mahony, J., Lavigne, R., & Coffey, A. (2018). Novel N4-like bacteriophages of pectobacterium atrosepticum. *Pharmaceuticals*, *11*(2). https://doi.org/10.3390/ph11020045

Caldwell, D., Kim, B. S., & Iyer-Pascuzzi, A. S. (2017). Ralstonia solanacearum differentially colonizes roots of resistant and susceptible tomato plants. *Phytopathology*, *107*(5), 528–536. https://doi.org/10.1094/PHYTO-09-16-0353-R

Campos, M. D., Félix, M. do R., Patanita, M., Materatski, P., & Varanda, C. (2021). High throughput sequencing unravels tomato-pathogen interactions towards a sustainable plant breeding. *Horticulture Research*, *8*(1). https://doi.org/10.1038/s41438-021-00607-x

Chen, J., Mao, S., Xu, Z., & Ding, W. (2019). Various antibacterial mechanisms of biosynthesized copper oxide nanoparticles against soilborne Ralstonia solanacearum. *RSC Advances*, *9*(7), 3788–3799. https://doi.org/10.1039/c8ra09186b

Cox, D. B. T., Gootenberg, J. S., Abudayyeh, O. O., Franklin, B., Kellner, M. J., Joung, J., & Zhang, F. (2017). RNA editing with CRISPR-Cas13. *Science*, *358*(6366), 1019–1027. https://doi.org/10.1126/science.aaq0180

Cronin, D., Moënne-Loccoz, Y., Fenton, A., Dunne, C., Dowling, D. N., & O'Gara, F. (1997). Ecological interaction of a biocontrol Pseudomonas fluorescens strain producing 2,4-diacetylphloroglucinol with the soft rot potato pathogen Erwinia carotovora subsp. atroseptica. *FEMS Microbiology Ecology*, *23*(2), 95–106. https://doi.org/10.1111/j.1574-6941.1997.tb00394.x

De Silva, N. I., Brooks, S., Lumyong, S., & Hyde, K. D. (2019). Use of endophytes as biocontrol agents. *Fungal Biology Reviews*, *33*(2), 133–148. https://doi.org/10.1016/j.fbr.2018.10.001

Dennehy, J. J., & Abedon, S. T. (2020). Phage infection and lysis. In David R. Harper, Stephen T. Abedon, Benjamin H. Burrowes, & Malcolm L. McConville (Eds.), *Bacteriophages* (pp. 1–43). Springer. https://doi.org/10.1007/978-3-319-40598-8_53-1

Desai, S., Kumar, G. P., Daniel Amalraj, L., Bagyaraj, D. J., & Ashwin, R. (2016). Exploiting PGPR and AMF biodiversity for plant health management. In Dhananjaya Pratap Singh, Harikesh Bahadur Singh, Ratna Prabha (Eds.), *Microbial inoculants in sustainable agricultural productivity:* Research perspectives (vol. 1, pp. 145–160). Springer.

Dimkić, I., Janakiev, T., Petrović, M., Degrassi, G., & Fira, D. (2022). Plant-associated Bacillus and Pseudomonas antimicrobial activities in plant disease suppression via biological control mechanisms: A review. In Physiological and molecular plant pathology (p. 117). Academic Press. https://doi.org/10.1016/j.pmpp.2021.101754

Doerflinger, M., Forsyth, W., Ebert, G., Pellegrini, M., Herold, M., & Hall, E. (2016). CRISPR/Cas9: The ultimate weapon to battle infectious diseases? *Cellular Microbiology*, *19*(2). https://doi.org/10.1111/cmi.12693

Dong, O. X., & Ronald, P. C. (2019). Genetic engineering for disease resistance in plants: Recent progress and future perspectives. *Plant Physiology*, *180*(1), 26–38. https://doi.org/10.1104/pp.18.01224

Doolotkeldieva, T., & Bobusheva, S. (2016). Fire blight disease caused by Erwinia amylovora on rosaceae plants in Kyrgyzstan and biological agents to control this disease. *Advances in Microbiology*, *6*(11), 831–851. https://doi.org/10.4236/aim.2016.611080

Doolotkeldieva, T., Bobusheva, S., & Suleymankisi, A. (2016). Biological control of Erwinia carotovora ssp. carotovora by Streptomyces species. *Advances in Microbiology*, *06*(02), 104–114. https://doi.org/10.4236/aim.2016.62011

Dy, R. L., Rigano, L. A., & Fineran, P. C. (2018). Phage-based biocontrol strategies and their application in agriculture and aquaculture. *Biochemical Society Transactions*, *46*(6), 1605–1613. https://doi.org/10.1042/BST20180178

Eichmeier, A., Kiss, T., Necas, T., Penazova, E., Tekielska, D., Bohunicka, M., Valentova, L., Cmejla, R., Morais, D., & Baldrian, P. (2019). High-throughput sequencing analysis of the bacterial community in stone fruit phloem tissues infected by "candidatus phytoplasma prunorum". *Microbial Ecology*, *77*(3), 664–675. https://doi.org/10.1007/s00248-018-1250-9

Elhalag, K., Eldin, M. N., Hussien, A., & Ahmad, A. (2018). Potential use of soilborne lytic podoviridae phage as a biocontrol agent against Ralstonia solanacearum. *Journal of Basic Microbiology*, *58*(8), 658–669. https://doi.org/10.1002/jobm.201800039

El-shakh, A. S. A., Kakar, K. U., Wang, X., Almoneafy, A. A., Ojaghian, M. R., Li, B., Anjum, S. I., & Xie, G. L. (2015). Controlling bacterial leaf blight of rice and enhancing the plant growth with endophytic and rhizobacterial Bacillus strains. *Toxicological and Environmental Chemistry*, *97*(6), 766–785. https://doi.org/10.1080/02772248.2015.1066176

Ezrari, S., Legrifi, I., Taoussi, M., Khadiri, M., Belabess, Z., & Lahlali, R. (2023). Plant–Pathogen Interactions and Global Food Security. *Plant Pathogen Interaction*, 11–52. https://doi.org/10.1007/978-981-99-4890-1_2

Fira, D., Dimkić, I., Berić, T., Lozo, J., & Stanković, S. (2018). Biological control of plant pathogens by Bacillus species. *Journal of Biotechnology*, *285*, 44–55. https://doi.org/10.1016/j.jbiotec.2018.07.044

Gao, H., Qi, G., Yin, R., Zhang, H., Li, C., & Zhao, X. (2016). Bacillus cereus strain S2 shows high nematicidal activity against Meloidogyne incognita by producing sphingosine. *Scientific Reports*, *6*. https://doi.org/10.1038/srep28756

Gao, W., Thamphiwatana, S., Angsantikul, P., & Zhang, L. (2014). Nanoparticle approaches against bacterial infections. *Wiley Interdisciplinary Reviews: Nanomedicine and Nanobiotechnology*, *6*(6), 532–547. https://doi.org/10.1002/WNAN.1282

Gašić, K., Kuzmanović, N., Ivanović, M., Prokić, A., Šević, M., & Obradović, A. (2018). Complete genome of the Xanthomonas euvesicatoria specific bacteriophage KΦ1, its survival and potential in control of pepper bacterial spot. *Frontiers in Microbiology*, *9*(AUG). https://doi.org/10.3389/fmicb.2018.02021

Gebily, D. A. S., Ghanem, G. A. M., Ragab, M. M., Ali, A. M., Soliman, N. E. Dink., & Abd El-Moity, T. H. (2021). Characterization and potential antifungal activities of three Streptomyces spp. as biocontrol agents against Sclerotinia sclerotiorum (Lib.) de Bary infecting green bean. *Egyptian Journal of Biological Pest Control*, *31*(1). https://doi.org/10.1186/s41938-021-00373-x

Ghorbanpour, M., Omidvari, M., Abbaszadeh-Dahaji, P., Omidvar, R., & Kariman, K. (2018). Mechanisms underlying the protective effects of beneficial fungi against plant diseases. *Biological Control*, *117*, 147–157. https://doi.org/10.1016/j.meegid.2014.01.011

Ghosh, M., Sharma, N., Singh, A. K., Gera, M., Pulicherla, K. K., & Jeong, D. K. (2018). Transformation of animal genomics by next-generation sequencing technologies: A decade of challenges and their impact on genetic architecture. *Critical Reviews in Biotechnology*, *38*(8), 1157–1175. https://doi.org/10.1080/07388551.2018.1451819

Gowdar, S. B., Deepa, H., & Amaresh, Y. S. (2018). A brief review on biocontrol potential and PGPR traits of Streptomyces sp. for the management of plant diseases. *Journal of Pharmacognosy and Phytochemistry*, *7*(5): 03–07.

Guo, Y., Fan, Z., Yi, X., Zhang, Y., Khan, R. A. A., & Zhou, Z. (2021). Sustainable management of soil-borne bacterium ralstonia solanacearum in vitro and in vivo through fungal metabolites of different trichoderma spp. *Sustainability (Switzerland)*, *13*(3), 1–14. https://doi.org/10.3390/su13031491

Gupta, C., Salgotra, R. K., & Mahajan, G. (2020). Rediscovery of genetic and genomic resources for future food security. In R. K. Salgotra, & S. M. Zargar (Eds.), *Rediscovery of Genetic and Genomic Resources for Future Food Security*. Springer. https://doi.org/10.1007/978-981-15-0156-2

Hamdache, A., Azarken, R., Lamarti, A., Aleu, J., & Collado, I. G. (2013). Comparative genome analysis of Bacillus spp. and its relationship with bioactive nonribosomal peptide production. *Phytochemistry Reviews*, *12*(4), 685–716. https://doi.org/10.1007/s11101-013-9278-4

Hernández-Díaz, J. A., Garza-García, J. J. O., Zamudio-Ojeda, A., León-Morales, J. M., López-Velázquez, J. C., & García-Morales, S. (2021). Plant-mediated synthesis of nanoparticles and their antimicrobial activity against phytopathogens. *Journal of the Science of Food and Agriculture*, *101*(4), 1270–1287. https://doi.org/10.1002/jsfa.10767

Idowu, O., Olawole, O., Idumu, O., & Salami, A. (2016). Bio-control effect of Trichoderma asperellum (Samuels) Lieckf. and Glomus intraradices Schenk on Okra seedlings infected with Pythium aphanidermatum (Edson) Fitzp and Erwinia carotovora (Jones). *American Journal of Experimental Agriculture*, *10*(4), 1–12. https://doi.org/10.9734/ajea/2016/21348

Imada, K., Sakai, S., Kajihara, H., Tanaka, S., & Ito, S. (2016). Magnesium oxide nanoparticles induce systemic resistance in tomato against bacterial wilt disease. *Plant Pathology*, *65*(4), 551–560. https://doi.org/10.1111/ppa.12443

Jamal, M., Khan, A. W., Muhammad, S., Ullah, A., Bukhari, S., Andleeb, S., & Ali, M. (2018). Bacteriophages: An overview of the control strategies against multiple bacterial infections in different fields. *Wiley Online Library*, *59*(2), 123–133. https://doi.org/10.1002/jobm.201800412

Jaroszuk-ściseł, J., Tyśkiewicz, R., Nowak, A., Ozimek, E., Majewska, M., Hanaka, A., Tyśkiewicz, K., Pawlik, A., & Janusz, G. (2019). Phytohormones (Auxin, gibberellin) and ACC deaminase in vitro synthesized by the mycoparasitic trichoderma DEMTKZ3A0 strain and changes in the level of auxin and plant resistance markers in wheat seedlings inoculated with this strain conidia. *International Journal of Molecular Sciences*, *20*(19). https://doi.org/10.3390/ijms20194923

Jaganathan, D., Ramasamy, K., Sellamuthu, G., Jayabalan, S., & Venkataraman, G. (2018). CRISPR for crop improvement: An update review. *Frontiers in plant science*, 9, 985–1002. https:// doi.org/ 10.3389/fpls.2018.00985

Jiang, C., Fan, Z., Li, Z., Niu, D., Li, Y., Zheng, M., Wang, Q., Jin, H., & Guo, J. (2020). Bacillus cereus AR156 triggers induced systemic resistance against Pseudomonas syringae pv. tomato DC3000 by suppressing miR472 and activating CNLs-mediated basal immunity in Arabidopsis. *Molecular Plant Pathology*, *21*(6), 854–870. https://doi.org/10.1111/mpp.12935

Kariuki, C. K., Mutitu, E. W., & Muiru, W. M. (2020). Effect of Bacillus and Trichoderma species in the management of the bacterial wilt of tomato (Lycopersicum esculentum) in the field. *Egyptian Journal of Biological Pest Control*, *30*(1). https://doi.org/10.1186/s41938-020-00310-4

Khan, M. (2014). Nanotechnology: Scope and application in plant disease management. https://doi.org/10.3923/ppj.2014.214.231

Kim, D., Hong, S., Kim, Y. T., Ryu, S., Kim, H. B., & Lee, J. H. (2018). Metagenomic approach to identifying foodborne pathogens on chinese cabbage. *Journal of Microbiology and Biotechnology*, *28*(2), 227–235. https://doi.org/10.4014/jmb.1710.10021

Knight, R., Vrbanac, A., Taylor, B. C., Aksenov, A., Callewaert, C., Debelius, J., Gonzalez, A., Kosciolek, T., Mccall, L.-I., Mcdonald, D., Melnik, A. V, Morton, J. T., Navas, J., Quinn, R. A., Sanders, J. G., Swafford, A. D., Thompson, L. R., Tripathi, A., Xu, Z. Z., … Dorrestein, P. C. (2018). Best practices for analyzing microbiomes. *Nature Reviews Microbiology*, *16*(7), 410–422. https://doi.org/10.1038/s41579-018-0029-9

Konappa, N., Krishnamurthy, S., Arakere, U. C., Chowdappa, S., & Ramachandrappa, N. S. (2020). Efficacy of indigenous plant growth-promoting rhizobacteria and Trichoderma strains in eliciting resistance against bacterial wilt in a tomato. *Egyptian Journal of Biological Pest Control*, *30*(1). https://doi.org/10.1186/s41938-020-00303-3

Kosicki, M., Tomberg, K., & Bradley, A. (2018). Repair of double-strand breaks induced by CRISPR–Cas9 leads to large deletions and complex rearrangements. *Nature Biotechnology*, *36*(8), 765–771. https://doi.org/10.1038/nbt.4192

Kushalappa, A. C., Yogendra, K. N., & Karre, S. (2016). Plant innate immune response: Qualitative and quantitative resistance. *Critical Reviews in Plant Sciences*, *35*(1), 38–55. https://doi.org/10.1080/07352689.2016.1148980

Legeay, J., Husson, C., Cordier, T., Vacher, C., Marcais, B., & Buée, M. (2019). Comparison and validation of Oomycetes metabarcoding primers for Phytophthora high throughput sequencing. *Journal of Plant Pathology*, *101*(3), 743–748. https://doi.org/10.1007/s42161-019-00276-9

Li, H. Y., Luo, Y., Zhang, X. S., Shi, W. L., Gong, Z. T., Shi, M., Chen, L. L., Chen, X. L., Zhang, Y. Z., & Song, X. Y. (2014). Trichokonins from *Trichoderma pseudokoningii* SMF2 induce resistance against Gram-negative Pectobacterium carotovorum subsp. carotovorum in Chinese cabbage. *FEMS Microbiology Letters*, *354*(1), 75–82. https://doi.org/10.1111/1574-6968.12427

López-Gresa, M. P., Payá, C., Ozáez, M., Rodrigo, I., Conejero, V., Klee, H., Bellés, J. M., & Lisón, P. (2018). A new role for green leaf volatile esters in tomato stomatal defense against pseudomonas syringe pv. Tomato. *Frontiers in Plant Science*, *871*. https://doi.org/10.3389/fpls.2018.01855

Luiz, C., da Rocha Neto, A. C., Franco, P. O., & Di Piero, R. M. (2017). Emulsions of essential oils and aloe polysaccharides: Antimicrobial activity and resistance inducer potential against Xanthomonas fragariae. *Tropical Plant Pathology*, *42*(5), 370–381. https://doi.org/10.1007/s40858-017-0153-5

Lundborg, L., Sampedro, L., Borg-Karlson, A. K., & Zas, R. (2019). Effects of methyl jasmonate on the concentration of volatile terpenes in tissues of Maritime pine and Monterey pine and its relation to pine weevil feeding. *Trees - Structure and Function*, *33*(1), 53–62. https://doi.org/10.1007/s00468-018-1757-1

Lyra, M., L., L., ACC., P., PD., P., TL., B., D., B. A., & Faivovich, J. (2020). High throughput DNA sequencing of museum specimens sheds light on a species that has been missing for the last 55 years: Bokermannohyla claresignata (Anura: Hylidae: Cophomantini). *Zoological Journal of the Linnean Society*. *190*(4), 1235–1255.

Maji, S., & Chakrabartty, P. K. (2014). Biocontrol of bacterial wilt of tomato caused by Ralstonia solanacearum by isolates of plant growth promoting rhizobacteria. *AJCS*, *8*(2), 208–214.

McKernan, K., Smith, D. R., Spangler, J., Helbert, Y., Lynch, R. C., Devitt-Lee, A., Zhang, L., Orphe, W., Warner, J., Foss, T., Hudalla, C. J., & Silva, M. (2016). Metagenomic analysis of medicinal Cannabis samples; pathogenic bacteria, toxigenic fungi, and beneficial microbes grow in culture-based yeast and mold tests. *F1000Research*, *5*. https://doi.org/10.12688/f1000research.9662.1

Mirik, M., Aysan, Y., & Cinar, O . (2008). Biological control of bacterial spot disease of pepper with bacillus strains. *Turkish Journal of Agriculture and Forestry*, *32*(5).

Mohammed, A. F., Oloyede, A. R., & Odeseye, A. O. (2020). Biological control of bacterial wilt of tomato caused by Ralstonia solanacearum using Pseudomonas species isolated from the rhizosphere of tomato plants. *Archives of Phytopathology and Plant Protection*, *53*(1–2), 1–16. https://doi.org/10.1080/03235408.2020.1715756

Monteiro, L., De Lima, R., Mariano, R., & Souto-Maior, A. M. (2005). Brazilian archives of biology and technology antagonism of Bacillus spp. Against Xanthomonas campestris pv. campestris. *Brazilian Archives of Biology and Technology*, *48*(1), 23–29.

Morrow, J. L., Om, N., Beattie, G. A. C., Chambers, G. A., Donovan, N. J., Liefting, L. W., Riegler, M., & Holford, P. (2020). Characterization of the bacterial communities of psyllids associated with Rutaceae in Bhutan by high throughput sequencing. *BMC Microbiology*, *20*(1). https://doi.org/10.1186/s12866-020-01895-4

Mundt, C. C. (2014). Durable resistance: A key to sustainable management of pathogens and pests. *Infection, Genetics and Evolution*, *27*, 446–455. https://doi.org/10.1016/j.meegid.2014.01.011

Hassan Abd-El-Khair. (2019). Biological Control of Phyto-pathogenic Bacteria. In El-Wakeil, N., Saleh, M., & Abu-hashim, M. (Eds.), *Cottage Industry of Biocontrol Agents and Their Applications Practical Aspects to Deal Biologically with Pests and Stresses Facing Strategic Crops* (pp. 299–336). Springer.

Nagai, H., Miyake, N., Kato, S., Maekawa, D., Inoue, Y., & Takikawa, Y. (2017). Improved control of black rot of broccoli caused by Xanthomonas campestris pv. campestris using a bacteriophage and a non-pathogenic Xanthomonas sp. strain. *Journal of General Plant Pathology*, *83*(6), 373–381. https://doi.org/10.1007/s10327-017-0745-4

Nelson, R., Wiesner-Hanks, T., Wisser, R., & Balint-Kurti, P. (2018). Navigating complexity to breed disease-resistant crops. *Nature Reviews Genetics*, *19*(1), 21–33. https://doi.org/10.1038/nrg.2017.82

Niu, D., Xia, J., Jiang, C., Qi, B., Ling, X., Lin, S., Zhang, W., Guo, J., Jin, H., & Zhao, H. (2016). Bacillus cereus AR156 primes induced systemic resistance by suppressing miR825/825* and activating defense-related genes in Arabidopsis. *Journal of Integrative Plant Biology*, *58*(4), 426–439. https://doi.org/10.1111/jipb.12446

Nizamani, M. M., Zhang, Q., Muhae-Ud-Din, G., & Wang, Y. (2023). High-throughput sequencing in plant disease management: A comprehensive review of benefits, challenges, and future perspectives. *Phytopathology Research*, *5*(1). https://doi.org/10.1186/s42483-023-00199-5

Orgwsundin, G. E., Isafcastiblanco, L. U., Yuan, X., & Dching-hongyang, A. N. (2016). Bacterial disease management: Challenges, experience, innovation and future prospects: Challenges in bacterial molecular plant pathology. *Molecular Plant Pathology*, *17*(9), 1506–1518. https://doi.org/10.1111/mpp.12436

Ortigosa, A., Gimenez-Ibanez, S., Leonhardt, N., & Solano, R. (2019). Design of a bacterial speck resistant tomato by CRISPR/Cas9-mediated editing of SlJAZ2. *Plant Biotechnology Journal*, *17*(3), 665–673. https://doi.org/10.1111/pbi.13006

Pacios-Michelena, S., Aguilar González, C. N., Alvarez-Perez, O. B., Rodriguez-Herrera, R., Chávez-González, M., Arredondo Valdés, R., Ascacio Valdés, J. A., Govea Salas, M., & Ilyina, A. (2021). Application of streptomyces antimicrobial compounds for the control of phytopathogens. *Frontiers in Sustainable Food Systems*, *5*. https://doi.org/10.3389/fsufs.2021.696518

Pandey, P., Irulappan, V., Bagavathiannan, M. V., & Senthil-Kumar, M. (2017). Impact of combined abiotic and biotic stresses on plant growth and avenues for crop improvement by exploiting physio-morphological traits. *Frontiers in Plant Science*, *8*. https://doi.org/10.3389/fpls.2017.00537

Park, J., Lee, G. M., Kim, D., Park, D. H., & Oh, C. S. (2018). Characterization of the lytic bacteriophage phiEaP-8 effective against both Erwinia amylovora and Erwinia pyrifoliae causing severe diseases in apple and pear. *Plant Pathology Journal*, *34*(5), 445–450. https://doi.org/10.5423/PPJ.NT.06.2018.0100

Pinheiro, L. A. M., Pereira, C., Barreal, M. E., Gallego, P. P., Balcão, V. M., & Almeida, A. (2020). Use of phage ϕ6 to inactivate Pseudomonas syringae pv. actinidiae in kiwifruit plants: In vitro and ex vivo experiments. *Applied Microbiology and Biotechnology*, *104*(3), 1319–1330. https://doi.org/10.1007/s00253-019-10301-7

Piombo, E., Abdelfattah, A., Droby, S., Wisniewski, M., Spadaro, D., & Schena, L. (2021). Metagenomics approaches for the detection and surveillance of emerging and recurrent plant pathogens. *Microorganisms*, *9*(1), 1–19. https://doi.org/10.3390/microorganisms9010188

Poland, J., & Rutkoski, J. (2016). Advances and challenges in genomic selection for disease resistance. *Annual Review of Phytopathology*, *54*, 79–98. https://doi.org/10.1146/annurev-phyto-080615-100056

Porter, T. M., & Hajibabaei, M. (2018). Scaling up: A guide to high-throughput genomic approaches for biodiversity analysis. *Molecular Ecology*, *27*(2), 313–338. https://doi.org/10.1111/mec.14478

Poveda, J., Roeschlin, R. A., Marano, M. R., & Favaro, M. A. (2021). Microorganisms as biocontrol agents against bacterial citrus diseases. *Biological Control*, *158*. https://doi.org/10.1016/j.biocontrol.2021.104602

Pradhan, D., Kumar, A., Singh, H., & Agrawal, U. (2019). High-throughput sequencing. In *Data processing handbook for complex biological data sources* (pp. 39–52). Elsevier. https://doi.org/10.1016/B978-0-12-816548-5.00004-6

Prashar, P., Kapoor, N., & Sachdeva, S. (2014). Rhizosphere: Its structure, bacterial diversity and significance. *Reviews in Environmental Science and Biotechnology*, *13*(1), 63–77. https://doi.org/10.1007/s11157-013-9317-z

Ramírez, M., Neuman, B. W., & Ramírez, C. A. (2020). Bacteriophages as promising agents for the biological control of Moko disease (Ralstonia solanacearum) of banana. *Biological Control*, *149*. https://doi.org/10.1016/j.biocontrol.2020.104238

Riedlmeier, M., Ghirardo, A., Wenig, M., Knappe, C., Koch, K., Georgii, E., Dey, S., Parker, J. E., Schnitzler, J. P., & Vlot, A. C. (2017). Monoterpenes support systemic acquired resistance within and between plants. *Plant Cell*, *29*(6), 1440–1459. https://doi.org/10.1105/tpc.16.00898

Rinaldo, A. R., & Ayliffe, M. (2015). Gene targeting and editing in crop plants: A new era of precision opportunities. *Molecular Breeding*, *35*(1), 1–15. https://doi.org/10.1007/s11032-015-0210-z

Román-Reyna, V., Dupas, E., Cesbron, S., Marchi, G., Campigli, S., Hansen, M. A., Bush, E., Prarat, M., Shiplett, K., Ivey, M. L. L., Pierzynski, J., Miller, S. A., Peduto Hand, F., Jacques, M.-A., & Jacobs, J. M. (2021). Metagenomic sequencing for identification of xylella fastidiosa from leaf samples. *MSystems*, *6*(5). https://doi.org/10.1128/msystems.00591-21

Rombouts, S., Volckaert, A., Venneman, S., Declercq, B., Vandenheuvel, D., Allonsius, C. N., Van Malderghem, C., Jang, H. B., Briers, Y., Noben, J. P., Klumpp, J., Van Vaerenbergh, J., Maes, M., & Lavigne, R. (2016). Characterization of novel bacteriophages for biocontrol of bacterial blight in leek caused by Pseudomonas syringae pv. porri. *Frontiers in Microbiology*, *7*(March). https://doi.org/10.3389/fmicb.2016.00279

Sanz-Gaitero, M. & Van, R. M. J. (2021). Structure and function of bacteriophages. In D. R. Harper, S. T. Abedon, B. H. Burrowes, & M. L. McConville (Eds.), *Bacteriophages: Biology, technology, therapy* (pp. 19–91). Springer.

Savale, A., Ghotekar, S., Pansambal, S., & Pardeshi, O. (2017). Green synthesis of fluorescent CdO nanoparticles using Leucaena leucocephala L. extract and their biological activities. *Journal of Bacteriology & Mycology: Open Access*, *5*(5). https://doi.org/10.15406/jbmoa.2017.05.00148

Sharifi, R., & Ryu, C. M. (2016). Are bacterial volatile compounds poisonous odors to a fungal pathogen Botrytis cinerea, alarm signals to arabidopsis seedlings for eliciting induced resistance, or both? *Frontiers in Microbiology*, *7*. https://doi.org/10.3389/fmicb.2016.00196

Shen, Q., Liu, L., Wang, L., & Wang, Q. (2018). Indole primes plant defense against necrotrophic fungal pathogen infection. *PLoS One*, *13*(11). https://doi.org/10.1371/JOURNAL.PONE.0207607

Shende, S. S., Gaikwad, N. D., & Bansod, S. D. (2016). Synthesis and evaluation of antimicrobial potential of copper nanoparticle against agriculturally important Phytopathogens. *International Journal of Biology Research*. https://api.semanticscholar.org/CorpusID:99932497

Singh, A., Gautam, P. K., Verma, A., Singh, V., Shivapriya, P. M., Shivalkar, S., & Samanta, S. K. (2020). Green synthesis of metallic nanoparticles as effective alternatives to treat antibiotics resistant bacterial infections: A review. *Biotechnology Reports*, *25*, E00427. https://www.sciencedirect.com/science/article/pii/S2215017X19305934

Singh, P., Singh, R. K., Guo, D. J., Sharma, A., Singh, R. N., Li, D. P., Malviya, M. K., Song, X. P., Lakshmanan, P., Yang, L. T., & Li, Y. R. (2021). Whole genome analysis of sugarcane root-associated endophyte pseudomonas aeruginosa B18: A plant growth-promoting bacterium with antagonistic potential against sporisorium scitamineum. *Frontiers in Microbiology*, *12*. https://doi.org/10.3389/fmicb.2021.628376

Song, G. C., & Ryu, C. M. (2018). Evidence for volatile memory in plants: Boosting defence priming through the recurrent application of plant volatiles. *Molecules and Cells*, *41*(8), 724–732. https://doi.org/10.14348/molcells.2018.0104

Studholme, D. J. (2011). Application of high-throughput genome sequencing to intrapathovar variation in Pseudomonas syringae. *Molecular Plant Pathology*, *12*(8), 829–838. https://doi.org/10.1111/j.1364-3703.2011.00713.x

Suminda, G. G. D., Bhandari, S., Won, Y., Goutam, U., Kanth Pulicherla, K., Son, Y. O., & Ghosh, M. (2022). High-throughput sequencing technologies in the detection of livestock pathogens, diagnosis, and zoonotic surveillance. *Computational and Structural Biotechnology Journal*, *20*, 5378–5392. https://doi.org/10.1016/j.csbj.2022.09.028

Suza, W. P., & Lee, D. J. (2021). *Genetics, agriculture, and biotechnology*. Iowa State University.

Syed Ab Rahman, S. F., Singh, E., Pieterse, C. M. J., & Schenk, P. M. (2018). Emerging microbial biocontrol strategies for plant pathogens. *Plant Science*, *267*, 102–111. https://doi.org/10.1016/j.plantsci.2017.11.012

Tahir, H. A. S., Gu, Q., Wu, H., Niu, Y., Huo, R., & Gao, X. (2017). Bacillus volatiles adversely affect the physiology and ultra-structure of Ralstonia solanacearum and induce systemic resistance in tobacco against bacterial wilt. *Scientific Reports*, *7*. https://doi.org/10.1038/srep40481

Tedersoo, L., Albertsen, M., Anslan, S., & Callahan, B. (2021). Perspectives and benefits of high-throughput long-read sequencing in microbial ecology. *Applied and Environmental Microbiology*, *87*(17), 1–19. https://doi.org/10.1128/AEM.00626-21

Thambugala, K. M., Daranagama, D. A., Phillips, A. J. L., Kannangara, S. D., & Promputtha, I. (2020). Fungi vs. fungi in biocontrol: An overview of fungal antagonists applied against fungal plant pathogens. *Frontiers in Cellular and Infection Microbiology*, *10*. https://doi.org/10.3389/fcimb.2020.604923

Tkacz, A., Hortala, M., & Poole, P. S. (2018). Absolute quantitation of microbiota abundance in environmental samples. *Microbiome*, *6*(1). https://doi.org/10.1186/s40168-018-0491-7

Triwidodo, A. (2014). Biological control of plant diseases caused by bacteria. *Jurnal Perlindungan Tanaman Indonesia*, *18*(1), 1–12. https://doi.org/10.22146/jpti.15352

Van Hop, D., Thi Phuong Hoa, P., Duc Quang, N., Huu Ton, P., Hoang, T. H., Van Hung, N., Thi Van, N., Van Hai, T., Thi Kim Quy, N., Thi Anh Dao, N., & Thi Thom, V. (2014). Biological control of xanthomonas oryzae pv. oryzae causing rice bacterial blight disease by streptomyces toxytricini vn08-a-12, isolated from soil and leaf-litter samples in Vietnam. *Biocontrol Science*, *19*, 3.

Vanneste, J. L. (2017). The scientific, economic, and social impacts of the New Zealand outbreak of bacterial canker of kiwifruit (Pseudomonas syringae pv. actinidiae). *Annual Review of Phytopathology*, *55*(1), 377–399. https://doi.org/10.1146/annurev-phyto-080516-035530

Vanti, G. L., Nargund, V. B., Basavesha K., N., Vanarchi, R., Kurjogi, M., Mulla, S. I., Tubaki, S., & Patil, R. R. (2019). Synthesis of Gossypium hirsutum-derived silver nanoparticles and their antibacterial efficacy against plant pathogens. *Applied Organometallic Chemistry*, *33*(1). https://doi.org/10.1002/aoc.4630

Voisard, C., Keel, C., Haas, D., & Dèfago, G. (1989). Cyanide production by Pseudomonas fluorescens helps suppress black root rot of tobacco under gnotobiotic conditions. *The EMBO Journal*, *8*(2), 351–358. https://doi.org/10.1002/j.1460-2075.1989.tb03384.x

Vu, N. T., & Oh, C. S. (2020). Bacteriophage usage for bacterial disease management and diagnosis in plants. *Plant Pathology Journal*, *36*(3), 204–217. https://doi.org/10.5423/PPJ.RW.04.2020.0074

Wada, N., Ueta, R., Osakabe, Y., & Osakabe, K. (2020). Precision genome editing in plants: State-of-the-art in CRISPR/Cas9-based genome engineering. *BMC Plant Biology*, *20*(1). https://doi.org/10.1186/s12870-020-02385-5

Walker, T. S., Bais, H. P., Déziel, E., Schweizer, H. P., Rahme, L. G., Fall, R., & Vivanco, J. M. (2004). Pseudomonas aeruginosa-plant root interactions: Pathogenicity, biofilm formation, and root exudation. *Plant Physiology*, *134*(1), 320–331. https://doi.org/10.1104/pp.103.027888

Wang, X., Wei, Z., Yang, K., Wang, J., Jousset, A., Xu, Y., Shen, Q., & Friman, V. P. (2019). Phage combination therapies for bacterial wilt disease in tomato. *Nature Biotechnology*, *37*(12), 1513–1520. https://doi.org/10.1038/s41587-019-0328-3

Wei, C., Liu, J., Maina, A. N., Mwaura, F. B., Yu, J., Yan, C., Zhang, R., & Wei, H. (2017). Developing a bacteriophage cocktail for biocontrol of potato bacterial wilt. *Virologica Sinica*, *32*(6), 476–484. https://doi.org/10.1007/s12250-017-3987-6

Wiesner-Hanks, T., & Nelson, R. (2016). Multiple disease resistance in plants. *Annual Review of Phytopathology*, *54*, 229–252. https://doi.org/10.1146/annurev-phyto-080615-100037

Wu, G., Liu, Y., Xu, Y., Zhang, G., Shen, Q., & Zhang, R. (2018). Exploring elicitors of the beneficial rhizobacterium Bacillus amyloliquefaciens SQR9 to induce plant systemic resistance and their interactions with plant signaling pathways. *Molecular Plant-Microbe Interactions*, *31*(5), 560–567. https://doi.org/10.1094/MPMI-11-17-0273-R

Xie, S., Jiang, H., Ding, T., Xu, Q., Chai, W., & Cheng, B. (2018). Bacillus amyloliquefaciens FZB42 represses plant miR846 to induce systemic resistance via a jasmonic acid-dependent signalling pathway. *Molecular Plant Pathology*, *19*(7), 1612–1623. https://doi.org/10.1111/mpp.12634

Yadav, S., Yadav, S., Kaushik, R., Saxena, A. K., & Arora, D. K. (2014). Genetic and functional diversity of fluorescent Pseudomonas from rhizospheric soils of wheat crop. *Journal of Basic Microbiology*, *54*(5), 425–437. https://doi.org/10.1002/jobm.201200384

Yu, Y., Gui, Y., Li, Z., Jiang, C., Guo, J., & Niu, D. (2022). Induced systemic resistance for improving plant immunity by beneficial microbes. *Plants*, *11*(3). https://doi.org/10.3390/plants11030386

Zaczek-Moczydłowska, Maja A., Young, G. K., Trudgett, J., Fleming, C. C., Campbell, K., & O'Hanlon, R. (2020a). Genomic characterization, formulation and efficacy in planta of a siphoviridae and podoviridae protection cocktail against the bacterial plant pathogens pectobacterium spp. *Viruses*, *12*(2). https://doi.org/10.3390/v12020150

Zaczek-Moczydłowska, Maja A., Young, G. K., Trudgett, J., Plahe, C., Fleming, C. C., Campbell, K., & O'Hanlon, R. (2020b). Phage cocktail containing Podoviridae and Myoviridae bacteriophages inhibits the growth of Pectobacterium spp. under in vitro and in vivo conditions. *PLoS One*, *15*(4). https://doi.org/10.1371/journal.pone.0230842

Zhang, H., & Wang, S. (2013). Rice versus Xanthomonas oryzae pv. oryzae: A unique pathosystem. *Current Opinion in Plant Biology*, *16*(2), 188–195. https://doi.org/10.1016/j.pbi.2013.02.008

Zubair, M., Farzand, A., Mumtaz, F., Khan, A. R., Sheikh, T. M. M., Haider, M. S., Yu, C., Wang, Y., Ayaz, M., Gu, Q., Gao, X., & Wu, H. (2021). Novel genetic dysregulations and oxidative damage in fusarium graminearum induced by plant defense eliciting psychrophilic bacillus atrophaeus ts1. *International Journal of Molecular Sciences*, *22*(22). https://doi.org/10.3390/ijms222212094

15 Beneficial Microbes for Phytonematode Control in Vegetable Crops

Salah-Eddine Laasli, Najwa Seddiqi Kallali, Ikram Leghrifi, Fouad Mokrini, Abdelfattah Dababat, and Rachid Lahlali

15.1 INTRODUCTION

Plant-parasitic nematodes (PPNs) are microscopic roundworms that infect the roots and other parts of various vegetable crops, causing significant yield losses and quality reduction worldwide (Ansari and Saleem, 2023). Some of the most damaging PPN species include root-knot nematodes (RKNs) (*Meloidogyne* spp.), cyst nematodes (*Heterodera* and *Globodera* spp.), and lesion nematodes (*Pratylenchus* spp.) (Sorribas et al., 2020). PPNs have a complex life cycle that involves egg, juvenile, and adult stages, and they can survive in various environmental conditions and soil types (Kumar and Yadav, 2020). PPNs can also interact with other soil-borne pathogens, such as fungi and bacteria, and form disease complexes that exacerbate damage to host plants (Siddiqui and Aziz, 2024).

The conventional method of PPN control uses chemical nematicides, which are synthetic compounds that kill or inhibit nematodes. However, chemical nematicides have several limitations and drawbacks, such as high cost, toxicity to non-target organisms, environmental pollution, and resistance to nematodes (Sivasubramaniam et al., 2020). Moreover, the use of chemical nematicides is restricted or banned in many countries due to the health and safety concerns of consumers and farmers (Ebone et al., 2019). Therefore, alternative and sustainable methods of PPN control that are effective, safe, and eco-friendly are needed.

One of the promising methods of PPN control is the use of beneficial microbes, which are living organisms that can suppress or reduce the nematode population and damage by various mechanisms (Silva et al., 2018; Abd-Elgawad, 2021). Beneficial microbes include fungi, bacteria, and actinomycetes that can act as biological control agents (BCAs) of PPNs (Lahlali et al., 2022). BCAs can directly affect the nematodes by parasitizing, predating, competing, or producing toxins or indirectly by inducing plant resistance, enhancing plant growth, or altering the soil microbiome (Al-Ani et al., 2020). The use of beneficial microbes for PPN control has several advantages over chemical nematicides, such as specificity, compatibility, biodegradability, and low cost (Mwamula et al., 2022).

The main objectives of this chapter are to review the current status and challenges of using beneficial microbes for PPN control in vegetable crops and to discuss the opportunities and future directions for improving the efficacy and consistency of microbial biocontrol of PPN.

15.2 BENEFICIAL MICROBES FOR NEMATODE CONTROL

15.2.1 Beneficial Bacteria

The discovery of beneficial microorganisms for regulating nematode numbers in crops represents a significant step toward more ecologically friendly agricultural methods. The complicated interactions between microbes and plant health can help manage nematode infestations. Beneficial bacteria

 DOI: 10.1201/9781003543237-15

use various strategies to treat agricultural nematode infections, including parasitism, antibiosis, competition, and the induction of systemic resistance (ISR). These strategies not only immediately combat nematodes but also prepare plants to survive future attacks, offering a holistic and permanent solution to nematode control (Ney et al., 2019; Singh et al., 2019) (Figure 15.1).

The parasitic association of bacteria, particularly those from *Pasteuria* species, with nematodes is a strong biocontrol approach with substantial agricultural implications. *Pasteuria* spp. are noted for their peculiar predatory method of action, in which they cling to the cuticle of a nematode, pierce it, and finally cause its death (Topalović et al., 2020).

In the antibiosis process, an important concept in the control of soil pathogens such as nematodes is that bacteria such as *Bacillus* and *Pseudomonas* produce harmful substances against them (Afriyie Boakye et al., 2022). These bacteria also synthesize a great variety of bioactive compounds, such as antibiotics, enzymes, and toxins, that interfere with the normal physiological processes in nematodes to reduce their infectivity and survival rates. By doing this, it is possible to reduce the extent of plant damage caused by nematodes, thus indicating that these bacterial groups can be used as natural enemies of nematode (Afriyie Boakye et al., 2022; Pandey et al., 2023). *Bacillus* spp. are known for producing diverse antimicrobial compounds, including lipopeptides (surfactins, iturins, fengycins), which are characterized by their strong antifungal and antibacterial activities that could alter the environment or directly affect the vital status of nematodes (Nadeem et al., 2021; Raaijmakers et al., 2010). Similarly, various secondary metabolites are produced by *Pseudomonas* spp., including phenazines, pyrrolnitrin, and 2,4-diacetyl phloroglucinol (Kenawy et al., 2019; Suresh et al., 2021). In various ways, these active biocompounds modify nematode physiology by interfering with cuticle structure, digestive metabolism, and reproductive potential. They not only attack nematodes but also make it difficult for other plant pathogens to live in the area; hence, the compounds are all-rounders in the plant protection scheme (Afriyie Boakye et al., 2022; Pandey et al., 2023).

Moreover, the competition between beneficial bacteria and nematodes for vital resources in the rhizosphere is another powerful strategy for biological control in agriculture. This arises from their need to use and use nutrients and space necessary for their growth, survival, and propagation. Nematodes are affected by the rapid growth of beneficial bacteria that outcompete them by absorbing more nutrients. Some bacterial strains increase the ability of plants to absorb nutrients while also depriving nematodes of food. *Bacillus* spp. produce phosphorus and other essential nutrients that indirectly affect nematode populations by altering nutrient dynamics in the rhizosphere

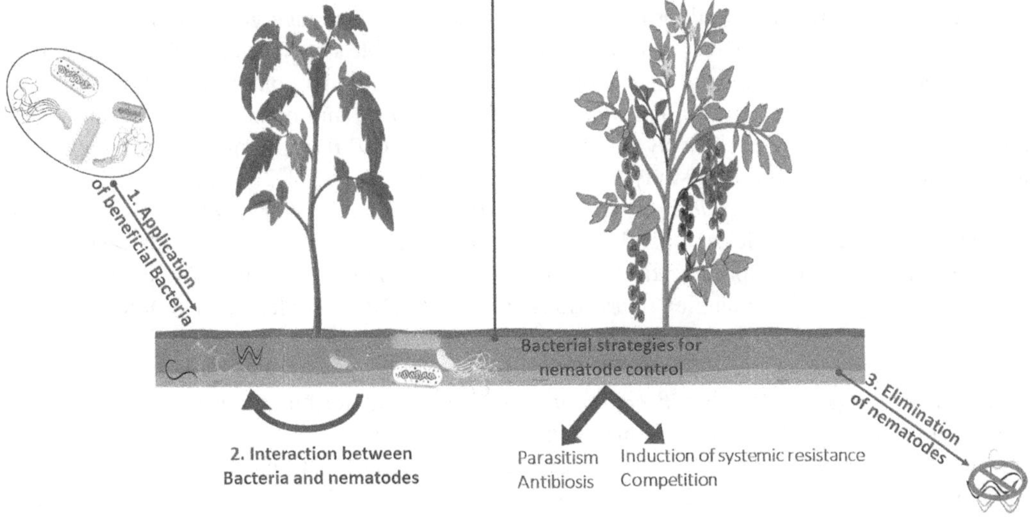

FIGURE 15.1 Bacterial strategies for nematode control in vegetable crops.

(Mendes et al., 2011). Additionally, *Pseudomonas fluorescens* has been reported to occupy root surfaces, thereby outcompeting PPNs in a way that reduces nematode infection rates (Lugtenberg & Kamilova, 2009). However, the presence of assisting microbes can also lead to a situation in which nematodes are not tolerated by their competing populations. One way is through the production of siderophores that bind iron in the soil, rendering it inaccessible to nematodes and other pathogens and retarding their growth (Ansari et al., 2017; Prabhakar, 2020). The above-mentioned process exemplifies other means by which helpful microbial organisms could help control nematode infections in farmlands (Aioub et al., 2022).

In addition, the induction of systemically resistant plants by beneficial microbes such as *Bacillus*, *Pseudomonas*, and *Rhizobium* strains represents a sophisticated biological defense strategy against nematode attacks (Topalović et al., 2020). The process involves the activation of the plant's innate defense mechanisms via hormone pathways and the production of defensive compounds, which in turn reduce the plant's vulnerability to nematode damage and overall resilience. The rhizosphere selects its microbial antagonists through root exudates released by plants, which are vital for host–microbe communication (Yildiz et al., 2023). This can be achieved by exploiting the natural interactions between plants and beneficial microbes, which can enhance plant defenses against nematodes and offer a sustainable alternative to chemical nematicides. The activation of ISR through microbial consortia not only contributes to the direct suppression of nematodes but also supports the overall health and productivity of crops, illustrating the potential of integrating beneficial microbes into crop management practices for improved plant resilience and yield (Topalović et al., 2020).

15.2.2 Successful Application of Beneficial Bacteria

Meloidogyne incognita is a nematode that causes root-knot disease in tomato plants. In a study conducted by Anwar ul Haq et al. (2022), the PGPR–*Pasteuria penetrans* interaction for *M. incognita* management in tomatoes was studied. The application of bacterial suspensions consisting of *Bacillus subtilis*, *P. fluorescens*, and *Enterobacter cloacae* along with *P. penetrans* significantly inhibited nematode reproduction while promoting plant growth. The combination of *P. penetrans* with *E. cloacae* was also effective at remarkably reducing egg masses, galls, females, and galling index. This study suggests that the symbiotic association between PGPRs and *P. penetrans* can be a promising alternative to chemical nematicides and that it accentuates the synergistic interactions among bacterial antagonists against nematodes.

In addition, to identify microbial agents for biological control, Qu et al. (2023) examined the microbial communities in the rhizosphere soil and roots of sponge gourd (*Luffa cylindrica*) infected and uninfected with *M. incognita*. Thirty-two cultivable strains of bacteria and eight fungal strains were identified, along with thousands of operational taxonomic units of bacteria and fungi. In particular, *Bacillus* sp., *Pseudomonas nitroreducens*, and other bacterial strains were found to be dominant or exclusively present under certain conditions, suggesting their potential for nematode suppression. The study found that *Streptomyces* sp. TR27 exhibited nematocidal activity, whereas *Bacillus amyloliquefaciens*, *Bacillus* sp. P35, and *Microbacterium azadirachta* exhibited repellent potential against *M. incognita* juveniles. These results provide insight into novel nematicides and emphasize the shift toward environmentally friendly biological control methods, highlighting the complex interactions between RKNs, host plants, and microorganisms (Table 15.1).

In addition, the objective achieved by Gowda et al. (2022) was to determine the bioefficacy of *Bacillus pumilus* in treating *Pseudomonas putida* liquid formulations against *M. incognita* in okra. The results indicated that seed treatment with *P. putida* and soil application of enriched farmyard manure significantly reduced nematode populations and egg mass production compared with the respective controls. This makes it obvious that the use of *P. putida* and *B. pumilus* will be a breakthrough in the development of biological nematicides that are environmentally friendly.

TABLE 15.1
Key Beneficial Bacteria for Agricultural Nematode Control

Bacteria Species	Controlled Nematode Species	Crop	References
Bacillus megaterium, *Pseudomonas fluorescens*	*Meloidogyne* spp., *Rotylenchulus reniformis*, *Radopholus similis*, *Helicotylenchus multicinctus*	Various	Nagachandrabose and Baidoo (2021)
Azospirillum spp.			
Bacillus spp., *Rhizobium* spp.	Root-knot nematode	*Phaseolus vulgaris*	Karanja et al. (2007)
Bacillus velezensis, *Pseudomonas aeruginosa*	*Macrophomina phaseolina*, *Meloidogyne hapla*	Strawberry	Camacho et al. (2023)
P. aeruginosa, *Bacillus cereus*	*Meloidogyne incognita*	Various	Mohamed et al. (2021)
Pasteuria thornei	*Pratylenchus brachyurus*, *Pratylenchus zeae*	Soybean, maize	De Souza Confort and Inomoto (2018)
Pasteuria spp.	Various plant-parasitic nematodes	Various	Devi (2023)
Bacillus sp., *Paenibacillus* sp., *Xanthomonas* sp.	*Meloidogyne graminicola*	Rice	Bui et al. (2020)
Pseudomonas diminuta, *Bacillus subtilis*	*Pratylenchus coffee*	Coffee	Hindersah et al. (2022)

15.2.3 Beneficial Fungi

The study of the dynamic relationship between soil fungi and nematodes leads to a deep knowledge of how the complex mechanisms that fungi use to trap, infect, and prey on them work. Nematode-trapping fungi are an interesting subgroup of fungi that have come up with various strategies to survive in nutrient-poor conditions by preying on nematodes. These include the development of complex traps, endoparasitism, and switching between a saprophytic and predatory existence (Zhang et al., 2020).

The omnivorous way of life of some mushrooms, such as the ones found in the Orbiliaceae family, is characterized by their activeness in hunting and eating nematodes using specialized trapping structures. They have developed various kinds of trapping mechanisms that include adhesive networks and constricting rings because of adaptation for prey capture, which was shaped by natural selection. The origin of these traps from one ancestor provides an interesting example of how fungi diversified their predatory strategies to facilitate nematode capture and consumption in different ecological settings (Zhang et al., 2020).

Moreover, the fungus that traps nematodes in natural conditions and turns into the so-called traps to trap, kill, and consume them is so amazing. Trap formation is a dynamic change of mode from saprophytism to predatism that signals entry into the carnivorous phase of these fungi. Investigations have recently brought to light some regulatory mechanisms underlying the formation of this equipment, which include pathways of signaling, small molecule compounds, worn body, peroxisome, autophagy, and pH-sensing receptors. Numerous findings suggest an intricate web of biological processes controlling predatory activities by nematode-trapping fungi, which can be exploited for enhanced biocontrol purposes (Zhu et al., 2022).

Nematophagous fungi have also a sophisticated biological interaction known as endoparasitism, where fungi infect and consume nematodes from the inside using some special structures of infectivity. In this way, it is self-evident that these fungi have evolved to exploit nematodes as their nutritive source. This procedure entails several principal stages, facilitated by certain specific fungal adaptations and enzymatic activities (Rahman et al., 2023).

In the process of endoparasitism, three key steps involve specific adaptations in fungi, which cause a variety of enzymatic activities. The initial phase in endoparasitism is the penetration of the nematode cuticle. To break through this protective barrier, some nematophagous fungi like *Nematophthora* species use mechanical pressure and enzymatic degradation. Several extracellular enzymes including chitinases and serine proteases produced by these organisms degrade cuticle components structurally; such enzyme breakdown is important because it enables the fungus to gain access to the worm's interior, establish infection, and initiate consumption from within (Lebrigand et al., 2016; Zhang et al., 2016).

After penetrating the cuticle, the fungi rely on special infectious structures to tether their bodies inside the nematode. These structures, which are different in each fungus, were made to enable increased absorption of nutrients from a host nematode. Such structural adaptations are important for evolution as they help maximize the ingestion of nematodes by fungi through intracellular means. Moreover, these fungi secrete a range of enzymes that further decompose the internal structures of nematodes, enabling nutrient assimilation into the fungi. The fungal mycelium moves and multiplies within a nematode's body, thereby engulfing nutrients and finally killing the pathogen. This notion is essential in showing how predatory and efficient such fungi can be in controlling various populations of nematodes within different settings when fed internally (Lebrigand et al., 2016; Zhang et al., 2016).

In the infection process, serine proteases and chitinases are of central importance. At the very beginning, enzymes take part in puncturing the cuticle and then they continue disintegrating the nematode's body parts. Acting on the chemical components found in eggshell and nematode cuticle, these enzymes act as crucial factors that enable virulence in fungi, leading to the consumption of nematodes (Rahman et al., 2023).

15.2.4 Successful Applications of Beneficial Fungi

A study has shed light on the use of fungi as a means of nematode control. It was found that several species of fungi from the genera *Arthrobotrys*, *Cylindrocarpon*, *Fusarium*, *Monacrosporium*, *Paecilomyces*, *Phoma*, *Plectosphaerella*, and *Verticillium* naturally inhabit the cysts and egg masses of several nematode species, such as *Globodera rostochiensis*, *G. pallida*, *Heterodera schachtii*, and *Meloidogyne hapla*. This is important due to the potential role of these fungi as natural enemies of nematodes, providing an ecological alternative to chemical control methods. Among others, *Fusarium oxysporum* was the most abundant fungal species among those studied, indicating that it could be a strong antagonist against nematodes based on one finding. The presence of these fungi on nematode cysts and egg masses suggests their involvement in the natural regulation of populations, which can be used to suppress PPNs in agriculture, thus reducing crop damage caused by them (Al-Ani et al., 2022) (Table 15.2).

TABLE 15.2
Key Beneficial Fungi Used for Agricultural Nematode Control

Fungi Species	Controlled Nematode Species	Crop	References
Arthrobotrys, *Monacrosporium*, *Nematophthora*, *Hirsutella*, *Verticillium*, *Paecilomyces*	various	Vegetable crops	Saxena (2004)
Nematophagous fungi	*Nacobbus aberrans*	Pepper	Bernardo et al. (2023)
Drechslerella brochopaga	*Meloidogyne hapla*	Horticultural and field crops	Hastuti et al. (2023)
Purporeocillum lilacinum	*Meloidogyne incognita*	Tomato	Niño-Arteaga et al. (2023)
Pochonia chlamydosporia	various	Various crops	Lopez-Nuñez et al. (2022)
Beauveria spp., *Arthrobotrys* spp., *Pochonia chlamydosporia*, *Metarhizium anisopliae*	*Pratylenchus goodeyi*, *Helicotylenchus* spp.	Banana	Ciancio et al. (2022)
Rhizoglomus fasciculatum, *Paecilomyces lilacinus*	*M. incognita*	*Capsicum annuum* (bell pepper)	Giri et al. (2022)

Furthermore, a study carried out in Kenya by Wachira (2015) was conducted to identify where and how many nematode-destroying fungi (NTF) exist in selected key vegetable growing areas to use them as BCAs against PPNs. This research, which was necessitated by the widespread use of nematicides for controlling PPNs, is praised as a milestone toward sustainable agriculture, considering that due to nematode infestations, some Kenyan vegetable crops can be lost at a rate of 80% in the worst cases and there are health and environmental concerns associated with chemicals used for the purpose. To achieve this objective, soil samples were collected from Kibwezi, Kinare, Kabete, Athi-river, and Machakos where vegetables are grown extensively and NTF isolated through the sprinkle method. The results showed that 171 fungal isolates had the potential of antagonizing nematodes while *Arthrobotrys* was the most prevalent genus. *Arthrobotrys* has an important reputation for its role in trapping nematodes, making it one of the most effective biocontrol agents applied across farms. These fungi were distributed differently across various surveyed areas suggesting that NTF effectiveness against nematode populations could be affected by indigenous environmental conditions and farming practices as well (Wachira, 2015).

15.2.5 Actinomycetes

Many biocontrol agents are created from actinomycetes, particularly those in the genus *Streptomyces*, to fight against RKNs, which severely attack the root systems of major crops leading to huge losses. *Streptomyces* sp. is such a potential BCA through its control of root-knot disease on chili plants in greenhouse conditions, resulting in 33.4% control efficacy (Ruanpanun, 2011). The actinomycetes produce bioactive compounds that give rise to this effectiveness, which is an alternative way of controlling PPNs. There are several different mechanisms by which actinomycetes control nematodes. These include producing chemicals that can kill nematodes, having antifungal effects, and producing plant growth-promoting substances such as indole-3-acetic acid and siderophores. They also increase the plant's resistance to pest attacks and thus promote plant health. This is a multimodal strategy for managing nematodes in crops (Ruanpanun, 2011).

Moreover, the usage of microbial antagonists such as actinomycetes in the control of PPNs emphasizes the necessity of understanding interplay within the host plant–nematode–target soil–microbial control agent–environment structure. In addition, there are organic soil amendments that stimulate antagonistic microorganisms to PPNs with the decomposition of organic matter, leading to the accumulation of nematocidal compounds in soils. This means a more extensive approach to nematode management by considering both soil health and microbial activity that improves BCAs like actinomycetes (Silva et al., 2022).

15.2.6 Successful Applications of Actinomycetes

In this context, relevant research has shown that actinomycetes can effectively control nematodes. For instance, *Streptomyces* species produce secondary metabolites such as spectinabilin that cause direct toxicity to nematodes. A recent study reported that 100% lethality of *Caenorhabditis elegans*, a model organism for studying nematodes, was caused by spectinabilin, showing its strong nematicidal feature. The reason for this direct toxicity is that the compound interferes with essential physiological processes in the nematode (Sun et al., 2023).

Also, in this regard, Ran et al. (2022) have also pointed out that actinomycetes produce compounds that can effectively inhibit the hatching of nematode eggs, which is an important step in controlling these worm populations. For instance, several *Micromonospora* sp. WH06-derived metabolites like benzene-propanoic acid were found to be potent inhibitors of *M. incognita* egg hatching. This function ensures that new nematodes do not emerge and therefore reduces the existing population and the damage they inflict on agriculture.

Moreover, Kang et al. (2022) emphasized the fact that compounds such as aureothin and alloaureothin, obtained from endophytic *Streptomyces* sp., can inhibit the growth, multiplication, and behavior of *Bursaphelenchus xylophilus*. These compounds disrupt the nematode's development and reproductive potentialities, thus lowering its viability and fertility levels. Stopping this interference is also important in sustainable nematode control because it prevents quick nematode population recovery, which results in further damage to agricultural systems.

15.3 CHALLENGES AND OPPORTUNITIES FOR MICROBIAL BIOCONTROL OF NEMATODES

The use of beneficial microbes for PPN control in vegetable crops is not a new concept, and several studies have reported the potential and effectiveness of various microbial BCAs against different PPN species and hosts. However, the application of beneficial microbes for PPN control in the field is still limited and inconsistent, and many challenges and gaps need to be addressed and overcome (Kozacki et al., 2024).

15.3.1 Challenges for Microbial PPN Management

One of the main challenges for microbial biocontrol of PPN is the variability and unpredictability of the performance and persistence of beneficial microbes in the field (Tóthné Bogdányi et al., 2021). The performance and persistence of beneficial microbes depend on several biotic and abiotic factors, such as soil properties, climate, crop management, nematode species and density, plant genotype and phenotype, and interactions with other soil microorganisms (Abdul Rahman et al., 2021; Malgioglio et al., 2022). These factors can influence the survival, growth, activity, distribution of beneficial microbes in the soil and the rhizosphere, and, consequently, their ability to control the nematodes.

Some of the soil properties that can affect the performance and persistence of beneficial microbes are pH, moisture, temperature, organic matter, nutrients, texture, and structure (Hartmann and Six, 2023). For instance, soil pH can affect the availability and solubility of nutrients and toxins and the activity and diversity of soil microorganisms (Gondal et al., 2021). Soil moisture and temperature can affect the water potential, oxygen availability, and metabolic rate of beneficial microbes and nematodes (Furtak and Wolińska, 2023). Soil texture and structure can affect the pore size, water movement, and air diffusion in the soil and thus the mobility and distribution of beneficial microbes and nematodes (Erktan et al., 2020).

Several plant-related factors can also affect the overall action of beneficial microbes including plant genotype and phenotype, growth stage and condition, and root exudates and signals (Nadarajah and Abdul Rahman, 2021). For instance, plant genotype and phenotype can affect the genetic and morphological traits of the plants, such as resistance, tolerance, and attractiveness to the nematodes, and also the compatibility and colonization of the beneficial microbes (Nerva et al., 2022). Plant growth stage and condition can affect the physiological and biochemical status of the plants, such as nutrient uptake, photosynthesis, and stress response and also the growth and activity of the beneficial microbes (Santoyo et al., 2021). Furthermore, plant root exudates and signals can affect the chemical and biological composition of the rhizosphere and also the attraction and communication of these microbes (Vives-Peris et al., 2020).

15.3.2 Opportunities for Microbial PPN Management

The management of PPNs in vegetables presents several opportunities, particularly in the realm of microbial interventions. One such approach is the use of BCAs, which have shown promise in reducing PPN populations and increasing crop yields (Abd-Elgawad, 2020). BCAs are especially beneficial in sustainable agriculture and should be incorporated into comprehensive pest management systems (Curk and Trdan, 2024).

Another innovative method for PPN management is the practice of grafting vegetable crops (Phani et al., 2023). Grafting, which involves the union of two plant varieties, can potentially reduce phytonematode damage. This technique relies on several factors, including plant responses mediated by resistance R genes, efficient crosstalk of defensive genes, alteration of plant metabolism and nutrient mobility, changes in plant vigor and physiological attributes, and modification of the rhizosphere microbial community (Phani et al., 2023). Moreover, the use of resistant rootstocks in both open-field conditions and greenhouses has been identified as a viable strategy for managing nematode incidence (Phani et al., 2021; 2023). This approach is part of a broader shift toward sustainability in agriculture, which includes organic and protected farming of high-value vegetable crops.

15.4 CONCLUSION

Beneficial microbes have been explored as a feasible alternative to chemical nematicides in crops against nematodes, thus revealing possibilities of increasing food production without causing harm to the environment. There is enormous potential for the use of bacteria, fungi, and actinomycetes in the biological control of nematodes through various mechanisms including parasitism, antibiosis, competition-induced resistance, trapping endoparasitism, and secretion of nematicidal compounds. Not only do such microbial agents provide an answer to some of these problems associated with chemical nematicides such as the development of resistance and environmental degradation but they are also compatible with principles of sustainable agriculture and integrated pest management.

However, this is seen to affect the performance and consistency of helpful beneficial microbes in the environment due to various factors such as soil properties, climate conditions, and crop management practices. As a result, novel approaches in formulation, delivery, inoculation, and integration with existing agricultural systems need to be explored to enhance the performance and persistence of microbial biocontrol agents. To unravel complex interactions between beneficial microbes, nematodes, and plant hosts, future studies should apply genomics analysis, molecular techniques, and bioinformatics tools. Additionally, the portrayal of microbial biocontrol agents in nematode management programs requires a multidisciplinary approach that includes not only biological or ecological research but also socio-economic considerations for farmers' acceptance and practicality. Therefore, it is important that more efforts continue being made on understanding modes of action; optimizing application methods; and determining long-term effects on soil health and crop productivity caused by microbial biocontrol agents. Researchers, farmers, and extension officers should work together to mix the beneficial microbes with nematode management practices. There is a need for more research on how microbial consortia interact with other formulations of microbes and their advantages to make biocontrol agents more effective and dependable. Finally, the tactical utilization of favorable microorganisms for managing nematodes has immense potential in sustainable vegetable production, food security, and environmental conservation.

ACKNOWLEDGMENTS

This study received financial support from the Phytopathology Unit within the Department of Plant Protection and Environment at the Ecole Nationale d'Agriculture de Meknes, Morocco.

REFERENCES

Abd-Elgawad, M. M. (2020). Biological control agents in the integrated nematode management of potato in Egypt. *Egyptian Journal of Biological Pest Control*, *30*, 1–13.

Abd-Elgawad, M. M. (2021). Optimizing safe approaches to manage plant-parasitic nematodes. *Plants*, *10*(9), 1911.

Abdul Rahman, N. S. N., Abdul Hamid, N. W., & Nadarajah, K. (2021). Effects of abiotic stress on soil microbiome. *International Journal of Molecular Sciences*, *22*(16), 9036.

Afriyie Boakye, T., Kwadwo Anning, D., Li, H., Osei, R., Boamah, S., & Anim, S. (2022). Mechanism of antagonistic bioagents in controlling root-knot nematodes (Meloidogyne sp.): A review. *Asian Research Journal of Agriculture*, 27–44. https://doi.org/10.9734/arja/2022/v15i230153

Aioub, A. A. A., Elesawy, A. E., & Ammar, E. E. (2022). Plant growth promoting rhizobacteria (PGPR) and their role in plant-parasitic nematodes control: A fresh look at an old issue. *Journal of Plant Diseases and Protection*, *129*(6), 1305–1321. https://doi.org/10.1007/s41348-022-00642-3

Al-Ani, L. K. T., Aguilar-Marcelino, L., Fiorotti, J., Sharma, V., Sarker, M. S., Furtado, E. L., ... & Herrera-Estrella, A. (2020). Biological control agents and their importance for the plant health. In J.S. Singh & S.R. Vimal (Eds.), *Microbial services in restoration ecology* (pp. 13–36). Elsevier.

Al-Ani, L. K. T., Soares, F. E. de F., Sharma, A., de los Santos-Villalobos, S., Valdivia-Padilla, A. V., & Aguilar-Marcelino, L. (2022). Strategy of nematophagous fungi in determining the activity of plant parasitic nematodes and their prospective role in sustainable agriculture. *Frontiers in Fungal Biology*, *3*. https://doi.org/10.3389/ffunb.2022.863198

Ansari, T., & Saleem, M. (2023). Plant parasitic nematodes: A silent threat to agricultural output and sustainable approaches for their management. In Hasanuzzaman, M. (Ed.), *Climate-Resilient Agriculture*, Vol 1. Springer, Cham. https://doi.org/10.1007/978-3-031-37424-1_36.

Ansari, R. A., Mahmood, I., Rizvi, R., Sumbul, A., & Safiuddin. (2017). Siderophores: Augmentation of soil health and crop productivity. In Kumar, V., Kumar, M., Sharma, S., & Prasad, R. (Eds.), *Probiotics in Agroecosystem*. Springer, Singapore. https://doi.org/10.1007/978-981-10-4059-7_15

Anwar ul Haq, M., Shahid, M., Zeeshan Niaz, M., Mahmood, K., Talal, M., Adil, Z., & Manzoor Ahmad, H. (2022). Biological management of root knot nematode, meloidogyne incognita through bacterial antagonists infecting tomato. *Journal of Agriculture and Veterinary Science*, *1*(2), 71–79. https://doi.org/10.55627/agrivet.01.02.0255

Bernardo, V. F., Gonzalez, M. A., Garita, S. A., Lopez, R. O., Arango, M. C., Saparrat, M. C. N., & Ruscitti, M. F. (2023). Evaluation of two nematophagous fungi for the control of false root-knot nematode Nacobbus aberrans in pepper crops. *Journal of General Plant Pathology*, *89*(6), 339–346. https://doi.org/10.1007/s10327-023-01145-1

Bui, H. X., Hadi, B. A. R., Oliva, R., & Schroeder, N. E. (2020). Beneficial bacterial volatile compounds for the control of root-knot nematode and bacterial leaf blight on rice. *Crop Protection*, *135*. https://doi.org/10.1016/j.cropro.2019.04.016

Camacho, M., de los Santos, B., Vela, M. D., & Talavera, M. (2023). Use of bacteria isolated from berry rhizospheres as biocontrol agents for charcoal rot and root-knot nematode strawberry diseases. *Horticulturae*, *9*(3). https://doi.org/10.3390/horticulturae9030346

Ciancio, A., Rosso, L. C., Lopez-Cepero, J., & Colagiero, M. (2022). Rhizosphere 16S-ITS metabarcoding profiles in banana crops are affected by nematodes, cultivation, and local climatic variations. *Frontiers in Microbiology*, *13*. https://doi.org/10.3389/fmicb.2022.855110

Compant, S., Duffy, B., Nowak, J., Clément, C., & Barka, E. A. (2005). Use of plant growth-promoting bacteria for biocontrol of plant diseases: Principles, mechanisms of action, and future prospects. *Applied and Environmental Microbiology*, *71*(9), 4951–4959). https://doi.org/10.1128/AEM.71.9.4951-4959.2005

Curk, M., & Trdan, S. (2024). Benefiting from complexity: Exploring enhanced biological control effectiveness via the simultaneous use of various methods for combating pest pressure in agriculture. *Agronomy*, *14*(1), 199. https://doi.org/10.3390/agronomy14010199

De Souza Confort, P. M., & Inomoto, M. M. (2018). Pasteuria thornei, a novel biological seed treatment for Pratylenchus brachyurus control in soybean. *Nematology*, *20*(6), 519–523. https://doi.org/10.1163/15685411-00003156

Devi, G. (2023). Pasteuria spp. as biocontrol agent against plant-parasitic nematodes: An overview. *International Journal of Plant & Soil Science*, *35*(20), 134–144. https://doi.org/10.9734/ijpss/2023/v35i203793

Ebone, L. A., Kovaleski, M., & Deuner, C. C. (2019). Nematicides: History, mode, and mechanism action. *Plant Science Today*, *6*(2), 91–97.

Erktan, A., Or, D., & Scheu, S. (2020). The physical structure of soil: Determinant and consequence of trophic interactions. *Soil Biology and Biochemistry*, *148*, 107876.

Furtak, K., & Wolińska, A. (2023). The impact of extreme weather events as a consequence of climate change on the soil moisture and on the quality of the soil environment and agriculture: A review. *Catena*, *231*, 107378.

Giri, B., Rawat, R., Saxena, G., Manchanda, P., Wu, Q. S., & Sharma, A. (2022). Effect of Rhizoglomus fasciculatum and Paecilomyces lilacinus in the biocontrol of root-knot nematode, Meloidogyne incognita in Capsicum annuum L. *Communicative and Integrative Biology*, *15*(1), 75–87. https://doi.org/10.1080/19420889.2021.2025195

Gondal, A. H., Hussain, I., Ijaz, A. B., Zafar, A., Ch, B. I., Zafar, H., ... & Usama, M. (2021). Influence of soil pH and microbes on mineral solubility and plant nutrition: A review. *International Journal of Agriculture and Biological Sciences*, *5*(1), 71–81.

Gowda, M. T., Sellaperumal, C., Reddy, R., & Pandey, K. (2022). Antagonistic potential of rhizobacterial liquid formulations against Meloidogyne incognita infecting okra under field conditions. *Vegetable Science*, *49*(02), 241–247. https://doi.org/10.61180/vegsci.2022.v49.i2.16

Haas, D., & Défago, G. (2005). Biological control of soil-borne pathogens by fluorescent pseudomonads. *Nature Reviews Microbiology*, *3*(4), 307–319). https://doi.org/10.1038/nrmicro1129

Hartmann, M., & Six, J. (2023). Soil structure and microbiome functions in agroecosystems. *Nature Reviews Earth & Environment*, *4*(1), 4–18.

Hastuti, L., Sari, R. W., Fauzi, F., Naibaho, D. C., Purba, R. T., & Putri, Q. A. (2023). Nematophagous fungi isolated from municipal waste-contaminated soil in medan city, north sumatera: Morphological identification, phylogeny analysis and assessment as root-knot nematodes biocontrol. *Yüzüncü Yıl Üniversitesi Tarım Bilimleri Dergisi*, *33*(4), 522–533. https://doi.org/10.29133/yyutbd.1230261

Hindersah, R., Asyiah, I. N., Harni, R., Rahayu, D. S., & Fitriatin, B. N. (2022). Formulation of soil beneficial microbes solid inoculant for controlling nematode in coffee. *Jurnal Penelitian Pertanian Terapan*, *22*(1), 58–66. https://doi.org/10.25181/jppt.v22i1.2126

Kang, M. K., Kim, J. H., Liu, M. J., Jin, C. Z., Park, D. J., Kim, J., Sung, B. H., Kim, C. J., & Son, K. H. (2022). New discovery on the nematode activity of aureothin and alloaureothin isolated from endophytic bacteria Streptomyces sp. AE170020. *Scientific Reports*, *12*(1). https://doi.org/10.1038/s41598-022-07879-w

Karanja, N. K., Mutua, G. K., & Kimenju, J. W. (2007). Evaluating the effect of Bacillus and Rhizobium. bi-inoculant on nodulation and nematode control in Phaseolus vulgaris L. In *Advances in Integrated Soil Fertility Management in sub-Saharan Africa: Challenges and Opportunities* (pp. 865–872). https://doi.org/10.1007/978-1-4020-5760-1_82

Kenawy, A., Dailin, D. J., Abo-Zaid, G. A., Malek, R. A., Ambehabati, K. K., Zakaria, K. H. N., Sayyed, R. Z., & El Enshasy, H. A. (2019). *Biosynthesis of antibiotics by PGPR and their roles in biocontrol of plant diseases* (pp. 1–35). https://doi.org/10.1007/978-981-13-6986-5_1

Kozacki, D., Soika, G., Skwiercz, A., & Malusà, E. (2024). Microbial-based products and soil management practices to control nematodes in organic horticultural crops. In Chaudhary, K. K., Meghvansi, M. K., & Siddiqui, S. (Eds.), *Sustainable Management of Nematodes in Agriculture, Vol.2: Role of Microbes-Assisted Strategies. Sustainability in Plant and Crop Protection, vol 19*. Springer, Cham. https://doi.org/10.1007/978-3-031-52557-5_1.

Kumar, Y., & Yadav, B. C. (2020). Plant-parasitic nematodes: Nature's most successful plant parasite. *International Journal of Research and Review*, *7*(3), 379–386.

Lahlali, R., Ezrari, S., Radouane, N., Kenfaoui, J., Esmaeel, Q., El Hamss, H., ... & Barka, E. A. (2022). Biological control of plant pathogens: A global perspective. *Microorganisms*, *10*(3), 596.

Lebrigand, K., He, L. D., Thakur, N., Arguel, M. J., Polanowska, J., Henrissat, B., Record, E., Magdelenat, G., Barbe, V., Raffaele, S., Barbry, P., & Ewbank, J. J. (2016). Comparative genomic analysis of drechmeria coniospora reveals core and specific genetic requirements for fungal endoparasitism of nematodes. *PLoS Genetics*, *12*(5). https://doi.org/10.1371/journal.pgen.1006017

Lopez-Nuñez, R., Suarez-Fernandez, M., Lopez-Moya, F., & Lopez-Llorca, L. V. (2022). Chitosan and nematophagous fungi for sustainable management of nematode pests. *Frontiers in Fungal Biology*, *3*. https://doi.org/10.3389/ffunb.2022.980341

Lugtenberg, B., & Kamilova, F. (2009). Plant-growth-promoting rhizobacteria. *Annual Review of Microbiology*, *63*, 541–556. https://doi.org/10.1146/annurev.micro.62.081307.162918

Malgioglio, G., Rizzo, G. F., Nigro, S., Lefebvre du Prey, V., Herforth-Rahmé, J., Catara, V., & Branca, F. (2022). Plant-microbe interaction in sustainable agriculture: The factors that may influence the efficacy of PGPM application. *Sustainability*, *14*(4), 2253.

Mendes, R., Kruijt, M., De Bruijn, I., Dekkers, E., Van Der Voort, M., Schneider, J. H. M., Piceno, Y. M., DeSantis, T. Z., Andersen, G. L., Bakker, P. A. H. M., & Raaijmakers, J. M. (2011). Deciphering the rhizosphere microbiome for disease-suppressive bacteria. *Science*, *332*(6033), 1097–1100. https://doi.org/10.1126/science.1203980

Mohamed, S. A., Ameen, H. H., Elkelany, U. S., El-Wakeel, M. A., Hammam, M. M., & Soliman, G. M. (2021). Genetic improvement of Pseudomonas aeruginosa and Bacillus cereus for controlling root knot nematode and two weeds under laboratory conditions. *Jordan Journal of Biological Sciences*, *14*(4), 859–865. https://doi.org/10.54319/jjbs/140428

Mwamula, A. O., Kabir, M. F., & Lee, D. (2022). A review of the potency of plant extracts and compounds from key families as an alternative to synthetic nematicides: History, efficacy, and current developments. *The Plant Pathology Journal*, *38*(2), 53.

Nadarajah, K., & Abdul Rahman, N. S. N. (2021). Plant–microbe interaction: Aboveground to belowground, from the good to the bad. *International Journal of Molecular Sciences*, *22*(19), 10388.

Nadeem, H., Niazi, P., Asif, M., Kaskavalci, G., & Ahmad, F. (2021). Bacterial strains integrated with surfactin molecules of Bacillus subtilis MTCC441 enrich nematocidal activity against Meloidogyne incognita. *Plant Biology*, *23*(6), 1027–1036. https://doi.org/10.1111/plb.13301

Nagachandrabose, S., & Baidoo, R. (2021). Humic acid – a potential bioresource for nematode control. *Nematology*, *24*(1), 1–10). https://doi.org/10.1163/15685411-bja10116

Nerva, L., Sandrini, M., Moffa, L., Velasco, R., Balestrini, R., & Chitarra, W. (2022). Breeding toward improved ecological plant-microbiome interactions. Trends in Plant Science, 27(11), 1134–1143. https://doi.org/10.1016/j.tplants.2022.06.004

Ney, L., Franklin, D., Mahmud, K., Cabrera, M., Hancock, D., Habteselassie, M., Newcomer, Q., Dahal, S., & Subedi, A. (2019). Sensitivity of nematode community analysis to agricultural management practices and inoculation with local effective microorganisms in the Southeastern United States. *Soil Systems*, *3*(2), 1–16. https://doi.org/10.3390/soilsystems3020041

Niño-Arteaga, D. M., Macías-López, B. C., Olivas-García, J. M., Alonso-Gómez, M. A., Sánchez, E., & Muñoz-Márquez, E. (2023). Biocontrol of the root-knot nematode Meloidogyne incognita with Purporeocillum lilacinum and liquid bio-formulates in tomato (Solanum lycopersicum). *Notulae Botanicae Horti Agrobotanici Cluj-Napoca*, *51*(1). https://doi.org/10.15835/nbha51113010

Pandey, P., Tripathi, A., Dwivedi, S., Lal, K., & Jhang, T. (2023). Deciphering the mechanisms, hormonal signaling, and potential applications of endophytic microbes to mediate stress tolerance in medicinal plants. *Frontiers in Plant Science*, *14*. https://doi.org/10.3389/fpls.2023.1250020

Phani, V., Gowda, M. T., & Dutta, T. K. (2023). Grafting vegetable crops to manage plant-parasitic nematodes: A review. *Journal of Pest Science*, *97*, 539–560.

Phani, V., Khan, M. R., & Dutta, T. K. (2021). Plant-parasitic nematodes as a potential threat to protected agriculture: Current status and management options. *Crop Protection*, *144*, 105573.

Prabhakar, P. K. (2020). Bacterial siderophores and their potential applications: A review. *Current Molecular Pharmacology*, *13*(4), 295–305. https://doi.org/10.2174/1874467213666200518094445

Qu, L., Dai, K., Wang, J., Cao, L., Rao, Z., & Han, R. (2023). Microbial landscapes of the rhizosphere soils and roots of Luffa cylindrica plant associated with Meloidogyne incognita. *Frontiers in Microbiology*, *14*. https://doi.org/10.3389/fmicb.2023.1168179

Raaijmakers, J. M., de Bruijn, I., Nybroe, O., & Ongena, M. (2010). Natural functions of lipopeptides from Bacillus and Pseudomonas: More than surfactants and antibiotics. *FEMS Microbiology Reviews*, *34*(6), 1037–1062). https://doi.org/10.1111/j.1574-6976.2010.00221.x

Rahman, M. U., Chen, P., Zhang, X., & Fan, B. (2023). Predacious strategies of nematophagous fungi as bio-control agents. *Agronomy*, *13*(11), 2685). https://doi.org/10.3390/agronomy13112685

Ran, Y., Zhang, Y., Wang, X., & Li, G. (2022). Nematicidal metabolites from the actinomycete micromonospora sp. WH06. *Microorganisms*, *10*(11). https://doi.org/10.3390/microorganisms10112274

Rodriguez-Kabana, R., Morgan-Jones, G., & Chet, I. (1987). Biological control of nematodes: Soil amendments and microbial antagonists. *Plant and Soil*, *100(1–3)*, 237–247. https://doi.org/10.1007/BF02370944

Ruanpanun, P. (2011). Diversity of actinomycetes and fungi from plant-parasitic nematode infested rhizosphere soils and their biocontrol potential on meloidogyne incognita. Thesis Dissertation. Chiang Mai: Graduate School, Chiang Mai University.

Santoyo, G., Guzmán-Guzmán, P., Parra-Cota, F. I., Santos-Villalobos, S. D. L., Orozco-Mosqueda, M. D. C., & Glick, B. R. (2021). Plant growth stimulation by microbial consortia. *Agronomy*, *11*(2), 219.

Saxena, G. (2004). Biocontrol of nematode-borne diseases in vegetable crops. In *Fruit and vegetable diseases* (pp. 397–450). https://doi.org/10.1007/0-306-48575-3_11

Silva, J. C. P. D., Medeiros, F. H. V. D., & Campos, V. P. (2018). Building soil suppressiveness against plant-parasitic nematodes. *Biocontrol Science and Technology*, *28*(5), 423–445.

Silva, G. da C., Kitano, I. T., Ribeiro, I. A. de F., & Lacava, P. T. (2022). The potential use of actinomycetes as microbial inoculants and biopesticides in agriculture. *Frontiers in Soil Science*, *2*. https://doi.org/10.3389/fsoil.2022.833181

Singh, A., Sharma, P., Kumari, A., Kumar, R., & Pathak, D. V. (2019). Management of root-knot nematode in different crops using microorganisms. In *Plant biotic interactions: State of the art*. https://doi.org/10.1007/978-3-030-26657-8_6

Sivasubramaniam, N., Hariharan, G., & Zakeel, M. C. M. (2020). Sustainable management of plant-parasitic nematodes: An overview from conventional practices to modern techniques. In Ansari, R., Rizvi, R., & Mahmood, I. (Eds.), *Management of Phytonematodes: Recent Advances and Future Challenges*. Springer, Singapore. https://doi.org/10.1007/978-981-15-4087-5_16

Sorribas, F. J., Djian-Caporalino, C., & Mateille, T. (2020). Nematodes. In R. Albajes, M. Lodovica Gullino, J. C. Lenteren, & Y. Elad (Eds.), *Integrated pest and disease management in greenhouse crops* (pp. 147–174). Springer Dordrecht.

Sun, Y., Xie, J., Tang, L., Odiba, A. S., Chen, Y., Fang, W., Wu, X., & Wang, B. (2023). Isolation, identification and molecular mechanism analysis of the nematicidal compound spectinabilin from newly isolated streptomyces sp. DT10. *Molecules*, *28*(11). https://doi.org/10.3390/molecules28114365

Suresh, P., Varathraju, G., Shanmugaiah, V., Almaary, K. S., Elbadawi, Y. B., & Mubarak, A. (2021). Partial purification and characterization of 2, 4-diacetylphloroglucinol producing Pseudomonas fluorescens VSMKU3054 against bacterial wilt disease of tomato. *Saudi Journal of Biological Sciences*, *28*(4), 2155–2167. https://doi.org/10.1016/j.sjbs.2021.02.073

Topalović, O., Hussain, M., & Heuer, H. (2020). Plants and associated soil microbiota cooperatively suppress plant-parasitic nematodes. *Frontiers in Microbiology*, *11*. https://doi.org/10.3389/fmicb.2020.00313

Tóthné Bogdányi, F., Boziné Pullai, K., Doshi, P., Erdős, E., Gilián, L. D., Lajos, K., ... Tóth, F. (2021). Composted municipal green waste infused with biocontrol agents to control plant parasitic nematodes: A review. *Microorganisms*, *9*(10), 2130.

Vives-Peris, V., De Ollas, C., Gómez-Cadenas, A., & Pérez-Clemente, R. M. (2020). Root exudates: From plant to rhizosphere and beyond. *Plant Cell Reports*, *39*, 3–17.

Wachira, M. P. (2015). Survey of nematode destroying fungi from selected vegetable growing areas in Kenya. *Agriculture, Forestry and Fisheries*, *4*(4), 159. https://doi.org/10.11648/j.aff.20150404.12

Yang, Y., Yang, E., An, Z., & Liu, X. (2007). Evolution of nematode-trapping cells of predatory fungi of the Orbiliaceae based on evidence from rRNA-encoding DNA and multiprotein sequences. Proceedings of the National Academy of Sciences of the United States of America, *104*(20), 8379–8384. https://doi.org/10.1073/pnas.0702770104

Yildiz, H. N., Altinok, H. H., Dikilitas, M., Günacti, H., & Ay, T. (2023). Suppression of tomato bacterial speck disease (Pseudomonas syringae pv. tomato (Okabe) Young, Dye, & Wilkie) via induced systemic resistance by Pseudomonas and Bacillus strains. *Botany*, *101*(9), 391–399. https://doi.org/10.1139/cjb-2022-0066

Yu, Q., & Josef, C. (1998). Fungi associated with cysts of Globodera rostochiensis, G. pallida, and Heterodera schachtii; and egg masses and females of Meloidogyne hapla in Belgium. *Phytoprotection*, *79*(2), 63–69. https://doi.org/10.7202/706135ar

Zhang, Y., Li, S., Li, H., Wang, R., Zhang, K. Q., & Xu, J. (2020). Fungi–nematode interactions: Diversity, ecology, and biocontrol prospects in agriculture. *Journal of Fungi*, *6*(4), 1–24). https://doi.org/10.3390/jof6040206

Zhang, L., Zhou, Z., Guo, Q., Fokkens, L., Miskei, M., Pócsi, I., Zhang, W., Chen, M., Wang, L., Sun, Y., Donzelli, B. G., Gibson, D. M., Nelson, D. R., Luo, J. G., Rep, M., Liu, H., Yang, S., Wang, J., Krasnoff, S. B., Xu, Y., … Lin, M. (2016). Insights into adaptations to a near-obligate nematode endoparasitic lifestyle from the finished genome of Drechmeria Coniospora. *Scientific Reports*, *6*, 23122. https://doi.org/10.1038/srep23122

Zhu, M. C., Li, X. M., Zhao, N., Yang, L., Zhang, K. Q., & Yang, J. K. (2022). Regulatory mechanism of trap formation in the nematode-trapping fungi. *Journal of Fungi*, *8*(4). https://doi.org/10.3390/jof8040406

16 The Role of Microbial Consortia in Controlling Soil-Borne Pathogens

Oumaima Benaissa, Ikram Leghrifi, Salah-Eddine Laasli, Kaoutar Bourak, Tourya Sagouti, Zineb Belabess, and Rachid Lahlali

16.1 INTRODUCTION

The complex interaction between plants and soil-borne pathogens poses an ongoing challenge in agricultural ecosystems, as these microorganisms have the potential to harm crop health and impose significant economic losses (Ezrari et al., 2023). Soil-borne pathogens, like fungi, bacteria, and oomycetes, inhabit the soil, posing a constant threat to food security worldwide (Katan, 2017; Legrifi et al., 2022). The consequences of their presence extend beyond visible disease symptoms, impacting both crop health and economic viability. Crop yields suffer, leading to reduced agricultural productivity and compromising the overall quality of harvested produce (Dixon & Tilston, 2010). The economic impact is substantial, with farmers contending not only with reduced income but also increased costs related to disease management and potential crop losses. These effects reverberate through the entire agricultural supply chain, affecting food availability, pricing, and global trade (Delgado-Baquerizo et al., 2020).

Historically, conventional disease control methods heavily relied on chemical pesticides to combat phytopathogens (St. Martin & Brathwaite, 2012; Turkolmez & Dervis, 2017). While these interventions demonstrated immediate efficacy in disease suppression, their use is laden with limitations and adverse consequences (Leadbeater & Gisi, 2009). The persistent use of pesticides has resulted in environmental degradation, including soil and water contamination, and raised concerns about the emergence of pesticide-resistant pathogen strains (Gisi & Sierotzki, 2015; Sánchez-Torres, 2021). Additionally, residues left by these chemicals in agricultural ecosystems pose risks to human health. The inherent shortcomings of conventional disease control methods necessitate a critical reevaluation of strategies that balance effectiveness (Purkayastha et al., 2016). The transition toward sustainable and eco-friendly disease management practices in agriculture is imperative, considering the limitations of conventional approaches (Ahirwar et al., 2020; N. S. Verma et al., 2023). Sustainable agriculture, grounded in ecological balance and resource conservation, is increasingly acknowledged as vital for the long-term resilience of agroecosystems (Stirling, 2018). Within this framework, the exploration of alternative strategies addressing the complexities of soil-borne pathogens becomes crucial.

The application of microbial consortia appears to be a promising solution in the search for sustainable disease management, as they offer a comprehensive and ecological approach (Vishwakarma et al., 2020). The interactions within these consortia, made up of various biological control agents (BCAs), offer an effective way of harnessing the power of nature to combat pathogens present in the soil (Lim et al., 2009). Utilizing microbial consortia offers a range of benefits, strategically addressing the biological limitations inherent in individual microorganisms (Verma et al., 2022). One notable advantage lies in the enhanced stability achieved by combining multiple microbes, a characteristic proven effective across a broader spectrum of environmental and ecological conditions

 DOI: 10.1201/9781003543237-16

(Woo et al., 2018). This stability is particularly advantageous when considering microorganisms with distinct modes of action, varied nutritional demands, and diverse growth characteristics, as it fosters synergistic interactions among the microbial community. Other benefits include decreased application costs and increased microbial biodiversity in the plant rhizosphere (Xu et al., 2011; Hassani et al., 2018). Despite these advantages, the successful introduction of microbial consortia faces several challenges.

This chapter seeks to unravel the potential of microbial consortia, examining their composition, ecological interactions, applications in agriculture, and the challenges that necessitate a nuanced understanding for their successful deployment in sustainable disease management strategies.

16.2 MICROBIAL CONSORTIA: COMPOSITION AND ASSEMBLY

16.2.1 Microbial Consortia

The quest for ecologically benign substitutes for chemical pesticides and fertilizers has led to increased interest in the application of beneficial microorganisms to manage plant diseases (Niu et al., 2020; Fira et al., 2018; Syed Ab Rahman et al., 2018a). Over the years, a wide range of fungal and bacterial species, as well as viruses, insects, and nematodes, have served as BCAs to combat soil-borne crop pathogens (Cazorla et al., 2016). However, the translation of promising results obtained in controlled laboratories or greenhouses to field crop production has remained a persistent obstacle in agricultural practices (Xu et al., 2011). While various microbial strains have demonstrated efficacy against pathogens under controlled environments, the limited success observed in commercial fields raises doubts about the practical application of these BCAs (Mazzola & Freilich, 2017).

One of the main hurdles lies in the inadequate colonization of the host rhizosphere, coupled with ineffective inhibition of the growth of soil-borne pathogens (Sarma et al., 2015). The rhizosphere serves as a critical interface for microbial interactions (Mazzola & Freilich, 2017), and successful biocontrol relies on the establishment and proliferation of beneficial microbes in this environment (Syed Ab Rahman et al., 2018a). However, factors including soil type, environmental conditions, and the presence of competing microorganisms can hinder the efficient colonization of BCAs (Woo & Pepe, 2018b). Moreover, the issue is exacerbated by the infections' inadequate growth suppression in the soil (Sarma et al., 2015). While laboratory studies may highlight the antagonistic potential of microbial strains against specific pathogens, the dynamic and complex nature of field conditions presents a different set of challenges (Xu et al., 2011). These include variations in temperature, humidity levels, and the presence of diverse microbial communities, all of which can affect the effectiveness of BCAs in suppressing pathogen proliferation (Mazzola & Freilich, 2017).

The use of multi-strain BCAs (MSBCAs) is a promising approach to overcoming the challenges associated with single-strain biocontrol (Sarma et al., 2015). By assembling different beneficial microbial strains into consortia, this strategy aims to enhance the stability and efficacy of disease inhibition in agricultural settings (Vorholt et al., 2017), taking into account the complexity of the soil environment and the dynamic nature of pathogen populations. This approach offers a more comprehensive solution to the limitations observed in single-strain applications (Bhatia et al., 2018), with various bacterial strains interacting synergistically to provide distinct capabilities to the consortium (Woo & Pepe, 2018b). The possible advantages of utilizing a variety of microbial consortia for biocontrol applications include increased flexibility, broader range disease resistance, and enhanced resilience to environmental fluctuations (Mazzola & Freilich, 2017; Sarma et al., 2015; Vorholt et al., 2017).

Moreover, the benefits of MSBCAs extend beyond their combined efficacy against infections (Pliego et al., 2008), as the various metabolic processes carried out by different strains in a consortium may support plant growth, soil health, and nutrient cycling (Sarma et al., 2015). This multipurpose strategy promotes overall ecosystem resilience and production in addition

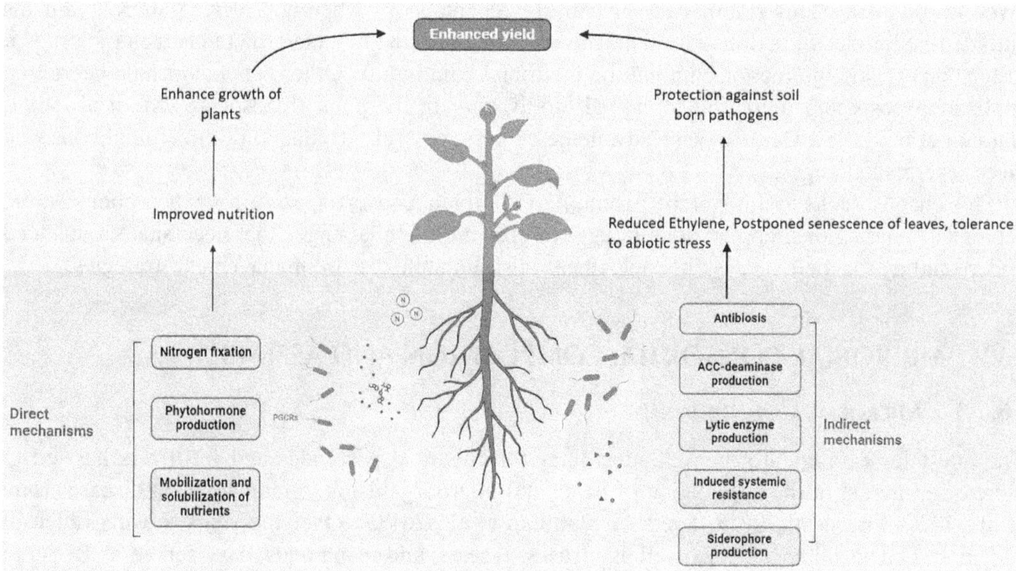

FIGURE 16.1 Direct and indirect mechanisms of the action of microbial consortia.

to addressing the pressing issue of disease management, which is in line with the larger objectives of sustainable agriculture (Thomloudi et al., 2019). Combining two or more antagonists can improve the potential or minimize the variability of biocontrol treatments (Haggag & Nofal, 2006). Recently, emphasis has been placed on combining BCAs with various disease control mechanisms to improve disease management and overcome variability in the performance of introduced BCAs. Efforts were also made to utilize a consortium of BCAs to achieve sustainable management of phytopathogens (Gujri, 2011). Certain antagonists have shown promise in preventing the growth of Fusarium wilt, hence reducing the severity of the illness (Gordon, 2017) and integrating chemical pesticides with biological control strategies may reduce environmental harm (Driouich et al., 2013).

16.2.2 Fusarium Wilt

Fusarium wilt is a devastating plant disease caused by the soil-borne fungus *Fusarium oxysporum* (Altinok, 2013). It affects various plants, including many economically important crops like tomatoes, bananas, and cucumbers (Aoki et al., 2014). Managing Fusarium wilt presents challenges, with one promising approach involving the use of antagonistic organisms to confer protection against the pathogen (Bonanomi et al., 2010). With over 100 host-specific strains (formae speciales) found worldwide, *F. oxysporum* encompasses various spores, including macroconidia and chlamydospores, all containing nuclei formed during hyphal mitosis. To effectively combat Fusarium wilt, the concept of a microbial consortium has emerged as an innovative strategy, entailing the deliberate introduction of diverse beneficial microorganisms into the soil. These microorganisms deploy unique mechanisms to combat pathogenic *Fusarium* species, creating an inhospitable environment through synergistic interactions among bacteria, fungi, and other beneficial microbes. Notably, certain microbial members like *Trichoderma* spp., *Pseudomonas* spp., and *Bacillus* spp. engage in antagonistic activities, producing antifungal compounds and competing for resources. Furthermore, the consortium can induce systemic resistance in host plants, enhancing resilience against Fusarium wilt. BCAs operate through mechanisms such as nutrient cycling, biofilm formation on plant roots, and overall improvement of soil health. Successful consortium implementation requires careful consideration of factors like application timing, compatibility testing, and ongoing monitoring, offering a sustainable and environmentally friendly solution for Fusarium wilt control. Moreover, research demonstrates that *Bacillus subtilis* spores, when used with glucose as a carrier, exhibit prolonged viability compared to other carriers. The treatment of seed using these spores proves to be an efficient delivery method, effectively suppressing vascular wilt disease in lentils and holding promise for practical application in commercial fields (El-Hassan & Gowen, 2006). Under in vitro conditions, *Collimonas arenae* strain Cal35 demonstrated effective antagonism against *F. oxysporum* f.sp. *pisi*. When combined with specific strains, this bacterial strain showed increased activity, leading to a reduction in disease severity and *F. oxysporum* levels in the endosphere of field tomato plant roots (Raza et al., 2024).

16.2.3 Pythium Root Rot

Soil pathogens, such as *Pythium* spp., while no longer classified as fungi, exhibit life cycles resembling those of true fungi, comprising vegetative (mycelium), asexual (sporangia and zoospores), and sexual (oospores) phases (H Ho, 2018). *Pythium* species are globally distributed (Rusagara Nzungize et al., 2012), posing a significant to agricultural productivity worldwide (Dissanayake et al., 1998). In regions like Kenya and Rwanda, traditional local bean varieties have incurred yield losses of up to 70% due to this disease, with hydroponically grown lettuces experiencing losses ranging from 35% to 54% at 18°C and 12% to 17% at 28°C (Namayanja et al., 2014). Over the past 20 years, *Pythium* root rot has gained prominence in several East and Central African countries, such as Burundi, the Democratic Republic of Congo, Kenya, and Uganda (Nzungize et al., 2011). This plant disease, caused by various *Pythium* species, aquatic molds within the oomycete group (Arora et al., 2021), thrives in moist, poorly drained soils (Lookabaugh et al., 2017), affecting a wide array of plants, including vegetables, ornamentals, and lawns (H Ho, 2018). Early detection of *Pythium* is crucial for implementing effective control measures (Pandey & Pandey, 2005), as different species exhibit varying sensitivities to control strategies, necessitating accurate species identification to optimize resource allocation (Wu et al., 2020). Various carbohydrate-active enzymes (CAZymes), including carbohydrate esterases, polysaccharide lyases, glycoside hydrolases, and proteases, contribute to *Pythium* pathogenesis by facilitating plant cell wall penetration and colonization (Lévesque et al., 2010; Zerillo et al., 2013).

Controlling *Pythium* spp. poses significant challenges due to their highly contagious nature, a broad range of hosts, and long lifespan, leading to increased fungicide application costs (Al-Mawaali et al., 2012; Krajaejun et al., 2018; Li et al., 2007; Okubara et al., 2014; Wheeler et al., 2016). Once infected, hosts exhibit various symptoms, including plant wilting, root rot due to excess moisture, crown rot, and leaf damage, often resulting in plant mortality (Deng et al., 2005; Higginbotham & Kidwell, 2004; Kageyama Kageyama et al., 2002; Lin et al., 2020).

Crop rotation, ineffective against *Pythium* due to its host diversity (Wu et al., 2020), is further complicated by the pathogen's ability to persist in the soil as resilient oospores even in the absence of a host (Matthiesen & Robertson, 2021). Consequently, managing *Pythium* effectively is imperative to mitigate substantial crop losses (Ho, 2018), though concerns over the environmental and health impacts of commercial fungicides, as well as the potential for *Pythium* resistance, necessitate careful consideration (Lookabaugh et al., 2017; Mares et al., 2004).

Given *Pythium*'s prevalence during summer periods and under high humidity conditions, researchers are exploring temperature and humidity-based control methods (Huzar-Novakowiski & Dorrance, 2018; Matthiesen & Robertson, 2021; Wheeler et al., 2016). Moreover, biocontrol is attracting growing interest due to the lower risks it poses to the environment and human health (Mavrodi et al., 2012; Van Os & Van Ginkel, 2001). Effective management of *Pythium* is challenging, considering factors such as its aggressiveness, the efficacy of fungicides, the applicability of BCAs, and the suitability of disease-resistant plants (Okubara et al., 2014) further complicated by environmental variables like air temperature, water pH, and soil type (Wu et al., 2020).

16.2.4 *Phytophthora* Root Rot

Phytophthora, despite its fungal-like morphology, belongs to Oomycota, a diverse group of microorganisms with the Stramenopiles. With over 100 species grouped into 10 distinct clades (Kroon et al., 2012; Martin et al., 2012), *Phytophthora* is characterized by cell walls containing cellulose instead of chitin, filamentous hyphae devoid of transverse septa and the production of sexual spores called oospores and motile zoospores (Giachero et al., 2022). *Phytophthora* root rot, caused by various *Phytophthora* species, poses a serious threat to various crops and ornamental plants worldwide (Tyler & Gijzen, 2014). This destructive disease mainly targets the roots of susceptible plants, leading to reduced vigor, wilting, and eventual plant death (Dorrance, 2018). Symptoms of *Phytophthora* root rot vary depending on the host plant and environmental conditions but generally include wilting, yellowing or browning of foliage, stunted growth, and, ultimately, plant death (Legrifi et al., 2023). Infected plants often exhibit rotting roots, which become dark and mushy as the disease progresses (Fry, 2008). In some cases, infection can spread to the stem, causing cankers and lesions (Hansen et al., 2012). *Phytophthora* root rot spreads via the movement of water, through either the soil or irrigation systems, facilitating pathogen access to susceptible plant roots (Fry, 2008). Once established in the soil, *Phytophthora* spp. can survive for extended periods, making it difficult to manage and eradicate (Martin et al., 2012). Therefore, *Phytophthora* root rot management generally involves using resistant cultivars, improving soil drainage, and treating seeds with fungicides (Dorrance, 2018). However, concerns regarding the adverse effects of conventional methods on the environment and pathogen resistance have prompted the exploration of alternative biocontrol strategies.

Biocontrol tactics utilizing microorganisms, ions, minerals, and plant extracts offer promising alternatives for addressing pathogen resistance and environmental considerations (Siah et al., 2018; Zehra et al., 2021). Recently developed metal chelates, particularly calcium chelates, show promise in controlling *Phytophthora* species, offering an alternative to phosphite, which, while effective, can lead to phytotoxicity and pathogen tolerance (Khdiar et al., 2023). Glasshouse trials have demonstrated a significant reduction in root lesion development when calcium chelate or phosphite was applied individually, with calcium chelate exhibiting superior performance at equivalent concentrations (Burgess et al., 2021; Khdiar et al., 2023). Additionally, the mixture of calcium chelates and phosphite showed the most potential in reducing root lesion development (Chai et al., 2017). These findings suggest calcium chelate's potential as a fungicide for managing *Phytophthora* infections, highlighting a promising avenue for disease control in both natural ecosystems and agriculture (Khdiar et al., 2023).

16.2.5 Rhizoctonia Root Rot

Rhizoctonia root rot, brought on by the fungus *Rhizoctonia solani*, is a major plant disease affecting a wide range of crops worldwide (Sumner & Bell, 1982). This pathogen infects the roots and lower stems of plants, inducing symptoms such as root rot, damping-off, and stunted growth (Gonzalez et al., 2011; Rahman et al., 2020). Symptoms of the disease generally appear in early summer, characterized by wilted or dead scattered across the field or clustered in certain areas (Gondal et al., 2019). The fungus can cause lesions on the hypocotyls, seed rots, and root rots (S. Behiry et al., 2023), with seedling damping-off occurring when germinating seedlings are infected before emergence from the soil (Das et al., 2019).

R. solani exerts a considerable global impact, leading to severe losses in a variety of crops. Sclerotia serves as the primary means of reproduction for *R. solani*, contrasting with the production of asexual spores (conidia) (Rubayet & Bhuiyan, 2016). The excessive use of chemical fungicides by farmers to combat *R. solani* not only poses significant risks to the environment and human health but also fosters the emergence of resistant disease strains (Behiry et al., 2022). For these reasons, biological control techniques are becoming increasingly popular as practical substitutes for plant disease control (Das et al., 2019). However, due to its broad host range, controlling this disease is difficult (Okubara et al., 2014). Although conventional fungicides demonstrate effectiveness, their expense and potential risks to human and animal well-being pose significant concerns (Marimuthu et al., 2013). In addition, the intensive application of chemical pesticides and fertilizers in contemporary agriculture has led to a number of drawbacks, such as pesticide resistance, the return of pests, and damage to beneficial creatures such as ladybugs and bees. Concerns are also being raised about chemical residues from the application of these products, which end up in food, feed, and forage (Ahmad et al., 2019; Bhuiyan et al., 2019; Das et al., 2019; Rubayet & Bhuiyan, 2016).

16.2.6 *Sclerotinia* Stem Rot

Throughout the world, rapeseed-growing areas are affected by *Sclerotinia* stem rot (SSR), a fungal infection caused by *Sclerotinia sclerotiorum* (Lib.) De Bary (Zhao & Meng, 2003). This disease is not specific to any particular host and capable of infecting over 600 types of plants, including members of various plant families, including Fabaceae, Amaranthaceae, Apiaceae, Brassicaceae, and Solanaceae, resulting in substantial economic losses annually (Liang & Rollins, 2018). Infected host plant stems develop large, white, or gray lesions throughout the infection, impeding seed growth and diminishing yields (Pressete et al., 2019). The production of long-lasting sclerotia by *S. sclerotiorum* enables its inoculum to persist in the soil for extended periods (Derbyshire & Denton-Giles, 2016). Current strategies for SSR management primarily revolve around cultural interventions and fungicide application (Pressete et al., 2019). Cultural practices aim to reduce soil sclerotia levels or create an inhospitable environment for SSR proliferation, although crop rotation's efficacy is limited due to sclerotia's prolonged survival and the pathogen's wide host range (Gondal et al., 2019). Consequently, the development of *Sclerotinia*-resistant rapeseed cultivars is essential for effective disease management in the future (Ding et al., 2021). Present strategies for managing SSR predominantly hinge on cultural control methods and the application of fungicides (Derbyshire & Denton-Giles, 2016). The majority of cultivation activities aim to reduce the amount of *S. sclerotiorum* inoculum in the soil or to create an environment that is not conducive to SSR growth. Breeding for SSR resistance is fraught with difficulties, despite the important advances in our comprehension of resistance pathways and sequencing and bioinformatics technologies. SSR resistance is a complex quantitative trait regulated by several genes (Ding et al., 2021) (Table 16.1).

TABLE 16.1
Various Soil-Borne Pathogens That Contribute to the Occurrence of Root Rot in Plants

Soil-Borne Pathogen	Strain	Plant	Disease	References
Armillaria	*A. cepistipes*	Oaks Ornamental Trees Shrubs	Root rot	Menge et al. (2012) Thomidis and Exadaktylou (2012)
Fusarium	*F. avenaceum* *F. graminearum* *F. oxysporum* *F. tricinctum* *F. graminearum*	Soybean	Root rot	Zhang et al. (2011)
	F. oxysporum f. sp. radicis cucumerinum	Cucumber	Root and stem rot	Kannangara et al. (2000)
	Fusarium spp.	Cassava	Root rot	Bandyopadhyay et al. (2006)

(Continued)

TABLE 16.1 (*Continued*)
Various Soil-Borne Pathogens That Contribute to the Occurrence of Root Rot in Plants

Soil-Borne Pathogen	Strain	Plant	Disease	References
Pythium	*Pythium* spp.	Bean	Root rot	Heyman et al. (2013)
	P. graminicola	Corn	Root rot	Thaines Bodah (2017)
	P. diclinum	Tobacco	Root rot	Graham et al. (2011)
	P. aphanidermatum	Ornamentals	Root rot	Henricot et al. (2014)
	P. ultimum	Cucumber		
	P. arrhenomanes	Sugarcane	Root rot	Dissanayake et al. (1998)
	Pythium spp.	Wheat	Root rot	Petkowski et al. (2013)
Phytophthora	*Phytophthora* spp.	Potato and tomato	Late blight stem rots	Thaines Bodah (2017)
	P. cambivora	Pea and fava bean	Root rot	Heyman et al. (2013)
	P. pisi sp. nov			
	P. citricola	*Aucuba japonica*	Root rot	Henricot et al. (2014)
	P. palmivora	Papaya	Root rot	Thaines Bodah (2017)
	Phytophthora spp.	Red raspberries	Root rot	Kurbetli and Ozan (2013)
		Chickpeas		Orlikowski et al. (2011)
		Pepper soybean		Truong et al. (2012)
		Ash trees		Du et al. (2013)
		Kiwi fruit avocado		Graham et al. (2011)
Phoma	*P. betae*		Leaf spot	Thaines Bodah (2017)
	P. terrestris	Onion	Pink root rot	Serafin et al. (2013)
	P. terrestris	Corn	Red root rot	Du et al. (2013)
	P. sclerotioides	Alfalfa	Brown root rot	Manjunatha et al. (2013)
Rhizoctonia	*R. solani*	*Glycine max* L.	Damping-off and root rot	Rahman et al. (2020)
	R.zeae	Corn	Damping-off, root rot, sunken lesions	Sumner and Bell (1982)
	R. solani	Tomato	Seedling damping-off and root rot	S. I. Behiry et al. (2022)
	R. solani	Tobacco	Leaf spot and root rot	Gonzalez et al. (2011)

16.3 ECOLOGICAL INTERACTION: MECHANISMS OF DISEASE SUPPRESSION

Microbial consortia exhibit a rich array of ecological interactions that underpin their effectiveness in suppressing soil pathogens (Figure 16.2). Understanding these mechanisms is fundamental to exploring the full effect of microbial consortia for sustainable agriculture.

16.3.1 Competition for Resources

Competition for resources is a fundamental and intricate strategy employed by BCAs to safeguard plants against pathogens (Iftikhar et al., 2020). This ecological phenomenon is marked by the proficient use of limited resources near host plants, strategically inhibiting plant pathogen growth (Prajapati et al., 2020). The rhizosphere microbiome, a pivotal component of nutrient cycling, relies heavily on the availability of plant exudates on the root surface and in the surrounding rhizosphere. Notably, the successful suppression of pathogens is contingent on the competitive utilization of nutrients inherent in root exudates, initiating a sophisticated interplay between BCAs and soil pathogens (Niu et al., 2020b).

The dynamics within microbial consortia further amplify the complexity of this competition. For instance, the synergistic combination of five strains of *Ralstonia* (non-virulent) and eight *Pseudomonas* strains, carefully structured into biocontrol consortia used to combat *Ralstonia*

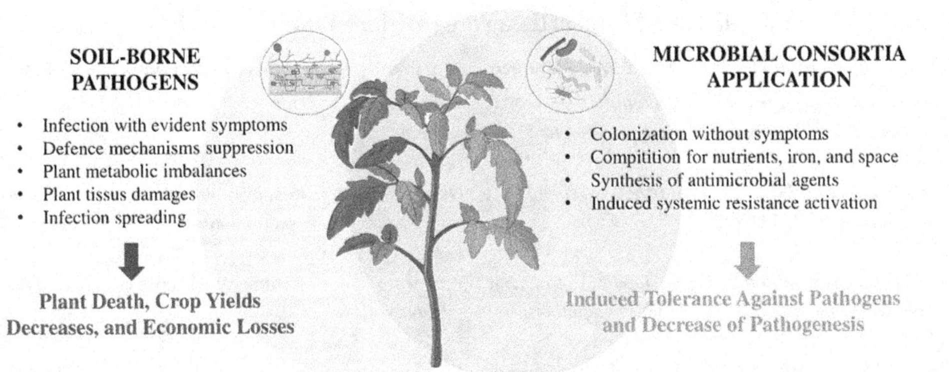

FIGURE 16.2 Mechanisms employed by microbial consortia for the control of attacks by soil-borne pathogens.

solanacearum, the causative agent of tomato wilt, exhibited significantly more potent inhibitory effects on the density of *R. solanacearum* population in comparison with to the individual strains (Hu et al., 2016). This strengthening of inhibition capacities arises from the expanded niche overlap exerted by these consortia on the targeted pathogens, illustrating the strategic advantage of such cooperative microbial interactions. The concept of niche overlap, especially in the context of decomposing 48 distinct single-carbon resources present in tomato root exudates, emerges as a critical determinant in understanding the effectiveness of biocontrol strategies (Wei et al., 2015).

In the rhizosphere, the persistent scarcity of assimilable iron resources is a direct consequence of the ongoing competition between soil-borne pathogens and disease-suppressing microorganisms (Gu et al., 2020). The recovery of iron by many soil microbes involves the secretion of siderophores, which are chemically diverse secondary metabolites exhibiting a pronounced preference for iron. Furthermore, siderophore production is subject to positive regulation through interspecies interactions among soil microbes (Kramer et al., 2020). Notably, the interaction of *Streptomyces coelicolor* with five other soil actinobacteria results in an enhanced siderophore diversity, demonstrating the adaptive nature of microbial communities in response to competition for limited resources (Traxler et al., 2013).

16.3.2 Antagonistic Traits: Production of Antimicrobial Compounds

Microorganisms exhibit a remarkable capacity to synthesize a diverse array of antimicrobial compounds, presenting a pivotal mode of action for directly inhibiting or inducing lethality in microbial adversaries within various environments (Syed Ab Rahman et al., 2018b; Tiku, 2018). A multitude of studies have elucidated the antimicrobial properties of compounds produced by beneficial microorganisms (BCAs) (Table 16.2), underscoring their suppressive effects on the proliferation of plant pathogens. These investigations predominantly focus on biocontrol strains, particularly those associated with the *Trichoderma*, *Pseudomonas*, and *Bacillus* genera, known for synthesizing antibiotics such as chitinase, lipopeptides, bacteriocins, 2,4-diacetylphloroglucinol (DAPG), phenazines, and polyketides (Caulier et al., 2019; Mingmongkolchai & Panbangred, 2018; Negi et al., 2017; Urbina-Salazar et al., 2019).

Certain inhibitory metabolites may initially present in low concentrations or remain dormant within pure cultures, yet they have the potential to be upregulated within a community setting through interactions among antagonistic bacterial and fungal organisms within the community. Instances such as the coexistence of chitinase-producing *Trichoderma atroviride* P1 and DAPG-producing

TABLE 16.2
Mode of Action of Microbial Consortia against Different Soil-Borne Pathogens

Microbial Consortium	Soil-Borne Pathogen	Plant	Mode of Action	References
Pseudomonas fluorescens EBC5 and *P. fluorescens* EBC6	*Pythium aphanidermatum*	*Capsicum annum*	ISR	Muthukumar et al. (2010)
Trichoderma viride **NCC 34 and** ***P. fluorescens*** **MTCC 1749**	*Rhizoctonia solani*	*Oryza sativa*	ISR; plant growth promotion	Mathivanan et al. (2005)
Gliocladium catenulatum and Trichoderma harzianum **DB11**	*Phytophthora fragariae* and *Phytophthora cactorum*	*Fragaria* x *ananassa*	Antimicrobial activities	Vestberg et al. (2004)
Pseudomonas **sp. Sp14** ***and T. harzianum*** **Tr6**	*Fusarium oxysporum f.* sp. *radicis cucumerinum*	*Cucumis sativus*	Expression of various defense-related genes	Alizadeh et al. (2013)
P. fluorescens NRI-N and *P. fluorescens* NBRI-N6	*Sclerotium rolfsii*	*Piper betel*	ISR	Singh et al. (2003)
P. fluorescens **Pf1** ***and Azospirillum*** **sp.** AZ204	*Rhizoctonia ataticola*	*Gossypium hirsutum*	Plant growth promotion	Marimuthu et al. (2013a)
Pseudomonas putida RE8 **and** ***P. putida WCS358***	*F. oxysporum*	*Raphanus sativus*	Mutual increase in population; ISR; siderophore	de Boer et al. (2003)
P. fluorescens **F113** ***and Stentrophomonas maltophila*** **W81**	*Pythium* spp.	*Beta vulgaris*	ISR	Dunne et al. (1998)
Bacillus sp. strains (SE49, IN937b, T4, and INRN)	*Sclerotium rolfsii*	*Cucumis sativus*	ISR	Jetiyanon (2007)
Glomus intraradices*, *Rhizobium **sp., and** ***P. straita***	*Macrophomina phaseolina*	*Cicer arietinum*	Increased NPK uptake; increased plant biomass	Akhtar and Siddiqui (2008)

Pseudomonas fluorescens exemplify this phenomenon. It has been observed that their interaction enhances biocontrol gene expression, indicating that upregulation can indeed occur in mixed antagonist environments (Lutz et al., 2004). Co-cultivation of *B. subtilis* with *Fusarium tricinctum* (endophytic fungus) resulted in a substantial enhancement in the accumulation of secondary metabolites, including those with antimicrobial properties (Ola et al., 2013). Thus, specific interactions between organisms across different taxonomic domains can amplify the production of antimicrobial compounds. Furthermore, these interactions have the potential to activate the biosynthesis of previously undiscovered compounds with antimicrobial proprieties. An illustrative case is the antibiotic amycomicin, whose production is contingent upon the interaction between two soil-dwelling actinobacteria: *Amycolatopsis* sp. AA4 as the producing strain and *S. coelicolor* M145 as the inducer (Pishchany et al., 2018). These instances underscore that microbial interactions, both inter- and intra-domain, play a key role in stimulating or activating the synthesis of antimicrobial compounds.

The influence of microbial interactions extends beyond the realm of antimicrobial synthesis to modulate competition for resources (Raaijmakers et al., 2002). This dynamic interplay can significantly lead to the heightened inhibition of pathogen development. Consequently, employing BCAs in multi-layer mixtures holds promise in enhancing the competitive effect of biocontrol microorganisms against soil-borne pathogens for the essential resources required for rhizosphere survival. Moreover, these beneficial microorganisms may induce enhanced niche overlap and synthesis of new antimicrobial molecules, amplifying the spectrum of plant pathogens targeted for suppression (Niu et al., 2020b).

16.3.3 Induced Systemic Resistance: Signaling Pathways and Plant Defense Activation

The plant immune system significantly influences the composition of microorganisms in the proximity of plant roots. This dynamic interplay between the plant and its root-associated microbiome not only widens the scope of plant immunity but also acts as a supplemented defensive barrier against plant pathogens (Teixeira et al., 2019). The interaction of beneficial microorganisms with plant roots can induce systemic host resistance to pathogens, possibly by activating induced systemic resistance (ISR) (Table 16.2) (Pieterse et al., 2014). The initiation of ISR begins with the recognition of specific microbial signals by the plant host. Microorganisms within the microbial consortia release molecular patterns or molecules that serve as cues for the plant to prepare for potential pathogenic encounters. These signals, often termed microbe-associated molecular patterns (MAMPs) or effectors, are known by plant receptors, initiating a cascade of events that lead to the activation of defense responses (Abdul Malik et al., 2020). Multiple investigations have outlined the multifaceted role of microbial consortia beyond enhancing rhizosphere colonization and eliminating soil-borne pathogen proliferation. The diversity of microbial signals within the consortium contributes to the robustness of ISR (Yin et al., 2021). Different strains within the consortium may produce distinct MAMPs, allowing for a broad spectrum of recognition by plant receptors (Buscaill & van der Hoorn, 2021). This diversity not only broadens the range of pathogens that can be targeted but also enhances the adaptability of the plant's defense system to various environmental conditions, leading to their heightened efficacy in controlling diseases (Emonet et al., 2021). Nevertheless, the precise mechanisms underlying how interactions among microbial consortia members enhance specific systemic resistance to soil-borne pathogens are not yet fully understood. One hypothesis suggests that the microbe–microbe interplay within these biocontrol consortia may lead to the increased production of specific elicitors and potent compounds, potentially more efficiently triggering ISR (Niu et al., 2020b).

16.4 APPLICATIONS IN AGRICULTURE

To ensure sustainable agricultural production in the face of global warming and growing concerns about the environmental effects of chemical fertilizers, it is essential to make use of symbiotic interactions between plants and plant growth-promoting microorganisms (Abhilash et al., 2016). Genetically mediated invasions of plant pathogens have a significant impact on many aspects of plant physiology and development (Agbodjato et al., 2016). Furthermore, when considering the multitude of advantages provided by consortia of plant-beneficial microorganisms, such as heightened crop stress tolerance, enhanced soil health, biological pest control, improved nutrient absorption, and the remediation of contaminated soils, integrating them into agricultural methods emerges as a potent strategy. This approach not only fosters healthier ecosystems but also optimizes agricultural productivity while mitigating environmental harm (Adesemoye et al., 2009; Ahmed et al., 2016).

16.4.1 Successful Implementation of Microbial Consortia

Many successful studies illustrate the effectiveness of microbial consortia in agricultural plant protection. In California, a mixture of beneficial fungi and bacteria, like *Bacillus* spp., *Pseudomonas* spp., and *Trichoderma* spp., was used to control soil-borne pathogens in tomato crops, improving plant health and yield while reducing the use of chemical fungicides (Ma et al., 2023). Furthermore, in Brazil, a combination of nitrogen-fixing bacteria and mycorrhizal fungi applied to corn seeds boosted nutrient uptake, enhanced pest resistance, and increased yields while reducing dependence on fertilizers and pesticides (de Novais et al., 2020). In Spain, foliar spraying and soil immersion of *Trichoderma* spp, *Bacillus* spp., and *Streptomyces* spp. mitigated citrus pathogens in organic orchards, promoting tree health and improving fruit quality without synthetic fungicides (Pylak et al., 2019). India's adoption of entomopathogenic fungi, plant growth-promoting rhizobacteria, and bacteria in rice cultivation orchestrated effective pest management against stem borers and leaf insects, resulting in increased cereal yields and sustainable farming paradigms with reduced reliance on chemical pesticides (Goyal & Manoharachary, n.d.; Shilev et al., 2019). These researches reveled the essential role of microbial consortia in fortifying agricultural ecosystems, building resilience, and promoting environmentally friendly practices.

16.4.2 Advantages of Using Microbial Consortia over Chemical Pesticides

The use of microbial strains in the form of axenic cultures has been frequently used in various sectors. In the natural environment, microbes exist as multi-species and outperform monocultures (Bhatt et al., 2021). Utilizing microbial consortia in agriculture offers numerous advantages over chemical pesticides (Huang et al., 2019). First, these consortia are environmentally friendly, consisting of naturally occurring microorganisms that are non-toxic and biodegradable, thus minimizing harm to non-target organisms and promoting soil and water quality (Qian et al., 2020). Additionally, microbial consortia can be tailored to target specific pests, pathogens, or plant diseases, reducing the risk of harming beneficial insects and pollinators (Brune & Bayer, 2012; Lee et al., 2018). Moreover, they are less prone to resistance development compared to chemical pesticides due to their multiple mechanisms of action (Bhatia et al., 2018). Furthermore, many microbial consortia contribute to soil health by enhancing fertility and nutrient uptake through symbiotic relationships with plants (Kaeberlein et al., 2002). Their compatibility with organic farming practices makes them suitable alternatives to synthetic chemicals, supporting sustainable agriculture (Rodríguez-Castillo et al., 2019). By integrating microbial consortia into integrated pest management (IPM) programs, growers can maximize pest control efficacy while minimizing environmental risks, thereby contributing to long-term agricultural sustainability (Raimondo et al., 2020).

16.4.3 Integration of Microbial Consortia in IPM Strategies

The integration of microbial consortia within IPM strategies represents a holistic approach to sustainable pest control in agriculture (Jaiswal et al., 2022). By incorporating microbial consortia alongside other IPM components such as cultural practices, BCAs, and selective pesticide applications, growers can enhance pest management efficacy while minimizing environmental impact (Baker et al., 2019; Raimondo et al., 2020). Microbial consortia offer multiple benefits within an IPM framework, including targeted management of pests and pathogens, reduced risk of resistance development, and promotion of soil health (Berini et al., 2018). In an IPM program, microbial consortia can be applied preventively to establish beneficial microorganisms in the agroecosystem, thereby providing long-term protection against pests and diseases (Qian et al., 2020). Additionally, microbial consortia can complement other IPM tactics by filling gaps in pest control where chemical pesticides may be less effective or unsustainable (Bhatia et al., 2018). By integrating microbial consortia into IPM strategies, growers can achieve sustainable pest management practices that prioritize environmental stewardship, minimize pesticide use, and promote the health and resilience of agricultural ecosystems (Raimondo et al., 2020).

16.5 CHALLENGES AND LIMITATIONS

Biopesticides and biofertilizers derived from natural or microbiological sources offer important advantages over synthetic alternatives, but they also present certain challenges (Aamir et al., 2020). First, they are generally less expensive than synthetic pesticides, which can reduce agricultural costs if properly managed. Second, they present less risk to the environment than traditional herbicides, as they are less likely to harm non-target organisms (Ahkami et al., 2017). However, continuous monitoring is essential to prevent microbial pesticides from developing harmful effects on non-target organisms, including humans (Adesemoye & Kloepper, 2009). In addition, organic biofertilizers are generally used in smaller quantities than conventional fertilizers, which can lead to a more efficient utilization of resources (Colla et al., 2015). However, the manufacture and maintenance of biofertilizers require skilled personnel, underlining the need for specialized knowledge and expertise in their production and application (Egamberdieva et al., 2019). Meeting these challenges will be essential to maximizing the benefits of biofertilizers and biopesticides in sustainable agricultural practices (Aamir et al., 2020).

16.6 CONCLUSIONS

In the search for sustainable and effective solutions to control pathogens, the role of microbial consortia appears to be an effective tool in the field of plant pathology. This review highlights the potential of microbial consortia to open up new avenues for formulations used in agriculture, particularly for the management of soil-borne pathogens. A crucial aspect of the development of microbial consortia as BCAs involves a comprehensive examination of their biocontrol activities, mode of action, environmental and human safety considerations, commercial viability, and formulation stability. These factors are essential in formulation and production processes to ensure the efficiency and safety of microbial consortia used for biocontrol.

In light of these considerations, it is imperative that the scientific community and industry actively support and encourage the application of microbial consortia. Rigorous, scientifically based studies on this subject are essential if we are to achieve a sound scientific basis and develop industrialized technologies capable of competing with traditional agrochemicals. Promoting these objectives promises to establish sustainable and effective alternatives for disease management in agriculture. The inherent issues associated with the utilization of agrochemicals further underscore the urgency of embracing microbial consortia. The latter has been implicated in various environmental issues, posing threats to both human health and terrestrial ecosystems. Therefore, the transition toward microbial consortia-based BCAs represents not only a scientific advancement but also a conscientious effort to address the pressing environmental and health challenges associated with conventional agricultural practice.

ACKNOWLEDGMENTS

This study received financial support from the Phytopathology Unit within the Department of Plant Protection and Environment at the Ecole Nationale d'Agriculture de Meknes, Morocco.

REFERENCES

Aamir, M., Rai, K. K., Zehra, A., Dubey, M. K., Kumar, S., Shukla, V., & Upadhyay, R. S. (2020). Microbial bioformulation-based plant biostimulants: A plausible approach toward next generation of sustainable agriculture. In *Microbial endophytes: functional biology and applications* (pp. 195–225). Elsevier. https://doi.org/10.1016/B978-0-12-819654-0.00008-9.

Abdul Malik, N. A., Kumar, I. S., & Nadarajah, K. (2020). Elicitor and receptor molecules: Orchestrators of plant defense and immunity. *International Journal of Molecular Sciences*, *21*(3), 963.

Abhilash, P. C., Tripathi, V., Edrisi, S. A., Dubey, R. K., Bakshi, M., Dubey, P. K., Singh, H. B., & Ebbs, S. D. (2016). Sustainability of crop production from polluted lands. *Energy, Ecology and Environment*, *1*(1), 54–65. https://doi.org/10.1007/s40974-016-0007-x

Adesemoye, A. O., & Kloepper, J. W. (2009). Plant-microbes interactions in enhanced fertilizer-use efficiency. *Applied Microbiology and Biotechnology*, *85*(1), 1–12). https://doi.org/10.1007/s00253-009-2196-0

Adesemoye, A. O., Torbert, H. A., & Kloepper, J. W. (2009). Plant growth-promoting rhizobacteria allow reduced application rates of chemical fertilizers. *Microbial Ecology*, *58*(4), 921–929. https://doi.org/10.1007/s00248-009-9531-y

Agbodjato, N. A., Noumavo, P. A., Adjanohoun, A., Agbessi, L., & Baba-Moussa, L. (2016). Synergistic effects of plant growth promoting rhizobacteria and chitosan on in vitro seeds germination, greenhouse growth, and nutrient uptake of maize (Zea mays L.). *Biotechnology Research International*, 1–11. https://doi.org/10.1155/2016/7830182

Ahirwar, N. K., Singh, R., Chaurasia, S., Chandra, R., & Ramana, S. (2020). Effective role of beneficial microbes in achieving the sustainable agriculture and eco-friendly environment development goals: A review. *Front. Microbiol*, *5*, 111–123.

Ahkami, A. H., Allen White, R., Handakumbura, P. P., & Jansson, C. (2017). Rhizosphere engineering: Enhancing sustainable plant ecosystem productivity. *Rhizosphere*, *3*, 233–243. https://doi.org/10.1016/j.rhisph.2017.04.012

Ahmad, A. G. M., Attia, A. Z. G., Mohamed, M. S., & Elsayed, H. E. (2019). Fermentation, formulation and evaluation of PGPR Bacillus subtilis isolate as a bioagent for reducing occurrence of peanut soil-borne diseases. *Journal of Integrative Agriculture*, *18*(9), 2080–2092. https://doi.org/10.1016/S2095-3119(19)62578-5

Ahmed, S. M., El-Zemity, S. R., Selim, R., Ahmed, S. M., Selim, R. E., & Kassem, F. A. (2016). A potential elicitor of green alga (ulva lactuca) and commercial algae products against late blight disease of solanum tuberosum L. *Asian Journal of Agriculture and Food Sciences*. https://www.researchgate.net/publication/334636356

Akhtar, M. S., & Siddiqui, Z. A. (2008). Biocontrol of a root-rot disease complex of chickpea by Glomus intraradices, Rhizobium sp. and Pseudomonas straita. *Crop Protection*, *27*(3–5), 410–417.

Alizadeh, H., Behboudi, K., Ahmadzadeh, M., Javan-Nikkhah, M., Zamioudis, C., Pieterse, C. M. J., & Bakker, P. A. H. M. (2013). Induced systemic resistance in cucumber and Arabidopsis thaliana by the combination of Trichoderma harzianum Tr6 and Pseudomonas sp. Ps14. *Biological Control*, *65*(1), 14–23.

Al-Mawaali, Q. S., Al-Sadi, A. M., Khan, A. J., Al-Hasani, H. D., & Deadman, M. L. (2012). Response of cucurbit rootstocks to Pythium aphanidermatum. *Crop Protection*, *42*, 64–68. https://doi.org/10.1016/j.cropro.2012.07.017

Altinok, H. H. (2013). Fusarium species isolated from common weeds in eggplant fields and symptomless hosts of fusarium oxysporum f. sp. melongenae in Turkey. *Journal of Phytopathology*, *161*(5), 335–340. https://doi.org/10.1111/jph.12074

Aoki, T., O'Donnell, K., & Geiser, D. M. (2014). Systematics of key phytopathogenic Fusarium species: Current status and future challenges. *Journal of General Plant Pathology*, *80*(3), 189–201). https://doi.org/10.1007/s10327-014-0509-3

Arora, H., Sharma, A., Sharma, S., Haron, F. F., Gafur, A., Sayyed, R. Z., & Datta, R. (2021). Pythium damping-off and root rot of capsicum annuum l.: Impacts, diagnosis, and management. *Microorganisms*, *9*(4). https://doi.org/10.3390/microorganisms9040823

Baker, B. P., Green, T. A., Loker, A. J., & Baker, B. (2019). Biological Control and Integrated Pest Management in Organic and Conventional Systems. *Biological Control*, *140*, 104095.

Bandyopadhyay, R., Mwangi, M., Aigbe, S. O., & Leslie, J. F. (2006). Fusarium species from the cassava root rot complex in West Africa. *Phytopathology*, *96*(6), 673–676. https://doi.org/10.1094/PHYTO-96-0673

Behiry, S. I., Al-Askar, A. A., Soliman, S. A., Alotibi, F. O., Basile, A., Abdelkhalek, A., Elsharkawy, M. M., Salem, M. Z. M., Hafez, E. E., & Heflish, A. A. (2022). Plantago lagopus extract as a green fungicide induces systemic resistance against Rhizoctonia root rot disease in tomato plants. *Frontiers in Plant Science*, *13*. https://doi.org/10.3389/fpls.2022.966929

Behiry, S., Soliman, S. A., Massoud, M. A., Abdelbary, M., Kordy, A. M., Abdelkhalek, A., & Heflish, A. (2023). Trichoderma pubescens Elicit Induced Systemic Resistance in Tomato Challenged by Rhizoctonia solani. *Journal of Fungi*, *9*(2). https://doi.org/10.3390/jof9020167

Berini, F., Katz, C., Gruzdev, N., Casartelli, M., Tettamanti, G., & Marinelli, F. (2018). Microbial and viral chitinases: Attractive biopesticides for integrated pest management. In *Biotechnology Advances*, *36*(3), 818–838). https://doi.org/10.1016/j.biotechadv.2018.01.002

Bhatia, S. K., Bhatia, R. K., Choi, Y. K., Kan, E., Kim, Y. G., & Yang, Y. H. (2018). Biotechnological potential of microbial consortia and future perspectives. *Critical Reviews in Biotechnology*, *38*(8), 1209–1229). https://doi.org/10.1080/07388551.2018.1471445

Bhatt, P., Bhatt, K., Sharma, A., Zhang, W., Mishra, S., & Chen, S. (2021). Biotechnological basis of microbial consortia for the removal of pesticides from the environment. *Critical Reviews in Biotechnology*, *41*(3), 317–338). https://doi.org/10.1080/07388551.2020.1853032

Bhuiyan, K., Sheikh, B., & Rahman, M. (2019). *Efficacy of Trichoderma-fortified compost in controlling soil-borne diseases of bush bean (Phaseolus vulgaris L.) and sustainable crop production*. https://www.researchgate.net/publication/335169097

Bonanomi, G., Antignani, V., Capodilupo, M., & Scala, F. (2010). Identifying the characteristics of organic soil amendments that suppress soilborne plant diseases. *Soil Biology and Biochemistry*, *42*(2), 136–144). https://doi.org/10.1016/j.soilbio.2009.10.012

Brune, K. D., & Bayer, T. S. (2012). Engineering microbial consortia to enhance biomining and bioremediation. *Frontiers in Microbiology*, *3*. https://doi.org/10.3389/fmicb.2012.00203

Burgess, T. I., Edwards, J., Drenth, A., Massenbauer, T., Cunnington, J., Mostowfizadeh-Ghalamfarsa, R., Dinh, Q., Liew, E. C. Y., White, D., Scott, P., Barber, P. A., O'gara, E., Ciampini, J., McDougall, K. L., & Tan, Y. P. (2021). Current status of Phytophthora in Australia. *Persoonia: Molecular Phylogeny and Evolution of Fungi*, *47*, 151–177. https://doi.org/10.3767/persoonia.2021.47.05

Buscaill, P., & van der Hoorn, R. A. L. (2021). Defeated by the nines: Nine extracellular strategies to avoid microbe-associated molecular patterns recognition in plants. *The Plant Cell*, *33*(7), 2116–2130.

Caulier, S., Nannan, C., Gillis, A., Licciardi, F., Bragard, C., & Mahillon, J. (2019). Overview of the antimicrobial compounds produced by members of the Bacillus subtilis group. *Frontiers in Microbiology*, *10*, 302.

Cazorla, F. M., Mercado-Blanco, J., Pidal, M., & Del Obispo, A. (2016). *Biological control of tree and woody plant diseases: An impossible task?. BioControl*, *61*(3), 233–242.

Chai, L. Q., Zhang, K. Y., Tang, L. J., Zhang, J. Y., & Zhang, H. S. (2017). Two mono- and dinuclear Ni(II) complexes constructed from quinazoline-type ligands: Synthesis, X-ray structures, spectroscopic, electrochemical, thermal, and antimicrobial studies. *Polyhedron*, *130*, 100–107. https://doi.org/10.1016/j.poly.2017.04.010

Colla, G., Rouphael, Y., Di Mattia, E., El-Nakhel, C., & Cardarelli, M. (2015). Co-inoculation of Glomus intraradices and Trichoderma atroviride acts as a biostimulant to promote growth, yield and nutrient uptake of vegetable crops. *Journal of the Science of Food and Agriculture*, *95*(8), 1706–1715. https://doi.org/10.1002/jsfa.6875

Das, I. R., Bhuiyan, M. K. A., Jannat, R., Kayesh, E., Tanbir Rubayet, M., Arefin, M. N., & Rubayet, M. T. (2019). Effect of bio-fortified compost in controlling soil-borne diseases of lentil (Lens culinaris l.) and enhance the crop growth and yield. *Advances in Biology & Earth Sciences*, *4*(2), 93–106.

de Boer, M., Bom, P., Kindt, F., Keurentjes, J. J. B., van der Sluis, I., Van Loon, L. C., & Bakker, P. A. H. M. (2003). Control of Fusarium wilt of radish by combining Pseudomonas putida strains that have different disease-suppressive mechanisms. *Phytopathology*, *93*(5), 626–632.

de Novais, C. B., Sbrana, C., da Conceição Jesus, E., Rouws, L. F. M., Giovannetti, M., Avio, L., Siqueira, J. O., Saggin Júnior, O. J., da Silva, E. M. R., & de Faria, S. M. (2020). Mycorrhizal networks facilitate the colonization of legume roots by a symbiotic nitrogen-fixing bacterium. *Mycorrhiza*, *30*(2–3), 389–396. https://doi.org/10.1007/s00572-020-00948-w

Delgado-Baquerizo, M., Guerra, C. A., Cano-Díaz, C., Egidi, E., Wang, J.-T., Eisenhauer, N., Singh, B. K., & Maestre, F. T. (2020). The proportion of soil-borne pathogens increases with warming at the global scale. *Nature Climate Change*, *10*(6), 550–554.

Deng, Z., Harbaugh, B. K., Kelly, R. O., Seijo, T., & Mcgovern, R. J. (2005). Screening for resistance to pythium root rot among twenty-three caladium cultivars. *HortTechnology*, *15*, 631–634

Derbyshire, M. C., & Denton-Giles, M. (2016). The control of sclerotinia stem rot on oilseed rape (Brassica napus): Current practices and future opportunities. *Plant Pathology*, *65*(6), 859–877). https://doi.org/10.1111/ppa.12517

Ding, L. N., Li, T., Guo, X. J., Li, M., Liu, X. Y., Cao, J., & Tan, X. L. (2021). Sclerotinia stem rot resistance in rapeseed: Recent progress and future prospects. *Journal of Agricultural and Food Chemistry*, *69*(10), 2965–297. https://doi.org/10.1021/acs.jafc.0c07351

Dissanayake, N., Hoy, J. W., & Griffin, J. L. (1998). Herbicide Effects on Sugarcane Growth, Pythium Root Rot, and Pythium arrhenomanes. *Phytopathology*, *88*, 530–535.

Dixon, G. R., & Tilston, E. L. (2010). Soil-borne pathogens and their interactions with the soil environment. *Soil Microbiology and Sustainable Crop Production*, *1*, 197–271.

Dorrance, A. E. (2018). Management of Phytophthora sojae of soybean: A review and future perspectives. *Canadian Journal of Plant Pathology*, *40*(2), 210–219. https://doi.org/10.1080/07060661.2018.1445127

Driouich, A., Follet-Gueye, M. L., Vicré-Gibouin, M., & Hawes, M. (2013). Root border cells and secretions as critical elements in plant host defense. *Current Opinion in Plant Biology*, *16*(4), 489–495. https://doi.org/10.1016/j.pbi.2013.06.010

Du, W., Zhao, X., Raju, T., Davies, P., & Trethowan, R. (2013). Studies on the resistance of some Australasian chickpeas (Cicer arietinum L.) to Phytophthora root rot disease. *Australian Journal of Crop Science*, *7*(6), 794–800.

Dunne, C., Moënne-Loccoz, Y., McCarthy, J., Higgins, P., Powell, J., Dowling, D. N., & O'gara, F. (1998). Combining proteolytic and phloroglucinol-producing bacteria for improved biocontrol of Pythium-mediated damping-off of sugar beet. *Plant Pathology*, *47*(3), 299–307.

Egamberdieva, D., Wirth, S., Bellingrath-Kimura, S. D., Mishra, J., & Arora, N. K. (2019). Salt-tolerant plant growth promoting rhizobacteria for enhancing crop productivity of saline soils. *Frontiers in Microbiology, 10*. https://doi.org/10.3389/fmicb.2019.02791

Emonet, A., Zhou, F., Vacheron, J., Heiman, C. M., Tendon, V. D., Ma, K.-W., Schulze-Lefert, P., Keel, C., & Geldner, N. (2021). Spatially restricted immune responses are required for maintaining root meristematic activity upon detection of bacteria. *Current Biology*, *31*(5), 1012–1028.

Ezrari, S., Legrifi, I., Taoussi, M., Khadiri, M., Belabess, Z., & Lahlali, R. (2023). Plant–pathogen interactions and global food security. In *Plant pathogen interaction* (pp. 11–52). Springer. https://doi.org/10.1007/978-981-99-4890-1_2

Fira, D., Dimkić, I., Berić, T., Lozo, J., & Stanković, S. (2018). Biological control of plant pathogens by Bacillus species. *Journal of Biotechnology*, *285*, 44–55. https://doi.org/10.1016/j.jbiotec.2018.07.044

Fry, W. (2008). Phytophthora infestans: The plant (and R gene) destroyer. In *Molecular Plant Pathology*, *9*(3), 385–402). https://doi.org/10.1111/j.1364-3703.2007.00465.x

Giachero, M. L., Declerck, S., & Marquez, N. (2022). Phytophthora root rot: Importance of the disease, current and novel methods of control. *Agronomy*, *12*(3). https://doi.org/10.3390/agronomy12030610

Gisi, U., & Sierotzki, H. (2015). Fungicide Resistance in Plant Pathogens. *Fungicide Resistance in Plant Pathogens*. https://doi.org/10.1007/978-4-431-55642-8

Gondal, A. S., Rauf, A., & Naz, F. (2019). Anastomosis Groups of Rhizoctonia solani associated with tomato foot rot in Pothohar Region of Pakistan. *Scientific Reports*, *9*(1). https://doi.org/10.1038/s41598-019-40043-5

Gonzalez, M., Pujol, M., Metraux, J. P., Gonzalez-Garcia, V., Bolton, M. D., & Borrás-Hidalgo, O. (2011). Tobacco leaf spot and root rot caused by Rhizoctonia solani Kühn. *Molecular Plant Pathology*, *12*(3), 209–216. https://doi.org/10.1111/j.1364-3703.2010.00664.x

Gordon, T. R. (2017). *Fusarium oxysporum and the Fusarium Wilt Syndrome. Annual Review of Phytopathology, 55*, 23–39. https://doi.org/10.1146/annurev-phyto-080615-095919

Goyal, A., & Manoharachary, C. (n.d.). *Future challenges in crop protection against fungal pathogens*. https://www.springer.com/series/11224

Graham, J., Hackett, C. A., Smith, K., Woodhead, M., MacKenzie, K., Tierney, I., Cooke, D., Bayer, M., & Jennings, N. (2011). Towards an understanding of the nature of resistance to Phytophthora root rot in red raspberry. *Theoretical and Applied Genetics*, *123*(4), 585–601. https://doi.org/10.1007/s00122-011-1609-5

Gu, S., Wei, Z., Shao, Z., Friman, V.-P., Cao, K., Yang, T., Kramer, J., Wang, X., Li, M., & Mei, X. (2020). Competition for iron drives phytopathogen control by natural rhizosphere microbiomes. *Nature Microbiology*, *5*(8), 1002–1010.

Gujri, M. (2011). Evaluation of disease control and plant growth promotion potential of biocontrol agents on Pisum sativum and comparison of their activity with popular chemical control agent-carbendazim Poornima Sharma. *Journal of Toxicology and Environmental Health Sciences*, *3*(5), 127–138. https://www.academicjournals.org/JTEHS

Haggag, W. M., & Nofal, M. A. (2006). Improving the biological control of Botryodiplodia disease on some Annona cultivars using single or multi-bioagents in Egypt. *Biological Control*, *38*(3), 341–349. https://doi.org/10.1016/j.biocontrol.2006.02.010

Hansen, E. M., Reeser, P. W., & Sutton, W. (2012). Phytophthora beyond agriculture. *Annual Review of Phytopathology*, *50*), 359–378). https://doi.org/10.1146/annurev-phyto-081211-172946

Hassani, M. A., Durán, P., & Hacquard, S. (2018). Microbial interactions within the plant holobiont. *Microbiome*, *6*, 1–17.

Henricot, B., Pérez Sierra, A., & Jung, T. (2014). Phytophthora pachypleura sp. nov., a new species causing root rot of Aucuba japonica and other ornamentals in the United Kingdom. *Plant Pathology*, *63*(5), 1095–1109. https://doi.org/10.1111/ppa.12194

Heyman, F., Blair, J. E., Persson, L., & Wikström, M. (2013). Root rot of pea and faba bean in southern Sweden caused by Phytophthora pisi sp. nov. *Plant Disease*, *97*(4), 461–471. https://doi.org/10.1094/PDIS-09-12-0823-RE

Higginbotham, R. W., & Kidwell, K. K. (2004). *Virulence of Pythium Species Isolated from Wheat Fields in Eastern Washington. Plant Disease*, *88*(9), 1021–1026.

Ho, H. (2018). The taxonomy and biology of phytophthora and pythium. *Journal of Bacteriology & Mycology: Open Access*, *6*(1). https://doi.org/10.15406/jbmoa.2018.06.00174

Hu, J., Wei, Z., Friman, V.-P., Gu, S., Wang, X., Eisenhauer, N., Yang, T., Ma, J., Shen, Q., & Xu, Y. (2016). Probiotic diversity enhances rhizosphere microbiome function and plant disease suppression. *MBio*, *7*(6), 10–1128.

Huang, Y., Zhan, H., Bhatt, P., & Chen, S. (2019). Paraquat Degradation From Contaminated Environments: Current Achievements and Perspectives. *Frontiers in Microbiology*, *10*. https://doi.org/10.3389/fmicb.2019.01754

Huzar-Novakowiski, J., & Dorrance, A. E. (2018). Genetic diversity and population structure of pythium irregulare from soybean and corn production fields in Ohio. *Plant Disease*, *102*(10), 1989–2000. https://doi.org/10.1094/PDIS-11-17-1725-RE

Iftikhar, Y., Sajid, A., Shakeel, Q., Ahmad, Z., & Ul Haq, Z. (2020). Biological antagonism: A safe and sustainable way to manage plant diseases. In Ul Haq, I., & Ijaz, S. (Eds.), *Plant disease management strategies for sustainable agriculture through traditional and modern approaches* (pp. 83–109). *Sustainability in Plant and Crop Protection*, vol 13. Springer, Cham. https://doi.org/10.1007/978-3-030-35955-3_5

Jaiswal, D. K., Gawande, S. J., Soumia, P. S., Krishna, R., Vaishnav, A., & Ade, A. B. (2022). Biocontrol strategies: An eco-smart tool for integrated pest and diseases management. *BMC Microbiology*, *22*(1). https://doi.org/10.1186/s12866-022-02744-2

Jetiyanon, K. (2007). Defensive-related enzyme response in plants treated with a mixture of Bacillus strains (IN937a and IN937b) against different pathogens. *Biological Control*, *42*(2), 178–185.

Kaeberlein, T., Lewis, K., & Epstein, S. S. (2002). *Isolating 'Uncultivable' Microorganisms in Pure Culture in a Simulated Natural Environment*. www.sciencemag.org/cgi/content/

Kageyama Kageyama, K. K., Suzuki Suzuki, M. M., Priyatmojo Priyatmojo, A. A., Oto Oto, Y. Y., Ishiguro Ishiguro, K. K., Suga Suga, H. H., Aoyagi Aoyagi, T. T., & Fukui Fukui, H. H. (2002). Characterization and Identification of Asexual Strains of Pythium Associated with Root Rot of Rose in Japan. *Journal of Phytopathology*, *151*, 485–49. https://doi.org/10.1046/j.1439-0434.2003.00757.x

Kannangara, T., Utkhede, R., Paul, J., & Punja, Z. (2000). Effects of mesophilic and thermophilic composts on suppression of Fusarium root and stem rot of greenhouse cucumber. *Canadian Journal of Microbiology*, *46*, 1021–8.

Katan, J. (2017). Diseases caused by soilborne pathogens: Biology, management and challenges. *Journal of Plant Pathology*, *99*(2), 305–315.

Khdiar, M. Y., Burgess, T. I., Barber, P. A., & Hardy, G. E. S. J. (2023). Calcium chelate is as effective as phosphite in controlling Phytophthora root rot in glasshouse trials. *Plant Pathology*, *72*(1), 112–119. https://doi.org/10.1111/ppa.13642

Krajaejun, T., Lohnoo, T., Jittorntam, P., Srimongkol, A., Kumsang, Y., Yingyong, W., Rujirawat, T., Reamtong, O., & Mangmee, S. (2018). Assessment of matrix-assisted laser desorption ionization-time of flight mass spectrometry for identification and biotyping of the pathogenic oomycete Pythium insidiosum. *International Journal of Infectious Diseases*, *77*, 61–67. https://doi.org/10.1016/j.ijid.2018.09.006

Kramer, J., Özkaya, Ö., & Kümmerli, R. (2020). Bacterial siderophores in community and host interactions. *Nature Reviews Microbiology*, *18*(3), 152–163.

Kroon, L. P. N. M., Brouwer, H., De Cock, A. W. A. M., & Govers, F. (2012). The genus Phytophthora anno 2012. *Phytopathology*, *102*(4), 348–364. https://doi.org/10.1094/PHYTO-01-11-0025

Kurbetli, I., & Ozan, S. (2013). Occurrence of phytophthora root and stem rot of kiwifruit in Turkey. *Journal of Phytopathology*, *161(11–12)*, 887–889. https://doi.org/10.1111/jph.12145

Leadbeater, A., & Gisi, U. (2009). The challenges of chemical control of plant diseases. In Gisi, U., Chet, I., & Gullino, M. (Eds.), *Recent Developments in Management of Plant Diseases. Plant Pathology in the 21st Century* (pp. 3–17), vol 1. Springer, Dordrecht. https://doi.org/10.1007/978-1-4020-8804-9_1

Lee, Y., Jeong, S. E., Hur, M., Ko, S., & Jeon, C. O. (2018). Construction and evaluation of a Korean native microbial consortium for the bioremediation of diesel fuel-contaminated soil in Korea. *Frontiers in Microbiology*, *9*. https://doi.org/10.3389/fmicb.2018.02594

Legrifi, I., Al Figuigui, J., Radouane, N., Ezrari, S., Belabess, Z., Tahiri, A., Amiri, S., & Lahlali, R. (2022). First report of Pythium schmitthenneri on olive trees and in Morocco. *Australasian Plant Disease Notes*, 1–5. https://doi.org/10.1007/s13314-022-00449-z

Legrifi, I., Taoussi, M., Al Figuigui, J., Abderrahim, L., Hussain, T., & Lahlali, R. (2023). Oomycetes root rot caused by Pythium spp. and Phytophthora spp.: Host range, detection, and management strategies, special case of olive trees. *Gesunde Pflanzen*, 1–29. https://doi.org/10.1007/s10343-023-00946-w

Lévesque, C., Brouwer, H., Cano, L., Hamilton, J. P., Holt, C., Huitema, E., Raffaele, S., Robideau, G. P., Thines, M., Win, J., Zerillo, M. M., Beakes, G. W., Boore, J. L., Busam, D., Dumas, B., Ferriera, S., Fuerstenberg, S. I., Gachon, C. M., Gaulin, E., … Robin Buell, C. (2010). Genome sequence of the necrotrophic plant pathogen Pythium ultimum reveals original pathogenicity mechanisms and effector repertoire. *Genome Biology*, *11*(7), R73 https://genomebiology.biomedcentral.com/articles/10.1186/gb-2010-11-7-r73

Li, B., Ravnskov, S., Xie, G., & Larsen, J. (2007). Biocontrol of Pythium damping-off in cucumber by arbuscular mycorrhiza-associated bacteria from the genus Paenibacillus. *BioControl*, *52*(6), 863–875. https://doi.org/10.1007/s10526-007-9076-2

Liang, X., & Rollins, J. A. (2018). Mechanisms of broad host range necrotrophic pathogenesis in sclerotinia sclerotiorum. *Phytopathology*, *108*(10), 1128–1140. https://doi.org/10.1094/PHYTO-06-18-0197-RVW

Lim, J.-H., Jung, H.-Y., & Kim, S.-D. (2009). Development of the microbial consortium for the environmental friendly agriculture by the antagonistic rhizobacteria. *Journal of Applied Biological Chemistry*, *52*(3), 116–120.

Lin, F., Wani, S. H., Collins, P. J., Wen, Z., Li, W., Zhang, N., McCoy, A. G., Bi, Y., Tan, R., Zhang, S., Gu, C., Chilvers, M. I., & Wang, D. (2020). QTL mapping and GWAS for identification of loci conferring partial resistance to Pythium sylvaticum in soybean (Glycine max (L.) Merr). *Molecular Breeding*, *40*(6). https://doi.org/10.1007/s11032-020-01133-9

Lookabaugh, E. C., Whipker, B., & Shew, B. B. (2017). Evaluation of poinsettia cultivars for resistance to pythium root rot caused by pythium aphanidermatum. *HortTechnology*, *27*(6), 805–812. https://doi.org/10.21273/HORTTECH03817-17

Lutz, M. P., Wenger, S., Maurhofer, M., Défago, G., & Duffy, B. (2004). Signaling between bacterial and fungal biocontrol agents in a strain mixture. *FEMS Microbiology Ecology*, *48*(3), 447–455.

Ma, M., Taylor, P. W. J., Chen, D., Vaghefi, N., & He, J. Z. (2023). Major soilborne pathogens of field processing tomatoes and management strategies. *Microorganisms*, *11*(2). https://doi.org/10.3390/microorganisms11020263

Manjunatha, S. V., Naik, M. K., Khan, M. F. R., & Goswami, R. S. (2013). Evaluation of bio-control agents for management of dry root rot of chickpea caused by Macrophomina phaseolina. *Crop Protection*, *45*, 147–150. https://doi.org/10.1016/j.cropro.2012.09.003

Mares, D., Tosi, B., Poli, F., Andreotti, E., & Romagnoli, C. (2004). Antifungal activity of Tagetes patula extracts on some phytopathogenic fungi: Ultrastructural evidence on Pythium ultimum. *Microbiological Research*, *159*(3), 295–304. https://doi.org/10.1016/j.micres.2004.06.001

Marimuthu, S., Ramamoorthy, V., Samiyappan, R., & Subbian, P. (2013a). Intercropping system with combined application of a zospirillum and p seudomonas fluorescens reduces root rot incidence caused by r hizoctonia bataticola and increases seed cotton yield. *Journal of Phytopathology*, *161*(6), 405–411.

Marimuthu, S., Ramamoorthy, V., Samiyappan, R., & Subbian, P. (2013b). Intercropping system with combined application of azospirillum and pseudomonas fluorescens reduces root rot incidence caused by rhizoctonia bataticola and increases seed cotton yield. *Journal of Phytopathology*, *161*(6), 405–411. https://doi.org/10.1111/jph.12084

Martin, F. N., Gloria Abad, Z., Balci, Y., & Ivors, K. (2012). Identification and detection of phytophthora: Reviewing our progress, identifying our needs. *Plant Disease*, *96*(8), 1080–1103. https://doi.org/10.1094/PDIS-12-11-1036-FE

Mathivanan, N., Prabavathy, V. R., & Vijayanandraj, V. R. (2005). Application of talc formulations of Pseudomonas fluorescens Migula and Trichoderma viride Pers. ex SF Gray decrease the sheath blight disease and enhance the plant growth and yield in rice. *Journal of Phytopathology*, *153*(11–12), 697–701.

Matthiesen, R. L., & Robertson, A. E. (2021). Comparison of aggressiveness and fungicide sensitivity of four Pythium spp. that cause damping-off of soybean in the United States. *Canadian Journal of Plant Pathology*, *43*(5), 769–782. https://doi.org/10.1080/07060661.2021.1881162

Mavrodi, O. V., Walter, N., Elateek, S., Taylor, C. G., & Okubara, P. A. (2012). Suppression of Rhizoctonia and Pythium root rot of wheat by new strains of Pseudomonas. *Biological Control*, *62*(2), 93–102. https://doi.org/10.1016/j.biocontrol.2012.03.013

Mazzola, M., & Freilich, S. (2017). Prospects for biological soilborne disease control: Application of indigenous versus synthetic microbiomes. *Phytopathology*, *107*(3), 256–263. https://doi.org/10.1094/PHYTO-09-16-0330-RVW

Menge, J. A., Douhan, G. W., Mckee, B., Pond, E., Bender, G. S., & Faber, B. (2012). Three new avocado rootstock cultivars tolerant to phytophthora root rot: 'Zentmyer', 'Uzi', and 'Steddom'. *HortScience*, *47*(8), 1191–1194. https://doi.org/10.21273/HORTSCI.47.8.1191

Mingmongkolchai, S., & Panbangred, W. (2018). Bacillus probiotics: An alternative to antibiotics for livestock production. *Journal of Applied Microbiology*, *124*(6), 1334–1346.

Muthukumar, A., Bhaskaran, R., & Sanjeevkumar, K. (2010). Efficacy of endophytic Pseudomonas fluorescens (Trevisan) migula against chilli damping-off. *Journal of Biopesticides*, *3*(Special Issue), 105.

Namayanja, A., Msolla, S. N., Buruchara, R., & Namusoke, A. (2014). Genetic analysis of resistance to pythium root rot disease in common bean (Phaseolus Vulgaris L.) genotypes. *Journal of Crop Improvement*, *28*(2), 184–202. https://doi.org/10.1080/15427528.2013.863815

Negi, Y. K., Prabha, D., Garg, S. K., & Kumar, J. (2017). Biological control of ragi blast disease by chitinase producing fluorescent Pseudomonas isolates. *Organic Agriculture*, *7*, 63–71.

Niu, B., Wang, W., Yuan, Z., Sederoff, R. R., Sederoff, H., Chiang, V. L., & Borriss, R. (2020a). Microbial interactions within multiple-strain biological control agents impact soil-borne plant disease. *Frontiers in Microbiology*, *11*. https://doi.org/10.3389/fmicb.2020.585404

Niu, B., Wang, W., Yuan, Z., Sederoff, R. R., Sederoff, H., Chiang, V. L., & Borriss, R. (2020b). Microbial interactions within multiple-strain biological control agents impact soil-borne plant disease. *Frontiers in Microbiology*, *11*, 1–16. https://doi.org/10.3389/fmicb.2020.585404

Nützmann, H.-W., Reyes-Dominguez, Y., Scherlach, K., Schroeckh, V., Horn, F., Gacek, A., Schümann, J., Hertweck, C., Strauss, J., & Brakhage, A. A. (2011). Bacteria-induced natural product formation in the fungus Aspergillus nidulans requires Saga/Ada-mediated histone acetylation. *Proceedings of the National Academy of Sciences*, *108*(34), 14282–14287.

Nzungize, J., Gepts, P., Buruchara, R., Male, A., Ragama, P., Busogoro, J. P., & Baudoin, J. P. (2011). Introgression of Pythium root rot resistance gene into Rwandan susceptible common bean cultivars. *African Journal of Plant Science*, *5*(3), 193–200. https://www.academicjournals.org/ajps

Okubara, P. A., Dickman, M. B., & Blechl, A. E. (2014). Molecular and genetic aspects of controlling the soilborne necrotrophic pathogens Rhizoctonia and Pythium. *Plant Science*, *228*, 61–70. https://doi.org/10.1016/j.plantsci.2014.02.001

Ola, A. R. B., Thomy, D., Lai, D., Brötz-Oesterhelt, H., & Proksch, P. (2013). Inducing secondary metabolite production by the endophytic fungus Fusarium tricinctum through coculture with Bacillus subtilis. *Journal of Natural Products*, *76*(11), 2094–2099.

Orlikowski, L. B., Ptaszek, M., Rodziewicz, A., Nechwatal, J., Thinggaard, K., & Jung, T. (2011). Phytophthora root and collar rot of mature Fraxinus excelsior in forest stands in Poland and Denmark. *Forest Pathology*, *41*(6), 510–519. https://doi.org/10.1111/j.1439-0329.2011.00714.x

Pandey, K.K. & Pandey, P.K. (2005). Differential response of biocontrol agents against soil pathogens on tomato, chilli and brinjal. *Indian Phytopath*, *58*(3), 329–331.

Petkowski, J. E., De Boer, R. F., Norng, S., Thomson, F., & Minchinton, E. J. (2013). Pythium species associated with root rot complex in winter-grown parsnip and parsley crops in south eastern Australia. *Australasian Plant Pathology*, *42*(4), 403–411. https://doi.org/10.1007/s13313-013-0211-5

Pieterse, C. M. J., Zamioudis, C., Berendsen, R. L., Weller, D. M., Van Wees, S., & Bakker, P. A. H. M. (2014). Induced systemic resistance by beneficial microbes. *Annual Review of Phytopathology*, *52*, 347–375.

Pishchany, G., Mevers, E., Ndousse-Fetter, S., Horvath Jr, D. J., Paludo, C. R., Silva-Junior, E. A., Koren, S., Skaar, E. P., Clardy, J., & Kolter, R. (2018). Amycomicin is a potent and specific antibiotic discovered with a targeted interaction screen. *Proceedings of the National Academy of Sciences*, *115*(40), 10124–10129.

Pliego, C., De Weert, S., Lamers, G., De Vicente, A., Bloemberg, G., Cazorla, F. M., & Ramos, C. (2008). Two similar enhanced root-colonizing Pseudomonas strains differ largely in their colonization strategies of avocado roots and Rosellinia necatrix hyphae. *Environmental Microbiology*, *10*(12), 3295–3304. https://doi.org/10.1111/j.1462-2920.2008.01721.x

Prajapati, S., Kumar, N., Kumar, S., & Maurya, S. (2020). Biological control a sustainable approach for plant diseases management: A review. *Journal of Pharmacognosy and Phytochemistry*, *9*(2), 1514–1523.

Pressete, C. G., Giannini, L. S. V., de Paula, D. A. C., do Carmo, M. A. V., Assis, D. M., Santos, M. F. C., Machado, J. da C., Marques, M. J., Soares, M. G., & Azevedo, L. (2019). Sclerotinia sclerotiorum (white mold): Cytotoxic, mutagenic, and antimalarial effects in vivo and in vitro. *Journal of Food Science*, *84*(12), 3866–3875. https://doi.org/10.1111/1750-3841.14910

Purkayastha, J., Arora, R., & Singh, L. (2016). Sustainable and novel eco-friendly approaches towards integrated disease and vector management. In Vijay Veer, & Gopalakrishnan, R. (Eds.), *Herbal Insecticides, Repellents and Biomedicines: Effectiveness and Commercialization*. Springer, New Delhi. https://doi.org/10.1007/978-81-322-2704-5_2

Pylak, M., Oszust, K., & Frąc, M. (2019). Review report on the role of bioproducts, biopreparations, biostimulants and microbial inoculants in organic production of fruit. *Reviews in Environmental Science and Biotechnology*, *18*(3), 597–616). https://doi.org/10.1007/s11157-019-09500-5

Qian, X., Chen, L., Sui, Y., Chen, C., Zhang, W., Zhou, J., Dong, W., Jiang, M., Xin, F., & Ochsenreither, K. (2020). Biotechnological potential and applications of microbial consortia. *Biotechnology Advances*, *40*. https://doi.org/10.1016/j.biotechadv.2019.107500

Raaijmakers, J. M., Vlami, M., & De Souza, J. T. (2002). Antibiotic production by bacterial biocontrol agents. *Antonie van Leeuwenhoek*, *81*, 537–547.

Rahman, M. T., Rubayet, M. T., & Bhuiyan, M. K. A. (2020). Integrated management of rhizoctonia root rot disease of soybean caused by Rhizoctonia solani. *Nippon Journal of Environmental Science*, *1*(7), 1018.

Raimondo, E. E., Aparicio, J. D., Bigliardo, A. L., Fuentes, M. S., & Benimeli, C. S. (2020). Enhanced bioremediation of lindane-contaminated soils through microbial bioaugmentation assisted by biostimulation with sugarcane filter cake. *Ecotoxicology and Environmental Safety*, *190*. https://doi.org/10.1016/j.ecoenv.2019.110143

Raza, A., Hassan, A., Akram, W., Anjum, T., Aftab, Z. e . H., & Ali, B. (2024). Seed coating with the synthetic consortium of beneficial Bacillus microbes improves seedling growth and manages Fusarium wilt disease. *Scientia Horticulturae*, *325*. https://doi.org/10.1016/j.scienta.2023.112645

Rodríguez-Castillo, G., Molina-Rodríguez, M., Cambronero-Heinrichs, J. C., Quirós-Fournier, J. P., Lizano-Fallas, V., Jiménez-Rojas, C., Masís-Mora, M., Castro-Gutiérrez, V., Mata-Araya, I., & Rodríguez-Rodríguez, C. E. (2019). Simultaneous removal of neonicotinoid insecticides by a microbial degrading consortium: Detoxification at reactor scale. *Chemosphere*, *235*, 1097–1106. https://doi.org/10.1016/j.chemosphere.2019.07.004

Rubayet, M. T., & Bhuiyan, M. K. A. (2016). Integrated management of stem rot of potato caused by sclerotium rolfsii. *Bangladesh Journal of Plant Pathology*, *32*(1&2), 7–14

Rusagara Nzungize, J., Lyumugabe, F., Busogoro, J.-P., & Baudoin, J.-P. (2012). B A Pythium root rot of common bean: Biology and control methods. A review. *Biotechnology, Agronomy, Society and Environment*, *16*(3), 405–413

S A El-Hassan, & S R Gowen. (2006). Formulation and Delivery of the Bacterial Antagonist Bacillus subtilis for Management of Lentil Vascular Wilt Caused by Fusarium oxysporum f. sp. lentis. *Journal of Phytopathology*, *154*, 148–155. https://doi.org/10.1111/j.1439-0434.2006.01075.x

Sánchez-Torres, P. (2021). Molecular mechanisms underlying fungicide resistance in citrus postharvest green mold. *Journal of Fungi*, *7*(9), 1–18. https://doi.org/10.3390/jof7090783

Sarma, B. K., Yadav, S. K., Singh, S., & Singh, H. B. (2015). Microbial consortium-mediated plant defense against phytopathogens: Readdressing for enhancing efficacy. *Soil Biology and Biochemistry*, *87*, 25–33. https://doi.org/10.1016/j.soilbio.2015.04.001

Serafin, L., Simpfendorfer, S., Gardner, M., & McMullen, G. (2013). *Northern Grains Region Trial Results Autumn*. www.dpi.nsw.gov.au

Shilev, S., Azaizeh, H., Vassilev, N., Georgiev, D., & Babrikova, I. (2019). Interactions in soil-microbe-plant system: Adaptation to stressed agriculture. In *Microbial interventions in agriculture and environment: research trends, priorities and prospects* (Vol. 1, pp. 131–171). Springer. https://doi.org/10.1007/978-981-13-8391-5_6

Siah, A., Magnin-Robert, M., Randoux, B., Choma, C., Rivière, C., Halama, P., & Reignault, P. (2018). *Natural agents inducing plant resistance against pests and diseases* (pp. 121–159). https://doi.org/10.1007/978-3-319-67045-4_6

Singh, A., Mehta, S., Singh, H. B., & Nautiyal, C. S. (2003). Biocontrol of collar rot disease of betelvine (Piper betle L.) caused by Sclerotium rolfsii by using rhizosphere-competent Pseudomonas fluorescens NBRI-N6 and P. fluorescens NBRI-N. *Current Microbiology*, *47*, 153–158.

St. Martin, C. C. G., & Brathwaite, R. A. I. (2012). Compost and compost tea: Principles and prospects as substrates and soil-borne disease management strategies in soil-less vegetable production. *Biological Agriculture and Horticulture*, *28*(1), 1–33. https://doi.org/10.1080/01448765.2012.671516

Stirling, G. R. (1988). Biological control of plant-parasitic nematodes. In G. O. Poinar & H.- B. Jansson (Eds.), *Diseases of nematodes*. CRC Press.

Sumner, D. R., & Bell, D. K. (1982). Root diseases induced in corn by Rhizoctonia solani and Rhizoctonia zeae. *Phytopathology*, *72*, 86–91.

Syed Ab Rahman, S. F., Singh, E., Pieterse, C. M. J., & Schenk, P. M. (2018a). Emerging microbial biocontrol strategies for plant pathogens. *Plant Science*, *267*, 102–111. https://doi.org/10.1016/j.plantsci.2017.11.012

Syed Ab Rahman, S. F., Singh, E., Pieterse, C. M. J., & Schenk, P. M. (2018b). Emerging microbial biocontrol strategies for plant pathogens. *Plant Science*, *267*, 102–111. https://doi.org/10.1016/j.plantsci.2017.11.012

Teixeira, P. J. P. L., Colaianni, N. R., Fitzpatrick, C. R., & Dangl, J. L. (2019). Beyond pathogens: Microbiota interactions with the plant immune system. *Current Opinion in Microbiology*, *49*, 7–17.

Thaines Bodah, E. (2017). Root rot diseases in plants: A review of common causal agents and management strategies. *Agricultural Research & Technology: Open Access Journal*, *5*(3). https://doi.org/10.19080/artoaj.2017.05.555661

Thomidis, T., & Exadaktylou, E. (2012). Effectiveness of cyproconazole to control Armillaria root rot of apple, walnut and kiwifruit. *Crop Protection*, *36*, 49–51. https://doi.org/10.1016/j.cropro.2012.02.003

Thomloudi, E. E., Tsalgatidou, P. C., Douka, D., Spantidos, T. N., Dimou, M., Venieraki, A., & Katinakis, P. (2019). Multistrain versus single-strain plant growth promoting microbial inoculants-The compatibility issue. *Hellenic Plant Protection Journal*, *12*(2), 61–77. https://doi.org/10.2478/hppj-2019-0007

Tiku, A. R. (2018). Antimicrobial compounds and their role in plant defense. In Singh, A., & Singh, I. (Eds.), *Molecular Aspects of Plant-Pathogen Interaction*. Springer, Singapore. https://doi.org/10.1007/978-981-10-7371-7_13.

Traxler, M. F., Watrous, J. D., Alexandrov, T., Dorrestein, P. C., & Kolter, R. (2013). Interspecies interactions stimulate diversification of the Streptomyces coelicolor secreted metabolome. *MBio*, *4*(4), 10–1128.

Truong, H. T. H., Kim, K. T., Kim, D. W., Kim, S., Chae, Y., Park, J. H., Oh, D. G., & Cho, M. C. (2012). Identification of isolate-specific resistance QTLs to phytophthora root rot using an intraspecific recombinant inbred line population of pepper (Capsicum annuum). *Plant Pathology*, *61*(1), 48–56. https://doi.org/10.1111/j.1365-3059.2011.02483.x

Turkolmez, S., & Dervis, S. (2017). Activity of metalaxyl-M plus mancozeb, fosetyl-Al, and phosphorous acid against Phytophthora crown and root rot of apricot and cherry caused by Phytophthora palmivora. *Plant Protection Science*, *53*(4), 216–225.

Tyler, B. M., & Gijzen, M. (2014). The Phytophthora sojae genome sequence: Foundation for a revolution. In *Genomics of plant-associated fungi and oomycetes: Dicot pathogens* (pp. 133–157). Springer. https://doi.org/10.1007/978-3-662-44056-8_7

Urbina-Salazar, A. del R., Inca-Torres, A. R., Falcón-García, G., Carbonero-Aguilar, P., Rodríguez-Morgado, B., del Campo, J. A., Parrado, J., & Bautista, J. (2019). Chitinase production by Trichoderma harzianum grown on a chitin-rich mushroom byproduct formulated medium. *Waste and Biomass Valorization*, *10*, 2915–2923.

Van Os, G. J., & Van Ginkel, J. H. (2001). *Suppression of Pythium root rot in bulbous Iris in relation to biomass and activity of the soil microflora*. www.elsevier.com/locate/soilbio.

Verma, H., Kant, C., Singh, S. K., White, J. F., Kumar, A., & Droby, S. (2022). Biocontrol potential of microbial consortia: Approaches in food security and disease management. In Kumar, A. (Ed.), *Microbial Biocontrol: Sustainable Agriculture and Phytopathogen Management*. Springer, Cham. https://doi.org/10.1007/978-3-030-87512-1_7

Verma, N. S., Kuldeep, D. K., Chouhan, M., Prajapati, R., & Singh, S. K. (2023). A review on eco-friendly pesticides and their rising importance in sustainable plant protection practices. *International Journal of Plant & Soil Science*, *35*(22), 200–214.

Vestberg, M., Kukkonen, S., Saari, K., Parikka, P., Huttunen, J., Tainio, L., Devos, N., Weekers, F., Kevers, C., & Thonart, P. (2004). Microbial inoculation for improving the growth and health of micropropagated strawberry. *Applied Soil Ecology*, *27*(3), 243–258.

Vishwakarma, K., Kumar, N., Shandilya, C., Mohapatra, S., Bhayana, S., & Varma, A. (2020). Revisiting plant–microbe interactions and microbial consortia application for enhancing sustainable agriculture: A review. *Frontiers in Microbiology*, *11*, 560406.

Vorholt, J. A., Vogel, C., Carlström, C. I., & Müller, D. B. (2017). Establishing causality: Opportunities of synthetic communities for plant microbiome research. *Cell Host and Microbe*, *22*(2), 142–155. https://doi.org/10.1016/j.chom.2017.07.004

Wei, Z., Yang, T., Friman, V.-P., Xu, Y., Shen, Q., & Jousset, A. (2015). Trophic network architecture of root-associated bacterial communities determines pathogen invasion and plant health. *Nature Communications*, *6*(1), 8413.

Wheeler, T. A., Russell, S. A., Anderson, M. G., Serrato-Diaz, L. M., French-Monar, R. D., & Woodward, J. E. (2016). Management of peanut pod rot I: Disease dynamics and sampling. *Crop Protection*, *79*, 135–142. https://doi.org/10.1016/j.cropro.2015.10.010

Woo, S. L., & Pepe, O. (2018a). Microbial consortia: Promising probiotics as plant biostimulants for sustainable agriculture. *Frontiers in Plant Science*, *9*, 1801.

Woo, S. L., & Pepe, O. (2018b). Microbial consortia: Promising probiotics as plant biostimulants for sustainable agriculture. *Frontiers in Plant Science*, *9*. https://doi.org/10.3389/fpls.2018.01801

Wu, W., Ogawa, F., Ochiai, M., Yamada, K., & Fukui, H. (2020). Common strategies to control pythium disease. *Reviews in Agricultural Science*, *8*, 58–69. https://doi.org/10.7831/ras.8.0_58

Xia, S., Xu, Y., Hoy, R., Zhang, J., Qin, L., & Li, X. (2020). The notorious soilborne pathogenic fungus sclerotinia sclerotiorum: An update on genes studied with mutant analysis. *Pathogens*, *9*(1). https://doi.org/10.3390/pathogens9010027

Xu, X. M., Jeffries, P., Pautasso, M., & Jeger, M. J. (2011). Combined use of biocontrol agents to manage plant diseases in theory and practice. *Phytopathology*, *101*(9), 1024–1031). https://doi.org/10.1094/PHYTO-08-10-0216

Xu, X.-M., Jeffries, P., Pautasso, M., & Jeger, M. J. (2011). Combined use of biocontrol agents to manage plant diseases in theory and practice. *Phytopathology*, *101*(9), 1024–1031.

Yin, C., Casa Vargas, J. M., Schlatter, D. C., Hagerty, C. H., Hulbert, S. H., & Paulitz, T. C. (2021). Rhizosphere community selection reveals bacteria associated with reduced root disease. *Microbiome*, *9*, 1–18.

Zehra, A., Raytekar, N. A., Meena, M., & Swapnil, P. (2021). Efficiency of microbial bio-agents as elicitors in plant defense mechanism under biotic stress: A review. *Current Research in Microbial Sciences*, *2*. https://doi.org/10.1016/j.crmicr.2021.100054

Zerillo, M. M., Adhikari, B. N., Hamilton, J. P., Buell, C. R., Lévesque, C. A., & Tisserat, N. (2013). Carbohydrate-active enzymes in pythium and their role in plant cell wall and storage polysaccharide degradation. *PLoS One*, *8*(9). https://doi.org/10.1371/journal.pone.0072572

Zhang, J. X., Xue, A. G., Zhang, H. J., Nagasawa, A. E., Tambong, J. T., Zhang, J. Z., & And Tambong, J. T. (2011). Response of soybean cultivars to root rot caused by Fusarium species. *Canadian Journal of Plant Science*, *90*, 767–776.

Zhao, J., & Meng, J. (2003). Genetic analysis of loci associated with partial resistance to Sclerotinia sclerotiorum in rapeseed (Brassica napus L.). *Theoretical and Applied Genetics*, *106*(4), 759–764. https://doi.org/10.1007/s00122-002-1171-2

17 Molecular Mechanisms and Breeding Strategies for Insect Pest Resistance in Fruit Crops

Priscilla Amponsah, Augustine Antwi-Boasiako, Justice Frimpong Amankwah, and Muhammad Khuram Razzaq

17.1 INTRODUCTION

Insects possess dangerous diseases and are the main rivals for food and fiber (Hill, 2012). Global agricultural production losses from pests and illnesses are estimated to reach 37%, of which 13% are attributable to insects (Savary et al., 2019). Because they directly cause damage such as fouling, deformations, necrosis, and the spread of diseases, insect pests have a major negative impact on agricultural production. By consuming and degrading natural fibers, they also hasten the process of degradation. Herbivorous insects destroy a fifth of the world's crop yield every year (Deutsch et al., 2018).

Since the mid-1990s, the utilization of chemical insecticides has decreased, and sustainable pest management techniques have been promoted by the introduction of insect-resistant plants, which are modified to express specific proteins poisonous to particular insects (Abbas, 2018). Genetically engineered plants, release chemicals that are toxic to insects, lowering the need for pesticides while increasing crop yields and promoting sustainable agricultural methods (Rao et al., 2017). These plants have undergone genetic modification to create toxins that are harmful to insects and are generated by the DNA of the plant (Abbas, 2018). The most prevalent type of plant resistant to insects is the *Bacillus thuringiensis* (Bt) plant (Xiao & Wu, 2019). These plants contribute to a decrease in pesticide use, which enhances crop yields. Examples of crops that have proven successful in controlling pests are maize and cotton, which are resistant to insects (Blanco et al., 2016). Research aiming at boosting the expression of insect resistance genes and resolving associated hazards has focused on the production of genetically engineered poplar that is resistant to insects in countries like China (Song, 2013).

The link between insect pests and their plant hosts is a dynamic interplay impacted by different elements such as insect traits and plant allelochemicals (Tanda, 2022). The intricate physiological systems of phytophagous insects are sculpted by their interactions with their host plants, which is the outcome of a protracted and ongoing evolutionary process that produced an astounding diversity of insects. Conventionally, plant–insect interactions are thought to be mutualistic, antagonistic, or commensalistic (Tanda, 2022). Mutualism is defined by the assistance given by both parties, with neither suffering damage and both profiting (Tripathy & Pradhan, 2018). In antagonistic partnerships, the other party suffers while the other party gains an advantage. Examples of this include phytophagy by insects (e.g., pest insects) and insectivory by plants (e.g., carnivorous plants). When two species coexist together, neither party suffers (as in the case of monarch butterfly larvae and some species of milkweed, which accumulate cardiac glycosides for defense) (Lawson, 2023). Insects and plants have an extremely diverse host relationship, with feeding habits ranging from highly specialized to generalist (Ali et al., 2012). The majority of insects, especially pests that affect crops, are specialists that have evolved to solely eat specific plants especially those in a single

DOI: 10.1201/9781003543237-17

family (Capinera, 2005). When insects consume plants, a practice known as phytophagy, they usually do it in an antagonistic manner, where the insect gains an advantage over the plant (Pérez-Hedo et al., 2022). This may cause the plant to experience a variety of consequences, including slower development, damage to the leaves, and a lower production of crops. Relationships between insects and plants are not confined to a straightforward exchange between two species; rather, they frequently entail intricate interactions with the plant microbiome (Sharma et al., 2021), which includes the endo- and exo-microbiomes (organisms that reside inside insects) and the microbiome (organisms that live on the outside of insects).

To lessen the need for insecticides, insect-resistant crop varieties are genetically modified to have inherent resistance to common pests and insects. By doing so, a more sustainable environment is encouraged and soil fertility is maintained (Alemu, 2020). A healthier soil can be maintained by avoiding insecticides because they include dangerous compounds that affect plant growth and soil quality (Baweja et al., 2020). The economy benefits from fewer pesticide applications as well because pesticides can be costly and raise farmer profits while lowering consumer costs. Furthermore, as they protect useful creatures like pollinators and beneficial insects, insect-resistant crop types can benefit the environment and help maintain a more sustainable and balanced ecosystem (Ali et al., 2023). Because of these main advantages, crop types that are resistant to insects are a useful tool in agriculture. They provide benefits for the economy, ecology, and environment in addition to lowering production costs and preserving soil fertility (Roberts & Mattoo, 2018). From an economic standpoint, they prevent pest damage to crop yields and lessen the requirement for costly insecticides. In terms of ecology, they encourage greater species diversity and improve the stability of ecosystems, resulting in a more sustainable system (Ratnadass et al., 2012). In terms of the environment, the decreased usage of pesticides lessens their effects on both the natural world and human health. Cultivars that are resistant to insects combine the benefits of biological, cultural, and natural methods to suppress insect pests, making them suitable for use with insecticides (Sharma et al., 2004). Plant resistance to insects is long-lasting, specialized, and only affects the intended pest (Mookiah et al., 2021).

Plants have developed physical traits and natural compounds, such as cuticle thickness and leaf hair, that influence their vulnerability to pests (Abebe, 2021). It has been determined that more than 30,000 plant compounds contribute to a plant's resistance to insects (Anulika et al., 2016). Plants have the ability to produce defense chemicals in response to herbivory through molecular mechanisms. These substances may be generated constitutively or under the influence of herbivores (Aljbory & Chen, 2018). Through genetic engineering, it is possible to create agricultural cultivars with increased resistance through induced resistance. Utilizing insecticidal genes and modifying gene expression patterns are two examples of the new biotechnological technologies being utilized to produce novel plant-resistant traits (Kumari et al., 2022). By reducing the need for insecticides and creating crop varieties with increased resistance, these resistance mechanisms can be utilized for pest management.

Plants have developed various resistance mechanisms to deter feeding and damage by insect pests in fruit crops. These mechanisms include morphological, biochemical, and molecular responses. Plants utilize these strategies to adapt to herbivory (Belete, 2018). These defenses are dynamic and can be generated constitutively or in reaction to injury to plants, which impacts herbivore development, feeding, and survival (War et al., 2012). Additionally, plants emit volatile chemical substances that draw herbivores' natural enemies (Aljbory & Chen, 2018). One way to make use of induced resistance as a pest management technique is to create crop cultivars that, in response to a mild infestation, will induce a response (Douglas, 2018). The creation of harmful substances that either kill or slow the growth of herbivores are examples of direct defenses. Alternatively, plants may use mechanical protection on their surface. Indirect defenses can be strengthened by offering food and shelter to herbivores or by releasing volatiles that attract herbivores' natural enemies (Tiku, 2021).

17.2 HOST PLANT RESISTANCE

17.2.1 Plant Morphological Barriers/Features

Plants employ a range of defense strategies, including chemical substances and structural barriers, to fend off insect pests. Morphological characteristics that pose a threat to insect survival include the length and quantity of internodes, the thickness of the leaves, rinds, and pith; and the existence of trichomes (Santiago Carabelos et al., 2011). Additional characteristics of morphology include the amount of silica present, anatomical alterations, tissue hardness, thickness of the cell wall and cuticle, surface waxes, form, color, and size. All of these traits work together to hinder insects' ability to proliferate, attract potential mates, reproduce, consume, and transmit disease. Certain defenses, such as morphological traits, are constant, but plants can produce chemicals to repel insects or draw in natural predators when necessary.

17.2.1.1 Color and Shape

Some insects exhibit preferences for specific plant types based on characteristics such as color and shape. This preference arises from the varying appeal of hues to different insects. A study emphasized the impact of leaf structure, color, and size on the choice of the host plant, with red-leafed plants like broccoli and cabbage proving less attractive to the imported cabbageworm. A similar effect is observed in red lettuce cultivars, which are less susceptible to cucumber beetle damage (Nagata et al., 2004). Furthermore, Norris's (1980) study reveals a preference for red plumage in certain bird species over green while turnips with bigger roots are less vulnerable to turnip maggot damage. According to Yepsen (1984), onion cultivars with densely packed leaves are attractive to thrips. These findings highlight the complex interplay between some plant visual traits and interactions with insects, herbivores, and mammals. Aphids that appear late in the growth cycle usually target leaves that reflect about 500 nm color, which is considered an excellent growth stage (Craigie, 2022).

17.2.1.2 Thickening of Cell Walls

Plants, being sessile organisms, have developed diverse defense strategies to fend off insect pests. Hardening plant tissues is one such tactic that has attracted a lot of attention in the realm of plant–insect exchanges. This natural defense mechanism is crucial for thwarting insect attacks and deterring pests from feeding within the plant tissues. One method which plants accomplish this task is by hardening their tissues. Plant tissue hardening, sometimes referred to as fortifying plant cells and structures, is a dynamic process that makes plants more resistant to insect attacks (Islam et al., 2020). Because their tissues are stronger, plants create a formidable barrier that keeps insects from feasting on them. This mechanism is remarkable since it shows itself in different ways in different plant species. Cultivars of rice offer a notable example of tissue hardening in action. Studies have indicated that rice plants possessing thicker hypodermal layers exhibit greater resistance against pest insects such as the Asiatic rice borer (*Chilo suppressalis*) and yellow stem borer (*Scirpophaga incertulas*) (Chavan and Patel, 2016).

17.2.1.3 Structure and Texture of Tissues

Insect–plant interactions are greatly influenced by the hardness and thickness of plant stems. For instance, a plant's ability to withstand a specific stem borer is closely linked to the characteristics of its stem tissues (Reynolds et al., 2009). Wheat-stem sawfly (*Cephus cinctus*) resistance is associated with solid stems (Talbert et al., 2014). According to Wallace et al. (1974), a certain amount of stem stiffness can cause eggs to become cracked and dry, which will make it harder for the larvae to migrate. In a similar vein, the degree of rind hardness and stalk fiber content dictate how well sugarcane resists *Saccharalis diatrea* (Santiago et al., 2013). Culminate plants with strong, woody stems that have strongly spaced vascular bundles are the best defense against the squash vine borer, *Melittia cucurbita*. The potato aphid, *Macrosiphum euphorbiae*, cannot penetrate the thick cortex of wild tomato cousins like *Lycopersicum hirsutum*, preventing it from entering their vascular tissue (Quiros et al., 1977). Compact tissue genotypes in cotton are positively associated with resistance against insect pests (Karar et al., 2017).

17.2.1.4 Surface Waxes

All naturally occurring waxy substances extracted from plants fall under the category of wax. From a chemical standpoint, wax is described as an ester that is created when a high molecular weight aliphatic alcohol and a long-chain fatty acid combine (Jetter et al., 2006; Peng, 2012). Wax content in plants can range from a tiny percentage to several percentages of their dry weight. Most vascular plants have a thin coating of material covering their outermost layer, called the cuticle, which is mostly made up of hydrophobic substances. Usually complex blends, these cuticular waxes are mostly made up of n-alkanes (C7–C62), alcohols, and acids combined (Peng, 2012). In addition to having substances that prevent insect attacks, the waxy coating of plant leaves plays a crucial part in the methods by which the plant regulates its water intake (Norris, 1980). The plant is shielded from problems including infections, insect feeding, and desiccation by these surface waxes on the epicuticle. Epicuticular waxes affect insects' eating behavior; they can either stimulate or discourage feeding (Reifenrath et al., 2005).

17.2.1.5 Trichomes

Trichomes are projections that emerge from the leaf, stem, and root epidermal layers (Fambrini and Pugliesi, 2019). Together, these extensions make up the trichome cover, or pubescence, that covers the surface of a plant. These projections that resemble hairs are called trichomes or hairs, and they come in a variety of sizes, shapes, and forms (Fambrini and Pugliesi, 2019). In plants, they fulfill a variety of vital physiological and ecological roles. Trichomes' mechanical impacts are mostly determined by attributes including density, erectness, length, and form (Akwu et al., 2023). There are two primary categories of trichomes: glandular and non-glandular. Insect-toxic compounds are released by glandular trichomes (Giuliani et al., 2021). These secretions serve as deterrents and repellents to insects by obstructing their movement on plant surfaces. In addition to interfering with food, glandular trichomes affect insect survival, development, and reproduction (Giuliani et al., 2021). Depending on whether the secretions produced by glandular trichomes are sticky or comprise allelochemicals (Johnson, 2011), the actions of the trichomes can differ.

17.2.1.6 Anatomical Adoption of Organs

Plants' suitability for herbivores might vary depending on even little changes to their morphological structure. These changes may have an impact on additional elements that raise insect mortality. In sorghum, the presence of both erect and lingular hairs, along with the tightness of the leaf sheath that forms an acute angle, inhibited the development of *Chilo partellus* larvae (Kishore, 1991). According to Patanakamjorn and Pathak (1967), some rice varieties that had tightly wound leaf sheaths showed less susceptibility to *C. suppressalis*, the striped stem borer. Spring wheat genotypes having tight leaf sheaths, dense hairs at the base of the leaves, and a thick waxy film covering the leaves showed resistance to *Hydrellia griseola* (Zhu, 1981). *Avenae Sitobion* and other grain aphids are warded off by wheat awns (Acreman & Dixon, 1986). According to Agarwal (1969), internode borer *Chilo sacchariphagus* indicus was more likely to affect sugarcane plants with loosely fitting leaf sheaths than those with tightly fitting sheaths. Mealybugs were found to exhibit the opposite behavior (Mehta et al., 1981). The cotton ovary surface's gossypol glands provided protection against bollworms (*Helicoverpa armigera*) (Rajarajeswari and Subbarao, 1997).

17.2.2 Biochemical Features of Host Plant Resistance

Chemical-based resistance is one of plants' primary defense strategies against herbivores and insect pests. The result of several interactions between plants and insects that affect the ultimate stage of plant population establishment by insects is plant susceptibility or resistance. According to Whittaker and Feeny (1971), several allelochemicals have the ability to repel, prevent feeding, or have other negative impacts on insects. A plant's resistance to insect pests may be explained by its ability to disrupt any one of several biological factors. These disruptions could be brought on by a plant's biochemical or biophysical traits. Plants that are resistant to insects often do so because of their chemical makeup rather than because of environmental or dietary variables (Carmona et al., 2011).

17.2.2.1 Non-Protein Amino Acids

The amino acids that are not found in proteins are known as non-protein amino acids or odd amino acids. Vertebrates cannot grow because of many amino acids that are not proteins. It is known that a few non-protein amino acids have bioactivities against insects, including canavanine, cyanoalanine, canaline, diaminobutyric acid, propionic acid, ß-Pyrozole-1-yl-L-alanine, mimosine, albizziine, and 3,4-dopa. Non-protein amino acids have a protective effect because they share structural similarities with amino acids that are frequently encountered (Huang et al., 2011). The negative effects of non-protein amino acids on insects are exacerbated by the analog molecule's integration into the protein synthesis process. By not incorporating these compounds into proteins, plants that contain them avoid becoming auto-toxic (Rosenthal, 1991). A few amino acids that aren't proteins can also hinder certain enzymes. Canavanine has strong insecticidal effects and shields plant seeds and tender foliage from insect infestations (Tanda, 2022). The fifth instar tobacco hornworm, *Manduca sexta*, was fed to a tiny dose of canavanine (88 ppm), which significantly reduced the overall weight of the resulting moth (Rosenthal & Dahlman, 1975). It decreased *M. sexta's* fertility as well as fecundity (Palumbo & Dahlman, 1978). In comparison to control plants, *M. sexta* larvae grew more slowly on tobacco plants treated with canavanine. Due to their reduced growth rate, the insects were more susceptible to environmental stressors and their natural adversaries. The amino acid β-cyano-L-alanine (BCNA) present in *Vicia sativa* was toxic to locusts at physiological concentrations in the diet. After a day of receiving BCNA, the hemolymph volume of mature locusts dramatically dropped, and within five days, they perished. The locust died because of a decreased ability to reabsorb water from the rectal lumen into the hemolymph (Schlesinger et al., 1976).

17.2.2.2 Lectins

Phytohaemagglutinins (HA), or lectins, are non-immunoglobulin proteins that have the ability to bind to the carbohydrate moieties of complex carbohydrates in a selective and reversible manner without altering the covalent structure of any known glycosyl residues (Prachi Vibhute et al., 2021). Boyd and Shapleigh (1954) gave the name 'lectin' its initial use. Lectins are found in a wide range of plant species, the majority of which are members of the Leguminosae family. Legumeaceous lectins are abundant in seeds. Roots, stems, and leaves of young red kidney bean and soybean seedlings have low lectin contents (Lis & Sharon, 1981). Lectins are also present in Graminaceous and Solanaceous plants. Cereal grains include lectins, which are mostly found in the embryo and are absent in the endosperm. Most legumes have cotyledons that are abundant in lectins. Plant vegetative parts also contain lectins, albeit to a lesser degree. On the contrary, lectins in vegetative portions are distinct from those in seeds.

17.2.2.3 Cyanogenic Glycosides

Despite being produced in the cytoplasm, cyanogenic glycosides (CGs) are intermediately polar nitrogenous secondary metabolites that are commonly stored in plant vacuoles central (Pagare et al., 2015). A variety of scholars have produced a number of publications regarding cynogenic substances. According to Malagon and Garrido (1990), CGS can function as phagostimulants or as a feeding deterrent. When exposed to CG dhurrin, which is present in sorghum, *Atherigona soccata* and *Chilo partellus* get activated to oviposit (Alborn et al., 1992). Bellotti and Arias (1993) reported that CG increased the cassava tubers' resistance to the bug *Cyrtomenus bergi*. The production of CGS involves three specific enzymes: two multifunctional cytochrome P450s and the third enzyme, a-VDPG-glycosyltransferase. By using gene technology, it is feasible to introduce the CG synthesis pathway into acynogenic plants (Casas et al., 1993). By preventing the synthesis of CG biosynthesis, antisense or co-suppression techniques can also be used to lower the level of CGS in plants (Estruch et al., 1997).

17.2.2.4 Glucosinolates

These are a uniform class of secondary plant products made of amino acids with sulfate and thioglucose moiety that are hydrophilic, non-volatile, and soluble in water (Dhruv et al., 2022). Glucosinolates seem to have a role in the chemical defenses that work well against most non-adapted phytophagous insects (Gols, 2014). Certain insects that have adapted to feed on crucifers can use glucosinolates in their host-seeking or host-recognition behaviors. Moreover, some glucosinolate concentrations encourage larval feeding in species that have evolved. Generally, when glucosinolate content rises, herbivorous insect damage tends to decrease. Plants have yielded over 100 different glucosinolates, all of which share the same general structure (Mayer, 2004). Since these substances are non-volatile, insects can only identify them through direct contact. These substances function as defense mechanisms for plants against a variety of generalist insects, such as aphids and grasshoppers. In addition, they interfere with larval chemoreceptors, impair egg hatchability and oviposition, and have a major effect on beetles in both their larval and adult stages. They also operate as a fumigant, nematicidal, and parasite attractant (Chun and Schoonhoven, 1973).

17.2.2.5 Alkaloids

Alkaloids are a complex and diverse collection of secondary plant metabolites, of which over 12,000 have been identified (Dey et al., 2020). The term 'alkaloid' designates a very simple substance that can have a variety of forms, making its description difficult to pin down. Alkaloids, which are constantly expanding, are mostly simple molecules containing one or more nitrogen atoms that usually belong to a cyclic system (Gurib-Fakim, 2006). Certain alkaloids disrupt the acetylcholine transmission pathway in insects' nervous systems, making them poisonous or repulsive to them (Shivanandappa & Rajashekar, 2014). Insect feeding is inhibited by alkaloids such as methylxanthines, pyrrolizidines, quinolizidines, indole alkaloids, benzylisoquinolines, and steroid alkaloids.

17.2.2.6 Phenolics

Aromatic rings with a hydroxyl group are known as phenols, and they include functional derivatives such methylethers, glycosides, and esters (Ho, 1992; Zeb, 2021). Plants all over the kingdom contain these distinct-purpose chemicals and polymers. The lignin found in the cell walls of ferns, gymnosperms, and angiosperms is one of the several types of polyphenols found in vascular plants (Espiñeira et al., 2011). Prominent phenolics include tannins, isoflavonoids, xanthones, biflavonoids, anthocyanins, and flavonoids. For the adult Japanese beetle, *Popilla japonica*, quercetin and rutin functioned as stimulants, while phorridzin, phloretin, naringenin, and catechin as feeding deterrents (Fulcher et al., 1998). Pest insect larvae, such as those of the tobacco budworm, cotton bollworm, pink bollworm, and greenbug, were negatively impacted by rutin and other flavonoids.

17.2.2.7 Terpenoids/Isoprenoids

A structurally diverse category of chemicals known as plant terpenoids includes saponins, triterpinoids, resquiterpenes, monoterpenes, and several other steroids. These substances have a variety of allomonic effects that have been demonstrated. Changes in insect behavior, metabolism, sensory physiology, development rates, metamorphosis, fertility, and longevity are among the effects. The basic metabolic, biochemical, physiological, and behavioral processes of insects may be disrupted by monoterpenoids (Finetti et al., 2021). Acute toxicity is a property of some terpenoids; others serve as attractants, repellents, or antifeedants; yet others have an impact on insect growth, development, and reproduction. The main line of defense against insect invasion is provided by plants since they contain repellents. Eucalyptus oils are used to repel insects, and common commercial insect repellents are citronella and camphor oil. Terpenoids have the ability to repel insects in addition to drawing vast numbers of them. Terpenoids are present in insecticides that have repelling properties that can be used to mask or entice, and certain terpenoids even work as attractants for beneficial insects (Ali & Agrwal, 2023). Monoterpenoids are essential to a plant's defensive system because of their long-term or sublethal effects. Plants have a variety of feeding deterrents that prevent insects from absorbing and using food, which slows their development and growth (Boncan et al., 2020).

17.2.2.8 Epicuticular Lipids

According to Eigenbrode and Espelie (1991), the protective cuticle that envelops plants' aerial sections is composed of a lipid polymer and a combination of lipid molecules referred to as epicuticular lipids or surface waxes. Because of these compounds' unique physical and chemical characteristics, plants are better able to withstand a variety of pressures, such as biological stress from insect pests, and they also have an impact on the uptake and effectiveness of pesticides (Eigenbrode and Espelie, 1995). Insect behavior is impacted by the cuticular chemicals, which disrupt eating, oviposition, and population growth. *Plutella xylostella* and *Delia radicum* increased their oviposition on *Brassica oleracea* leaves as a result of artificially disrupting or removing the epicuticular lipids using mechanical polishing or detergents (Eigenbrode and Espelie, 1995).

17.2.2.9 Tannins

Tannins are phenolic polymers that have a long history of being connected to the leather-making process of animal hides (Falcão and Araújo, 2018). These polyphenols have free phenolic groups that have the ability to attach to proteins. Tannins have a variety of properties, including the ability to attach to other materials, coagulate proteins, and block enzymes (Das et al., 2020). When it comes to insects, the usual results are lower food intake, less weight gain, and lower food usage efficiency. When it comes to cotton bolls, tannin content helps to protect them from pink bollworm (*Pectinophora gossypiiella*) infestation (Legé, 1998).

17.2.2.10 Proline

Cotton jassid and pink bollworm incidence were found to be lower in cotton types with high proline concentration in their leaves (Waghmare, 2022). For most insects, sugars are vital components of their diet and their primary source of sustenance. In general, insects need sugars in their diet, and a higher sugar content promotes better growth (Levin et al., 2017). Insect phagostimulants include sucrose, fructose, and glucose, among other sugars. Total free sugars and the jassid population exhibited a negative connection (Kaur, 2008). Similarly, non-reducing sugars and leafhopper egg production in cotton were found to be negatively correlated (Halder et al., 2017). According to Kumar et al. (2014), there was a positive association between cotton sensitivity to whiteflies and the amount of sugar in the leaves.

17.2.3 Phytohormones and Herbivore-Induced Plant Volatiles

Phytohormones and herbivore-induced plant volatiles (HIPVs) are integral components of plant communication networks, shaping responses to environmental cues and herbivore threats (Peleg and Blumwald, 2011). Phytohormones like jasmonic acid (JA), salicylic acid (SA), and ethylene orchestrate diverse defense mechanisms in plants, from secondary metabolite synthesis to physical barrier induction (Wasternack and Hause, 2013). Concurrently, HIPVs emitted in response to herbivore attacks serve as airborne signals with ecological roles, acting as repellents, attractants for natural enemies, and warnings for neighboring plants (Karban and Shiojiri, 2009). Recent research explored the defensive responses induced by seven specific HIPVs in tomatoes, unveiling their potential repellent effects against pests and attraction of natural enemies (Pérez-Hedo et al., 2021).

17.2.3.1 Biosynthesis of HIPVs

HIPVs are organic volatile molecules that plants release in reaction to injury from herbivorous insects. Various plant metabolites are used as precursors in complex enzymatic pathways that biosynthesize HIPVs (Dudareva et al., 2006). Several biosynthetic pathways are associated with HIPVs. These include: the terpenoid pathway, phenylpropanoid and benzenoid pathway, amino acid pathway, and lipoxygenase (lox) pathway.

Terpenoid pathways are derived from isoprene units. Terpenoids play a vital role in plant defense systems. They belong to several classes, including hemiterpenes, monoterpenes, sesquiterpenes, homoterpenes, and some diterpenes. Sesquiterpenes are generated in the cytosol by the mevalonic acid pathway, whereas monoterpenes and diterpenes are formed in plastids via the methylerythritol phosphate pathway. JA and its precursors are essential for inducing the synthesis of terpenoid compounds and activating certain synthesis-related genes (Ament et al., 2004). Phenylpropanoid and benzenoid pathway is produced from L-phenylalanine and is catalyzed by enzymatic reactions, such as trans-cinnamic acid production via the deamination of L-phenylalanine ammonia lyase (PAL). The amino acid pathway breaks down amino acids, such as isoleucine, into keto acids and aldehydes. The following amino acid degradation routes yield a variety of volatile chemicals as precursors: methionine, valine, leucine, isoleucine, and alanine. These volatiles, which are produced by the breakdown of amino acids, are vital to indirect plant defense because they attract herbivores' natural enemies (Dudareva et al., 2006). Lipoxygenase (LOX) pathway involves the oxygenation of polyunsaturated fatty acids, namely, linolenic and linoleic acid, by LOX enzymes. This results in the production of hydroperoxides, which are then transformed into volatile alcohols and aldehydes, such as green-leaf volatiles (Allmann & Baldwin, 2010).

17.2.3.2 Genetic Manipulation of HIPVs

An important development in plant defense against insect pests is the genetic engineering of plants to increase their volatile emission. In an attempt to manipulate insect behavior, plants have been genetically modified to emit target volatiles more readily. For example, plants that have undergone metabolic engineering on tobacco have shown higher production of terpenoids, especially the diterpene cembratriene-ol, which has changed the responses of insects (Lücker et al., 2004). In *Arabidopsis*, overexpression of the terpene synthase strawberry nerolidol synthase produced the sesquiterpene alcohol (3S)-(E)-nerolidol, which attracted *P. persimilis*, a predatory mite (Kappers et al., 2005). Similar to this, (E)-β-farnesene, (E)-α-bergamotene, and other herbivore-induced sesquiterpene hydrocarbons have been created to be released by transgenic maize plants that overexpress the corn *TPS10* gene. Another noteworthy development is the production of linalool by transgenic lines of *Arabidopsis*, which effectively repels the aphid *Myzus persicae* (Aharoni et al., 2003).

17.2.3.3 Phytohormones and HIPVs

Phytohormones and HIPVs are two important defense mechanisms that plants have evolved as stationary creatures against herbivore attacks. Plant activities are regulated by phytohormones, which are messenger molecules, and HIPVs, which are aerial signals that enable complex interactions among plants. Phytohormones, especially JA, work in concert with HIPVs to create a sophisticated communication system that allows plants to react to herbivore threats dynamically. A crucial phytohormone called JA increases in response to a herbivore attack, starting signaling cascades that trigger defensive mechanisms like the production of secondary metabolites and the activation of genes linked to defense. In the face of biotrophic infections, SA interacts with JA to enhance defense (Horiuchi et al., 2001). Lima bean plants provide an excellent example of how phytohormones are involved in plant defense and HIPV emission. When JA is applied, a volatile mixture that resembles what plants with spider mite infestations emit is released. Methyl salicylate highlights the delicate balance between phytohormones and HIPVs by making the spider mite-induced mix more appealing to *Phyoseiulus persimilis*, the predatory mite (Dicke et al., 1999).

17.2.4 Plant-Defensive Proteins

Plants produce enzymes and secondary metabolites as part of their defensive mechanism in reaction to stressors such as chemical attacks, heavy metals, pollution, and unfavorable environmental conditions (Aguirre-Becerra et al., 2021). Early identification of crop and forest damage is made easier by the molecular indicators for plant diseases that are provided by the makeup and abundance of plant defense proteins. Plant defense proteins comprise a wide variety of compounds, such as cyclotides, pore-forming toxins, and arcelins (Kocyigit et al., 2023). These proteins have antiviral, antifungal, antibacterial, and anti-insect characteristics. Plants use physical defenses like spines and thorns as well as chemical defenses like production and secretion to protect themselves against a variety of stresses (Hanley et al., 2007). When surface receptors attach to signaling molecules (elicitors) produced by pathogens or plants, signaling pathways are started, triggering defense responses in response to the pathogen (Abdul Malik et al., 2020). The upkeep of plant health, agricultural output, and ecosystem sustainability depend on these defense mechanisms

17.2.4.1 Plant Thionins

Thionins are distinguished by their distinct structure that is rich in cysteines, which form disulfide bonds to form a compact and stable three-dimensional shape (Stec, 2006). One important feature of thionins is that they are essential for a plant's defense against many diseases. The thionins that are generated function as a first line of defense by preventing the growth and colonization of incoming microbes. This quick and focused reaction highlights how important thionins are to the plant's intrinsic defense mechanism. Alpha and beta forms of thionins are very poisonous to animal cells, causing leakage of cells and the release of ions, proteins, and nucleotides (Kocyigit et al., 2023). These thionins exhibit toxicity against a variety of taxa, such as insect larvae, mammalian cells, fungi, and bacteria (Stec, 2006).

17.2.4.2 Ribosome-Inactivating Proteins

By depurinating ribosomal RNA, ribosome-inactivating proteins (RIPs) interfere with the mechanism of cellular protein synthesis and prevent the production of vital proteins in target cells, including those of invasive pathogens. Plants have a powerful defense mechanism in place when pathogen multiplication is hindered by this interference with protein synthesis. A well-researched example of a plant-derived RIP that exhibits activity against ribosomes of animals, plants, yeast, and bacteria is ricin derived from *Ricinus communis* seeds (Puri et al., 2012). Furthermore, adenine can be removed from nucleic acid substrates by some RIPs, while others perform functions similar to those of chitinase, lipase, and deoxyribonucleases (Peumans et al., 2001).

17.2.4.3 Cyclotides

Cyclotides' unique circular shape and resilience make them essential to plant defense. Their ability to rupture pathogen cell membranes confers antimicrobial capabilities (Nawrot et al., 2014). Furthermore, they can obstruct herbivores and pest cellular processes, rendering them adaptable protectors against a range of plant dangers. These peptides, which have three disulfide linkages and 28–37 amino acid residues together to form a cyclic cystine knot motif, show remarkable structural stability in the face of environmental stresses such as biological agents, chemicals, and high temperatures (Colgrave and Craik, 2004). Cyclotides are found in many plant groups, including the Poaceae, Fabaceae, Violaceae, Rubiaceae, and Cucurbitaceae (Tammineni et al., 2020) They have been linked to several documented actions, such as insecticidal, antiviral, and anticancer effects.

17.2.4.5 Pore-Forming Toxins

Toxins known as pore-forming toxins (PFTs) cause cell lysis by piercing the target cells' membranes (Dalla Serra and Martínez, 2011). By undermining the integrity of microbial membranes, these toxins help plants defend themselves against diseases by allowing cellular contents to seep out and eventually kill the invasive pathogen (Etxaniz et al., 2018). They play a key role in forming pores in cell membranes that are filled with water. With a molecular weight of 54.8 kDa, enterolobin, which is obtained from *Enterolobium contortisiliquum* seeds, is a well-researched example of a plant PFT. It has been shown that this cytolytic protein is harmful to *Callosobruchus maculatus* beetle larvae (Sousa et al., 1993).

17.2.4.6 Arcelins

Some leguminous plants include seeds that contain proteins called arcelins, which have antibacterial qualities (Jain et al., 2022). These proteins are essential to plant defense because they impede the establishment of pests and diseases, especially in the concentrated seed environment. The antibacterial properties of seeds help protect them against pathogens during the germination and early growth stages. Eight different arcelin protein variants have been found; their molecular weights range roughly from 27 to 42 kDa. Research has shown that in the guts of insects, arcelins interact with glycoproteins and components of cell membranes (Ferreira et al., 2007). Certain arcelin variations show resistance to proteolysis, which causes insect larvae to starve to death.

17.2.4.7 Canatoxin and Urease Enzymes

Enzymes called urease and canatoxin are essential to plants' defensive mechanisms against herbivores. The digestion processes of herbivores are disrupted by aflatoxin, which prevents protein synthesis, while urease enzymes speed up the breakdown of urea into ammonia, which is harmful to herbivores and affects how they consume and absorb nutrients (Bertram et al., 2013). These protective elements include insecticidal and antifungal qualities. They include canatoxins and different ureases, which are present in soybean and pigeon pea seeds. Pepcanatox, a 10-kDa peptide released by canatoxin during insect digestion, causes membranes to become porous, increasing permeability (Dang and Van Damme, 2015). The insecticidal and antifungal properties of ureases and canatoxins are attributed to this pore-forming activity.

17.2.4.8 Plant α-Amylase Inhibitors and Protease Inhibitors

These inhibitors interfere with the digestive systems of herbivores. Protease inhibitors target enzymes involved in protein breakdown, while α-amylase inhibitors block the function of this enzyme, which is essential for the digestion of carbohydrates (Franco et al., 2002). These inhibitors serve as a defense mechanism against herbivorous attacks by hindering the ability of herbivores to digest food. Six categories comprise α-amylase inhibitors: Kunitz-like from cereals like wheat, barley, and rice, knottin-like from *Amaranthus hypochondriacus* seeds, and cereal-type from cereals inhibiting α-amylases in birds, mammals, and insects *c-purothionin*-like from *Sorghum bicolour*, lectin-like from *Proteus vulgaris* inhibiting α-amylases in *Callosbruchus maculatus* and *Callosobruchus chinensis*, and thaumatin-like from *Zea mays* inhibiting digesting amylase enzymes in insects such as *Sitophilus zeamais* and *Rizopherta dominica* (Jain et al., 2022).

17.2.5 Lectins

Lectins play a pivotal role in plant defense by identifying and binding to carbohydrates present on the surface of microbial cells (Naithani et al., 2021). Pathogens may be recognized and grouped thanks to this interaction, which prevents them from attaching themselves to plant tissues and infiltrating them. Lectins play a vital role in strengthening the plant's defenses against infections by obstructing the adhesion of pathogens to host cells (Naithani et al., 2021). Their particular type and three-dimensional architectures influence the large range of biological functions they display. Some plant lectins are poisonous, but others have antiviral, antibacterial, antifungal, and anticancer properties.

17.2.5.1 Plant Lectins

In the intricate dynamics of plant defense mechanisms, one class of proteins stands out for its versatile role in confronting the persistent challenge posed by various phytophagous insects – plant lectins (Michiels et al., 2010). Lectins are a diverse group of proteins known for their unique biological function: the ability to recognize and adhere to specific carbohydrate combinations. They are an integral part of the plant defense system. Reversible attachment is a defining characteristic of this recognition system, which is important in the intricate relationship between plants and insects (Oguri, 2022). Plant lectins form a heterogeneous family, encompassing proteins with distinct structural and functional attributes. Their significance in plant biology extends beyond defense, as lectins participate in a spectrum of physiological processes, ranging from seed development to response mechanisms against pathogens. Through extensive genomic and transcriptomic analyses, the prevalence of lectin sequences has been revealed across the plant kingdom, emphasizing their ubiquitous nature. A wide variety of plant species have produced lectins that have been discovered and purified (Oguri, 2022). They are frequently present in a wide range of foods, including vegetables, cereals, and legumes (Liu et al., 2010).

17.2.5.2 Plant Lectins as Defense Proteins

There has been a longstanding hypothesis suggesting that plant lectins function defensively against pathogens and herbivores. Substantial evidence now corroborates this notion, indicating that these proteins, which bind to carbohydrates, contribute to the innate immune system of plants. They operate through both direct toxicity and signaling pathways to defend against invading organisms. Many plant lectins exhibit binding specificity toward the glycoproteins and glycolipids found on the cell membranes of fungi, bacteria, viruses, nematodes, and insects (Peumans and Van Damme, 2001). By recognizing foreign carbohydrate patterns, lectins can mediate pathogen recognition and subsequent signaling to launch other plant defenses (Lannoo and Van Damme, 2014). Additionally, some lectins directly disrupt the metabolism and physiology of attackers. For example, chitin-binding lectins interfere with fungal growth and insect digestion through interactions with chitin on cell walls and the peritrophic matrix (PM) lining insect midguts (Mishra et al., 2019). High toxicity across taxa suggests that lectins have broadly adapted as part of non-host defenses in plants (Lannoo and Van Damme, 2014).

17.2.5.3 Mechanisms of Plant Lectin Toxicity against Insects

Many insects have a Peritrophic membrane (PM), a physical barrier secreted by certain epithelial cells that protect the midgut epithelium (Hegedűs et al., 2009). The PM comprises a grid-like network with glycan structures filling interstitial spaces, acting as a molecular sieve. It is made up of a chitinous lattice held together by chitin-binding glycoproteins such as peritrophins (Hegedűs et al., 2009). The PM becomes a prime target for lectins because of the abundance of glycoproteins and chitin fibrils in it. Significant deviations in PM development and disturbances in microvilli structures have been found in a number of studies investigating the impact of plant lectins on the ultrastructural organization of the insect midgut. To demonstrate the effect of lectins, the analysis of cotton leafworm larvae fed on a meal containing Gleheda showed significant abnormalities in the dissected PM. According to Harper et al. (2001), studies conducted on the midgut structures of the European corn borer (*O. nubilalis*) following the ingestion of a meal containing wheat germ aglutinin (WGA) revealed hypersecretion of disordered PM layers into the midgut lumen and the presence of disintegrating microvilli. Furthermore, once WGA was consumed, the midgut microvilli of *Drosophila* larvae showed notable morphological alterations (Li et al., 2009).

17.2.6 Plant-Defensive Enzymes

Plants serve as the primary source within ecosystems, exerting both direct and indirect influences on human life. However, their susceptibility to pathogens, pests, parasites, and interference by insects contributes substantially to losses in yield and productivity, posing a significant obstacle to crop distribution.

In response to these challenges, plants activate intricate immune response pathways, functioning as resilient organisms capable of withstanding various stresses. In certain instances, plants release enzymes induced either naturally or through exogenous triggers, collectively known as plant defense enzymes (Kant et al., 2015).

Defense enzymes like β-1,3-glucanase, PAL, peroxidase, chitinase, chitosanase, and polyphenol oxidase are closely linked to plant resistance. As part of plant resistance against fungal diseases, chitinases (EC 3.2.1.14) and β-1,3-glucanase (EC 3.2.1.39) target chitin and β-1,3 glucan, the major polysaccharides in fungal cell walls. Plants include a large variety of chitinases and chitosanases (EC 3.2.1.132), which are categorized as Pathenogenesis related protein (PR) proteins (El Ouakfaoui & Asselin, 1992; Neuhaus, 1999). Chitosanases hydrolyze chitosan, whereas chitinases degrade chitin.

17.3 BREEDING STRATEGIES FOR INSECT PEST RESISTANCE IN FRUIT CROPS

17.3.1 Conventional Breeding Techniques

Developing fruit crops for resistance to insect pests forms an essential element of managing pests (Mitchell et al., 2016). A fruit crop with inherit potential to tolerant attack is termed to be resistant. These stresses imposed by fruit crops by insect pests are challenging especially in monoculture farming and particularly under continuous farming on a piece of land. The ability to enhance fruit crop resistance to insect pest tolerance is centered on the fundamentals of crop improvement, heredity, and entomology. Hence, this led to the development of several strategies for breeding for insect resistance among fruit crops. This section will only focus on the following strategies involving phenotypic selection and hybridization forming the major component of conventional breeding techniques.

Selection of genetic materials from the germplasm pool offers the chance to identify a fruit genotype conferring resistance (R) to insect pests. Usually, such R genes detected in the germplasm emerge via the evolutionary process of plants and insects that are involved in reciprocal selection. The exchange of R genes identified could be introduced into the cultivar quickly (within a year) or slowly (over 10 years). Evaluation of germplasm could help identify genotype materials that confer resistance from fruit crops to insect pest resistance. For instance, phenotypic screening of mango and tomato germplasm led to the identification of genotypes resistance to fruit fly and leaf miners, respectively (Mahmoud et al., 2022; Sison et al., 2020) (Table 17.1). In search for resistance in apple plants against woolly apple aphids via phenotypic screening assisted in identifying cultivars such as Golden Delicious, Delbarestivale, and Red Miracle that are immune and resistant genotypes (Stark Gold, Raritan, Early Gold, and Argi Gala) (Saeid & Ateyyat, 2014; Sandanayaka et al., 2005)

Breeding for insect pest resistance is the top priority of fruit breeders (Campos et al., 2017). A number of researchers have utilized hybridization coupled with marker-assisted selection (MAS) in an attempt to develop bananas resistant to weevils (Kiggundu et al., 2003; Uwimana et al., 2021). Integration of MAS into conventional breeding strategies could limit the breeding cycle (Seifi et al., 2013); however, introgression of key traits from unrelated species remains a challenge. For instance, integrating hybridization by crossing triploid and tetraploid bananas with the application of markers resulted in identifying quantitative trait loci (QTLs) for weevil resistance (*Cosmopolites sordidus* Germar) (Uwimana et al., 2021).

17.4 NOVEL APPROACHES THROUGH GENOMICS

17.4.1 Genome Editing Strategies for Engineering Insect Pest Resistance

A vital technique in modern biology is genome editing, which enables scientists to change an organism's DNA. Genome editing techniques have considerably improved over the past 20 years,

TABLE 17.1
Application of Conventional Breeding Strategies in Identifying Fruit Crops Resistance to Insect Pest

Pest	Common Name	Fruit	Strategy	No. of Cultivars/ Hybrids	Conclusion	References
Bactrocera dorsalis (Hendel)	Mango fruit fly	Mango	Phenotypic screening	33	12-209, 12-127, 12-103, and 16-010 exhibited promising resistance ratings	Sison et al. (2020)
Tuta absoluta (Meyrick)	Tomato leaf miner	Tomato	Phenotypic screening	36 wild and cultivated	Hab-7 is the most resistant	Mahmoud et al. (2022)
Bemisia tabaci (Gennadius)	Tomato whitefly	Tomato	Hybridization		Cultivars responded differentially	Silva et al. (2014)
Cosmopolites sordidus (Germar)	Banana root borer/weevil	Banana	Hybridization	45 clones	Ehanced resistance	Kiggundu et al. (2003)
C. sordidus (Germar)	Banana weevi	Banana	Hybridization and markers	138	Identified QTLs	Uwimana et al. (2021)
Aphis gossypii (Glover)	Melon aphid	melon	Phenotypic screening	23	Identified three genotypes that could help manage the pest	KIRIŞIK et al. (2020)
Eriosoma lanigerum (Hausmann)	Woolly apple aphid	Apple	Phenotypic screening	22	Cultivars responded differentially and detected immune types	Saeid et al. (2014)
E. lanigerum (Hausmann)	Woolly apple aphid	Apple	Phenotypic screening	16	Identified resistant and suscetipble cultivars	Sandanayaka et al. (2005)

employing nucleases to cut particular target regions of the genome. Promising methods for creating insect pest resistance in agriculture include the CRISPR/Cas9 system, RNA interference, and gene drive technology (Khalil, 2020). The use of genome editing tools like ZFNs, TALENs, or CRISPR/Cas9 can help create de novo resistance in crops with insufficient resistant genes, offering a potential solution for effective insect pest management (Komal et al., 2023). Genome editing has shown high specificity, efficiency, and cost-effectiveness, making it a promising tool for combating insect pests in agriculture (Wani et al., 2022). These methods have the power to transform pest control and advance sustainable agricultural methods.

An efficient two-phase strategy for insect genome editing entails altering the target insects and reintroducing them into their natural environment (Babu et al., 2003). Since insects have specialized detoxification enzymes that enable them to get past chemical defense responses in a variety of plant species, targeting detoxification genes in polyphagous pests may be a practical tactic (Xia et al., 2021). Nevertheless, a species may entirely disappear due to this 'super-Mendelian inheritance' in as few as 15–20 generations, upsetting the delicate ecological balance.

To apply genome editing techniques, it is necessary to pinpoint the precise insect pest that has to be eradicated, comprehend the resistance characteristics that wild relatives of cultivated crops carry, identify the genes causing these characteristics, and create the editing plan (Nair, 2019). The availability of target genes, efficiency, specificity, and convenience of use are among the criteria that influence the choice of genome editing tool. To make the appropriate alterations, the genome editing tool must enter the plant cells, which can be accomplished in a number of ways (Sedeek et al., 2019). The modified plants must be checked for successful gene editing once the editing tool is injected into the plant cells (Ding et al., 2016). Following verification of the editing process' success, field tests or bioassays can be used to assess the plant's resistance to the intended insect pest (Zhang et al., 2011) (Table 17.2).

TABLE 17.2
Successful Application of Genome Editing Strategies for Enhanced Pest Resistance in Fruit Crops

Plant	Method	Target Gene	Trait	References
Citrus (*Citrus sinensis*)	CRISPR/Cas9	*CsLOB1*	Disease susceptibility gene for citrus bacterial canker, enhanced resistance to citrus canker disease	Moon et al. (2022); Jia et al. (2022)
Apple (*Malus domestica*)	(CRISPR/Cas9-FLP/FRT gene editing system)	*MdDIP1*/ *MdDIPM4*	Resistance to apple proliferation disease, enhanced pest resistance	Pompili et al. (2020)
Grapevine, (*Vitis vinifera*)	CRISPR/Cas9	*MLO-7*, *VvWRKY52*	Enhanced resistance to powdery mildew	Karavolias et al. (2021)
Kiwi (*Actinidia chinensis*)	CRISPR/Cas9	*AcWRKY40*	Enhanced resistance to *Botrytis cinerea*	Long et al. (2023)
Grape (*Vitis*)	Polyethylene glycol-mediated delivery	*MLO7*	Enhanced biotic resistance against powdery mildew	Martín-Valmaseda et al. (2023)
Orange (*Citrus sinensis*)	Polyethylene glycol-mediated delivery	*CsNPR3*	Induced biotic stress tolerance	Martín-Valmaseda et al. (2023)
Apple (*Malus pumila*)	Polyethylene glycol-mediated delivery	*DIPM1*, *DIPM2*, *DIPM4*	Improved resistance against *Erwinia amylovora*	Wang et al. (2019)

17.4.2 RNAi Approach for Insect Resistance

A useful technique for controlling insects in a species-specific way is RNA interference (RNAi), a highly conserved sequence-specific technique that inhibits the production of a targeted gene (Kim & Zhang, 2022). In this biological process, tiny RNA molecules either direct inhibitory chromatin changes or reduce the stability or translation capacity of the targeted mRNA to silence gene expression at the transcriptional or post-transcriptional level (Dykes & Emanueli, 2017). Double-stranded RNA (dsRNA), which can originate from an external source like a virus or transgene, or an endogenous genomic site, is the catalyst for gene silence (Sadeq et al., 2021). RNAi is a more effective, adaptable, precise, stable, and user-friendly method of controlling pests. The infestation of insects and pests has a significant impact on agricultural production worldwide and is expected to worsen in the future years (War et al., 2016). The use of pesticides may become ineffective as a result of environmental toxicity and insect resistance. Targeted insect genes have been silenced by genetic engineering techniques with the introduction of host-delivered RNA interference, commonly referred to as host-induced gene silencing (Ghag, 2017; Saakre et al., 2023). This has provided a unique insect pest resistance strategy for battling a variety of agricultural pests. RNAi has the potential to be a very valuable and successful plant protection technique for achieving long-lasting insect pest resistance in crops, provided that it is well-designed and delivered (Joga et al., 2021). In laboratory settings, RNAi-based crop protection techniques have been routinely employed to control insects, despite an insufficient understanding of RNAi mechanisms (Liu et al., 2020). Out of all the methods that have been tried, the direct application of dsRNA as an insecticide and the expression of dsRNA in transgenic crop plants both seem promising for usage in the field (Nitnavare et al., 2021). However, further research is needed in a number of useful areas of dsRNA application.

17.4.3 QTL Mapping and Candidate Genes for Insect Resistance Gene against Pests of Fruit Crops

The process of QTL mapping involves locating genomic areas called QTLs that are linked to particular traits or phenotypes (Khan, 2015). A main QTL (Wf-1) that controls resistance against two distinct whitefly species was found in a tomato study on host plant resistance (Vosman et al., 2019). According to the study, Wf-1 controls glandular trichome type IV production, which permits the buildup of bioactive metabolites (Vosman et al., 2019). In tomato plants, a number of QTLs have been linked to aphid resistance; these include two in cowpea, five in barley, four additive and two pairs of epistatic QTLs in melon, and one significant candidate QTL on chromosome 7 that regulates foxglove aphid resistance in soybean (Liang et al., 2016). Two QTLs for resistance to the green peach aphid were found, and a putative QTL in apples for resistance to the pink apple aphid and another for resistance to the green apple aphid have been localized (Han et al., 2021). Using a large-scale methodology, researchers examined data from metabolomics analysis, trichome phenotyping, and whitefly resistance testing (Vosman et al., 2019). They most likely used a population of recombinant inbred lines (RILs) that resulted from a hybrid between the wild-related *Solanum galapagense* and the cultivated tomato species *Solanum lycopersicum* (Vosman et al., 2019). Tests for whitefly resistance were used to evaluate the degree of resistance in the RIL population, and trichome density, size, and other properties were probably phenotyped (Vosman et al., 2019).

17.4.4 MAS for Identification of Insect Resistance Gene against Pests of Fruit Crops

A technique known as MAS uses DNA molecular marker patterns, either in addition to or instead of trait values, to help breeding systems identify suitable individuals. It combines conventional genetics with molecular biology to enable the identification of genes governing desired features (Hasan et al., 2021). When choosing organisms for desirable qualities like color, meat quality, or disease resistance, MAS has shown to be an invaluable tool (Farokhzadeh & Fakheri, 2014). A valuable tool in crop breeding, molecular markers are categorized according to their relationship with QTLs or genetic maps. This allows for the selection of desired qualities, including stress tolerance, without requiring phenotypic assessment (Hasan et al., 2021). Abiotic stress tolerance, disease resistance, and salt tolerance are examples of stress tolerance traits that can be successfully bred for using MAS thanks to recent developments in marker technology and the creation of nested association mapping populations (Wani et al., 2018). In addition, MAS can be used to choose parents for breeding programs by identifying people with desired genetic features and evaluating genetic diversity. Additionally, it can be applied to marker-assisted pyramiding, which is the process of breeding several genes or traits into a single plant, and marker-assisted introgression, which is the process of moving particular genes or traits from one plant to another (Mekonnen et al., 2017). MAS is a vital technique in agricultural biotechnology, having drastically reduced the time required to produce new crop types from nearly 25 years to 7–10 years (Mekonnen et al., 2017). It entails locating and utilizing molecular or genetic markers, which are DNA sequences containing nucleic acids that are close to the desired gene's DNA sequence (Nadeem et al., 2018). Instead of choosing based on the feature itself, these markers are then utilized to indirectly select the desired attribute, such as disease resistance, productivity, or quality (Gupta et al., 2010) (Table 17.3).

TABLE 17.3
List of Fruit Resistance Genes against Major Economic Insects

Fruit	Resistance Gene	Target Insect	Reference
Cucumber (*Cucumis sativus*)	*NBS-LRR, lectin (GNA, ASAL)*	Aphids	Liang et al. (2015)
Apple (*Malus domestica*)	*Er1, Er3, and Er4* (*Eriosoma lanigerum* resistance gene 1, 3, and 4)	Woolly apple aphid (WAA; *Eriosoma lanigerum hausm*)	Bus et al. (2008)
Apple (*M. domestica*), Kiwi (*Actinidia deliciosa*)	*Bacillus thuringiensis*, *Bt* genes	Leaf roller	Wearing et al. (1994)

17.5 CONCLUSION AND FUTURE PERSPECTIVES

In summary, the creation of insect-resistant plants via cutting-edge genomics techniques offers a viable path toward sustainable agriculture. Through the utilization of biotechnology, scientists are advancing environmentally sustainable farming practices by fortifying plant defenses against insect pests. These genetically engineered crops could boost crop yields, decrease the need for pesticides, and improve global food security. In addition to supporting a more resilient and sustainable food system, ongoing research and innovation in this area show significant promise for solving global agricultural concerns.

REFERENCES

Abbas, M. S. T. (2018). Genetically engineered (modified) crops (Bacillus thuringiensis crops) and the world controversy on their safety. *Egyptian Journal of Biological Pest Control*, *28*(1), 1–12

Abdul Malik, N. A., Kumar, I. S., & Nadarajah, K. (2020). Elicitor and receptor molecules: Orchestrators of plant defense and immunity. *International Journal of Molecular Sciences*, *21*(3). https://doi.org/10.3390/ijms21030963

Abebe, W. (2021). Review on plant defense mechanisms against insect pests. *International Journal of Novel Research in Interdisciplinary Studies*, *8*, 15–39.

Acreman, T., & Dixon, A. (1986). The role of awns in the resistance of cereals to the grain aphid, Sitobion avenae. *Annals of Applied Biology*, *109*(2), 375–381.

Agarwal, R. (1969). Morphological characteristics of sugarcane and insect resistance. *Entomologia experimentalis et applicata, 12*(5), 767–776.

Aguirre-Becerra, H., Vazquez-Hernandez, M. C., Saenz de la O, D., Alvarado-Mariana, A., Guevara-Gonzalez, R. G., Garcia-Trejo, J. F., & Feregrino-Perez, A. A. (2021). Role of stress and defense in plant secondary metabolites production. In D. Pal & A. K. Nayak (Eds.), *Bioactive natural products for pharmaceutical applications* (pp. 151–195). Springer. https://doi.org/10.1007/978-3-030-54027-2_5

Aharoni, A., Giri, A. P., Deuerlein, S., Griepink, F., de Kogel, W.-J., Verstappen, F. W. A., Verhoeven, H. A., Jongsma, M. A., Schwab, W., & Bouwmeester, H. J. (2003). Terpenoid metabolism in wild-type and transgenic arabidopsis plants. *The Plant Cell, 15*(12), 2866–2884. https://doi.org/10.1105/tpc.016253

Akwu, N. A., Naidoo, Y., Singh, M., Dewir, Y. H., Magyar-Tábori, K., Lekhooa, M., & Aremu, A. O. (2023). Development and biomechanics of Grewia lasiocarpa E. Mey. Ex Harv. trichomes exudate. *Plants*, *12*(11). https://doi.org/10.3390/plants12112198

Alborn, H., Stenhagen, G., & Leuschner, K. (1992). Biochemical selection of sorghum crop varieties resistant to sorghum shoot fly (*Atherigona soccata*) and stem borer (*Chilo partellus*): role of allelochemicals. In: Rizvi, S.J.H., & Rizvi, V. (Eds.), *Allelopathy*. Springer, Dordrecht.

Alemu, M. (2020). Trend of biotechnology applications in pest management: A review. *International Journal of Applied Sciences and Biotechnology*, *8*(2), 108–131.

Ali, J. G., & Agrawal, A. A. (2012). Specialist versus generalist insect herbivores and plant defense. *Trends in plant science*, *17*(5), 293–302.

Ali, M. A., Abdellah, I. M., & Eletmany, M. R. (2023). Towards sustainable management of insect pests: Protecting food security through ecological intensification. *International Journal of Chemical and Biochemical Sciences*, *24*(4), 386–394.

Ali, M. Y., Naseem, T., Holopainen, J. K., Liu, T., Zhang, J., & Zhang, F. (2023). Tritrophic interactions among arthropod natural enemies, herbivores and plants considering volatile blends at different scale levels. *Cells*, *12*(2), 251.

Aljbory, Z., & Chen, M. S. (2018). Indirect plant defense against insect herbivores: A review. *Insect Science*, *25*(1), 2–23.

Allmann, S., & Baldwin, I. (2010). Insects betray themselves in nature to predators by rapid isomerization of green leaf volatiles. *Science (New York, N.Y.)*, *329*, 1075–1078. https://doi.org/10.1126/science.1191634

Ament, K., Kant, M., Sabelis, M., Haring, M., & Schuurink, R. (2004). Jasmonic acid is a key regulator of spider mite-induced volatile terpenoid and methyl salicylate emission in tomato. *Plant Physiology*, *135*, 2025–2037. https://doi.org/10.1104/pp.104.048694

Anulika, N. P., Ignatius, E. O., Raymond, E. S., Osasere, O.-I., & Abiola, A. H. (2016). The chemistry of natural product: Plant secondary metabolites. *International Journal of Technology Enhancements and Emerging Engineering Research*, *4*(8), 1–9.

Babu, R. M., Sajeena, A., Seetharaman, K., & Reddy, M. (2003). Advances in genetically engineered (transgenic) plants in pest management: An over view. *Crop Protection*, *22*(9), 1071–1086.

Baweja, P., Kumar, S., & Kumar, G. (2020). Fertilizers and pesticides: Their impact on soil health and environment. *Soil Health*, 59, 265–285).

Belete, T. (2018). Defense mechanisms of plants to insect pests: From morphological to biochemical approach. *Trend in Technical & Scientific Research*, *2*(2), 30–38.

Bellotti, Anthony C., & Arias V. B. (1993). The possible role of HCN in the biology and feeding behavior of the cassava burrowing bug (Cyrtomenus bergi Froeschner: Cydnidae: Hemiptera). In Roca, William M.; Thro, & Ann Marie (Eds.), *International Scientific Meeting Cassava Biotechnology Network (1, 1992, Cartagena de Indias, Colombia). Proceedings. Centro Internacional de Agricultura Tropical (CIAT), Cali, CO*. pp. 406–409. (Working document no. 123).

Bertram, T. A., Ludlow, J. W., Basu, J., & Muthupalani, S. (2013). Digestive tract. In W. M. Haschek, C. G. Rousseaux, & M. A. Wallig (Eds.), *Haschek and Rousseaux's handbook of toxicologic pathology* (3rd edn, pp. 2277–2359). Academic Press. https://doi.org/10.1016/B978-0-12-415759-0.00056-X

Blanco, C. A., Chiaravalle, W., Dalla-Rizza, M., Farias, J. R., García-Degano, M. F., Gastaminza, G., Mota-Sánchez, D., Murúa, M. G., Omoto, C., Pieralisi, B. K., Rodríguez, J., Rodríguez-Maciel, J. C., Terán-Santofimio, H., Terán-Vargas, A. P., Valencia, S. J., & Willink, E. (2016). Current situation of pests targeted by Bt crops in Latin America. *Current Opinion in Insect Science*, *15*, 131–138. https://doi.org/10.1016/j.cois.2016.04.012.

Boncan, D. A. T., Tsang, S. S. K., Li, C., Lee, I. H. T., Lam, H.-M., Chan, T.-F., & Hui, J. H. L. (2020). Terpenes and terpenoids in plants: Interactions with environment and insects. *International Journal of Molecular Sciences*, *21*(19). https://doi.org/10.3390/ijms21197382

Boyd, W. C., & Shapleigh, E. (1954). Antigenic relations of blood group antigens as suggested by tests with lectins. *The Journal of Immunology*, *73*(4), 226–231.

Bus, V., Chagné, D., Bassett, H., Bowatte, D., Calenge, F., Celton, J.-M., …, Ranatunga, A. (2008). Genome mapping of three major resistance genes to woolly apple aphid (Eriosoma lanigerum Hausm.). *Tree Genetics & Genomes*, *4*, 223–236.

Campos, H., Caligari, P. D., Brown, A., Tumuhimbise, R., Amah, D., Uwimana, B., Nyine, M., Mduma, H., Talengera, D., & Karamura, D. (2017). Bananas and plantains (Musa spp.). In H. Campos & O. Ortiz (Eds.), *Genetic improvement of tropical crops* (pp. 219–240). Springer.

Capinera, J. L. (2005). Relationships between insect pests and weeds: An evolutionary perspective. *Weed Science*, *53*(6), 892–901.

Carmona, D., Lajeunesse, M. J., & Johnson, M. T. J. (2011). Plant traits that predict resistance to herbivores. *Functional Ecology*, *25*(2), 358–367. https://doi.org/10.1111/j.1365-2435.2010.01794.x

Casas, A. M., Kononowicz, A. K., Zehr, U. B., Tomes, D. T., Axtell, J. D., Butler, L. G., Bressan, R. A., & Hasegawa, P. M. (1993). Transgenic sorghum plants via microprojectile bombardment. *Proceedings of the National Academy of Sciences*, *90*(23), 11212–11216.

Chavan, S., & Patel, K. (2016). A review on morphological and biochemical basis of host plant resistance against paddy stem borer. *Advances in Life Sciences*, *5*(20), 8900–8906.

Chun, M. W., & Schoonhoven, L. (1973). Tarsal contact chemosensory hairs of the large white butterfly Pieris brassicae and their possible role in oviposition behaviour. *Entomologia Experimentalis et Applicata*, *16*(3), 343–357.

Colgrave, M. L., & Craik, D. J. (2004). Thermal, chemical, and enzymatic stability of the cyclotide kalata B1: The importance of the cyclic cystine knot. *Biochemistry*, *43*(20), 5965–5975.

Craigie, G. (2022). *Detecting sugarcane aphid (Melanaphis sacchari) infestation in grain sorghum (Sorghum bicolor) using leaf spectral response* (Doctoral dissertation). Kansas State University. https://hdl.handle.net/2097/41811

Dalla Serra, M., & Martínez, M. T. (2011). Pore-forming toxins. *eLS*. https://doi.org/10.1002/9780470015902.a0002655.pub2

Dang, L., & Van Damme, E. J. M. (2015). Toxic proteins in plants. *Phytochemistry*, *117*, 51–64. https://doi.org/10.1016/j.phytochem.2015.05.020

Das, A. K., Islam, M. N., Faruk, M. O., Ashaduzzaman, M., & Dungani, R. (2020). Review on tannins: Extraction processes, applications and possibilities. *South African Journal of Botany*, *135*, 58–70.

Deutsch, C. A., Tewksbury, J. J., Tigchelaar, M., Battisti, D. S., Merrill, S. C., Huey, R. B., & Naylor, R. L. (2018). Increase in crop losses to insect pests in a warming climate. *Science*, *361*(6405), 916–919.

Dey, P., Kundu, A., Kumar, A., Gupta, M., Lee, B. M., Bhakta, T., Dash, S., & Kim, H. S. (2020). Analysis of alkaloids (indole alkaloids, isoquinoline alkaloids, tropane alkaloids). In A. Sanches Silva, S. F. Nabavi, M. Saeedi, & S. M. Nabavi (Eds.), *Recent advances in natural products analysis* (pp. 505–567). Elsevier. https://doi.org/10.1016/B978-0-12-816455-6.00015-9

Dhruv, J., Dobaria, J., & Shukla, Y. (2022). Plant secondary metabolites in stress: An overview. *Indian Journal of Agricultural Biochemistry*, *35*(2), 120–132.

Dicke, M., & Baldwin, I. T. (2010). The evolutionary context for herbivore-induced plant volatiles: Beyond the 'cry for help.' *Induced Biogenic Volatile Organic Compounds from Plants*, *15*(3), 167–175. https://doi.org/10.1016/j.tplants.2009.12.002

Ding, Y., Li, H., Chen, L.-L., & Xie, K. (2016). Recent advances in genome editing using CRISPR/Cas9. *Frontiers in Plant Science*, *7*, 703.

Douglas, A. E. (2018). Strategies for enhanced crop resistance to insect pests. *Annual Review of Plant Biology*, *69*, 637–660.

Dudareva, N., Negre, F., Nagegowda, D. A., & Orlova, I. (2006). Plant volatiles: Recent advances and future perspectives. *Critical Reviews in Plant Sciences*, *25*, 417–440.

Dykes, I. M., & Emanueli, C. (2017). Transcriptional and post-transcriptional gene regulation by long non-coding RNA. *Genomics, Proteomics & Bioinformatics*, *15*(3), 177–186.

Eigenbrode, S. D., & Espelie, K. E. (1995). Effects of plant epicuticular lipids on insect herbivores. *Annual Review of Entomology*, *40*(1), 171–194. https://doi.org/10.1146/annurev.en.40.010195.001131

Eigenbrode, S. D., & Shelton, A. M. (1990). Behavior of neonate diamondback moth larvae (Lepidoptera: Plutellidae) on glossy-leafed resistant Brassica oleracea L. *Environmental Entomology*, *19*(5), 1566–1571.

El Ouakfaoui, S., & Asselin, A. (1992). Multiple forms of chitosanase activities. *Phytochemistry*, *31*(5), 1513–1518.

Espiñeira, J. M., Novo Uzal, E., Gómez Ros, L. V., Carrión, J. S., Merino, F., Ros Barceló, A., & Pomar, F. (2011). Distribution of lignin monomers and the evolution of lignification among lower plants. *Plant Biology*, *13*(1), 59–68. https://doi.org/10.1111/j.1438-8677.2010.00345.x

Estruch, J. J., Carozzi, N. B., Desai, N., Duck, N. B., Warren, G. W., & Koziel, M. G. (1997). Transgenic plants: An emerging approach to pest control. *Nature Biotechnology*, *15*(2), 137–141.

Etxaniz, A., González-Bullón, D., Martín, C., & Ostolaza, H. (2018). Membrane repair mechanisms against permeabilization by pore-forming toxins. *Toxins*, *10*(6), 234.

Falcão, L., & Araújo, M. E. M. (2018). Vegetable tannins used in the manufacture of historic leathers. *Molecules*, *23*(5), 1081.

Fambrini, M., & Pugliesi, C. (2019). The dynamic genetic-hormonal regulatory network controlling the trichome development in leaves. *Plants*, *8*(8). https://doi.org/10.3390/plants8080253

Farokhzadeh, S., & Fakheri, B. (2014). Marker-assisted selection for disease resistance: Applications in breeding. *International Journal of Agriculture and Crop Sciences (IJACS)*, *7*(14), 1392–1405.

Ferreira, R. B., Monteiro, S., freitas, R., Santos, C. N., Chen, Z., Batista, L. M., Duarte, J., Borges, A., & Teixeira, A. R. (2007). The role of plant defence proteins in fungal pathogenesis. *Molecular Plant Pathology*, *8*(5), 677–700. https://doi.org/10.1111/j.1364-3703.2007.00419.x

Finetti, L., Tiedemann, L., Zhang, X., Civolani, S., Bernacchia, G., & Roeder, T. (2021). Monoterpenes alter TAR1-driven physiology in Drosophila species. *Journal of Experimental Biology*, *224*(1), jeb232116. https://doi.org/10.1242/jeb.232116

Franco, O. L., Rigden, D. J., Melo, F. R., & Grossi-de-Sá, M. F. (2002). Plant α-amylase inhibitors and their interaction with insect α-amylases. *European Journal of Biochemistry*, *269*(2), 397–412. https://doi.org/10.1046/j.0014-2956.2001.02656.x

Fulcher, A. F., Ranney, T. G., Burton, J. D., Walgenbach, J. F., & Danehower, D. A. (1998). Role of foliar phenolics in host plant resistance of Malus taxa to adult Japanese beetles. *HortScience, 33*(5), 862–865. https://doi.org/10.21273/HORTSCI.33.5.862.

Ghag, S. B. (2017). Host induced gene silencing, an emerging science to engineer crop resistance against harmful plant pathogens. *Physiological and Molecular Plant Pathology*, *100*, 242–254.

Giuliani, C., Bottoni, M., Ascrizzi, R., Milani, F., Spada, A., Flamini, G., & Fico, G. (2021). Morphology and phytochemistry of Teucrium chamaedrys L. (Lamiaceae) cultivated at the Ghirardi Botanic Garden (Lombardy, Northern Italy). *Flora*, *282*, 151898. https://doi.org/10.1016/j.flora.2021.151898

GOLS, R. (2014). Direct and indirect chemical defences against insects in a multitrophic framework. *Plant, Cell & Environment*, *37*(8), 1741–1752. https://doi.org/10.1111/pce.12318

Gupta, P., Kumar, J., Mir, R., & Kumar, A. (2010). 4 Marker-assisted selection as a component of conventional plant breeding. *Plant Breeding Reviews*, *33*(4), 145–217.

Gurib-Fakim, A. (2006). Medicinal plants: Traditions of yesterday and drugs of tomorrow. *Molecular Aspects of Medicine*, *27*(1), 1–93. https://doi.org/10.1016/j.mam.2005.07.008

Halder, J., Kushwaha, D., Deb, D., Karmakar, P., Krishna, R., Nagendran, K., Rai, A.B., & Singh, B. (2017). Assessment of susceptibility of *Ocimum tenuiflorum* to lace wing bug [*Cochlochila bullita* (Stal.)]: Host's physico-morphic and biochemical properties. *Journal of Entomology and Zoology Studies*, *5*(3), 814–821.

Han, Y., & Korban, S.S. (2021). Genetic and physical mapping of the apple genome. In S.S. Korban (Eds.), *The Apple Genome. Compendium of Plant Genomes*. Springer, Cham. https://doi.org/10.1007/978-3-030-74682-7_7.

Hanley, M. E., Lamont, B. B., Fairbanks, M. M., & Rafferty, C. M. (2007). Plant structural traits and their role in anti-herbivore defence. *Perspectives in Plant Ecology, Evolution and Systematics*, *8*(4), 157–178. https://doi.org/10.1016/j.ppees.2007.01.001

Harper, E. P., Baldwin, R. G., Gansneder, B. G., & Chronister, J. L. (2001). Full-time women faculty off the tenure track: Profile and practice. *The Review of Higher Education*, *24*(3), 237–257.

Hasan, N., Choudhary, S., Naaz, N., Sharma, N., & Laskar, R. A. (2021). Recent advancements in molecular marker-assisted selection and applications in plant breeding programmes. *Journal of Genetic Engineering and Biotechnology*, *19*(1), 1–26.

Hegedűs, B., Viharos, L., Gervain, M., & Gálfi, M. (2009). The effect of low-level laser in knee osteoarthritis: A double-blind, randomized, placebo-controlled trial. *Photomedicine and Laser Surgery*, *27*(4), 577–584.

Hill, D. S. (2012). *The economic importance of insects*. Springer.

Ho, C.-T. (1992). Phenolic compounds in food. In *Phenolic compounds in food and their effects on health II* (Vol. 507, pp. 2–7). American Chemical Society. https://doi.org/10.1021/bk-1992-0507.ch001

Horiuchi, J., Arimura, G., Ozawa, R., Shimoda, T., Takabayashi, J., & Nishioka, T. (2001). Exogenous ACC enhances volatiles production mediated by jasmonic acid in lima bean leaves. *FEBS Letters*, *509*(2), 332–336. https://doi.org/10.1016/S0014-5793(01)03194-5

Huang, T., Jander, G., & de Vos, M. (2011). Non-protein amino acids in plant defense against insect herbivores: Representative cases and opportunities for further functional analysis. *Plant-Insect Interactions*, *72*(13), 1531–1537. https://doi.org/10.1016/j.phytochem.2011.03.019

Islam, W., Tayyab, M., Khalil, F., Hua, Z., Huang, Z., & Chen, H. Y. (2020). Silicon-mediated plant defense against pathogens and insect pests. *Pesticide Biochemistry and Physiology*, *168*, 104641.

Jain, M., Amera, G. M., Muthukumaran, J., & Singh, A. K. (2022). Insights into biological role of plant defense proteins: A review. *Biocatalysis and Agricultural Biotechnology*, *40*, 102293.

Jetter, R., Kunst, L., & Samuels, A. L. (2006). Composition of plant cuticular waxes. *Annual Plant Reviews: Biology of the Plant Cuticle*, *23*, 145–181.

Jia, H., Wang, Y., Su, H., Huang, X., & Wang, N. (2022). LbCas12a-D156R efficiently edits LOB1 effector binding elements to generate canker-resistant citrus plants. *Cells*, *11*(3), 315.

Joga, M. R., Mogilicherla, K., Smagghe, G., & Roy, A. (2021). RNA interference-based forest protection products (FPPs) against wood-boring coleopterans: Hope or hype? *Frontiers in Plant Science*, *12*, 733608.

Johnson, M. T. (2011). *Evolutionary ecology of plant defences against herbivores* (Vol. 25, pp. 305–311): Wiley Online Library.

Kant, M., Jonckheere, W., Knegt, B., Lemos, F., Liu, J., Schimmel, B., Villarroel, C., Ataide, L., Dermauw, W., & Glas, J. (2015). Mechanisms and ecological consequences of plant defence induction and suppression in herbivore communities. *Annals of Botany*, *115*(7), 1015–1051.

Kappers, I. F., Verstappen, F. W. A., Luckerhoff, L. L. P., Bouwmeester, H. J., & Dicke, M. (2010). Genetic variation in jasmonic acid- and spider mite-induced plant volatile emission of cucumber accessions and attraction of the predator Phytoseiulus persimilis. *Journal of Chemical Ecology*, *36*(5), 500–512. https://doi.org/10.1007/s10886-010-9782-6

Karar, H., Abbas, Q., Ali, S., & Shahid, M. R. (2017). *Comparative evaluation of some novel vs commercial cotton genotypes against insect pest complex for seed cotton yield and fibre traits*. https://api.semanticscholar.org/CorpusID:232238219

Karavolias, N. G., Horner, W., Abugu, M. N., & Evanega, S. N. (2021). Application of gene editing for climate change in agriculture. *Frontiers in Sustainable Food Systems*, *5*, 685801.

Karban, R., & Shiojiri, K. (2009). Self-recognition affects plant communication and defense. *Ecology Letters*, *12*, 502–506. https://doi.org/10.1111/j.1461-0248.2009.01313.x

Kaur, N. (2008). *Developing artificial rearing techniques for hemlock woolly adelgid, Adelges tsugae and balsam woolly adelgid, Adelges piceae; artificial infestation and epicuticular wax study of Carolina hemlock, Tsuga caroliniana, provenances* (Master's thesis, North Carolina State University). NC State University. http://www.lib.ncsu.edu/resolver/1840.16/590.

Khalil, A. M. (2020). The genome editing revolution. *Journal of genetic engineering and biotechnology*, *18*(1), 1–16.

Khan, S. (2015). QTL mapping: A tool for improvement in crop plants. *Research Journal of Recent Sciences*, *2277*, 2502.

Kiggundu, A., Gold, C. S., Labuschagne, M. T., Vuylsteke, D., & Louw, S. (2003). Levels of host plant resistance to banana weevil, Cosmopolites sordidus (Germar)(Coleoptera: Curculionidae), in African Musa germplasm. *Euphytica*, *133*, 267–277.

Kim, D. S., & Zhang, J. (2023). Strategies to improve the efficiency of RNAi-mediated crop protection for pest control. *Entomologia Generalis*, *43*(1), 5–19.

KIRIŞIK, M., ÜNLÜ, M., & Erler, F. (2020). Evaluation of some melon lines for their resistance against melon aphid, Aphis gossypii Glover. *Turkish Journal of Agriculture and Forestry*, *44*(3), 322–329.

Kishore, P. (1991). Morphological factors responsible for conferring resistance in sorghum cultivars to the stem borer, Chilo partellus (Swinhoe). *Journal of Entomological Research*, *15*(3), 163–168.

Kocyigit, E., Kocaadam-Bozkurt, B., Bozkurt, O., Ağagündüz, D., & Capasso, R. (2023). Plant toxic proteins: Their biological activities, mechanism of action and removal strategies. *Toxins*, *15*(6). https://doi.org/10.3390/toxins15060356

Komal, J., Desai, H., Samal, I., Mastinu, A., Patel, R., Kumar, P. D., …, Bhoi, T. K. (2023). Unveiling the genetic symphony: Harnessing crispr-cas genome editing for effective insect pest management. *Plants*, *12*(23), 3961.

Kumari, P., Jasrotia, P., Kumar, D., Kashyap, P. L., Kumar, S., Mishra, C. N., …, Singh, G. P. (2022). Biotechnological approaches for host plant resistance to insect pests. *Frontiers in Genetics*, *13*, 914029.

Lannoo, N., & Van Damme, E. J. M. (2014). Lectin domains at the frontiers of plant defense. *Frontiers in Plant Science*, *5*. https://doi.org/10.3389/fpls.2014.00397

Lawson, N. (2023).*Wildscape: Trilling Chipmunks, Beckoning Blooms, Salty Butterflies, and Other Sensory Wonders of Nature*. Chronicle Books. Princeton Architectural Press. ISBN: 978-1-79722-247-9.

Legé, K. E. (1998). Tannins in cotton. In Bajaj, Y.P.S. (Ed.), *Cotton. Biotechnology in Agriculture and Forestry*, vol 42. Springer, Berlin, Heidelberg. https://doi.org/10.1007/978-3-642-80373-4_23

Levin, E., McCue, M. D., & Davidowitz, G. (2017). More than just sugar: Allocation of nectar amino acids and fatty acids in a Lepidopteran. *Proceedings of the Royal Society B: Biological Sciences*, *284*(1848), 20162126.

Li, H., Sun, L., Mittapalli, O., Muir, W., Xie, J., Wu, J., Schemerhorn, B., Sun, W., Pittendrigh, B., & Murdock, L. (2009). Transcriptional signatures in response to wheat germ agglutinin and starvation in Drosophila melanogaster larval midgut. *Insect Molecular Biology*, *18*(1), 21–31.

Liang, D., Chen, M., Qi, X., Xu, Q., Zhou, F., & Chen, X. (2016). QTL mapping by SLAF-seq and expression analysis of candidate genes for aphid resistance in cucumber. *Frontiers in Plant Science*, *7*, 1000.

Liang, D., Liu, M., Hu, Q., He, M., Qi, X., Xu, Q., …, Chen, X. (2015). Identification of differentially expressed genes related to aphid resistance in cucumber (Cucumis sativus L.). *Scientific Reports*, *5*(1), 9645.

Lis, H., & Sharon, N. (1981). Lectins in higher plants. In A. Marcus (Ed.), *Proteins and nucleic acids* (pp. 371–447). Academic Press. https://doi.org/10.1016/B978-0-12-675406-3.50016-5.

Liu, S., Jaouannet, M., Dempsey, D. M. A., Imani, J., Coustau, C., & Kogel, K.-H. (2020). RNA-based technologies for insect control in plant production. *Biotechnology Advances*, *39*, 107463.

Long, L., Gu, L., Wang, S., Cai, H., Wu, J., Wang, J., & Yang, M. (2023). Progress in the understanding of WRKY transcription factors in woody plants. *International Journal of Biological Macromolecules*, *242*(1), 124379. https://doi.org/10.1016/j.ijbiomac.2023.124379

Lücker, J., Bowen, P., & Bohlmann, J. (2004). Vitis vinifera terpenoid cyclases: Functional identification of two sesquiterpene synthase cDNAs encoding (+)-valencene synthase and (–)-germacrene D synthase and expression of mono- and sesquiterpene synthases in grapevine flowers and berries. *Phytochemistry*, *65*(19), 2649–2659. https://doi.org/10.1016/j.phytochem.2004.08.017

Mahmoud, A. M. A., Ebadah, I. M. A., Moawad, S. S., Mahmoud, Y. A., Abd El-Hameid, A. R., & Sadek, H. (2022). Tuta absoluta (Meyrick) resistance in tomato (Solanum lycopersicum L.) accessions and species. *International Journal of Pest Management*, *68*(1), 1–13. https://doi.org/10.1080/09670874.2022.2055195.

Malagon, J., & Garrido, A. (1990). Relation between cyanogenic glycosides content and the resistance to Capnodis tenebrionis (L.) in stone fruits. *Boletín de Sanidad Vegetal, Plagas*, *16*(2), 499–503.

Martín-Valmaseda, M., Devin, S. R., Ortuño-Hernández, G., Pérez-Caselles, C., Mahdavi, S. M. E., Bujdoso, G., …, Alburquerque, N. (2023). CRISPR/Cas as a genome-editing technique in fruit tree breeding. *International Journal of Molecular Sciences*, *24*(23), 16656.

Mayer, A. M. (2004). Resistance to herbivores and fungal pathogens: Variations on a common theme? A review comparing the effect of secondary metabolites, induced and constitutive, on herbivores and fungal pathogens. *Israel Journal of Plant Sciences*, *52*(4), 279–292.

Mehta, U., Jayanthi, R., & David, H. (1981). Occurrence of pink mealybug on certain early maturing sugarcane varieties. *Pestology*, *5*, 9–10.

Mekonnen, T., Haileselassie, T., & Tesfaye, K. (2017). Identification, mapping and pyramiding of genes/quantitative trait loci (QTLs) for durable resistance of crops to biotic stresses. *Journal of Plant Pathology and Microbiology*, *8*, 412. doi: 10.4172/2157-7471.1000412.

Michiels, K., Van Damme, E. J., & Smagghe, G. (2010). Plant-insect interactions: what can we learn from plant lectins? *Archives of Insect Biochemistry and Physiology: Published in Collaboration with the Entomological Society of America, 73*(4), 193–212.

Mishra, A., Behura, A., Mawatwal, S., Kumar, A., Naik, L., Mohanty, S. S., Manna, D., Dokania, P., Mishra, A., & Patra, S. K. (2019). Structure-function and application of plant lectins in disease biology and immunity. *Food and Chemical Toxicology*, *134*, 110827.

Mitchell, C., Brennan, R. M., Graham, J., & Karley, A. J. (2016). Plant defense against herbivorous pests: Exploiting resistance and tolerance traits for sustainable crop protection. *Frontiers in Plant Science*, *7*, 1132.

Mookiah, S., Sivasubramaniam, B., Thangaraj, T., & Govindaraj, S. (2021). Host plant resistance. In Omkar (Ed.), *Molecular Approaches for Sustainable Insect Pest Management* (pp. 1–56). Springer Singapore. https://doi.org/10.1007/978-981-16-3591-5_1

Moon, T. T., Maliha, I. J., Khan, A. A. M., Chakraborty, M., Uddin, M. S., Amin, M. R., & Islam, T. (2022). CRISPR-Cas genome editing for insect pest stress management in crop plants. *Stresses*, *2*(4), 493–514.

Nadeem, M. A., Nawaz, M. A., Shahid, M. Q., Doğan, Y., Comertpay, G., Yıldız, M., …, Labhane, N. (2018). DNA molecular markers in plant breeding: Current status and recent advancements in genomic selection and genome editing. *Biotechnology & Biotechnological Equipment*, *32*(2), 261–285.

Nagata, R., Nuessly, G., & McAuslane, H. (2004). Host plant resistance in valmaine cos lettuce to the banded cucumber beetle. *HortScience*, *39*(4), 765D – 765.

Nair, K. P. (2019). Utilizing crop wild relatives to combat global warming. *Advances in Agronomy*, *153*, 175–258.

Naithani, S., Komath, S. S., Nonomura, A., & Govindjee, G. (2021). Plant lectins and their many roles: Carbohydrate-binding and beyond. *Journal of Plant Physiology*, *266*, 153531.

Nawrot, R., Barylski, J., Nowicki, G., Broniarczyk, J., Buchwald, W., & Goździcka-Józefiak, A. (2014). Plant antimicrobial peptides. *Folia Microbiologica*, *59*, 181–196.

Neuhaus, J. M. (1999). Bias and efficiency loss due to misclassified responses in binary regression. *Biometrika*, *86*(4), 843–855.

Nitnavare, R. B., Bhattacharya, J., Singh, S., Kour, A., Hawkesford, M. J., & Arora, N. (2021). Next generation dsRNA-based insect control: Success so far and challenges. *Frontiers in Plant Science*, *12*, 673576.

Norris, D. M., & Kogan, M. (1980). Biochemical and morphological bases of resistance. In F. G. Maxwell & P. R. Jennings (Eds.), *Breeding Plants Resistant to Insects* (pp. 22–61). Wiley.

Oguri, S. (2022). Structure and function of plant chitin-binding lectins and tomato lectin. *Trends in Glycoscience and Glycotechnology*, *34*(200), E75–E80.

Pagare, S., Bhatia, M., Tripathi, N., Pagare, S., & Bansal, Y. (2015). Secondary metabolites of plants and their role: Overview. *Current Trends in Biotechnology and Pharmacy*, *9*(3), 293–304.

Palumbo, R., & Dahlman, D. (1978). Reduction of manduca sexta fecundity and fertility by L-canavanine. *Journal of Economic Entomology*, *71*(4), 674–676.

Patanakamjorn, S., & Pathak, M. (1967). Varietal resistance of rice to the Asiatic rice borer, Chilo suppressalis (Lepidoptera: Crambidae), and its association with various plant characters. *Annals of the Entomological Society of America*, *60*(2), 287–292.

Peleg, Z., & Blumwald, E. (2011). Hormone balance and abiotic stress tolerance in crop plants. *Current Opinion in Plant Biology*, *14*(3), 290–295. https://doi.org/10.1016/j.pbi.2011.02.001

Peng, C. (2012). *Synthesis of very-long-chain bifunctional and isotope-labeled compounds for biochemical investigations into novel compounds in plant cuticular waxes* (T). University of British Columbia. Retrieved from https://open.library.ubc.ca/collections/ubctheses/24/items/1.0073469

Peumans, W. J., Hao, Q., & Van Damme, E. J. (2001). Ribosome-inactivating proteins from plants: More than RNA N-glycosidases? *The FASEB Journal*, *15*(9), 1493–1506.

Pompili, V., Dalla Costa, L., Piazza, S., Pindo, M., & Malnoy, M. (2020). Reduced fire blight susceptibility in apple cultivars using a high-efficiency CRISPR/Cas9-FLP/FRT-based gene editing system. *Plant Biotechnology Journal*, *18*(3), 845–858.

Prachi Vibhute, M. S., Jaabir, M., Sangeetha Bharath, S., & Sivakamavalli, J. (2021). Overview of lectins. In P. Elumalai & S. Lakshmi (Eds.), *Lectins: Innate immune defense and therapeutics* (pp. 1–25). Springer. https://doi.org/10.1007/978-981-16-7462-4_1

Puri, M., Sharma, D., & Barrow, C. J. (2012). Enzyme-assisted extraction of bioactives from plants. *Trends in Biotechnology*, *30*(1), 37–44.

Pérez-Hedo, M., Alonso-Valiente, M., Vacas, S., Gallego, C., Rambla, J. L., Navarro-Llopis, V., Granell, A., & Urbaneja, A. (2021). Eliciting tomato plant defenses by exposure to herbivore induced plant volatiles. *Entomologia Generalis*, *41*(3), 209–218.

Pérez-Hedo, M., Bouagga, S., Zhang, N. X., Moerkens, R., Messelink, G., Jaques, J. A., …, Pappas, M. L. (2022). Induction of plant defenses: The added value of zoophytophagous predators. *Journal of Pest Science*, *95*(4), 1501–1517.

Quiros, C. F., Stevens, M. A., Rick, C. M., & Kok-Yokomi, M. (1977). Resistance in tomato to the pink form of the potato aphid (macrosiphum euphorbiae thomas): The role of anatomy, epidermal hairs, and foliage composition. *Journal of the American Society for Horticultural Science*, *102*(2), 166–171.

Rajarajeswari, V., & Subbarao, I. (1997). Gossypol glands in relation to resistance to bollworm (Helicoverpa armigera) in upland cotton (Gossypium hirsutum). *The Indian Journal of Agricultural Sciences*, *67*(7).

Rao, D. Easwar, Divya, K., Prathyusha, I.V.S.N., Rama Krishna, Ch., & Chaitanya, K.V. (2017). Insect-Resistant Plants. In Suresh Kumar Dubey, Ashok Pandey, & Rajender Singh Sangwan (Eds.), *Current Developments in Biotechnology and Bioengineering* (pp. 47–74). Elsevier. https://doi.org/10.1016/B978-0-444-63661-4.00003-7.

Ratnadass, A., Fernandes, P., Avelino, J., & Habib, R. (2012). Plant species diversity for sustainable management of crop pests and diseases in agroecosystems: A review. *Agronomy for Sustainable Development*, *32*, 273–303.

Reifenrath, K., Riederer, M., & Müller, C. (2005). Leaf surface wax layers of Brassicaceae lack feeding stimulants for Phaedon cochleariae. *Entomologia Experimentalis et Applicata*, *115*(1), 41–50.

Reynolds, O. L., Keeping, M. G., & Meyer, J. H. (2009). Silicon-augmented resistance of plants to herbivorous insects: A review. *Annals of Applied Biology*, *155*(2), 171–186.

RH, W. (1971). Allelochemics: Chemical interactions between species. *Science*, *171*, 757–770.

Roberts, D. P., & Mattoo, A. K. (2018). Sustainable agriculture – Enhancing environmental benefits, food nutritional quality and building crop resilience to abiotic and biotic stresses. *Agriculture*, *8*(1), 8.

Rosenthal, G. A. (1991). Nonprotein amino acids as protective allelochemicals. In G. A. Rosenthal & M. R. Berenbaum (Eds.), *Herbivores: Their Interactions with Secondary Plant Metabolites* (pp. 1–34). Academic Press

Rosenthal, G. A., & Dahlman, D. (1975). Non-protein amino acid-insect interactions – II. Effects of canaline-urea cycle amino acids on growth and development of the tobacco hornworm, Manduca Sexta L. (Sphingidae). *Comparative Biochemistry and Physiology Part A: Physiology*, *52*(1), 105–108.

Saakre M, Jaiwal S, Rathinam M, Raman KV, Tilgam J , Paul K, Sreevathsa R, Pattanayak D. (2024). Host-delivered RNA interference for durable pest resistance in plants: Advanced methods, challenges, and applications. *Molecular Biotechnology*, *66*, 1786–1805. https://doi.org/10.1007/s12033-023-00833-9.

Sadeq, S., Al-Hashimi, S., Cusack, C. M., & Werner, A. (2021). Endogenous double-stranded RNA. *Non-Coding RNA*, *7*(1), 15.

Saeid, A.-R., & Ateyyat, M. (2014). Phenotypic and molecular screening of apple genotypes to woolly apple aphid resistance. *Notulae Botanicae Horti Agrobotanici Cluj-Napoca*, *42*(1), 99–103.

Sandanayaka, W., Bus, V., & Connolly, P. (2005). Mechanisms of woolly aphid [Eriosoma lanigerum (Hausm.)] resistance in apple. *Journal of Applied Entomology*, *129(9–10)*, 534–541.

Santiago, R., Barros-Rios, J., & Malvar, R. A. (2013). Impact of cell wall composition on maize resistance to pests and diseases. *International Journal of Molecular Sciences*, *14*(4), 6960–6980.

Santiago Carabelos, R., Butrón Gómez, A. M., Revilla Temiño, P., & Malvar Pintos, R. A. (2011). Is the basal area of maize internodes involved in borer resistance? *BMC Plant Biology*, 11, 137. doi: 10.1186/1471-2229-11-137.

Santiago, R., Souto, X., Sotelo, J., Butrón, A., & Malvar, R. (2003). Relationship between maize stem structural characteristics and resistance to pink stem borer (Lepidoptera: Noctuidae) attack. *Journal of Economic Entomology*, *96*(5), 1563–1570.

Savary, S., Willocquet, L., Pethybridge, S. J., Esker, P., McRoberts, N., & Nelson, A. (2019). The global burden of pathogens and pests on major food crops. *Nature Ecology & Evolution*, *3*(3), 430–439.

Schlesinger, H., Applebaum, S., & Birk, Y. (1976). Effect of β-cyano-l-alanine on the water balance of Locusta migratoria. *Journal of Insect Physiology*, *22*(10), 1421–1425.

Sedeek, K. E., Mahas, A., & Mahfouz, M. (2019). Plant genome engineering for targeted improvement of crop traits. *Frontiers in Plant Science, 10*, 114.

Seifi, A., Visser, R., & Bai, Y. (2013). How to effectively deploy plant resistances to pests and pathogens in crop breeding. *Euphytica, 190*, 321–334.

Sharma, G., Malthankar, P. A., & Mathur, V. (2021). Insect–plant interactions: A multilayered relationship. *Annals of the Entomological Society of America, 114*(1), 1–16.

Sharma, H., & Ortiz, R. (2002). Host plant resistance to insects: An eco-friendly approach for pest management and environment conservation. *Journal of Environmental Biology, 23*(2), 111–135.

Sharma, H. C., Sharma, K. K., & Crouch, J. H. (2004). Genetic transformation of crops for insect resistance: potential and limitations. *Critical Reviews in Plant Sciences, 23*(1), 47–72.

Shivanandappa, T., & Rajashekar, Y. (2014). Mode of action of plant-derived natural insecticides. In D. Singh (Ed.), *Advances in plant biopesticides* (pp. 323–345). Springer. https://doi.org/10.1007/978-81-322-2006-0_16

Silva, K. F., Michereff-Filho, M., Fonseca, M. E., Silva-Filho, J. G., Texeira, A. C., Moita, A. W., Torres, J. B., Fernández-Muñoz, R., & Boiteux, L. S. (2014). Resistance to B emisia tabaci biotype B of S olanum pimpinellifolium is associated with higher densities of type IV glandular trichomes and acylsugar accumulation. *Entomologia Experimentalis Et Applicata, 151*(3), 218–230.

Sison, M. L. J., Cortaga, C. Q., Barcos, A. K. S., Laurel, N. R., & Mateo, J. M. C. (2020). Resistance screening and influence of fruit physico-biochemical properties of "Carabao" and other mango varieties against Oriental fruit fly, *Bactrocera dorsalis* (Diptera: Tephritidae), in the Philippines. *Philippine Journal of Science, 149*(4), 1175–1187.

Song, X. (2013). *Genetically modified poplars in China* [Graduating Project, University of British Columbia]. University of British Columbia Open Collections. doi: http://dx.doi.org/10.14288/1.0075635.

Sousa, M., Morhy, L., Richardson, M., Hilder, V., & Gatehouse, A. (1993). Effects of the cytolytic seed protein enterolobin on coleopteran and lepidopteran insect larvae. *Entomologia experimentalis et applicata, 69*(3), 231–238.

Stec, B. (2006). Plant thionins – the structural perspective. *Cellular and Molecular Life Sciences CMLS, 63*(12), 1370–1385. https://doi.org/10.1007/s00018-005-5574-5

Talbert, L. E., Sherman, J. D., Hofland, M. L., Lanning, S. P., Blake, N. K., Grabbe, R., Lamb, P. F., Martin, J. M., & Weaver, D. K. (2014). Resistance to Cephus cinctus Norton, the wheat stem sawfly, in a recombinant inbred line population of wheat derived from two resistance sources. *Plant Breeding, 133*(4), 427–432.

Tammineni, R., Gulati, P., Kumar, S., & Mohanty, A. (2020). An overview of acyclotides: Past, present and future. *Phytochemistry, 170*, 112215.

Tanda, A. S. (2022). Mutualistic plant plants associations associations related to insect resistance insect resistance molecular advances in insect resistance of field crops. In *Modern and applied approaches* (pp. 1–42): Springer.

Tanda, A.S. (2022). Mutualistic Plant Associations Related to Insect Resistance. In Tanda, A.S. (Ed.), *Molecular Advances in Insect Resistance of Field Crops*. Springer, Cham. https://doi.org/10.1007/978-3-030-92152-1_1

Tiku, A.R. (2021). Direct and Indirect Defence Against Insects. In Singh, I.K., & Singh, A. (Eds.), *Plant-Pest Interactions: From Molecular Mechanisms to Chemical Ecology*. Springer, Singapore. https://doi.org/10.1007/978-981-15-2467-7_8.

Tripathy, A., & Pradhan, R. K. (2018). Symbiotic interactions, law of purposive association and the+/+ nature of all co-evolution. *Indian Journal of Science and Technology, 11*, 43.

Uwimana, B., Mwanje, G., Batte, M., Akech, V., Shah, T., Vuylsteke, M., & Swennen, R. (2021). Continuous mapping identifies loci associated with weevil resistance [Cosmopolites sordidus (Germar)] in a triploid banana population. *Frontiers in Plant Science, 12*, 753241.

Vosman, B., Kashaninia, A., van't Westende, W., Meijer-Dekens, F., van Eekelen, H., Visser, R. G., …, Voorrips, R. E. (2019). QTL mapping of insect resistance components of Solanum galapagense. *Theoretical and Applied Genetics, 132*, 531–541.

Waghmare, V.N. (2022). Cotton Breeding. In Yadava, D.K., Dikshit, H.K., Mishra, G.P., & Tripathi, S. (Eds.), *Fundamentals of Field Crop Breeding*. Springer, Singapore. https://doi.org/10.1007/978-981-16-9257-4_11

Wallace, L., McNeal, F., & Berg, M. (1974). Resistance to both Oulema melanopus and Cephus cinctus in pubescent-leaved and solid-stemmed wheat selections. *Journal of Economic Entomology, 67*(1), 105–107.

Wang, T., Zhang, H., & Zhu, H. (2019). CRISPR technology is revolutionizing the improvement of tomato and other fruit crops. *Horticulture Research, 6*, Article 77. https://doi.org/10.1038/s41438-019-0159-x.

Wani, S. H., Choudhary, M., Barmukh, R., Bagaria, P. K., Samantara, K., Razzaq, A., …, Varshney, R. K. (2022). Molecular mechanisms, genetic mapping, and genome editing for insect pest resistance in field crops. *Theoretical and Applied Genetics*, *135*(11), 3875–3895.

Wani, S. H., & Gosal, S. (Eds.). (2018). *Marker-assisted breeding for abiotic stress tolerance in crop plants. In Biotechnologies of crop improvement* (Vol. 3, pp. 1–20). https://doi.org/10.1007/978-3-319-94746-4_1.

War, A. R., Paulraj, M. G., Ahmad, T., Buhroo, A. A., Hussain, B., Ignacimuthu, S., & Sharma, H. C. (2012). Mechanisms of plant defense against insect herbivores. *Plant Signaling & Behavior*, *7*(10), 1306–1320.

War, A. R., Taggar, G. K., War, M. Y., & Hussain, B. (2016). Impact of climate change on insect pests, plant chemical ecology, tritrophic interactions and food production. *International Journal of Clinical and Biological Sciences*, *1*(02), 16–29.

Wasternack, C., & Hause, B. (2013). Jasmonates: Biosynthesis, perception, signal transduction and action in plant stress response, growth and development. An update to the 2007 review in annals of botany. *Annals of Botany*, *111*(6), 1021–1058. https://doi.org/10.1093/aob/mct067

Wearing, C., & Hokkanen, H. (1994). Pest resistance to Bacillus thuringiensis: Case studies of ecological crop assessment for Bt gene incorporation and strategies of management. *Biocontrol Science and Technology*, *4*(4), 573–590.

Whittaker, R. H., & Feeny, P. P. (1971). Allelochemics: Chemical Interactions between Species: Chemical agents are of major significance in the adaptation of species and organization of communities. *Science*, *171*(3973), 757–770.

Xia, J., Guo, Z., Yang, Z., Han, H., Wang, S., Xu, H., …, Xie, W. (2021). Whitefly hijacks a plant detoxification gene that neutralizes plant toxins. *Cell*, *184*(7), 1693–1705. e1617.

Xiao, Y., & Wu, K. (2019). Recent progress on the interaction between insects and Bacillus thuringiensis crops. *Philosophical Transactions of the Royal Society B*, *374*(1767), 20180316.

Yepsen, R. B. (1984). *The encyclopedia of natural insect and disease control: The most comprehensive guide to protecting plants-vegetables, fruit, flowers, trees, and lawns-without toxic chemicals*. Rodale Press.

Zeb, A. (2021). Chemistry of phenolic antioxidants. In A. Zeb (Ed.), *Phenolic antioxidants in foods: Chemistry, biochemistry and analysis* (pp. 25–87). Springer. https://doi.org/10.1007/978-3-030-74768-8_2

Zhang, H., Yin, W., Zhao, J., Jin, L., Yang, Y., Wu, S., …, Wu, Y. (2011). Early warning of cotton bollworm resistance associated with intensive planting of Bt cotton in China. *PLoS One*, *6*(8), e22874.

Zhu, J. (1981). Preliminary observations on the resistance of varieties of spring wheat to Hydrellia griseola (Fallen). *Insect Knowledge (Kunchong Zhishi)*, *18*(5), 213–214.

18 Physiological and Molecular Mechanism of Insect Herbivory Tolerance in Plants

A Potential Tool for Resistance Breeding

Kaushik Pramanik, Amit Layek, Durga Mahalanobish, Debashis Roy, and Debjyoti Majumder

18.1 INTRODUCTION

The emergence of new variations of disease-causing agents and insect types is increasingly threatening the production of crops. These pathogens are responsible for around 15% of global food production losses and present a significant obstacle in developing crops that can resist them (Sanghera et al., 2011). As genetic polymorphism is observed in these pathogenic agents and insect populations, alterations in the climatic factors are believed to further this polymorphism, causing the evolution of violent biotypes or strains (Anderson et al., 2004). This evolution will change how the interaction between hosts and these pathogens plays out. As a result, outbreaks of diseases or insect pests are likely to persist and even worsen by spreading to new regions where they were not previously established. This situation significantly impacts the available methods for managing these issues. While a combination of methods offers greater reliability, in areas where resources are limited – such as small-scale farming – plant breeders must maximize the use of various genetic variations for disease and pest resistance found in cultivated crops and their related wild plants. Therefore, it's crucial to study how these tolerance and/or resistance variations work to leverage them for enhancing the quality of the most commonly used crop varieties, which are crucial for the livelihoods of many impoverished rural communities.

Despite its attractive characteristics, it appears that tolerance has received less attention than the other two types of plant resistance. Tolerance may be the ideal strategy for managing insect pests, as it aligns with a fundamental principle of integrated pest management (IPM), which involves accepting a certain level of pest damage. By enhancing a plant's ability to withstand injury, we can effectively achieve target-based pest management tactics that exert minimal selective pressure, thereby reducing the risk of pests developing resistance. Unlike antixenosis and antibiosis, tolerance does not influence insect biology or behavior. Consequently, pests cannot develop resistance to plants that exhibit tolerance. So, it's clear that the intangible benefits of tolerance in the realm of plant resistance should not be underestimated. This paper delves into the concept of tolerance and the factors that have constrained its adoption in plant resistance and IPM.

18.2 THE ALTERATIONS IN PLANT'S SYSTEM TOWARD PATHOGENS AND HERBIVORES

Autotrophic organisms must develop complex defense mechanisms to thrive in a world encompassed by heterotrophic creatures. Encounters with heterotrophs can range from tiny microbes to massive

DOI: 10.1201/9781003543237-18

creatures like moose, necessitating plants to have comprehensive protective strategies. When it comes to defense against microbial threats, plants employ effective mechanisms on a small scale, exemplified by the hypersensitive response (HR). This process entails the immediate death of cells surrounding an infection site, which then become filled with antimicrobial compounds to halt pathogen spread (Lam et al., 2001). While HR is a potent defense, it incurs substantial costs and expects an erudite recognition system to prevent unwarranted activation. Consequently, the HR, often preceded by a burst of reactive oxygen species (ROS), that may contribute to cell death, gets triggered when the surveillance system plant, including R-genes that encode receptor proteins, detects specific cues from striking pathogens, such as lipids, peptides, proteins, and polysaccharides (Van Breusegem et al., 2001). While the HR can deter sedentary herbivores that target specific plant tissues, like phloemfeeders, it proves ineffective against most free-roaming herbivory insects, as they evade it by simply moving to a different feeding location. This independence of herbivory insects from their hosts, along with their physiological autonomy, significantly extends the spatial scale of plant-herbivore interactions, influencing both responses within individual plants and the dynamics of the surrounding ecological community.

Many insect herbivores naturally approach plants after their parents have meticulously chosen and, in certain situations, influenced the host plant through actions like microbial inoculations (e.g., bark beetles) or modifications in the plant's source-sink interactions (e.g., aphids) (Larson and Whitham, 1997). These parental efforts guarantee that the juvenile herbivore is developmentally and nutritionally primed for its first assault. With its strong mandibles, maxillae, and other eating organs, the young herbivore can overcome the plant's natural defense. They also possessed an efficient digestive system to break down and absorb plant materials in an environment controlled by the herbivore itself, as well as the mobility and sensory systems necessary to adapt to variations in the plant's suitability (Renwick, 2001). This blend of physiological and behavioral independence plays a pivotal role in explaining the fundamental distinctions in how plants react to herbivores compared to pathogens and how the act of herbivore predation is perceived.

Compared to the attack by pathogens, herbivore infestations often involve physical damage to the plant, and the detection of herbivore attacks typically triggers alterations in the plant's response to wounds. Furthermore, the self-sufficiency of herbivores, both in terms of their physiology and behavior, allows plants to employ defense mechanisms that would be unproductive against pathogens (Figure 18.1). For instance, plants produce secondary (2°) compounds that precisely target the

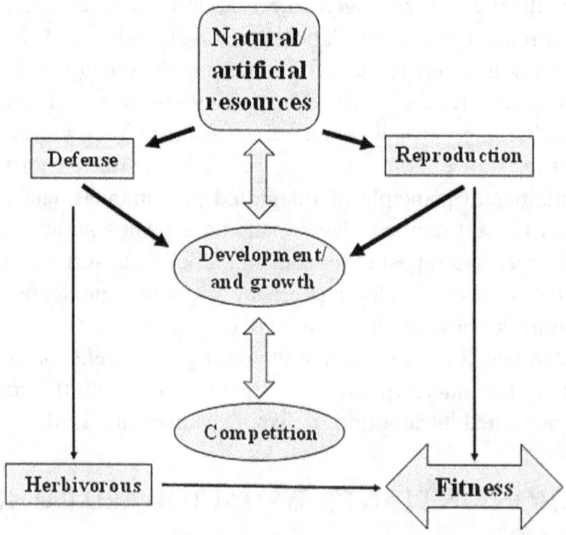

FIGURE 18.1 A visual representation depicting how a plant's overall fitness is determined by the interaction of resource distribution among different plant functions, ecological relationships, and trade-offs based on resources.

specific organs of herbivores, such as the nervous and digestive systems (Rosenthal and Berenbaum, 1992). Additionally, they utilize interactions with higher trophic levels as a defensive strategy by conveying data or providing nutritional incentives to the NEs of herbivores. This recruitment of herbivory NEs significantly expands the spatial scale of the associations (Donath and Boland, 1994). The world of plant-herbivore interactions has been extensively explored in the ecological and entomological literature, with the recent integration of molecular techniques. To assist readers with a molecular focus in navigating this extensive body of work, we have chosen recent reviews covering various aspects of these interactions and provided an overview of key concepts and terminology commonly used to describe them. Following this, we offer an examination of recent literature that highlights advancements in our molecular comprehension of how plants recognize herbivore attacks and the mechanisms of resistance that have been confirmed through insect bioassays.

18.3 LIMITATIONS IMPOSED ON THE PROGRESS AND APPLICATION OF TOLERANCE

18.3.1 It Is Challenging to Identify Tolerance, and Our Understanding of the Underlying Mechanisms Is Limited

The dominant use of antixenosis and antibiosis in plant resistance can be largely attributed to their practicality. The mechanisms of antibiosis have often been quite straightforward to detect and enhance through breeding. This is primarily because they are frequently governed by a major gene or a single gene, and their impacts on herbivory insects are easily observable. We acknowledge that the accurate chemical pathways for antibiosis in many systems remain unknown. This lack of understanding has not impeded the identification of antibiosis and the successful breeding of wheat varieties resistant to *Cephus cinctus*, even though the particular mechanisms triggering larval mortality have not yet been pinpointed (Varella et al., 2015). Whereas, antixenosis mechanisms may not be as easily distinguishable as antibiosis and they are still more discernible than mechanisms of tolerance. It is due to antixenotic mechanisms often involving visible plant characteristics that can be detected by sight, and insect responses can typically be noticed and quantified. For instance, features like glandular trichomes in alfalfa and the frego bract in cotton, which inhibit larval feeding and egg-laying, are easily noticeable and effective (Ranger and Hower, 2001). Even mechanisms that are less visually conspicuous, such as chemical deterrents, tissue thickness, and surface waxes, can be instantly detected and tested (Weaver et al., 2009; Chamarthi et al., 2011).

Compared to antibiosis and antixenosis, our understanding of tolerance is relatively limited. Though, tolerance to damage from arthropods has been recognized in numerous plants, including barley, sorghum, alfalfa, rice, wheat, maize, cowpea, cotton, okra, turnip, muskmelon, tea, Spanish cedar, tall fescue, northern red oak, and *Brassica rapa*. It has also been identified in sugarcane, lentil, soybean, switchgrass, cacao, potato, tomato, strawberry, and cassava (Schuster et al., 1980; Strauss and Agrawal, 1999; Koch et al., 2016). In some of these crops, tolerance is a crucial aspect of their resistance. For instance, sorghum's resistance to the *Schizaphis graminum* relies on the ability of saplings to survive the damage caused by feeding. This is evidently a tolerance response, as resistant lines have no impact on the behavior or biology of the *S. graminum*. In the case of barley, the discovery of virulent populations of *Diuraphis noxia*, impelled the growth of tolerant lines like 'Sydney' and 'Stoneham' to lower the selection load on the *D. noxia*, thus enhancing the stability of these genotypes (Mornhinweg et al., 2009; Marimuthu and Smith, 2012). Despite its successful application in some crops, there is still a lack of comprehensive knowledge involving the underlying root of tolerance.

Morphological aspects related to tolerance might involve features like protected meristems, their numbers, and evolving flexibility. At the molecular level, only a small number of specific substances, such as peroxidases, SNF1-related catalases, and kinases, have been recognized as playing a role in herbivory tolerance via resource distribution (Schwachtje et al., 2006) or the detoxification of ROS (Ramm et al., 2013). It's important to recognize that the mechanisms contributing to tolerance can vary depending on factors such as the herbivore's specialization (e.g., generalists or specialists) (Delaney et al., 2009; Foyer et al., 2015), feeding behavior (e.g., sucking or chewing), the plant's mutual associations (e.g., some milkweed species become more tolerant to the herbivores when they have vesicular-arbuscular-mycorrhizal fungi), and environmental conditions (Wise and Abrahamson, 2007; Tao et al., 2016). These various factors make it more complex to identify and understand tolerance mechanisms. Additionally, some of these mechanisms are consistently active, while others are triggered in response to specific conditions. Assessing germplasm that displays induced tolerance often requires evaluating it in the incidence of pest infestations, which can be more complicated due to seasonal variations in pest infestations at specific locations. Several crops that exhibited tolerance have been detected by chance. Typically, the development of resistant cultivars has resulted from broad screenings to identify any form of resistance. For instance, some alfalfa cultivars such as 'Team', tolerant to weevil damage, came about through extensive germplasm testing. More than 2 million plants were subjected to weevil infestation over a ten-year breeding period to identify any resistance. 'Team' was eventually released in 1970 and is thought to possess all three types of resistance, with tolerance being the predominant factor. It's worth noting that the primary aim of the scientists was not to understand the hidden mechanisms but to create a resistant plant variety. Large-scale inspections focused solely on plant tolerance have also yielded productive results (Dunn et al., 2011).

18.3.2 We Have Yet to Uncover the Genetics Behind Tolerance for Most of the Cases

The capacity to anticipate the physical traits of a plant based on its genetic makeup is crucial for speeding up the enhancement of crops. This is primarily because it enhances the efficiency and accuracy of screening and selecting germplasm. However, comprehending the genetic basis of plant tolerance to herbivores requires identifying varying alleles and the shuffling or separation of these alleles. To fulfill these needs, creating and assessing large breeding populations is necessary. The lack of insight regarding the mechanisms behind tolerance poses challenges for accurately characterizing plants and identifying genetic variations. Despite these difficulties, genetic diversity related to tolerance to leaf herbivory insects has been explained in both crop and non-crops (Punnuri et al., 2013). Similar to antixenosis and antibiosis, it appears that multiple genetic loci and their relations mainly influence tolerance. Although researchers have identified quantitative trait loci (QTLs) correlated with tolerance to herbivores, no genes related to this trait have been cloned till now. Consequently, additional investigation should aim to improve the genetic precision of these target QTLs, with the ultimate goal of identifying and cloning the genes responsible for tolerance.

18.3.3 There Are Still Significant Barriers Preventing the Establishment of Efficient Methods for Widespread Screening of Tolerance, Especially on a Large Scale

A significant challenge in breeding for insect tolerance lies in the struggle to locate distinct characteristics that can be readily, accurately, and consistently measured when plants are subjected to natural or artificial insect pressure. While screening techniques are labor-intensive or onerous for researchers, they are generally impractical for assessing the numerous plant lines routinely examined in breeding programs. For instance, when assessing wheat tolerance to *Rhopalosiphum padi*, a range of methods can be employed that target various plant traits, such as biomass ratio, test weight, grain yield, and the improvement of shoots and roots (Lamb and MacKay, 1995; Dunn et al., 2011). However, not all these methods are conducive to swiftly evaluating thousands of plants. Therefore, the development of high-performance phenotyping techniques that enable the precise analysis of numerous plant lines will significantly aid in creating tolerant crop varieties. Challenges related to executing high-quantity phenotyping in plant propagation programs include the facilities, software licenses, and costs of apparatus needed for experiments, shortage of skilled manpower to handle large datasets, and a need for established values for data analysis (Goggin et al., 2015).

18.3.4 Entomologists Have Focused Less on Tolerance Because Their Primary Interests, Research Guidance, and Funding Prospects Are Predominantly Oriented Toward Mechanisms That Influence Pest Biology Rather Than Plant Biology

Historically, entomologists have played a central role in studying plant resistance to insects, and their perspective has often dominated the field. As a result, rather than focusing on how plants respond to insect damage, entomologists have frequently used plants to implement pest control measures. This approach highlights a drawback in training scientists, where they may become overly specialized in plant biology. Very few entomologists have received official hands-on training in various attributes of plant biology, such as anatomy, metabolism, photosynthesis, and water relations (Jacobs and Frickel, 2009; Rhoten, 2004). Entomologists who are experts in considering the plant's role in interactions with insects could possibly enhance the research and development of tolerant plant varieties. Furthermore, ambidextrous research teams with combined knowledge of plant physiology, pest biology, and agronomy have the potential to create tolerant cultivars (Peterson and Higley, 2001). However, this type of research is still in its early stages.

18.3.5 The Emphasis in Plant Resistance Efforts Continues to be on Regulating Pest Infestations Instead of Addressing Plant Stress

Our perspective is that research on plant resistance while appearing progressive and aligned with IPM principles has mainly taken a one-sided method to pest control, similar to the historical use of insecticides in the 1950s and 1960s. Resistant cultivars, particularly those employing antixenosis and antibiosis mechanisms, essentially act to suppress insect infestations. Rather than insecticide use, the ecological impact of employing resistant lines is relatively minimal. In this sense, resistant lines align with one of IPM's goals, minimizing negative environmental consequences. However, cultivars with antibiosis mechanisms can exert selective pressure on insect biomass, potentially promoting resistance enhancement. Even though resistant lines may be a more favorable pest control strategy, they do not essentially represent a truly ecological pest management approach. New methods for introducing resistance in plants may also face sustainability challenges. For instance, plants that produce the *Bt* toxin (Tabashnik et al., 2008). have resulted in the selection of resistance, even though the number of pests was not at economically damaging levels.

The distinction between 'control' and 'management' within IPM is crucial in understanding why tolerance is not as widely emphasized in plant resistance strategies. In the context of pest management, these terms have been extensively discussed (Menalled et al., 2016). In summary, 'control' refers to approaches that primarily target pests and often involve eradicating pests, thereby exerting significant selection pressure for resistance. This focus on pest eradication includes highly effective strategies like the use of antibiotics in crops, as exemplified by *Bt* crops. In contrast, 'management' involves a strategy that centers on the careful and strategic use of various methods to achieve a desired outcome (Pedigo and Higley, 1996). Tolerance, as a form of plant resistance, aligns well with the concept of management.

18.3.6 Other Biological Factors

In theory, tolerance is a highly appealing concept for inclusion in IPM programs. However, due to the limited research and understanding of tolerance, we cannot predict whether and to what extent specific environmental factors, like heat or drought stress, might make a tolerant plant variety highly vulnerable to pest damage. This concern becomes particularly significant in the context of changing climate and the increasing occurrence of drought-prone regions. Research on non-crop species has demonstrated that drought can reduce a plant's capacity to withstand herbivore attacks. Even among closely related plant species, responses to herbivory can vary under drought conditions. Consequently (Shibel and Heard, 2016), the influence of environmental factors on a plant's capacity to tolerate herbivory insects may need to be evaluated on a case-by-case basis for each crop variety and cultivar. In various agricultural ecosystems, certain arthropods transit between different crop species throughout their life cycle. For instance, *Helicoverpa zea*, in North Carolina, can cause damage to tobacco, corn, wild host plants, soybean, and cotton. If an area has only one crop species that is tolerant to corn earworm damage, it could lead to undesirable populations of this pest for the other crop species in that area.

18.3.7 Economic and Social Issues

In the United States, farmers often prioritize pest control to minimize risks, possibly even more than they do to maximize crop yields (Higley, 2006). The traditional mindset that 'the only good bug is a dead bug' still holds strong in modern agriculture systems. Furthermore, private agricultural companies may be hesitant to promote tolerant crop varieties because they don't want their customers to question whether their products are genuinely resistant to pests. Therefore, it's crucial to prioritize education about tolerance and tolerant crop varieties if this resistance approach is to succeed. Tolerant crop varieties need to possess desirable agronomic traits. For instance, Nguessan and Quisenberry (1994) detected various rice lines that could tolerate damage from *Sitophilus oryzae*. However, these lines did not meet the desired agronomic standards. This represents a significant challenge in integrating tolerance into crop plants and is a matter that researchers must address. Another substantial restriction is that tolerant crops might be more susceptible to superficial damage compared to crops that exhibit different types of resistance. Consumer preferences for the appearance of vegetables and fruits are influenced by how the products look. Therefore, consumers' preference for unblemished food items could restrict the adoption of tolerance in certain crop species.

18.4 SPECIFIC PLANT'S RESPONSE TO HERBIVORES

Plants employ various types of defense mechanisms in response to insect attacks, including constitutive, induced, direct, and indirect defenses. They have established a multi-layered defense system to combat insect pests. This diverse array of defense strategies ranges from external characters to the production of chemical 2° metabolites and adaptive responses. In some cases, when the direct defense involving the creation of toxic 2°metabolites to deter insects is ineffective, plants release substances that attract natural predators of the damaging insects (Erb and Reymond, 2019). Throughout this indirect defense, plants provide habitat and sustenance to these predators, reducing the impact of herbivory pressure.

A plant's classification as tolerant or resistant depends on its capacity to adapt strategies for self-defense within a given context. Fundamentally, if a plant's physical and chemical attributes can deter insects and reduce the harm done by herbivores, it is categorized as resistant. Resistant plants aim to discourage insect development and sustenance, often exerting significant selective pressure on the herbivores that attack them. Conversely, tolerant plants may not deter insects, but they can mitigate the negative impacts of herbivory. Tolerant plants typically impose minimal selective pressure, allowing insects to thrive without significant hindrance. Insect activities like herbivory, egg-laying, and colonization can trigger a range of plant defense mechanisms. These responses include fortifying or altering physical barriers, producing compounds that counteract insects, emitting herbivore-induced plant volatiles (HIPVs), generating HR/ROS, activating defense signaling pathways, and expressing defense-related genes (Blenn et al., 2012). Moreover, prepping the plant's defenses against insect attacks enables a swift initiation of these protective responses. Consequently, plants employ various individual or combined strategies to reduce or discourage herbivores (Table 18.1).

TABLE 18.1
Plant-Protecting Proteins against Herbivores

Defense Protein	Observed Plant Species	Target Insect Pests	Sources
LOXs	*Sitobion avenae*	*Triticum aestivum*	Zhao et al. (2009)
	Bemisia tabaci	*Nicotiana attenuata*	Kempema et al. (2007)
	Myzus persicae; *Macrosiphium euphorbiae*	Tomato	Fidantsef et al.(1999)
PPOs	*Helicoverpa armigera*	Tomato	Johnson et al. (2009)
	Spodoptera frugiperda	Buffalo-grass	Heng-Moss et al. (2004)
SOD	*Aphis medicaginis*	*Medicago sativa*	Huang et al. (2007)
PIs	*Schizaphis graminum*	*Sorghum bicolor*	Zhu-Salzman et al. (2004)
	Plutella xylostella	Transgenic *Arabidopsis*/oil seed rape	De Leo et al. (2007)
	Spodoptera littoralis	*Gossypium hirsutum*	Dunse et al. (2010)
	Spodoptera exigua	*Nicotiana attenuata*	De Leo et al. (2007)
Chitinases	*Schizaphis graminum*	*Sorghum bicolor*	Zhu-Salzman et al. (2004)
Peroxidases	*Populus* sp.	*Lymantria dispar*	Chen et al. (2005)
	Spodoptera frugiperda	*Oryza sativa*	Stout et al. (2009)
	Bemisia tabaci	*Arabidopsis*	Kempema et al. (2007)
	Spodoptera littoralis	*Zea mays*	Chen et al. (2009)
Chitinases	*Schizaphis graminum*	*Sorghum bicolor*	Zhu-Salzman et al. (2004)

18.4.1 Herbivory Elicitors

Any substance originating from herbivores that have a cellular-level interaction with a plant is considered a possible elicitor. To date, we have identified herbivore-specific elicitors in the oral exudations of lepidopteran larvae and the oviposition fluid of weevils. These fluids frequently come into contact with plant injuries. Furthermore, there is some indication that microorganisms residing in the digestive systems of herbivores play a role in generating elicitors observed in oral exudations (Spiteller et al, 2000). The first category of lepidopteran elicitors consists of lytic enzymes (glucosidase) derived from *Plasmodiophora brassicae* in cabbage, which prompts the release of terpenoid volatiles (Mattiacci et al., 1995). Similar lytic enzymes have been discovered in the secretions of various other lepidopterans (glucose oxidases) in the saliva of *H. zea*, as well as in sucking insects like whitefly (*B. tabaci*) (Funk, 2001), where alkaline phosphatase is found in their saliva. Additionally, a wide range of watery digestive enzymes are present in the saliva of aphids (Miles, 1999), but it is important to note that their functions as elicitors of defense responses have yet to be definitively confirmed.

The second group of elicitors consists of fatty acid–amino acid conjugates (FACs) and has been identified as the regurgitant of larval Geometridae, Noctuidae, and Sphingidae (Halitschke et al., 2001). One specific FAC, called FAC-volicitin, was discovered in *S. exigua* and has the ability to trigger the damaged maize plants to emit the same combination of volatile indole and terpenoid odors that are released when the plants are attacked by caterpillars (Alborn et al., 1997). But according to a recent study on unharmed maize plants, volicitin only caused plants to release VOCs around midnight, and the amount released by injured seedlings was much more than that of unharmed plants. Moreover, volicitin was discovered to be less effective than jasmonic acid (JA) at evoking volatile organic compound (VOC) emission (Schmelz et al., 2001). Interestingly, it's worth noting that none of the isolated enantiomers of volicitin displayed elicitor activity in the lima bean, a host for the beet armyworm (Spiteller et al., 2001). This suggests that there is a need for further investigation into the role of volicitin as an elicitor.

In recent times, a new category of specific elicitors of herbivores has been identified, and they trigger a unique form of defense response in pea plants. These particular elicitors are referred to as bruchins and are long-chain diols that have been diesterified with 3-hydroxypropanoic acid (3-HP). They are so named since they are expelled alongside the oviposition fluid of cowpea and pea weevils belonging to the family Bruchidae. In a few pea plant varieties, bruchins prompt the development of neoplastic growth on the pods (Doss et al., 2000). This growth raises the eggs out of their oviposition site, effectively blocking larval entry into the pods. This neoplastic growth has the added benefit of exposing the larvae to potential threats like parasites, predators, and desiccation, making it function as an indirect defense mechanism. Equally, when elm leaf beetles, scientifically known as *Xanthogaleruca luteola*, lay their eggs, it triggers the release of specific VOCs from the field elm trees they inhabit. These VOCs, induced by the act of oviposition, serve to attract parasitic wasps that prey on the beetle eggs (Meiners and Hilker, 2000). However, the exact biochemical root of this elicitation process remains a mystery.

The initial stages of the process by which herbivores trigger plant responses are not yet fully understood. Unlike the well-defined R-gene products involved in pathogen recognition, no equivalent receptors have been identified for recognizing herbivore elicitors. Some evidence suggests that glucose oxidases may enhance the production of hydrogen peroxide (H_2O_2) at the site of the attack, potentially amplifying the elicitation by generating ROS. Another proposed mechanism, put forth by Engelberth et al. (2000), involves a channel-forming peptide called alamethicin, which is generated by *Trichoderma viridae*. Similar to the volatile chemicals released when the salicylic acid (SA) cascade antagonizes the JA signaling pathway, this peptide also causes the lima bean to release them. On the contrary, it is noteworthy that these channel-forming peptides have yet to be found in insect herbivore exudate. To date, the only system that has been able to establish a connection from the elicitor to the successive signaling processes is the *Nicotiana attenuata-Manduca* system. In this system, FACs found in the oral exudations of two *Manduca* species trigger significant releases of JA and ethylene (ET). These released compounds play a role in modifying the wound responses of *N. attenuate* (Schittko et al., 2000; Winz and Baldwin, 2001).

18.4.2 Octadecanoids

Researchers have confirmed the importance of octadecanoids in plant–insect interactions by manipulating three enzymes in the pathway: JA carboxyl methyltransferase (JMT), lipoxygenase (LOX), and hydroperoxide lyase (HPL). The enzyme LOX, which is composed of nonheme iron, aids in the process of oxygenating linolenic acid, leading to the production of 9-and 13-hydroperoxides. These hydroperoxides are then transformed into oxoacids and aldehydes. The products derived from 13-hydroperoxy linolenic acid can undergo additional enzymatic processes involving reduction, cyclization, and β-oxidation, ultimately leading to the production of JAs. LOX proteins have various roles in plants, including influencing growth, development, maturation, senescence, and metabolic reactions to pathogen attacks and injuries. These responses are believed to be resolved through the synthesis of oxylipin signals such as oxo-phytodienoic acid (OPDA) (ortho phenylene diamine) and JA (Creelman and Mullet, 1997; Schaller, 2001). Researches indicate that different forms of LOX enzymes serve various functions. In the case of *Arabidopsis thaliana*, reducing the expression of a specific LOX isoform through co-suppression led to decreased levels of JA induced by wounds (Bell et al., 1995), while the basal JA levels remained unaffected. Multiple LOX isoforms are coded by gene families, with at least three such isoforms identified in potatoes. When transgenic potatoes were created without the LOX-H3 isoform through antisense technology, they did not produce proteinase inhibitors (PIs) in response to damage. Interestingly, when the larvae of beet armyworm and Colorado potato beetle were fed on these antisense plants, they became larger compared to those fed on normal plants. What's puzzling is that the impact of LOX-H3 on resistance did not seem to be related to its role in the JA increase following wounds. Both antisense and normal plants had similar JA levels after injury, and even adding extra JA couldn't restore normal PI levels in

LOX-H3 antisense plants (Royo et al., 1999). These findings imply that LOX-H3 likely generates a signal independent of JA through various mechanisms. It could directly influence a response in *pin* genes, be part of a regulatory network affecting the appearance of other LOX genes, or impact the amount of ET in response to injury, which is needed, along with JA, for optimal *pin2* expression in tomatoes (O'Donnell et al., 1996). An interesting observation is that LOX-H3 antisense potatoes produce more flowers and about 20% more tubers on average than normal plants, implying that the resistance associated with this LOX isoform may come at a cost in terms of plant fitness when not required (Royo et al., 1999).

A resultant product of LOX enzymes, namely 13(*S*)-HPOT, serves as a substrate for numerous other enzymes that perform a crucial role in communications between insects and plants. This includes the enzyme HPL, which breaks down 13-hydroperoxide into two types of compounds: C_6-aldehydes and C12-oxoacids (Schaller, 2001). The C_6-aldehydes are believed to serve multiple functions, such as protecting the area of plant damage from microbial infections, acting as a direct defense against certain leaf herbivores, and potentially triggering the expression of genes associated with plant defense. On the contrary, the C_{12}-products, which comprise traumatic acid and traumatin, may be tangled in the process of wound healing (Noordermeer et al., 2001). Additionally, nowadays, it has been advised that C_6-aldehydes may also play a role in directly stimulating the expression of genes related to defense, participate in systemin-based signaling, and facilitate communication between plants (Farmer, 2001).

Recent research involving potato plants with reduced HPL activity has revealed that the expression of the HPL gene is controlled by the plant's developmental stage, and the levels of its activity are regulated after the transcription of the gene, which aligns with findings from similar research on tomatoes (Vancanneyt et al., 2001). Furthermore, these HPL-depleted plants displayed normal gene expression in response to wounds, including genes like LOX-H1, allene oxide synthase (AOS), LOX-H3, *Pin2*, prosystemin, and LAP, even though they produced lower levels of hexenal and 3-hexenal. This suggests that these specific chemical compounds do not serve as signaling molecules. Notably, aphids, *Myzus persicae* did more damage on plants with reduced HPL activity than on wild plants, indicating that the C_6-aldehydes produced by HPL may play a significant role in the plant's direct defense mechanisms (Vancanneyt et al., 2001). When the LOX enzyme produces 13(*S*)-HPOT, and it doesn't get broken down by HPL, it can serve as a substrate for AOS, which generates an unstable compound called allene oxide. This allene oxide is then converted into a chemically pure form of 9*S*,13*S*-OPDA through cyclization by allene oxide cyclase (Schaller, 2001). Subsequent processes, including β-oxidation and reduction, lead to the production of JA. Both JA and OPDA dynamically regulate the appearance of genes linked to plant defense and trigger resistance mechanisms.

Many of the genes associated with the biosynthesis of JA are activated in response to physical injury, and some of these genes, such as LOX, AOS, and 12-OPDA reductase, are also stimulated when JA is applied externally. This suggests that there is a positive feedback mechanism in place. In one particular case, introducing extra AOS from flax into potato plants increased the levels of JA even without external stimulation. However, this didn't impact the expression of *pin2* genes, implying that there might be a signal for *pin*-gene expression that is independent of JA. While most octadecanoids are known to be involved in responses that affect the entire plant, their role as a mobile signaling system hasn't been definitively proven. Experiments using labeled 14C-JA applied to *Nicotiana sylvestris* plant leaves provided evidence that JA can travel from wounded leaves to the roots, as shown by the temporary increase in JA levels in the roots after leaf injury. This suggests that JA can serve as a signal transmitted from injured leaves to the plant's roots (Zhang et al., 1997).

In contrast, experiments using glucuronidase reporter genes fused with the AOS promoter from *N. tabacum* and *A. thaliana* showed local and systemic stimulation of AOS occurred upon wounding, but this induction was limited to the site where OPDA or JA was applied. This suggests that these two octadecanoids were not carried throughout the transformed plants. Conversely, a gene responsible for converting JA into methyl jasmonate (MJ), known as S-adenosyl-L-methionine–JMT, was recently identified in *A. thaliana*. JMT gene expression increased both locally and systemically after injury and MJ treatments. When plants were genetically modified to overexpress JMT, they had higher MJ levels and constantly stated JA-responsive genes like *PDF1. 2* and *VSP*, even though their JA levels were normal. This led to enhanced resistance against *Botrytis cinerea*, implying that MJ can control the expression of defense genes and potentially function as a mobile signal within and between plants. Earlier research had also suggested that MJ could act as an airborne signal for interplant communication during defense responses (Seo et al., 2001).

18.4.3 Wound Signals

Both physical damage and herbivory feeding trigger two types of plant responses that spread throughout the entire plant or are confined to the site of injury. The systemic responses rely on portable signals, which can take the form of electrical (Herde et al., 1999), hydraulic (Malone and Alarcon, 1995), or chemical (Ryan, 2000) signals. In the case of tomatoes, the production of PIs is stimulated by signals carried through both the xylem and phloem. When the petiole is steam-girdled, effectively killing the phloem while leaving the xylem intact, this does not impede the systemic induction of PIs following severe wounding, indicating the presence of a xylem-transported signal. However, small injuries do not trigger a signal transported through the xylem; instead, they lead to PI production in distant organs that still have intact phloem connections (Rhodes et al., 1999). Furthermore, the expression of the *Pin2* (PI II) gene, which codes for a PI, can be brought by electrical indicators correlated with wounding. The response and transmission of wound-induced electrical signals seem to depend on the intact abscisic acid (ABA) signaling pathways, as they are absent in tomato mutants, and deficient in ABA (Herde et al., 1999).

Immediately after injury, plants produce ROS (containing the O^{2-} within the wounded tissue) and hydrogen peroxide (H_2O_2) in the damaged area and systemically throughout the entire plant. Oligogalacturonides, which are also produced in response to injury, are believed to serve as a key signal indicating tissue damage, as they can transiently induce ROS production. Additionally, both oligogalacturonides and chitosan derived from fungi can trigger the generation of ROS and the appearance of PI genes that respond to wounds. However, oligogalacturonides are also brought about by systemins and have limited mobility. As such, they probably represent a local, transitory stage in the signaling process after system formation rather than serving as the main mobile signal. Systemin, an oligopeptide with phytohormonal properties, is believed to be the crucial signal for injury in certain solanaceous plants like black nightshade, pepper, tomato, and tobacco. It is derived from prosystemin and is likely moved through the plant's phloem. Systemin is highly active even at very low concentrations (femtomolar levels) and is the most well-validated mobile signal (Ryan, 2000). In tomatoes that have been genetically modified to express the prosystemin cDNA in an antisense orientation and cannot express the PI I and II after wounding, rendering them more vulnerable to *M. sexta* (Orozco-Cardenas et al., 1993). Conversely, plants engineered to overexpress the prosystemin gene demonstrate constantly activated responses to injury and increased expression of PI transcripts.

18.4.4 Wound-Elicited Responses

Wound-induced resistance in plants is largely governed by compounds originating from the 'octadecanoid' pathway, primarily involving C_{18}fatty acids. The formation of numerous defense-related substances, such as antinutritive factors, toxins, and antidigestive enzymes, relies on signaling by octadecanoids, including 12-OPDA, JAs, and MJ, all of which are derivatives of linolenic acid (Creelman and Mullet, 1997). It is not entirely clear how herbivore-specific elicitors affect the local and systemic signaling pathways that spread the signals produced at the site of damage and alter the defense-related gene expression throughout the plant. A promising number of studies employ herbivore bioassays to assess the effects of various manipulations, especially those related to signal transduction pathways. Additionally, there is a surge in research employing insect attack and transcriptional profiling to recognize new signaling routes.

18.5 ROLE OF PLANT'S SECONDARY (2°) METABOLITES IN INSECT DEFENSE

Plants produce a diverse array of bioactive molecules and 2°metabolites that counteract insects. These naturally occurring plant chemicals serve to not only diminish insect damage but also restrict the scope of herbivory injury and influence successive defense mechanisms. While 2°metabolites primarily play a role in direct defense, they can also participate in indirect defense by providing a habitat for natural predators of particular insects (Erb and Reymond, 2019). Many bioactive compounds created by plants function as toxins for insects, impacting their digestive and nervous systems, which can lead to slowed growth or even mortality. Additionally, these compounds play a crucial role in controlling the taste, smell, and appearance of a plant or its various parts (Kessler et al., 2006). Alkaloids are present in nearly all plants and perform a vital role in defending insect pests. The presence of endophytic fungi, *Acremonium coenophialum* in *Festuca arundinacea*, limits aphid feeding through the synthesis of pyrrolizidine alkaloids (PAs) (Johnson, 2011). These PAs are converted into toxic forms when they enter the basic digestive tracts of insects, making them powerful anti-feeders that primarily deter aphids and other pests. Moreover, modified forms of glucosinolates, like indole glucosinolate in *A. thaliana*, provide increased resistance against *M. persicae* (Erb and Reymond, 2019). Plant terpenoids act as repellents, toxins, and deterrents, which inhibit insect attacks in plants. Terpenoids are found in volatile compounds and resins and are essential elements of plant essential oils, all of which have a negative impact on insect attacks.

Plant peptides and various components contribute to plant defense by hindering the digestive abilities of insects. Substances like chitinases, lectins, and α-amylase inhibitors (α-AI) function as antidigestive proteins, disrupting the assimilation of consumed plant parts, or as antinutritive agents, reducing herbivore intake of plant parts. Plants produce PIs that serve to impede the digestive system of insects. Insects from the Hemiptera class possess vital enzymes like cysteine proteases and metalloproteinases; inhibiting these enzymes within the insect's digestive system can lead to severe mortality of these insects. Similarly, plant α-AI hinders the α-amylases in attacking herbivory insects, thereby limiting the breakdown of starch in their diet. In the case of wheat, α-AIs have been observed to deter beetles, mealworms, and weevils from targeting the grains and the plants (Fürstenberg-Hägg et al., 2013). The introduction of bean α-AIslinto *Pisum sativum* has provided resistance against the weevil, *Bruchus pisorum* (Morton et al., 2000). Transgenic tomato varieties containing poplar chitinase have shown resistance against *Leptinotarsa decemlineata* by preventing their development (Lawrence and Novak, 2006).

Similarly, polyphenol oxidase (PPO) enzymes also play a role in plant defense, often triggered by mechanical injury or wounds. When insects feed on plants, they damage plant cells, leading to the release of PPOs that, in turn, generate ROS (Mahanil et al., 2008). Elevating the expression of tomato, PPO has been shown to enhance resistance to *S. litura* by impeding its growth and enhancing mortality rates. Many plants have resin and laticifers ducts that store resins and latex. When insects attack, these ducts rupture, causing latex to be released at the site of the injury, deterring or trapping the insects. In addition to being adhesive to capture insects, the latex of certain plants can also contain toxic substances. For example, *A. cannabinum*'s latex includes alkaloids, phenolics, PIs, and terpenoids, which serve as anti-nutrients or toxins when consumed by herbivores (Erb and Reymond, 2019).

18.6 PLANTS' DEFENSE AGAINST INSECTS THROUGH SPECIFIC CHEMICALS

Plants produce various specialized compounds, such as VOCs, nectars, and food bodies, which serve to support, attract, and provide shelter to scavengers that prey on insects, thus indirectly contributing to the defense against insects. VOCs are primarily generated in roots and flowers to draw in insects, pollinators, and scavengers. Maize plants generate over 30 volatile compounds, such as sesquiterpenes and (E)-α-bergamotene, when subjected to damage from the leaf worm, *S. littoralis* (Erb et al., 2012). Remarkably, VOCs, or HIPVs, also play a role in communication among plants within and between them and in priming defense responses against insects. For instance, *Artemisia tridentata* and *N. attenuata* plants have been observed to engage in this behavior, where injured *A. tridentata* plants release VOCs that induce chemical defense responses in *N. attenuata*(Kessler et al., 2006). Similarly, when undamaged leaves are exposed to volatiles released from wounded leaves, it triggers heightened defensive reactions against larval feeding by the *Lymantria dispar* (Maffei et al., 2012). Plants generate nutrient-rich food bodies (FBs) as a way to establish a mutually beneficial relationship with other organisms, which, in turn, provide safety against insects. An example of this mutualism is seen in the relationship between *Pheidole bicornis* and *Piper fimbriulatum* ants, where the ants defend the plant from various insects while benefiting from the FBs as a food source (Fischer et al., 2002). Many plants also produce nectars to attract pollinating insects, parasitosis, and predators, significantly contributing to indirect defense (Erb and Reymond, 2019). While nectars are typically found in flowers, extrafloral nectaries (EFNs) are created and placed on plants' leaves and shoots. The emission of these EFNs increases when the plant is under attack by insects.

18.7 STRUCTURAL AND PHYSICAL ADAPTATION IN INSECT DEFENSE

Plants obtained modified physical characteristics like waxy coating, trichomes, hard cuticular layers, and spines to counteract insects with different feeding habits, providing a robust defense against insect attachment, herbivory, and egg-laying. Forming a protective film or crystalline layer through epicuticular waxes, plants prevent insects from landing, feeding, or depositing eggs on them (Blenn et al., 2012). When subjected to insect-induced stress, plants can alter the production and composition of these waxes from their natural state. For instance, when the cabbage white butterfly lays eggs on *A. thaliana*, it triggers alterations in the wax concentration, increasing the amount of tetratriacontanoic acid (C34) and reducing the tetracosanoic acid (C24) (Bricchi et al., 2012). This changing composition serves to attract egg parasitoids like *Trichogramma brassicae*. Likewise, plants employ strategies to deter herbivory by increasing the toughness of their leaves and roots. The reinforcement of the epidermal structures of plants discourages insect feeding, making it difficult for insects to consume them. Root toughness is often improved by depositing lignin polymers, which act as a deterrent against insect attacks. Additionally, the deposition of substances such as silica, suberin, callose, and cellulose reinforces cell walls, effectively limiting insect feeding by creating a formidable barrier.

Plants have additional strategies for defense against herbivores, including stimulating significant root regrowth when under attack, leading to increased root numbers and higher root density. Alterations to plant leaves, such as the development of spines and thorns and the expansion of adaptive features like trichomes, also serve as protective measures against insects. These characteristics deter larger leaf herbivory from feeding on the plants and impede their mobility and attachment (Aljbory and Chen, 2018). The trichomes, with or without glands, also play a major role in pest management. In the case of *N. attenuata*, the glandular trichomes generate a substantial amount of O-acyl sugars. These sugars tortuously prompt the larvae of *Manduca sexta* to release volatile compounds, which, in a cascading effect, attract predators of the larvae (Cheng et al., 2013b). Likewise, a higher density of trichomes on raspberry leaves acts as a deterrent, preventing or decreasing egg-laying by the mite, *Tetranychus urticae*(Karley et al., 2016).

18.8 SIGNALING AND PERCEPTION IN PLANT–INSECT INTERACTIONS

In addition to physical and biochemical defenses, plants depend on various internal functions to detect and respond to herbivory attacks. These processes include recognizing specific compounds produced by insects, maintaining ion gradients across the plasma membrane, triggering bursts of calcium ions (Ca^{2+}), generating ROS, and initiating complex signaling pathways (Erb et al., 2012). The stimulation of these signaling events successively leads to the accumulation of hormones, and the activation of defense-related genes and genes involved in the production of phytohormones

(Blenn et al., 2012). Typically, the plant detects the presence of herbivores at the place of attack, but this perception can extend to neighboring cells and through the entire plant, triggering a systemic defense pathway. In many cases, insects that attack plants secrete oral secretions (OSs) or substances known as elicitors. Specific receptors are present on the plant cell membrane to detect these elicitors (Maffei et al., 2012). The initial signals triggered by the leaf herbivory include changes in the electrical charge of the plasma membrane (depolarization), movements of ions in or out of the cell, and bursts of ROS or Ca^{2+}. A significant portion of insect OS consists of compounds called FACs, which can serve as potent elicitors for activating defense responses in the plant. For example, when cotton leafworms feed on lima beans and release their OS, it rapidly depolarizes the plasma membrane, initializing defensive behavior against the insect. The depolarization of specific membrane proteins and membrane potentials, such as glutamate receptor-like (GLRs) proteins, play a role in regulating the expression of genes made by JAs and transmitting signals related to plant injuries (Mousavi et al., 2013). These FACs and OS, in turn, trigger kinase signaling pathways that are essential for modulating the plant's defense against pest attacks. Applying FACs derived from *Manduca sexta* to injured leaves of *N. attenuate*, triggered the initiation of salicylic acid-induced protein kinase, MAPKs (mitogen-activated protein kinases), and wound-induced protein kinase resulted in the gathering of hormones like JAs, SA, and ET (Wu et al., 2007). Likewise, when BPH is fed on rice, it leads to the expression of multiple *Oryza sativa* mitogen-activated protein kinases (Nanda et al., 2018).

Again, the respiratory-burst oxidase homolog (RBOH) has the ability to interact with calcium ions (Ca^{2+}) and MAPK signaling in addition to generating ROS, indicating their crucial role in herbivory-induced defense responses. The combination of ROS production and Ca^{2+} signaling has been observed in *A. thaliana*, where the presence of Ca^{2+} effectively activates RBOH (Drerup et al., 2013). The plant's ROS-mediated defense signaling against sap-suckers is typically achieved over the accretion of hydrogen peroxide (H_2O_2) and increased activity of enzymes like peroxidases (POD), and catalases (CAT), which is a common phenomenon across various plant species (Kerchev et al., 2012). Among various ion species, calcium ions (Ca^{2+}) play a crucial role in changing cell membrane potentials and transmitting signals during herbivory feeding. Under usual conditions, the concentration of Ca^{2+} ions in the apoplast and cytoplasm typically remains in the micromolar and nanomolar ranges, respectively. However, when insects feed on the plant, the balance of Ca^{2+} ions is disrupted, resulting in a surge of Ca^{2+} ions entering the cytosol. This shift in ionic concentrations due to the Ca^{2+} surge triggers subsequent signaling cascades for defense responses (Dodd et al., 2010). Phytohormones have a crucial role in precisely adjusting plant defense mechanisms. JAs serves as a central player in fine-tuning plant defenses when confronted with insect attacks, promoting both direct and indirect defense strategies (Yang et al., 2019). When plant parts are damaged by herbivory or physical injury, the intracellular concentration of JA increases and accumulates in plant cells. This buildup of JA triggers a range of defense responses, which can encompass physical changes like the development of trichomes, as well as chemical reactions such as the release of VOCs, the emission of EFNs, the accumulation of 2° metabolites, and the expression of JA-responsive genes (Wasternack et al., 2013).

In tomatoes, SA provides defense against *H. armigera* by promoting the generation of ROS (Peng et al., 2004). SA-induced accumulation of hydrogen peroxide (H_2O_2) also deters the herbivore from feeding on the plant, as higher concentrations of H_2O_2 have negative effects on the insect's digestive system and its development (Maffei et al., 2007). In a similar manner, ET signaling collaborates with SA and JA to either trigger or inhibit particular aspects of the plant's defense system against insect attacks. For example, the production of ET aided the growth of the fall armyworm (*S. fugiparda*) in maize (Harfouche et al., 2006). Conversely, disrupting the ET signaling pathway resulted in a reduced aphid population in tomatoes and increased resistance to *S. littoralis* in *A. thaliana* (Mantelin et al., 2009). Ethylene also plays a role in inducing the release of various VOCs during interactions between plants and insects, as detected in lima beans, maize, and European alder plants (Erb and Reymond, 2019).

18.9 MOLECULAR ADVANCES IN DIRECT DEFENSE

It is possible to modify almost any characteristic in a plant to serve as a direct defense mechanism. Nevertheless, several research in this field have concentrated on a wide range of 2° metabolites that act as toxins, substances that reduce digestibility, or repellents. While various potential direct defense traits have been identified, concrete evidence of a particular metabolite's defensive role is largely limited to a few cases where we thoroughly understand how it is produced and can manipulate its expression. In the ongoing competition among plants and insects, plants are estimated to adapt and make innovative defenses as insects develop resistance to the primary set of defenses (Ehrlich and Raven, 1964). Therefore, the mechanisms behind the initiation of flexibility in direct defenses are of great interest. Most of the molecular research in this area has concentrated on model plant systems from the Solanaceae and Brassicaceae families, and, as a result, these systems have seen the most significant advancements in our understanding of direct defenses. Because PIs and other protective proteins are direct products of plants, researchers have assessed their defensive effects by genetically modifying various plant species. For instance, they have recently introduced an *A. thaliana* cysteine PI gene (*Atcys*) into a white poplar, which rendered the plant resistant to *Chrysomela populi*. This resistance was achieved by inhibiting the digestive proteinase activity of the beetles from the family Chrysomellidae (Delledonne et al., 2001). To stay ahead of herbivores that are constantly modifying themselves, plants have developed modern pathways for producing many new active PIs. Among these, few may be capable of retaining their protective function in the ever-changing chemical ecosystems of a plant as it grows and ages. The arrangement of certain PIs, which may have multiple domains, allows a plant to generate inhibitors, effective for a wide range of proteases and maintaining their protecting function in diverse chemical surroundings.

Throughout evolutionary history, genes responsible for these PIs have accumulated numerous mutations in their reactive sites. This is a pattern we would expect to see in an ongoing evolutionary competition where there's pressure to diversify the properties of the PIs, possibly as a response to the changing strategies of herbivores. *Arabidopsis* plants produce a variety of glucosinolates, which serve to protect them from generalist herbivores and pathogens. However, these compounds can also act as attractants for herbivores and oviposition specialists in the Brassicaceae family. This diversity in metabolite profiles within populations is likely maintained due to different selection pressures. A model explaining how the diversity of glucosinolate profiles, both in terms of types and amounts, is preserved was suggested by Kliebenstein et al. (2001). They looked at 39 different *Arabidopsis* ecotypes and found that 14 different leaf glucosinolate profiles, or 34 different structures, could be produced by variations in just five specific genetic loci, each of which is responsible for different branching points in the glucosinolate biosynthesis pathway. Most of these glucosinolates are produced through the extension of the methionine molecule. Furthermore, it was observed that a single locus had a substantial influence on a significant portion (about 75%) of the quantitative variation in glucosinolate levels. Researchers have employed genetic transformation to investigate the significance of both the types and amounts of glucosinolates in *Arabidopsis* plants. For instance, when they introduced the *CYP79A1* gene from *Sorghum bicolor* into *Arabidopsis* plants, this gene, which typically alters L-tyrosine into p-hydroxyphenylacetaldoxime during the production of cyanogenic glycosides, caused the production of p-hydroxybenzylglucosinolate (specifically, sinalbin, a glucosinolate not usually observed in *Arabidopsis*). This genetic alteration resulted in a fourfold rise in total glucosinolate levels, primarily due to the higher production of sinalbin (Nielsen et al., 2001). However, despite this significant increase in glucosinolate investment, the acceptance of these transformed plants by two specialist flea beetles from the Brassicaceae family (*Phyllotreta nemorum* and *P. cruciferae*) was not affected when tested in choice experiments. This indicates that even with a substantial increase in glucosinolate production, the plant is not likely to experience reduced herbivore pressure from these specialist insects.

To eliminate a specialist herbivore community from a plant, it may be necessary for the plant to develop an entirely new defense mechanism. Recent advancements in plant transformation techniques have enabled researchers to simulate such an evolutionary event. For example, they have successfully introduced the entire biosynthetic pathway for dhurrin, a type of cyanogenic glucoside not typically found in Brassicaceae plants, into *A. thaliana*. This was achieved by expressing two versatile microsomal P_{450} enzymes (CYP71E1 and CYP79A1) and a solvable UDPG-glucosyltransferase (sbHMNGT) (Tattersall et al., 2001). Remarkably, these three enzymes naturally unified themselves into a functional complex that effectively facilitated the transfer of reaction products. As a result, the genetically modified plants could produce and store significant quantities of dhurin. These plants exhibited complete resistance to *P. nemorum*, a specialist herbivore of the Brassicaceae family. This breakthrough in transferring entire 2° metabolite pathways between different plant species empowers researchers to investigate an elementary principle put forth by Ehrlich and Raven's (1964) theory of plant-herbivore coevolution. This theory suggests that the diversity of 2°metabolites serves as a limitation on the advancement of herbivore-host selection. In *Arabidopsis* plants, their resistance to herbivores has been linked to glucosinolate production and breakdown (specifically by myrosinase enzymes), as well as the presence of trichomes, which are hair-like structures on the plant's surface (Mauricio, 1998). Yet, a recent genetic analysis showed a specific genetic locus that couldn't be linked to any of the formerly known traits associated with resistance.

Jander et al. (2001) performed crosses between the Columbia ecotype, which showed much better resilience, and the Landsberg *erecta* ecotype, which was sensitive to the larvae of the generalist nocturnal insect *Trichoplusia ni*. This genetic region on chromosome 1 known as the '*TASTY*' locus was linked to this vulnerability. This locus was particularly noticeable in genes affecting trichome density, blooming duration, disease resistance, and glucosinolate levels. It was, however, close to a recently identified locus known as '*esp*', which is responsible for generating epithionitriles throughout the hydrolysis of glucosinolates instead of isothiocyanates (Lambrix et al., 2001). Glucosinolates are compounds that are broken down during injury and herbivore attacks, making them important for plant defense. This outcome underscores the significance of studying direct defense mechanisms not only in plants but also within the digestive systems of insects that feed on these plants. Herbivores consume plants within an ecosystem that they can biochemically influence, and they adapt their detoxification and digestive systems to counteract the effects of the plant's direct defenses. Investigating these mechanisms and the biochemical processes that occur in the guts of insects could offer valuable understandings of the plant characteristics that serve as direct defenses.

However, it's essential to broaden the scope beyond just the insect gut and comprise aspects like insect activities and the NEs of insects. For instance, when *Brassica napus* were genetically modified with a gene from potatoes that codes for a PI, diamondback moth (DBM) larvae responded to their reduced digestive efficacy by consuming more plants (Winterer and Bergelson, 2001). Such compensatory behaviors indicate that many direct defenses, especially those that hinder herbivory development without killing them, may not be effective as defenses if they are not combined with indirect defenses. Indirect defenses are plant attributes that enhance the foraging efficacy of NEs of leaf herbivores. By employing a combination of defenses that hinder herbivore development and increase the likelihood of their mortality before they can reproduce, plant populations can significantly curb herbivore populations, thus creating a more effective defense strategy (Moran and Hamilton, 1980).

18.10 ADVANCEMENTS IN MOLECULAR MECHANISMS OF INDIRECT DEFENSIVE STRATEGIES

When assessing the costs associated with direct defenses, it becomes clear that certain defense mechanisms can place significant metabolic burdens on plants and reduce their overall fitness, especially when plants are under additional stress, like competition with other plants (Baldwin, 2001). As a result, plants have evolved defensive partnerships where they invest minimal resources in releasing information-containing VOCs or offering nutritional rewards to attract insects from higher trophic levels to help meet their defensive needs. This mutualistic arrangement benefits both plant genotypes that use traits, improving the efficiency of NE's genotypes capable of utilizing these plant traits (Dicke and van Loon, 2000). Many plants display similar defensive features inadvertently to herbivores, and many predators make good use of these characteristics to increase the efficiency of their foraging.

While there is growing evidence of the efficiency of these defenses in the ecosystems, there is still much to learn about the underlying mechanisms governing the manifestation of these attributes.

18.10.1 Volatile Organic Compounds

Researchers have analyzed the release of VOCs from over 15 plant species involved in various interactions, including plant-caterpillar-parasitoid, plant-caterpillar-predatory bug, and plant-spider mite-predatory mite interactions (Meiners and Hilker; 2000; Kessler and Baldwin, 2001). In all of these systems, the host plant releases VOCs that are triggered by damage and herbivory, serving as signals in interactions involving three trophic levels. A closer examination of the VOCs emitted by these plants has revealed the presence of many common signaling compounds shared among different plants. However, several compounds, specific to particular plant species and are released in response to cues from specific herbivores (Dicke and van Loon, 2000; Halitschke et al., 2001). Volatile compounds come mostly from three different biosynthetic routes. First, there are the 'green-leaf volatiles' (GLVs), which are C_6-aldehydes and -alcohols generated from α- linoleic acid and linolenic acid through their corresponding hydroperoxides (Noordermeer et al., 2001). Some of these GLVs may serve as direct defenses, as shown in a trial involving transgenic potatoes with reduced hydroperoxide lyase (HPL), which demonstrated reduced resistance to aphid populations. GLVs also act as kairomones, which are signals conveying information. For example, *Cis*-3-hexen-1-ol is often noticed in the air around plants after herbivory damage, and in a field study, increasing its release from plants attracted a generalist predator (Halitschke et al., 2000; Vancanneyt et al., 2001). Another green-leaf volatile, trans-2-hexenal, possesses biocidal properties and is commonly emitted by plants following herbivore injury. It triggers the production of sesquiterpenoid phytoalexins in damaged *Arabidopsis* and cotton, signifying its potential role in signaling within and between plants. The electrophilic α, β-unsaturated carbonyl group found in many *GLV* may enable them to bring defense and stress responses in plants, although these possible functions require further confirmation (Bate and Rothstein, 1998; Farmer, 2001).

The second group of volatile compounds originates from two isoprenoid pathways- mevalonate and nonmevalonate. Mono- and sesquiterpenes derived from these pathways have important roles as kairomones, fascinating parasitoids, and predators to plants under attack, and they also serve as phytoalexins, which are substances defending against pathogens (Turlings and Benrey, 1998). Following arthropod damage, many terpenoids are produced momentarily and systemically. Preliminary data suggests that the transcriptional level may be involved in the regulation of some terpenoids' release. Notably, cytosolic sesquiterpene synthases and plastidial monoterpene synthases have a high level of structure resemblance and similar reaction pathways, which has made it challenging to create functional knockouts for these genes. Although HMGR (hydroxymethyl glutaryl-CoA reductase) transcript levels rapidly increase in response to wounding and are further developed by herbivore regurgitants (Korth and Dixon, 1997) and C_6 aldehydes (Bate and Rothstein, 1998), no mutant plants lacking this gene have been developed. A maize mutant was identified with an *Ac* insertion (a transpositional mutation) in the sesquiterpene cyclase, *stc1*. Normally, this cyclase is instigated 15- to 30-fold by herbivory damage, insect OSs, and refined volicitin in wild plants. However, in the mutant plants, *stc1* was not incorporated, and analysis of emitted volatile compounds indicated that *stc1* codes for a synthase responsible for producing a naphthalene-based sesquiterpene.

The third pathway occupies a set of volatile mixtures released by plants when they are damaged by herbivores. These compounds originate from the shikimate pathway, which connects carbohydrate metabolism to the production of aromatic substances in plants and microorganisms (Herrmann and

Weaver, 1999). Methyl salicylate is developed from this pathway and emitted when plants are subjected to herbivore damage, but not when they experience mechanical injury, as seen in wild tobacco and lima beans. In experiments, when methyl salicylate was employed on lima beans, it attracted predatory mites searching for prey. However, in a field study involving wild tobacco, predatory bugs were not drawn to methyl salicylate (Kessler and Baldwin, 2001). Another metabolite arising from the shikimate pathway is indole, released by maize plants following injury by beet armyworms, but not in response to manual or mechanical injury. The combination of indole and terpenoids released by maize plants is appealing to the *Cotesia marginiventris*, which parasitizes several lepidopteran larvae (Turlings and Benrey, 1998). The IGL (indole-3-glycerol phosphate lyase) enzyme promotes the creation of free indole, and, similar to naphthalene-comprising terpenoids triggered by volicitin, it is triggered by caterpillar regurgitant, serving as an elicitor for these chemical responses. VOCs serve an ecological role by fascinating NEs to reduce herbivore pressure on plants, and this role has been observed in natural settings. However, there is currently no concrete evidence that plants relying on this indirect defense mechanism experience tangible fitness benefits (Kessler, 2001). To establish this, it is essential to gain a deep understanding of the mechanisms governing the release of herbivore-induced VOCs and to manipulate these mechanisms in real field conditions. Doing so will not only provide the necessary evidence but will also assess the potential agricultural value of this defense mechanism. Plants that emit VOC against herbivory attacks could open the door to a novel era in ecological pest control in agriculture.

18.10.2 Extrafloral Nectar

Plants employ various strategies, including extrafloral nectar production, to entice the NEs of herbivores. This is well-developed in ant plants, which generate extrafloral nectar and structures known as Mullerian and Pearl FBs to attract ants (Agrawal and Rutter, 1998). These ant mutualists produce effective defense, to the point that over evolutionary time, some ant-housing acacias have seemingly lost their biochemical defenses. EFNs are observed in more or less 66 plant families and have been shown to serve a defensive role by enticing ants and wasps that prey on the herbivores, thus reducing herbivore injury. While findings have demonstrated that the number of EFNs and the amino acid amount in the nectar increases the herbivore attacks, yet little was known about the mechanisms underlying this response. Research on *Macaranga tanarius* trees revealed that EFN secretion enhanced not only after herbivore feeding but also in response to mechanical leaf injury and the application of exogenous JAs. Moreover, the use of phenidone, an antagonist JA synthesis, hampered wound-induced extrafloral nectar secretion (Heil, 2001). It was found that both the transient rise in JA levels and the increase in nectar flow following damage were strongly linked to the extent of herbivore damage. Increased nectar secretion led to more visitors, including defenders, which significantly reduced herbivore damage. This indirect defense mechanism is reliant on the JA signaling cascade, similar to its role in herbivore-induced VOC release. Ecological research has revealed that the value of indirect defenses facilitated by plant traits, such as VOC emission and nectar production, primarily lies in the spatial information they offer to predators. These plant traits assist small predators in locating actively feeding herbivores within the considerably greater spatial dimensions of host plants. Consequently, for the successful implementation of indirect defenses in crops, it is crucial to maintain the spatial information conveyed by these signals.

18.11 INTRICATE INTERPLAY AND ORGANIZED SYNCHRONIZATION OF INSECT-TRIGGERED REACTIONS

The alterations in plant metabolism triggered by pathogen or herbivory attacks are likely controlled by complex alterations in gene expression. These changes encompass genes responsible for both 1° and 2° metabolism. Various methods, such as subtractive libraries, microarrays, DDRT-PCR, and AFLP-cDNA display, enable scientists to investigate modifications in the 'transcriptome', which represent the genes that are activated in response to herbivore attacks. These methods also help identify variations in gene expression between different plant genotypes with varying levels of resistance. Initial findings from the application of these techniques to study plant–insect relations indicate that the changes induced by herbivory attacks are similar in terms of scope and extent to those seen in response to pathogen attacks (Maleck et al., 2000; Schenk et al., 2000). The initial microarray investigation of plant–insect relations examined the dynamics, timing, and control of the countenance of 150 genes in *A. thaliana* that are activated in response to injury (Reymond et al., 2000). When assessing the responses to wounding over a specific period, different gene classes with similar behaviors were identified, one of which was connected to the arrival of cues from the jasmonate cascade, including JA, dinor-12-oxo-phytodienoic acid (dnOPDA), and 12-OPDA. However, not all genes within this cluster relied on JA signaling, as evident when comparing the responses of the JA-insensitive *coi1-1* mutant with those of wild-type plants. Furthermore, comparing the gene expression designs in plants injured by mechanical means and those damaged by *Pieris* caterpillars showed significant differences, predominantly in the expression of genes associated with water stress, which were diminished in insect-damaged plants. It's important to note that the timing and the extent of injury caused by insects were not perfectly replicated in the mechanically wounded plants, making it challenging to determine if the absence of drought-related gene expression was a definite retort to *Pieris* caterpillar feeding. This microarray analysis was conducted using a customized chip with a preselected set of genes for this particular plant–insect interaction. With the accessibility of microarrays that obscure the entire genome of *A. thaliana*, we eagerly anticipate the first truly impartial examinations of the transcriptional modifications prompted by leaf herbivory attacks in this model plant.

18.12 CONCLUSION

The plant's ability to tolerate biological stresses is influenced by a combination of its internal functions, physical structure, genetics, biochemistry, growth, and evolutionary factors. Researchers have gathered extensive data on genetic markers, specific genes, proteins, and substances linked to how plants defend themselves. This chapter points out these elements comprehensively, offering a foundation for understanding the diverse ways by which plants defend themselves. It also touches upon some genes and techniques that hold the potential for enhancing plant defense. The interaction between plants and pathogens or insects is intricate, involving multiple communication pathways that monitor and control the entry of these agents. Discovering and coordinating an efficient signaling pathway for defense, leading to an effective defense strategy, remains challenging due to the numerous genes and proteins expressed during these interactions. Nonetheless, certain components of tolerance show promise, though further research is needed to understand the exact signaling patterns involved in which these components participate. However, finding more efficient receptors is crucial to combat the elusive behaviors commonly seen in pathogens and pests. This is essential for ensuring that signals can effectively reach the subsequent components. Further research is required to explore how defense genes or QTLs can adapt to evolving biological threats and changing environmental conditions.

REFERENCES

Agrawal, A. A., & Rutter, M. T. (1998). Dynamic anti-herbivore defense in ant-plants: The role of induced responses. *Oikos*, 227–236. https://doi.org/10.2307/3546834

Alborn, H. T., Turlings, T. C. J., Jones, T. H., Stenhagen, G., Loughrin, J. H., & Tumlinson, J. H. (1997). An elicitor of plant volatiles from beet armyworm oral secretion. *Science*, *276*(5314), 945–949. https://doi.org/10.1126/science.276.5314.945

Aljbory, Z., & Chen, M. S. (2018). Indirect plant defense against insect herbivores: A review. *Insect Science*, *25*(1), 2–23. https://doi.org/10.1111/1744-7917.12436

Anderson, P. K., Cunningham, A. A., Patel, N. G., Morales, F. J., Epstein, P. R., & Daszak, P. (2004). Emerging infectious diseases of plants: Pathogen pollution, climate change and agrotechnology drivers. *Trends in Ecology & Evolution*, *19*(10), 535–544. https://doi.org/10.1016/j.tree.2004.07.021

Baldwin, I. T. (2001). An ecologically motivated analysis of plant-herbivore interactions in native tobacco. *Plant Physiology*, *127*(4), 1449–1458. https://doi.org/10.1104/pp.010762

Bate, N. J., & Rothstein, S. J. (1998). C6-volatiles derived from the lipoxygenase pathway induce a subset of defense-related genes. *The Plant Journal*, *16*(5), 561–569. https://doi.org/10.1046/j.1365-313x.1998.00324.x

Bell, E., Creelman, R. A., & Mullet, J. E. (1995). A chloroplast lipoxygenase is required for wound-induced jasmonic acid accumulation in Arabidopsis. *Proceedings of the National Academy of Sciences*, *92*(19), 8675–8679. https://doi.org/10.1073/pnas.92.19.8675

Blenn, B., Bandoly, M., Küffner, A., Otte, T., Geiselhardt, S., Fatouros, N. E., & Hilker, M. (2012). Insect egg deposition induces indirect defense and epicuticular wax changes in *Arabidopsis thaliana*. *Journal of Chemical Ecology*, *38*, 882–892. https://doi.org/10.1007/s10886-012-0132-8

Bricchi, I., Bertea, C. M., Occhipinti, A., Paponov, I. A., & Maffei, M. E. (2012). Dynamics of membrane potential variation and gene expression induced by *Spodoptera littoralis*, *Myzus persicae*, and *Pseudomonas syringae* in Arabidopsis. *PLoS One*, *7*(10), e46673. https://doi.org/10.1371/journal.pone.0046673

Chamarthi, S. K., Sharma, H. C., Vijay, P. M., & Narasu, M. L. (2011). Leaf surface chemistry of sorghum seedlings influencing expression of resistance to sorghum shoot fly, *Atherigona soccata*. *Journal of Plant Biochemistry and Biotechnology*, *20*, 211–216. https://doi.org/10.1007/s13562-011-0048-3

Chen, H., Wilkerson, C. G., Kuchar, J. A., Phinney, B. S., & Howe, G. A. (2005). Jasmonate-inducible plant enzymes degrade essential amino acids in the herbivore midgut. *Proceedings of the National Academy of Sciences*, *102*(52), 19237–19242. https://dx.doi.org/10.1073/pnas.0509026102

Chen, Y., Ni, X., & Buntin, G. D. (2009). Physiological, nutritional, and biochemical bases of corn resistance to foliage-feeding fall armyworm. *Journal of Chemical Ecology*, *35*, 297–306. https://dx.doi.org/10.1007/s10886-009-9600-1.

Cheng, X., Zhu, L., & He, G. (2013). Towards understanding of molecular interactions between rice and the brown planthopper. *Molecular plant*, *6*(3), 621–634.

Creelman, R. A., & Mullet, J. E. (1997). Biosynthesis and action of jasmonates in plants. *Annual Review of Plant Biology*, *48*(1), 355–381. https://doi.org/10.1146/annurev.arplant.48.1.355

De Leo, F., Bonadé-Bottino, M., Ceci, L. R., Gallerani, R., & Jouanin, L. (2001). Effects of a mustard trypsin inhibitor expressed in different plants on three lepidopteran pests. *Insect Biochemistry and Molecular Biology*, *31*(6–7), 593–602. https://dx.doi.org/10.1016/S0965-1748(00)00164-8

Delaney, K. J., Haile, F. J., Peterson, R. K., & Higley, L. G. (2009). Seasonal patterns of leaf photosynthesis after insect herbivory on common milkweed, *Asclepias syriaca*: Reflection of a physiological cost of reproduction, not defense? *The American Midland Naturalist*, *162*(2), 224–238. https://doi.org/10.1674/0003-0031-162.2.224

Delledonne, M., Allegro, G., Belenghi, B., Balestrazzi, A., Picco, F., Levine, A., Zelasco, S., Calligai,P., & Confalonieri, M. (2001). Transformation of white poplar (*Populus alba* L.) with a novel *Arabidopsis thaliana* cysteine proteinase inhibitor and analysis of insect pest resistance. *Molecular Breeding*, *7*, 35–42. https://doi.org/10.1023/A:1009605001253

Dicke, M., & van Loon, J. J. (2000). Multitrophic effects of herbivore-induced plant volatiles in an evolutionary context. *Entomologia Experimentalis et Applicata*, *97*(3), 237–249. https://doi.org/10.1046/j.1570-7458.2000.00736.x

Dodd, A. N., Kudla, J., & Sanders, D. (2010). The language of calcium signaling. *Annual Review of Plant Biology*, *61*, 593–620. https://doi.org/10.1146/annurev-arplant-070109-104628

Donath, J., & Boland, W. (1994). Biosynthesis of acyclic homoterpenes in higher plants parallels steroid hormone metabolism. *Journal of Plant Physiology*, *143*(4–5), 473–478. https://doi.org/10.1016/S0176-1617(11)81809-2

Doss, R. P., Oliver, J. E., Proebsting, W. M., Potter, S. W., Kuy, S., Clement, S. L., Williamson, R. T., Carney, J. R., & DeVilbiss, E. D. (2000). Bruchins: Insect-derived plant regulators that stimulate neoplasm formation. *Proceedings of the National Academy of Sciences*, *97*(11), 6218–6223. https://doi.org/10.1073/pnas.110054697

Drerup, M. M., Schlücking, K., Hashimoto, K., Manishankar, P., Steinhorst, L., Kuchitsu, K., & Kudla, J. (2013). The calcineurin B-like calcium sensors CBL1 and CBL9 together with their interacting protein kinase CIPK26 regulate the Arabidopsis NADPH oxidase RBOHF. *Molecular Plant*, *6*(2), 559–569. https://doi.org/10.1093/mp/sst009

Dunn, B. L., Porter, D. R., Baker, C. A., & Carver, B. F. (2011). Screening USDA-ARS wheat germplasm for bird cherry-oat aphid tolerance. *Journal of Crop Improvement*, *25*(2), 176–182. https://doi.org/10.1080/15427528.2011.547092

Dunse, K. M., Stevens, J. A., Lay, F. T., Gaspar, Y. M., Heath, R. L., & Anderson, M. A. (2010). Coexpression of potato type I and II proteinase inhibitors gives cotton plants protection against insect damage in the field. *Proceedings of the National Academy of Sciences*, *107*(34), 15011–15015. https://dx.doi.org/10.1073/pnas.1009241107

Ehrlich, P. R., & Raven, P. H. (1964). Butterflies and plants: A study in coevolution. *Evolution*, *18*(4), 586–608. https://doi.org/10.2307/2406212

Engelberth, J., Koch, T., Schüler, G., Bachmann, N., Rechtenbach, J., & Boland, W. (2001). Ion channel-forming alamethicin is a potent elicitor of volatile biosynthesis and tendril coiling. Cross talk between jasmonate and salicylate signaling in lima bean. *Plant Physiology*, *125*(1), 369–377. https://doi.org/10.1104/pp.125.1.369

Erb, M., & Reymond, P. (2019). Molecular interactions between plants and insect herbivores. *Annual Review of Plant Biology*, *70*, 527–557. https://doi.org/10.1146/annurev-arplant-050718-095910

Erb, M., Meldau, S., & Howe, G. A. (2012). Role of phytohormones in insect-specific plant reactions. *Trends in Plant Science*, *17*(5), 250–259. https://doi.org/10.1016/j.tplants.2012.01.003

Farmer, E. E. (2001). Surface-to-air signals. *Nature*, *411*(6839), 854–856. https://doi.org/10.1038/35081189

Fidantsef, A. L., Stout, M. J., Thaler, J. S., Duffey, S. S., & Bostock, R. M. (1999). Signal interactions in pathogen and insect attack: Expression of lipoxygenase, proteinase inhibitor II, and pathogenesis-related protein P4 in the tomato, *Lycopersicon esculentum*. *Physiological and Molecular Plant Pathology*, *54*(3–4), 97–114. https://dx.doi.org/10.1006/pmpp.1998.0192

Fischer, R. C., Richter, A., Wanek, W., & Mayer, V. (2002). Plants feed ants: Food bodies of myrmecophytic Piper and their significance for the interaction with *Pheidole bicornis* ants. *Oecologia*, *133*, 186–192. https://doi.org/10.1007/s00442-002-1000-y

Foyer, C. H., Verrall, S. R., & Hancock, R. D. (2015). Systematic analysis of phloem-feeding insect-induced transcriptional reprogramming in Arabidopsis highlights common features and reveals distinct responses to specialist and generalist insects. *Journal of Experimental Botany*, *66*(2), 495–512. https://doi.org/10.1093/jxb/eru491

Funk, C. J. (2001). Alkaline phosphatase activity in whitefly salivary glands and saliva. *Archives of Insect Biochemistry and Physiology: Published in Collaboration with the Entomological Society of America*, *46*(4), 165–174. https://doi.org/10.1002/arch.1026

Fürstenberg-Hägg, J., Zagrobelny, M., & Bak, S. (2013). Plant defense against insect herbivores. *International Journal of Molecular Sciences*, *14*(5), 10242–10297. https://doi.org/10.3390/ijms140510242

Goggin, F. L., Lorence, A., & Topp, C. N. (2015). Applying high-throughput phenotyping to plant–insect interactions: Picturing more resistant crops. *Current Opinion in Insect Science*, *9*, 69–76. https://doi.org/10.1016/j.cois.2015.03.002

Halitschke, R., Keßler, A., Kahl, J., Lorenz, A., & Baldwin, I. T. (2000). Ecophysiological comparison of direct and indirect defenses in *Nicotiana attenuata*. *Oecologia*, *124*, 408–417. https://doi.org/10.1007/s004420000389

Halitschke, R., Schittko, U., Pohnert, G., Boland, W., & Baldwin, I. T. (2001). Molecular interactions between the specialist herbivore *Manduca sexta* (Lepidoptera, Sphingidae) and its natural host *Nicotiana attenuata*. III. Fatty acid-amino acid conjugates in herbivore oral secretions are necessary and sufficient for herbivore-specific plant responses. *Plant Physiology*, *125*(2), 711–717. https://doi.org/10.1104/pp.125.2.711

Harfouche, A. L., Shivaji, R., Stocker, R., Williams, P. W., & Luthe, D. S. (2006). Ethylene signaling mediates a maize defense response to insect herbivory. *Molecular Plant-Microbe Interactions*, *19*(2), 189–199. https://doi.org/10.1094/MPMI-19-0189

Heil, M., Koch, T., Hilpert, A., Fiala, B., Boland, W., & Linsenmair, K. E. (2001). Extrafloral nectar production of the ant-associated plant, *Macaranga tanarius*, is an induced, indirect, defensive response elicited by jasmonic acid. *Proceedings of the National Academy of Sciences*, *98*(3), 1083–1088. https://doi.org/10.1073/pnas.98.3.1083

Heng-Moss, T., Sarath, G., Baxendale, F., Novak, D., Bose, S., Ni, X., & Quisenberry, S. (2004). Characterization of oxidative enzyme changes in buffalo-grasses challenged by *Blissus occiduus*. *Journal of Economic Entomology*, *97*(3), 1086–1095. https://doi.org/10.1093/jee/97.3.1086

Herde, O., Peña Cortés, H., Wasternack, C., Willmitzer, L., & Fisahn, J. (1999). Electric signaling and Pin2 gene expression on different abiotic stimuli depend on a distinct threshold level of endogenous abscisic acid in several abscisic acid-deficient tomato mutants. *Plant Physiology*, *119*(1), 213–218. https://doi.org/10.1104/pp.119.1.213

Herrmann, K. M., & Weaver, L. M. (1999). The shikimate pathway. *Annual Review of Plant Biology*, *50*(1), 473–503. https://doi.org/10.1146/annurev.arplant.50.1.473

Higley L. G.. (2006). The devil and Leon Higley: An IPM story. In: M. Gray (Ed.), *Illinois Crop Protection Technology Conference* (pp. 132–134). University of Illinois Extension.

Jacobs, J. A., & Frickel, S. (2009). Interdisciplinarity: A critical assessment. *Annual Review of Sociology*, *35*, 43–65. https://doi.org/10.1146/annurev-soc-070308-115954

Jander, G., Cui, J., Nhan, B., Pierce, N. E., & Ausubel, F. M. (2001). The *TASTY* locus on chromosome 1 of Arabidopsis affects feeding of the insect herbivore *Trichoplusia ni*. *Plant Physiology*, *126*(2), 890–898. https://doi.org/10.1104/pp.126.2.890

Johnson, M. T. (2011). Evolutionary ecology of plant defences against herbivores. *Functional Ecology*, *25*(2), 305–311. https://doi.org/10.1111/j.1365-2435.2011.01838.x

Johnson, M. T., Smith, S. D., & Rausher, M. D. (2009). Plant sex and the evolution of plant defenses against herbivores. *Proceedings of the National Academy of Sciences*, *106*(43), 18079–18084. https://dx.doi.org/10.1073/pnas.0904695106

Karley, A. J., Mitchell, C., Brookes, C., McNicol, J., O'neill, T., Roberts, H., Graham, J., & Johnson, S. N. (2016). Exploiting physical defence traits for crop protection: Leaf trichomes of *Rubus idaeus* have deterrent effects on spider mites but not aphids. *Annals of Applied Biology*, *168*(2), 159–172. https://doi.org/10.1111/aab.12252

Kempema, L. A., Cui, X., Holzer, F. M., & Walling, L. L. (2007). Arabidopsis transcriptome changes in response to phloem-feeding silverleaf whitefly nymphs. Similarities and distinctions in responses to aphids. *Plant Physiology*, *143*(2), 849–865. https://dx.doi.org/10.1104/pp.106.090662

Kerchev, P. I., Fenton, B., Foyer, C. H., & Hancock, R. D. (2012). Plant responses to insect herbivory: Interactions between photosynthesis, reactive oxygen species and hormonal signalling pathways. *Plant, Cell & Environment*, *35*(2), 441–453. https://doi.org/10.1111/j.1365-3040.2011.02399.x

Kessler, A., & Baldwin, I. T. (2001). Defensive function of herbivore-induced plant volatile emissions in nature. *Science*, *291*(5511), 2141–2144. https://doi.org/10.1126/science.291.5511.2141

Kessler, A., Halitschke, R., Diezel, C., & Baldwin, I. T. (2006). Priming of plant defense responses in nature by airborne signaling between *Artemisia tridentata* and *Nicotiana attenuata*. *Oecologia*, *148*, 280–292. https://doi.org/10.1007/s00442-006-0365-8

Kliebenstein, D. J., Kroymann, J., Brown, P., Figuth, A., Pedersen, D., Gershenzon, J., & Mitchell-Olds, T. (2001). Genetic control of natural variation in Arabidopsis glucosinolate accumulation. *Plant Physiology*, *126*(2), 811–825. https://doi.org/10.1104/pp.126.2.811

Koch, K. G., Chapman, K., Louis, J., Heng-Moss, T., & Sarath, G. (2016). Plant tolerance: A unique approach to control hemipteran pests. *Frontiers in Plant Science*, *7*, 1363. https://doi.org/10.3389/fpls.2016.01363

Korth, K. L., & Dixon, R. A. (1997). Evidence for chewing insect-specific molecular events distinct from a general wound response in leaves. *Plant Physiology*, *115*(4), 1299–1305. https://doi.org/10.1104/pp.115.4.1299

Lam, E., Kato, N., & Lawton, M. (2001). Programmed cell death, mitochondria and the plant hypersensitive response. *Nature*, *411*(6839), 848–853. https://doi.org/10.1038/35081184

Lamb, R. J., & Mackay, P. A. (1995). Tolerance of antibiotic and susceptible cereal seedlings to the aphids *Metopolophium dirhodum* and *Rhopalosiphum padi*. *Annals of Applied Biology*, *127*(3), 573–583. https://doi.org/10.1111/j.1744-7348.1995.tb07613.x

Lambrix, V., Reichelt, M., Mitchell-Olds, T., Kliebenstein, D. J., & Gershenzon, J. (2001). The Arabidopsis epithiospecifier protein promotes the hydrolysis of glucosinolates to nitriles and influences *Trichoplusia ni* herbivory. *The Plant Cell*, *13*(12), 2793–2807. https://doi.org/10.1105/tpc.010261

Larson, K. C., & Whitham, T. G. (1997). Competition between gall aphids and natural plant sinks: Plant architecture affects resistance to galling. *Oecologia*, *109*, 575–582. https://doi.org/10.1007/s004420050119

Lawrence, S. D., & Novak, N. G. (2006). Expression of poplar chitinase in tomato leads to inhibition of development in Colorado potato beetle. *Biotechnology Letters*, *28*, 593–599. https://doi.org/10.1007/s10529-006-0022-7

León, J., Rojo, E., & Sánchez-Serrano, J. J. (2001). Wound signalling in plants. *Journal of Experimental Botany*, *52*(354), 1–9. https://doi.org/10.1093/jexbot/52.354.1

Maffei, M. E., Arimura, G. I., & Mithöfer, A. (2012). Natural elicitors, effectors and modulators of plant responses. *Natural Product Reports*, *29*(11), 1288–1303. https://doi.org/10.1039/C2NP20053H

Maffei, M. E., Mithöfer, A., & Boland, W. (2007). Insects feeding on plants: Rapid signals and responses preceding the induction of phytochemical release. *Phytochemistry*, *68*(22–24), 2946–2959. https://doi.org/10.1016/j.phytochem.2007.07.016

Mahanil, S., Attajarusit, J., Stout, M. J., & Thipyapong, P. (2008). Overexpression of tomato polyphenol oxidase increases resistance to common cutworm. *Plant Science*, *174*(4), 456–466. https://doi.org/10.1016/j.plantsci.2008.01.006

Maleck, K., Levine, A., Eulgem, T., Morgan, A., Schmid, J., Lawton, K. A., Dangl, J. L., & Dietrich, R. A. (2000). The transcriptome of *Arabidopsis thaliana* during systemic acquired resistance. *Nature Genetics*, *26*(4), 403–410. https://doi.org/10.1038/82521

Malone, M., & Alarcon, J. J. (1995). Only xylem-borne factors can account for systemic wound signalling in the tomato plant. *Planta*, *196*(4), 740–746. https://doi.org/10.1007/BF00197340

Mantelin, S., Bhattarai, K. K., & Kaloshian, I. (2009). Ethylene contributes to potato aphid susceptibility in a compatible tomato host. *New Phytologist*, *183*(2), 444–456. https://doi.org/10.1111/j.1469-8137.2009.02870.x

Marimuthu, M., & Smith, C. M. (2012). Barley tolerance of Russian wheat aphid (Hemiptera: Aphididae) biotype 2 herbivory involves expression of defense response and developmental genes. *Plant Signaling & Behavior*, *7*(3), 382–391. https://doi.org/10.4161/psb.19139

Mattiacci, L., Dicke, M., & Posthumus, M. A. (1995). Beta-Glucosidase: An elicitor of herbivore-induced plant odor that attracts host-searching parasitic wasps. *Proceedings of the National Academy of Sciences*, *92*(6), 2036–2040. https://doi.org/10.1073/pnas.92.6.2036

Mauricio, R. (1998). Costs of resistance to natural enemies in field populations of the annual plant *Arabidopsis thaliana*. *The American Naturalist*, *151*(1), 20–28. https://doi.org/10.1086/286099

Meiners, T., & Hilker, M. (2000). Induction of plant synomones by oviposition of a phytophagous insect. *Journal of Chemical Ecology*, *26*, 221–232. https://doi.org/10.1023/A:1005453830961

Menalled, F. D., Peterson, R. K., Smith, R. G., Curran, W. S., Páez, D. J., & Maxwell, B. D. (2016). The eco-evolutionary imperative: Revisiting weed management in the midst of an herbicide resistance crisis. *Sustainability*, *8*(12), 1297. https://doi.org/10.3390/su8121297

Miles, P. W. (1999). Aphid saliva. *Biological Reviews*, *74*(1), 41–85. https://doi.org/10.1017/S0006323198005271

Moran, N., & Hamilton, W. D. (1980). Low nutritive quality as defense against herbivores. *Journal of Theoretical Biology*, *86*(2), 247–254. https://doi.org/10.1016/0022-5193(80)90004-1

Mornhinweg, D. W., Bregitzer, P. P., Porter, D. R., Peairs, F. B., Baltensperger, D. D., Hein, G. L., Randolph, T. A., Koch, M., & Walker, T. (2009). Registration of 'Sidney'spring feed barley resistant to Russian wheat aphid. *Journal of Plant Registrations*, *3*(3), 214–218. https://doi.org/10.3198/jpr2009.04.0205crc

Morton, R. L., Schroeder, H. E., Bateman, K. S., Chrispeels, M. J., Armstrong, E., & Higgins, T. J. (2000). Bean α-amylase inhibitor 1 in transgenic peas (*Pisum sativum*) provides complete protection from pea weevil (*Bruchus pisorum*) under field conditions. *Proceedings of the National Academy of Sciences*, *97*(8), 3820–3825. https://doi.org/10.1073/pnas.070054597

Mousavi, S. A., Chauvin, A., Pascaud, F., Kellenberger, S., & Farmer, E. E. (2013). GLUTAMATE RECEPTOR-LIKE genes mediate leaf-to-leaf wound signalling. *Nature*, *500*(7463), 422–426. https://doi.org/10.1038/nature12478

Nanda, S., Wan, P. J., Yuan, S. Y., Lai, F. X., Wang, W. X., & Fu, Q. (2018). Differential responses of osMPK s in IR56 rice to two BPH populations of different virulence levels. *International Journal of Molecular Sciences*, *19*(12), 4030. https://doi.org/10.3390/ijms19124030

N'guessan, F. K., & Quisenberry, S. S. (1994). Screening selected rice lines for resistance to the rice water weevil (Coleoptera: Curculionidae). *Environmental Entomology*, *23*(3), 665–675. https://doi.org/10.1093/ee/23.3.665

Nielsen, J. K., Hansen, M. L., Agerbirk, N., Petersen, B. L., & Halkier, B. A. (2001). Responses of the flea beetles *Phyllotreta nemorum* and *P. cruciferae* to metabolically engineered *Arabidopsis thaliana* with an altered glucosinolate profile. *Chemoecology*, *11*, 75–83. https://doi.org/10.1007/PL00001835

Noordermeer, M. A., Veldink, G. A., & Vliegenthart, J. F. (2001). Fatty acid hydroperoxide lyase: A plant cytochrome P450 enzyme involved in wound healing and pest resistance. *Chembiochem*, *2*(7–8), 494–504. https://doi.org/10.1002/1439-7633(20010803)2:7/8%3C494::AID-CBIC494%3E3.0.CO;2-1

O'donnell, P. J., Calvert, C., Atzorn, R., Wasternack, C. H. M. O., Leyser, H. M. O., & Bowles, D. J. (1996). Ethylene as a signal mediating the wound response of tomato plants. *Science*, *274*(5294), 1914–1917. https://doi.org/10.1126/science.274.5294.1914

Orozco-Cardenas, M., McGurl, B., & Ryan, C. A. (1993). Expression of an antisense prosystemin gene in tomato plants reduces resistance toward *Manduca sexta* larvae. *Proceedings of the National Academy of Sciences*, *90*(17), 8273–8276. https://doi.org/10.1073/pnas.90.17.8273

Pedigo, L. P. (1996). *Entomology and pest management* (2nd ed.). Prentice-Hall Inc.

Pedigo, L. P., & Higley, L. G. (1996). Introduction to pest management and thresholds. In *Economic thresholds for integrated pest management* (pp. 3–8). University of Nebraska Press.

Peng, J., Deng, X., Jia, S., Huang, J., Miao, X., & Huang, Y. (2004). Role of salicylic acid in tomato defense against cotton bollworm, *Helicoverpa armigera* Hubner. *Zeitschrift für Naturforschung C*, *59*(11–12), 856–862. https://doi.org/10.1515/znc-2004-11-1215

Peterson, R. K., & Higley, L. G. (2001). Photosynthesis, yield loss, and injury guilds. In *Biotic Stress and Yield Loss* (pp. 83–97). CRC Press.

Punnuri, S., Huang, Y., Steets, J., & Wu, Y. (2013). Developing new markers and QTL mapping for greenbug resistance in sorghum [*Sorghum bicolor* (L.) Moench]. *Euphytica*, *191*, 191–203. https://doi.org/10.1007/s10681-012-0755-4

Ramm, C., Saathoff, A., Donze, T., Heng-Moss, T., Baxendale, F., Twigg, P., Baird, L., & Amundsen, K. (2013). Expression profiling of four defense-related buffalograss transcripts in response to chinch bug (Hemiptera: Blissidae) feeding. *Journal of Economic Entomology*, *106*(6), 2568–2576. https://doi.org/10.1603/EC13267

Ranger, C. M., & Hower, A. A. (2001). Role of the glandular trichomes in resistance of perennial alfalfa to the potato leafhopper (Homoptera: Cicadellidae). *Journal of Economic Entomology*, *94*(4), 950–957. https://doi.org/10.1603/0022-0493-94.4.950

Renwick, J. A. A. (2001). Variable diets and changing taste in plant–insect relationships. *Journal of Chemical Ecology*, *27*, 1063–1076. https://doi.org/10.1023/A:1010381509601

Reymond, P., Weber, H., Damond, M., & Farmer, E. E. (2000). Differential gene expression in response to mechanical wounding and insect feeding in Arabidopsis. *The Plant Cell*, *12*(5), 707–719. https://doi.org/10.1105/tpc.12.5.707

Rhodes, J. D., THAIN, J. F., & WILDON, D. C. (1999). Evidence for physically distinct systemic signalling pathways in the wounded tomato plant. *Annals of Botany*, *84*(1), 109–116. https://doi.org/10.1006/anbo.1999.0900

Rhoten, D. (2004). Interdisciplinary research: Trend or transition. *Items and Issues*, *5*(1–2), 6–11.

Rosenthal, G. A., & Berenbaum, M. R. (2012). *Herbivores: their interactions with secondary plant metabolites: Ecological and evolutionary processes* (Vol. 2). Academic press.

Royo, J., León, J., Vancanneyt, G., Albar, J. P., Rosahl, S., Ortego, F., Castanera, P., & Sánchez-Serrano, J. J. (1999). Antisense-mediated depletion of a potato lipoxygenase reduces wound induction of proteinase inhibitors and increases weight gain of insect pests. *Proceedings of the National Academy of Sciences*, *96*(3), 1146–1151. https://doi.org/10.1073/pnas.96.3.1146

Ryan, C. A. (2000). The systemin signaling pathway: Differential activation of plant defensive genes. *Biochimica et Biophysica Acta (BBA)-Protein Structure and Molecular Enzymology*, 1477(1–2), 112–121. https://doi.org/10.1016/S0167-4838(99)00269-1

Sanghera, G. S., Wani, S. H., Singh, G., Kashyap, P. L., & Singh, N. B. (2011). Designing crop plants for biotic stresses using transgenic approach. *Vegetos*, *24*(1), 1–25.

Schaller, F. (2001). Enzymes of the biosynthesis of octadecanoid-derived signalling molecules. *Journal of Experimental Botany*, *52*(354), 11–23. https://doi.org/10.1093/jexbot/52.354.11

Schenk, P. M., Kazan, K., Wilson, I., Anderson, J. P., Richmond, T., Somerville, S. C., & Manners, J. M. (2000). Coordinated plant defense responses in Arabidopsis revealed by microarray analysis. *Proceedings of the National Academy of Sciences*, *97*(21), 11655–11660. https://doi.org/10.1073/pnas.97.21.11655

Schittko, U., Preston, C. A., & Baldwin, I. T. (2000). Eating the evidence? *Manduca sexta* larvae cannot disrupt specific jasmonate induction in *Nicotiana attenuata* by rapid consumption. *Planta*, *210*, 343–346. https://doi.org/10.1007/PL00008143

Schmelz, E. A., Alborn, H. T., & Tumlinson, J. H. (2001). The influence of intact-plant and excised-leaf bioassay designs on volicitin-and jasmonic acid-induced sesquiterpene volatile release in *Zea mays*. *Planta*, *214*, 171–179. https://doi.org/10.1007/s004250100603

Schuster, D. J., Price, J. F., Martin, F. G., Howard, C. M., & Albregts, E. E. (1980). Tolerance of strawberry cultivars to two spotted spider mites in Florida. *Journal of Economic Entomology*, *73*(1), 52–54. https://doi.org/10.1093/jee/73.1.52

Schwachtje, J., Minchin, P. E., Jahnke, S., van Dongen, J. T., Schittko, U., & Baldwin, I. T. (2006). SNF1-related kinases allow plants to tolerate herbivory by allocating carbon to roots. *Proceedings of the National Academy of Sciences*, *103*(34), 12935–12940. https://doi.org/10.1073/pnas.0602316103

Seo, H. S., Song, J. T., Cheong, J. J., Lee, Y. H., Lee, Y. W., Hwang, I., Lee, J. S., & Choi, Y. D. (2001). Jasmonic acid carboxyl methyltransferase: A key enzyme for jasmonate-regulated plant responses. *Proceedings of the National Academy of Sciences*, *98*(8), 4788–4793. https://doi.org/10.1073/pnas.081557298

Shibel, Z., & Heard, S. B. (2016). Synergistic and additive effects of drought stress and simulated herbivory on two goldenrods, *Solidago altissima* and *S. gigantea*. *Botany*, *94*(8), 635–642. https://doi.org/10.1139/cjb-2016-0060

Smith, C. M. (Ed.). (2005). *Plant resistance to arthropods: Molecular and conventional approaches*. Springer. https://doi.org/10.1007/1-4020-3702-3_4

Spiteller, D., Dettner, K., & Boland, W. (2000). Gut bacteria may be involved in interactions between plants, herbivores and their predators: Microbial biosynthesis of N-acylglutamine surfactants as elicitors of plant volatiles. *Biological Chemistry*, *381*(8), 755–762. https://doi.org/10.1515/BC.2000.096

Spiteller, D., Pohnert, G., & Boland, W. (2001). Absolute configuration of volicitin, an elicitor of plant volatile biosynthesis from lepidopteran larvae. *Tetrahedron Letters*, *42*(8), 1483–1485. https://doi.org/10.1016/S0040-4039(00)02290-5

Stout, M. J., Riggio, M. R., & Yang, Y. (2009). Direct induced resistance in *Oryza sativa* to *Spodoptera frugiperda*. *Environmental Entomology*, *38*(4), 1174–1181. https://dx.doi.org/10.1603/022.038.0426

Strauss, S. Y., & Agrawal, A. A. (1999). The ecology and evolution of plant tolerance to herbivory. *Trends in Ecology & Evolution*, *14*(5), 179–185. https://doi.org/10.1016/S0169-5347(98)01576-6

Tabashnik, B. E., Gassmann, A. J., Crowder, D. W., & Carrière, Y. (2008). Insect resistance to *Bt* crops: Evidence versus theory. *Nature Biotechnology*, *26*(2), 199–202. https://doi.org/10.1038/nbt1382

Tao, L., Ahmad, A., de Roode, J. C., & Hunter, M. D. (2016). Arbuscular mycorrhizal fungi affect plant tolerance and chemical defences to herbivory through different mechanisms. *Journal of Ecology*, *104*(2), 561–571. https://doi.org/10.1111/1365-2745.12535

Tattersall, D. B., Bak, S., Jones, P. R., Olsen, C. E., Nielsen, J. K., Hansen, M. L., Høj, P. B., & Møller, B. L. (2001). Resistance to an herbivore through engineered cyanogenic glucoside synthesis. *Science*, *293*(5536), 1826–1828. https://doi.org/10.1126/science.1062249

Turlings, T. C., & Benrey, B. (1998). Effects of plant metabolites on the behavior and development of parasitic wasps. *Ecoscience*, *5*(3), 321–333. https://doi.org/10.1080/11956860.1998.11682472

Van Breusegem, F., Vranová, E., Dat, J. F., & Inzé, D. (2001). The role of active oxygen species in plant signal transduction. *Plant Science*, *161*(3), 405–414. https://doi.org/10.1016/S0168-9452(01)00452-6

Vancanneyt, G., Sanz, C., Farmaki, T., Paneque, M., Ortego, F., Castañera, P., & Sánchez-Serrano, J. J. (2001). Hydroperoxide lyase depletion in transgenic potato plants leads to an increase in aphid performance. *Proceedings of the National Academy of Sciences*, *98*(14), 8139–8144. https://doi.org/10.1073/pnas.141079498

Varella, A. C., Weaver, D. K., Sherman, J. D., Blake, N. K., Heo, H. Y., Kalous, J. R., Chao, S., Hofland, M. L., Martin, J. M., Kephart, K.D., & Talbert, L. E. (2015). Association analysis of stem solidness and wheat stem sawfly resistance in a panel of North American spring wheat germplasm. *Crop Science*, *55*(5), 2046–2055. https://doi.org/10.2135/cropsci2014.12.0852

Wasternack, C., & Hause, B. (2013). Jasmonates: Biosynthesis, perception, signal transduction and action in plant stress response, growth and development. An update to the 2007 review in Annals of Botany. *Annals of Botany*, *111*(6), 1021–1058. https://doi.org/10.1093/aob/mct067

Weaver, D. K., Buteler, M., Hofland, M. L., Runyon, J. B., Nansen, C., Talbert, L. E., Lamb, P., & Carlson, G. R. (2009). Cultivar preferences of ovipositing wheat stem sawflies as influenced by the amount of volatile attractant. *Journal of Economic Entomology*, *102*(3), 1009–1017. https://doi.org/10.1603/029.102.0320

Wei, H., Zhikuan, J., & Qingfang, H. (2007). Effects of herbivore stress by *Aphis medicaginis* Koch on the malondialdehyde contents and the activities of protective enzymes in different alfalfa varieties. *Acta Ecologica Sinica*, *27*(6), 2177–2183. https://dx.doi.org/10.1016/S1872-2032(07)60048-1

Winterer, J., & Bergelson, J. (2001). Diamondback moth compensatory consumption of protease inhibitor-transformed plants. *Molecular Ecology*, *10*(4), 1069–1074. https://doi.org/10.1046/j.1365-294X.2001.01239.x

Winz, R. A., & Baldwin, I. T. (2001). Molecular interactions between the specialist herbivore *Manduca sexta* (Lepidoptera, Sphingidae) and its natural host *Nicotiana attenuata*. IV. Insect-induced ethylene reduces jasmonate-induced nicotine accumulation by regulating putrescine N-methyltransferase transcripts. *Plant Physiology*, *125*(4), 2189–2202. https://doi.org/10.1104/pp.125.4.2189

Wise, M. J., & Abrahamson, W. G. (2007). Effects of resource availability on tolerance of herbivory: A review and assessment of three opposing models. *The American Naturalist*, *169*(4), 443–454. https://doi.org/10.1086/512044

Wu, J., Hettenhausen, C., Meldau, S., & Baldwin, I. T. (2007). Herbivory rapidly activates MAPK signaling in attacked and unattacked leaf regions but not between leaves of *Nicotiana attenuata*. *The Plant Cell*, *19*(3), 1096–1122. https://doi.org/10.1105/tpc.106.049353

Yang, J., Duan, G., Li, C., Liu, L., Han, G., Zhang, Y., & Wang, C. (2019). The crosstalks between jasmonic acid and other plant hormone signaling highlight the involvement of jasmonic acid as a core component in plant response to biotic and abiotic stresses. *Frontiers in Plant Science*, *10*, 1349. https://doi.org/10.3389/fpls.2019.01349

Zhang, Z. P., & Baldwin, I. T. (1997). Transport of [2–14 C] jasmonic acid from leaves to roots mimics wound-induced changes in endogenous jasmonic acid pools in *Nicotiana sylvestris*. *Planta*, *203*, 436–441. https://doi.org/10.1007/s004250050211

Zhao, L. Y., Chen, J. L., Cheng, D. F., Sun, J. R., Liu, Y., & Tian, Z. (2009). Biochemical and molecular characterizations of *Sitobion avenae*-induced wheat defense responses. *Crop Protection*, *28*(5), 435–442. https://dx.doi.org/10.1016/j.cropro.2009.01.005

Zhu-Salzman, K., Salzman, R. A., Ahn, J. E., & Koiwa, H. (2004). Transcriptional regulation of sorghum defense determinants against a phloem-feeding aphid. *Plant Physiology*, *134*(1), 420–431. https://dx.doi.org/10.1104/pp.103.028324

Index

Note: **Bold** page numbers refer to tables and *italic* page numbers refer to figures.

For Product Safety Concerns and Information please contact our EU
representative GPSR@taylorandfrancis.com
Taylor & Francis Verlag GmbH, Kaufingerstraße 24, 80331 München, Germany

www.ingramcontent.com/pod-product-compliance
Lightning Source LLC
LaVergne TN
LVHW080342011026
849161LV00009B/2157
* 9 7 8 1 0 3 2 8 9 5 1 7 8 *